Analog Integrated Circuit Design by Simulation

About the Author

Uğur Çilingiroğlu received his M.S. in electrical engineering from Istanbul Technical University in 1973, and Ph.D. in microelectronics from Southampton University in 1978. He currently holds a faculty position at Yeditepe University. Dr. Çilingiroğlu has authored numerous papers and two previous books.

Analog Integrated Circuit Design by Simulation

Techniques, Tools, and Methods

Uğur Çilingiroğlu

Yeditepe University
Istanbul, Turkey

McGraw Hill Education

New York Chicago San Francisco
Athens London Madrid
Mexico City Milan New Delhi
Singapore Sydney Toronto

Library of Congress Control Number: 2019931446

McGraw-Hill Education books are available at special quantity discounts to use as premiums and sales promotions or for use in corporate training programs. To contact a representative, please visit the Contact Us page at www.mhprofessional.com.

Analog Integrated Circuit Design by Simulation:
Techniques, Tools, and Methods

1 2 3 4 5 6 7 8 9 QVS 24 23 22 21 20 19

ISBN 978-1-260-44145-1
MHID 1-260-44145-8

Sponsoring Editor
Lara Zoble

Editorial Supervisor
Stephen M. Smith

Production Supervisor
Pamela A. Pelton

Acquisitions Coordinator
Elizabeth M. Houde

Project Manager
Tania Andrabi, Cenveo® Publisher Services

Copy Editor
Tulsi Rawat, Cenveo Publisher Services

Proofreader
Linda Manis Leggio

Indexer
Robert Swanson

Art Director, Cover
Jeff Weeks

Composition
Cenveo Publisher Services

Contents

CHAPTER 1
Preliminaries

This book is about linear analog integrated circuits. An analog circuit is qualified as linear if it exhibits a linear relationship between its input and output variables. Regulators, amplifiers, buffers, and filters are typical examples of such circuits. They are used mostly for regulating power supplies, conditioning sensor signals prior to analog-to-digital conversion, or driving an actuator/transducer after digital-to-analog conversion. The last two cases are exemplified in Fig. 1.1 with two simple applications. Figure 1.1(a) shows a voltage amplifier matching up the voltage-mode signal of a temperature-sensing diode to the input range of an analog-to-digital converter (ADC). Figure 1.1(b) shows a transimpedance amplifier converting and filtering the current-mode output signal of a digital-to-analog converter (DAC) into a voltage-mode signal to drive an actuator/transducer load. Both examples exhibit a feature common to almost all linear integrated-circuit applications: Their architecture is in the form of a *closed-loop configuration* consisting of (a) an *operational amplifier* (opamp), and (b) an *external network* built usually with passive components. In a closed-loop configuration, the opamp itself acts as an inaccurate, imprecise, and nonlinear supplier of large voltage gain while the negative feedback provided by the external network trades off the excessive voltage gain for accuracy, precision, and linearity, which are the three fundamental attributes of linear analog signal processing.

Integrated-circuit design in general is a two-stage process. The first stage, called *electrical design*, delivers a complete circuit schematic of the closed-loop configuration annotated with the dimensions and other physical parameters of (a) all external-network components and (b) all opamp internal devices. The second stage, called *physical design*, converts this schematic into an artwork defining the layout to be fabricated. This book is primarily on electrical design; physical design is covered as much as it pertains to electrical design.

Electrical design is a top-down process going through three layers of abstraction as illustrated with a block diagram in Fig. 1.2. Electrical performance of the application, as specified in terms of what we call closed-loop metrics, is the input of the topmost *application layer*. In this layer, design activity begins with the selection of a closed-loop configuration that can handle the application, and moves to the translation of the specified closed-loop metrics into (a) open-loop metrics, which define the target performance of the opamp itself, and (b) external-network device metrics, which define the target performance of external-network components. The intermediate *circuit layer* of design is based on open-loop metrics. Its outcome is twofold: (a) adopting an opamp circuit topology that can satisfy these open-loop metrics, and (b) specifying the target performance of opamp internal devices in terms of device metrics. In the final *device layer* of design, the device metrics specified in application and circuit layers are translated into bias conditions and physical parameters for all opamp devices and external-network components.

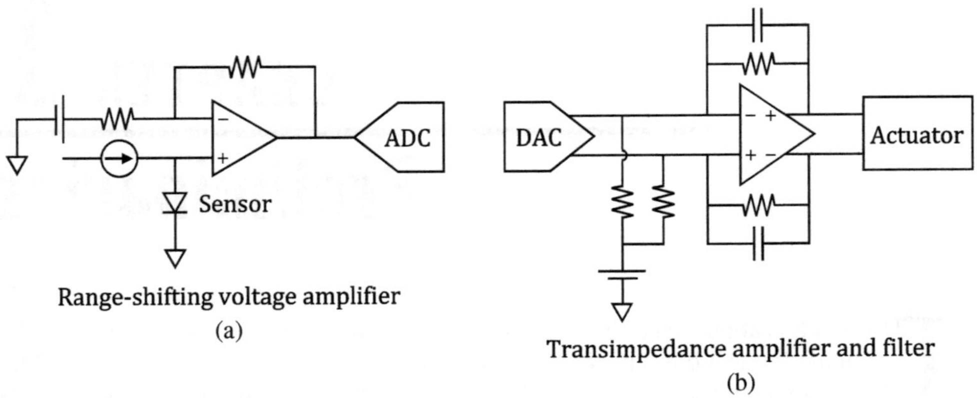

FIGURE 1.1 Two linear integrated-circuit application examples. (a) A single-ended voltage amplifier for conditioning a thermal-sensor signal prior to analog-to-digital conversion. (b) A fully-differential transimpedance amplifier for driving an actuator.

Although the block diagram shown in Fig. 1.2 indicates a one-way flow, it frequently becomes necessary to loop back and redesign as overconstrained or underconstrained design spaces are encountered along the way. However, this does not alter the definitions indicated in Fig. 1.2 for the input metrics or outcomes of any of the three layers.

The main body of this book is divided into three chapters, each dealing with one of the three layers of design. The application layer is covered in Chapter 2. The device layer comes next in Chapter 3. Finally, in Chapter 4, we study the circuit layer and also work on complete linear integrated-circuit design examples involving the contents of the preceding chapters.

The methodology practiced in this book for design and verification is based on Spice simulations, which are run on version-27 of the open-source platform *Ngspice*. The reader is referred to the companion website of this book, www.mhprofessional.com/AICDS, for instructions regarding its download and installation.

Circuit description in Ngspice is based on direct netlist entry whereas commercial simulation platforms generate netlists from schematic entries. Schematic entry is generally faster, easier, and foolproof, which are valuable assets in professional use because they minimize the time and attention needed for circuit description. However, the intricacies of circuit description are better learned by paying attention to details, which is why netlist entry is preferred in this book. Input files describing the netlists of all simulation examples can be downloaded from this book's website. Each of the folders named `chap2`, `chap3`, and `chap4` contains the files used in one particular chapter. The first step of running a simulation with any of these files is to open the Ngspice executable. Next, the folder containing the input file is defined as the current directory if it is not already so. For a file residing in, for example, `c:\spice\chap3` of an MS Windows installation, this is done by entering

```
cd c:\spice\chap3
```

at the prompt as shown at the top of Fig. 1.3. Next, the filename itself is entered as exemplified in the middle part of Fig. 1.3 for an input file `ex3-16.sp`. Finally, entering the command `run` as shown at the bottom of Fig. 1.3 initiates the simulation. As the

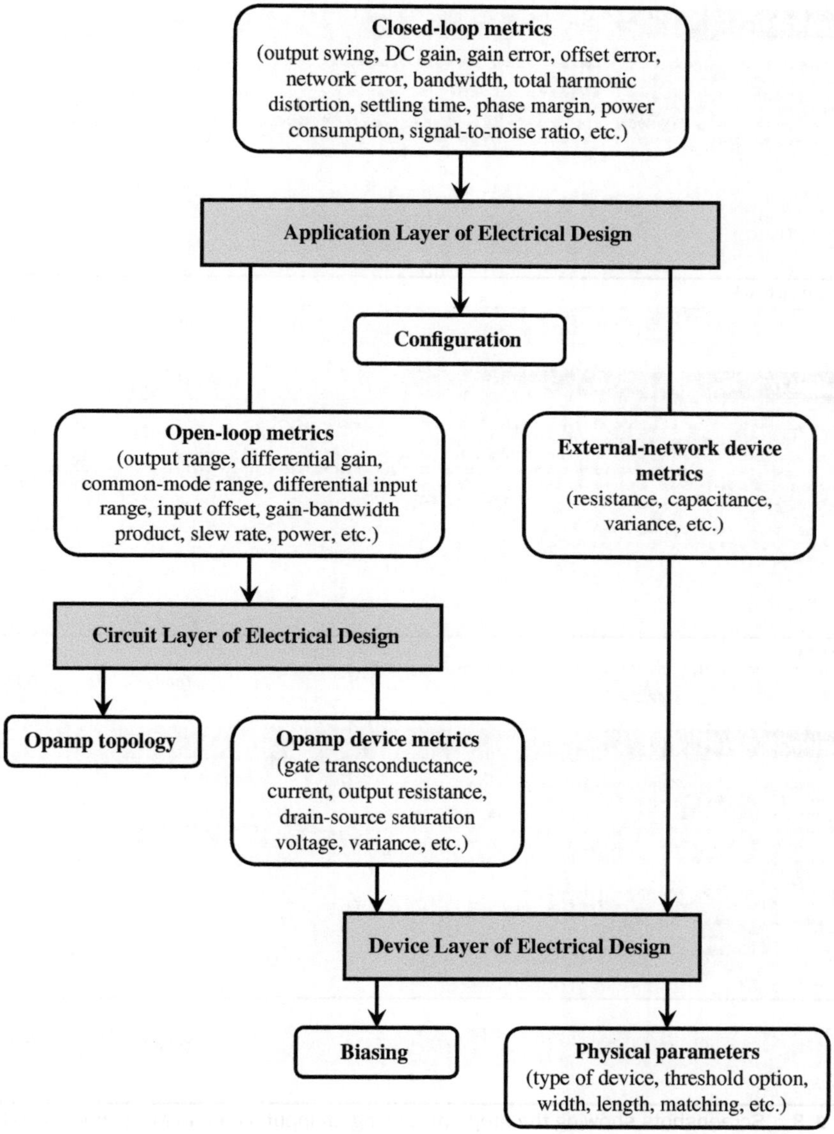

FIGURE 1.2 Layers of abstraction in electrical design.

reader will experience initially in Example 2.9 of Subsection 2.4.2, it is possible to start a simulation without entering a `run` command provided that the input file is properly scripted. As to reading out or plotting the outcomes of a simulation, the reader will find the necessary descriptions inside the examples presented throughout the following three chapters.

Input files are prepared with a text editor, and saved with extension `.sp`. Each begins with a title line and ends with a `.end` line. The lines in between are expected to describe

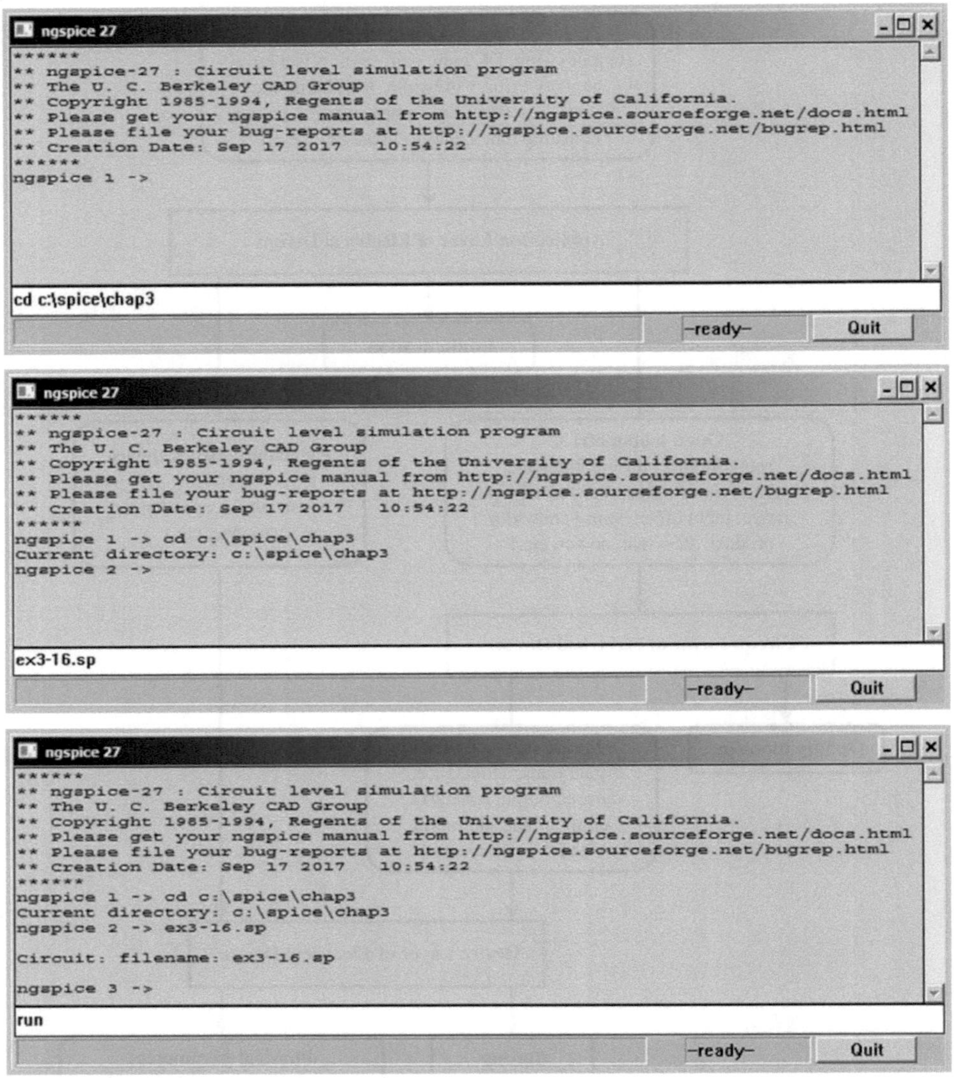

FIGURE 1.3 Screenshots showing the steps of running an input file in an MS Windows installation. *Top:* Changing the directory. *Middle:* Declaring the filename. *Bottom:* Initiating the simulation.

at least the network of components, their parameters, and the type of analysis to be performed. For an example, the reader is referred to file `ex2-1.sp`, which netlists the circuit schematic shown in Fig. 2.5. Its line-by-line description is as follows:

Line 1 Filename is typed in the title line. Since Spice does not consider this line as a declaration, any text can be included without affecting the outcome of the simulation.

Line 2 This line quantifies certain parameters. Typographically, it spans over two lines as indicated by the plus sign appearing at the beginning of the extension. This is how a long line can be accommodated in an input file.

Line 3 Another file `opmacro1.sp` is called by this line. This additional file actually describes the netlist of the opamp `x1` declared in line 4 as a *subcircuit*. If a circuit is involved in multiple simulations, or multiple instances of a circuit are involved in a simulation, we prefer to describe such a circuit as a subcircuit and simply include its instances whenever or wherever needed.

Line 4 This is an element line where an instance of the subcircuit `opamp` is called. Note that a subcircuit instance is identified with a name beginning with `x`. The three numbers appearing after the name are the node numbers of the top-level circuit in which the subcircuit is embedded.

Lines 5 and 6 These lines begin with `r`, and therefore describe the instances of resistors. Each contains two node identifiers and the value of the resistance.

Lines 7 and 8 Independent voltage sources are defined in these lines. They begin with `v`, and continue with two node identifiers. The first between the two is the reference node. The third number `0.54` appearing in line 7 is the voltage imposed onto the reference node with respect to the ground node. Note that 0 is the universal identifier of ground node. The voltage of the source `vd` is undefined in line 8 because it is changed as a part of the analysis defined next in line 9.

Line 9 Beginning with a dot, this line describes a command. Specifically, it defines a dc simulation as indicated by `.dc`. The voltage of the independent source `vd` is the variable of this simulation. It varies between 0.476 and 0.723 V in 1-mV increments.

The end of the input file is declared in the very last line.

Although the examples included in the following chapters will help the reader master the basics of simulation with Ngspice, it will be much more beneficial at this stage to read the specific sections of the user manual, which is included in the downloaded Ngspice package.

CHAPTER 2
Application Layer

2.1 Introduction

As stated in Chapter 1, the application layer of electrical design involves (a) selection of an appropriate closed-loop configuration for the application, and (b) translation of the specified closed-loop metrics into open-loop metrics for the opamp and device metrics for the external network. Available closed-loop amplifier configurations are well documented in related literature. No attempt is made in this book to present their inventory but most popular configurations are presented in examples throughout this chapter. The main emphasis of the chapter is on the analytical techniques and tools used in translating closed-loop metrics into open-loop metrics and external-network device metrics. These techniques and tools are generalized into four major architectural classes depending on (a) whether amplification is performed in continuous time or in discrete time, and (b) whether signal representation is single ended or fully differential. The case of continuous-time/single-ended amplifier configurations, being the most fundamental of the four, is covered in the first nine sections of this chapter. Based on the foundation thus established, coverage is then extended to continuous-time/fully-differential configurations in Section 2.10, to discrete-time/single-ended configurations in Section 2.11, and finally to discrete-time/fully-differential configurations in Section 2.12.

Most of the concepts presented in this chapter are supported with Spice simulations conducted with a parameterized single-ended or fully-differential opamp macromodel defined as a subcircuit. The single-ended version is netlisted in the input file `opmacro1.sp`. As shown in Fig. 2.1(a), this macro interfaces with the closed-loop configuration through the following three pins:

1. `ninv`: Noninverting input.
2. `inv`: Inverting input.
3. `out`: Output.

The user-defined parameters of the single-ended opamp macro are described in the following list for future reference.

- `vdd`: Positive power-supply voltage. Introduced in Subsection 2.2.1.
- `vss`: Negative power-supply voltage (zero for a single-supply opamp). Introduced in Subsection 2.2.1.
- `voh`: Higher limit V_{OH} of output range. Introduced in Subsection 2.2.1.

FIGURE 2.1 Symbols and pin configurations of the opamp macromodels defined as subcircuits. (a) Single-ended `opmacro1.sp`. (b) Fully-differential `opmacro2.sp`.

- `vol`: Lower limit V_{OL} of output range. Introduced in Subsection 2.2.1.
- `a0`: DC differential gain A_0 at the midpoint of the output range. Introduced in Subsection 2.4.1.
- `fgbw`: Gain-bandwidth product f_{GBW}. Introduced in Subsection 2.5.1.
- `vdir`: Differential-input range V_{DIR}. Introduced in Subsection 2.5.2.
- `fnd`: Nondominant pole frequency f_{nd}. Introduced in Section 2.7.
- `vos` at `vout`: Offset voltage V_{os} defined at an output bias voltage V_O. Introduced in Subsection 2.4.1. Unlike the preceding parameters, these two are set by default to `vos`= 0 and `vout`= (`voh`+`vol`)/2 but can be reset by adding the following lines to the main input file that contains the opamp subcircuit `opmacro1.sp`:

```
.control
alter v.<subcircuit identifier>.vos=<value>
alter v.<subcircuit identifier>.vout=<value>
.endc
```

  where `<subcircuit identifier>` is the name of the opamp instance, such as `x1`, and `<value>` is the reset value.

The input file netlisting the fully-differential opamp macro is `opmacro2.sp`. As shown in Fig. 2.1(b), this macro interfaces with the closed-loop configuration through the following five pins:

1. `ninv`: Noninverting input.
2. `inv`: Inverting input.
3. `outp`: Noninverting output. Described in Subsection 2.10.1.
4. `outn`: Inverting output. Described in Subsection 2.10.1.
5. `ocm`: This is the pin where the output common-mode voltage V_{OCM} is externally applied as described in Subsection 2.10.1.

The user-defined parameters of the single-ended opamp macro described previously apply also to this opamp with one exception. The exception is the replacement of the single-ended open-loop dc gain parameter `a0` with the differential open-loop dc gain parameter `adif0` which is defined in Subsection 2.10.3.

Another adjustable opamp parameter is the offset voltage `vos`, whose default value is zero. The user may alter it by adding the following line to the main input file containing the opamp subcircuit `opmacro2.sp`:

```
.control
alter v.<subcircuit identifier>.vos=<value>
.endc
```

where `<subcircuit identifier>` is the name of the opamp instance, such as `x1`, and `<value>` is the reset value. The offset voltage of a fully-differential opamp is described in Subsection 2.10.3.

2.2 First-Order DC Response

DC response is of prime importance in design because most applications operate with dc bias and are supposed to handle signals whose frequency spectra include dc. Furthermore, a closed-loop configuration treats a time-varying signal not much differently than a pure dc as long as its bandwidth is wider than the frequency spectrum of the signal. The voltage amplifier shown in Fig. 1.1(a) is a good example. It is biased with a dc voltage as well as a dc current, and is driven by a signal of thermal origin which usually remains constant for most of the time and varies slowly otherwise.

In this section, we present an analysis of closed-loop dc response using an ideal opamp model based on the so-called *virtual-short approximation*. Despite its ultimate simplicity, the model is still sufficiently accurate for specifying the resistive components of the external network and open-loop dc range metrics of the opamp. In some applications, the designer has to consider also the second-order features of closed-loop dc response. Those are discussed in Section 2.4 on the basis of a more accurate opamp model.

2.2.1 First-Order Open-Loop DC Transfer Characteristic and Range Limitations

Shown in Fig. 2.2(a) is the symbol convention we generally use for opamp terminal voltages regardless of the type of response being analyzed. v_O is the total instantaneous output voltage. v_{IN} and v_{IP} represent the total instantaneous voltages of the inverting and noninverting input terminals, respectively. In steady state, the opamp does not draw any dc current from these input terminals because they are capacitively terminated inside the opamp. All of these voltages are defined with respect to system ground. The difference between the two input voltages, as defined by

$$v_{ID} \doteq v_{IP} - v_{IN}, \tag{2.1}$$

is called *differential-input voltage*. The average of the two defined by

$$v_{CM} \doteq \frac{v_{IP} + v_{IN}}{2} \tag{2.2}$$

is called *common-mode voltage*.

In this book, the relationship imposed by the opamp between v_O and v_{ID} is generally called *open-loop transfer function* if expressed analytically, or *open-loop transfer characteristic* if displayed graphically. The dc form of the open-loop transfer characteristic is shown in Fig. 2.2(b) for an ideal single-supply opamp powered by a positive supply-voltage V_{DD} with respect to system ground. The ideal features of the characteristic are associated with its segment located between the two output-voltage levels $v_O = V_{OL}$ and $v_O = V_{OH}$. In a

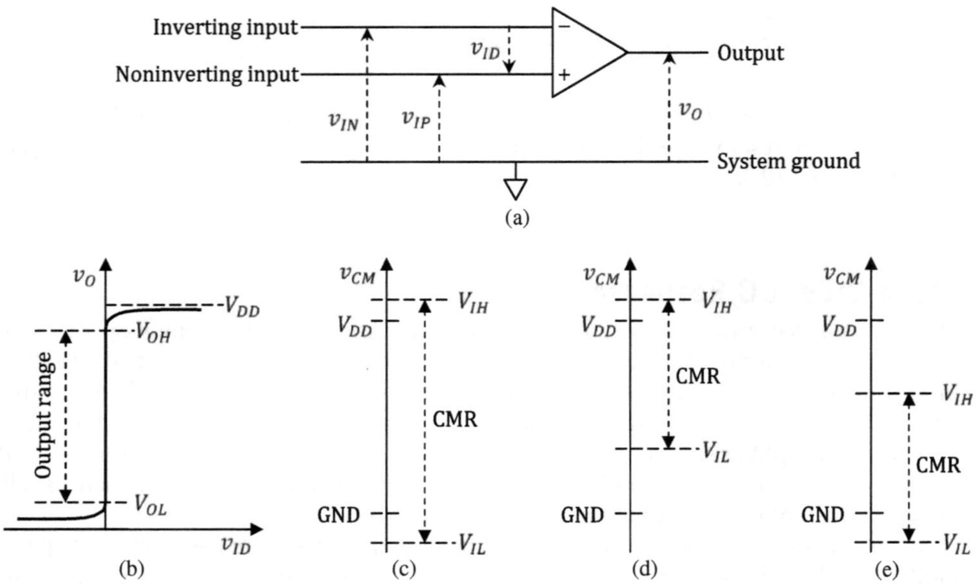

FIGURE 2.2 (a) Symbol convention for opamp input and output variables. (b) Open-loop dc transfer characteristic and output range in an ideal single-supply opamp. (c) A rail-to-rail common-mode range. (d) A common-mode range excluding ground. (e) A common-mode range excluding the supply voltage V_{DD}.

real opamp, this segment is nonlinear with a large but finite and varying slope as will be discussed in detail in Section 2.4. In first-order dc modeling, we assume it to be perfectly vertical and located at

$$v_{ID} = 0. \tag{2.3}$$

When functioning properly, the negative feedback established by the external network of the closed-loop configuration keeps the operation point of the opamp on this ideally vertical segment. The range between V_{OL} and V_{OH} is an open-loop metric called *output range*. It must be made wide enough by design to safely accommodate the so-called *output swing*, which is a closed-loop metric defined as the range in which v_O is expected to vary. This design constraint can be stated as

$$V_{OL} < V_{O(\mathrm{min})}, \tag{2.4}$$

and

$$V_{OH} > V_{O(\mathrm{max})}, \tag{2.5}$$

where $V_{O(\mathrm{min})}$ and $V_{O(\mathrm{max})}$ are the limits of the output swing.

According to (2.1) and (2.2), the common-mode voltage of an ideal opamp complying with (2.3) is given by

$$v_{CM} = v_{IP} = v_{IN}, \tag{2.6}$$

which enables us to interpret the common-mode voltage as the common value of the identical v_{IP} and v_{IN}. Ideally, an opamp is expected to be insensitive to v_{CM}, that is, v_O should remain unchanged even when v_{IP} and v_{IN} commonly change in time. In reality, all opamps are somewhat sensitive to v_{CM}, but the sensitivity is, to a first-order approximation, negligible as long as v_{CM} remains within a particular range known as *common-mode range* (CMR). This range is an open-loop metric. It must be made sufficiently wide by design to accommodate the minimum and maximum values $V_{CM(\mathrm{min})}$ and $V_{CM(\mathrm{max})}$ of the common-mode voltage. Representing the lower and upper limits of CMR with V_{IL} and V_{IH}, this design constraint can be expressed with

$$V_{IL} < V_{CM(\mathrm{min})}, \tag{2.7}$$

and

$$V_{IH} > V_{CM(\mathrm{max})}. \tag{2.8}$$

The range $V_{CM(\mathrm{min})} \leq v_{CM} \leq V_{CM(\mathrm{max})}$ is called *common-mode swing* in this book. It is a closed-loop metric.

Depending on the topology of the opamp, V_{IL} may be lower or higher than ground. Likewise, V_{IH} may be lower or higher than V_{DD}. If CMR includes both ground and V_{DD}, as shown in Fig. 2.2(c), the opamp is said to have a *rail-to-rail input*. Yet, some other topologies offer a CMR excluding ground or V_{DD}, as depicted in Figs. 2.2(d) and (e), respectively.

In the past, amplifiers used to be built mostly with *split-supply* opamps which are powered symmetrically by a positive supply V_{DD} and a negative supply V_{SS}. The main benefit of split-supply powering is that it enables an amplifier to handle bipolar signals[1] without necessitating any bias. Over the years, however, the diversification of signal-processing applications has introduced many other forms of signals which can't be processed without bias anyway. More importantly, the cost of incorporating two separate power supplies is prohibitively high for the rapidly expanding application area of battery-operated portable equipment. These trends have diminished the popularity of split-supply opamps. Still, it is worth knowing that the only effect the split-supply scheme has on the first-order open-loop dc transfer characteristic is to shift it down as shown in Fig. 2.3(a). Since the output range now includes $v_O = 0$, the opamp can put out a signal centered around the ground. As shown in Figs. 2.3(b) to (d), the CMR also shifts down to include the ground, which is why a split-supply opamp can also handle a common-mode signal swinging around ground.

2.2.2 DC Analysis with Virtual-Short Approximation

In order to specify V_{IL} and V_{IH} of the common-mode range and V_{OL} and V_{OH} of the output range for a given application, one has to determine first the limits $V_{CM(\mathrm{min})}$ and $V_{CM(\mathrm{max})}$ of the common-mode swing and $V_{O(\mathrm{min})}$ and $V_{O(\mathrm{max})}$ of the output swing. In many cases, the limits of the output swing are readily available from closed-loop specifications. In other cases, they need to be determined from a dc analysis of the closed-loop configuration. The limits of the common-mode swing are also derived from the same analysis. A simple first-order analysis based on (2.3) is deemed sufficient for this purpose despite its

[1] Swinging symmetrically around ground.

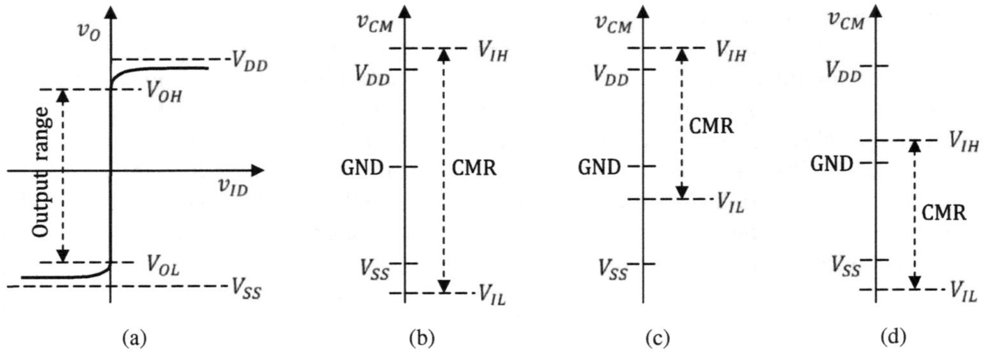

FIGURE 2.3 DC range metrics in an ideal split-supply opamp. (a) Open-loop dc transfer characteristic and output range. (b) A rail-to-rail common-mode range. (c) A common-mode range excluding the negative supply voltage V_{SS}. (d) A common-mode range excluding the positive supply voltage V_{DD}.

inaccuracies because the safety margins by which these dc range metrics are quantified from (2.4), (2.5), (2.7), and (2.8) are reasonably wide anyway.

Known as *virtual-short approximation,* (2.3) enables us to relate v_O and v_{CM} to the input variables of a closed-loop configuration directly from the external network while treating the opamp as an instrument that virtually equates v_{IN} to v_{IP}. The following examples introduce the reader to its use in first-order dc closed-loop analysis and in specifying open-loop dc range metrics.

Example 2.1 An annotated schematic of the temperature-sensor application of Fig. 1.1(a) is shown in Fig. 2.4(a). It is built with a 1.8-V single-supply opamp, a sensor diode biased by a dc current I, and a feedback network comprising a bias source V_B and resistors R_1 and R_2. This closed-loop configuration is expected to amplify and shift the temperature-dependent voltage v_D of the sensor diode into an output swing between $V_{O(\min)} = 0.150$ V and $V_{O(\max)} = 1.650$ V corresponding to the temperature range $-40°\text{C} \leq T \leq 125°\text{C}$. The voltage v_D across the diode equals $V_D = 626$ mV at $T = 25°\text{C}$, and exhibits a temperature sensitivity $\theta \doteq \partial v_D/\partial T = -1.5$ mV/°C within the specified temperature range. We will

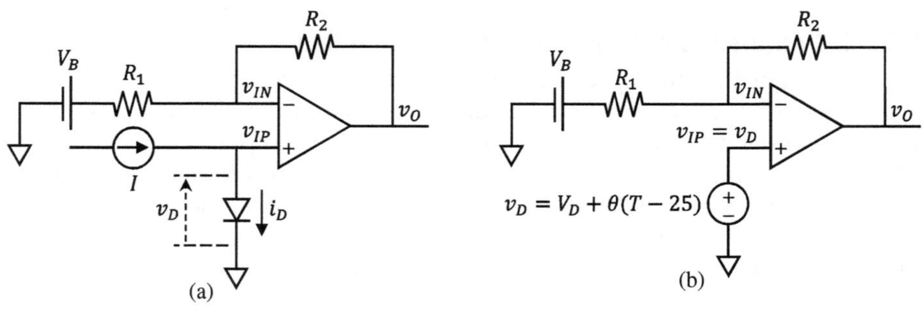

FIGURE 2.4 (a) Annotated schematic of the temperature-sensor application. (b) Equivalent circuit.

determine (a) R_2/R_1 and V_B for the external network, and (b) the limits of output range and common-mode range for the opamp.

Work The sensor diode and its bias current source I can be represented with a single dc voltage source v_D modeled by

$$v_D = V_D + \theta(T - 25)$$

and shown in Fig. 2.4(b). Considering the specified values of V_D, θ, and T, we identify the following range for v_D:

$$0.476\,\text{V} \leq v_D \leq 0.723\,\text{V}. \tag{2.9}$$

We now analyze the closed-loop configuration shown in Fig. 2.4(b) with a view to determining the relationship between its output voltage v_O and input variables V_B and v_D. This relationship will be used for specifying the dc range specifications and design constraints. Note that the analytical relationship between the output voltage and input variables of a closed-loop configuration is generally called *closed-loop transfer function* in this book. For all linear configurations, it assumes the form

$$v_O = \sum_{j=1}^{n} A_{cl,j} x_j, \tag{2.10}$$

where $x_j, j = 1, \ldots, n$ are the input variables of the configuration, and $A_{cl,j}$ is the *closed-loop gain* for input variable x_j.

Let's begin with the observation $v_{IP} = v_D$ in Fig. 2.4(b). Since $v_{IN} = v_{IP}$ due to virtual short, we therefore write

$$v_{IN} = v_D. \tag{2.11}$$

Additionally, the currents flowing in R_1 and R_2 are identical, hence

$$\frac{v_{IN} - V_B}{R_1} = \frac{v_O - v_{IN}}{R_2}. \tag{2.12}$$

Canceling v_{IN} between (2.11) and (2.12) and solving for v_O, we obtain the following first-order dc transfer function of the closed-loop configuration

$$v_O = \left(1 + \frac{R_2}{R_1}\right) v_D - \frac{R_2}{R_1} V_B, \tag{2.13}$$

which indeed complies with the general form (2.10), and indicates $A_{cl,D} = 1 + (R_2/R_1)$ for v_D and $A_{cl,B} = -R_2/R_1$ for V_B.

We now use (2.13) for determining the external-network device metrics R_2/R_1 and V_B. To this end, we rewrite this equation once for $v_O = V_{O(\min)} = 0.150$ V and $v_D = V_{D(\min)} = 0.476$ V, and next for $v_O = V_{O(\max)} = 1.650$ V and $v_D = V_{D(\max)} = 0.723$ V. Only V_B and R_2/R_1 are unknown in the resulting two equations. A simultaneous solution of the two yields $V_B = 0.540$ V and $R_2/R_1 = 5.05$. Note that an additional constraint is needed for resolving the ratio R_2/R_1 into individual values of R_1 and R_2. If not provided by closed-loop performance specifications as in the present case, the designer creates one in accordance with best design principles. It may be related to such concerns as accuracy and precision, power dissipation, footprint, etc.

Next, we determine the specifications of the dc range metrics of the opamp. Since $V_{O(\min)}$ and $V_{O(\max)}$ are already specified for this application, the limits of the opamp output range are to comply with $v_{OL} < 0.150$ V and $v_{OH} > 1.650$ V in accordance with (2.4) and (2.5). As to the limits of the common-mode range, we refer to (2.7) and (2.8) after determining the limits $V_{CM(\min)} = 0.476$ V and $V_{CM(\max)} = 0.723$ V of common-mode swing from (2.11) and (2.6) for two extreme values of v_D.

Finally, we verify the performance of the equivalent circuit of Fig. 2.4(b) with a Spice simulation. This can be done with an operation-point (`.op`) analysis or a dc analysis (`.dc`). In the case of an operation-point analysis, each independent source is assigned a fixed value entered by the user. The simulator determines the dc voltage of each and every node and the dc current of each and every independent voltage source, and prints a table of these variables. The dc analysis differs from the operation-point analysis in that one or two independent sources are swept over a specified range. The corresponding values of node voltages and independent voltage-source currents can be plotted as functions of those sources swept in the dc analysis. In the present example, we must either run two separate operation-point analyses for $V_{D(\min)}$ and $V_{D(\max)}$, or run a single dc analysis sweeping v_D between these two extreme values. We prefer the latter. The Spice schematic of the equivalent circuit of Fig. 2.4(b) is shown in Fig. 2.5. The input file `ex2-1.sp` containing its netlist is as follows:

```
filename: ex2-1.sp              line 1
.param vdd=1.8 vss=0 voh=1.65 vol=0.15 a0=1meg ; line 2
+  vdir=0.25 fgbw=10meg fnd=10g   ; line 2 continuation
.incl opmacro1.sp               ; line 3
x1 1 2 3 opamp                  ; line 4
r1 4 2 10k                      ; line 5
r2 2 3 50.5k                    ; line 6
vb 4 0 0.54                     ; line 7
vd 1 0                          ; line 8
.dc vd 0.476 0.723 1m           ; line 9
.end
```

The second line of the file specifies the parameters of the opamp. As indicated by `vdd=1.8 vss=0`, a single-supply opamp is used. Also note from `voh=1.65` and `vol=0.15` that the output range of the opamp is just as wide as the output swing.

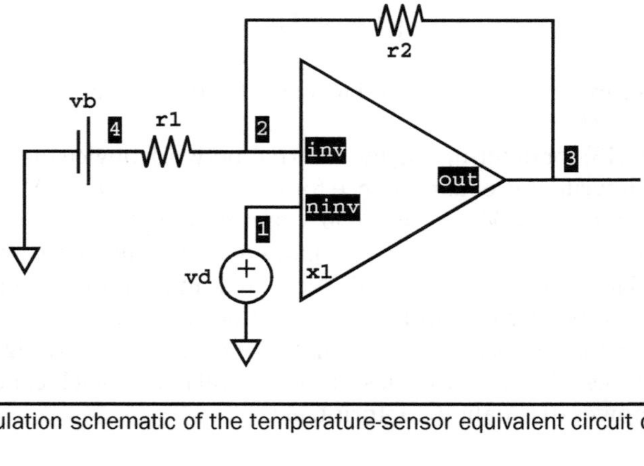

FIGURE 2.5 Simulation schematic of the temperature-sensor equivalent circuit of Example 2.1.

Among the remaining four parameters in line 2, only `a0` is relevant to a first-order dc analysis. This parameter represents the slope of the dc transfer characteristic within the output range. In the ideal opamp assumed throughout the present section this slope should be infinite. The specified `a0=1meg` adequately approximates this assumption. The final two parameters `fgbw` and `fnd` are of no significance in an operation-point analysis or dc analysis. Their values are set arbitrarily.

The third line of the input file includes the opamp macro whose instance is defined in line 4. Lines 5 and 6 define the feedback resistors. We arbitrarily adopt 10-kΩ resistance for R_1. This leads to $R_2 = 50.5$ kΩ since R_2/R_1 has to be 5.05. `vb` and `vd` appearing in lines 7 and 8 define the bias source V_B and diode voltage v_D, respectively.

Also notice the instruction of a dc analysis in line 9. It sweeps `vd` between the expected minimum and maximum levels in 1-mV increments.

For running the simulation, first, change the directory to the folder where `ex2-1.sp` resides. Assuming `c:\spice\chap2` to be the proper folder, enter

```
cd c:\spice\chap2
```

to do that. Next, load the input file by entering

```
ex2-1.sp
```

and start the simulation by entering `run`. Upon completion, entering

```
plot v(3) v(2) v(1)
```

will plot v_O, v_{IN}, and v_{IP}. Notice the virtual equality between v_{IN} and v_{IP}, and verify the limits of the output swing and common-mode swing. ▲

Example 2.2 A closed-loop configuration is qualified as an *inverting* configuration for any input variable that is processed with a negative dc closed-loop gain. If an input variable is processed with a positive dc closed-loop gain, then, the configuration is said to be *noninverting* for that particular input variable. For example, as one can ascertain from (2.13), the configuration shown in Fig. 2.4(b) is inverting for V_B but noninverting for v_D.

Shown in Fig. 2.6 is the most basic noninverting configuration for amplifying a voltage-mode signal source v_i whose source resistance is represented by r_i. V_B is a dc bias-voltage source.

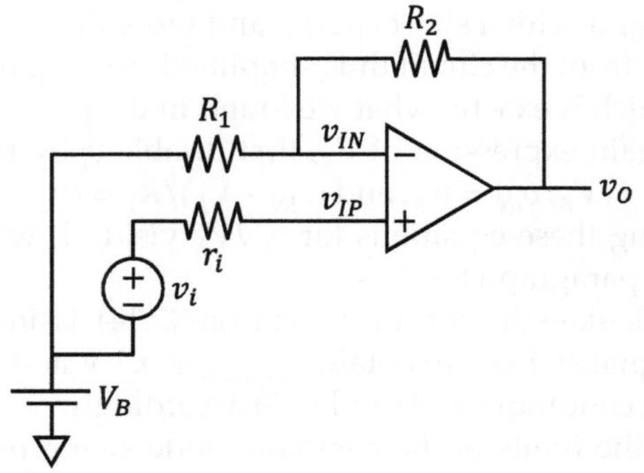

FIGURE 2.6 The noninverting amplifier of Example 2.2.

1. Derive closed-loop dc gain expressions for v_i and V_B.
2. This amplifier is supposed to provide a closed-loop dc gain $A_{cl,i} = 3$ V/V for v_i whose swing is specified by $V_{i(\min)} = -0.2$ V, $V_{i(\max)} = 0.2$ V. Assuming $V_B = 0.9$ V, determine the output-range and common-mode range specifications of the opamp.

Work We can work out part 1 by (a) recognizing $v_{IP} = V_B + v_i$ from the circuit, (b) assuming $v_{IN} = v_{IP}$ as implied by virtual short, and (c) equating the currents of R_1 and R_2 as

$$\frac{v_{IN} - V_B}{R_1} = \frac{v_O - v_{IN}}{R_2}.$$

Solving these equations for v_O yields the closed-loop dc transfer function

$$v_O = \left(1 + \frac{R_2}{R_1}\right) v_i + V_B, \tag{2.14}$$

which indicates closed-loop dc gains $A_{cl,i} = 1 + (R_2/R_1)$ for v_i and $A_{cl,B} = 1$ for V_B. Incidentally, notice the absence of the source resistance r_i of the signal source in gain expressions. This is due to the fact that the purely capacitive internal loading of the noninverting opamp terminal does not admit any dc current through this resistance. Since source resistances are more inaccurate and more imprecise than those of the resistors used intentionally in the external network, we indeed strive for minimizing their effect on closed-loop gain. Apparently, the noninverting configuration offers a perfect solution in this respect.

We now take the opportunity provided by part 1 of this example to expose the reader to the use of *superposition* in the derivation of a closed-loop gain expression and in the construction of the entire closed-loop transfer function whenever necessary. Superposition, which is applicable to any linear circuit of multiple independent sources, offers a simpler and faster derivation in closed-loop dc analysis. It is applied by disabling all input variables except one, and determining the output variable by analyzing the simplified configuration. This process is repeated for all input variables one by one. Summing up the expressions thus derived for all output variables yields the actual output variable in the presence of all input variables. Recall that an electrical variable is disabled by replacing its source with a zero, that is, short-circuiting a voltage source or open-circuiting a current source. For the closed-loop dc gain due to v_i in Fig. 2.6 for example, disable V_B by replacing it with a short circuit, and write directly $v_{IP} = v_i$, $v_{IN} = v_{IP}$ and $v_{IN}/R_1 = (v_O - v_{IN})/R_2$ from the circuit thus simplified. Solving these equations for v_O/v_i yields $1 + (R_2/R_1)$, which is exactly what we found in the preceding paragraph for $A_{cl,i}$. For determining the gain expression of V_B, first disable v_i by replacing it with a short circuit. Then, write $v_{IP} = V_B$, $v_{IN} = v_{IP}$, and $(v_{IN} - V_B)/R_1 = (v_O - v_{IN})/R_2$ from the circuit thus simplified. Solving these equations for v_O/V_B yields 1, which indeed is what we found in the previous paragraph for $A_{cl,B}$.

In part 2, we first calculate the output swing from (2.14). Using the specified values of $A_{cl,i}$, V_B, and input-signal swing, we obtain $V_{O(\min)} = 0.3$ V and $V_{O(\max)} = 1.5$ V. Output range must satisfy the conditions (2.4) and (2.5) accordingly.

Next, we calculate the limits of the common-mode swing from $v_{CM} = v_{IN} = v_{IP} = V_B + v_i$ as $V_{CM(\min)} = 0.7$ V and $V_{CM(\max)} = 1.1$ V. The common-mode range of the opamp must satisfy (2.7) and (2.8) accordingly.

The reader is now invited to verify the output swing and common-mode swing with a dc simulation. Assume a single-supply ideal opamp of $V_{DD} = 1.8$ V; set V_{OL} and V_{OH} to satisfy the output range determined above; and deploy $r_i = 500\ \Omega$, $R_1 = 10\ \text{k}\Omega$, and $R_2 = 20\ \text{k}\Omega$. ▲

Example 2.3 The *unity-gain buffer* configuration shown in Fig. 2.7 is to be built with an opamp of the following common-mode range and output-range specifications: $V_{IL} = 0.7$ V, $V_{IH} = 2$ V, $V_{OL} = 0.2$ V, $V_{OH} = 1.6$ V. The input signal v_i varies between 0 and 0.5 V. Determine the viable range of bias voltage V_B for this application.

Work Unity-gain buffer is a special case of the noninverting configuration of Fig. 2.6 for $R_2 = 0$ and $R_1 = \infty$. Notice $v_O = v_{IN}$ from the schematic, and $v_{IN} = v_{IP} = V_B + v_i$ due to virtual short. Therefore

$$v_{CM} = v_O = v_i + V_B,$$

which indicates a closed-loop gain of unity for both v_i and V_B. Considering the specified limits of v_i, this equation leads us to

$$V_{CM(\text{min})} = V_{O(\text{min})} = V_B,$$

and

$$V_{CM(\text{max})} = V_{O(\text{max})} = 0.5 + V_B.$$

Using these expressions in (2.7) and (2.8) for the common-mode range and in (2.4) and (2.5) for the output range, and considering the specified limits of the two range metrics, we obtain

$$V_{IL} = 0.7\,\text{V} < V_{CM(\text{min})} = V_B,$$

$$V_{IH} = 2\,\text{V} > V_{CM(\text{max})} = 0.5 + V_B,$$

$$V_{OL} = 0.2\,\text{V} < V_{O(\text{min})} = V_B,$$

and

$$V_{OH} = 1.6\,\text{V} > V_{O(\text{max})} = 0.5 + V_B.$$

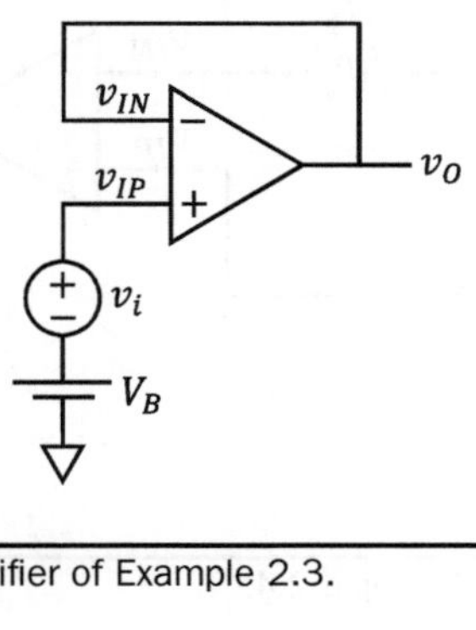

FIGURE 2.7 Unity-gain buffer amplifier of Example 2.3.

Therefore, V_B is constrained by the common-mode range through

$$0.7\,\text{V} < V_B < 1.5\,\text{V},$$

and by the output range through

$$0.2\,\text{V} < V_B < 1.1\,\text{V}.$$

The viable range of V_B satisfying both constraints is therefore described by $0.7\,\text{V} < V_B < 1.1\,\text{V}$.

The reader is now invited to verify the result with dc simulations. Assume a single-supply ideal opamp of $V_{DD} = 1.8$ V; set V_{OL} and V_{OH} to their specified values; and, repeat the simulation for different values of V_B inside and outside the viable range determined above. Also verify the unity closed-loop gain. ▲

Example 2.4 Shown in Fig. 2.8 is the most basic inverting configuration for amplifying a voltage-mode signal v_i. r_i represents the source resistance of the signal source. V_B is a dc bias-voltage source.

1. By applying superposition, derive closed-loop dc gain expressions for v_i and V_B and construct the closed-loop dc transfer function.
2. Determine the output-range and common-mode range specifications of the opamp for the following specifications: $R_2 = 10$ kΩ, $R_1 = 2.5$ kΩ, $r_i = 100$ Ω, $V_B = 0.9$ V, $V_{i(\text{min})} = -0.15$ V, $V_{i(\text{max})} = 0.15$ V.

Work First, consider the gain for v_i. By shorting out V_B, assuming virtual short, and equating the current of R_2 to that of R_1 and r_i, we obtain

$$\frac{0 - v_i}{R_1 + r_i} = \frac{v_O - 0}{R_2},$$

whose solution for v_O yields the closed-loop gain $-R_2/(R_1 + r_i)$ for v_i. Next, we consider the gain for V_B. By shorting out v_i, assuming virtual short, and equating the current of

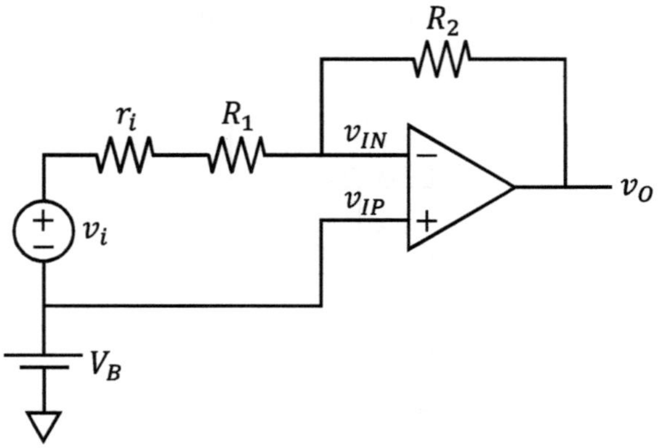

FIGURE 2.8 Inverting amplifier of Example 2.4.

R_2 to that of R_1 and r_i, we obtain

$$\frac{V_B - V_B}{R_1 + r_i} = \frac{v_O - V_B}{R_2},$$

whose solution for v_O yields the closed-loop gain 1 for V_B. Superposition leads to the following closed-loop dc transfer function:

$$v_O = -\frac{R_2}{R_1 + r_i} v_i + V_B. \tag{2.15}$$

In part 2, we calculate $V_{O(\min)} = 0.323$ V and $V_{O(\max)} = 1.48$ V from (2.15). As usual, output range must comply with (2.4) and (2.5) for these limits of the output swing. Next, notice from the schematic that the common-mode voltage is fixed at $v_{CM} = V_B$ regardless of the value of v_i. Any common-mode range including $V_B = 0.9$ V is therefore good enough for this application.

As a final note, notice the presence of r_i in the closed-loop gain expression for v_i. While a dependence on a source resistance is inevitable in an inverting configuration, we can reduce its effect by selecting large external-network resistances R_1 and R_2.

Simulated verification is again left to the reader as an exercise. ▲

Example 2.5 The transresistance amplifier shown in Fig. 2.9 is to provide a closed-loop gain of $A_{cl,i} = -50$ kΩ for the current-mode signal i_i varying between 0 and 10 μA. V_B is specified as 1 V. Determine the resistance R as well as the common-mode and output ranges.

Work Since $v_{IP} = V_B = 1$ V, virtual short implies $v_{IN} = 1$ V, hence $v_{CM} = 1$ V. Therefore, any common-mode range including 1 V is acceptable for this application. Now equate the signal current to the current flowing in R:

$$i_i = \frac{v_{IN} - v_O}{R}.$$

Since $v_{IN} = V_B$, this expression yields the closed-loop dc transfer function

$$v_O = -Ri_i + V_B, \tag{2.16}$$

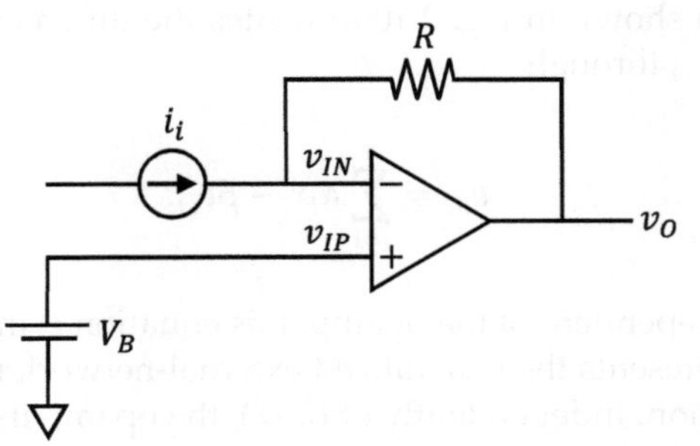

FIGURE 2.9 Transresistance amplifier of Example 2.5.

which indicates a closed-loop dc gain $A_{cl,i} = -R$ for i_i, and a closed-loop dc gain $A_{cl,B} = 1$ for V_B. The specified value of $A_{cl,i}$ implies $R = 50$ kΩ. Considering also the specified value of V_B and the limits of i_i, (2.16) yields $V_{O(\max)} = 1$ V and $V_{O(\min)} = 0.5$ V. Output range should be wide enough to accommodate these two extremes of the output swing.

The reader will benefit from simulating this application too. Of course, the upstream node of the current source should be tied to ground or any dc voltage-source such as V_B. ▲

2.3 Unified Closed-Loop Model

Virtual-short approximation is a very a powerful tool enabling us to analyze the first-order dc response of any closed-loop configuration. Regardless of the complexity of the external network, the analytical procedure is simple, quick, and error-free even for a novice designer. However, this approximation is invalid in the cases of accurate dc response, frequency response, and step response. Modeling of these more sophisticated responses necessitates more complex opamp models, and therefore involves tedious and potentially error-prone circuit analysis. Fortunately, we don't need to face this complexity each time we deal with a specific application because all closed-loop configurations can be represented with one unified closed-loop transfer function for each of these responses. Unification is done by combining a generalized external-network model with a response-specific open-loop transfer function. In this section, we present the generalized external-network model and apply it to the unification of the first-order closed-loop dc transfer function as a simple (and useful) example. Extension to accurate dc response, frequency response, and step response will be presented in Sections 2.4, 2.5, and 2.6, respectively.

2.3.1 A Generalized External-Network Model

Single-ended continuous-time closed-loop configurations deploying linear external-network components can be modeled by the unified block diagram shown in Fig. 2.10(a), where v_O stands for the output voltage and x_j, $j = 1, \ldots, n$, denote the voltage or current sources of input signals or bias. The reader is referred to [1] for the origins of this model. In addition to the opamp, the diagram includes two types of blocks representing the *feedback factor* β and *input factors* α_j, $j = 1, \ldots, n$. All of these factors are defined solely by the external network. β has to be positive in order to establish a negative-feedback loop. α_j may be positive or negative depending on the type of the configuration.

The block diagram shown in Fig. 2.10(a) relates the differential-input voltage v_{ID} of the opamp to x_j and v_O through

$$v_{ID} = \sum_{j=1}^{n} \alpha_j x_j - \beta v_O. \tag{2.17}$$

Since β and α_j are independent of the opamp, this equation is imposed solely by the external network. It represents the generalized external-network model mentioned at the beginning of this section. Independently of (2.17), the opamp itself also relates v_O to v_{ID} through its own open-loop transfer function. By eliminating v_{ID} between the open-loop transfer function and (2.17), we therefore obtain a unified closed-loop transfer function

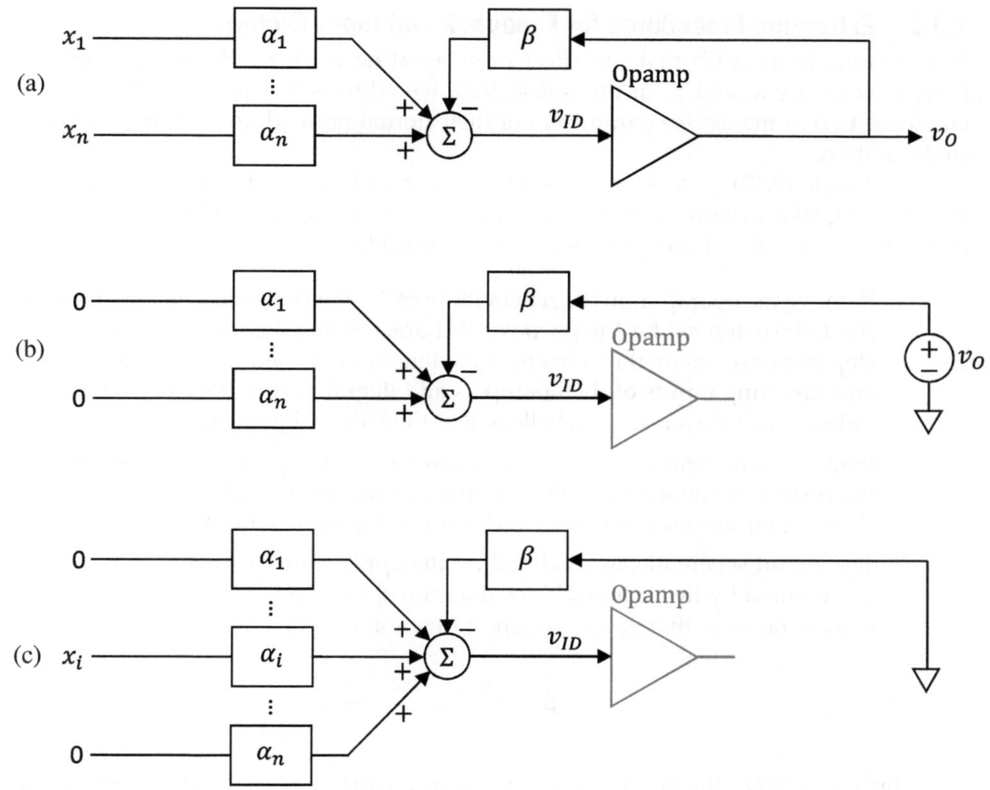

FIGURE 2.10 (a) Unified block diagram of single-ended closed-loop configurations. (b) Characterization of feedback factor β with short-circuited voltage-mode input sources and open-circuited current-mode input sources. (c) Characterization of an input factor α_i with the short-circuited output port.

relating v_O to x_j for all single-ended continuous-time closed-loop configurations. However, as mentioned earlier in this section, open-loop transfer function is response specific. We will develop one in Section 2.4 specifically for the accurate modeling of dc response, and another in Section 2.5 specifically for modeling the frequency and step responses.

Equation (2.17), being so general, is of course applicable also to the simplest case of first-order dc response. In this case, the open-loop transfer function to be combined with (2.17) is (2.3) of the virtual-short approximation. This combination leads to the unified closed-loop dc transfer function

$$v_O = \sum_{j=1}^{n} \frac{\alpha_j}{\beta} x_j, \tag{2.18}$$

which indeed yields (2.10), and indicates a closed-loop dc gain

$$A_{cl,i} = \frac{\alpha_i}{\beta} \tag{2.19}$$

for an input variable x_i.

2.3.2 Extraction Procedures for Feedback and Input Factors

It is obvious from (2.17) that the effect of external network on closed-loop behavior is represented by α_j and β. In this subsection, we address the question of how to relate these two to the device parameters of the external network in a given closed-loop configuration.

According to (2.17), β can be extracted from the circuit schematic of the configuration by (a) setting all x_j to zero, (b) applying a voltage v_O to the output, and (c) evaluating the ratio $-v_{ID}/v_O$. Details of this procedure are explained below.

1. Remove the opamp from the schematic because it has no effect on β. If, however, β is to be extracted for the purpose of characterizing the frequency response or step response, retain the parasitic impedances associated with the noninverting and inverting inputs of the opamp as an integral part of the external network unless you have a reason to believe that they are insignificant.
2. Force an independent test voltage source v_O to the port where the output of the opamp is connected in the original configuration, and set all x_j to zero by shorting input-voltage sources and opening input-current sources.
3. The circuit schematic, as modified by the application of steps 1 and 2, is now represented by the reduced block diagram shown in Fig. 2.10(b). Analyze it to determine $-v_{ID}$, that is, $v_{IN} - v_{IP}$, as a function of v_O. β is given by

$$\beta = -\frac{v_{ID}}{v_O} = \frac{v_{IN} - v_{IP}}{v_O}. \tag{2.20}$$

According to (2.17), the input factor α_i of an input variable x_i can be extracted from the circuit schematic of the configuration by setting (a) all x_j except x_i to zero, (b) setting v_O to zero, and (c) evaluating the ratio v_{ID}/x_i, as detailed below.

1. Remove the opamp from the schematic because it has no effect on α_i. If, however, α_i is to be extracted for the purpose of characterizing the frequency response or step response, retain the parasitic impedances associated with the noninverting and inverting inputs of the opamp as an integral part of the external network unless you have a reason to believe that they are insignificant.
2. Short-circuit the port where the output of the opamp is connected in the original configuration, and set all x_j except x_i to zero by shorting voltage sources and opening current sources. Retain x_i as an input test source.
3. The circuit schematic, as modified by the application of steps 1 and 2, is now representable by the reduced block diagram shown in Fig. 2.10(c). Analyze it to determine v_{ID}, that is, $v_{IP} - v_{IN}$ as a function of x_i. α_i is given by

$$\alpha_i \doteq \frac{v_{ID}}{x_i} = \frac{v_{IP} - v_{IN}}{x_i}. \tag{2.21}$$

Example 2.6 Extract the feedback and input factors of the temperature sensor of Fig. 2.4(b), noninverting amplifier of Fig. 2.6, unity-gain buffer of Fig. 2.7, inverting amplifier of Fig. 2.8, and transresistance amplifier of Fig. 2.9, and construct the closed-loop transfer function of each for first-order dc response.

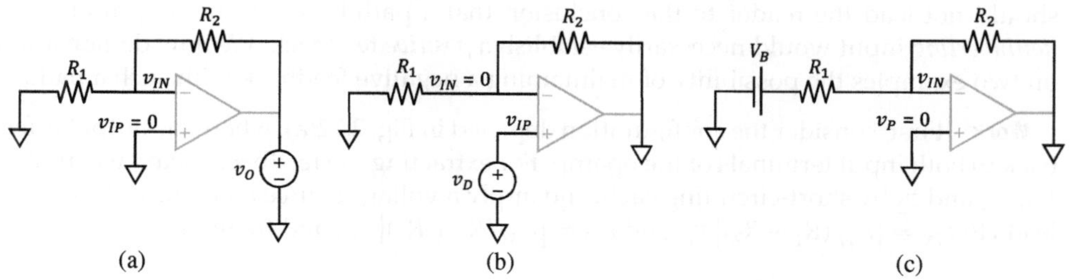

FIGURE 2.11 Reduced schematics for extracting the feedback and input factors of the temperature sensor of Fig. 2.4. (a) Feedback factor. (b) Input factor of v_D. (c) Input factor of V_B.

Work

Temperature Sensor The reduced schematic for β-extraction is shown in Fig. 2.11(a). Compare with Fig. 2.4(b), and notice the following differences: (a) opamp is removed, (b) each voltage source is replaced by a short circuit, and (c) a voltage source v_O is applied to the output port.

$v_{IP} = 0$ is immediately obvious from Fig. 2.11(a). v_{IN} is determined by applying the voltage divider rule to R_1 and R_2. The outcome $v_{IN} = [R_1/(R_1 + R_2)]v_O$ in addition to $v_{IP} = 0$ implies $v_{ID} = -[R_1/(R_1 + R_2)]v_O$, which yields $\beta = R_1/(R_1 + R_2)$ from (2.20).

The reduced schematic for extracting the input factor α_D of the voltage source v_D is shown in Fig. 2.11(b). Notice the shorting of V_B and the output port. Since $v_{IP} = v_D$ and $v_{IN} = 0$, (2.21) yields $\alpha_D = 1$.

The reduced schematic for extracting the input factor α_B of the voltage source V_B is shown in Fig. 2.11(c). Notice the shorting of v_D as well as the output port. Since $v_{IP} = 0$ and $v_{IN} = [R_2/(R_1 + R_2)]V_B$, (2.21) yields $\alpha_B = -R_2/(R_1 + R_2)$.

Closed-loop transfer function for first-order dc response: By substituting the feedback and input factors found from Fig. 2.11 into (2.18), we arrive at the very same closed-loop transfer function described earlier by (2.13).

Noninverting Amplifier Feedback factor: $\beta = R_1/(R_1 + R_2)$. Input factor for v_i: $\alpha_i = 1$. Input factor for V_B: $\alpha_B = R_1/(R_1+R_2)$. By substituting these feedback and input factors into (2.18), we obtain the very same transfer function described by (2.14).

Unity-Gain Buffer Feedback factor: $\beta = 1$. Input factor for v_i: $\alpha_i = 1$. Input factor for V_B: $\alpha_B = 1$. Substituting these feedback and input factors into (2.18) yields the transfer function $v_O = v_i + V_B$.

Inverting Amplifier Feedback factor: $\beta = (R_1+r_i)/(R_1+r_i+R_2)$. Input factor for v_i: $\alpha_i = -R_2/(R_1+r_i+R_2)$. Input factor for V_B: $\alpha_B = (R_1+r_i)/(R_1+r_i+R_2)$. By substituting these feedback and input factors into (2.18), we arrive at the very same transfer function described by (2.15).

Transresistance Amplifier Feedback factor: $\beta = 1$. Input factor for i_i: $\alpha_i = -R$. Input factor for V_B: $\alpha_B = 1$. By substituting these feedback and input factors into (2.18), we obtain the very same transfer function described by (2.16). ▲

Example 2.7 The feedback loop of the configurations we worked on in Example 2.6 is closed by a path between the output and inverting input of the opamp. This observation should not lead the reader to the conclusion that a path between the output and the *noninverting* input would necessarily establish a *positive* feedback. We now demonstrate on two examples the possibility of maintaining a negative feedback with such a path.

Work First, consider the configuration depicted in Fig. 2.12(a), where the output is fed back to both input terminals of the opamp. For extracting the feedback factor, we disable V_B, v_1, and v_2 by short-circuiting each, and apply a voltage source v_O to the output. This leads to $v_{IN} = [R_1/(R_1 + R_2)]\, v_O$ and $v_{IP} = [R_3/(R_3 + R_4)]\, v_O$, and therefore

$$\beta = \frac{R_1R_4 - R_2R_3}{(R_1 + R_2)(R_3 + R_4)}.$$

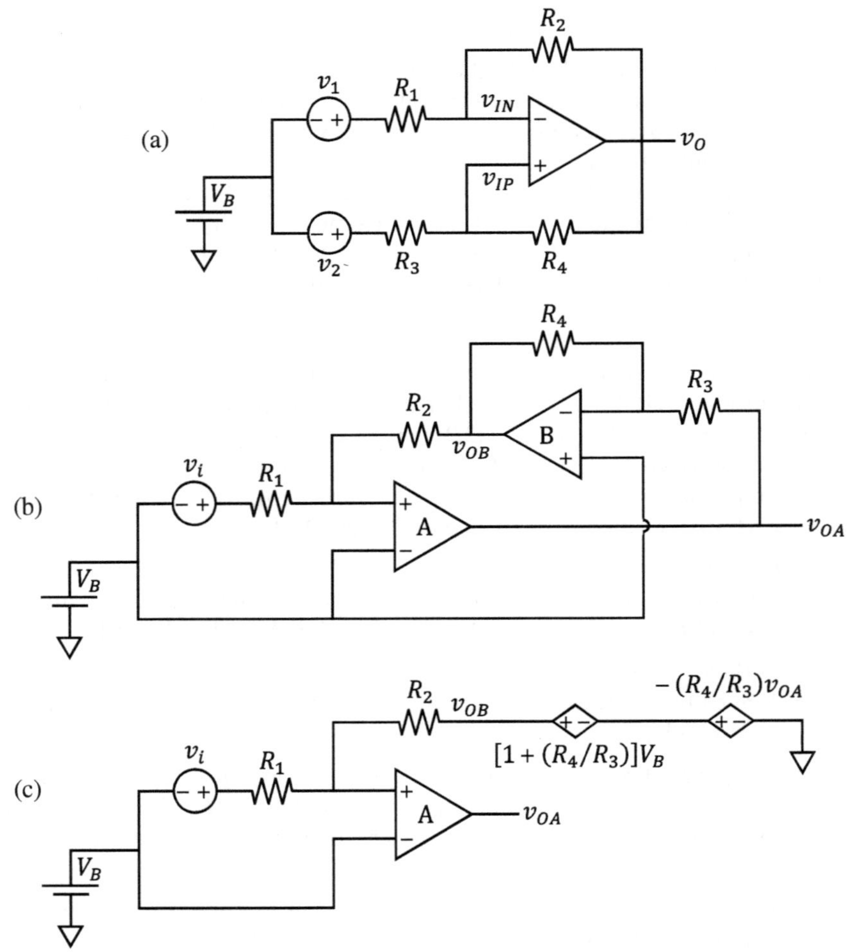

FIGURE 2.12 Closed-loop configurations of Example 2.7. (a) A configuration with feedback to both inputs of the opamp. (b) A configuration with an active feedback loop driving the noninverting input of opamp-A. (c) Replacing the inverting amplifier of part (b) by an equivalent circuit.

The feedback polarity of this configuration is negative provided that the condition $R_1R_4 > R_2R_3$ of a positive β is satisfied. The reader is invited to extract also the input factors, and arrive at

$$v_O = \frac{R_4(R_1 + R_2)}{R_1R_4 - R_2R_3}v_2 - \frac{R_2(R_3 + R_4)}{R_1R_4 - R_2R_3}v_1 + V_B$$

for the dc transfer function.

Next, consider the configuration depicted in Fig. 2.12(b), and recognize the combination of opamp-B and resistances R_3 and R_4 as an inverting amplifier. It is also apparent that this amplifier and resistances R_1 and R_2 form an *active* feedback path for opamp-A. This path starts at the output terminal of voltage v_{OA} and ends up at the *noninverting* input terminal. Yet, it is expected to yield a negative feedback due to the inclusion of the inverting amplifier. We now test this expectation by extracting the factor β_A of the loop across opamp-A. The first step we take to this end is to relate the output voltage v_{OB} of the inverting amplifier to v_{OA}. We do this by considering v_{OA} and V_B as the input variables and v_{OB} as the output voltage of this amplifier. Extracting the feedback and input factors, and combining in (2.18), we obtain

$$v_{OB} = \left(1 + \frac{R_4}{R_3}\right)V_B - \frac{R_4}{R_3}v_{OA}.$$

In the main schematic of Fig. 2.12(b), we now replace the inverting amplifier by an equivalent circuit representing this relationship, and therefore arrive at the schematic shown in Fig. 2.12(c). The feedback factor β_A is extracted from this circuit by (a) short-circuiting v_i and V_B, and (b) applying a voltage source v_{OA} to the output. Since the V_B-dependent voltage source inside the feedback path is also replaced by a short circuit due to $V_B = 0$, the reduced schematic yields

$$\beta_A = \frac{R_1}{R_1 + R_2}\frac{R_4}{R_3}$$

for the feedback factor of opamp-A. Since β_A is positive, opamp-A indeed operates with a negative feedback. Figure 2.12(c) also yields the input factors

$$\alpha_B = \frac{R_1}{R_1 + R_2}\frac{R_4}{R_3},$$

$$\alpha_i = \frac{R_2}{R_1 + R_2},$$

and therefore enables us to construct the following expression of v_{OA} from (2.18):

$$v_{OA} = V_B + \frac{R_2R_3}{R_1R_4}v_i. \quad \blacktriangle$$

2.4 Accurate Modeling of DC Response

As we observed in Example 2.6, (2.18) relates the dc output voltage of a closed-loop configuration to input variables through external-network resistances only. Designers rely on this equation for determining the target values of these resistances from the specified values of closed-loop dc gain metrics. The resistance values thus determined are then used as device metrics in resistor design for the external network. However, no single fabricated copy of the closed-loop configuration built with these resistors will deliver the specified values of the dc gain metrics exactly. One obvious reason for this deviation is the simplifying assumption of virtual short in (2.18). Secondly, since it is impossible to fabricate a resistor with perfect accuracy and precision, a deviation is inevitable regardless of how faithful the opamp model used in design is. There exist many applications, most notably in analog-to-digital conversion, in which this deviation should be confined by design into specified limits. This calls for an accurate modeling of the dc response of closed-loop configurations, and a detailed understanding of the limits of accuracy and precision available from external-network components.

We characterize the deviation of the closed-loop dc response from its target with what we call *dc error* throughout this book. It is defined as the difference between the measured and targeted values of the dc output voltage for a specified set of input variables. The mean value of this error, as determined statistically over a large number of fabricated copies of the closed-loop configuration, is a closed-loop metric by which the *accuracy* of the application is quantified. The standard deviation of its distribution is yet another closed-loop metric; it quantifies the *precision* of the application. In this section, we develop a model that accounts for the imperfections of the dc response, and use it as a tool for controlling the dc error by design.

2.4.1 Open-Loop and Closed-Loop DC Transfer Functions

A realistic open-loop dc transfer characteristic is shown in Fig. 2.13. The horizontal axis is greatly scaled up in comparison with the vertical axis in order to highlight the imperfections of the characteristic. We observe three such imperfections within the output range

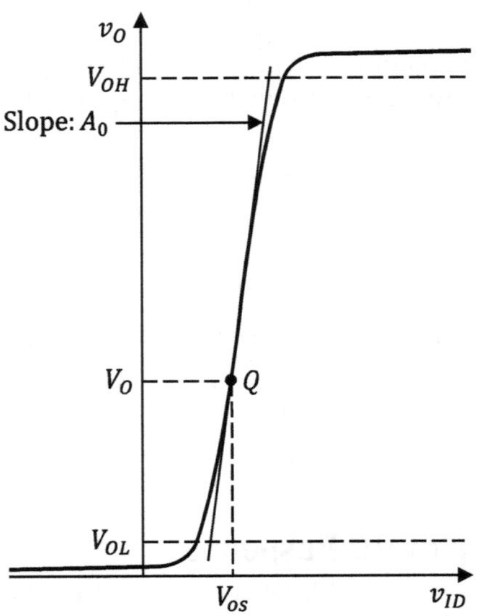

FIGURE 2.13 A realistic open-loop dc transfer characteristic.

between V_{OL} and V_{OH}. Namely, a mild nonlinearity, a finite slope, and a shift along the horizontal axis. Considering the mildness of the nonlinearity, we presently approximate the characteristic with a linear segment of slope A_0 but will return to nonlinearity in Subsection 2.5.3 where the ac concept of harmonic distortion will be discussed. A_0 is called *dc differential gain* throughout this book. Due to the shift along the horizontal axis, the output bias voltage V_O is established for $v_{ID} = V_{os}$ instead of $v_{ID} = 0$. V_{os} is called *input offset voltage*. As explained in Section 2.1, V_{os} and the expected output bias V_O for which it is specified are represented with parameters `vos` and `vout` in the input file `opmacro1.sp` of our single-ended opamp macro.[2]

Considering the imperfections described above, we can express the open-loop dc transfer function as

$$v_O = V_O + A_0\left(v_{ID} - V_{os}\right). \tag{2.22}$$

We now eliminate v_{ID} between (2.22) and the generalized external-network expression (2.17), and thus obtain the following accurate form of the closed-loop dc transfer function

$$v_O = \frac{\beta A_0}{1+\beta A_0}\sum_{j=1}^{n}\frac{\alpha_j}{\beta}x_j + \frac{V_O}{1+\beta A_0} - \frac{A_0 V_{os}}{1+\beta A_0}. \tag{2.23}$$

Note that if we replace A_0 with infinity and V_{os} with zero, and therefore eliminate opamp imperfections, this equation will indeed turn into (2.18) of the first-order dc response.

2.4.2 DC Error

Suppose that an application is designed for a feedback factor β and input factors α_j. Its target output voltage for a given set x_j, $j = 1, \ldots, n$, of dc input variables is described by (2.18). Now, consider a fabricated copy of the application whose feedback and input factors turn out to be β' and α'_j. The actual output voltage v'_O of this copy is obtained from (2.23) after replacing v_O with v'_O, β with β', and α_j with α'_j. By subtracting (2.18) from the resulting equation, we obtain the following expression of dc error:

$$\Delta v_O \doteq v'_O - v_O = \frac{\beta' A_0}{1+\beta' A_0}\sum_{j=1}^{n}\left(\frac{\alpha'_j}{\beta'} - \frac{\alpha_j}{\beta}\right)x_j - \frac{v_O - V_O}{1+\beta' A_0} - \frac{A_0 V_{os}}{1+\beta' A_0}.$$

We now introduce two approximations to make this equation more design friendly. First, we replace $1+\beta' A_0$ with $\beta' A_0$ because the latter is many orders of magnitude greater than unity. Next, β' appearing in the denominator of the second and third error components

[2] As also stated in Section 2.1, unless specifically reset by the user, the default values of `vos` and `vout` in `opmacro1.sp` are `vos=` 0 and `vout=` `(voh+vol)`/2. If an offset-free opamp is to be simulated for an expected output bias voltage different from the default, then, the user should reset only `vout`. If an offset is to be added to an opamp that is expected to operate at the default output bias, then, only `vos` should be reset. If an offset has to be specified at an output bias other than the default, then, both `vos` and `vout` should be reset. However, in an application where the dc differential gain parameter `a0` is larger than several tens of thousands, there is no need to reset `vout` because the dc transfer characteristic will be steep enough to practically eliminate the sensitivity of V_{os} to V_O.

is replaced by β at the expense of a relative error $(\beta' - \beta)/\beta$ which does not exceed a few percent in practice. We also recognize the difference $v_O - V_O$ as the target value of the *output signal*, and represent it with v_o. Finally, we represent α'_j/β' and α_j/β with the closed-loop gain symbols $A'_{cl,j}$ and $A_{cl,j}$, respectively. These transformations lead to

$$\Delta v_O = E_{\text{net}} + E_{\text{gain}} + E_{os}, \tag{2.24}$$

where

$$E_{\text{net}} \doteq \sum_{j=1}^{n} \left(A'_{cl,j} - A_{cl,j}\right) x_j \tag{2.25}$$

is called *external-network error*,

$$E_{\text{gain}} \doteq -\frac{v_o}{\beta A_0} \tag{2.26}$$

is called *gain error*, and

$$E_{os} \doteq -\frac{V_{os}}{\beta} \tag{2.27}$$

is called *offset error*. Next, we elaborate on each of these three components with a view to develop a methodology by which we can design for a specified limit on dc error.

External-Network Error

External-network error is due to the deviation of closed-loop dc gains $A'_{cl,j}$, $j = 1, \ldots, n$, from their respective design targets $A_{cl,j}$. Since gains are determined solely by external-network resistances, the root cause of this error component is the imperfection of integrated resistors. Indeed, the resistance measured on the fabricated copies of a resistor will exhibit a Gaussian random distribution of variance $\sigma^2[R]$ around a mean value $\overline{R}$, and generally speaking, even this mean value will shift from the resistance R targeted in design. Both $\overline{R}$ and $\sigma^2[R]$ are predictable, and, as to be detailed in Section 3.10, there exist various electrical and physical design techniques by which we can control them. Assuming an external network of m resistors of resistance R_k, $k = 1, \ldots, m$, the mean value $\overline{E_{\text{net}}}$ and variance $\sigma^2[E_{\text{net}}]$ of the external-network error can be related to the mean values and variances of these resistances by applying the *delta method* of statistics. This method approximates the variance $\sigma^2[y]$ of a function $y = f(x_1, \ldots, x_m)$ of m uncorrelated random variables $x_1, \ldots, x_m$ with

$$\sigma^2[y] \approx \left(\frac{\partial y}{\partial x_1}\right)^2 \sigma^2[x_1] + \cdots + \left(\frac{\partial y}{\partial x_m}\right)^2 \sigma^2[x_m], \tag{2.28}$$

where $\sigma^2[x_1], \ldots, \sigma^2[x_m]$ are the variances of the variables. Its application to (2.25) yields

$$\sigma^2[E_{net}] \approx \sum_{k=1}^{m} \left[\frac{\partial}{\partial R_k} \sum_{j=1}^{n} A'_{cl,j} x_j \right]^2 \sigma^2[R_k], \tag{2.29}$$

which prescribes the following procedure for determining $\sigma^2[E_{net}]$:

1. Construct the first-order dc transfer function (2.18) in terms of external-network resistances.
2. Differentiate the transfer function with respect to one of the resistances, and multiply the square of the derivative by the variance of the resistance.
3. Repeat step 2 for each of the remaining resistances.
4. The sum of the outcomes of steps 2 and 3 equals the variance of the external-network error.

The mean value $\overline{E_{net}}$ of the network error, on the other hand, is given by

$$\overline{E_{net}} = \sum_{j=1}^{n} \left(\overline{A'_{cl,j}} - A_{cl,j} \right) x_j, \tag{2.30}$$

where the mean value $\overline{A'_{cl,j}}$ of closed-loop gain for input x_j is well approximated by substituting $\overline{R_k}$ for R_k in the expression of $A_{cl,j}$.

We now exemplify the use of (2.29) and (2.30) on three of the closed-loop configurations involved in previous examples.

First, consider the transresistance amplifier of Example 2.5, whose first-order closed-loop dc transfer function is described by (2.16). The variance and mean of its external-network error are determined from (2.29) and (2.30) as $\sigma^2[E_{net}] = i_i^2\sigma^2[R]$ and $\overline{E_{net}} = -(\overline{R} - R)i_i$. Note that the mean value of the external-network error is proportional to the shift in the mean value of the resistance, which is typically on the order of tens of a percent in relative terms. Considering also the presence of a distribution around the mean, we conclude that a pretty large external-network error may be associated with a transresistance amplifier.

Next, consider the noninverting amplifier of Example 2.2. Its first-order closed-loop dc transfer function, given by (2.14), yields

$$\sigma^2[E_{net}] \approx \left(\frac{R_2}{R_1} \right)^2 \left(\frac{\sigma^2[R_2]}{R_2^2} + \frac{\sigma^2[R_1]}{R_1^2} \right) v_i^2 \tag{2.31}$$

from (2.29), and

$$\overline{E_{net}} = \left[\overline{\left(\frac{R_2}{R_1} \right)} - \frac{R_2}{R_1} \right] v_i \tag{2.32}$$

from (2.30). As will be discussed in Subsection 3.10.2, there exist a number of physical design techniques by which the mean value of the ratio of two resistances can be implemented very accurately. The mean value of the network error can thus be virtually eliminated in a noninverting amplifier. Obviously, any closed-loop configuration whose gains are functions of resistance ratios enjoys this property. The inverting amplifier of Example 2.4 is no exception provided that the source resistance r_i is negligible in comparison with R_1.

Finally, consider the unity-gain buffer of Example 2.3, which is characterized by the first-order closed-loop dc transfer function $v_o = v_i + V_B$. As expected from an external network containing only a short circuit, (2.29) and (2.30) yield $\sigma^2[E_{net}] = 0$ and $\overline{E_{net}} = 0$, hence no external-network error.

Gain Error

Gain error E_{gain} is due to the fact that the dc differential gain A_0 of the open-loop dc transfer characteristic is finite. Actually, A_0 is also randomly distributed among the fabricated copies of the opamp, but we ignore this distribution because its effect on E_{gain} is secondary to the finite value of A_0. By adopting the values targeted in design for β and A_0, we therefore treat E_{gain} as a deterministic entity. Since the two are positive-valued parameters, the polarity indicated by (2.26) for E_{gain} is definitely the opposite of the polarity of the output signal v_o. Also note that, since β is usually constrained by the target value of the closed-loop gain through (2.19), A_0 is left as the sole instrument by which this error can be controlled by design. Obviously, the larger the A_0, the smaller the gain error.

Offset Error

Offset error E_{os} arises from a shift V_{os} of the open-loop transfer characteristic along the differential-input voltage axis at the operation point. V_{os} exhibits a Gaussian distribution of variance $\sigma^2[V_{os}]$ around a mean value $\overline{V_{os}}$. The extent of the resulting distribution in offset error is characterized by variance $\sigma^2[E_{os}]$, which can be related to $\sigma^2[V_{os}]$ by applying (2.28) to (2.27):

$$\sigma^2[E_{os}] = \frac{\sigma^2[V_{os}]}{\beta^2}. \tag{2.33}$$

The mean value $\overline{E_{os}}$ of offset error is described by

$$\overline{E_{os}} = -\frac{\overline{V_{os}}}{\beta}, \tag{2.34}$$

which, very much like the external-network error, can be virtually eliminated by following certain principles in the electrical and physical design of the opamp.

Designing for a Specified DC Error

The dc error described by (2.24) exhibits a random distribution around a mean value among all fabricated copies of an application. In some copies, it definitely will exceed what the application can tolerate. Since such copies are disqualified, production yield is inevitably lesser than 100%. Excessive yield loss can be prevented by design but this calls for an understanding of the relationship between the yield and dc-error parameters. Next, we develop a model for this relationship.

Since yield Y is given by the probability of Δv_O being inside a tolerable range $[-a, a]$, we can quantify it with

$$Y = \int_{-a}^{a} f(\Delta v_O)\, \mathrm{d}\Delta v_O, \tag{2.35}$$

where $f(\Delta v_O)$ denotes the probability-density function of the dc error. Assuming a Gaussian distribution for the dc error, and representing its standard deviation and mean with $\sigma[\Delta v_O]$ and $\overline{\Delta v_O}$, respectively, we can describe $f(\Delta v_O)$ with

$$f(\Delta v_O) = \frac{1}{\sigma[\Delta v_O]\sqrt{2\pi}} \exp\left\{-\frac{1}{2}\left[\frac{\Delta v_O - \overline{\Delta v_O}}{\sigma[\Delta v_O]}\right]^2\right\},$$

which is plotted qualitatively in Fig. 2.14(a). Integrating this function in accordance with (2.35) yields the area under the curve between $\Delta v_O = -a$ and $\Delta v_O = a$, which is analytically described by

$$Y = \frac{1}{2}\mathrm{erf}\left\{\frac{a - \overline{\Delta v_O}}{\sqrt{2}\ \sigma[\Delta v_O]}\right\} - \frac{1}{2}\mathrm{erf}\left\{\frac{-a - \overline{\Delta v_O}}{\sqrt{2}\ \sigma[\Delta v_O]}\right\}, \tag{2.36}$$

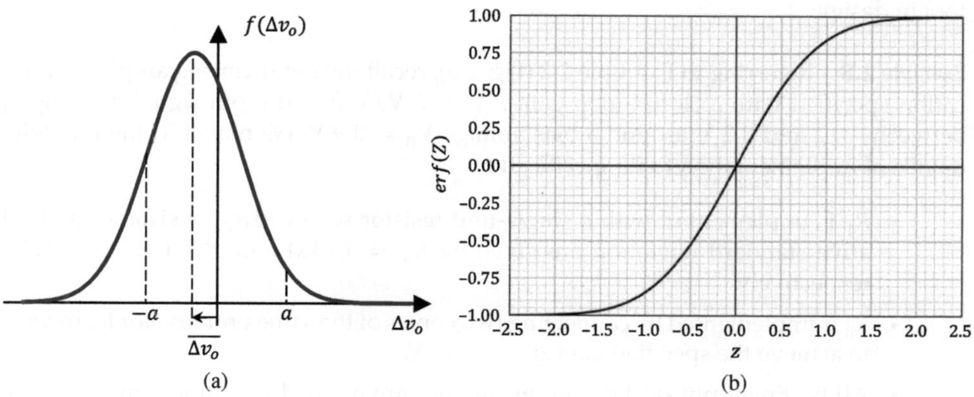

FIGURE 2.14 (a) Gaussian probability-density function of total output-signal error. $\overline{\Delta v_O}$ represents mean error due to systematic causes. The probability of total output signal being inside the interval $[-a, a]$ is given by the integral of the curve from $-a$ to a. (b) Error function.

where

$$\text{erf}(z) \doteq \frac{2}{\sqrt{\pi}} \int_0^z \exp\left(-t^2\right) dt,$$

is the error function whose variation with the argument z is shown in Fig. 2.14(b). It is obvious from (2.36) that Y is determined by three parameters, namely, the mean value $\overline{\Delta v_O}$ and standard deviation $\sigma[\Delta v_O]$ of dc error, and the specified limit a of tolerable error.

$\overline{\Delta v_O}$ is a metric by which we quantify the *accuracy* of an application. We model it by summing up the mean values of the three error components appearing in (2.24), that is,

$$\overline{\Delta v_O} = \overline{E_{\text{net}}} + E_{\text{gain}} + \overline{E_{os}}, \tag{2.37}$$

where the three terms on the right-hand side are described by (2.30), (2.26), and (2.34), respectively.

$\sigma[\Delta v_O]$ is a metric by which we quantify the *precision* of an application. Since gain error has no random component, this metric is due to network and offset errors only. Considering (2.28), it can be modeled by

$$\sigma[\Delta v_O] = \sqrt{\sigma^2[E_{\text{net}}] + \sigma^2[E_{os}]}, \tag{2.38}$$

where the terms on the right-hand side are determined from (2.31) and (2.34), respectively.

As concluded earlier from (2.36), production yield is controlled by $\overline{\Delta v_O}$, $\sigma[\Delta v_O]$, and a. While the last of these three is usually derived from upper-level specifications of the application in hand, the first two are controllable by design. For a given error limit, the designer can manipulate these two variables to guarantee the specified yield. Shown in Fig. 2.15 is a graphical design tool relating Y to these two variables as implied by (2.36) but notice that both variables are normalized with respect to the error-limit parameter a in order to make the tool universal. Next, in Example 2.8, we exemplify the use of this tool in design.

Example 2.8 Referring to Example 2.2 (Fig. 2.6), recall the noninverting amplifier application specified with a closed-loop gain $A_{cl,i} = 3$ V/V for an input signal v_i swinging between −0.2 and 0.2 V around a bias voltage $V_B = 0.9$ V. We now add the following details of implementation to its specification:

- R_1 is implemented with a single-unit resistor whose target resistance and relative standard deviation are given by $R_u = 10$ kΩ and $\sigma[R_u]/R_u = 0.325\%$, respectively.
- R_2 is implemented by connecting two copies of the same unit resistor R_u in series to achieve the specified gain $A_{cl,i} = 3$ V/V.
- All pertinent physical design techniques are applied for eliminating the mean values of network and offset errors.
- The standard deviation of offset is expected to be $\sigma[V_{os}] = 800$ μV.

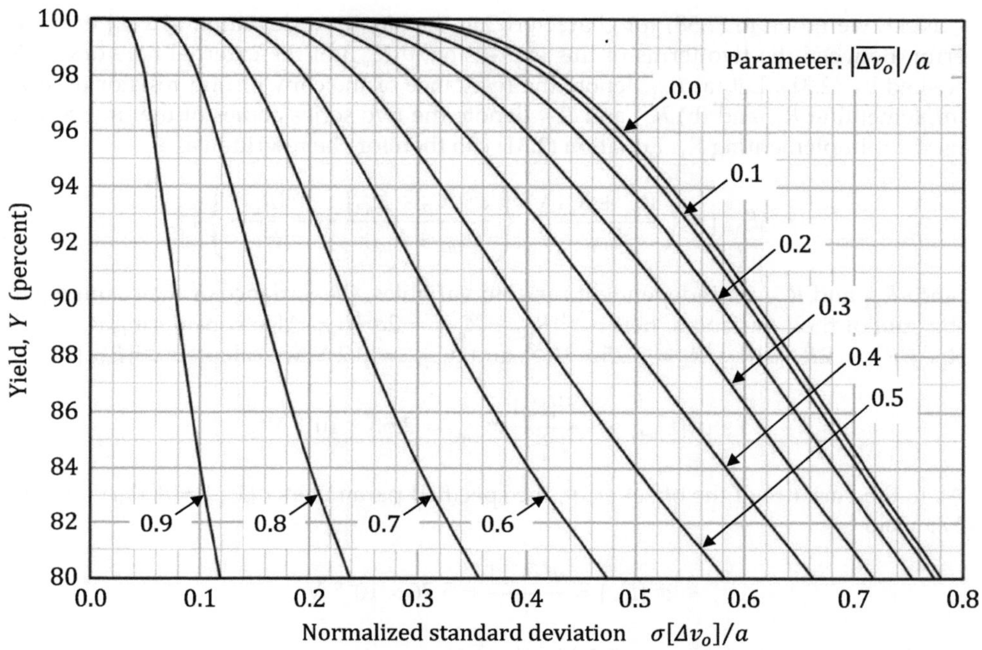

FIGURE 2.15 Universal curves of yield Y versus normalized standard deviation $\sigma[\Delta v_O]/a$ and normalized mean $\overline{\Delta v_O}/a$.

Determine (a) the minimum value of the dc differential gain A_0 needed for achieving a yield $Y \geq 90\%$ within the dc-error range -5 mV $\leq \Delta v_O \leq 5$ mV, and (b) the theoretical upper limit of yield for $A_0 \to \infty$.

Work For part a, consider the fact that the dc error is affected by A_0 only through the gain-error component E_{gain} and the effect manifests itself only through the mean value $\overline{\Delta v_O}$. Since $\overline{E_{net}}$ and $\overline{E_{os}}$ are negligible in this design, (2.37) and (2.26) yield

$$|\overline{\Delta v_O}| = \frac{|v_o|_{max}}{\beta A_0},$$

for the largest absolute value of the mean dc error in terms of the largest absolute value of v_o. The latter is determined from the specified closed-loop gain and input signal swing as $|v_o|_{max} = 3 \times 0.2 = 0.6$ V. Also considering $\beta = R_1/(R_1 + R_2) = 1/3$ as found for this application in Example 2.6, we arrive at

$$|\overline{\Delta v_O}|_{max} = \frac{1.8}{A_0}. \tag{2.39}$$

Next, we move to (2.38) for characterizing the standard deviation $\sigma[\Delta v_O]$ of the dc error. Between the two terms of this expression, $\sigma[E_{net}]$ of the external network is described by (2.31). Let (a) R_{u1} denote the resistance of the copy of unit resistor used for implementing R_1, and (b) R_{u2} and R_{u3} denote the two series copies of the unit resistor used for implementing R_1. Equation (2.31) can therefore be rewritten as

$$\sigma^2[E_{net}] \approx \left(\frac{R_{u2}+R_{u3}}{R_{u1}}\right)^2 \left\{\frac{\sigma^2[R_{u2}+R_{u3}]}{(R_{u2}+R_{u3})^2} + \frac{\sigma^2[R_{u1}]}{R_{u1}^2}\right\} v_i^2.$$

Since R_{u2} and R_{u3} are independent random variables, the variance of their sum equals the sum of their variances, hence $\sigma^2[R_{u2}+R_{u3}] = 2\sigma^2[R_u]$. Also considering $R_{u1} = R_u$, $R_{u2}+R_{u3} = 2R_u$, $\sigma[R_u]/R_u = 3.25 \times 10^{-3}$, and $v_{i(max)} = 0.2$ V, we calculate $\sigma^2[E_{net}]$ as

$$\sigma^2[E_{net}] \approx \frac{6\sigma^2[R_u]}{R_u^2} v_{i(max)}^2 = 2.53 \times 10^{-6}\ \text{V}^2.$$

As to the variance of the offset error, the specified parameters $\sigma[V_{os}] = 800$ μV and $\beta = 1/3$ yield

$$\sigma^2[E_{os}] = \frac{\sigma^2[V_{os}]}{\beta^2} = 5.76 \times 10^{-6}\ \text{V}^2.$$

Combining these variances in (2.38) leads us to

$$\sigma[\Delta v_o] = 2.88 \times 10^{-3}\ \text{V}.$$

Considering $a = 5 \times 10^{-3}$ V, we arrive at $\sigma[\Delta v_o]/a = 0.58$. According to Fig. 2.15, $Y = 90\%$ is possible for this much normalized standard deviation if the normalized mean does not exceed $|\overline{\Delta v_O}|/a = 0.18$. Using (2.39) and $a = 5$ mV in this equation, and solving for A_0, we conclude that the latter should be no less than 2000 V/V.

As to part b of the problem, first, note from (2.39) that $\overline{\Delta v_O}$ would vanish if we could make A_0 infinitely large. Also considering $\sigma[\Delta v_o]/a = 0.58$, we read 91.6% from Fig. 2.15 for this ideal case. ▲

Example 2.9 The robustness of an integrated-circuit design against random parameter variations is verified with Monte Carlo simulations. In this example, we perform an operation-point Monte Carlo analysis to verify the yield targeted in Example 2.8. The schematic of the noninverting amplifier is originally given in Fig. 2.6. It is redrawn in Fig. 2.16 after excluding the ineffective source resistance r_i but including the node identifiers of the following input file `ex2-9.sp` used in simulation:

```
filename: ex2-9.sp ; line 1
.param vdd=1.8 vss=0 voh=1.6 vol=0.2 a0=2000 vdir=0.25 ; line 2
+  fgbw=10meg fnd=10g ; line-2 continuation
.incl opmacro1.sp ; line 3
x1 1 2 3 opamp ; line 4
ru2 2 5 10k ; line 5
ru3 5 3 10k ; line 6
ru1 2 4 10k ; line 7
vb 4 0 0.9 ; line 8
vi 1 4 0.2 ; line 9
```

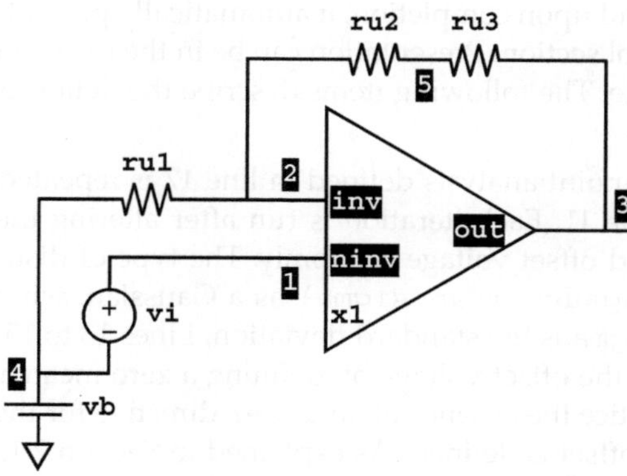

FIGURE 2.16 Simulation schematic of the noninverting amplifier of Example 2.9.

```
.control ; line 10
repeat 200 ; line 11
define agauss(mean,sigma) (mean+sigma*sgauss(0)) ; line 12
alter ru2=agauss(10k,32.5) ; line 13
alter ru3=agauss(10k,32.5) ; line 14
alter ru1=agauss(10k,32.5) ; line 15
alter v.x1.vos=agauss(0,0.8m) ; line 16
op ; line 17
set appendwrite ; line 18
wrdata mc v(3) ; line 19
end ; line 20
.endc ; line 21
.end ; line 22
```

Notice the following key features of the file:

- An instance of the single-ended opamp macro presented in Section 2.1 is defined in line 3. Its dc transfer characteristic is qualitatively similar to what is shown in Fig. 2.13. It has a mild nonlinearity whose inflection point is located midway between V_{OL} and V_{OH}. The slope at this point is defined by the user with the parameter `a0` to represent the dc differential gain A_0. Line 2 of the input file is where V_{OL}, V_{OH}, and A_0 are defined. The same line also includes non-dc parameters `vdir`, `fgbw`, and `fnd`, which, being irrelevant to the present example, are specified with arbitrary but positive values.
- R_2 is constructed with two series 10-kΩ resistances `ru2` and `ru3`. R_1 is a single 10-kΩ resistance labeled `ru1`. These are defined in lines 5 to 7.
- Input signal is set to the specified maximum 0.2 V in line 9.
- The file includes a *control section* starting in line 10 and ending in line 21. A control section generally includes a sequence of commands that define all simulation requirements regarding the type of analysis as well as the manipulation and presentation of data. A simulation defined by a control section doesn't necessitate a `run` command for running. It starts running automatically after the file has

been called, and upon completion, it automatically presents the data defined inside the control section. Presentation can be in the form of plotting, printing, or storing in a file. The following items describe the details of the present control section.

- An operation-point analysis defined in line 17 is repeated for 200 times as instructed in line 11. Each iteration is run after altering the values of the three resistances and offset voltage randomly. The type of distribution is defined in line 12 with `agauss(mean,sigma)` as a Gaussian, where `mean` is the mean value, and `sigma` is the standard deviation. Lines 13 to 15 alter the resistances. Line 16 alters the offset voltage by defining a zero mean and 0.8-mV standard deviation. Notice the absence of an `alter` directive for the output bias voltage at which the offset is defined. As explained in Section 2.1, the default value of this voltage equals the average of `voh` and `vol`, hence 0.9 V, which is indeed what the output voltage of the present application is supposed to be when there is no signal.
- Lines 18 and 19 create a text file `mc.data` in the form of a list, where the second column reports the value of the output voltage `v(3)` computed in each of the 200 runs.[3]

Since the target value of the maximum output voltage is specified as 1.5 V and the tolerable error is 5 mV, any output voltage smaller than 1.495 V or larger than 1.505 V will indicate a disqualified copy of the amplifier. The yield target being 90%, we expect 20 out of 200 runs to be disqualified. Now, run the simulation, copy the contents of the text file `mc.data` to a spreadsheet, sort the second column, and count the disqualified runs. Is the number consistent with the expectation? Repeat the simulation a couple of more times to affirm the level of consistency but don't forget to delete the text file `mc` before running each new simulation. Otherwise the results of the present simulation will be appended to those of the previous one. ▲

2.5 Frequency Response

The variation of an open-loop or closed-loop metric with the frequency of a sinusoidal input source in steady state is called *frequency response* in this book. Although, in practice, a pure sinusoidal signal is an exception rather than a rule, any signal of practical significance can be represented through Fourier transform with a collection of sinusoidal components distributed over a frequency spectrum. This is why frequency response is qualified as a very general tool of invaluable utility in the study of opamps and closed-loop configurations.

In Subsection 2.5.1, we present a unified model for the frequency response of closed-loop configurations operating under small-signal conditions, and determine the factors controlling the output spectrum. Frequency response to a large signal may be corrupted by what we call *slewing*, which is an open-loop imperfection arising from opamp input-range limitations. We consider the consequences of this effect in Subsection 2.5.2.

---

[3] Incidentally, the first column of `mc.data` reports the variable of the very last `alter` line, that is, the offset voltage `vos`.

In the first two subsections, we ignore the effect of opamp nonlinearity on frequency response and thus assume a constant slope A_0 for the open-loop dc transfer characteristic. The effect of nonlinearity, as manifested in the form of *harmonic distortion*, is the subject of Subsection 2.5.3.

2.5.1 Small-Signal Frequency Response

For modeling the small-signal frequency response, first, consider the unified closed-loop configuration of Fig. 2.10(b), and suppose that all inputs other than an x_i are tied to dc sources whereas x_i is a pure sinewave with zero mean. This input composition will create a dc component and a sinusoidal component for each of v_{ID} and v_O. The dc components can be characterized with the methods presented in Section 2.2, 2.3, or 2.4. Here we consider only the sinusoidal components. Representing these components with v_{id} and v_o, the generalized model equation (2.17) yields $v_{id} = \alpha_i x_i - \beta v_o$ for the relationship imposed by the external network between these components. v_o and v_{id} are further related by $v_o = A v_{id}$, where A represents the differential gain of the opamp itself. By eliminating v_{id} between these two expressions, we obtain the following closed-loop gain expression relating the sinusoidal component v_o of the output voltage to the small-signal sinusoidal input source x_i:

$$A_{cl,i} \doteq \frac{v_o}{x_i} = \frac{\beta A}{1+\beta A}\frac{\alpha_i}{\beta}, \tag{2.40}$$

where β, α_i, and A are generally dependent on frequency due to the capacitive components or parasitics associated with the opamp and external network. Equation (2.40) is the basis of the small-signal frequency-response analysis of a closed-loop configuration.

Open-Loop Transfer Function

The dependence of the differential gain A of an opamp on frequency is represented approximately by the single-pole transfer function

$$A \doteq \frac{v_o}{v_{id}} = \frac{A_0}{1+\frac{s}{2\pi f_o}}, \tag{2.41}$$

where A_0 is the dc differential gain introduced in Subsection 2.4.1, f_o (Hz) denotes the pole frequency known as *open-loop bandwidth*, $s = j2\pi f$ is the complex-frequency variable, and f (Hz) is the frequency of the sinusoidal excitation. According to (2.41), the variation of the magnitude $|A|$ with frequency, also known as *magnitude response*, is described by

$$|A| = \frac{A_0}{\sqrt{1+\left(\frac{f}{f_o}\right)^2}}, \tag{2.42}$$

and plotted qualitatively in Fig. 2.17. Notice the logarithmic scaling of the frequency and magnitude axes. For frequencies lower than f_o, the magnitude asymptotically approaches

$$|A| \approx A_0 \quad \text{for } f \ll f_o.$$

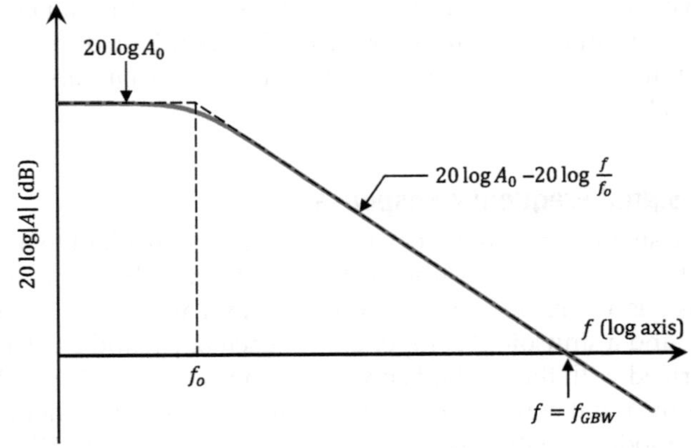

FIGURE 2.17 Magnitude response of a single-pole open-loop transfer function A.

For $f = f_o$, it equals $A_0/\sqrt{2}$. For frequencies higher than f_o, it becomes a decreasing function of f according to

$$|A| \approx \frac{f_{GBW}}{f} \quad \text{for } f \gg f_o,$$

where

$$f_{GBW} \doteq A_0 f_o \tag{2.43}$$

is an open-loop metric of outstanding design utility called *gain-bandwidth product*. As observed in Fig. 2.17, $|A|$ drops to 1 V/V (0 dB) for $f = f_{GBW}$.

Closed-Loop Gain: The Case of a Frequency-Independent External Network

By substituting the single-pole open-loop gain function (2.41) into (2.40) for A, we obtain

$$A_{cl,i} = \frac{\beta A_0}{1 + \beta A_0} \frac{1}{1 + \frac{s}{2\pi(1+\beta A_0)f_o}} \frac{\alpha_i}{\beta}. \tag{2.44}$$

Considering the fact that βA_0 is much larger than unity, we can simplify this equation into

$$A_{cl,i} = \frac{1}{1 + \frac{s}{2\pi\beta f_{GBW}}} \frac{\alpha_i}{\beta}, \tag{2.45}$$

where the product $A_0 f_o$ is replaced by f_{GBW} as per (2.43).

In amplifier applications, closed-loop configuration is built with an external network of resistive components only. As a consequence, neither α_i nor β depends on frequency in such applications. Filter applications, however, rely not only on resistive components but also on capacitive ones in their external network. This makes α_i and β frequency dependent.

We first consider amplifier applications, for which, none of the parameters appearing in (2.45) varies with frequency except the complex-frequency variable s itself. This leads to a single-pole low-pass transfer function

$$A_{cl,i} = \frac{A_{cl0,i}}{1 + \frac{s}{2\pi f_{cl}}}, \tag{2.46}$$

where

$$A_{cl0,i} \doteq \frac{\alpha_i}{\beta} \tag{2.47}$$

is dc closed-loop gain, and

$$f_{cl} \doteq \beta f_{GBW} \tag{2.48}$$

is closed-loop small-signal bandwidth. The magnitude of this transfer function is described by

$$|A_{cl,i}| = \frac{|A_{cl0,i}|}{\sqrt{1 + \left(\frac{f}{f_{cl}}\right)^2}}, \tag{2.49}$$

where $|A_{cl0,i}|$ is the absolute value of the dc closed-loop gain. Note that β in a frequency-independent external network is unconditionally positive. Input factor α_i, however, may be positive or negative, as exemplified in several different cases in Example 2.6. This is why only α_i needs to be represented with its absolute value in $|A_{cl0,i}|$.

As mentioned at the beginning of Section 2.4, the closed-loop gain expression (2.19) derived by assuming a virtual short is the basis of external-network design. Equations (2.49) and (2.47) indeed confirm (2.19) for dc ($f = 0$) but, as expected from a single-pole low-pass transfer function, they also predict a gradually decreasing closed-loop gain as f approaches f_{cl}. At $f = f_{cl}$ the magnitude is expected to drop to $A_{cl0,i}/\sqrt{2}$, and thereafter become inversely proportional to frequency. These predictions are illustrated graphically in Fig. 2.18 with a typical magnitude-response plot on logarithmic frequency and gain axes.

The relatively slow variation of the magnitude of closed-loop gain with frequency in between dc and f_{cl} is not a concern in many amplifier applications. In such applications, f_{cl} is set equal to the highest frequency f_{max} to be amplified. Yet, there are other applications where a better flatness is needed for the magnitude response within the frequency band $0 \leq f \leq f_{max}$ of operation. This calls for an f_{cl} wider than f_{max}. The bandwidth needed in such cases depends on how much deterioration can be tolerated in the amplitude of the output signal within the range $0 \leq f \leq f_{max}$. Representing the output amplitude with $V_o(f_{max})$ for $f = f_{max}$ and with $V_o(0)$ for $f \rightarrow 0$, this difference is quantified with

$$\Delta V_o \doteq V_o(f_{max}) - V_o(0), \tag{2.50}$$

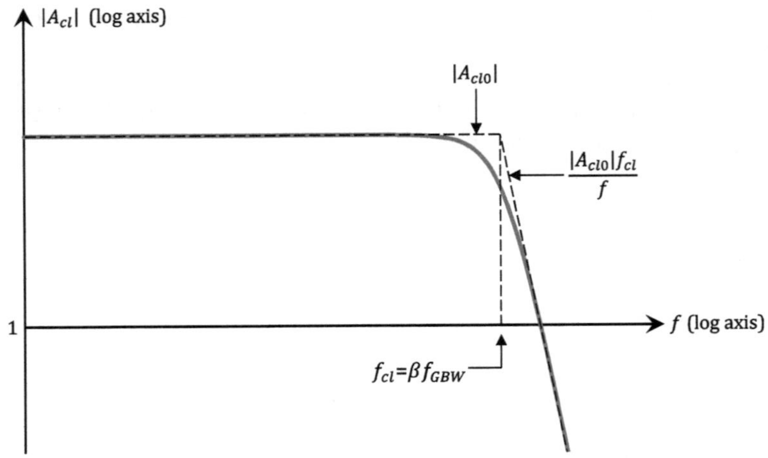

FIGURE 2.18 Magnitude response of closed-loop configurations with a frequency-independent external network.

which is called *magnitude error*. Representing the input amplitude with X_i, and considering $V_o(f_{\max}) = |A_{cl,i}(f_{\max})|X_i$ and $V_o(0) = A_{cl0,i}X_i$, we obtain the following expression from (2.50) and (2.49):

$$\Delta V_o = \left[\frac{1}{\sqrt{1+\left(\frac{f_{\max}}{f_{cl}}\right)^2}} - 1\right] V_o(0), \tag{2.51}$$

whose solution for f_{cl} yields the required closed-loop small-signal bandwidth for a given set of $f_{\max}$, $V_o(0)$, and $\Delta V_o(0)$.

Closed-Loop Gain: The Case of a Frequency-Dependent External Network

As mentioned previously, closed-loop applications often necessitate filtering too. In such cases, we match the closed-loop gain $A_{cl,i}$ to the desired filter characteristic through α_i/β by deploying an external network of resistors and capacitors. The effect of the relatively inaccurate and imprecise open-loop gain A on frequency response is minimized by setting the gain-bandwidth product f_{GBW} of the opamp much higher than the highest characteristic frequency of the target filter response. This practice works by keeping the magnitude of $\beta A/(1+\beta A)$ of (2.40) close to unity within the frequency band of interest.

For a frequency-dependent external network of single-pole response, (2.49) is still a valid description of the magnitude of the closed-loop transfer function α_i/β. Equation (2.47) of $A_{cl0,i}$ is also valid provided that its right-hand side is interpreted as the dc value of $|\alpha_i|/\beta$. For f_{cl}, however, (2.48) is invalid; it has to be replaced by the pole frequency of α_i/β.

Example 2.10 Determine the necessary gain-bandwidth product for achieving a closed-loop small-signal bandwidth $f_{cl} = 3$ MHz in the following amplifier configurations:

1. The noninverting amplifier of Fig. 2.6 for $A_{cl0,i} = 3$ V/V with respect to v_i (see Examples 2.2 and 2.6).
2. The unity-gain buffer amplifier of Fig. 2.7 (see Examples 2.3 and 2.6).
3. The inverting amplifier of Fig. 2.8 for $A_{cl0,i} = -3$ V/V with respect to v_i (see Examples 2.4 and 2.6).
4. The transresistance amplifier of Fig. 2.9 for $A_{cl0,i} = -50$ kΩ (see Examples 2.5 and 2.6).

Verify the result by running an ac simulation for each configuration. Adopt $V_{DD} = 1.8$ V, $V_{SS} = 0$, $V_B = 0.9$ V, $V_{OH} = 1.65$ V, $V_{OL} = 0.15$ V, and $A_0 = 1000$ V/V for the dc parameters. Set f_{GBW} to whatever value you have determined for the application. The remaining two parameters are irrelevant to this example; you may set them to `vdir=0.25` and `fnd=2g`.

Work

1. Noninverting amplifier: $A_{cl0,i} = 1 + (R_2/R_1) = 3$ V/V implies $\beta = R_1/(R_1 + R_2) = 1/3$, for which, (2.48) yields $f_{GBW} = 9$ MHz.
2. Unity-gain buffer amplifier: Since $\beta = 1$, (2.48) yields $f_{GBW} = 3$ MHz.
3. Inverting amplifier: $A_{cl0,i} = -R_2/(R_1 + r_i) = -3$ V/V implies $\beta = (R_1 + r_i)/(R_1 + r_i + R_2) = 1/4$, for which, (2.48) yields $f_{GBW} = 12$ MHz.
4. Transresistance amplifier: Since $\beta = 1$, (2.48) yields $f_{GBW} = 3$ MHz.

An input file (`ex2-10a.sp`) suitable for verifying the closed-loop bandwidth performance of the noninverting amplifier is presented below:

```
filename: ex2-10a.sp
.param vdd=1.8 vss=0 voh=1.65 vol=0.15 a0=1000 vdir=0.25
+  fgbw=9meg fnd=2g
.incl opmacro1.sp
x1 1 2 3 opamp
r1 2 4 10k
r2 2 3 20k
vi 1 4 dc 0 ac 1m
vb 4 0 dc 0.9
.ac dec 1000 1k 50meg  ;  line 9
.end
```

Note the following properties:

- Gain-bandwidth product is set to 9 MHz as the solution of the problem indicates.
- R_1 is set arbitrarily to 10 kΩ. R_2 is twice as large as needed for a closed-loop dc gain of 3 V/V. The source resistance r_i is omitted since it is ineffective in this configuration.

- Input signal magnitude is set to 1 mV but this selection is not critical because what we are interested in is small-signal gain, hence the ratio of output and input signal magnitudes.
- AC analysis is defined in line 9. It involves 1000 different frequencies in each decade between 1 kHz and 50 MHz.

Plot the magnitude response with the command

```
plot vdb(3)-vdb(1)
```

The plot looks like Fig. 2.18 where the gain axis is calibrated in decibel units. First, zoom in on the low-frequency segment of the plot, and read the dc gain $A_{cl0,i}$. It turns out to be 9.5164 dB whereas the specified 3 V/V corresponds to 9.5424 dB. The difference between the two is due to gain error, which has been ignored in frequency-response modeling as a result of the assumption $\beta A_0 \to \infty$ made in arriving at (2.45). For determining the simulated closed-loop bandwidth, zoom in around the target bandwidth 3 MHz, and read the frequency at which the gain drops from its low-frequency value by 3 to 6.5164 dB. The unity-gain and inverting amplifiers are simulated along the same lines. As to the transresistance amplifier, you can use the following input file (`ex2-10b.sp`):

```
filename: ex2-10b
.param vdd=1.8 vss=0 voh=1.65 vol=0.15 a0=1000 vdir=0.25
+  fgbw=3meg fnd=2g
.incl opmacro1.sp
x1 1 2 3 opamp
r 2 3 50k
ii 0 4 dc 0 ac 1m
vs 4 2 0
vb 1 0 dc 0.9
.ac dec 1000 1k 50meg
.end
```

Notice the short-circuit voltage source `vs` inserted in series with the current-mode signal source. This source acts as an ammeter, and enables us to characterize the current of the signal source. Signal source has a 1-mA magnitude but again this is not critical because what we will plot is the ratio of the magnitudes of the output and input signals. Plotting with

```
plot mag(v(3))/mag(i(vs))
```

yields the magnitude of closed-loop transresistance gain A_{cl} in units of ohms along a linear axis. Nominally, it should read 50 kΩ at dc and low frequencies, and drop by a factor of $\sqrt{2} = 1.41$ at the closed-loop bandwidth but the actual dc value will be slightly different due to gain error as pointed out in connection with the simulation of the noninverting amplifier. ▲

Example 2.11 The noninverting amplifier considered in Example 2.10 can tolerate a maximum magnitude error −5 mV within the frequency band $0 \leq f \leq 500$ kHz when driven by an ac signal of 0.2 V amplitude. Assuming small-signal operation, determine the minimum gain-bandwidth product needed for this application, and verify the specified error with an ac simulation.

Work The specifications $A_{cl0,i} = 3$ V/V and $V_i = 0.2$ V of Example 2.10 yield $V_o(0) = 0.6$ V. Substituting this together with $\Delta V_o = -5$ mV and $f_{\max} = 500$ kHz into (2.51), and solving for the bandwidth, we obtain $f_{cl} = 3.85$ MHz as a minimum. For this much bandwidth and $\beta = 1/3$ of the amplifier, the necessary gain-bandwidth product is calculated from (2.48) as $f_{GBW} = 11.5$ MHz.

Simulation can be performed with the input file `ex2-10a.sp` used in Example 2.10 but don't forget to set `fgbw=11.5meg`. Also make sure that the magnitude of the input signal is defined exactly as 200 mV because you are not supposed to plot a ratio of magnitudes but the magnitude of the output signal only. Plot by entering the command

```
plot vm(3)
```

and read the output magnitude for (a) $f = 1$ kHz as an approximation to actual dc magnitude, and (b) $f = 500$ kHz as the magnitude at the maximum operation frequency. Magnitude error is the difference between the two. Note that the dc magnitude will be somewhat different from the expected 600 mV due to the dc gain error. ▲

Example 2.12 Shown in Fig. 2.19(a) is the simulation schematic of a low-pass filter application. External-network components are specified as $R_1 = 15$ kΩ, $R_2 = 150$ kΩ, and $C = 10$ pF. The gain-bandwidth product of the opamp is 3 MHz. Determine the frequency response of the voltage gain with respect to signal source v_i, and verify with a simulation.

Work The input factor α_i of v_i and the feedback factor β are determined directly from the schematic as follows:

$$\alpha_i = -\frac{R_2}{R_1 + R_2 + sCR_1R_2}, \tag{2.52}$$

$$\beta = \frac{R_1(1 + sCR_2)}{R_1 + R_2 + sCR_1R_2}. \tag{2.53}$$

Substituting these two into (2.47) yields

$$A_{cl,i} = \frac{\alpha_i}{\beta} = -\frac{R_2}{R_1(1 + sCR_2)}, \tag{2.54}$$

from which, we predict a dc gain $A_{cl0,i} = -R_2/R_1 = -10$ V/V, and a corner frequency $f_P = 1/2\pi CR_2 = 106$ kHz, at which, the gain drops by 3 dB.

The following input file `ex2-12.sp` is appropriate for verification:

```
filename: ex2-12.sp
.param vdd=1.8 vss=0 voh=1.65 vol=0.15 a0=10000 vdir=0.25
+  fgbw=3meg fnd=2g
.incl opmacro1.sp
x1 1 2 3 opamp
r1 2 4 15k
r2 2 3 150k
c 2 3 10p
vi 4 1 dc 0 ac 1m
```

```
vb 1 0 dc 0.9
.ac dec 1000 100 100meg
.meas ac bw when vm(3)=7.07m ; line 11
.end
```

Note that the command `.meas` is introduced in line 11 for letting the simulator measure and put out the value of the corner frequency. On this line, `ac` specifies the type of simulation, and `bw` represents the name we arbitrarily assign to corner frequency. The rest "`when vm(3)=7.07m`" instructs the simulator to measure the specific frequency, at which, the magnitude of the ac output signal `vm(3)` equals 7.07 mV. Considering the 1-mV amplitude of the input signal and the 10-V/V dc gain of the closed-loop configuration, the frequency corresponding to this much ac output signal will indeed be the corner frequency. However, the simulator puts out `bw=7.77464e+004`, which is considerably

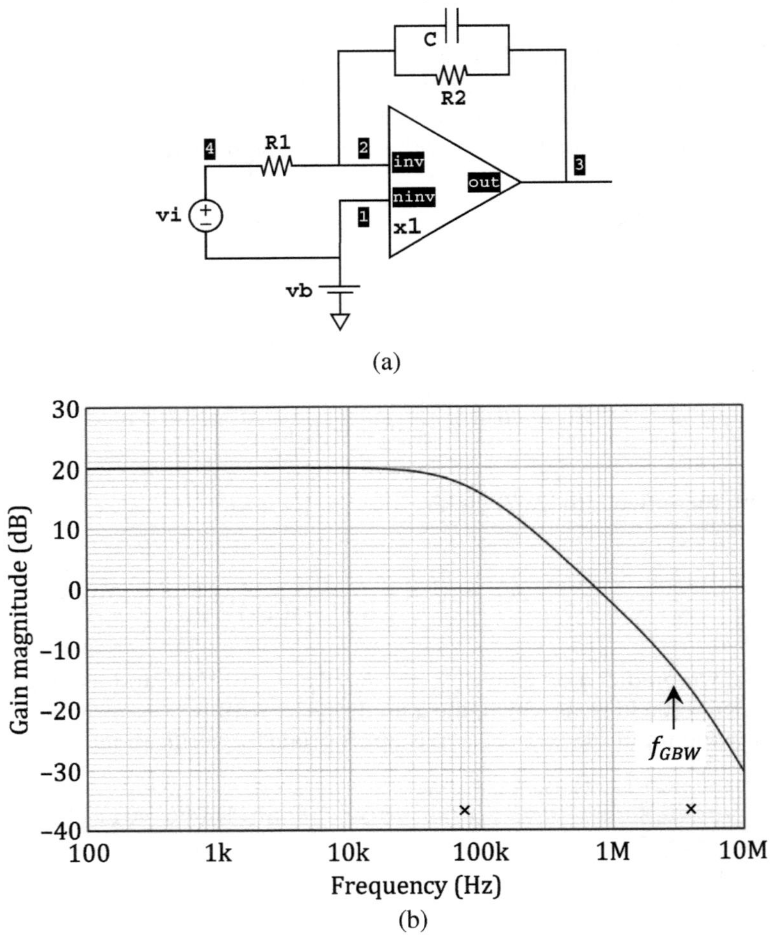

FIGURE 2.19 (a) Simulation schematic of the filter application of Example 2.12. (b) Magnitude response of the closed-loop gain for v_i, where pole frequencies are marked with crosses.

lower than the prediction $f_P = 106$ kHz made above. The reason for this large discrepancy is the inadequacy of the 3-MHz gain-bandwidth product.

For a quantitative assessment of the effect of f_{GBW} on the corner frequency, we need to rely on an exact closed-loop transfer function unlike the approximate (2.54). To this end, we customize (2.44) for this particular application by substituting (2.52) for α_i and (2.53) for β. The outcome exhibits two poles at the frequencies

$$f_{P1,P2} = \frac{1}{2}\left(f_{GBW} + \frac{f_P}{\beta_0}\right) \pm \sqrt{\frac{1}{4}\left(f_{GBW} + \frac{f_P}{\beta_0}\right)^2 - f_{GBW} f_P}, \tag{2.55}$$

where f_P is the nominal pole frequency $f_P = 1/2\pi CR_2$ associated with the approximate transfer function (2.54), and $\beta_0 = R_1/(R_1 + R_2)$ is the dc value of the feedback factor β. A numerical evaluation of (2.55) yields $f_{P1} = 77.7$ kHz and $f_{P2} = 4.09$ MHz. The former of the two is obviously a perfect match to the corner frequency measured by the simulator.

The simulated magnitude response can be plotted by entering

```
plot vdb(3)-vdb(4)
```

The outcome, displayed in Fig. 2.19(b), clearly exhibits the effect of both poles, whose frequencies are marked by crosses. The reader is now invited to repeat the simulation with, say, a tenfold higher f_{GBW}, and observe the change in the frequencies. As a matter of fact, for $f_{GBW} \gg f_P$, we expect $f_{P1} = f_P$ and $f_{P2} = f_{GBW}$ from (2.55). ▲

2.5.2 Slew-Limited Bandwidth

As we will discuss in Chapter 4, each internal amplifier stage of an opamp functions as a transconductor generating a current proportional to its voltage-mode input signal. Whenever the absolute value $|v_{ID}|$ of the differential-input voltage of an opamp becomes too large for one of these stages to sustain its transconductor functionality, the opamp starts *slewing*. Since a slewing opamp can no more respond proportionately to the variation of v_{ID}, the negative-feedback loop of the external network is broken. The configuration is thus forced to operate in open loop as long as slewing continues. Slewing is undesirable because it reduces the effective bandwidth of frequency response and slows down the speed of step response in closed-loop configurations. In this section, we discuss and model the effect of slewing on the effective bandwidth of amplifier configurations in particular. Its effect on the step response of amplifiers is discussed in Subsection 2.6.2.

Although, in practice, the transconductance of an opamp stage deteriorates gradually as a result of an increasing differential-input voltage, we approximate this fact by assuming an abrupt transition into slewing whenever $|v_{ID}|$ exceeds an open-loop metric to be called *differential-input range* and represented with V_{DIR} in this book.[4] The condition of slew-free operation is therefore defined quite generally by

$$|v_{ID}| < V_{DIR} \tag{2.56}$$

[4] In most texts on opamps, particularly in opamp data sheets, the term "differential-input range" refers to the maximum voltage that can be applied between the two inputs before the reliability of the opamp is jeopardized. It usually appears in the maximum-rating list, and should not be confused with what we define as V_{DIR} in this text.

Now suppose that one of the inputs of a closed-loop configuration receives a sinusoidal signal of amplitude X_i and frequency f around a possible dc bias, and let the amplitude of the corresponding sinusoidal differential-input signal of the opamp be V_{id}. No matter how large X_i may be, V_{id} is virtually zero for $f \to 0$ as long as CMR or output-range limitations are not breached. This is what the virtual-short approximation implies. Therefore, the condition (2.56) of slew-free operation is readily satisfied for low-frequency excitation regardless of how large X_i may be. However, the condition is challenged at high frequencies because V_{id} is an increasing function of not only X_i but also f as explained next.

Consider the general relationship $v_{id} = v_o/A$ between the output signal v_o of a closed-loop configuration and the differential-input signal v_{id} of the opamp. Since v_o is related to the input signal x_i by $v_o = A_{cl,i}x_i$, this expression can be rewritten quite generally as $v_{id} = (A_{cl,i}/A)x_i$. V_{id} is therefore related to X_i by the magnitude of $A_{cl,i}/A$ as follows

$$V_{id} = \left|\frac{A_{cl,i}}{A}\right| X_i.$$

For a given input signal amplitude X_i, V_{id} can be evaluated from this expression by substituting (2.41) for A, and (2.46) for $A_{cl,i}$. Considering also (2.47) for $A_{cl0,i}$, and assuming $f_o \ll f_{cl}$, the outcome is as follows

$$V_{id} = \frac{|\alpha_i|}{\beta A_0} \frac{\sqrt{1 + \left(\frac{f}{f_o}\right)^2 + \left(\frac{f^2}{f_o f_{cl}}\right)^2}}{1 + \left(\frac{f}{f_{cl}}\right)^2} X_i. \tag{2.57}$$

The variation of V_{id} with f as predicted by this equation for a fixed X_i is plotted qualitatively in Fig. 2.20. Notice how V_{id} increases with f starting from a vanishingly small

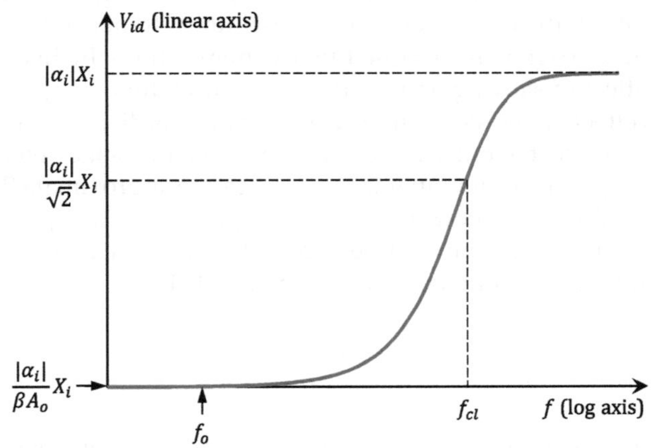

FIGURE 2.20 Variation of the amplitude V_{id} of differential-input voltage with frequency for a given closed-loop input signal amplitude X_i.

$V_{id} = |\alpha_i|X_i/\beta A_o$ for dc, reaching $|\alpha_i|X_i/\sqrt{2}$ for $f = f_{cl}$,[5] and finally saturating at $V_{id} = |\alpha_i|X_i$ for $f \to \infty$.

According to these results and (2.56), a slew-free operation within the closed-loop small-signal bandwidth f_{cl} is possible for any amplitude X_i that satisfies the condition

$$X_i < \frac{\sqrt{2}V_{DIR}}{|\alpha_i|}. \tag{2.58}$$

If X_i does not satisfy (2.58), then, slewing will prevail at a frequency lesser than f_{cl} and thus impose a narrower bandwidth.

The bandwidth observed for the maximum input-signal amplitude $X_i = X_{i(\max)}$ is called *full-power bandwidth* regardless of whether it equals f_{cl} due to slew-free operation or falls short of f_{cl} due to slew-limited operation. In order to formulate this closed-loop metric, we solve (2.57) for f under the assumption $f \gg f_o$, and substitute V_{DIR} and $X_{i(\max)}$ for V_{id} and X_i, respectively. Denoting the full-power bandwidth with f_{fp}, the outcome is given below alongside the slew-free case:

$$f_{fp} = f_{cl} \qquad \text{for } X_{i(\max)} \le \sqrt{2}V_{DIR}/\,|\alpha_i|\,, \tag{2.59a}$$

$$f_{fp} = \frac{f_{cl}}{\sqrt{\left(\frac{|\alpha_i|X_{i(\max)}}{V_{DIR}}\right)^2 - 1}} \qquad \text{for } X_{i(\max)} \ge \sqrt{2}V_{DIR}/\,|\alpha_i|\,. \tag{2.59b}$$

In an application where f_{fp} is slew limited, we substitute the specified f_{fp} into (2.59b), and solve the equation for f_{cl} for determining the corresponding small-signal bandwidth. Equation (2.48) will then provide us with the value of f_{GBW} to be targeted in design for achieving the specified f_{fp}.

Finally, note that analytically estimating the magnitude error in the case of a slew-limited bandwidth is impractical if not impossible. The reader is advised to rely on iterative simulation with our opamp macro to determine the necessary f_{GBW} if the closed-loop metric of concern is the magnitude error instead of the bandwidth.

Example 2.13 An opamp of $V_{DIR} = 0.15$ V is to be used in the following four amplifier applications, where a sinusoidal input signal is to be processed with 3-MHz full-power bandwidth. Magnitude error is of no concern.

- The noninverting amplifier of Fig. 2.6 for $A_{cl0,i} = 3$ V/V and $V_{i(\max)} = 0.25$ V.
- The unity-gain buffer amplifier of Fig. 2.7 for $V_{i(\max)} = 0.3$ V.
- The inverting amplifier of Fig. 2.8 for $A_{cl0,i} = -3$ V/V, $r_i = 0$, and $V_{i(\max)} = 0.25$ V.
- The transresistance amplifier of Fig. 2.9 for $A_{cl0,i} = -50$ kΩ and $I_{i(\max)} = 10$ μA.

[5] Derivation of this particular value of V_{id} also involves (2.43) and (2.48).

1. Determine the minimum closed-loop small-signal bandwidth and the corresponding gain-bandwidth product needed by each application for the specified values of dc gain and maximum input-signal amplitude.
2. Run a transient simulation on the unity-gain buffer amplifier for each of the following values of V_{DIR}: (a) 500, 150, 100, and 50 mV. Set the gain-bandwidth product to the value calculated in the first part of the example. For the rest of the opamp parameters, use $V_{DD} = 1.8$ V, $V_{SS} = 0$, $V_{OH} = 1.65$ V, $V_{OL} = 0.15$ V, $A_0 = 1000$ V/V, and $f_{nd} = 2$ GHz. Set $V_B = 0.9$ V, and inspect the effect of V_{DIR} on the output-voltage waveform.

Work

- Noninverting amplifier: $A_{cl0,i} = 3$ V/V implies $\beta = 1/3$ and $\alpha_i = 1$. For the specified $V_{i(\max)} = 0.25$ V and $V_{DIR} = 0.15$ V, (2.58) is not satisfied. The bandwidth of this application is therefore slew limited. For the specified values of f_{fp}, α_i, V_i, and V_{DIR}, (2.59b) yields $f_{cl} = 4$ MHz, which implies $f_{GBW} = 12$ MHz from (2.48).
- Unity-gain buffer amplifier: $A_{cl0,i} = 1$ V/V implies $\beta = 1$ and $\alpha_i = 1$. For the specified $V_{i(\max)} = 0.3$ V and $V_{DIR} = 0.15$ V, (2.58) is not satisfied. The bandwidth of this application is also slew limited. Again using (2.59b) and (2.48), we arrive at $f_{cl} = 5.2$ MHz and $f_{GBW} = 5.2$ MHz.
- Inverting amplifier: $A_{cl0,i} = 3$ V/V and $R_i = 0$ imply $\beta = 1/4$ and $|\alpha_i| = 3/4$. For the specified $V_{i(\max)} = 0.25$ V and $V_{DIR} = 0.15$ V, (2.58) is satisfied. According to (2.59a), bandwidth is therefore defined by f_{cl}. We target $f_{cl} = 3$ MHz, for which, (2.48) yields $f_{GBW} = 12$ MHz.
- Transresistance amplifier: $A_{cl0,i} = 50$ kΩ implies $\beta = 1$ and $|\alpha_i| = 50$ kΩ. For the specified $I_{i(\max)} = 10$ μA and $V_{DIR} = 0.15$ V, (2.58) is not satisfied. Again a slew-limited bandwidth is involved. Substituting into (2.59b) $f_{fp} = 3$ MHz together with the specified values of α_i, V_i, and V_{DIR}, we obtain $f_{cl} = 9.54$ MHz, for which, (2.48) yields $f_{GBW} = 9.54$ MHz.

Verification of the unity-gain buffer is performed with the following input file `ex2-13.sp`:

```
filename: ex2-13.sp
.param vdd=1.8 vss=0 voh=1.65 vol=0.15 a0=10000 vdir=0.05
+  fgbw=5.2meg fnd=2g
.incl opmacro1.sp
x1 1 2 2 opamp
vb 3 0 0.9
vi 1 3 sin 0 300m 3meg
.tran 0.1n 3u
.end
```

Running the simulation for each of the specified values of V_{DIR}, and plotting the output voltage `v(2)`, we obtain the waveforms shown in Fig. 2.21. Notice the negligible effect of V_{DIR} on the waveform as long as it is greater than 150 mV, for which, the opamp is optimized for the specifications of the present example. For smaller values of V_{DIR}, however, $f_{GBW} = 5.2$ MHz becomes insufficient for handling the 3-MHz signal. In this case,

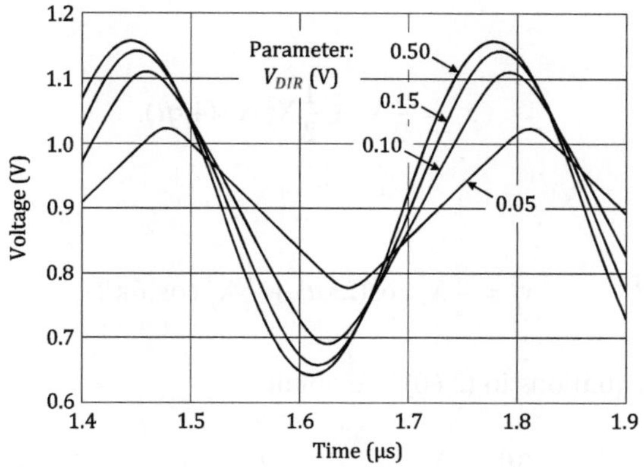

FIGURE 2.21 Output-voltage waveforms of the unity-gain buffer amplifier of Example 2.13.

the output voltage is rapidly transformed into a triangular waveform due to excessive slewing. ▲

Example 2.14 How much is the smallest value of V_{DIR} that prevents the amplifier described in Example 2.11 from slewing for a maximum signal amplitude $V_{i(\max)} = 0.2$ V?

Work V_{DIR} must be large enough to satisfy (2.58). Since $\alpha_i = 1$ and $V_{i(\max)} = 0.2$ V, this condition calls for $V_{DIR} > 0.141$ V. ▲

2.5.3 Harmonic Distortion

The nonlinearity of an open-loop transfer characteristic was depicted in Fig. 2.13 but ignored thereafter. Its effect on closed-loop dc response is indeed negligible as long as the output swing is confined to the output range. In frequency response, however, the nonlinearity leads to an undesirable effect called *harmonic distortion*, which manifests itself with spurious output-voltage components at the harmonics of the input frequency. In this section, we construct a model for this imperfection, and derive design guidelines for controlling its effect on the output signal.

The nonlinearity of the open-loop transfer function leads to a nonlinearity in the closed-loop transfer function. We can represent this nonlinearity by expanding the closed-loop transfer function into a Taylor series in terms of the related input signal x_i. In most cases, a third-order expansion adequately approximates the transfer function, that is,

$$v_o \approx A_{cl,i}x_i + B_2x_i^2 + B_3x_i^3, \tag{2.60}$$

where B_2 and B_3 are the coefficients of second-order nonlinearity and third-order nonlinearity, respectively. The coefficient of x_i is the closed-loop gain $A_{cl,i}$ because, after all, this equation must reduce to $v_o = A_{cl,i}x_i$ in the absence of nonlinearity ($B_2 = 0$, $B_3 = 0$). Next, we specify x_i as a pure sinusoidal signal of frequency f and amplitude X_i, that is,

$$x_i = X_i\cos(2\pi ft). \tag{2.61}$$

Therefore

$$x_i^2 = \frac{1}{2}X_i^2 + \frac{1}{2}X_i^2\cos(4\pi ft), \tag{2.62}$$

and

$$x_i^3 = \frac{3}{4}X_i^3\cos(2\pi ft) + \frac{1}{4}X_i^3\cos(6\pi ft). \tag{2.63}$$

Using these three equations in (2.60), we obtain

$$v_o = \frac{B_2}{2}X_i^2 + \left(A_{cl,i}X_i + \frac{3B_3}{4}X_i^3\right)\cos(2\pi ft) + \frac{B_2}{2}X_i^2\cos(4\pi ft) + \frac{B_3}{4}X_i^3\cos(6\pi ft). \tag{2.64}$$

The first term on the right-hand side of this equation is a small dc component caused by second-order nonlinearity. It will be ignored in the rest of the analysis due to its irrelevance to the frequency response. The second term of (2.64) represents the *fundamental* component of v_o. The part $3B_3X_i^3/4$ of the coefficient of this component is caused by third-order nonlinearity. It is usually negligible in comparison with the nominal amplitude $A_{cl,i}X_i$ of the output signal. For the purpose of harmonic-distortion modeling, we therefore simplify (2.64) into the following form:

$$v_o = \underbrace{A_{cl,i}X_i\cos(2\pi ft)}_{\textit{Fundamental}} + \underbrace{\frac{B_2}{2}X_i^2\cos(4\pi ft)}_{\textit{2nd harmonic}} + \underbrace{\frac{B_3}{4}X_i^3\cos(6\pi ft)}_{\textit{3rd harmonic}}, \tag{2.65}$$

where spurious output signals at frequencies $2f$ and $3f$ are labeled as *2nd harmonic* and *3rd harmonic*, respectively. Several different metrics are in use for quantifying the harmonic distortion of closed-loop configurations. In amplifier applications, the so-called *total harmonic distortion* (THD) finds widespread use. It is defined generally by

$$\text{THD} \doteq \sqrt{\text{HD}_2^2 + \text{HD}_3^2 + \cdots}, \tag{2.66}$$

where HD_k is defined by

$$\text{HD}_k \doteq \frac{k\text{-th harmonic amplitude}}{\text{Fundamental amplitude}},$$

and called *k-th order harmonic distortion*. Considering (2.65), we can express the second-order and third-order harmonic distortion metrics as

$$\text{HD}_2 \doteq \frac{\text{2nd harmonic amplitude}}{\text{Fundamental amplitude}} = \frac{B_2X_i}{2A_{cl,i}} = \frac{B_2V_o}{2A_{cl,i}^2}, \tag{2.67}$$

and

$$\text{HD}_3 \doteq \frac{\text{3rd harmonic amplitude}}{\text{Fundamental amplitude}} = \frac{B_3 X_i^2}{4A_{cl,i}} = \frac{B_3 V_o^2}{4A_{cl,i}^3}, \tag{2.68}$$

where $V_o = A_{cl,i}X_i$ is the amplitude of the ideal output signal. THD, as defined by (2.66), is usually multiplied by 100 and the outcome is expressed as a percentage. It is also common to express THD in units of decibels (dB) after taking the logarithm of (2.66), and multiplying the outcome by 20.

In order to derive design guidelines for controlling THD, we need to relate (2.67) and (2.68) to opamp properties. We do this by first expanding the open-loop transfer function $v_O = f(v_{ID})$ of an opamp into Taylor series around the dc operation point, and then, rearranging the series in terms of the output signal v_o and the differential-input signal v_{id}. This leads to

$$v_o = Av_{id} + A_2 v_{id}^2 + A_3 v_{id}^3 + \cdots, \tag{2.69}$$

where (a) A is the small-signal differential gain of the opamp, (b) A_2 is the coefficient of the second-order nonlinearity, and (c) A_3 is the coefficient of the third-order nonlinearity. As usual in a closed-loop configuration, v_{id} is set by the external network to

$$v_{id} = \alpha_i x_i - \beta v_o \tag{2.70}$$

in accordance with (2.17).

We now relate the coefficients B_2 and B_3 of closed-loop nonlinearity to the coefficients A, A_2, and A_3 of open-loop nonlinearity. To this end, we first substitute (a) (2.70) into (2.69) for v_{id}, and (b) (2.60) into the resulting equation for v_o. Ignoring higher-order components beyond the third, the outcome is an equation in the form $ax_i + bx_i^2 + cx_i^3 = 0$. By equating each of the three coefficients to zero, and solving, we arrive at:

$$A_{cl,i} = \frac{A\alpha_i}{1+\beta A} \approx \frac{\alpha_i}{\beta}, \tag{2.71}$$

$$B_2 = \frac{A_2\alpha_i^2}{(1+\beta A)^3} \approx \frac{A_2\alpha_i^2}{(\beta A)^3} = \frac{A_2}{\beta A^3}A_{cl,i}^2, \tag{2.72}$$

$$B_3 = \frac{A_3\alpha_i^3}{(1+\beta A)^4} - \frac{2A_2^2\alpha_i^3\beta}{(1+\beta A)^5} \approx \frac{A_3\alpha_i^3}{(\beta A)^4} - \frac{2A_2^2\alpha_i^3\beta}{(\beta A)^5} = \frac{A_3}{\beta A^4}A_{cl,i}^3 - \frac{2A_2^2}{\beta A^5}A_{cl,i}^3, \tag{2.73}$$

where approximations are due to $\beta A \gg 1$. Note that the second term of the final part of (2.73) is usually negligible in comparison with the first term; hence

$$B_3 \approx \frac{A_3}{\beta A^4}A_{cl,i}^3. \tag{2.74}$$

Finally, we use (2.72) and (2.74) in (2.67) and (2.68), respectively, to arrive at

$$\mathrm{HD}_2 = \frac{A_2}{2\beta A^3} V_o, \tag{2.75}$$

$$\mathrm{HD}_3 = \frac{A_3}{4\beta A^4} V_o^2. \tag{2.76}$$

Although not explicitly stated, we have ignored the frequency dependence of the coefficients of nonlinearity in the derivation of (2.75) and (2.76). As a consequence, theoretically, their validity is restricted to fundamental frequencies lesser than $f_o/3$ so that even the 3rd harmonic frequency remains inside the open-loop bandwidth of the opamp. Fortunately, we can extend their validity to higher fundamental frequencies simply by decomposing the third and fourth powers of A appearing in their denominators into $A^3 \rightarrow |A(f)|^2|A(2f)|$ and $A^4 \rightarrow |A(f)|^3|A(3f)|$, where $|A(f)|$, $|A(2f)|$, and $|A(3f)|$ represent the magnitude of the open-loop gain for the fundamental and harmonic frequencies. Otherwise, A_2, A_3, and V_o are evaluated at the fundamental frequency. The reader is referred to [2] for a more precise treatment of this extension.

We can now derive from (2.75) and (2.76) the design guidelines we need for controlling THD. Note that β and V_o cannot be freely manipulated for this purpose because the former is involved in numerous other closed-loop metrics, and the latter is dictated by the output-swing specification of the target application. Still, it is worth keeping in mind that THD decreases with β and increases with V_o. Next, consider A_2, which is proportional to the second derivative of the open-loop transfer characteristic of the opamp. If the dc operation point coincides with the inflection point of this characteristic where the second derivative vanishes, then, theoretically, A_2 and therefore HD_2 will also vanish. As will be discussed in Chapter 4, there exist certain design practices by which we can approximate this condition but meeting it exactly is virtually impossible due to the inaccuracy and imprecision associated with the open-loop transfer characteristic. A_3 is even more uncontrollable because it is proportional to the third derivative of the very same open-loop transfer characteristic. Although A_2 and A_3 are not amenable to manipulation, both are known to be functions of A. Most typically, the magnitude of A_2 is proportional to $|A(f)|^2$, and that of A_3 is proportional to $|A(f)|^3$. Considering these dependencies together with the decomposition presented in the preceding paragraph for A^3 and A^4, we arrive at $\mathrm{HD}_2 \propto 1/|A(2f)|$ and $\mathrm{HD}_3 \propto 1/|A(3f)|$. It is obvious from this conclusion that both harmonic distortion components hence THD can be reduced by increasing the differential gain. Considering the magnitude response depicted in Fig. 2.17, raising only the dc differential gain A_0 will suffice as long as the specified harmonic frequencies are lower than the open-loop bandwidth f_o. Otherwise, it is necessary to increase the gain-bandwidth product f_{GBW} in order to reduce the THD.

Example 2.15 We can simulate a circuit for THD by running a transient analysis with a sinusoidal input signal of the fundamental frequency. Consider the noninverting amplifier of Fig. 2.6 as specified in Example 2.13 with $A_{cl0,i} = 3$ V/V, $f_{fp} = 3$ MHz, and $V_{i(\max)} = 0.25$ V. Set opamp parameters to $V_{DD} = 1.8$ V, $V_{SS} = 0$, $V_{OH} = 1.65$ V, $V_{OL} = 0.15$ V, $A_0 = 1000$ V/V, $V_{DIR} = 0.15$ V, $f_{GBW} = 12$ MHz, and $f_{nd} = 2$ GHz. Assume $V_B = 0.9$ V, and determine by simulation the total harmonic distortion of this amplifier for the following values of the fundamental frequency: $f = 100$ Hz, $f = 1$ kHz, $f = 10$ kHz, $f = 100$ kHz, and $f = 1$ MHz.

Work Input file ex2-15.sp is as follows:

```
filename: ex2-15.sp
.param vdd=1.8 vss=0 voh=1.65 vol=0.15 a0=1000 vdir=0.15
+  fgbw=12meg fnd=2g
.incl opmacro1.sp
x1 1 2 3 opamp
r2 2 3 20k
r1 2 4 10k
vb 4 0 0.9
vin 1 4 sin 0 0.25 100 ; line 8
.tran 10u 100m
.end
```

Notice the sinusoidal input signal voltage of zero dc, 0.25-V amplitude, and 100-Hz frequency in line 8. Transient analysis is defined in line 9 with a time step $T_{\text{step}} = 10\ \mu\text{s}$ and duration $T_{\text{stop}} = 100$ ms. It is essential to define the duration no less than 10 times longer than the period of the input signal. This is how we can ensure settling into a periodic steady state well before the very last period of simulation begins. It is the last period during which the simulator performs the Fourier analysis needed for determining the harmonic components. It is also advisable not to define a time step longer than 1/100 of the fundamental period. This will minimize the harmonic content of the input signal itself. After the completion of the simulation, the result of the analysis is displayed with the command

```
fourier 100 v(3)
```

where `100` specifies the fundamental frequency, and `v(3)` defines the voltage where the harmonics are to be evaluated. The result is reported with the following table:

```
Fourier analysis for v(3):
No. Harmonics: 10, THD: 0.0115298 %, Gridsize: 200, Interpolation Degree: 1

Harmonic Frequency   Magnitude   Phase        Norm. Mag   Norm. Phase
-------- ---------   ---------   -----        ---------   -----------
 0       0           0.899996    0            0           0
 1       100         0.747594    -0.0015309   1           0
 2       200         6.33051e-06 90.9469      8.46783e-06 90.9485
 3       300         8.21587e-05 1.42176      0.000109897 1.42329
 4       400         2.97264e-06 -88.109      3.97628e-06 -88.107
 5       500         2.39973e-05 -177.63      3.20994e-05 -177.63
 6       600         1.19474e-06 92.8306      1.59812e-06 92.8321
 7       700         7.00963e-06 3.30472      9.37625e-06 3.30625
 8       800         4.43711e-07 -86.236      5.93518e-07 -86.235
 9       900         2.04751e-06 -175.76      2.7388e-06  -175.76
```

Notice that eight harmonics are evaluated in addition to dc and fundamental but we may prefer limiting the report to the first four lines since the harmonics of significance are usually the second and third. We can do this with the command

```
set nfreqs=4
```

before performing the Fourier analysis. The table reports THD as 0.0115%, which is predominantly due to HD_3 as we read from the column of normalized magnitude. This is a consequence of the odd symmetry arising from the coincidence of the dc operation point

$V_O = 0.9$ V with the inflection point $V_M = (V_{OL} + V_{OH})/2$ of the open-loop dc transfer characteristic of the opamp.

Repeating the simulation for the rest of the specified fundamental frequencies, we obtain the following THD figures: 0.0119% for 1 kHz, 0.03322% for 10 kHz, 0.3074% for 100 kHz, and 1.403% for 1 MHz. Notice the rapid rise in distortion as the fundamental frequency is raised above 10 kHz. The reason is that the harmonic frequencies of such a high fundamental are increasingly less suppressed due to the roll-off in $|A|$ beyond the open-loop bandwidth $f_o = f_{GBW}/A_0 = 12$ kHz.

The reader is invited to repeat these simulations for a smaller input amplitude, say, $V_i = 0.1$ V. This will test our prediction that THD should decrease with the decreasing amplitude.

It is also advisable to repeat the simulations after moving the operation point away from the inflection point of the transfer characteristic, and to examine the corresponding change in HD_2. This can be done by retaining the amplitude of the input signal but reducing the bias voltage V_B from 0.9 V to, say, 0.7 V. ▲

2.6 Step Response

Closed-loop time-domain response to a stepping input variable is of interest in many amplifier applications. We call it *step response* in this book. The most typical closed-loop performance metric associated with step response is the time spent in transient while the configuration readjusts its output voltage. This metric is called *settling time*. Modeling of step response and settling time depends on whether the opamp slews or not during the transient. These cases are considered separately in the following two subsections.

2.6.1 Slew-Free Step Response

An amplifier input variable $x_i(t)$ abruptly stepping from an initial dc level X_1 to a final dc level X_2 at $t = 0$ is shown in Fig. 2.22(a) for a rising step and in Fig. 2.22(b) for a falling step. In the following analysis, we designate the height $X_i \doteq X_2 - X_1$ of the step as *input signal*. Note that X_i is positive for a rising step and negative for a falling step. The amplifier responds to this input signal by changing its output voltage $v_O(t)$ from an initial dc level V_{O1} corresponding to X_1 to a final dc level V_{O2} corresponding to X_2. Its output signal, defined as $v_o(t) \doteq v_O(t) - V_{O1}$, gradually evolves in time from zero to a final value

$$V_{O2} - V_{O1} = \frac{\alpha_i}{\beta} X_i$$

as predicted by a first-order dc analysis. Since β is a nonnegative parameter, the polarity of $v_o(t)$ is determined by $\alpha_i X_i$. It is positive as depicted in Fig. 2.22(c) for $\alpha_i X_i > 0$, and negative as shown in Fig. 2.22(d) for $\alpha_i X_i < 0$.

Step response is governed by the closed-loop transfer function (2.45) as long as the opamp complies with the small-signal open-loop transfer function (2.41) and does not slew any time during the transient. Since (2.45) is a single-pole transfer function, the corresponding time-domain solution for the output signal is described by

$$v_o(t) = \left[1 - \exp\left(-2\pi f_{cl} t\right)\right] \frac{\alpha_i}{\beta} X_i, \tag{2.77}$$

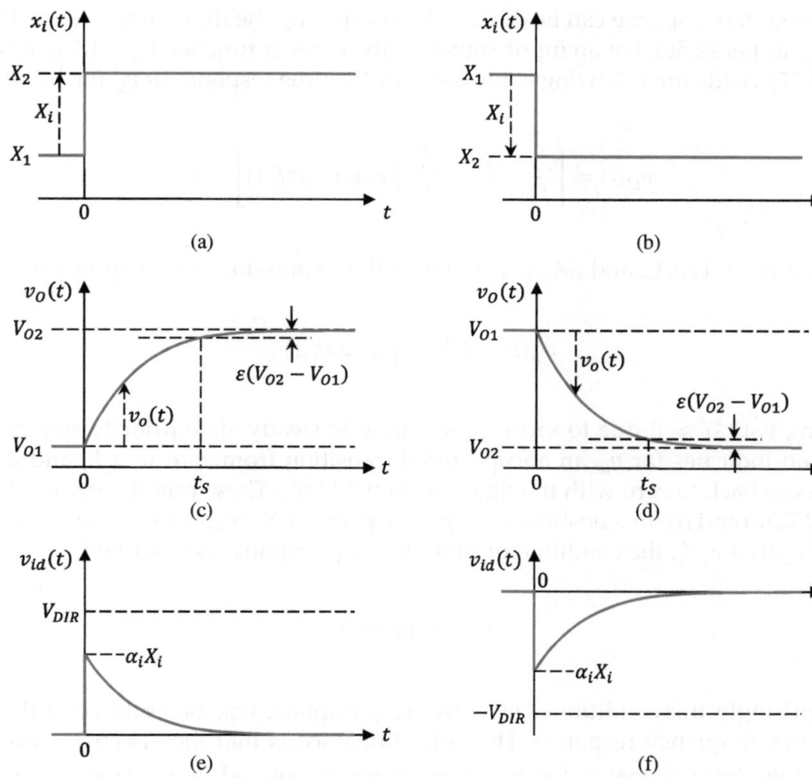

FIGURE 2.22 Small-signal step response of a closed-loop amplifier configuration. (a) A rising input step waveform. (b) A falling input step waveform. (c) Output voltage responding to a positive $\alpha_i X_i$. (d) Output voltage responding to a negative $\alpha_i X_i$. (e) Differential-input voltage responding to a positive $\alpha_i X_i$. (f) Differential-input voltage responding to a negative $\alpha_i X_i$.

which predicts an infinitely long time for $v_o(t)$ to reach its final value. In practice, we consider the transient to be complete once the difference between $v_o(t)$ and its final value has dropped to a specified fraction ϵ of the final value. This fraction, called *settling error*, is specified in accordance with the accuracy requirements of the application. The time needed for completing the transient in this sense is defined as settling time t_S. By substituting $(1 - \epsilon)\alpha_i X_i/\beta$ for v_o in (2.77), and solving the equation for time, we obtain the following expression for the settling time:

$$t_S = \frac{1}{2\pi f_{cl}} \ln\left(\frac{1}{\epsilon}\right), \tag{2.78}$$

according to which, a shorter settling time for a given settling error is possible only by increasing the small-signal bandwidth.

It is worth reemphasizing that (2.77), hence (2.78), are valid only if the opamp has a single-pole transfer function and does not slew any time during transition. The condition of slew-free step response can be derived by comparing the differential-input signal v_{id} with V_{DIR} as per (2.56). For an input-signal X_i, the transfer function $v_{id} = (A_{cl,i}/A)x_i$ leading to (2.57) yields the following expression for the time response of v_{id} for $t \geq 0^+$:

$$v_{id}(t) = \left[\frac{f_o}{f_{cl}} + \left(1 - \frac{f_o}{f_{cl}}\right)\exp\left(-2\pi f_{cl}t\right)\right]\alpha_i X_i,$$

but since $f_o/f_{cl} = 1/\beta A_o$, and βA_0 is very large, this expression is well approximated by

$$v_{id}(t) = \alpha_i X_i \exp\left(-2\pi f_{cl}t\right).$$

Assuming $v_{id}(0^-) \approx 0$ due to virtual short in a dc steady state prior to stepping, this expression indicates for v_{id} an abrupt initial transition from zero to $\alpha_i X_i$ and a subsequent decay back to zero with the time constant $1/2\pi f_{cl}$. These predictions are depicted in Figs. 2.22(e) and (f) for a positive $\alpha_i X_i$ and a negative $\alpha_i X_i$, respectively. Since the largest value of $v_{id}(t)$ is $\alpha_i X_i$, the condition of slew-free step response is described by

$$V_{DIR} > |\alpha_i||X_i|. \tag{2.79}$$

Not surprisingly, this condition of slew-free step response resembles the condition (2.58) of slew-free frequency response. The only difference is that the effective (rms) value $X_i/\sqrt{2}$ of the input signal in frequency response is replaced by the step height $|X_i|$ in step response.

Example 2.16 Assuming a slew-free step response, determine the necessary gain-bandwidth product for the unity-gain buffer amplifier of Example 2.3 (Fig. 2.7) for 0.1% settling error and 100-ns settling time. Verify the result with a transient simulation for $V_B = 0.9$ V and $A_0 = 1000$ V/V.

Work The gain-bandwidth product f_{GBW} of the unity-gain buffer amplifier equals the closed-loop bandwidth f_{cl} as indicated by (2.48) for $\beta = 1$. Substituting $\epsilon = 0.1 \times 10^{-2}$ and $t_S = 100 \times 10^{-9}$ s into (2.78), and solving for f_{cl}, we obtain $f_{GBW} = 11$ MHz.

Verification can be carried out with the following input file `ex2-16.sp`:

```
filename: ex2-16.sp
.param vdd=1.8 vss=0 voh=1.6 vol=0.2 a0=1000 vdir=2
+  fgbw=11meg fnd=2g
.incl opmacro1.sp
x1 1 2 2 opamp
vb 4 0 0.9
vi 1 4 pulse 0 0.5 0 0.1n
.tran 0.05n 1u
.meas tran ymin min v(2)
.meas tran ymax max v(2)
.meas tran teval when v(2)=(1.399478-0.9)*(1-0.001)+0.9
.end
```

Opamp macro parameters include $V_{OH} = 1.65$ V, $V_{OL} = 0.2$ V in compliance with the specifications of Example 2.3, and $V_B = 0.9$ V as an acceptable bias in accordance with the solution of Example 2.3. Also note the adoption of a large $V_{DIR} = 2$ V to preclude slewing, and $f_{GBW} = 11$ MHz as calculated above. As specified in Example 2.3, input signal is stepped from zero to 0.5 V. Stepping is completed in 0.1 ns after it starts at $t = 0$. Transient simulation has a maximum time step 0.05 ns, and lasts for 1 μs, which is long enough to enable full settling.

Although settling time can be extracted manually from the output-voltage plot, a quicker way is to let the simulator measure its value. The three measurement lines appearing in the file serve this purpose. The first two measure the initial value V_{O1} and final value V_{O2} of the output voltage `v(2)`, and put out the result as `ymin` and `ymax`, respectively. Since A_0 is not infinite as implicitly assumed in our step-response formulations, gain error makes these dc values generally different from what we predict on the basis of virtual short. Although the differences are usually small, they may still be (and frequently are) comparable to the specified settling error. This is why it is imperative to measure the exact values of the initial and final values of the output voltage before measuring the settling time accurately. The third measurement line is a command to extract the settling time t_S which is defined as the time for the output voltage to reach

$$v_O(t_S) = \left(V_{O2} - V_{O1}\right)(1 - \epsilon) + V_{O1},$$

but the variables on the right-hand side of this expression must be declared numerically. For this reason, we first run the simulation for determining V_{O1} and V_{O2} only. This initial run yields

```
ymax   =  1.399478e+000 at=  1.000000e-006
ymin   =  9.000000e-001 at=  0.000000e+000
```

Using these values, we numerically define the third measurement line of the input file and rerun the simulation. The result for the settling time is put out as[6]

```
teval   =   1.12430e-007
```
▲

2.6.2 Slew-Limited Step Response

A step input of height $|X_i|$ larger than $V_{DIR}/|\alpha_i|$ immediately forces the opamp into slewing because it breaches the condition (2.79). The way in which the output signal v_o and the differential-input signal v_{id} respond to such an excessive input step is illustrated in parts (c), (d), (e), and (f) of Fig. 2.23. Notice that $|v_{id}|$ exceeds V_{DIR} at $t = 0^+$ just after the step has been applied, and remains so until $t = t_{\text{slew}}$. The opamp slews during this period. However, a slewing opamp is insensitive to any part of $|v_{id}|$ in excess of V_{DIR} because of the collapsing transconductance of one of its amplifier stages. This has two important implications for the step response: (a) The negative-feedback loop is broken through the opamp because opamp is blind to the signal returning from the output, and (b) the opamp responds as if $|v_{id}|$ is stepped only by V_{DIR} because this is the maximum

[6] It is important to note that the minimum and maximum values obtained from the first two measurement lines can be used in the third line only if the step response does not exhibit any overshoot or undershoot. These transient events are described in Section 2.7.

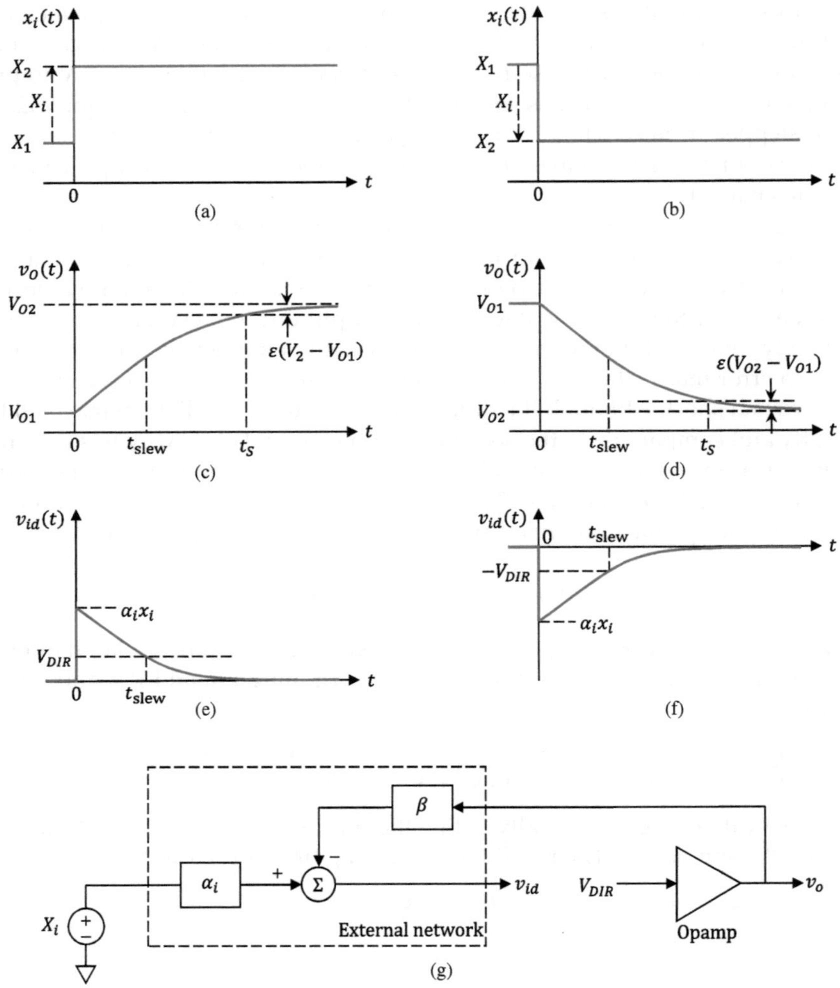

FIGURE 2.23 Step response of a closed-loop amplifier configuration with a slewing opamp. (a) A rising input step waveform. (b) A falling input step waveform. (c) Output voltage responding to a positive $\alpha_i X_i$. (d) Output voltage responding to a negative $\alpha_i X_i$. (e) Differential-input voltage responding to a positive $\alpha_i X_i$. (f) Differential-input voltage responding to a negative $\alpha_i X_i$. (g) Configuration during slewing.

it can sense at its differential input. The state of this open-loop configuration is depicted in Fig. 2.23(g). For analyzing the part of the step response for $0 \leq t \leq t_{\text{slew}}$, we therefore consider an opamp in open loop with a differential-input signal magnitude stepped by V_{DIR} at $t = 0$. Adopting the open-loop transfer function (2.41), the time evolution of the output signal is therefore described by

$$v_o(t) = \text{sgn}(\alpha_i X_i) A_0 V_{DIR} \left[1 - \exp\left(-2\pi f_o t\right)\right], \quad \text{for} \quad 0 \leq t \leq t_{\text{slew}},$$

where $\text{sgn}(\alpha_i X_i)$ is the sign of the product $\alpha_i X_i$. The time constant $1/2\pi f_o$ of the exponential variation indicated by this expression is many orders of magnitude longer than the time t_{slew} spent in slewing because the opamp bandwidth f_o is very narrow. This enables us to approximate the expression by adopting the first-order term of its power-series expansion around $t = 0$, and thus predict the following linear response:

$$v_o(t) = \text{sgn}(\alpha_i X_i) 2\pi A_0 f_0 V_{DIR} t.$$

By replacing $A_0 f_o$ with f_{GBW} in accordance with (2.43), we can formulate the constant time-rate of change of the output signal as

$$S_R \doteq \left| \frac{dv_o(t)}{dt} \right| = 2\pi \times f_{GBW} \times V_{DIR}, \tag{2.80}$$

and therefore represent the time evolution of v_o during slewing with

$$v_o(t) = \text{sgn}(\alpha_i X_i) S_R t, \quad 0 \le t \le t_{\text{slew}}. \tag{2.81}$$

Note that S_R is an open-loop performance metric known as *slew rate*. While $v_o(t)$ evolves in accordance with (2.81), $v_{id}(t)$ follows

$$v_{id}(t) = \alpha_i X_i - \beta v_o(t),$$

as implied by the model depicted in Fig. 2.23(g) during slewing. Substituting (2.81) for $v_o(t)$ in this expression, we obtain

$$v_{id}(t) = \alpha_i X_i \left(1 - \beta S_R t\right). \tag{2.82}$$

Slewing comes to an end at $t = t_{\text{slew}}$ when $|v_{id}|$ drops back to V_{DIR}. Substituting these values into (2.82), and solving for t_{slew}, we obtain

$$t_{\text{slew}} = \frac{|\alpha_i||X_i| - V_{DIR}}{\beta S_R}. \tag{2.83}$$

As expected, (2.83) yields a physically meaningful positive slewing time only if the condition $|\alpha_i||X_i| > V_{DIR}$ of slewing is satisfied.

The negative-feedback loop is closed again from $t = t_{\text{slew}}$ onward. The rest of the transient response is governed by the small-signal closed-loop transfer function (2.45), which implies the following variation for the output signal:

$$v_o(t) = \frac{\alpha_i X_i}{\beta} - \text{sgn}(\alpha_i X_i) \frac{V_{DIR}}{\beta} \exp\left[-2\pi f_{cl}(t - t_{\text{slew}})\right], \quad \text{for } t \ge t_{\text{slew}}. \tag{2.84}$$

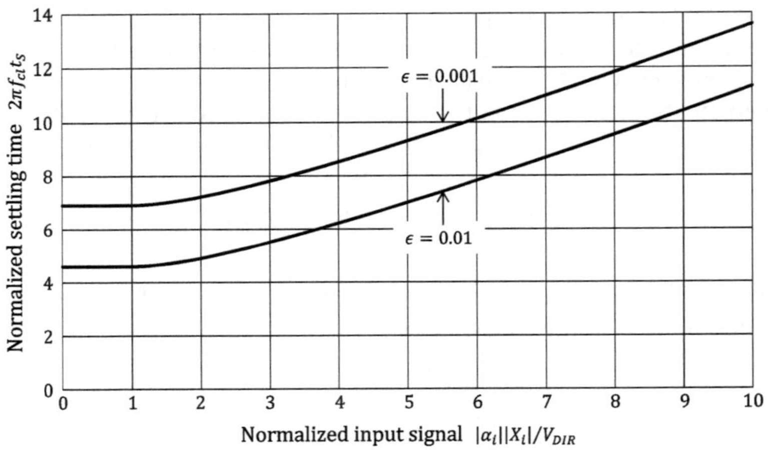

FIGURE 2.24 Variation of the normalized settling time $2\pi f_{cl}t_S$ with the normalized input signal $|\alpha_i||X_i|/V_{DIR}$ for $\epsilon = 0.1\%$ and $\epsilon = 1\%$. Slewing prevails for $|\alpha_i||X_i|/V_{DIR} > 1$.

Settling time t_S can be determined from (2.84) by substituting $(1 - \epsilon)\alpha_i X_i/\beta$ for v_o and (2.83) for t_{slew}, and solving the equation for time. This leads to

$$t_S = \frac{1}{2\pi f_{cl}}\left[\frac{|\alpha_i||X_i|}{V_{DIR}} + \ln\left(\frac{V_{DIR}}{|\alpha_i||X_i|}\right) - 1 + \ln\left(\frac{1}{\epsilon}\right)\right]. \tag{2.85}$$

This expression can be represented with a universal plot of $2\pi f_{cl}t_S$ versus $|\alpha_i||X_i|/V_{DIR}$ for a specific settling error ϵ. Note that $2\pi f_{cl}t_S$ represents the settling time as normalized to small-signal time constant $1/2\pi f_{cl}$, whereas $|\alpha_i||X_i|/V_{DIR}$ is input step height normalized to $V_{DIR}/|\alpha_i|$. The plots shown in Fig. 2.24 belong to the limits $\epsilon = 1\%$ and $\epsilon = 0.1\%$ most typically specified for settling error in practice. Recall that a step response is slew free for $|\alpha_i||X_i|/V_{DIR} \leq 1$, for which, the settling time is described by (2.78). Otherwise, the response is subject to slewing, for which (2.85) describes the settling time. Figure 2.24 covers both domains, and clearly shows how the settling time for a given small-signal closed-loop bandwidth is elongated by slewing.

Example 2.17 This example involves the application considered in Example 2.16 but the differential-input range is specified as $V_{DIR} = 0.1$ V. Determine the necessary gain-bandwidth product for 0.1% settling error and 100-ns settling time. Verify with simulation.

Work We first test the configuration for slewing. Since $\alpha_i = 1$ for the unity-gain buffer, and X_i is specified as 0.5 V, the product $|\alpha_i||X_i|$ equals 0.5 V. Step response will involve slewing because $V_{DIR} = 0.1$ V is smaller than $|\alpha_i||X_i|$. Equation (2.85) yields $f_{cl} = 14.8$ MHz. Since $\beta = 1$ for the configuration, we obtain $f_{GBW} = 14.8$ MHz from (2.48).

For verification, we run `ex2-16.sp` of Example 2.16 after setting V_{DIR} to 0.1 V and f_{GBW} to 14.8 MHz. Simulation yields $t_S = 113.3$ ns.

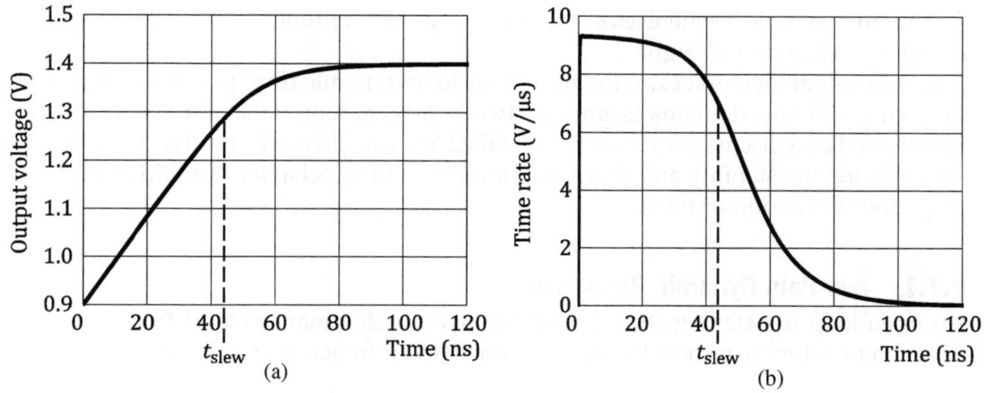

FIGURE 2.25 Waveforms of the unity-gain buffer amplifier of Example 2.17. (a) Output voltage. (b) Time derivative of output voltage. t_{slew} is the time slewing ends.

For the slew rate, we predict $SR = 9.3$ V/μs from (2.80). The corresponding slew time is calculated from (2.83) as $t_{slew} = 43$ ns. Simulated plots of the output voltage itself and its time derivative are given in Figs. 2.25(a) and (b), respectively. The calculated value of t_{slew} is also marked on these plots. Note from (2.81) that the derivative of the output voltage should be equal to the slew rate for $t \leq t_{slew}$. According to Fig. 2.25(b), the rate indeed closely matches the calculated value of the slew rate for the early part of slewing but, unlike the theoretical expectation, it gradually declines in the later part of slewing. This is due to the fact that the transition between slewing and small-signal operation is not as abrupt as assumed in (2.79). ▲

2.7 Multi-Pole Dynamic Response and Stability

Our treatment of small-signal closed-loop frequency response in Subsection 2.5.1 and step response in Subsection 2.6.1 is based on the single-pole open-loop transfer function described by (2.41). In reality, the open-loop transfer function of an opamp contains multiple poles, and, in some cases, also zeros. Among these poles, the one located at the lowest frequency is called the *dominant pole*. Others are collectively named as *nondominant poles*. Equation (2.41) represents an ideal case where the dominant pole is located at $f = f_o$ whereas all nondominant poles are at $f = \infty$. It nevertheless works well in predicting the frequency and step responses of an amplifier as long as the finite frequencies of the nondominant poles are at least a decade higher than the closed-loop bandwidth f_{cl}. This condition is usually met in simple opamp topologies. In complex topologies, however, nondominant pole frequencies may be much closer to f_{cl}, or even lower than the latter. Their impact on closed-loop frequency and step responses may be favorable or not depending on how many of them encroach how far upon the closed-loop bandwidth. For example, two or more nondominant poles located at frequencies lesser than f_{cl} may render the amplifier dysfunctional by causing oscillation. On the other hand, one nondominant pole whose frequency is a factor of 2 to 4 higher than f_{cl} helps minimize the settling time provided that other nondominant pole frequencies are much higher than

f_{cl}. Shaping the pole frequencies of the open-loop transfer function is therefore a critical design task in opamp design.

In this section, we first examine the closed-loop dynamic response of an amplifier in frequency and time domains assuming a two-pole open-loop transfer function. This will be followed by a discussion of how the so-called *loop gain* βA can be used as an instrument for verifying the stability and characterizing the settling behavior of multi-pole closed-loop amplifier configurations.

2.7.1 Two-Pole Dynamic Response

An open-loop transfer function containing one nondominant pole of frequency f_{nd} in addition to a dominant pole located at a much lower frequency f_o is described by

$$A(s) = \frac{A_0}{\left(1 + \frac{s}{2\pi f_o}\right)\left(1 + \frac{s}{2\pi f_{nd}}\right)}, \tag{2.86}$$

where A_0 denotes the dc differential gain as usual. Substituting this expression into (2.40), and assuming $\beta A_0 \gg 1$ and $f_o \ll f_{nd}$ as well as frequency-independent α_i and β as appropriate for amplifier applications, we obtain the following closed-loop gain expression:

$$A_{cl,i} = \frac{1}{1 + \frac{s}{2\pi f_{cl}} + \frac{s^2}{(2\pi)^2 f_{nd} f_{cl}}} \frac{\alpha_i}{\beta}, \tag{2.87}$$

Note that f_{cl} is still defined by (2.48) and will still be referred to as closed-loop bandwidth but, unlike the case of a single-pole response, it is no more necessarily the frequency at which the magnitude $|A_{cl,i}|$ drops by 3 dB with respect to $|A_{cl0,i}|$. Closed-loop dynamic response in frequency or time domain is now shaped up by the poles

$$s_{1,2} = -\pi f_{nd}\left(1 \pm \sqrt{1 - 4\frac{f_{cl}}{f_{nd}}}\right) \tag{2.88}$$

of (2.87). The shape of the response depends directly on the ratio f_{nd}/f_{cl} as we now discuss and exemplify with ac and transient simulations on the noninverting amplifier of Fig. 2.6 for the specifications $A_{cl,i} = 3$ V/V and $f_{cl} = 3$ MHz defined in Example 2.10. The input file (`twopole.sp`) to be used in these simulations is as follows:

```
filename: twopole.sp
.param vdd=1.8 vss=0 voh=1.65 vol=0.15 a0=1000 vdir=0.25
+  fgbw=9meg fnd=30meg
.incl opmacro1.sp
x1 1 2 3 opamp
r1 2 4 10k
r2 2 3 20k
vb 4 0 dc 0.9
```

```
********ac simulation*********
vi 1 4 dc 0 ac 1m
.ac dec 100 1k 50meg
*****transient simulation*****
*vi 1 4 pulse -100m 100m 0 0.1n 0.1n 1u
*.tran 0.5n 2u
******************************
.end
```

This file is created by supplementing the ac file `ex2-10a.sp` of Example 2.10 with a transient analysis. Notice the definition of the input signal of the transient analysis as a pulse of duration 1 μs switching from −100 to 100 mV at $t = 0$ and falling back at $t = 1$ μs. Since the circuit is biased with 0.9 V, and closed-loop gain is 3 V/V, the output voltage is expected to switch between 0.6 and 1.2 V in transient analysis. Also note that the nondominant pole frequency is set by the parameter `fnd` in line 3.

We begin our discussion with the case of $f_{nd}/f_{cl} > 4$. In this case, closed-loop dynamic response is said to be *overdamped*, for which, (2.88) indicates real and negative poles. For very large values of f_{nd}/f_{cl}, one of these poles is located at the bandwidth f_{cl}, and the other at the nondominant pole frequency f_{nd}. Since f_{nd} is much higher than the bandwidth, dynamic behavior is very similar to what is observed for single-pole response. This case is exemplified for $f_{nd}/f_{cl} = 10$ in Fig. 2.26(a) with simulated open-loop and closed-loop magnitude responses, and in Fig. 2.26(b) with the closed-loop step response.

Decreasing f_{nd}/f_{cl} toward 4 lowers the frequency of the high-frequency pole of (2.88) below f_{nd} and raises the lower-frequency pole above f_{cl}. As a result, the frequency at which the magnitude drops by 3 dB gradually increases while the step response exhibits a faster time evolution but, otherwise, the shape of both responses remains qualitatively similar to their respective single-pole counterparts.

For $f_{nd}/f_{cl} = 4$, (2.88) yields real but coincident poles $s_1 = s_2 = -4\pi f_{cl}$. In this case, dynamic response is said to be *critically damped*. As illustrated in Figs. 2.26(c) and (d), both magnitude and time responses still retain single-pole forms but now the frequency at which the magnitude drops by 3 dB is somewhat higher than f_{cl}. Also notice the markedly faster step response in comparison with the case of overdamped response.

The poles defined by (2.88) become complex conjugate as f_{nd}/f_{cl} drops below 4. In this case, the dynamic response is said to be *underdamped*. The frequency at which the magnitude drops by 3 dB continues to rise but, qualitatively, the magnitude response still retains its single-pole shape as long as $f_{nd}/f_{cl} \geq 2$. This is exemplified in Fig. 2.26(e), which shows the magnitude response for $f_{nd}/f_{cl} = 2$. Step response, on the other hand, undergoes a significant change for $f_{nd}/f_{cl} < 4$. As illustrated in Fig. 2.26(f) for $f_{nd}/f_{cl} = 2$, it starts to exhibit overshooting and undershooting episodes before eventually settling. This is what we call *ringing*.

For $f_{nd}/f_{cl} < 2$, ringing becomes more pronounced in step response as exemplified in Fig. 2.26(h) for $f_{nd}/f_{cl} = 1$ while the magnitude response significantly deviates from its single-pole form and starts exhibiting what we call *gain peaking* as exemplified in Fig. 2.26(g) for $f_{nd}/f_{cl} = 1$.

Our observations on the step response of the two-pole amplifier configuration raise the question as to which particular value of f_{nd}/f_{cl} should we target in design in order to minimize the settling time. For an answer, refer to Fig. 2.27, where we show the final phase of settling in all four cases considered in Fig. 2.26. The figure also shows two settling-error bands of width $2\epsilon_1 V_o$ and $2\epsilon_2 V_o$ around the target value of the output pulse. Since

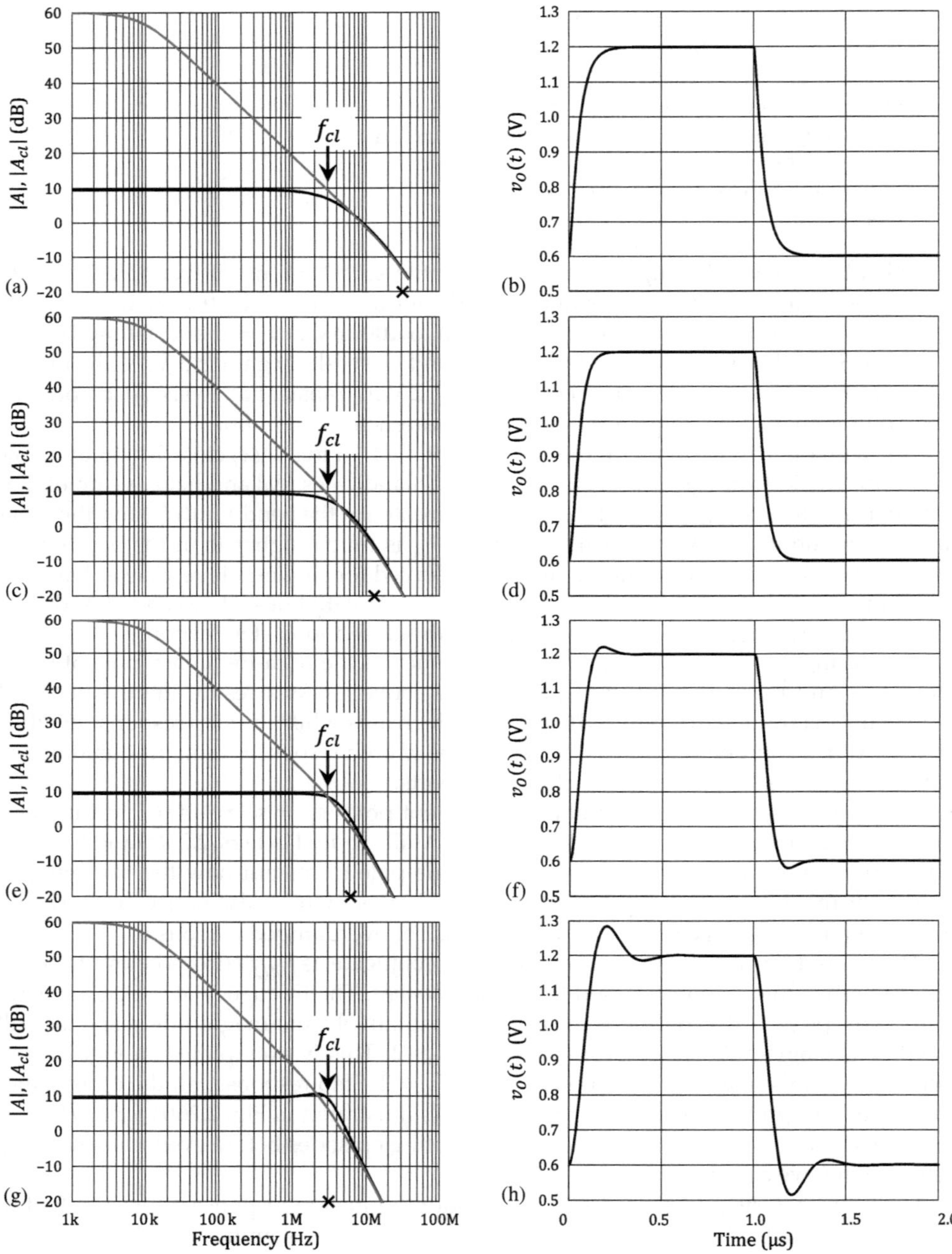

FIGURE 2.26 Simulated two-pole magnitude response and step response as obtained with `twopole.sp`. Opamp differential gain is plotted in gray. Nondominant pole frequency is marked with a cross. (a, b): $f_{nd}/f_{cl} = 10$. (c, d): $f_{nd}/f_{cl} = 4$. (e, f): $f_{nd}/f_{cl} = 2$. (g, h): $f_{nd}/f_{cl} = 1$.

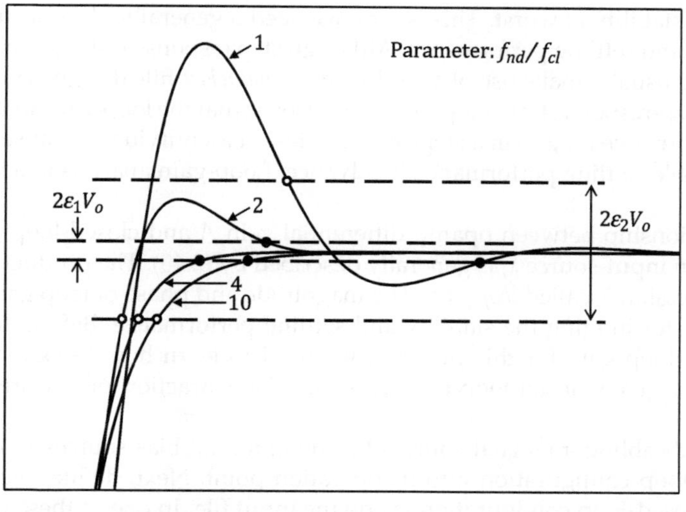

FIGURE 2.27 Final phase of settling for f_{nd}/f_{cl} = [1, 2, 4, 10].

a ringing output voltage can cross the boundaries of an error band more than once, we need to generally define the settling time t_S as the time to the very last crossing into the specified error band from outside. Settling times complying with this definition are marked in Fig. 2.27 with closed circles for the narrower band of settling error ϵ_1, and with open circles for the wider band of settling error ϵ_2. What we observe is that the shortest settling time is achieved with $f_{nd}/f_{cl} = 4$ for the narrower error band but with $f_{nd}/f_{cl} = 2$ for the wider one. This leads us to a rule of thumb that the settling time is minimized by placing the nondominant pole frequency in between $2f_{cl}$ of the slightly underdamped response and $4f_{cl}$ of the critically damped response. Where exactly the optimum lies depends on how large the specified settling error is. Obviously, an underdamped response is more suitable for a large-settling error, whereas a critically damped response is more desirable for a small-settling error.

The closed-loop step response exemplified in Figs. 2.26 and 2.27 is free of slewing because its specifications satisfy (2.79). In those cases where slewing is involved, the effect of the nondominant pole is observed during the final slew-free phase of time evolution. The rule of thumb stated above is still valid in such cases.

An important conclusion we can draw from two-pole dynamic response is that (2.78) and (2.85) somewhat overestimate the settling time if a nondominant pole creates a critically damped or mildly underdamped response. Trying to include these secondary phenomena would make these two equations unduly complex. Instead, we prefer keeping them as simple as they are, and using them as a first-order approximation without losing sight of their shortcomings.

2.7.2 Loop Gain, Phase Margin, and Stability

As we pointed out earlier, most opamps contain more than two poles and in some cases exhibit zeros as well. Although we ignore these excess poles and zeros while designing an opamp, they may affect the settling time of a closed-loop configuration at best, and threaten its stability at worst. This is why we need a general tool for testing the design for stability and settling performance. Although step response is the ultimate test for this purpose, we usually make use of an *ac loop-gain analysis* while design is in progress, and apply the step-response test at the end. The reason is that the loop-gain analysis proves to be much more revealing than a step-response test in identifying the causes of instability or undesirable settling performance in advance. Loop-gain analysis is what we discuss here.

The relationship between opamp differential gain A and closed-loop gain $A_{cl,i}$ with respect to an input-source x_i is generally described by (2.40). The product βA appearing in this expression is called *loop gain*. The magnitude and phase of loop gain are what we use as tools for judging the stability and settling performance. Before discussing how to interpret loop gain for this purpose, we need to learn how to extract it with simulation for a given closed-loop configuration. The extraction procedure is simple yet general.[3]

Start by disabling all signal sources but retaining all bias sources in order to bring the closed-loop configuration into its operation point. Next, define two copies of the resulting closed-loop configuration inside the input file. In one of these copies, insert a floating ac test voltage source v_x of zero dc value inside the feedback loop as exemplified in Fig. 2.28(a). Assign its reference polarity arbitrarily. Define A as the node where the signal flowing in the feedback loop enters this source, and B as the node where it exits. Represent the ac voltages of these nodes with v_a and v_b, respectively, and define

$$\mathcal{R}_V \doteq -\frac{v_a}{v_b} \tag{2.89}$$

as *voltage return ratio*. Now, take the second copy of the biased closed-loop configuration and insert between the very same nodes A and B an ac test current source i_x of zero dc level and two zero-volt voltage sources V_{S1} and V_{S2} acting as ammeters as exemplified in Fig. 2.28(b). Represent the ac currents flowing through these two ammeters in the direction of signal flow with i_a and i_b, and define

$$\mathcal{R}_I \doteq -\frac{i_a}{i_b} \tag{2.90}$$

as *current return ratio*. Finally, run an ac simulation on these two copies of the closed-loop configuration, and evaluate the loop gain βA as related to $\mathcal{R}_V$ and $\mathcal{R}_I$ through

$$\beta A = \frac{\mathcal{R}_V \mathcal{R}_I - 1}{2 + \mathcal{R}_V + \mathcal{R}_I}. \tag{2.91}$$

It is worth emphasizing that any wire inside the feedback loop can be split for the purpose of creating the nodes A and B for test-source insertion. In the case of Fig. 2.28, for example, the wires on both sides of the feedback resistor R_2 are suitable for source insertion. So is the wire connecting the common node of R_1 and R_2 to the inverting input of the opamp. Although $\mathcal{R}_V$ and $\mathcal{R}_I$ vary depending on the location selected for source insertion, βA itself will remain indifferent.

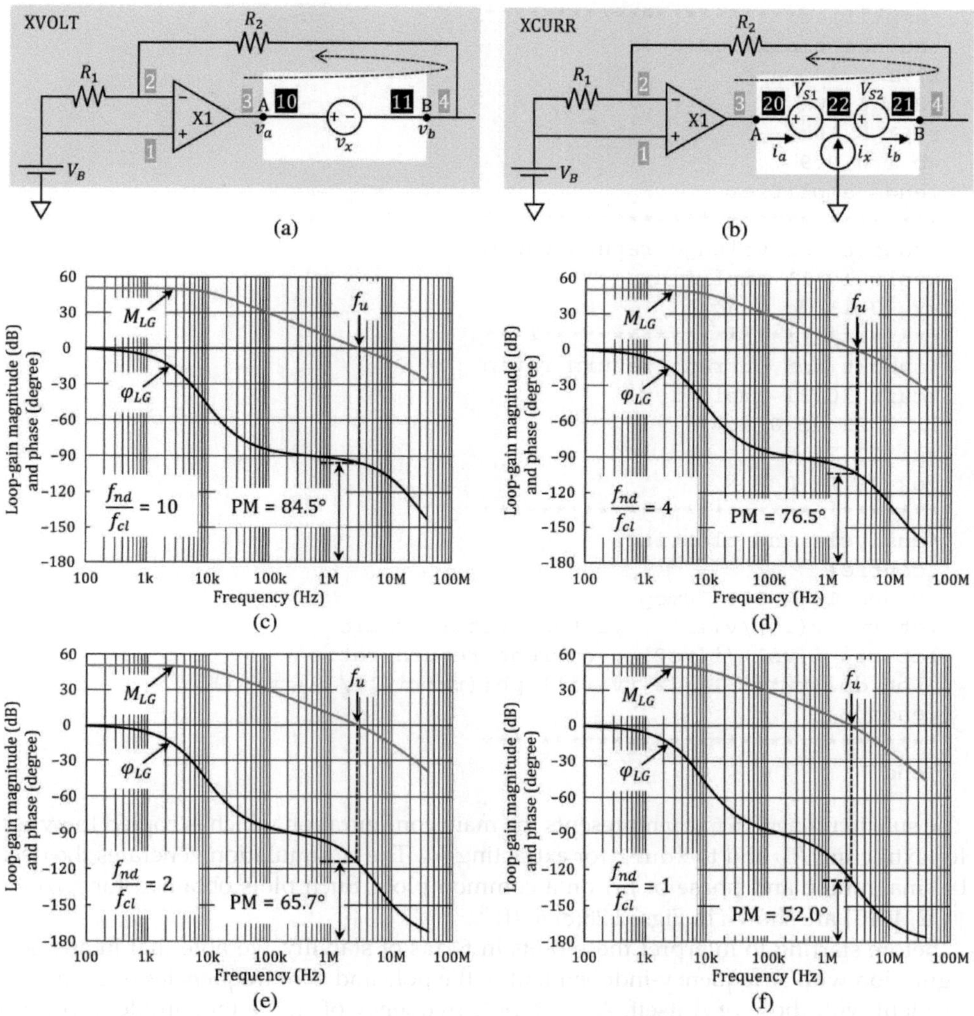

FIGURE 2.28 (a) Inserting an ac test voltage source v_x into the feedback loop for extracting the voltage return ratio $\mathcal{R}_V$ by simulation. The dashed arrow indicates the direction of signal flow. (b) Inserting an ac test current source i_x for extracting the current return ratio $\mathcal{R}_I$ by simulation. The dashed arrow indicates the direction of signal flow. (c–f) Magnitude M_{LG} and phase φ_{LG} plots of loop gain for four different values of f_{nd}/f_{cl}. PM indicates the phase margin.

As an example, we now apply the simulated loop-gain extraction procedure to the two-pole noninverting amplifier considered in Subsection 2.7.1. The circuit to be simulated is shown in Figs. 2.28(a) and (b). Its input file (`loopgain.sp`) is as follows:

```
filename: loopgain.sp
.param vdd=1.8 vss=0 voh=1.65 vol=0.15 a0=1000 vdir=0.25
+  fgbw=9meg fnd=3meg
.incl opmacro1.sp
```

```
*amplifier*********************
.subckt amplifier 3 4
x1 1 2 3 opamp
r2 2 4 20k
r1 1 2 10k
vb 1 0 0.9
.ends amplifier
*******************************
*subckt for voltage return ratio
xvolt 10 11 amplifier
vx 10 11 dc 0 ac 1
*******************************
*subckt for current return ratio
xcurr 20 21 amplifier
ix 0 22 dc 0 ac 1
vs1 20 22 0
vs2 22 21 0
*******************************
*analysis and plotting
.control
ac dec 1000 100 10meg
let rv=-v(10)/v(11) ; voltage return ratio
let ri=-i(vs1)/i(vs2) ; current return ratio
plot db((rv*ri-1)/(2+rv+ri)) ph((rv*ri-1)/(2+rv+ri))
.endc
*******************************
.end
```

The subcircuit `amplifier` represents the main configuration which is copied to `xvolt` for extracting $\mathcal{R}_V$ and to `xcurr` for extracting $\mathcal{R}_I$. The ac simulation generates the decibel magnitude and phase of βA on a common plot.[7] Such plots obtained for $f_{nd}/f_{cl} = \{1, 2, 4, 10\}$ are shown in Figs. 2.28(c) to (f).[8]

Before starting to interpret these plots in terms of stability, we note that in any configuration with a frequency-independent β, the pole and zero frequencies of βA are coincident with those of A itself. At each pole frequency of A, the magnitude of βA will therefore change its roll-off rate by -20 dB/dec while the phase will exhibit a $-45°$ rotation. Total phase rotation per pole is of course $-90°$, but this much rotation is completed

---

[7] Before running the simulation, set the unit of phase angle to degrees with command "`set units=degrees`."

[8] If the impedance seen by the test source through node A (where the loop signal flows into the source) is at least an order of magnitude smaller than the impedance seen through node B (where the loop signal flows out of the source), then, (2.91) is simplified into $\beta A \approx \mathcal{R}_V$. In that case, we don't need to involve the test current source i_x; a single copy of the configuration containing the test voltage source v_x will suffice for extracting the loop gain. The reader may prefer exploiting this opportunity if prior to simulation he/she can identify a loop wire where the impedances satisfy the necessary condition. The wire split in Fig. 2.28 is a typical example because the output impedance of the opamp seen by v_x through node A is negligibly small in comparison with the impedance this source sees beyond node B. The reader may wish to test this claim by plotting the magnitude and phase of $\mathcal{R}_V$ with the command "`plot db(-v(10)/v(11)) ph(-v(10)/v(11))`" after running the simulation, and comparing the result with what the simulation automatically generates.

in an approximately two-decade wide-frequency range centered at the pole frequency. The effect of a left half-plane zero is just the opposite of a pole. The roll-off rate changes by +20 dB/dec while the phase is rotated by +45° at its frequency. Total phase rotation equals +90°. There are cases where we encounter a right half-plane zero, which changes the roll-off rate by +20 dB/dec but rotates the phase by a total of −90°. These properties enable us to easily identify pole/zero frequencies on magnitude and phase characteristics. In the example of Fig. 2.28 we have no zeros but two poles, one located at $f = f_o$ and another at $f = f_{nd}$. Note that f_o is fixed in all four cases of Fig. 2.28 at 9 kHZ as calculated from (2.43) for f_{GBW} = 9 MHz and A_0 = 1000 V/V. f_{nd}, however, equals 30 MHz in Fig. 2.28(c), 12 MHz in (d), 6 MHz in (e), and 3 MHz in (f). Since (2.48) implies f_{cl} = 3 MHz for f_{GBW} = 9 MHz and $\beta = 1/3$, the values of f_{nd}/f_{cl} corresponding to these four cases are 10, 4, 2, and 1, respectively.

We can now move to the issue of interpreting the magnitude and phase plots for the extent of stability. First, note that the particular frequency at which the magnitude M_{LG} drops to 1 V/V (0 dB) is called *unity-gain frequency*. It is represented by f_u on the plots of Fig. 2.28. The phase $\varphi_{LG}(f_u)$ at this frequency is an indicator of whether the closed-loop configuration is definitely stable, and if so, how much the extent of damping is. To be specific, we evaluate the so-called *phase margin* for this purpose. It is defined by

$$\text{PM} \doteq \varphi_{LG}(f_u) - (-180°). \tag{2.92}$$

According to Barkhausen's stability criterion, a closed-loop configuration is stable if PM > 0. All of the four cases displayed in Fig. 2.28 exhibit a positive PM, hence stability, which is perfectly consistent with what we observed from the step responses simulated earlier and shown in Fig. 2.26. But notice from Fig. 2.28 how PM decreases as the nondominant pole frequency f_{nd} is lowered. In an amplifier built with a single-pole opamp, PM equals 90° because the total rotation of φ_{LG} is due to the single-pole f_o, and amounts to −90° only. The presence of the nondominant pole adds further rotation to the phase, and thus lowers the phase margin. But if its frequency f_{nd} is a decade or so higher than f_u, the effect is negligible as observed in Fig. 2.28(c) of the highly overdamped case of f_{nd}/f_{cl} = 10 where we read PM = 84.5°. In the critically damped case of Fig. 2.28(d), where f_{nd}/f_{cl} = 4, the effect of the nondominant pole on φ_{LG} is much more pronounced at $f = f_u$. As a result, PM is lowered to 76.5°. It is further lowered to 65.7° in the case of f_{nd}/f_{cl} = 2, and to 52.0° in that of f_{nd}/f_{cl} = 1, as observed in Figs. 2.28(e) and (f), respectively. Recall our conclusion in Subsection 2.7.1 that the fastest settling time is observed if f_{nd}/f_{cl} is in between 2 and 4. We now know that this range corresponds approximately to 65° ≤ PM ≤ 75°. This conclusion is important because damping specification is usually issued in terms of phase margin. The designer needs to translate this specification into f_{nd}/f_{cl} which happens to be the ultimate determinant of the extent of damping. The relationship between the two is discussed next.

The formal procedure of analytically modeling the phase margin of a given closed-loop configuration is as follows:

1. Write an s-domain expression for βA.
2. Replace each s with $j2\pi f$.
3. Determine the magnitude and phase expressions of the resulting complex function.
4. Equate the magnitude expression to zero, and solve for f to obtain f_u.

5. Substitute f_u into the phase expression to obtain $\varphi_{LG}(f_u)$.
6. Use $\varphi_{LG}(f_u)$ in (2.92) to obtain PM.

This procedure yields a simple and very useful expression in the special but significant case of amplifier configurations built around a two-pole opamp. Adopting (2.86) for A, and assuming a frequency-independent β, it leads us to the unity-gain frequency expression

$$f_u \approx \left\{ \left[\left(\frac{f_{cl}}{f_{nd}} \right)^2 + \frac{1}{4} \right]^{1/2} - \frac{1}{2} \right\}^{1/2} f_{nd}, \tag{2.93}$$

where f_{cl} is substituted for $\beta A_0 f_o$ as admitted by (2.43) and (2.48).[9]

Note that (2.93) indicates $f_u \approx f_{cl}$ for $f_{nd} \gg f_{cl}$. Although it also predicts f_u to fall increasingly short of f_{cl} as we decrease f_{nd}, the difference remains less than 10% of f_{cl} even for $f_{nd} = 2f_{cl}$. This fact, also verified by the plots of Fig. 2.28, enables us to assume $f_u \approx f_{cl}$ for the typical range of damping targeted in practice.

As to the phase at unity-gain frequency, the procedure yields

$$\varphi_{LG}(f_u) = -90° - \arctan\left(\frac{f_u}{f_{nd}}\right),$$

whose substitution into (2.92) leads to the following expression for the phase margin:

$$\text{PM} = \arctan\left(\frac{f_{nd}}{f_u}\right).$$

Considering $f_u \approx f_{cl}$ for all practical cases of damping, this expression can be turned into

$$\text{PM} \approx \arctan\left(\frac{f_{nd}}{f_{cl}}\right). \tag{2.94}$$

Equation (2.94) provides the designer with a very simple tool by which the phase-margin specification of an amplifier can be translated into the value of f_{nd} to be targeted in design.

2.8 Effect of External Network on Differential Gain

This chapter's coverage up to now may have given the reader the false impression that the differential gain A of an opamp is unconditionally independent of the properties of the external network. A actually depends on external-network parameters due to the nonzero output impedance of the opamp, and unless properly managed by design, this dependence may have undesirable consequences. For modeling this effect in the context

[9] This expression is subject to the assumption $f_o \ll f_u$ and $f_o \ll f_{nd}$, which are justified in all applications where the two-pole approximation is valid.

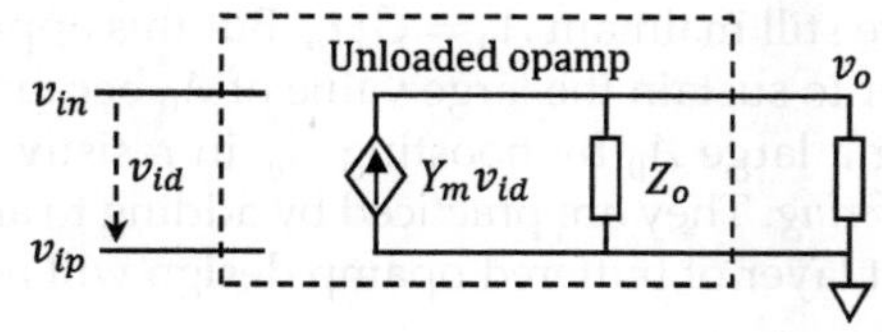

FIGURE 2.29 Equivalent circuit for the opamp block of Figs. 2.10(a) and (b).

of small-signal dc, frequency, and step responses, we ignore the nonlinearity and offset of the open-loop dc transfer characteristic, and represent the opamp block of Fig. 2.10 with the abstract Norton equivalent circuit shown in Fig. 2.29. This equivalent circuit comprises an *effective transadmittance* Y_m and an *output impedance* Z_o which are specific to the opamp, and are absolutely independent of the external network. Z_L, on the other hand, represents the load impedance imposed onto the opamp by external-network components as well as the circuit or device driven by the closed-loop configuration. This impedance is totally independent of opamp parameters. Since small-signal output signal v_o is related to the differential-input signal v_{id} by $v_o = Y_m(Z_o \parallel Z_L)v_{id}$, the open-loop transfer function $A \doteq v_o/v_{id}$ of the opamp is described by

$$A = \frac{Z_L}{Z_o + Z_L} Y_m Z_o, \tag{2.95}$$

where $Y_m Z_o$ represents the *unloaded* differential gain of the opamp. Indeed, if we disconnect the load impedance ($Z_L = \infty$), (2.95) will reduce to $A = Y_m Z_o$. Note that an ideal opamp with a vanishing output impedance ($Z_o = 0$) would also exhibit $A = Y_m Z_o$ regardless of Z_L. Obviously, what is responsible for the dependence of A on Z_L is the nonzero output impedance Z_o.

According to (2.95), the poles and possible zeros of A are contributed collectively by the opamp through Y_m and Z_o and its loading through Z_L. The dc form of this transfer function is obtained by replacing the impedances with the corresponding resistances, and Y_m with the transconductance G_m. This leads to

$$A_0 = \frac{R_L}{r_o + R_L} G_m r_o. \tag{2.96}$$

Note that R_L becomes infinite and therefore turns (2.96) into $A_0 = G_m r_o$ if (a) the external network is not resistive as in the case of unity-gain buffer amplifier or in the case of any discrete-time closed-loop configuration, and (b) the opamp does not drive a resistive circuit. In such a case, even a very large A_0 can be implemented with a simple opamp featuring a very large r_o and a moderate G_m. Such an opamp is usually referred to as *operational transconductance amplifier* (OTA) due to its high-output resistance. OTAs are very economical in terms of topological complexity, footprint, and design labor, and therefore much preferred in those applications involving nonresistive loading. In the case of resistive loading with a small R_L, however, the large r_o of an OTA can no more sustain a large A_0. If we maintain $r_o \gg R_L$, (2.96) will lead to $A_0 = R_L G_m$, and therefore indicate a necessity for boosting G_m in order to sustain the same A_0. Alternatively, we may reduce

r_o below R_L, and therefore still maintain $A_0 = G_m r_o$. But this approach will also necessitate a boosted G_m in order to sustain the large value of A_0 because r_o is now small. Both approaches to sustaining a large A_0 by boosting G_m in resistively loaded opamps are commonly known as *buffering*. They are practiced by adding to an OTA a special output stage called *buffer*. Circuit layer of buffered opamp design will be presented in Sections 4.9 and 4.10.

As pointed out in connection with (2.95), load impedance Z_L generally affects the pole and zero distribution of opamp differential gain A. Furthermore, its dc value R_L is a limiting factor of opamp dc differential gain A_0, as our discussion of (2.96) has revealed. This is why Z_L or R_L must be taken into consideration whenever the opamp is to be designed for a specified A or A_0. We do this by applying the following four-step procedure for extracting A from the circuit schematic of a closed-loop configuration:

Step 1 Disable all input sources.

Step 2 Separate the inverting and noninverting input terminals of the opamp from the rest of the configuration. However, attach the input impedances of these two terminals to the external network if deemed significant.

Step 3 Include the input impedance of the circuit driven by the closed-loop configuration.

Step 4 Apply an independent voltage-mode signal source v_{id} to the separated differential input of the opamp. By analyzing the resulting schematic, relate the output signal v_o to v_{id}. The ratio v_o/v_{id} is the differential gain A.

Example 2.18 Formulate the opamp differential gain A in terms of transadmittance Y_m and output impedance Z_o for (a) the noninverting amplifier of Fig. 2.6, (b) the unity-gain buffer of Fig. 2.7, (c) the transresistance amplifier of Fig. 2.9, (d) the inverting amplifier of Fig. 2.8, and (e) the filtering amplifier of Fig. 2.19. The circuit driven by each of these configurations has an input impedance Z_{in}.

Noninverting Amplifier It is obvious from the closed-loop schematic shown in Fig. 2.6 that the separation of the two inputs of the opamp from the rest of the circuit and disabling the sources will place the series combination of R_1 and R_1 across the output port. Since Z_{in} is also across the output port, the load impedance is described by $Z_L = (R_1 + R_2) \parallel Z_{in}$. The final schematic to be used for extracting A is shown in Fig. 2.30(a). It yields

$$A = \frac{(R_1 + R_2) \parallel Z_{in}}{Z_o + \left[(R_1 + R_2) \parallel Z_{in}\right]} Y_m Z_o$$

in accordance with (2.95).

Unity-Gain Buffer Referring to Fig. 2.7, note that the separation of the two inputs of the opamp and disabling the bias and signal sources leave no external-network impedance to load the output. Z_{in} being the sole load item, the final schematic to be used for extracting A is as shown in Fig. 2.30(b). A is therefore given by

$$A = \frac{Z_{in}}{Z_o + Z_{in}} Y_m Z_o.$$

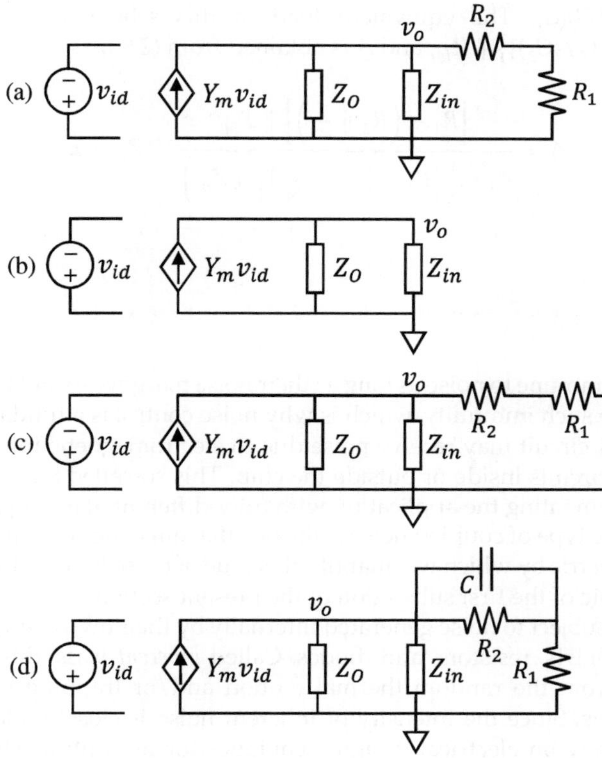

FIGURE 2.30 Equivalent circuits for extracting A in Example 2.18. (a) Noninverting amplifier. (b) Unity-gain buffer and transresistance amplifiers. (c) Inverting amplifier. (d) Filtering amplifier.

Transresistance Amplifier Referring to Fig. 2.9, we open-circuit the current source i_i, short-circuit the bias source V_B, and separate the inputs of the opamp from the rest of the circuit. Although R connects the output port with one of its terminals, it does not cause any loading because the other terminal floats. Only Z_{in} is present as a load. The final schematic for extracting A is the same as in Fig. 2.30(b). The expression of A derived for the unity-gain buffer is therefore valid for this amplifier as well.

Inverting Amplifier Separating the opamp inputs and shorting out the signal and bias sources in Fig. 2.8 yields the schematic shown in Fig. 2.30(c). The load impedance indicated by this schematic is $Z_L = (R_i + R_1 + R_2) \parallel Z_{in}$. A is therefore given by

$$A = \frac{(R_i + R_1 + R_2) \parallel Z_{in}}{Z_o + \left[(R_i + R_1 + R_2) \parallel Z_{in}\right]} Y_m Z_o.$$

Filtering Amplifier Referring to Fig. 2.19, we separate the inputs of the opamp from the rest of the circuit, and short-circuit the signal and bias sources to arrive at Fig. 2.30(d). The equivalent load in this schematic is described by $Z_L = \{R_1 + [R_2 \parallel (1/sC_2)]\} \parallel Z_{in}$, and A is obtained from (2.95) as

$$A = \frac{\left[R_1 + \left(R_2 \parallel \frac{1}{sC}\right)\right] \parallel Z_{in}}{Z_o + \left\{\left[R_1 + \left(R_2 \parallel \frac{1}{sC}\right)\right] \parallel Z_{in}\right\}} Y_m Z_o. \quad \blacktriangle$$

2.9 Noise

Digital circuits are immune to noise as long as their noise margins are not breached. Analog circuits have no such immunity, which is why noise control is a fundamental aspect of analog design. A circuit may receive noise due to electromagnetic interference emanating from other circuits inside or outside the chip. This so-called *coupled noise* is best controlled by implementing the application with fully-differential opamps and external networks. A specific type of coupled noise is the one that emanates directly from power-supply lines. The metric by which we quantify this type of noise in open loop and closed loop will be the topic of the first subsection of the present section.

Circuits are also subject to noise generated internally by their own conductive components such as MOSFETs, resistors, and diodes. Called *inherent noise*, this type of noise arises essentially from the random thermal motion and/or trapping of the current-conducting electrons. Since the intensity of inherent noise is closely related to device size and bias, we rely on electrical design techniques for its control. The second and third subsections of the present section are devoted to this type of noise.

2.9.1 Power Supply Noise

Power supply voltages fluctuate around the dc values V_{DD} and V_{SS} due to the temporal variation of the current flowing in the finite conductances of supply lines. This fluctuation leaks to the output as noise. The relative significance of this noise in open loop is quantified with a metric called *power supply rejection ratio (PSRR)*, which is defined as

$$\text{PSRR} \doteq 20\log\left(\frac{A}{A_{ps}}\right), \tag{2.97}$$

where A is the differential gain and A_{ps} is the small-signal voltage gain from the power line to the output of the opamp in open loop. PSRR is measured in units of decibel (dB).

Denoting the noise voltage on the power-supply line with v_{ps}, we can express the noise voltage measured at the output of the opamp in open loop as $A_{ps}v_{ps}$. When the loop is closed, this noise voltage appears in series with the output terminal of the opamp as depicted in Fig. 2.31. The closed-loop output noise voltage v_o can be related to v_{ps} by evaluating this block diagram, in which, all other voltage and current sources are disabled. The outcome is

$$v_o = \frac{A_{ps}v_{ps}}{1 + \beta A},$$

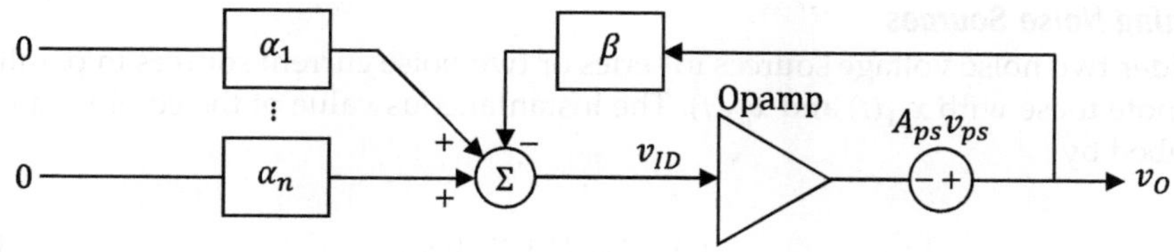

FIGURE 2.31 Block diagram for relating closed-loop output noise voltage v_o to power-supply noise voltage v_{ps}.

which, assuming $\beta A \gg 1$ and considering together with (2.97), yields

$$A_{cl,ps} \doteq \frac{v_o}{v_{ps}} = \frac{1}{\beta 10^{\mathrm{PSRR}/20}} \tag{2.98}$$

for the closed-loop gain from power-supply line to the output.

2.9.2 Fundamentals of Inherent Noise Analysis

As exemplified in Fig. 2.32 with a waveform $x_n(t)$, inherent noise is a randomly time-varying voltage or current of zero mean value. Its intensity is usually expressed in terms of *mean square* (ms) or *root mean square* (rms) values. These two values are denoted by X_n^2 and X_n, respectively, and are defined by

$$X_n^2 \doteq \lim_{T\to\infty}\left[\frac{1}{T}\int_0^T x_n^2(t)\,\mathrm{d}t\right], \tag{2.99}$$

and

$$X_n \doteq \sqrt{\lim_{T\to\infty}\left[\frac{1}{T}\int_0^T x_n^2(t)\,\mathrm{d}t\right]}. \tag{2.100}$$

In this subsection, we will review the techniques of noise analysis in analog circuits.

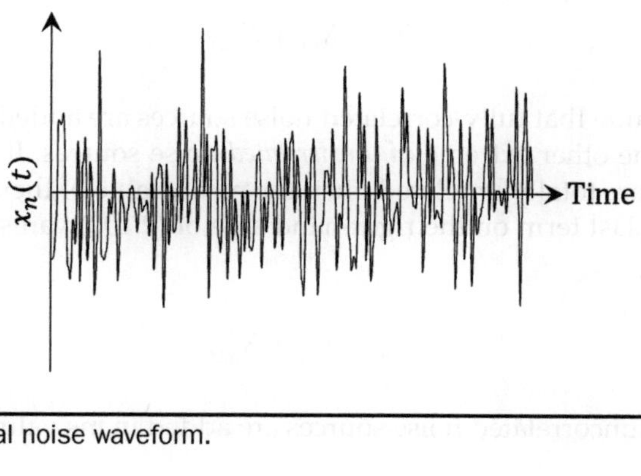

FIGURE 2.32 A typical noise waveform.

Summing Noise Sources

Consider two noise voltage sources in series or two noise current sources in parallel. Let us denote these with $x_{n1}(t)$ and $x_{n2}(t)$. The instantaneous value of the combined noise is described by

$$x_n(t) = x_{n1}(t) + x_{n2}(t). \tag{2.101}$$

The ms value of this combined noise is formulated by substituting (2.101) into (2.99):

$$\begin{aligned} X_n^2 &= \lim_{T\to\infty}\left\{\frac{1}{T}\int_0^T \left[x_{n1}(t)+x_{n2}(t)\right]^2 \,\mathrm{d}t\right\} \\ &= \lim_{T\to\infty}\left\{\frac{1}{T}\int_0^T \left[x_{n1}^2(t)+x_{n2}^2(t)+2x_{n1}(t)x_{n2}(t)\right] \,\mathrm{d}t\right\}, \end{aligned}$$

which yields

$$X_n^2 = X_{n1}^2 + X_{n2}^2 + 2\lim_{T\to\infty}\left[\frac{1}{T}\int_0^T x_{n1}(t)x_{n2}(t)\,\mathrm{d}t\right]. \tag{2.102}$$

Now, suppose that $x_{n1}(t)$ and $x_{n2}(t)$ share a common origin, so that their waveforms are identical with a possible difference in scale; that is, $x_{n2}(t) = ax_{n1}(t)$, where a is a constant. In this case of so-called *fully-correlated* noise sources, the last term on the right-hand side of (2.102) turns into

$$2a\lim_{T\to\infty}\left[\frac{1}{T}\int_0^T x_{n1}^2\,\mathrm{d}t\right] = 2aX_{n1}^2 = 2X_{n1}X_{n2}.$$

Therefore, (2.102) can be rewritten as follows

$$X_n^2 = X_{n1}^2 + X_{n2}^2 + 2X_{n1}X_{n2} = \left[X_{n1}+X_{n2}\right]^2,$$

which indicates

$$X_n = X_{n1} + X_{n2}. \tag{2.103}$$

We therefore conclude that fully-correlated noise sources are added in rms values.

Now consider the other extreme of *uncorrelated* noise sources. If the physical origins of $x_{n1}(t)$ and $x_{n2}(t)$ are totally unrelated, then the mean value of their product $x_{n1}(t)x_{n2}(t)$ and, therefore, the last term on the right-hand side of (2.102) vanish. This turns (2.102) into

$$X_n^2 = X_{n1}^2 + X_{n2}^2, \tag{2.104}$$

which tells us that uncorrelated noise sources are added in ms values.

Power Spectral Density and Noise Shaping

Inherent noise is a collection of infinitesimal sinewave components whose frequencies are spread over a continuous spectrum. The efficacy of these components is measured by the contribution they made to ms noise per unit frequency. This measure is called *power spectral density* (PSD), and is denoted by S (A^2/Hz or V^2/Hz) in this book. Total ms noise is determined by integrating PSD over frequency f as per

$$X_n^2 = \int_0^{\infty} S(f)\, df. \tag{2.105}$$

Now suppose that an input noise source of PSD $S_i(f)$ feeds noise to a circuit whose input-to-output transfer function is denoted by $A_n(f)$. The PSD $S_o(f)$ of the corresponding output noise is described by

$$S_o(f) \doteq |A_n(f)|^2\, S_i(f), \tag{2.106}$$

which indicates that the PSD is shaped by the magnitude of the transfer function as the noise passes from input to output. If the output receives noise from multiple uncorrelated sources, then, (2.106) is generalized into

$$S_o(f) = \sum_{k=1}^{m} |A_{nk}(f)|^2\, S_k(f), \tag{2.107}$$

where k is source index, m is source count, $A_{nk}(f)$ is the transfer function from source k to the output, and $S_k(f)$ is the PSD of source k.

The total inherent noise at the output of a given circuit is determined with (2.105) and (2.107). For a given circuit topology, we first represent the noise properties of each MOSFET or resistor or diode with a noise source of an appropriate $S_k(f)$, and determine $|A_k(f)|^2$ from source to output. The PSD $S_o(f)$ of the output noise is then calculated from (2.107). Next, the total ms output noise is quantified by integrating $S_o(f)$ over frequency in accordance with (2.105).

Resistor Noise

Resistor noise is mostly due to the random thermal motion of electrons. As shown in Fig. 2.33(a), this so-called *thermal noise* is modeled with a noise current source i_n placed

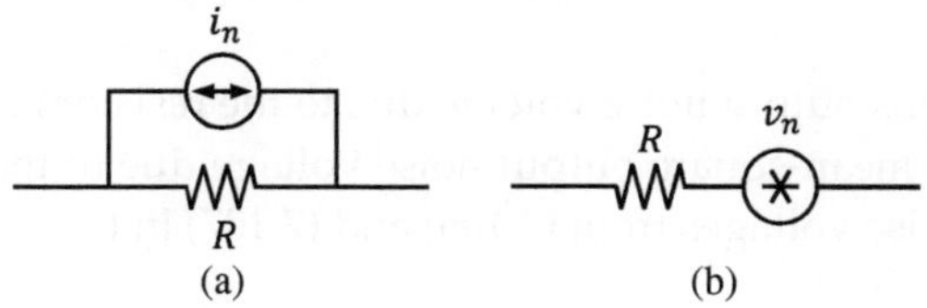

FIGURE 2.33 Representing the inherent noise of a resistor with (a) a current source, or (b) a voltage source.

in parallel with the resistor itself. The double-headed arrow used in the current-source symbol identifies it as a noise source. Its power spectral density, measured in units of A^2/Hz, is given by

$$S_i = 4\frac{E_{th}}{R}, \tag{2.108}$$

in which R (Ω) is the resistance of the resistor and E_{th} (J) is the *thermal energy* defined by $E_{th} = kT$, where $k = 1.38 \times 10^{-23}$ J/K is Boltzmann's constant and T (K) is absolute temperature. Note that E_{th} equals 4.14×10^{-21} J at room temperature $T = 300$ K.

As shown in Fig. 2.33(b), we can represent the resistor/noise-source combination with a Thévenin's equivalent, where the star symbol of the voltage source v_n identifies it as a noise source. The power spectral density S_v of this source, measured in V^2/Hz, is obtained by substituting into (2.106) R for A_n and (2.108) for $S(f)$. This leads to

$$S_v = 4E_{th}R. \tag{2.109}$$

2.9.3 Closed-Loop Noise Analysis

The noise performance of a closed-loop configuration is often specified by a metric called *signal-to-noise ratio* and abbreviated as SNR. It is defined as

$$\text{SNR} \doteq 10\log\frac{V^2_{o(\max)}}{V^2_{no}}, \tag{2.110}$$

where $V^2_{o(\max)}$ is the maximum ms output signal and V^2_{no} is the ms voltage of the output noise generated by the amplifier. An SNR specification obviously imposes a limit on the acceptable level of closed-loop output noise. Since noise is contributed collectively by opamp devices and external-network resistors, such a limit inevitably constrains the design of these components. In this subsection, we develop a general model for this design constraint by relating the closed-loop output noise to (a) external-network resistances and (b) opamp open-loop output noise.

Considering the fact that the noise sources associated with the opamp and external network are uncorrelated, (2.104) enables us to express the ms output noise voltage V^2_{no} of a closed-loop configuration quite generally as

$$V^2_{no} = V^2_{no(\text{ext})} + V^2_{no(\text{opamp})}, \tag{2.111}$$

where $V^2_{no(\text{ext})}$ is the ms output noise voltage due to the resistors of the external network and $V^2_{no(\text{opamp})}$ is the mean-square output noise voltage due to the opamp. We now formulate these two noise voltages from (2.105) and (2.107) in two separate paragraphs.

Output Noise due to External Network

Consider the resistors R_k, $k = 1, \ldots, m$ of the external network of a closed-loop configuration, and represent their noise voltages with v_{nk}. The ms output noise due to all these m

resistors is determined from (2.105), (2.107), and (2.109). Since the small-signal transfer function from v_{nk} to the output is the closed-loop gain $A_{cl,k}$ by definition, the outcome is as follows:

$$V^2_{no(\text{ext})} = 4E_{th} \int_0^{\infty} \sum_{k=1}^{m} |A_{cl,k}|^2 R_k \, df.$$

Assuming single-pole response for $A_{cl,k}$, and therefore substituting (2.49) into this equation and integrating, we arrive at

$$V^2_{no(\text{ext})} = 2\pi E_{th} f_{cl} \sum_{k=1}^{m} A^2_{cl0,k} R_k, \tag{2.112}$$

For $A_{cl0,k}$ and f_{cl}, the reader is referred to (2.47) and (2.48) if the external network is purely resistive. Otherwise, their description is given in the paragraph on frequency-dependent external networks in Subsection 2.5.1.

Example 2.19 Suppose that the noninverting and inverting amplifier configurations of Example 2.10 operate with $r_i = 0$ at the room temperature $T = 300$ K. Determine the maximum acceptable values of R_1 and R_2 that keep the rms output noise voltage contribution of their external network below 80 μV. Verify your design with noise simulations. Also determine the room-temperature rms output noise voltage contribution of the external network for the transresistance amplifier considered in the same example.

Noninverting Amplifier Consider (2.112), whose left-hand side is the mean-square value of the specified output noise. Among the parameters appearing on the right-hand side, f_{cl} equals 3 MHz as specified in Example 2.10. As to the extraction of $A_{cl0,k} = \alpha_k/\beta$, we need to construct an equivalent circuit by (a) including the noise voltage sources of R_1 and R_2 to the original circuit schematic presented in Fig. 2.6, and (b) disabling all other sources as well as r_i as suggested in problem statement. Since the square of α_k/β is needed in (2.112), the sign of α_k has no significance in noise calculations, which is why the reference terminal of a noise source v_{nk} can be assigned arbitrarily when determining α_k. The result, shown in Fig. 2.34(a), yields

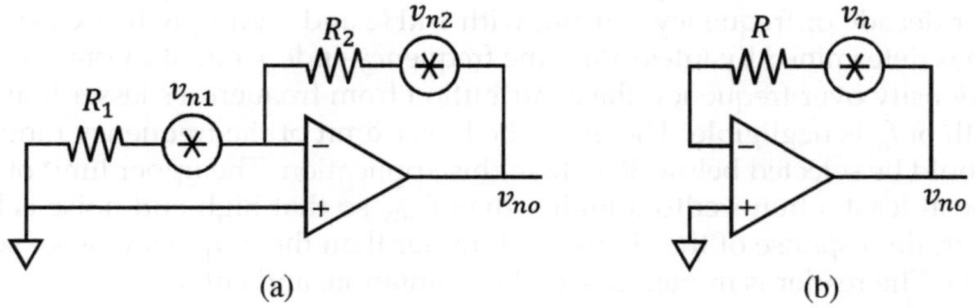

FIGURE 2.34 Equivalent circuits for extracting α_k/β in Example 2.19. (a) Noninverting and inverting amplifiers. (b) Transresistance amplifier.

$\alpha_1 = R_2/(R_1 + R_2)$, $\alpha_2 = R_1/(R_1 + R_2)$, and $\beta = R_1/(R_1 + R_2)$. Substituting these into (2.112), we arrive at

$$V^2_{no(\text{ext})} = 2\pi E_{th} f_{cl} \left(\frac{R_2}{R_1} + 1 \right) R_2,$$

where $f_{cl} = 3$ MHz, $E_{th} = 4.14 \times 10^{-21}$ J, $V_{no(\text{ext})} = 80$ μV, and $R_2/R_1 = 2$ as determined in Example 2.10. Solving this equation for R_2 yields 27.5 kΩ as the maximum permissible resistance for R_2. Considering $R_2/R_1 = 2$, R_1 should be 13.75 kΩ if the maximum is selected for R_2.

Inverting Amplifier Original schematic is shown in Fig. 2.8. In compliance with Example 2.10, we assume $f_{cl} = 3$ MHz and $R_2/R_1 = 3$. Ignoring r_i, the schematic for extracting α_k/β becomes identical to the one shown in Fig. 2.34(a). Calculation yields 20.6 kΩ for the largest permissible value for R_2. R_1 has to be one-third of R_2.

Transresistance Amplifier Original schematic is shown in Fig. 2.9. Equivalent circuit for noise analysis is given in Fig. 2.34(b). It yields $\alpha = 1$ and $\beta = 1$. Considering the original bandwidth specification $f_{cl} = 3$ MHz and resistance $R = 50$ kΩ as per Example 2.10, (2.112) yields $V_{no(\text{ext})} = 62.4$ μV.

The following input file `ex2-19.sp` is appropriate for verifying the noise performance of the noninverting amplifier:

```
filename: ex2-19.sp
.param vdd=1.8 vss=0 voh=1.65 vol=0.15 a0=1000 vdir=0.25
+   fgbw=9meg fnd=2g
.incl opmacro1.sp
x1 1 2 3 opamp
r1 2 4 13.75k
r2 2 3 27.5k
vi 1 4 dc 0 ac 1m
vb 4 0 dc 0.9
.noise v(3) vi dec 10 1k 1g ; line 9
.end
```

Notice (a) the opamp setting `fgbw=9meg` needed for $f_{cl} = 3$ MHz, (b) an ac input source v_i with an arbitrarily selected magnitude 1 mV, and (c) the noise analysis defined in line 9. In this line, `v(3)` specifies the voltage of the output port and `vi` represents the independent source of the ac input variable. The rest of the line specifies 10 simulation points per decade of frequency starting with 1 kHz and ending with 1 GHz. Since output noise is determined by integrating the frequency-independent thermal-noise power spectral density over frequency, the contribution from frequencies lesser than, say, one-hundredth of f_{cl} is negligible. Therefore the lower limit of the frequency range of simulation should be selected below 30 kHz in this application. The upper limit of frequency should be at least a hundredfold higher than f_{GBW} so that high-end noise is limited by the magnitude response of the circuit itself rather than the frequency range selected for simulation. The reader is invited to run the simulation, and enter

```
print onoise_total
```

to read $V_{no(\text{ext})}$. ▲

Output Noise due to Opamp

In closed loop, opamp contributes $V^2_{no(\text{opamp})}$ to ms output voltage as indicated in (2.111). We begin modeling this contribution by considering the noise properties of the opamp in open loop, where the PSD of the output noise voltage is described similarly to (2.107) by

$$S_{o(\text{opamp})}(f) = \sum_{k=1}^{m} |A_k(f)|^2 S_k(f) \tag{2.113}$$

where k is the index of the noise-generating internal device such as a MOSFET or a resistor, $|A_k(f)|$ is the transfer function from the noise source of device k to the output, and S_k is the PSD of the same noise source. Although multiple devices contribute to $S_{o(\text{opamp})}$, we prefer to represent all with a single fictitious noise voltage source v_{ni} generating the very same $S_{o(\text{opamp})}(f)$ at the output while placed in series with either the noninverting input terminal or the inverting input terminal of an otherwise noise-free opamp as shown in Figs. 2.35(a) and (b). Considering the fact that the transfer function from input to output is the open-loop differential gain $A(f)$ of the opamp, and representing the equivalent PSD of this *input-referred* noise voltage source with $S_{i(\text{opamp})}(f)$, we can rely on (2.106) to write

$$S_{o(\text{opamp})}(f) = |A(f)|^2 S_{i(\text{opamp})}(f). \tag{2.114}$$

Equating the right-hand sides of (2.113) and (2.114), and solving for $S_{i(\text{opamp})}(f)$, we obtain

$$S_{i(\text{opamp})}(f) = \sum_{k=1}^{m} \left|\frac{A_k(f)}{A(f)}\right|^2 S_k(f). \tag{2.115}$$

The input-referred noise source, when placed inside the loop together with the opamp, acts as any other input source of the closed-loop configuration. The ms value of the closed-loop output noise it generates is therefore given by

$$V^2_{no(\text{opamp})} = \int_0^{\infty} |A_{cl,ni}|^2 S_{i(\text{opamp})}(f)\, df, \tag{2.116}$$

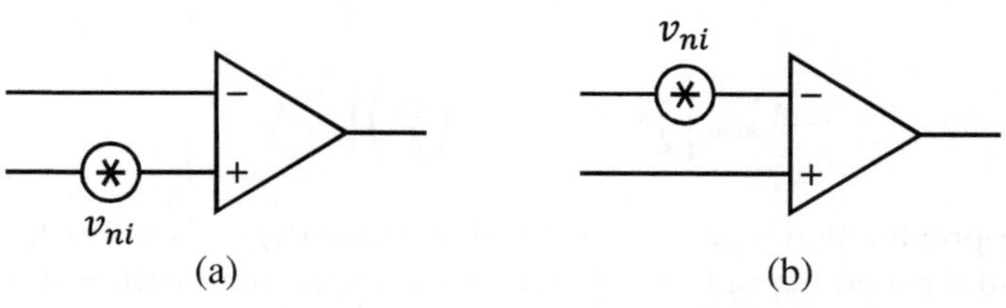

FIGURE 2.35 Representing the noise properties of an opamp with the combination of a noise-free opamp and input-referred noise voltage source v_{ni}. The latter is connected either to the noninverting terminal as in (a) or to the inverting terminal as in (b).

where $A_{cl,ni}$ is the closed-loop gain from v_{ni} to the output of the closed-loop configuration. Since v_{ni} drives the differential input of the opamp directly, its input factor is given by $\alpha_{ni} = 1$, which leads to

$$A_{cl,ni} = \frac{\beta A}{1+\beta A}\frac{1}{\beta} \tag{2.117}$$

from (2.40). Equation (2.116), together with (2.115) and (2.117), provides an ultimately general model for the contribution of the opamp to the output noise of the closed-loop configuration. The model is customized for any given application by (a) determining $A_k(f)$ and $A(f)$ from opamp topology in open loop, (b) determining $A_{cl,ni}$ from (2.117), (c) constructing $S_{i(\text{opamp})}(f)$ in accordance with (2.115), and (d) integrating as per (2.116).

For the usual case of single-pole frequency response, (2.117) yields

$$|A_{cl,ni}|^2 = \frac{A^2_{cl0,ni}}{1+(\frac{f}{f_{cl}})^2} \tag{2.118}$$

where $A_{cl0,ni}$ and f_{cl} are the dc closed-loop gain and closed-loop bandwidth, respectively. Furthermore, (2.115) yields

$$S_{i(\text{opamp})}(f) = k_{th} + \frac{k_{fl}}{f} \tag{2.119}$$

where k_{th} (V^2/Hz) is a constant determined by the thermal noise properties of the MOSFETs and resistors of the opamp, and k_{fl} (V^2) is another constant arising from the so-called *flicker noise* of MOSFETs. Several examples of extracting k_{th} and k_{fl} by applying (2.115) to a given opamp topology will be presented in Chapter 4. Modeling of $S_k(f)$ for this purpose will be presented earlier in Section 3.8. Integrating the product of (2.118) and (2.119) as per (2.116) while retaining the lower limit of the integral at $f = 0$ will obviously produce an infinite noise voltage for the flicker-noise contribution. In order to bypass this theoretical pitfall, we initially replace the lower limit of the integral with an arbitrary $f = f_1$, and consider the limiting case of $f_1 \rightarrow 0$ after evaluating the integral. The outcome is

$$\begin{aligned} V^2_{no(\text{opamp})} &= A^2_{cl0,ni}\left[k_{th}f_{cl}\left(\frac{\pi}{2} - \arctan\frac{f_1}{f_{cl}}\right) + k_{fl}\ln\sqrt{1+\left(\frac{f_{cl}}{f_1}\right)^2}\right] \\ &\approx A^2_{cl0,ni}\left[\frac{\pi}{2}k_{th}f_{cl} + k_{fl}\ln\left(\frac{f_{cl}}{f_1}\right)\right], \end{aligned} \tag{2.120}$$

where the approximation is justified by the limit value $(f_1/f_{cl}) \rightarrow 0$. The logarithmic form of the second term on the right-hand side of this expression renders the exact value of f_1 relatively inconsequential. To see why, consider $f_{cl} = 1$ MHz as an example, and suppose that the application is powered up only for 1 s. Since noise frequencies less than 1 Hz will be relatively ineffective because they have no chance to complete even a single cycle, $f_1 = 1$ Hz will be a reasonable lower limit leading to $\ln(f_{cl}/f_1) = 13.8$. If the very same application were operated continuously for, say, 30 years, then, the lowest noise frequency of relevance would be a nanohertz leading to $\ln(f_{cl}/f_1) = 34.5$, which is only a factor of 2.5 greater than the first application despite the billionfold difference between the values of f_1 involved.

2.10 Fully-Differential Continuous-Time Amplifiers

In this section, we will extend the techniques and tools of closed-loop modeling to fully-differential continuous-time amplifiers. Simulations will be conducted with the opamp macro defined in `opmacro2.sp` as described in Section 2.1.

2.10.1 First-Order DC Response and Range Limitations

What is unique about a fully-differential opamp is that it offers two complementary output terminals as depicted in Fig. 2.36(a). The output terminal marked with a minus sign is inverting. The one with a plus sign is noninverting. Instantaneous voltages defined between each of these output terminals and ground are represented with v_{ON} and v_{OP},

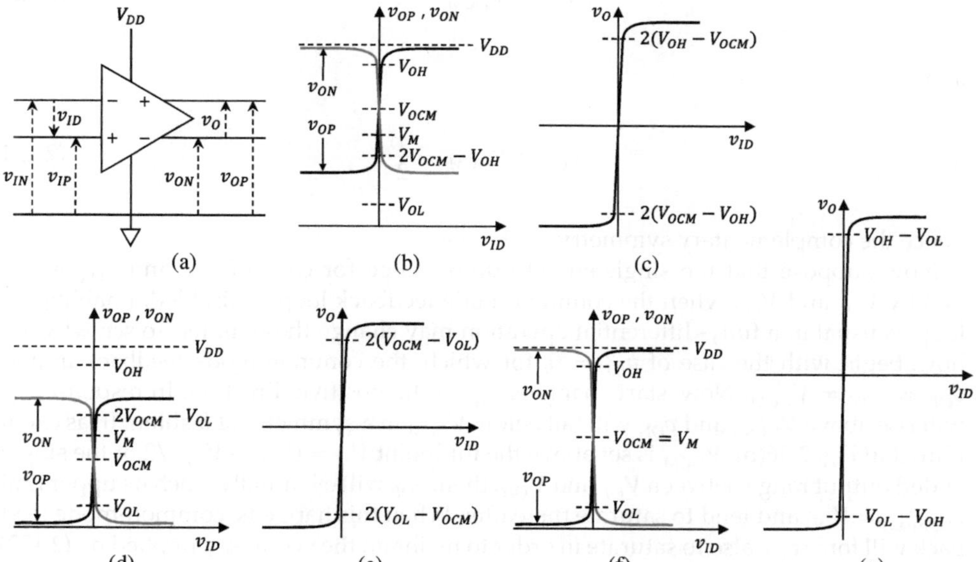

FIGURE 2.36 (a) Fully-differential opamp symbol and voltage definitions. (b) Single-ended open-loop dc transfer characteristics for $V_{OCM} > V_M$, where V_M is the average of the single-ended output range limits V_{OL} and V_{OH}. (c) Fully-differential open-loop dc transfer characteristic for $V_{OCM} > V_M$. (d) Single-ended open-loop dc transfer characteristics for $V_{OCM} < V_M$. (e) Fully-differential open-loop dc transfer characteristic for $V_{OCM} < V_M$. (f) Single-ended open-loop dc transfer characteristics for $V_{OCM} = V_M$. (g) Fully-differential open-loop dc transfer characteristic for $V_{OCM} = V_M$.

respectively. These two are the *single-ended output voltages* of the opamp. Their difference, defined by

$$v_O \doteq v_{OP} - v_{ON}, \tag{2.121}$$

is called *differential-output voltage*. It represents the output signal of the opamp.

Examples of open-loop dc transfer characteristics relating v_{OP} and v_{ON} individually to the differential-input voltage v_{ID} are shown qualitatively in Figs. 2.36(b), (d), and (f). Note that v_{OP} and v_{ON} are equalized at an externally defined *output common-mode voltage* V_{OCM} for $v_{ID} = 0$. Otherwise they vary steeply with v_{ID} in a complementary and symmetrical fashion around V_{OCM}. What maintains this complementary symmetry is an internal *common-mode feedback loop* (CMFB) circuit imposing the constraint

$$\frac{v_{OP} + v_{ON}}{2} = V_{OCM} \tag{2.122}$$

on the single-ended output voltages independently of v_{ID}. Combined with (2.121), this constraint indeed implies

$$v_{OP} = V_{OCM} + \frac{v_O}{2} \tag{2.123}$$

and

$$v_{ON} = V_{OCM} - \frac{v_O}{2}, \tag{2.124}$$

hence the complementary symmetry.

Now suppose that the single-ended output range for each of v_{OP} and v_{ON} is limited by V_{OL} and V_{OH} when the common-mode feedback loop is disabled. Enabling the loop as usual in a fully-differential operation may change these limits. To see why and how, begin with the case of $v_{ID} = 0$, for which, the common-mode feedback implies $v_{OP} = v_{ON} = V_{OCM}$. Now, start changing v_{ID} in the positive direction. In response, v_{OP} will rise above V_{OCM}, and v_{ON} will fall below V_{OCM} in a symmetrical fashion. If, as exemplified in Fig. 2.36(b), V_{OCM} is set above the midpoint $V_M = (V_{OL} + V_{OH})/2$ of the single-ended output range between V_{OL} and V_{OH}, then, v_{OP} will eventually reach its upper limit $v_{OP(\max)} = V_{OH}$ and tend to saturate thereafter. When this happens, common-mode feedback will force v_{ON} also to saturate in order to maintain the symmetry implied by (2.122). As a consequence, v_{ON} will be limited to $v_{ON(\min)} = 2V_{OCM} - V_{OH}$ instead of V_{OL}. In this case, the *differential-output range*, in which, the swing of v_O is confined, is given by

$$V_{OH(\text{dif})} \doteq v_{OP(\max)} - v_{ON(\min)} = V_{OH} - (2V_{OCM} - V_{OH}) = 2(V_{OH} - V_{OCM}). \tag{2.125}$$

If we change v_{ID} in the negative direction, v_{IN} will reach the upper limit $v_{ON(\max)} = V_{OH}$ before v_{IP} falls to V_{OL}, and, as a result, the common-mode feedback will limit $v_{IP(\min)}$ to

$2V_{OCM} - V_{OH}$. In this case, the lower limit of the differential-output range, defined as $V_{OL(\text{dif})} \doteq v_{OP(\min)} - v_{ON(\max)}$, is described by

$$V_{OL(\text{dif})} = 2(V_{OCM} - V_{OH}). \tag{2.126}$$

$V_{OH(\text{dif})}$ and $V_{OL(\text{dif})}$, as predicted by (2.125) and (2.126), are marked on the fully-differential open-loop dc transfer characteristic shown in Fig. 2.36(c).

The reader can easily conclude by following the analytical reasoning of the preceding paragraph that, if $V_{OCM} < V_M$, then, the single-ended output range is limited by V_{OL} at the bottom, and by $2V_{OCM} - V_{OL}$ at the top. The single-ended and fully-differential open-loop transfer characteristics of this case are shown in Figs. 2.36(d) and (e), respectively. Notice that the limits of the differential-output range in this case are described by

$$V_{OH(\text{dif})} = 2(V_{OCM} - V_{OL}), \tag{2.127}$$

and

$$V_{OL(\text{dif})} = 2(V_{OL} - V_{OCM}). \tag{2.128}$$

Finally, consider the case of $V_{OCM} = V_M$. In this case, single-ended output range will be limited by V_{OL} at the bottom, and by V_{OH} at the top. The corresponding single-ended and fully-differential open-loop transfer characteristics are shown in Figs. 2.36(f) and (g), respectively. Obviously, the limits of the differential-output range are given by

$$V_{OH(\text{dif})} = V_{OH} - V_{OL}, \tag{2.129}$$

and

$$V_{OL(\text{dif})} = V_{OL} - V_{OH}. \tag{2.130}$$

A comparison between (2.129) to (2.130) and (2.125) to (2.128) clearly shows that setting $V_{OCM} = V_M$ maximizes the differential-output range to $2(V_{OH} - V_{OL})$, which is twice as wide as that of a single-ended opamp. This is one of the outstanding benefits of using a fully-differential opamp instead of a single-ended one.

As observed in Figs. 2.36(c), (e), or (g), the differential-output range includes $v_O = 0$, which enables the opamp to put out a signal centered around the ground instead of a bias voltage. In this sense, a fully-differential opamp resembles a split-supply single-ended opamp but, unlike the latter, it offers this property in a single-supply configuration.

The output-range and common-mode range requirements of a fully-differential configuration are determined by applying the virtual-short approximation very much like the way in which we apply it to single-ended configurations. As exemplified next, first-order dc closed-loop transfer function is extracted also with the virtual-short approximation.

Example 2.20 Fully-differential voltage and transresistance amplifier applications are implemented by the generic closed-loop configurations shown in Figs. 2.37(a) and (b), respectively. We now analyze the first-order dc response of these configurations, and determine their output-range and common-mode range requirements as well as dc gain expressions.

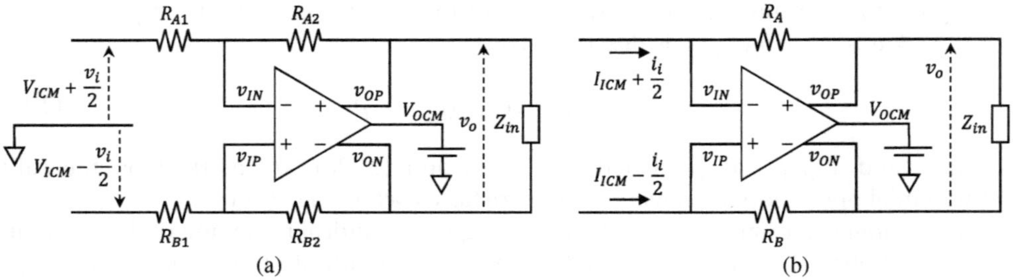

FIGURE 2.37 Generic fully-differential continuous-time amplifier configurations. (a) Voltage amplifier. (b) Transresistance amplifier.

Work First, consider the voltage amplifier shown in Fig. 2.37(a). Notice the presence of two feedback paths established by (a) R_{A1} and R_{A2} around the inverting-input and noninverting-output terminals of the opamp, and (b) R_{B1} and R_{B2} around the remaining input and output terminals. The corresponding resistors of the two paths are nominally identical, that is, $R_{B1} = R_{A1}$ and $R_{B2} = R_{A2}$. Also note that the input signal v_i is applied differentially around a dc *input common-mode voltage* V_{ICM}. Taking into consideration the equality of the corresponding resistances of the two loops, and thus dropping the letters A and B from the subscripts, we obtain the following two independent equations by equating the currents of the two resistors in each loop:

$$\frac{v_{OP} - v_{IN}}{R_2} = \frac{v_{IN} - [V_{ICM} + (v_i/2)]}{R_1},$$

$$\frac{v_{ON} - v_{IP}}{R_2} = \frac{v_{IP} - [V_{ICM} - (v_i/2)]}{R_1}.$$

By using (2.123) and (2.124) for v_{OP} and v_{ON} in these equations, and assuming virtual short $v_{IN} = v_{IP}$, we obtain

$$v_O = -\frac{R_2}{R_1} v_i \tag{2.131}$$

for the ideal dc differential-output voltage, and

$$v_{CM} = v_{IN} = v_{IP} = \frac{R_1 V_{OCM} + R_2 V_{ICM}}{R_1 + R_2} \tag{2.132}$$

for the common-mode voltage of the opamp.[10] Notice from (2.132) that v_{CM} remains in between V_{ICM} and V_{OCM} and does not change with the input signal. In the special but

[10] The reader is cautioned that the term "common-mode voltage" and its symbol v_{CM}, as defined by (2.2), describe the common-mode level of the inverting and noninverting input voltages of an opamp unless it is specified by the adjective "input" or "output" to imply V_{ICM} or V_{OCM}, respectively.

very common case of $V_{ICM} = V_{OCM}$ it equals V_{OCM}, which prevents the external network from conducting a dc current at the operation point.

Now consider (2.131). Obviously, the differential-output voltage v_O doesn't carry any bias and is proportional to the input signal v_i. It therefore represents the output signal v_o directly. Also note that the closed-loop dc voltage gain $A_{cl0} = -R_2/R_1$ is identical to the gain offered by a single-ended inverting amplifier configuration. But, if we define the output signal as $v_o \doteq v_{ON} - v_{OP}$ instead of $v_o \doteq v_{OP} - v_{ON}$, then, the closed-loop gain will be $A_{cl0} = R_2/R_1$, hence noninverting amplification. The very same configuration can therefore be utilized as inverting or noninverting amplifier simply by properly defining the reference convention of the output signal.

The differential-output range of a closed-loop configuration, as limited by $V_{OL(\mathrm{dif})}$ and $V_{OH(\mathrm{dif})}$, should be wide enough to accommodate the differential-output swing indicated by (2.131) for the minimum and maximum levels of v_i. Once $V_{OL(\mathrm{dif})}$ and $V_{OH(\mathrm{dif})}$ have thus been determined, we can transform them into single-ended output-range limits V_{OL} and V_{OH} through the appropriate equation pair among (2.125) to (2.130) for use as open-loop metrics in opamp design.

Now consider the generic fully-differential transresistance amplifier configuration shown in Fig. 2.37(b). The two feedback resistors R_A and R_B are nominally identical. Also notice the differential application of the current-mode input signal i_i around a dc input common-mode current I_{ICM}. Assuming $R_A = R_B$, and therefore dropping the subscripts, we can express the voltages across the two resistors by

$$v_{IN} - v_{OP} = R\left(I_{ICM} + \frac{i_i}{2}\right),$$

$$v_{IP} - v_{ON} = R\left(I_{ICM} - \frac{i_i}{2}\right).$$

By applying the virtual-short approximation $v_{IN} = v_{IP}$ to these two equations, we obtain

$$v_O \doteq v_{OP} - v_{ON} = -Ri_i \tag{2.133}$$

for the differential-output voltage hence the output signal. It is obvious from (2.133) that the configuration offers a transconductance gain $A_{cl0} = -R$ exactly like the single-ended version considered in Example 2.5. Considering $(v_{OP} + v_{ON})/2 = V_{OCM}$, we also obtain

$$v_{CM} = v_{IN} = v_{IP} = V_{OCM} + RI_{ICM} \tag{2.134}$$

for the common-mode voltage, which, obviously, is independent of the input signal. Also note $v_{CM} = V_{OCM}$ in the case of $I_{ICM} = 0$. Output range requirements for a given application are determined exactly like the voltage amplifier discussed previously. ▲

Example 2.21 The fully-differential amplifier configuration of Fig. 2.37(a) is built with $R_{A2} = R_{B2} = 30$ kΩ and $R_{A1} = R_{B1} = 10$ kΩ to provide a closed-loop dc gain of 3 V/V. The opamp operates with $V_{SS} = 0$ and $V_{DD} = 1.8$ V, and is specified by $V_{OL} = 0.15$ V, $V_{OH} = 1.65$ V, hence $V_M = 0.9$ V. The reader is invited to verify the dc range metrics with simulations.

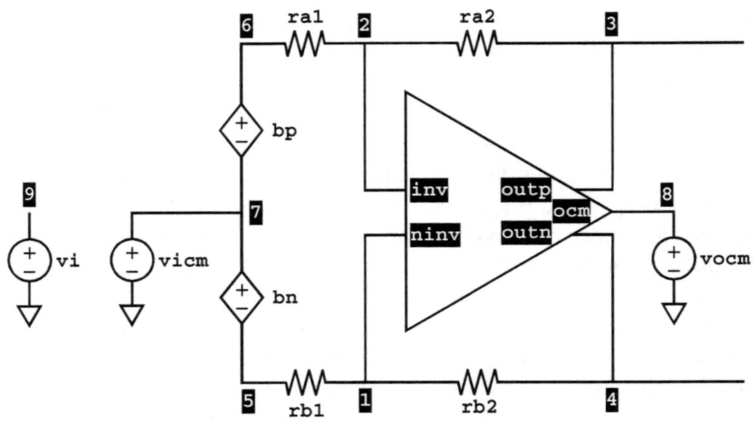

FIGURE 2.38 Simulation schematic used in Example 2.21.

The schematic to be used in simulations is shown in Fig. 2.38. Its input file `ex2-21.sp` is as follows:

```
filename: ex2-21.sp
.param vdd=1.8 vss=0 voh=1.65 vol=0.15 adif0=10k vdir=0.25
+  fgbw=10meg fnd=10g ; line 2
.incl opmacro2.sp
x1 1 2 3 4 8 fdopamp
ra2 2 3 30k
ra1 6 2 10k
rb2 1 4 30k
rb1 5 1 10k
vocm 8 0 0.9
vicm 7 0 0.9
bp 6 7 v=v(9)/2
bn 7 5 v=v(9)/2
vi 9 0
.dc vi -0.8 0.8 1m
.end
```

The opamp parameters declared in line 2 include a very large dc differential gain `adif0=10k` which approximates the slope of an ideal fully-differential open-loop dc transfer characteristic. Differential-input range, gain-bandwidth product, and nondominant pole frequency appearing in the same line are set to arbitrary values since they are irrelevant to a dc analysis. V_{OCM} is defined by the voltage source `vocm` to 0.9 V. V_{ICM} is also set to 0.9 V by the voltage source `vicm`. Input signal source v_i is represented by `vi`. It is applied differentially around V_{ICM} by dependent voltage sources `bp` and `bn` in compliance with the application depicted originally in Fig. 2.37(a) and is swept between −0.8 and 0.8 V.

Since $V_{OCM} = 0.9$ V is coincident with V_M, the differential-output range is maximized to ±1.5 V as indicated by (2.129) and (2.130). Considering the closed-loop gain of 3 V/V, the corresponding permissible limits of input-signal swing are ±0.5 V. Now, run the simulation and plot v(4,3) as the differential-output voltage v_O. Input signal voltage v_i, being the variable of sweep, will automatically appear on the horizontal axis. The result will probably come as a surprise because the output range will appear to extend between −1.8 and 1.8 V corresponding to an input signal swing ±0.6 V! In order to see the real picture, the reader is now invited to enter the following command to plot the closed-loop gain as represented by the derivative of v_O with respect to v_i:

```
plot deriv(v(4,3))
```

The closed-loop gain is expected to be 3 V/V. Zooming in on the uppermost segment of the plot window where the gain is in between 2.8 and 3 V/V will clearly show the gain dropping rapidly outside the range $-0.5\text{V} < v_i < 0.5\text{V}$.[11] The output range corresponding to this particular input-signal range is of course ±1.5 V as estimated above. This simulation shows us that an opamp can function even beyond the limits of the output range but the closed-loop gain deteriorates rapidly as these limits are breached.

Next, plot v_{IP} or v_{IN} as v_{CM}. Is the result consistent with (2.132)?

Finally, repeat the simulation for $V_{ICM} = V_{OCM} = 0.5$ V and again plot v_O and its derivative. Interpret your observations. ▲

2.10.2 Unified Closed-Loop Model

The unified block diagram of fully-differential continuous-time closed-loop configurations is shown in Fig. 2.39. Notice the presence of two external networks. The upper network generates the inverting input voltage v_{IN} of the opamp through the feedback factor β_A and the input factors α_{Aj} of input variables x_{Aj}, $j = 1, \ldots, n$. The lower network generates the noninverting input voltage v_{IP} through the feedback factor β_B and

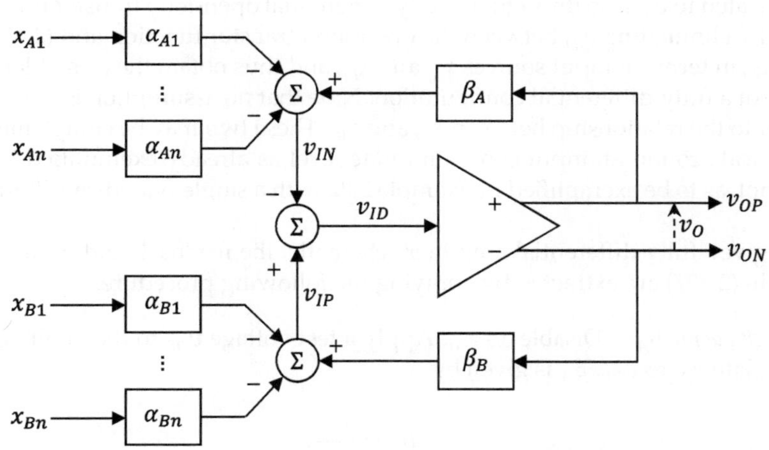

FIGURE 2.39 Unified block-diagram of fully-differential continuous-time closed-loop configurations.

---

[11] Incidentally, the maximum closed-loop gain observed within this range is slightly smaller than 3 V/V; why?

the input factors α_{Bj} of input variables x_{Bj}, $j = 1, \ldots, n$. v_{OP} and v_{ON} are the noninverting and inverting output voltages, which are interrelated by (2.122). The block diagram implies

$$v_{IN} = \beta_A v_{OP} - \sum_{j=1}^{n} \alpha_{Aj} x_{Aj}, \tag{2.135}$$

and

$$v_{IP} = \beta_B v_{ON} - \sum_{j=1}^{n} \alpha_{Bj} x_{Bj}. \tag{2.136}$$

Considering also (2.123) and (2.124), these two equations lead to the following expression for the differential-input voltage $v_{ID} \doteq v_{IP} - v_{IN}$:

$$v_{ID} = \sum_{j=1}^{n} \left(\alpha_{Aj} x_{Aj} - \alpha_{Bj} x_{Bj} \right) - \beta v_O + (\beta_B - \beta_A) V_{OCM}, \tag{2.137}$$

where

$$\beta \doteq \frac{\beta_A + \beta_B}{2}. \tag{2.138}$$

Since the feedback and input factors are independent of the opamp, (2.137) is imposed solely by the two external networks.

v_O is related to v_{ID} also through the fully-differential open-loop transfer function of the opamp. By eliminating v_{ID} between the open-loop transfer function and (2.137), we can express v_O in terms of input sources x_{Aj} and x_{Bj}, and thus obtain the closed-loop transfer function of a fully-differential configuration. Note that no assumption is made in (2.137) in regard to the relationship between x_{Aj} and x_{Bj}. These two may be complementary and symmetrical around an input common-mode level as already exemplified in Example 2.20, or not, as to be exemplified in Example 2.22 with a single-ended-to-differential converter.

For a given fully-differential amplifier schematic, the feedback and input factors appearing in (2.137) are extracted by applying the following procedure.

Feedback Factor β_A: Disable all x_{Aj}. Apply a test voltage v_{OP} to the noninverting output. Relate v_{IN} to v_{OP}. β_A is given by

$$\beta_A = \frac{v_{IN}}{v_{OP}}. \tag{2.139}$$

Input Factor α_{Ai}: Disable all x_{Aj} other than x_{Ai}, and short-circuit the noninverting output to ground ($v_{OP} = 0$). Apply a test voltage x_{Ai} to input-*i*. Relate v_{IN} to x_{Ai}. α_{Ai} is given by

$$\alpha_{Ai} = -\frac{v_{IN}}{x_{Ai}}. \tag{2.140}$$

Feedback Factor β_B: Disable all x_{Bj}. Apply a test voltage v_{ON} to the inverting output. Relate v_{IP} to v_{ON}. β_B is given by

$$\beta_B = \frac{v_{IP}}{v_{ON}}. \tag{2.141}$$

Input Factor α_{Bi}: Disable all x_{Bj} other than x_{Bi}, and short-circuit the inverting output to ground ($v_{ON} = 0$). Apply a test voltage x_{Bi} to input-*i*. Relate v_{IP} to x_{Bi}. α_{Bi} is given by

$$\alpha_{Bi} = -\frac{v_{IP}}{x_{Bi}}. \tag{2.142}$$

It is important to note that the differences between β_A and β_B, and between α_{Ai} and α_{Bi} are due to the limited accuracy and precision by which we can match the corresponding components of the upper and lower external networks in practice. For this reason, such differences are relevant only in the accurate dc analysis of a given configuration. In all other types of analysis, we usually assume the corresponding components to be perfectly matching, hence $\beta \doteq \beta_A = \beta_B$ and $\alpha_i \doteq \alpha_{Ai} = \alpha_{Bi}$, which enables us to consider the upper network only for extracting the factors. This assumption also simplifies (2.137) into the following fully-differential counterpart of (2.17):

$$v_{ID} = \sum_{j=1}^{n} \alpha_j \left(x_{Aj} - x_{Bj}\right) - \beta v_O. \tag{2.143}$$

In the case of a first-order dc response, we substitute zero for v_{ID} in (2.143), and thus obtain

$$v_O = \sum_{j=1}^{n} \frac{\alpha_j}{\beta} \left(x_{Aj} - x_{Bj}\right) \tag{2.144}$$

for the closed-loop dc transfer function. The closed-loop dc gain expression

$$A_{cl,i} = \frac{\alpha_i}{\beta}$$

indicated by (2.144) for the differential-input signal $x_{Ai} - x_{Bi}$ is exactly the same as (2.19) of a single-ended configuration.

Example 2.22 Determine β, α, and the first-order dc closed-loop transfer function of the fully-differential amplifier configurations of Fig. 2.37, and of the fully-differential summing amplifier and single-ended-to-differential converter shown in Figs. 2.40(a) and (b), respectively.

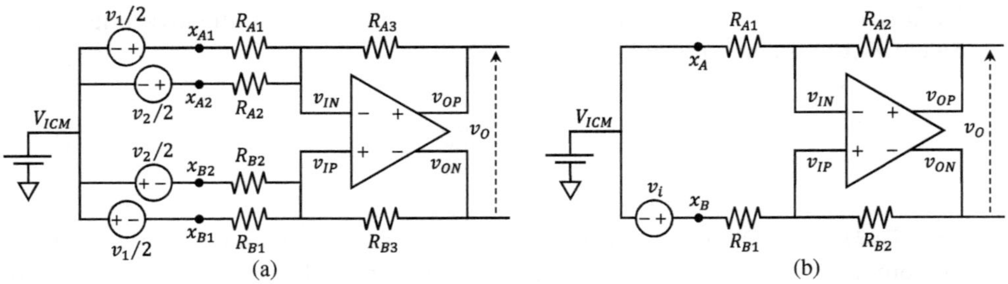

FIGURE 2.40 (a) Summing amplifier. (b) Single-ended-to-differential converter.

Voltage Amplifier, Fig. 2.37(a)

$$\beta = \frac{R_1}{R_1 + R_2},$$

$$\alpha = -\frac{R_2}{R_1 + R_2}.$$

Substituting these factors together with $x_A = V_{ICM} + (v_i/2)$ and $x_B = V_{ICM} - (v_i/2)$ into (2.144), we obtain the very same (2.131).

Transresistance Amplifier, Fig. 2.37(b) $\beta = 1$, $\alpha = -R$, $x_A = I_{ICM} + (i_i/2)$, and $x_B = I_{ICM} - (i_i/2)$ yield the very same (2.133).

Summing Amplifier, Fig. 2.40(a)

$$\beta = \frac{R_1 \parallel R_2}{R_3 + (R_1 \parallel R_2)},$$

$$\alpha_1 = -\frac{R_2 \parallel R_3}{R_1 + (R_2 \parallel R_3)},$$

$$\alpha_2 = -\frac{R_1 \parallel R_3}{R_2 + (R_1 \parallel R_3)}.$$

Also considering $x_{A1} = V_{ICM} + (v_1/2)$, $x_{A2} = V_{ICM} + (v_2/2)$, $x_{B1} = V_{ICM} - (v_1/2)$, and $x_{B2} = V_{ICM} - (v_2/2)$, we obtain

$$v_O = -\frac{R_3}{R_1}v_1 - \frac{R_3}{R_2}v_2$$

from (2.144).

Single-Ended-to-Differential Converter, Fig. 2.40(b) Notice (a) the presence of the common-mode voltage V_{ICM} at both inputs, and (b) the single-ended application of the signal v_i only to input B. Therefore, $x_A = V_{ICM}$ and $x_B = V_{ICM} + v_i$. β and α turn out to be exactly the same as those of the voltage amplifier. Therefore, (2.144) yields

$$v_O = \frac{R_2}{R_1} v_i. \quad \blacktriangle$$

2.10.3 Accurate Modeling of DC Response

For the accurate modeling of dc response, consider the fully-differential open-loop dc transfer characteristics shown in Figs. 2.36(c), (e), and (g). Representing the slope with A_{dif0} around $v_O = 0$ in any of them, and allowing for a possible offset V_{os}, we can approximate the associated transfer function with

$$v_O = A_{\text{dif0}} \left(v_{ID} - V_{os} \right), \tag{2.145}$$

which is the fully-differential counterpart of the single-ended accurate transfer function (2.22) presented in Subsection 2.4.1. Notice the absence of an output bias term such as V_O of (2.22). This is due to the fact that the differential-output of a fully-differential opamp is not biased.

Next, we cancel v_{ID} between (2.145) and (2.137) and obtain the following accurate expression for closed-loop dc transfer function

$$v_O = \frac{\beta A_{\text{dif0}}}{1 + \beta A_{\text{dif0}}} \sum_{j=1}^{n} \left(\frac{\alpha_{Aj}}{\beta} x_{Aj} - \frac{\alpha_{Bj}}{\beta} x_{Bj} \right) - \frac{A_{\text{dif0}}}{1 + \beta A_{\text{dif0}}} V_{os} + \frac{A_{\text{dif0}} (\beta_B - \beta_A)}{1 + \beta A_{\text{dif0}}} V_{OCM}, \tag{2.146}$$

which is the fully-differential counterpart of (2.23) derived in Subsection 2.4.1.

Suppose that a configuration is designed for a feedback factor β and input factors α_j. Its target output voltage for a given set x_{Aj} and x_{Bj} of input variables is described by (2.144). Now, consider a fabricated sample whose feedback and input factors turn out to be β'_A and α'_{Aj} for the upper network, and β'_B and α'_{Bj} for the lower network. The output voltage v'_O measured on this sample for the same set x_{Aj} and x_{Bj} of input variables is obtained from (2.146) after replacing (a) v_O with v'_O, (b) β_A with β'_A, (c) α_{Aj} with α'_{Aj}, (d) β_B with β'_B, and (e) α_{Bj} with α'_{Bj}. By subtracting (2.144) from the resulting equation, we obtain the following expression of dc error:

$$\Delta v_O \doteq v'_O - v_O = \frac{\beta' A_{\text{dif0}}}{1 + \beta' A_{\text{dif0}}} \sum_{j=1}^{n} \left[\left(\frac{\alpha'_{Aj}}{\beta'} - \frac{\alpha_j}{\beta} \right) x_{Aj} - \left(\frac{\alpha'_{Bj}}{\beta'} - \frac{\alpha_j}{\beta} \right) x_{Bj} \right]$$
$$- \frac{v_O}{1 + \beta' A_{\text{dif0}}} - \frac{A_{\text{dif0}} V_{os}}{1 + \beta' A_{\text{dif0}}} + \frac{A_{\text{dif0}} \left(\beta'_B - \beta'_A \right) V_{OCM}}{1 + \beta' A_{\text{dif0}}},$$

where β' is defined by

$$\beta' \doteq \frac{\beta'_A + \beta'_B}{2}. \tag{2.147}$$

Applying the usual approximation of replacing (a) $1+\beta' A_{\text{dif0}}$ with $\beta' A_{\text{dif0}}$ in all four error components, and (b) β' with β in the second and third error components, we turn this equation into

$$\Delta v_O = E_{\text{net}} + E_{\text{gain}} + E_{os} + E_{cm}, \tag{2.148}$$

where

$$E_{\text{net}} \doteq \sum_{j=1}^{n} \left[\underbrace{\left(\frac{\alpha'_{Aj}}{\beta'} - \frac{\alpha_j}{\beta} \right)}_{E_{Aj}} x_{Aj} - \underbrace{\left(\frac{\alpha'_{Bj}}{\beta'} - \frac{\alpha_j}{\beta} \right)}_{E_{Bj}} x_{Bj} \right] \tag{2.149}$$

is the external-network error,

$$E_{\text{gain}} \doteq -\frac{1}{\beta A_{\text{dif0}}} v_O \tag{2.150}$$

is the gain error,

$$E_{os} \doteq -\frac{1}{\beta} V_{os} \tag{2.151}$$

is the offset error, and

$$E_{cm} \doteq \frac{\beta'_B - \beta'_A}{\beta'} V_{OCM} \tag{2.152}$$

is the *common-mode error*. Comparing the components of (2.148) with those of its single-ended counterpart (2.24) reveals certain similarities and also differences. First, considering the fact that the output voltage v_O and the output signal v_o are identical in a fully-differential configuration, gain-error formulation is not different from its single-ended counterpart except for the replacement of A_0 by A_{dif0}. Offset error is also similar to its single-ended counterpart. As to the external-network error, there is no difference between the two variants except for the involvement of both upper and lower halves in the fully-differential configuration. Probably the most significant difference is the presence of a *common-mode error* in (2.148), which has no counterpart in (2.24). This error arises from the mismatch between the realized values of the feedback factors. In the voltage amplifier configurations of Figs. 2.37(a) and 2.40, this error actually cancels out the part of the external-network error associated with V_{ICM} provided that V_{ICM} and V_{OCM} are identical. In that case, (2.148) reduces to

$$\Delta v_O = \underbrace{\sum_{j=1}^{n} \left(E_{Aj} x_{aj} - E_{Bj} x_{bj} \right)}_{\substack{\text{External-}\\ \text{network}\\ \text{error}}} \underbrace{-\frac{1}{\beta A_{\text{dif0}}} v_O}_{\substack{\text{Gain}\\ \text{error}}} \underbrace{-\frac{1}{\beta} V_{os}}_{\substack{\text{Offset}\\ \text{error}}}, \tag{2.153}$$

in which, E_{Aj} and E_{Bj} are defined in (2.149), and x_{aj} and x_{bj} represent the signal components of x_{Aj} and x_{Bj}, respectively. The mean value $\overline{\Delta v_O}$ and the standard deviation $\sigma[v_O]$ of the dc error described by (2.153) are given by

$$\overline{\Delta v_O} = \sum_{j=1}^{n}\left(\overline{E_{Aj}}x_{aj} - \overline{E_{Bj}}x_{bj}\right) - \frac{1}{\beta A_{\text{dif0}}}v_O - \frac{1}{\beta}\overline{V_{os}}, \tag{2.154}$$

and

$$\sigma[\Delta v_O] = \sqrt{\sum_{j=1}^{n}\left(\sigma^2[E_{Aj}]x_{aj}^2 + \sigma^2[E_{Bj}]x_{bj}^2\right) + \frac{1}{\beta^2}\sigma^2[V_{os}]}. \tag{2.155}$$

These two equations are the fully-differential counterparts of (2.37) and (2.38). If their conditions of validity are not satisfied, the mean and standard deviation have to be formulated directly from (2.148).

Example 2.23 Consider the voltage-amplifier application of Example 2.21. Retain all parameter values other than the open-loop differential gain A_{dif0}. Assume 4000 V/V for the latter, and note the following specifications: (a) Resistance ratios don't exhibit any systematic deviation from their target values, and therefore the mean values of the resistance ratios satisfy $\overline{R_{A2}/R_{A1}} = \overline{R_{B2}/R_{B1}} = R_2/R_1$, where R_2 and R_1 are designed values. (b) The relative standard deviation of each resistance is specified by $\sigma[R]/R = 0.4\%$. (c) The mean value $\overline{V_{os}}$ of the offset is zero. (d) The standard deviation of the offset is specified by $\sigma[V_{os}] = 1$ mV.

How much will the mean value and standard deviation of dc error Δv_O be if the amplifier is driven by an input signal $v_i = 0.5$ V?

Work Since $V_{ICM} = V_{OCM}$ in this application, the mean value and the standard deviation of the dc error are described by (2.154) and (2.155), respectively. First, consider the external-network error components. For E_A, we refer to (2.149) for its definition, and take the following three steps:

1. For α'_A, substitute

$$\alpha'_A = -\frac{R_{A2}}{R_{A1} + R_{A2}}.$$

2. For β', adopt (2.138) and replace β_A and β_B with

$$\beta'_A = \frac{R_{A1}}{R_{A1} + R_{A2}},$$

$$\beta'_B = \frac{R_{B1}}{R_{B1} + R_{B2}}.$$

3. For α/β, substitute

$$\frac{\alpha}{\beta} = \frac{R_2}{R_1}.$$

These steps lead us to

$$E_A = -\frac{2\frac{R_{A2}}{R_{A1}}\left(1+\frac{R_{B2}}{R_{B1}}\right)}{2+\frac{R_{A2}}{R_{A1}}+\frac{R_{B2}}{R_{B1}}} + \frac{R_2}{R_1}. \tag{2.156}$$

For E_B, we follow a similar path and obtain

$$E_B = -\frac{2\frac{R_{B2}}{R_{B1}}\left(1+\frac{R_{A2}}{R_{A1}}\right)}{2+\frac{R_{B2}}{R_{B1}}+\frac{R_{A2}}{R_{A1}}} + \frac{R_2}{R_1}. \tag{2.157}$$

For determining the mean value of the external-network error, we evaluate (2.156) and (2.157) after replacing R_{A2}/R_{A1} with $\overline{R_{A2}/R_{A1}} = R_2/R_1$, and R_{B2}/R_{B1} with $\overline{R_{B2}/R_{B1}} = R_2/R_1$. The outcome is $\overline{E_A} = 0$ and $\overline{E_B} = 0$. Since $\overline{V_{os}}$ is also zero, (2.154) yields

$$\overline{\Delta v_O} = -\frac{1}{\beta A_{\text{dif0}}} v_O = -\frac{1}{\beta A_{\text{dif0}}} A_{cl0} v_i = -\frac{1}{\frac{1}{4}\times 4000} \times 3 \times 0.5 = -1.5 \text{ mV}.$$

We now move to the standard deviation described by (2.155). As to $\sigma^2[E_A]$ and $\sigma^2[E_B]$, note from (2.156) and (2.157) that E_A and E_B are functions of the uncorrelated random variables R_{A2}/R_{A1} and R_{B2}/R_{B1}. Applying (2.28) leads us to

$$\sigma^2[E_A] = \left[2\frac{\left(1+\frac{R_{B2}}{R_{B1}}\right)\left(2+\frac{R_{B2}}{R_{B1}}\right)}{\left(2+\frac{R_{A2}}{R_{A1}}+\frac{R_{B2}}{R_{B1}}\right)^2}\right]^2 \sigma^2\left[\frac{R_{A2}}{R_{A1}}\right] + \left[2\frac{\frac{R_{A_2}}{R_{A1}}\left(1+\frac{R_{A2}}{R_{A1}}\right)}{\left(2+\frac{R_{A2}}{R_{A1}}+\frac{R_{B2}}{R_{B1}}\right)^2}\right]^2 \sigma^2\left[\frac{R_{B2}}{R_{B1}}\right],$$

and

$$\sigma^2[E_B] = \left[2\frac{\left(1+\frac{R_{A2}}{R_{A1}}\right)\left(2+\frac{R_{A2}}{R_{A1}}\right)}{\left(2+\frac{R_{A2}}{R_{A1}}+\frac{R_{B2}}{R_{B1}}\right)^2}\right]^2 \sigma^2\left[\frac{R_{B2}}{R_{B1}}\right] + \left[2\frac{\frac{R_{B_2}}{R_{B1}}\left(1+\frac{R_{B2}}{R_{B1}}\right)}{\left(2+\frac{R_{A2}}{R_{A1}}+\frac{R_{B2}}{R_{B1}}\right)^2}\right]^2 \sigma^2\left[\frac{R_{A2}}{R_{A1}}\right].$$

The right-hand sides of these equations are substituted into (2.155) after replacing the resistance ratios inside the coefficients of the variances with the designed value R_2/R_1, and using the following equations for the variances themselves:

$$\sigma^2\left[\frac{R_{A2}}{R_{A1}}\right] = \left(\frac{R_{A2}}{R_{A1}}\right)^2\left(\frac{\sigma^2[R_{A2}]}{R_{A2}^2}+\frac{\sigma^2[R_{A1}]}{R_{A1}^2}\right) = 2\left(\frac{R_2}{R_1}\right)^2\frac{\sigma^2[R]}{R^2},$$

$$\sigma^2\left[\frac{R_{B2}}{R_{B1}}\right] = \left(\frac{R_{B2}}{R_{B1}}\right)^2\left(\frac{\sigma^2[R_{B2}]}{R_{B2}^2}+\frac{\sigma^2[R_{B1}]}{R_{B1}^2}\right) = 2\left(\frac{R_2}{R_1}\right)^2\frac{\sigma^2[R]}{R^2}.$$

We also substitute $v_i/2$ for x_a and $-v_i/2$ for x_b, and thus turn (2.155) into

$$\sigma[\Delta v_O] = \sqrt{\frac{\left(2+\frac{R_2}{R_1}\right)^2 + \left(\frac{R_2}{R_1}\right)^2}{\left(1+\frac{R_2}{R_1}\right)^2}\left(\frac{R_2}{R_1}\right)^2 \frac{\sigma^2[R]}{R^2}\left(\frac{v_i}{2}\right)^2 + \frac{1}{\beta^2}\sigma^2[V_{os}]}.$$

Plugging in the specified parameter values, this equation yields

$$\sigma[\Delta v_O] = \sqrt{\frac{(2+3)^2+3^2}{(1+3)^2} \times 3^2 \times 0.004^2 \times \left(\frac{0.5}{2}\right)^2 + \frac{1}{\left(\frac{1}{4}\right)^2} \times 0.001^2} = 5.93 \text{ mV}.$$

The reader is now invited to verify our predictions for $\overline{\Delta v_O}$ and $\sigma[\Delta v_O]$ by running a Monte Carlo analysis with the following input file `ex2-23.sp`:

```
filename: ex2-23.sp
.param vdd=1.8 vss=0 voh=1.65 vol=0.15 adif0=4000 vdir=0.25
+  fgbw=10meg fnd=10g
.incl opmacro2.sp
x1 1 2 3 4 8 fdopamp
ra2 2 3 30k
ra1 6 2 10k
rb2 1 4 30k
rb1 5 1 10k
vocm 8 0 0.9
vicm 7 0 0.9
bp 6 7 v=v(9)/2
bn 7 5 v=v(9)/2
vi 9 0 0.5
.control
repeat 200
define agauss(mean,sigma)  (mean+sigma*sgauss(0))
alter ra2=agauss(30k,120)
alter ra1=agauss(10k,40)
alter rb2=agauss(30k,120)
alter rb1=agauss(10k,40)
alter v.x1.vos=agauss(0,1m)
op
set appendwrite
wrdata mcdif v(4,3)
end
.endc
.end
```

The schematic is similar to that of Fig. 2.38. The reader may benefit from revisiting Example 2.9 for interpreting the part of the input file in between `.control` and `.endc`. The data of the differential-output voltage, defined as $v_O \doteq v_{ON} - v_{OP}$ by `v(4,3)`, accumulate in the second column of the file `mcdif`. Just copy it to a data-analysis spreadsheet and compute the mean value and standard deviation. ▲

2.10.4 Frequency Response and Step Response

Unless compromised by an inadequately slow common-mode feedback loop, the frequency and step responses of a fully-differential amplifier are described by the very same model expressions presented in Sections 2.5 and 2.6 for single-ended amplifiers. However, some of the symbols appearing in those expressions need to be reinterpreted as explained in the following list:

- x_i and v_o represent differential variables, that is, $x_i \doteq x_{Ai} - x_{Bi}$ and $v_o \doteq v_{OP} - v_{ON}$.
- α_i and β are defined for one-half of the external network, that is, $\alpha_i \doteq \alpha_{Ai} = \alpha_{Bi}$ and $\beta \doteq \beta_B = \beta_A$. Note that the two halves of the external network are assumed to be perfectly identical when analyzing the frequency response or step response. This assumption enables us to determine α_i and β from any half.
- A and A_0 are to be replaced by the differential counterparts A_{dif} and A_{dif0}.
- The odd symmetry of the fully-differential open-loop transfer characteristic around V_{OCM} suppresses the second-order harmonic distortion, and, therefore, implies $B_2 = 0$ in (2.60). This is yet another benefit of implementation with a fully-differential configuration.

Example 2.24 Suppose that the input signal v_i of the voltage amplifier of Example 2.21 is a sinewave of amplitude $V_i = 0.5$ V. Determine the necessary gain-bandwidth product for achieving a full-power bandwidth of 3 MHz for the cases of (a) $V_{DIR} = 0.5$ V, and (b) $V_{DIR} = 0.15$ V. Verify the predictions with ac and transient simulations for a sufficiently large A_{dif0} to avoid dc imperfections.

Work Considering $\alpha = 3/4$ and $V_i = 0.5$ V, we conclude from (2.58) that the frequency response of the amplifier will be slew free for $V_{DIR} = 0.5$ V but slew limited for $V_{DIR} = 0.15$ V. In the former case, we use $f_{cl} = 3$ MHz and $\beta = 1/4$ in (2.48), and obtain $f_{GBW} = 12$ MHz. In the slew-limited case, we first calculate $f_{cl} = 6.87$ MHz from (2.59b) for $f_{fp} = 3$ MHz, $\alpha_i = 3/4$, $V_{DIR} = 0.15$ V, and $X_{i(\text{max})} = 0.5$ V. Next, we use this result together with $\beta = 1/4$ in (2.48). This leads to $f_{GBW} = 27.5$ MHz.

Input file ex2-24.sp used for simulating the case of $V_{DIR} = 0.5$ V is as follows:

```
filename: ex2-24.sp
.param vdd=1.8 vss=0 voh=1.65 vol=0.15 adif0=10000 vdir=0.5
+  fgbw=12meg fnd=10g
.incl opmacro2.sp
x1 1 2 3 4 8 fdopamp
ra2 2 3 30k
ra1 6 2 10k
rb2 1 4 30k
rb1 5 1 10k
vocm 8 0 0.9
vicm 7 0 0.9
bp 6 7 v=v(9)/2
bn 7 5 v=v(9)/2
****ac analysis***********
*vi 9 0 dc 0 ac 1m
*.ac dec 1000 1k 100meg
****transient analysis****
```

```
vi 9 0 sin 0 0.5 3meg
.tran 0.5n 3u
.end
```

Run the ac analysis and plot `vdb(4,3)-vdb(9)` for small-signal magnitude response. Read f_{cl} as the frequency at which the magnitude drops by 3 dB. Does it confirm the prediction? Next, run the transient part of the file, where the differential input voltage is defined as a sinewave of 0.5-V amplitude and 3-MHz frequency. Since its frequency equals the designed value of f_{cl}, the amplitude of the differential-output voltage is expected to be $A_{cl0}V_i/\sqrt{2} = 1.061$ V. Plot the output signal `v(4,3)` and read its amplitude in one of the later cycles. It will turn out to be slightly lesser than expected. For an explanation of this difference, recall that the bandwidth model we developed in Subsection 2.5.2 is based on the assumption that the transition from small-signal behavior to slewing is abrupt around the threshold $v_{ID} = V_{DIR}$. As stated in the same subsection, transition is actually gradual, and therefore, a certain amount of transconductance deterioration should be expected unless v_{ID} is much smaller than V_{DIR}. Although $V_{DIR} = 0.5$ V of the present case is larger than $v_{ID} = \alpha V_i = (3/4)0.5 = 0.375$ V, the difference is not large enough to completely rule out transconductance deterioration. This leads to a smaller effective value for f_{cl}. The ac simulation does not show this effect because it relies on a linear analysis, and therefore, models only a pure small-signal behavior.

For the case of $V_{DIR} = 0.15$ V, use the same input file after resetting V_{DIR} and f_{GBW} but run only a transient simulation because the small-signal ac simulation doesn't take into account the consequences of slewing. ▲

Example 2.25 Assume $V_{DIR} = 0.15$ V for the amplifier specified in Example 2.24, and suppose that it is driven by a differential-input signal v_I stepping from −0.5 to 0.5 V. Determine the slew rate S_R, slewing time t_{slew}, and settling time t_S for $\epsilon = 1\%$. Verify with a transient simulation.

Work

- Use $V_{DIR} = 0.15$ V and $f_{GBW} = 27.5$ MHz in (2.80). Slew rate is calculated as $S_R = 25.9$ V/µs.
- Substitute into (2.83) the following: 3/4 for α_i, 1 V for X_i (step height), $V_{DIR} = 0.15$ V, $\beta = 1/4$, and $S_R = 2.59\times10^7$ V/s. Slew time is calculated as $t_{\text{slew}} = 92.7$ ns.
- Substitute into (2.85) the following: 3/4 for α_i, 1 V for X_i (step height), $V_{DIR} = 0.15$ V, $f_{cl} = 6.87$ MHz, and $\epsilon = 0.01$. Settling time is calculated as $t_S = 162$ ns.

The transient part of the input file `ex2-24.sp` can be used for verification after setting V_{DIR} to 0.15 V and defining the input source with

```
vi 9 0 pulse -0.5 0.5 10n 0 0 0.5u
```

Plotting `v(3,4)` will display the differential-output voltage waveform. Since this voltage is expected to change by 3 V, settling time should be determined as the difference in time between $t = 10$ ns when the input signal is stepped and the moment the differential-output voltage crosses $3(1 - 0.01) = 2.970$ mV. Measurement will yield 209 ns instead of the predicted 162 ns. The difference is due to the fact that the transconductance is recovered not abruptly but gradually after the opamp exits slewing for $t > t_{\text{slew}}$. As to the verification of the slew rate and slew time, simply plot the derivative of `v(3,4)`. ▲

2.10.5 Loop Gain, Differential Gain, and Noise

Loop Gain

The contents of Section 2.7 concerning the multi-pole dynamic response and stability of single-ended configurations are applicable also to fully-differential configurations. However, the simulation technique applied to the extraction of loop gain is modified to involve both halves of the differential configuration. Shown in Fig. 2.41(a) is the fully-differential counterpart of Fig. 2.28(a) for determining the voltage return ratio. Notice (a) the disabling of all signal sources but retaining the input and output common-mode voltage sources V_{ICM} and V_{OCM}, and (b) the insertion of two oppositely referenced replicas of the same ac test voltage source v_x inside the two differential loops. The voltage return ratio is defined by

$$\mathcal{R}_V \doteq -\frac{v_{an} - v_{ap}}{v_{bn} - v_{bp}}. \tag{2.158}$$

For extracting the current return ratio, we drive a copy of the same fully-differential configuration with an ac test current source i_x as shown in Fig. 2.41(b). V_{SP1}, V_{SP2}, V_{SN1},

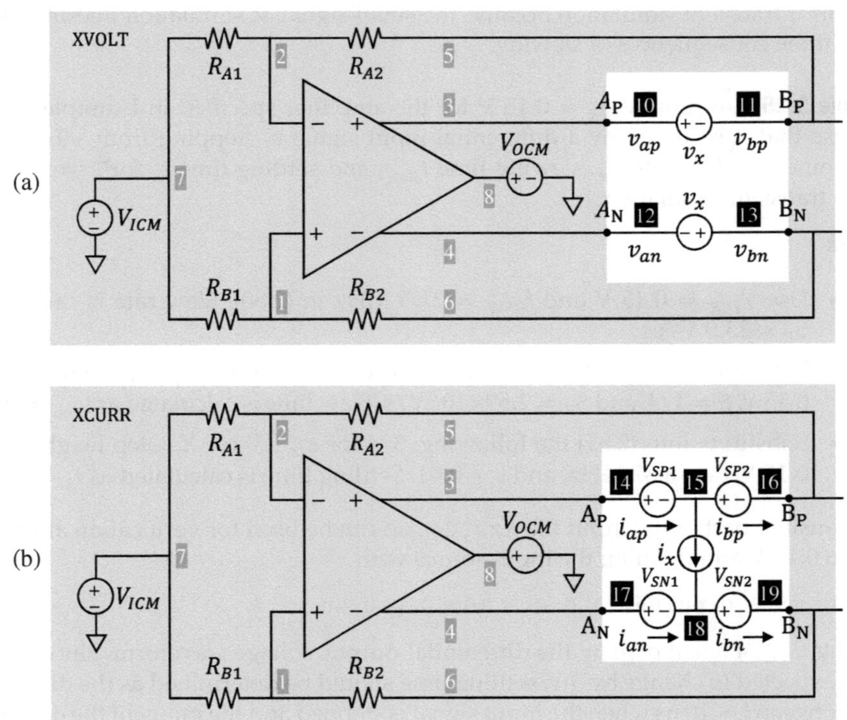

FIGURE 2.41 Inserting test sources into the differential-mode feedback loop for extracting the voltage return ratio as in (a), and current return ratio as in (b). Node numbers correspond to those of the input file "`fdloopgain.sp`".

and V_{SN2} are zero-volt voltage sources inserted for measuring the ac currents i_{ap}, i_{an}, i_{bp}, and i_{bn}, respectively. The current return ratio is defined by

$$\mathcal{R}_I \doteq -\frac{i_{an} - i_{ap}}{i_{bn} - i_{bp}}. \tag{2.159}$$

Using the outcomes of (2.158) and (2.159) in (2.91) yields the loop gain βA_{dif}. The following input file `fdloopgain.sp` extracts the magnitude and phase of the loop gain of the fully-differential amplifier example of Example 2.24. Node numbering is compliant with Fig. 2.41. The reader may use this file to verify the validity of (2.93) for unity-gain frequency, and of (2.94) for phase margin.

```
filename: fdloopgain.sp
.param vdd=1.8 vss=0 voh=1.65 vol=0.15 adif0=10000 vdir=0.15
+  fgbw=12meg fnd=6meg
.incl opmacro2.sp
*************amplifier************
.subckt amplifier 3 4 5 6
x1 1 2 3 4 8 fdopamp
ra2 2 5 30k
ra1 7 2 10k
rb2 1 6 30k
rb1 7 1 10k
vocm 8 0 0.9
vicm 7 0 0.9
.ends amplifier
**subckt for voltage return ratio**
xvolt 10 12 11 13 amplifier
e1 10 11 9 0 1
e2 13 12 9 0 1
vx 9 0 dc 0 ac 1
**subckt for current return ratio**
xcurr 14 17 16 19 amplifier
vsp1 14 15 0
vsp2 15 16 0
vsn1 17 18 0
vsn2 18 19 0
ix 15 18 dc 0 ac 1
*******analysis and plotting*******
.control
ac dec 1000 10 10meg
let rv=-(v(12,10)/v(13,11)) ; voltage rr
let ri=-(i(vsn1)-i(vsp1))/(i(vsn2)-i(vsp2)) ; current rr
plot db((rv*ri-1)/(2+rv+ri)) ph((rv*ri-1)/(2+rv+ri))
.endc
.end
```

The simulation-based loop-gain extraction technique described so far is applicable to the differential-mode loop closed over the external network. Each fully-differential configuration also has a common-mode loop closed internally for the purpose of equating the mean value of the two output voltages to V_{OCM}. This second loop can also affect the settling performance of the closed-loop configuration, and may even lead to instability if not designed properly, which is why we need to evaluate its gain too.

Common-mode loop gain is extracted by driving the loop with common-mode test sources as depicted in Fig. 2.42. Notice the dashed wires sensing the two output voltages and feeding them back to the opamp for common-mode control. Normally, these wires are not shown anywhere in application schematics because they sense the output voltages from within the opamp. Here we pull them out of the opamp symbol in order to explicitly show the reader that the test sources should be inserted within the common-mode feedback loop created by these wires. The schematic shown in Fig. 2.42(a) is for the extraction of voltage return ratio $\mathcal{R}_V$ defined by

$$\mathcal{R}_V \doteq -\frac{v_{an} + v_{ap}}{v_{bn} + v_{bp}}. \tag{2.160}$$

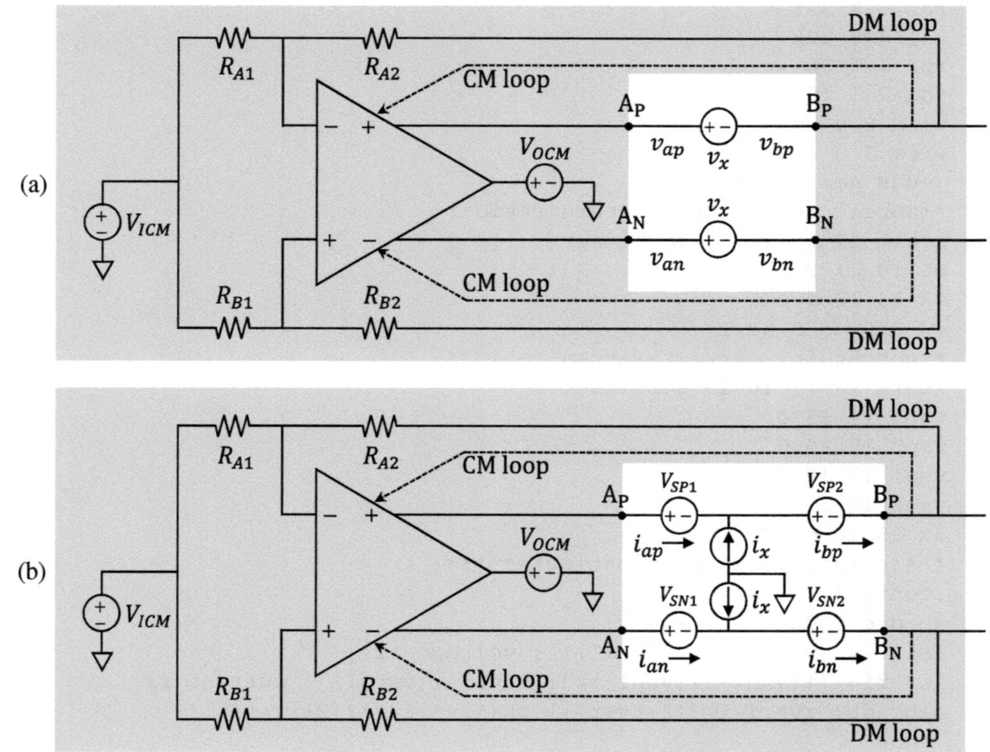

FIGURE 2.42 Inserting test sources into the common-mode (CM) feedback loop for extracting the voltage return ratio as in (a), and current return ratio as in (b). Differential-mode feedback loop is identified by DM.

Current return ratio $\mathcal{R}_I$ is extracted from the schematic shown in Fig. 2.42(b), and is defined by

$$\mathcal{R}_I \doteq -\frac{i_{an} + i_{ap}}{i_{bn} + i_{bp}}. \tag{2.161}$$

Once these two return ratios have been extracted, the corresponding common-mode loop gain is determined from (2.91).

The reader is cautioned that the fully-differential opamp macro `opmacro2.sp` is not suitable for applying the procedure of simulated return ratio extraction. Example 4.29 in Chapter 4 will provide an opportunity to apply the procedure on an actual device-level schematic of a fully-differential amplifier.

Differential Gain

We now move to the extraction of the differential gain A_{dif} of a fully-differential configuration. The Norton equivalent circuit of a fully-differential opamp is shown in Fig. 2.43. $v_{op} = v_o/2$ and $v_{on} = -v_o/2$ represent the single-ended signal voltages at the noninverting- and inverting-output terminals. G_m and Z_o are the single-ended transconductance and output-impedance parameters of the opamp. The load impedance, denoted by Z_L, is due to the parallel combination of (a) one-half $Z_{in}/2$ of the input impedance of the circuit driven differentially by the configuration, and (b) the external-network impedance seen from each single-ended output port.

The differential gain of the opamp is defined by

$$A_{\text{dif}} \doteq \frac{v_{op} - v_{on}}{v_{id}} = \frac{v_o}{v_{id}}. \tag{2.162}$$

Thanks to the symmetry of the fully-differential opamp, we can extract A_{dif} from one-half of the equivalent circuit. Upper half, for example, yields

$$A_{\text{dif}} = \frac{2v_{op}}{v_{id}} = 2\frac{Z_L}{Z_o + Z_L}G_m Z_o. \tag{2.163}$$

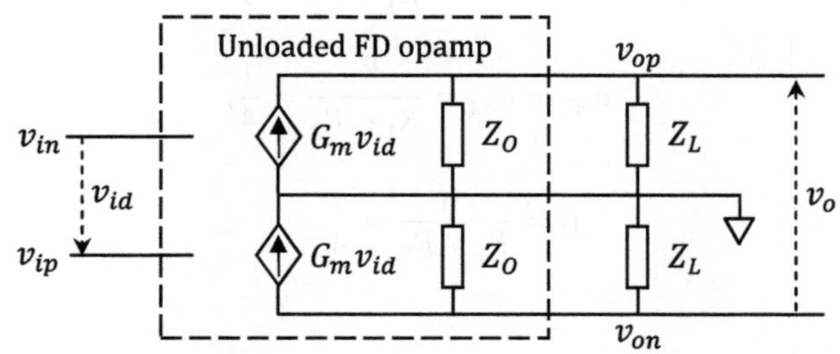

FIGURE 2.43 Equivalent circuit of a fully-differential opamp.

Comparing this expression with its single-ended counterpart (2.95), we observe a factor of 2 difference, which is due to the fact that all parameters appearing on the right-hand side are single ended whereas A_{dif} is defined differentially by (2.162). Equation (2.163) enables us to determine A_{dif} from any of the two halves of the opamp and external network.

Noise

The contents of Section 2.9 on noise are applicable to fully-differential configurations. However, when determining from (2.112) the thermal output noise due to external network, the reader should consider all resistors in the upper and lower halves of the external network. Due to the symmetry between the two halves, this will double the external-network component of mean-square output noise in comparison with a single-ended configuration built with similar resistors.

As far as coupled noise is concerned, fully-differential configurations offer much better immunity than single-ended configurations because if the upper and lower networks are properly laid out so that their corresponding features receive identical copies of the same noise, then, the configuration will reject the noise as a common-mode disturbance. This is yet another significant benefit of fully-differential configurations.

Example 2.26 Determine the room-temperature rms output noise generated by the external network of the noninverting amplifier of Fig. 2.37(a) for the opamp and external-network specifications declared in the input file `ex2-24.sp`. Verify the result with a noise simulation.

Work The mean-square value of the external-network noise is described by (2.112) where k is the resistor index. Therefore,

$$V^2_{no(\text{ext})} = 2\pi E_{th} f_{cl}\left[\left(\frac{\alpha_{A1}}{\beta}\right)^2 R_{A1} + \left(\frac{\alpha_{A2}}{\beta}\right)^2 R_{A2} + \left(\frac{\alpha_{B1}}{\beta}\right)^2 R_{B1} + \left(\frac{\alpha_{B2}}{\beta}\right)^2 R_{B2}\right].$$

Considering the symmetry of the configuration, the feedback factor, and input factors appearing in this expression can be extracted from one-half as

$$\alpha_{A1} = \alpha_{B1} = \frac{R_2}{R_1 + R_2} = \frac{3}{4},$$

$$\alpha_{A2} = \alpha_{B2} = \frac{R_1}{R_1 + R_2} = \frac{1}{4},$$

$$\beta = \frac{R_1}{R_1 + R_2} = \frac{1}{4}.$$

Also considering $f_{cl} = \beta f_{GBW} = (1/4) \times 12\,(\text{MHz}) = 3$ MHz and $E_{th} = 4.14 \times 10^{-21}$ J at room temperature, we obtain $V_{no(\text{ext})} = 136.8\ \mu\text{V}$.

Simulation is performed with the following input file ex2-26.sp:

```
filename: ex2-26.sp
.param vdd=1.8 vss=0 voh=1.65 vol=0.15 adif0=10000 vdir=0.5
+  fgbw=12meg fnd=10g
.incl opmacro2.sp
x1 1 2 3 4 8 fdopamp
ra2 2 3 30k
ra1 6 2 10k
rb2 1 4 30k
rb1 5 1 10k
vocm 8 0 0.9
vicm 7 0 0.9
bp 6 7 v=v(9)/2
bn 7 5 v=v(9)/2
vi 9 0 dc 0 ac 1m
.noise v(4,3) vi dec 100 1k 2g
.end
```

After running the simulation, simply print onoise_total to read the rms value of the output noise. ▲

2.11 Discrete-Time Amplifiers

Discrete-time amplifiers are built with an external network of capacitors and switches instead of resistors, which is why they are also known as *switched-capacitor* amplifiers. They perform amplification by sampling and storing the input variables across capacitors first, and then properly redistributing the corresponding electrical charge. This process is repeated in cycles at a frequency defined by a clock signal. Each cycle essentially contains two successive phases, one for sampling and storage, and another for amplification. Secondary phases may also be deployed for improving the performance.

Discrete-time amplifiers can be regarded as dc processors in the sense that their electrical variables should settle into a dc steady state in each and every phase in order to be considered valid. Prior to settling, however, a transient response is inevitably observed in each phase. The settling time of these transient events determines the minimum clock frequency, hence the maximum throughput of the amplifier. Our analytical interest in modeling a discrete-time amplifier is therefore focused usually on the dc and transient responses although frequency response may also be of interest in certain applications.

2.11.1 DC Analysis with Charge Conservation

Charge conservation is the physical basis of discrete-time operation. We now illustrate its use in first-order dc analysis on an example of sampled-analog amplifier whose closed-loop schematic is shown in Fig. 2.44(a). Notice the presence of five switches in addition to two capacitors. These switches are controlled by two-phase nonoverlapping clocks φ_1 and φ_2 whose timing diagram is also included in Fig. 2.44(a). We assume ideal switches closing when clock is high ($\varphi = 1$) and opening when clock is low ($\varphi = 0$). Since clock states $\varphi_1 = 1$ and $\varphi_2 = 1$ are nonoverlapping, switches controlled by φ_1 cannot be closed simultaneously with those controlled by φ_2.

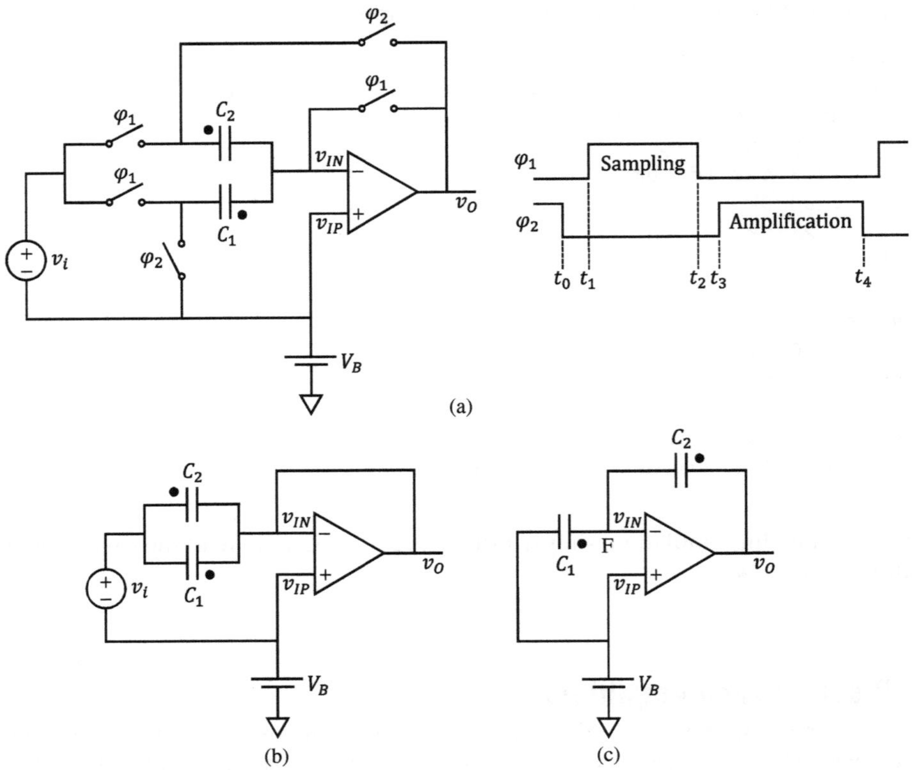

FIGURE 2.44 (a) Schematic and timing diagram of a discrete-time amplifier. (b) Sampling-phase schematic. (c) Amplification-phase schematic.

The period in between $t = t_1$ and $t = t_2$ is called *sampling phase*. Closing of the switches controlled by φ_1 and opening of those controlled by φ_2 lead to the *sampling-phase schematic* shown in Fig. 2.44(b). Following the expiration of the sampling phase at $t = t_2$, all switches remain open until $t = t_3$ when the switches controlled by φ_2 are closed to initiate the so-called *amplification phase*, during which, the amplifier is represented by the *amplification-phase schematic* shown in Fig. 2.44(c). This phase continues until $t = t_4$ when switches are opened again. Note that the input signal is actually sampled at $t = t_2$, and the corresponding amplified output signal is made available after the amplifier has settled into a dc steady state in the amplification phase. Assuming that the output is read out at $t = t_4$, closed-loop gain is defined by

$$A_{cl} \doteq \frac{v_o(t_4)}{v_i(t_2)}. \tag{2.164}$$

We now relate this gain to external-network parameters as a demonstration of the utility of the charge conservation principle in the first-order dc analysis of discrete-time configurations. As usual in any first-order dc analysis, we assume an offset-free opamp of infinite open-loop dc gain.

The first thing we do is to select arbitrarily one plate of each capacitor as a reference plate in the main schematic. The dots seen in Fig. 2.44(a) mark these reference plates. These reference marks are carried over to phase schematics as shown in Figs. 2.44(b) and (c). This practice enables us to identify the same plate in different phase schematics.

Next, we consider the sampling instant $t = t_2$ and the sampling-phase schematic shown in Fig. 2.44(b). Assuming that the opamp is in a dc steady state, and thus invoking the virtual-short approximation, we observe $v_{IN} = v_{IP} = V_B$. The output being short-circuited to the inverting input, we also observe $v_O = V_B$ regardless of the value $v_i(t_2)$ of the input signal. The charge established by these node voltages at the reference plates is given by

$$Q_{C1}(t_2) = C_1\{V_B - [V_B + v_i(t_2)]\} = -C_1 v_i(t_2) \tag{2.165}$$

for C_1, and by

$$Q_{C2}(t_2) = C_2\{[V_B + v_i(t_2)] - V_B\} = C_2 v_i(t_2) \tag{2.166}$$

for C_2. Unreferenced plates carry the same amount of charge with opposite polarity.

Opamp stays in open loop during the brief interval $t_2 < t < t_3$ between the end of the sampling phase and the beginning of the amplification phase. Also obvious for this interval is that both plates of both capacitors float because all of the switches are open. Barring certain secondary effects to be discussed in Subsection 2.11.3, the capacitive charges described by (2.165) and (2.166) are therefore retained throughout this interval, and are carried over to the amplification phase.

Now, consider the amplification phase starting at $t = t_3$ and the phase schematic depicted in Fig. 2.44(c). The capacitor plates connecting the node F remain floating throughout this phase. As a consequence, the sum of the charges established in the sampling phase at these particular plates is conserved. Since the floating plate happens to be the referenced plate in C_1 and unreferenced plate in C_2, the conserved charge Q equals $Q_{C1} - Q_{C2}$. According to (2.165) and (2.166), Q can therefore be expressed as

$$Q = -(C_1 + C_2)v_i(t_2). \tag{2.167}$$

Although this charge is conserved throughout the amplification phase, the voltages across the two capacitors differ from their respective sampling-phase values. Once the transient response caused by reconfiguration at $t = t_3$ has died away, and all voltages have settled to a dc steady state, virtual short will again impose $v_{IN} = v_{IP} = V_B$ until the amplification phase expires at $t = t_4$. Since the voltage across C_1 vanishes due to virtual short, no charge is left at the floating reference plate of this capacitor. On the other hand, the charge at the floating unreferenced plate of C_2 is now given by $C_2[V_B - v_O(t_4)]$, where $v_O(t_4)$ represents the output voltage in the steady-state period of the amplification phase. The charge conserved is therefore transferred in its entirety to C_2, and is described by

$$Q = C_2[V_B - v_O(t_4)]. \tag{2.168}$$

Equating the right-hand sides of (2.167) and (2.168), and solving for the output voltage, we obtain

$$v_O(t_4) = \left(1 + \frac{C_1}{C_2}\right) v_i(t_2) + V_B, \tag{2.169}$$

which indicates an output bias voltage $V_O = V_B$ and an output signal

$$v_o(t_4) = \left(1 + \frac{C_1}{C_2}\right) v_i(t_2). \tag{2.170}$$

According to (2.164), the closed-loop gain is described by

$$A_{cl} = 1 + \frac{C_1}{C_2}. \tag{2.171}$$

Example 2.27 The discrete-time amplifier configuration of Fig. 2.44(a) is specified by $V_B = 0.9$ V, $v_i = \pm 0.2$ V, and $A_{cl} = 3$ V/V. Determine the common-mode range and output-range specifications of its opamp.

Work Considering Fig. 2.44(b) for sampling phase, we notice $v_{IP} = v_{IN} = v_O = V_B = 0.9$ V. Therefore any common-mode range and output range that includes 0.9 V is suitable for this phase. Next, we consider the amplification phase schematic of Fig. 2.44(c), in which v_{IP} and v_{IN} are still equal to V_B hence 0.9 V but v_O swings between 0.3 and 1.5 V as implied by (2.169) for $V_B = 0.9$ V, $C_1/C_2 = 2$, and $v_i(t_2) = \pm 0.2$ V. Therefore, any common-mode range around 0.9 V is acceptable while the output range should comply with $V_{OL} < 0.3$ V and $V_{OH} > 1.5$ V. ▲

Example 2.28 Shown in Fig. 2.45(a) is the schematic and timing diagram of a discrete-time integrator known as *noninverting stray-insensitive switched-capacitor integrator*. Notice the following three separate phases of operation: (a) resetting and sampling (RS) phase defined by $\varphi_{RS} = 1$, $\varphi_1 = 1$, $\varphi_2 = 0$; (b) sampling phase defined by $\varphi_{RS} = 0$, $\varphi_1 = 1$, $\varphi_2 = 0$; and (c) integration phase defined by $\varphi_{RS} = 0$, $\varphi_1 = 0$, $\varphi_2 = 1$. Integration of the input signal v_i over time starts at the end $t = t_0$ of the RS phase, and continues over successive cycles of sampling and integration phases. Perform a first-order dc analysis with charge conservation to determine the output voltage v_O at the end of each of the two integration phases depicted in the timing diagram.

Work It is obvious from the phase schematics shown in Figs. 2.45(b) to (d) that the configuration has a floating node (node F) in sampling and integration phases but none in the RS phase. Note as a general rule that the charge conserved in a discrete-time configuration is defined during the particular phase in which none of the nodes float. When analyzing a configuration, it is most convenient to begin with that particular phase because its node voltages can be determined without referring back to previous phases. In the amplifier of Example 2.27, the sampling phase had this property, which is why we analyzed that phase first. For the same reason, we begin with the RS phase in the present configuration. Also note that the equation $v_{IN} = v_{IP} = V_B$ will be applied in all phases due to the validity of the virtual-short approximation in a dc analysis.

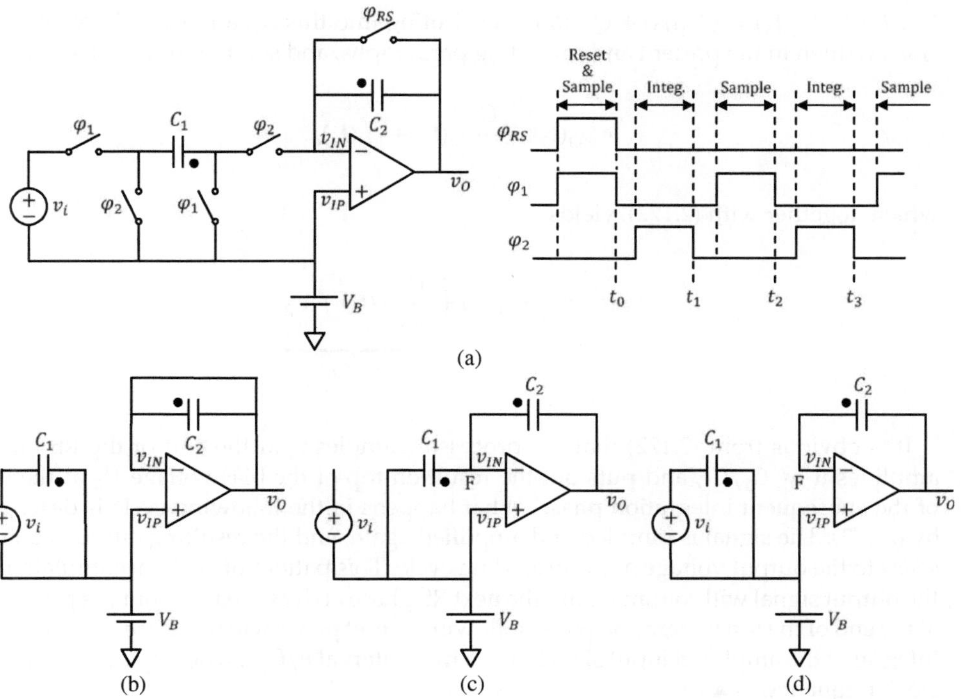

FIGURE 2.45 Discrete-time integrator schematics of Example 2.28. (a) Main schematic. (b) Reset-and-sample phase. (c) Integration phase. (d) Sampling phase.

For $t = t_0$, refer to Fig. 2.45(b), and observe $v_O(t_0) = v_{IN} = V_B$, which implies $Q_{C2}(t_0) = 0$ for the charge on the reference plate of C_2. Also obvious is $Q_{C1}(t_0) = -C_1 v_i(t_0)$ for the charge on the reference plate of C_1.

For $t = t_1$, refer to Fig. 2.45(c). Since $v_{IN} = V_B$, the reference-plate charges of C_1 and C_2 are described by $Q_{C1}(t_1) = 0$ and $Q_{C2}(t_1) = C_2[V_B - v_O(t_1)]$. By solving the charge conservation equation $Q_{C1}(t_1) + Q_{C2}(t_1) = Q_{C1}(t_0) + Q_{C2}(t_0)$ after substituting the charge expressions written so far, we obtain

$$v_O(t_1) = \frac{C_1}{C_2} v_i(t_0) + V_B. \tag{2.172}$$

For $t = t_2$, refer to Fig. 2.45(d). Since $v_{IN} = V_B$, the charge on the reference plate of C_2 is given by $Q_{C2}(t_2) = C_2[V_B - v_O(t_2)]$. Due to the floating state of the plate, this charge must be equal to $Q_{C2}(t_1)$. Referring to the preceding paragraph for the latter, the equality can be expressed as $Q_{C2}(t_2) = C_2[V_B - v_O(t_1)]$. What is also obvious from the phase schematic is $Q_{C1}(t_2) = -C_1 v_i(t_2)$.

For $t = t_3$, refer to Fig. 2.45(c). Considering $v_{IN} = V_B$, we write $Q_{C1}(t_3) = 0$ and $Q_{C2}(t_3) = C_2[V_B - v_O(t_3)]$ for the reference-plate charges. Charge conservation implies

$Q_{C1}(t_3) + Q_{C2}(t_3) = Q_{C1}(t_2) + Q_{C2}(t_2)$. Substituting into this equation the charge expressions written in the present and preceding paragraphs, and solving for $v_O(t_3)$, we obtain

$$v_O(t_3) = \frac{C_1}{C_2} v_i(t_2) + v_O(t_1),$$

which, together with (2.172), yields

$$v_O(t_3) = \frac{C_1}{C_2} v_i(t_2) + \underbrace{\frac{C_1}{C_2} v_i(t_0) + V_B}_{v_O(t_1)} \cdot \tag{2.173}$$

It is obvious from (2.172) that the processor samples v_i at the end of the RS phase, amplifies it by C_1/C_2, and puts out the result on top of the bias voltage V_B at the end of the subsequent integration phase. What happens in the following cycle is described by (2.173). The signal is sampled and amplified again, and the resulting output signal is *added* to the output voltage of the preceding cycle. This pattern of cyclic accumulation of the output signal will continue until the next RS phase arrives. Reading out the processor at the end of the nth integration phase following a reset phase will therefore yield the time integral of the amplified input signal over a time interval n/f_{CLK}, where f_{CLK} represent the clock frequency. ▲

2.11.2 Unified Closed-Loop Model

The unified closed-loop model presented in Fig. 2.10(a) of Section 2.3 is valid in any phase of discrete-time operation provided that (a) the opamp remains in closed loop throughout the phase, and (b) charge conservation is accounted for by properly complementing the capacitors with *initial-voltage* sources. In a dc analysis involving a floating node, we insert an initial-voltage source in series with each of the capacitors attached to the floating node, and set its value to the final voltage acquired by the capacitor in the preceding phase. In a transient analysis, we do this for all capacitors regardless of whether their nodes float or not. These initial-voltage sources become a part of the set of input variables x_j of the phase under analysis. Actually, a second type of input variable is also involved in accurate dc analysis. It comes in the form of an impulse-current source representing the parasitic *charge injection* and *clock feedthrough* effects observed in real MOSFET switches. A detailed modeling of these two parasitic effects will be presented in Chapter 3. Considering also the presence of input signals and bias sources, the unified model generally includes multiple sources. However, depending on the type of the response to be analyzed, some of these sources may be omitted. The types of sources involved in analyzing a discrete-time amplifier for first-order dc response, accurate dc response, and transient response are presented in Table 2.1 for future reference.

The unified closed-loop model indicated by Fig. 2.10(a) is analytically described by

$$v_{ID} \doteq v_{IP} - v_{IN} = \sum_{j=1}^{n} \alpha_j x_j - \beta v_O \tag{2.174}$$

Response	Signal and bias sources	Initial-voltage sources	Impulse-current sources
First-order dc	Yes	Yes*	No
Accurate dc	Yes	Yes*	Yes†
Transient	Yes	Yes	No

*Only for the capacitances attached to floating nodes.
†Only for the switches attached to floating nodes.

TABLE 2.1 Relevance of Sources in Discrete-Time Amplifier Analysis

for the differential-input voltage of the opamp, and by the opamp transfer function

$$v_O = f(v_{ID}) \tag{2.175}$$

for the output voltage. In the specific case of first-order dc analysis, (2.175) turns into $v_{ID} = 0$ due to virtual short, and (2.174) yields

$$v_O = \sum_{j=1}^{n} \frac{\alpha_j}{\beta} x_j \tag{2.176}$$

for the output voltage of the phase being analyzed.

Example 2.29 Using the unified model, determine the relationship between $v_O(t_4)$ and $v_i(t_2)$ by performing a first-order dc analysis on the amplifier of Fig. 2.44(a).

Work Since a first-order dc analysis is involved, we rely on (2.176) for each of the two phases. By the same token, we ignore charge-injection and clock-feedthrough effects and therefore omit impulse-current sources. As also indicated in Table 2.1, initial-voltage sources are needed for capacitors only if the phase to be analyzed contains a floating node. An inspection of the main circuit schematic given in Fig. 2.44(a) indicates the inverting-input node of the opamp as such in the amplification phase. None of the nodes is floating in the sampling phase. Therefore, the final voltages established across C_1 and C_2 at the end of the sampling phase will be carried over by initial-voltage sources to the analysis of the amplification phase.

As stated in Example 2.28, it is convenient to begin with a phase that doesn't contain any floating node. We therefore begin with the sampling phase, whose phase schematic is shown in Fig. 2.46(a). Final voltages across the capacitors are marked as $v_{C1}(t_2)$ and $v_{C2}(t_2)$. We carry over these two to the amplification-phase schematic shown in Fig. 2.46(b). Needless to say, the reference polarities of these signals must be preserved as we do this.

In the sampling-phase analysis, we first determine the feedback factor β and input factors α_i of v_i, and α_B of V_B. An inspection of Fig. 2.46(a) reveals $\beta = 1$, $\alpha_i = 0$, and $\alpha_B = 1$, for which, (2.176) indicates $v_O(t_2) = V_B$ for the output voltage at the end of the sampling phase.

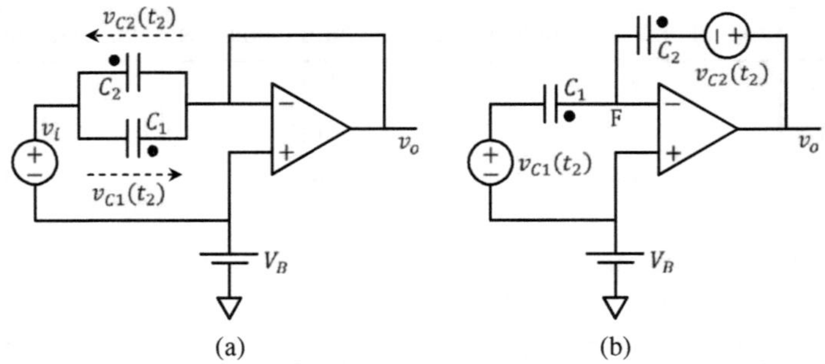

FIGURE 2.46 Discrete-time amplifier of Example 2.29. (a) Sampling-phase schematic. (b) Amplification-phase schematic.

Next, we determine the final values $v_{C1}(t_2)$ and $v_{C2}(t_2)$ of capacitive voltages at the end of the sampling phase. Considering $v_O(t_2) = V_B$, Fig. 2.46(a) indicates

$$v_{C1}(t_2) = -v_i(t_2),$$

and

$$v_{C2}(t_2) = v_i(t_2).$$

These two initial voltages are represented by voltage sources in the amplification-phase schematic shown in Fig. 2.46(b). Inspecting the schematic, we determine the input factors as

$$\alpha_{C1} = -\frac{\frac{1}{sC_2}}{\frac{1}{sC_1} + \frac{1}{sC_2}} = -\frac{C_1}{C_1 + C_2}, \tag{2.177}$$

$$\alpha_{C2} = \frac{\frac{1}{sC_1}}{\frac{1}{sC_1} + \frac{1}{sC_2}} = \frac{C_2}{C_1 + C_2}, \tag{2.178}$$

$$\alpha_B = 1 - \frac{\frac{1}{sC_2}}{\frac{1}{sC_1} + \frac{1}{sC_2}} = \frac{C_2}{C_1 + C_2}. \tag{2.179}$$

Inspection also yields

$$\beta = \frac{\frac{1}{sC_1}}{\frac{1}{sC_1} + \frac{1}{sC_2}} = \frac{C_2}{C_1 + C_2}. \tag{2.180}$$

By substituting these results into (2.176), we obtain the following expression for the output voltage to be observed at the end of the amplification phase:

$$v_O(t_4) = \left(1 + \frac{C_1}{C_2}\right) v_i(t_2) + V_B, \tag{2.181}$$

which, not surprisingly, indicates the same relationship as (2.169). ▲

Example 2.30 Repeat the analysis of Example 2.28 with the unified model.

Work Similar to what has been done in Example 2.29, we ignore impulse-current sources and rely on (2.176) for all phases. The main circuit schematic of Fig. 2.45(a) shows no floating node in the reset-and-sample (RS) phase. Analysis begins with that phase and proceeds through the following steps.

1. *Analysis for* $t = t_0$ The RS-phase schematic shown in Fig. 2.47(a) directly indicates $v_O(t_0) = v_{IN}(t_0) = v_{IP}(t_0) = V_B$. Since C_1 and C_2 float in the subsequent amplification phase, we need to determine the final voltages across these capacitors

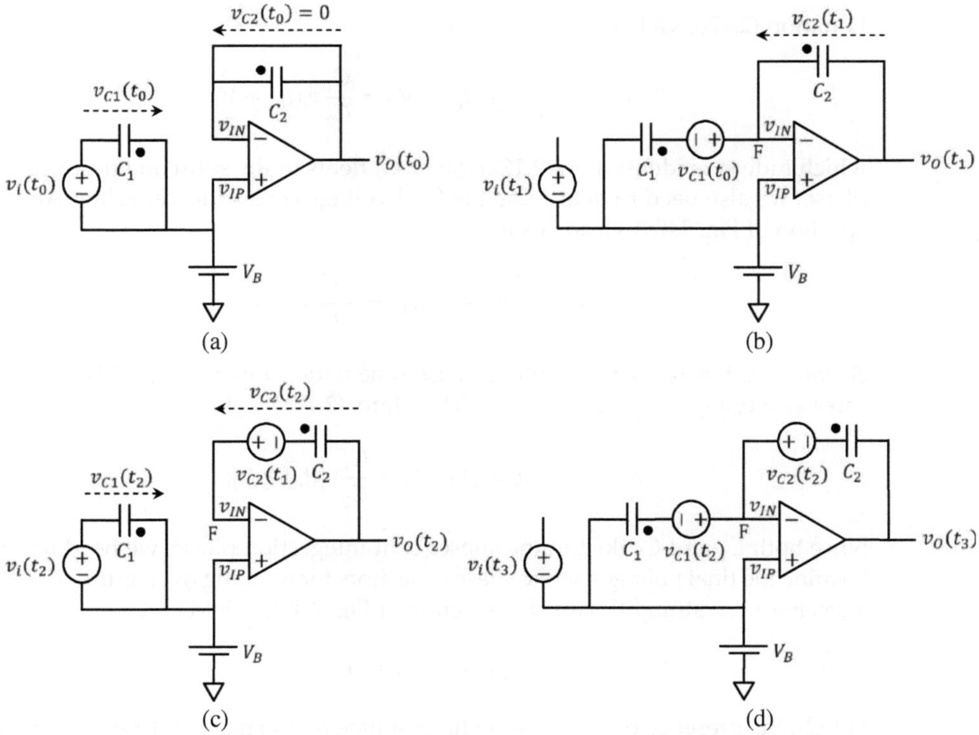

FIGURE 2.47 Phase schematics of Example 2.30. (a) Reset-and-sample phase. (b) First integration phase. (c) Sampling phase. (d) Second integration phase.

for carrying over as initial voltages. Considering the input signal across C_1 and the short-circuit across C_2, these voltages are described by

$$v_{C1}(t_0) = -v_i(t_0),$$

$$v_{C2}(t_0) = 0.$$

2. *Analysis for $t = t_1$* A schematic belonging to the first integration phase after reset and sample is given in Fig. 2.47(b). Notice the initial-voltage source $v_{C1}(t_0)$ in series with C_1. No voltage is carried over for C_2 because $v_{C2}(t_0) = 0$. We determine the following factors by inspection:

$$\alpha_i = 0,$$

$$\alpha_{C1} = -\frac{C_1}{C_1 + C_2},$$

$$\alpha_B = 1 - \frac{C_1}{C_1 + C_2} = \frac{C_2}{C_1 + C_2},$$

$$\beta = \frac{C_2}{C_1 + C_2}.$$

Equation (2.176) yields

$$v_O(t_1) = -\frac{C_1}{C_2} v_{C1}(t_0) + V_B = \frac{C_1}{C_2} v_i(t_0) + V_B,$$

which indeed is identical to (2.172). Since C_2 floats in the subsequent sampling phase, we also need to determine the final voltage across this capacitor. An inspection of Fig. 2.47(b) leads us to

$$v_{C2}(t_1) = V_B - v_O(t_1) = -\frac{C_1}{C_2} v_i(t_0).$$

3. *Analysis for $t = t_2$* The sampling-phase schematic shown in Fig. 2.47(c) indicates $\alpha_i = 0$, $\alpha_{C2} = -1$, $\alpha_B = 1$, $\beta = 1$. Therefore, (2.176) yields

$$v_O(t_2) = -v_{C2}(t_1) + V_B = \frac{C_1}{C_2} v_i(t_0) + V_B.$$

Since both C_1 and C_2 float in the subsequent integration phase, we need to determine the final voltages across these capacitors for carrying over as initial voltages. For C_1, a straightforward inspection of Fig. 2.47(c) yields

$$v_{C1}(t_2) = -v_i(t_2).$$

For C_2, the presence of the initial-voltage source $v_{C2}(t_1)$ may confuse the reader as to which two nodes to select for defining $v_{C2}(t_2)$. The general rule is that the nodes where the terminals of the capacitor are physically attached are to be selected. Since C_2 is physically in between the output node and node F, the arrow identifying $v_{C2}(t_2)$ in Fig. 2.47(c) is drawn exactly in between these two nodes. It obviously indicates

$$v_{C2}(t_2) = V_B - v_O(t_2) = v_{C2}(t_1) = -\frac{C_1}{C_2}v_i(t_0).$$

4. *Analysis for $t = t_3$* The schematic shown in Fig. 2.47(d) for the second integration phase indicates

$$\alpha_i = 0,$$

$$\alpha_{C1} = -\frac{C_1}{C_1 + C_2},$$

$$\alpha_{C2} = -\frac{C_2}{C_1 + C_2},$$

$$\alpha_B = \frac{C_2}{C_1 + C_2},$$

$$\beta = \frac{C_2}{C_1 + C_2},$$

whose substitution into (2.176) yields

$$v_O(t_3) = -\frac{C_1}{C_2}v_{C1}(t_2) - v_{C2}(t_2) + V_B = \frac{C_1}{C_2}v_i(t_2) + \frac{C_1}{C_2}v_i(t_0) + V_B.$$

This is exactly the same conclusion reached in Example 2.28 with the charge conservation principle and expressed by (2.173). ▲

2.11.3 Accurate Modeling of DC Response

The dc response of discrete-time amplifiers exhibits not only the very same gain, offset, and external-network errors observed in continuous-time amplifiers but also errors stemming from the imperfections of MOSFET switches used in their implementation. Every MOSFET switch stores mobile charge in its channel when closed, and releases part of this charge to the adjoining external-network capacitors when opened. Called *charge injection*, this effect causes a dc error by corrupting the voltage stored in any capacitor that becomes afloat due to switch opening. Another switch-related dc error is due to *clock feedthrough*, which also corrupts the voltage of a floating capacitor by coupling the clock pulse via extrinsic overlap capacitances of the MOSFET switch. Both effects can be modeled by attaching an impulse-current source to any node that becomes afloat when the switch is opened. We represent this current source with $Q\delta(t)$, where Q is the total charge transferred to the node due to these effects, and $\delta(t)$ is the Dirac delta function. Details of modeling Q in terms of MOSFET properties will be presented in Subsection 3.9.2.

For an accurate dc analysis of a discrete-time amplifier, we begin with the particular phase in which there is no floating node. Neither a capacitor initial-voltage source nor an impulse-current source is involved in this phase. We rely on (2.23) for determining the output voltage accurately, and use the outcome for determining the capacitor voltages that should be carried over to the following phase as initial-voltage sources. Next, we insert these sources into the schematic of the following phase together with the impulse-current sources representing charge-injection and clock feedthrough, and determine the output voltage accurately again with (2.23). We repeat this procedure for all subsequent phases including the final phase in which the output is read out. Equation (2.23), as applied to the final phase, yields the accurate value of the output voltage. We characterize the dc error by subtracting the first-order estimate of the output voltage from this accurate value.

Example 2.31 Formulate the dc error of the amplifier configuration of Fig. 2.44(a).

Work We begin with the sampling phase, in which, none of the nodes is afloat. Phase schematic, which is identical to that of Fig. 2.46(a), is presented in Fig. 2.48(a). Feedback and input factors, as already determined in Example 2.29, are $\beta = 1$, $\alpha_i = 0$, and $\alpha_B = 1$. After substituting these into (2.23) and assuming $V_O = V_B$ for offset definition, we obtain the following accurate description of the output voltage at the final instant $t = t_2$ of the sampling phase:

$$v_O(t_2) = -\frac{A_0 V_{os}}{1 + A_0} + V_B.$$

Figure 2.48(a) also indicates the following final voltages for the two capacitors:

$$v_{C1}(t_2) = v_O(t_2) - v_i(t_2) - V_B = -\frac{A_0 V_{os}}{1 + A_0} - v_i(t_2), \tag{2.182}$$

$$v_{C2}(t_2) = -v_O(t_2) + v_i(t_2) + V_B = \frac{A_0 V_{os}}{1 + A_0} + v_i(t_2). \tag{2.183}$$

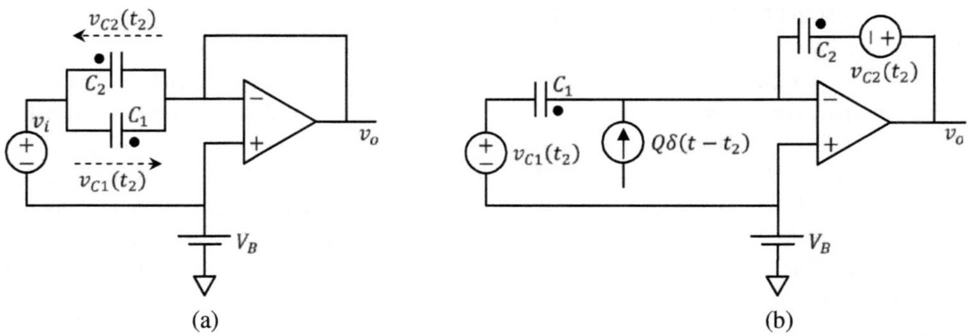

FIGURE 2.48 Phase schematics for the switched-capacitor amplifier of Example 2.31. (a) Sampling phase. (b) Amplification phase.

Shown in Fig. 2.48(b) is the amplification-phase schematic. Notice the inclusion of an impulse-current source $Q\delta(t-t_2)$ in addition to two capacitive initial-voltage sources carried over from the sampling phase. This impulse-current source represents the parasitic charge transferred by the opening feedback switch to the floating inverting-input node of the opamp. It is disabled by open-circuiting while extracting the feedback factor and the input factors associated with $v_{C1}(t_2)$, $v_{C2}(t_2)$, and V_B. These four factors are described by

$$\beta = \frac{C_2}{C_1 + C_2},$$

$$\alpha_{C1} = -\frac{C_1}{C_1 + C_2},$$

$$\alpha_{C2} = \frac{C_2}{C_1 + C_2},$$

$$\alpha_B = 1 - \frac{C_1}{C_1 + C_2} = \frac{C_2}{C_1 + C_2}.$$

As to the input factor of $Q\delta(t-t_2)$, we disable the three voltage sources by short-circuiting each, and observe $v_{IN} = Q/(C_1 + C_2)$ from the phase schematic because Q is the charge injected to the parallel equivalent of C_1 and C_2. This, together with $v_{IP} = 0$, leads us to

$$\alpha_Q = -\frac{1}{C_1 + C_2}.$$

Equation (2.23), rewritten explicitly in terms of the four amplification-phase input factors, yields

$$v_O(t_4) = \frac{\beta A_0}{1 + \beta A_0}\left[\frac{\alpha_{C1}}{\beta} v_{C1}(t_2) + \frac{\alpha_{C2}}{\beta} v_{C2}(t_2) + \frac{\alpha_B}{\beta} V_B + \frac{\alpha_Q}{\beta} Q\right]$$
$$+ \frac{V_B}{1 + \beta A_0} - \frac{A_0 V_{os}}{1 + \beta A_0}.$$

By determining all four ratios of α/β from previous equations and substituting them together with (2.182) and (2.183) into this equation, we obtain

$$v'_O(t_4) = \frac{\beta' A_0}{1 + \beta' A_0}\left(1 + \frac{C'_1}{C'_2}\right) v_i(t_2) - \frac{A_0 V_{os}}{(1 + \beta' A_0)(1 + A_0)} - \frac{\beta' A_0}{1 + \beta' A_0}\frac{Q}{C'_2} + V_B$$

for the output voltage observed at the end of the amplification phase on a fabricated sample of the amplifier. Note that the prime sign signifies the measured value of a parameter or variable. Subtracting from this equation the target output voltage described by (2.169) and using the output-signal symbol $v_o(t_4)$ for $[1+(C'_1/C'_2)]v_i(t_2)$ bring us to the following expression of dc error:

$$\Delta v_{O(dc)} \doteq v'_O(t_4) - v_O(t_4) = \left(\frac{C'_1}{C'_2} - \frac{C_1}{C_2}\right) v_i(t_2) - \frac{v_o(t_4)}{1+\beta' A_0}$$

$$- \frac{A_0 V_{os}}{(1+\beta' A_0)(1+A_0)} - \frac{\beta' A_0}{1+\beta' A_0} \frac{Q}{C'_2}.$$

We now simplify this equation by replacing (a) $1+A_0$ with A_0, (b) $1+\beta' A_0$ with $\beta' A_0$, (c) β' with β as we did in Subsection 2.4.2, and (d) C'_2 of the last term with the target value C_2. The outcome is as follows:

$$\Delta v_{O(dc)} = \underbrace{\left(\frac{C'_1}{C'_2} - \frac{C_1}{C_2}\right) v_i(t_2)}_{\substack{\text{External-}\\ \text{network}\\ \text{error}}} \underbrace{- \frac{v_o(t_4)}{\beta A_0}}_{\substack{\text{Gain}\\ \text{error}}} \underbrace{- \frac{V_{os}}{\beta A_0}}_{\substack{\text{Offset}\\ \text{error}}} \underbrace{- \frac{Q}{C_2}}_{\substack{\text{Charge}\\ \text{error}}} . \tag{2.184}$$

It is worth comparing the dc error performance of this discrete-time amplifier with that of the continuous-time counterpart shown in Fig. 2.6 and analyzed in Examples 2.2 and 2.6. The dc error of the continuous-time amplifier is described by (2.24) as follows:

$$\Delta v_O = \underbrace{\left(\frac{R'_2}{R'_1} - \frac{R_2}{R_1}\right) v_i}_{\substack{\text{External-}\\ \text{network}\\ \text{error}}} \underbrace{- \frac{v_o}{\beta A_0}}_{\substack{\text{Gain}\\ \text{error}}} \underbrace{- \frac{V_{os}}{\beta}}_{\substack{\text{Offset}\\ \text{error}}} ,$$

where β equals its counterpart in (2.184) for the same closed-loop gain. Obviously, external-network error depends on the difference between the measured and targeted ratios of two capacitances in the discrete-time amplifier and two resistances in the continuous-time amplifier. Since ratioed capacitors are usually more accurate and precise than ratioed resistors, the discrete-time amplifier is superior in this respect. Assuming identical opamps for both amplifiers, we observe a parity between the two in gain-error performance. An interesting disparity between the two amplifiers is observed in their offset-error performance. The discrete-time amplifier is a factor of A_0 better in this respect but this is a unique property of the particular configuration of this amplifier. Simpler discrete-time amplifier configurations may not offer this property. Finally, notice the unique charge error of the discrete-time amplifier, for which, no counterpart exists in the dc error of the continuous-time amplifier. The techniques applied for minimizing the charge error will be described in Subsection 3.9.2. ▲

2.11.4 Transient Response

The limited closed-loop bandwidth prevents a discrete-time configuration from instantaneously tracking a rapidly varying input signal in the sampling phase. The delay associated with this transient creates a *tracking error* manifesting itself as a difference between the sampled and actual values of the input signal. A discrete-time amplifier is inevitably forced into another type of transient state as its switches are reconfigured at the beginning of each new phase. Although this transient decays as the phase progresses, the final value of the output voltage still exhibits a *settling error* due to the finite-phase duration. Both of these transient errors can be reduced by designing the amplifier for a sufficiently wide closed-loop bandwidth f_{cl}. Before we discuss how f_{cl} can be controlled for this purpose, recall from Section 2.8 that a discrete-time closed-loop configuration is usually constructed with a simple unbuffered opamp called operational transconductance amplifier (OTA). A property common to most OTA topologies is that the frequency dependence of their open-loop transfer function (2.95) is primarily due to the external-network impedance Z_L instead of Y_m or Z_o, which enables us to assume $Y_m = G_m$ and $Z_o = r_o$. Z_L itself is almost purely capacitive as long as the conductance of a closed external-network switch is sufficiently high; hence $Z_L \approx 1/sC_L$.[12] These approximations turn (2.95) into the single-pole form of (2.41), where $A_0 = G_m r_o$ and $f_o = 1/2\pi r_o C_L$. According to (2.43), the gain-bandwidth product of a typical OTA is therefore described by

$$f_{GBW} = \frac{G_m}{2\pi C_L}, \tag{2.185}$$

which, together with (2.48) leads to the following expression for closed-loop bandwidth:

$$f_{cl} = \frac{\beta G_m}{2\pi C_L}. \tag{2.186}$$

According to this expression, a large bandwidth calls for a large transconductance and a small load capacitance.

The technique of analyzing a phase for transient response is presented next. As usual, we will consider the case of slew-free transient first, and then, extend the results to the case of a slewing transient.

Slew-Free Transient

As summarized in Table 2.1 of Subsection 2.11.2, analyzing a phase for transient response involves the voltages carried over from the preceding phase for all capacitances regardless of whether or not the present phase contains a floating node. Each of these voltages is represented by a step-voltage source placed in series with the capacitance it belongs to. Bias and input-signal sources are also included. However, the transient analysis technique we are about to describe is based on the assumption that the time variation of input signals in sampling phases is sufficiently slow to inhibit the tracking error. The technique is therefore applicable to the modeling of the settling error only.

[12] The effect of external-network switches on frequency response is negligible as long as the conductance of each switch is at least a factor of five larger than G_m. The conductances of signal and bias sources are also practically ineffective if they also are larger than the transconductance by at least the same factor.

In the case of a slew-free transient, the waveforms of the output and differential-input voltages of the OTA in any phase are described generally by

$$v_O(t) = v_O(\infty) - \left[v_O(\infty) - v_O(0)\right] \exp\left(-2\pi f_{cl} t\right), \tag{2.187}$$

$$v_{ID}(t) = v_{ID}(0) \exp\left(-2\pi f_{cl} t\right), \tag{2.188}$$

where $v_O(0)$ and $v_{ID}(0)$ represent the initial values at $t = 0$ when the phase begins, and $v_O(\infty)$ is the asymptotic value that would be reached by the output voltage in an infinitely long phase. $v_O(\infty)$ is determined by performing a dc analysis for the phase concerned. Also note that the condition of slew-free operation

$$V_{DIR} > \left|v_{ID}(0)\right| \tag{2.189}$$

is specified at the beginning of the phase because, as indicated by (2.188), $|v_{ID}|$ is maximized at that instant.

Equations (2.187) and (2.188) fully characterize the transient response provided that $v_O(0)$, $v_{ID}(0)$, $v_O(\infty)$, and f_{cl} are extracted from the phase schematic. The steps to be taken for this purpose are explained next.

1. *Initial Output Voltage* As long as switch resistances are negligibly small, the initial output voltage $v_O(0)$ is determined primarily by the combination of voltage sources and capacitive external network. Since the limited gain-bandwidth product prevents the OTA from responding instantaneously at $t = 0$, none of the OTA parameters other than the parasitic input and output capacitances is effective on $v_O(0)$. We therefore remove the OTA from the phase schematic but retain its input and output capacitances if considered significant, and determine the transfer function from each voltage source to the output. Denoting the transfer function of the voltage source $v_j(0)$ with γ_j, $v_O(0)$ is determined by superposition from

$$v_O(0) = \sum_{j=1}^{n} \gamma_j v_j(0). \tag{2.190}$$

   Note that the set of $v_j(0)$ includes (a) initial voltages carried over from the previous phase for all capacitors, (b) any input-signal source that may be driving the configuration in the present phase, and (c) any bias source.

2. *Initial Differential-Input Voltage* $v_{ID}(0)$ is determined from (2.174) for the specific value $v_O(0)$ found in step 1, that is,

$$v_{ID}(0) = \sum_{j=1}^{n} \alpha_j v_j(0) - \beta v_O(0). \tag{2.191}$$

   We extract α_j and β from the phase schematic, and use them in (2.191) together with $v_O(0)$ and $v_j(0)$ to determine $v_{ID}(0)$. We test for slew-free operation by comparing the result with V_{DIR} as per (2.189).

3. *Closed-Loop Bandwidth* Closed-loop bandwidth may change from one phase to the next. We calculate its value from (2.186) for the equivalent load capacitance C_L. The latter is extracted from the current phase schematic after disabling all sources.
4. *Asymptotic Output Voltage* $v_O(\infty)$ is what a dc analysis predicts for the output signal of the phase being analyzed. It is determined from the dc phase schematic.
5. *Final Output Voltage and Capacitor Voltages* The first four steps described so far enable us to characterize the transient response in terms of $v_O(t)$ and $v_{ID}(t)$ for the present phase. In order to determine the capacitor voltages to be carried over to the next phase, we calculate the final values $v_O(T)$ and $v_{ID}(T)$ of output and differential-input voltages by substituting phase duration T for t in (2.187) and (2.188), that is,

$$v_O(T) = v_O(\infty) - \left[v_O(\infty) - v_O(0)\right] \exp\left(-2\pi f_{cl} T\right), \tag{2.192}$$

$$v_{ID}(T) = v_{ID}(0) \exp\left(-2\pi f_{cl} T\right). \tag{2.193}$$

Next, by inspecting the schematic of the present phase we determine the final voltages of external-network capacitors for the specific values indicated by (2.192) and (2.193). As stated in Example 2.30, these voltages are defined across the nodes where the terminals of the capacitor are physically attached. Finally, we carry these voltages over to the transient analysis of the subsequent phase in the form of step sources representing initial capacitor voltages.

Example 2.32 Assuming slew-free transient response in all phases, we will derive an analytical settling-error expression for the amplifier of Fig. 2.44(a) loaded by a circuit of input capacitance C_I. Settling error is defined as the difference between (a) the output voltage observed at the end of the amplification phase, and (b) the asymptotic output voltage predicted by a dc analysis in the same phase. Timing diagram is copied to Fig. 2.49(a), where T_S and T_A represent the durations of the sampling and amplification phases, respectively. The analytically derived settling-error model will be verified by simulation for $C_I = 0.5$ pF, $C_1 = 2$ pF, $C_2 = 1$ pF, $G_m = 500$ μA/V, $r_o = 7$ MΩ, $V_B = 0.9$ V. Clock and input-signal waveforms are defined by `vi`, `vphi1`, and `vphi2` in the input file `ex2-32.sp`, and are displayed in Fig. 2.50.

Work We begin with the sampling phase, whose transient-phase schematic is shown in Fig. 2.49(b). Notice the inclusion of the capacitor voltage sources $v_{CI}(t_0)$ and $v_{C2}(t_0)$ carried over from the preceding amplification phase for C_I and C_2, respectively. Assuming complete settling in that phase and ignoring the dc error, we can model the former of these two voltages with (2.181) as

$$v_{CI}(t_0) = \left(1 + \frac{C_1}{C_2}\right) v_{i(\text{pre})} + V_B, \tag{2.194}$$

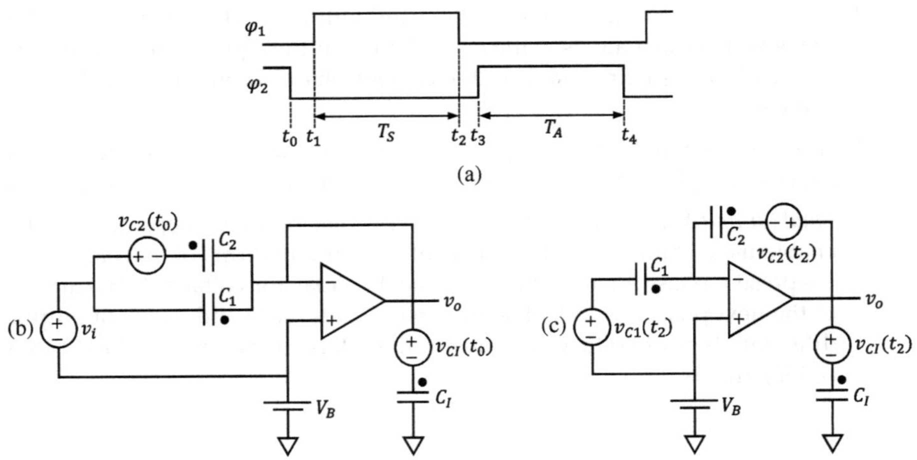

FIGURE 2.49 Example 2.32. (a) Timing diagram. (b) Sampling-phase transient schematic. (c) Amplification-phase transient schematic.

where $v_{i(\text{pre})}$ is the input signal sampled in the preceding cycle. For $v_{C2}(t_0)$, we can write

$$v_{C2}(t_0) = \left(1 + \frac{C_1}{C_2}\right) v_{i(\text{pre})} \tag{2.195}$$

because C_2 is placed between the output and and inverting-input terminals of the OTA in the preceding amplification phase. Also note from Fig. 2.49(b) that no initial-voltage source is inserted for C_1 because at the end of the preceding amplification phase this capacitor is placed between the virtually short-circuited input terminals of the OTA.

The steps of sampling-phase analysis are presented next.

1. *Initial Output Voltage in Sampling Phase* γ_j are determined by inspecting the phase schematic of Fig. 2.49(b) after removing the OTA. Parasitic capacitances of the OTA are assumed to be negligibly small, and therefore not included in the schematic.

   - For $v_{C2}(t_1)$: $\gamma_{C2} = -\frac{C_2}{C_1+C_2+C_I}$.
   - For $v_{CI}(t_1)$: $\gamma_{CI} = \frac{C_I}{C_1+C_2+C_I}$.
   - For v_i: $\gamma_i = \frac{C_1+C_2}{C_1+C_2+C_I}$.
   - For V_B: $\gamma_B = \frac{C_1+C_2}{C_1+C_2+C_I}$.

   By substituting into (2.190) these expressions as well as (2.194) and (2.195), we obtain

$$v_O(t_1) = \frac{C_1 + C_2}{C_1 + C_2 + C_I}\left[v_i - \left(1 - \frac{C_I}{C_2}\right) v_{i(\text{pre})}\right] + V_B.$$

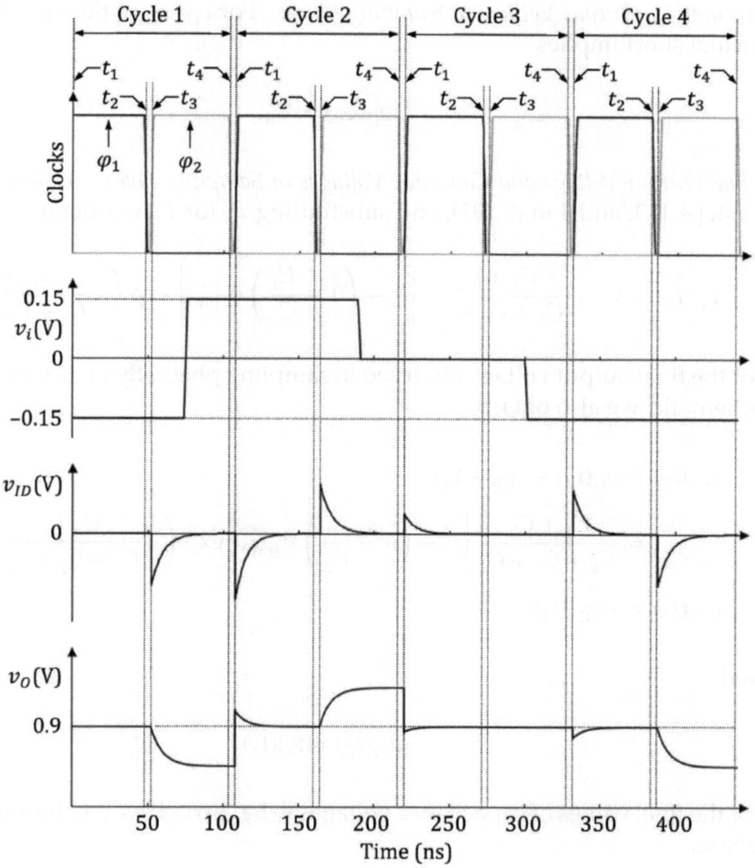

FIGURE 2.50 Simulated waveforms of Example 2.32.

2. *Initial Differential-Input Voltage in Sampling Phase* Although initial differential-input voltage is described generally by (2.191), our assumption of complete settling in the preceding amplification phase enables us to determine $v_{ID}(t_1)$ simply from the phase schematic shown in Fig. 2.49(b). Considering the short-circuit between the output and inverting-input terminals of the OTA, the outcome is

$$v_{ID}(t_1) = V_B - v_O(t_1).$$

3. *Closed-Loop Bandwidth in Sampling Phase* Inspecting the phase schematic after disabling the voltage sources, we observe the parallel combination of C_1, C_2, and C_I across the output port of the OTA. The load capacitance is therefore given by $C_L = C_1 + C_2 + C_I$. Also considering $\beta = 1$, (2.186) yields

$$f_{cl} = \frac{G_m}{2\pi(C_1 + C_2 + C_I)}.$$

4. *Asymptotic Output Voltage in Sampling Phase* For $v_O(\infty)$ of the sampling phase, virtual short implies

$$v_O(\infty) = V_B.$$

5. *Final Output Voltage and Capacitor Voltages in Sampling Phase* Using the results of steps 1, 3, and 4 in (2.192), and substituting T_S for T, we obtain

$$v_O(t_2) = V_B + \frac{C_1 + C_2}{C_1 + C_2 + C_I}\left[v_i - \left(1 - \frac{C_I}{C_2}\right) v_{i(\text{pre})}\right] \exp\left(-\frac{G_m T_S}{C_1 + C_2 + C_I}\right)$$

for the final output voltage observed in sampling phase. By inspecting the phase schematic, we also obtain

$$\begin{aligned} v_{C1}(t_2) &= v_O(t_2) - (V_B + v_i) \\ &= \frac{C_1 + C_2}{C_1 + C_2 + C_I}\left[v_i - \left(1 - \frac{C_I}{C_2}\right) v_{i(\text{pre})}\right] \exp\left(-\frac{G_m T_S}{C_1 + C_2 + C_I}\right) - v_i, \\ v_{C2}(t_2) &= -v_{C1}(t_2), \end{aligned}$$

and

$$v_{CI}(t_2) = v_O(t_2)$$

for the final values of capacitance voltages to be carried over to the amplification phase.

Now, we move to amplification phase, whose phase schematic is shown in Fig. 2.49(c).

1. *Initial Output Voltage in Amplification Phase* γ_j are determined by inspection as follows.

- For $v_{C1}(t_2)$: $\gamma_{C1} = \frac{C_1 C_2}{C_1 C_2 + C_1 C_I + C_2 C_I}$.
- For $v_{C2}(t_2)$: $\gamma_{C2} = \frac{C_1 C_2}{C_1 C_2 + C_1 C_I + C_2 C_I}$.
- For $v_{CI}(t_2)$: $\gamma_{CI} = \frac{C_1 C_I + C_2 C_I}{C_1 C_2 + C_1 C_I + C_2 C_I}$.
- For V_B: $\gamma_B = \frac{C_1 C_2}{C_1 C_2 + C_1 C_I + C_2 C_I}$.

Equation (2.190) therefore yields

$$\begin{aligned} v_O(t_3) = V_B &+ \frac{C_1 C_I + C_2 C_I}{C_1 C_2 + C_1 C_I + C_2 C_I} \frac{C_1 + C_2}{C_1 + C_2 + C_I}\left[v_i - \left(1 - \frac{C_I}{C_2}\right) v_{i(\text{pre})}\right] \\ &\times \exp\left(-\frac{G_m T_S}{C_1 + C_2 + C_I}\right). \end{aligned}$$

2. *Initial Differential-Input Voltage in Amplification Phase* $v_{ID}(t_3)$ is determined from (2.191) for the following factors obtained from phase schematic:

$$\beta = \frac{C_2}{C_1 + C_2},$$

$$\alpha_{C1} = -\frac{C_1}{C_1 + C_2},$$

$$\alpha_{C2} = \frac{C_2}{C_1 + C_2},$$

$$\alpha_{CI} = 0,$$

$$\alpha_B = 1 - \frac{C_1}{C_1 + C_2} = \frac{C_2}{C_1 + C_2}.$$

The result is

$$v_{ID}(t_3) = v_i - \frac{C_1C_2 + C_1C_I + 2C_2C_I}{C_1C_2 + C_1C_I + C_2C_I}\frac{C_1 + C_2}{C_1 + C_2 + C_I}\left[v_i - \left(1 - \frac{C_I}{C_2}\right)v_{i(\mathrm{pre})}\right] \times \exp\left(-\frac{G_m T_S}{C_1 + C_2 + C_I}\right).$$

However, since $v_{ID}(t_3)$ is needed only for determining whether the amplification-phase transient is slew free or not, the accuracy offered by this expression is not indispensable. Assuming a reasonable settling in the sampling phase, we can assume $T_S \to \infty$, and therefore simplify this expression into

$$v_{ID}(t_3) \approx v_i.$$

3. *Closed-Loop Bandwidth in Amplification Phase* Disabling all voltage sources of the phase schematic leaves C_I in parallel with the series combination of C_1 and C_2 at the output port. Therefore, $C_L = C_I + [C_1C_2/(C_1 + C_2)]$. Also considering $\beta = C_2/(C_1 + C_2)$, (2.186) yields

$$f_{cl} = \frac{C_2 G_m}{2\pi(C_1C_2 + C_1C_I + C_2C_I)}.$$

4. *Asymptotic Output Voltage in Amplification Phase* $v_O(\infty)$ is generally described by (2.176). For the feedback and input factors extracted in step 2 and the expressions of $v_{C1}(t_2)$ and $v_{C2}(t_2)$ derived in sampling-phase analysis, it yields

$$v_O(\infty) = V_B + \left(1 + \frac{C_1}{C_2}\right) \times \left\{v_i - \frac{C_1 + C_2}{C_1 + C_2 + C_I}\left[v_i - \left(1 - \frac{C_I}{C_2}\right)v_{i(\mathrm{pre})}\right]\exp\left(-\frac{G_m T_S}{C_1 + C_2 + C_I}\right)\right\}.$$

5. *Final Output Voltage in Amplification Phase* We can now determine $v_O(t_4)$ by using the results of steps 1, 3, and 4 of this phase in (2.192), and then replacing T with T_A. Ignoring the terms proportional to the negligibly small product

$$\exp\left(-\frac{G_m T_S}{C_1 + C_2 + C_I}\right) \exp\left(-\frac{C_2 G_m T_A}{C_1 C_2 + C_1 C_I + C_2 C_I}\right),$$

we obtain

$$v_O(t_4) = \left(1 + \frac{C_1}{C_2}\right) v_i + V_B$$
$$- \frac{(C_1 + C_2)^2}{C_2(C_1 + C_2 + C_I)} \left[v_i - \left(1 - \frac{C_I}{C_2}\right) v_{i(\text{pre})}\right] \exp\left(-\frac{G_m T_S}{C_1 + C_2 + C_I}\right)$$
$$- \frac{C_1 + C_2}{C_2} v_i \exp\left(-\frac{C_2 G_m T_A}{C_1 C_2 + C_1 C_I + C_2 C_I}\right).$$

Settling error $\Delta v_{O(\text{set})}$ is the difference between the final-output voltage described by this equation and its asymptotic value for $T_S \to \infty$ and $T_A \to \infty$. This definition leads us to

$$\Delta v_{O(\text{set})} = - \frac{(C_1 + C_2)^2}{C_2(C_1 + C_2 + C_I)} \left[v_i - \left(1 - \frac{C_I}{C_2}\right) v_{i(\text{pre})}\right] \exp\left(-\frac{G_m T_S}{C_1 + C_2 + C_I}\right)$$
$$- \frac{C_1 + C_2}{C_2} v_i \exp\left(-\frac{C_2 G_m T_A}{C_1 C_2 + C_1 C_I + C_2 C_I}\right). \quad (2.196)$$

The reader is reminded that the expressions used for describing $v_O(\infty)$ in the foregoing analysis are based on a first-order dc analysis. As a consequence, the dc error $\Delta v_{O(dc)}$ described by (2.184) is not included in (2.196). The difference observed at the end of an amplification phase between the actual output voltage $v'_O(t_4)$ of a fabricated copy and the first-order estimate $v_O(t_4)$ used in design equals the sum of the dc and settling errors, that is,

$$v'_O(t_4) = v_O(t_4) + \Delta v_{O(dc)} + \Delta v_{O(\text{set})}. \quad (2.197)$$

We now test the findings of the analysis by running a transient simulation with the following input file `ex2-32.sp`:

```
filename: ex2-32.sp
.param vdir=5 gm0=0.5m ro=7meg
*********************ota**************************
*ninv: noninverting input
*inv: inverting input
*out: output
.subckt ota ninv inv out
gm 0 out cur='gm0*vdir*tanh((v(ninv)-v(inv))/vdir)'
rout out 0 ro
bb 0 out i=0.9/ro
.ends ota
```

```
*******************switch**********************
.subckt switch in out cont
s in out cont 0 mod
cin in 0 25f
cout out 0 25f
.model mod sw vt=0.9 ron=1 roff=10g
.ends switch
******************amplifier********************
xota 1 2 3 ota
xs1 2 3 7 switch
xs2 6 4 7 switch
xs3 6 5 7 switch
xs4 4 3 8 switch
xs5 5 1 8 switch
c1 5 2 2p
c2 4 2 1p
ci 3 0 0.5p
vb 1 0 0.9
vi 6 1 pwl 0 -0.15 74n -0.15 76n 0.15 189n 0.15
+  191n 0 300n 0 302n -0.15 441 -0.15
vphi1 7 0 pulse 1.8 0 47n 2n 2n 58n 112n
vphi2 8 0 pulse 0 1.8 51n 2n 2n 50n 112n
.tran 10p 441n
.options trtol=1
.end
```

The subcircuit `ota` defines an OTA macromodel suitable for first-order dc, transient, and ac simulations. Its noninverting input, inverting input, and output terminals are accessed through pins `ninv`, `inv`, and `out`, respectively. Differential-input range, transconductance, and output resistance are represented by user-defined parameters `vdir`, `gm0`, and `ro`. It is a zero-offset OTA having 0.9-V output voltage for $v_{ID} = 0$. As explained earlier in this subsection, no internal pole is defined for the OTA.

The subcircuit `switch` defines an ideal switch whose resistance equals 1 Ω when closed and 10 GΩ when open. Its trip voltage, defined by `vt`, equals 0.9 V. It closes when the voltage at the control terminal `cont` is higher than the trip voltage. Two small capacitances `cin` and `cout` are added to the current-conducting terminals `in` and `out` for emulating the parasitic capacitances associated with real switches. These capacitances help remove certain transient artifacts that would otherwise be observed on simulated waveforms.

As to the dc error modeled generally by (2.184), neither an offset component nor a charge-error component will be observed in this simulation because no provision is made for these dc imperfections. External-network error will also be absent because no parametric variation is defined. However, the gain-error component will be present due to the finite dc open-loop gain $A_0 = 3500$ V/V as implied by $G_m = 0.5$ mA/V and $r_o = 7$ MΩ.

The nonoverlapping clocks are defined by `vphi1` and `vphi2`. Phase duration is 52 ns for each sampling and amplification. As shown in Fig. 2.50, simulation covers four successive cycles. Input signal `vi` varies only inside the amplification phases. Its sampled value equals −0.15 V in cycle 1, 0.15 V in cycle 2, 0 in cycle 3, and −0.15 V in cycle 4.

We now calculate the settling error $\Delta v_{O(\text{set})}$ in all four cycles, and compare the outcome with simulated values. Considering $V_B = 0.9$ V, $1 + (C_1/C_2) = 3$, and $v_i = -0.15$ V, the asymptotic output voltage $v_O(\infty)$ calculated from the first-order dc expression (2.169) for cycle 1 equals 0.45 V. The dc error of the same cycle, as obtained from the gain-error term of (2.184) for $v_o(t_4) = -0.45$ V, $\beta = 1/3$, and $A_0 = 3500$ V/V equals 387 μV. The settling error, as calculated from (2.196) for $v_{i(\text{pre})} = 0$, $G_m = 0.5$ mA/V, $C_1 = 2$ pF, $C_2 = 1$ pF, $C_I = 0.5$ pF, and $T_S = T_A = 52$ ns, equals 496 μV. According to (2.197), these numbers lead to a final output voltage $v'_O(t_4) = 0.450883$ V at $t = 104$ ns.

In cycle 2, considering $v_i = 0.15$ V, we obtain $v_O(\infty) = 1.35$ V from (2.169). Gain error, as calculated from (2.184) for $v_o(t_4) = 0.45$ V, equals −387 μV. For the settling error, (2.196) yields −611 μV for $v_i = 0.15$ V and $v_{i(\text{pre})} = -0.15$ V. Therefore, $v'_O(t_4)$ at $t = 216$ ns in cycle 2 is expected to be 1.349002 V.

In cycle 3 we expect $v_O(\infty) = 0.9$ V because $v_i = 0$. Equations (2.184) and (2.196) yield $\Delta v_{O(dc)} = 0$ and $\Delta v_{O(\text{set})} = 115$ μV, respectively, and therefore $v'_O(t_4) = 0.900115$ V at $t = 328$ ns.

In cycle 4, $v_i = -0.15$ V implies $v_O(\infty) = 0.45$ V. Considering $\Delta v_{O(dc)} = 387$ μV and $\Delta v_{O(\text{set})} = 496$ μV, we expect to observe $v'_O(t_4) = 0.450883$ V at $t = 440$ ns.

The reader is now invited to run the simulation and plot the output-voltage waveform by entering

```
plot v(3)
```

Zooming in on the waveform around the final instant of each amplification phase will enable the reader to compare the actual and predicted final values of the output voltage.

The reader will also benefit from comparing the calculated and simulated values of v_{ID} at $t = t_2$ in all cycles, and experimenting with (a) larger and smaller values of C_1, C_2, and G_m to affirm the predicted effect of f_{cl} on settling, (b) larger values of `ron` comparable to $1/G_m$ to see how switch-on resistance affects the settling behavior, and (c) a time-varying v_i in sampling phases, say, a sinewave of frequency 1 MHz or more, to observe the resulting tracking error. ▲

Slewing Transient

The OTA of a discrete-time amplifier initially exhibits a slew-limited transient response in any phase that doesn't satisfy the condition defined by (2.189). As discussed in Subsection 2.6.2, output voltage varies linearly with time during slewing. This variation is modeled by

$$v_O(t) = v_O(0) + \text{sgn}[v_{ID}(0)]S_R t, \quad 0 \le t \le t_{\text{slew}}, \tag{2.198}$$

where $v_O(0)$ is the initial value of the output voltage, S_R is the slew rate given by (2.80), and t_{slew} is the duration of slewing. Differential-input voltage also varies linearly with time as described by

$$v_{ID}(t) = v_{ID}(0) - \text{sgn}[v_{ID}(0)]\beta S_R t, \quad 0 \le t \le t_{\text{slew}}. \tag{2.199}$$

Slewing ends when $|v_{ID}|$ is reduced to the differential-input range V_{DIR} of the opamp. Substituting $\text{sgn}[v_{ID}(0)]V_{DIR}$ for $v_{ID}(t)$ in (2.199) and solving the resulting equation for time, we obtain the following expression for the duration of slewing:

$$t_{\text{slew}} = \frac{|v_{ID}(0)| - V_{DIR}}{\beta S_R}. \tag{2.200}$$

For $t \geq t_{\text{slew}}$, the loop is closed, and the voltages tend to their final values exponentially in accordance with

$$v_O(t) = v_O(\infty) - \left[v_O(\infty) - v_O(t_{\text{slew}})\right] \exp\left[-2\pi f_{cl}(t - t_{\text{slew}})\right], \quad t \geq t_{\text{slew}}, \tag{2.201}$$

$$v_{ID}(t) = \text{sgn}[v_{ID}(0)]V_{DIR} \exp\left[-2\pi f_{cl}(t - t_{\text{slew}})\right], \quad t \geq t_{\text{slew}}, \tag{2.202}$$

where $v_O(0)$, $v_{ID}(0)$, f_{cl}, and $v_O(\infty)$ are extracted by applying the first four steps of the procedure presented for the slew-free transient analysis, whereas $v_O(t_{\text{slew}})$ is determined by substituting t_{slew} for t in (2.198).

Any interested reader can use the OTA subcircuit described in Example 2.32 for simulating slewing transients. Setting a sufficiently small `vdir` will force the OTA to slew.

2.11.5 Loop-Gain Extraction

Needless to say, stability must be maintained with an adequate phase margin in all phases of discrete-time operation. This is verified by extracting the loop-gain of each phase separately by applying the simulation procedure described in Subsection 2.7.2. Before applying the procedure, however, the phase schematic has to be properly modified in order to avoid certain pitfalls of ac simulation with discrete-time configurations.

One reason for modification is that the switch model we have used thus far in simulations is not amenable to ac analysis. We overcome this problem by replacing each open switch with an open circuit, and each closed switch either with a short circuit or, better yet, with a resistance representing the switch parameter `ron`.

A need for modification arises also in the case of a phase schematic containing a floating node. This is because such a node may prevent the simulator from determining the dc operation point around which the ac simulation is to be performed. This problem may be encountered also in device-level schematics where switches are constructed with MOSFETs. Since a MOSFET usually exhibits an extremely large resistance in a cutoff state, a node isolated by MOSFET switches remains in an almost perfectly floating state, which may cause a simulation failure. We can solve this floating-node problem by imposing a dc voltage onto the floating node via a large inductance. This combination, while maintaining the floating state in high-frequency ac simulation, keeps the node at dc, and thus enables the simulator to perform an operation-point analysis.

How the dc voltage of the originally floating node should be set depends on whether the node is located at the input of the OTA or elsewhere in the phase schematic. In the latter case, we can set the dc voltage arbitrarily without being concerned with accuracy because it won't affect the operation point of other nodes. If, however, the floating node is at the input of the OTA, then, the dc voltage of this node must be set exactly to the value needed for keeping the dc output voltage at the particular level around which stability is to be assessed. One way of doing this is to use a fictitious dc feedback loop. For an explanation of how such a loop works, consider Fig. 2.51(a) where

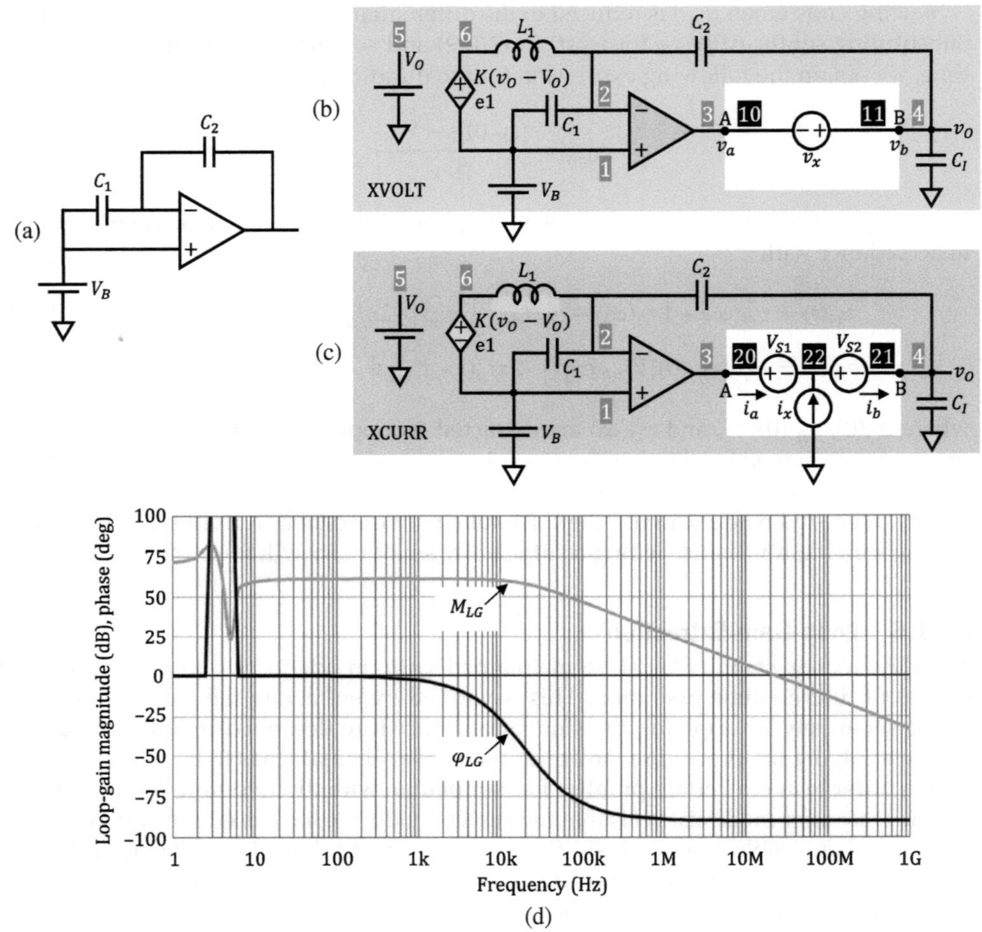

FIGURE 2.51 (a) Amplification-phase schematic of the amplifier of Fig. 2.44(a) where signal sources are disabled. (b) Simulation schematic for voltage return ratio extraction. (c) Simulation schematic for current return ratio extraction. (d) Extracted magnitude M_{LG} and phase ϕ_{LG} of loop gain.

the amplification-phase schematic of the amplifier of Fig. 2.44(a) is shown after short-circuiting the closed switches and open-circuiting the open switches. Notice the presence of a floating node at the inverting terminal, and suppose that we need to extract βA in this phase around a particular level V_O of the output voltage. The simulation schematic we construct for this purpose is shown in Fig. 2.51(b) for voltage return ratio and in Fig. 2.51(c) for current return ratio. In both, we insert a voltage-controlled voltage source (VCVS) `e1` sensing the difference between the actual output voltage v_O and the target output voltage V_O around which βA has to be extracted. The output of this VCVS drives the differential input of the OTA via a very large inductance `l1`. The negative-feedback loop thus created by the VCVS and OTA automatically sets the dc voltage of the inverting input to the value needed for equating v_O to V_O. The inductance effectively disables this feedback loop for the critical frequencies of loop-gain extraction. This scheme works for a VCVS gain *K* around 1 as long as the dc differential gain of the OTA is high.

The magnitude and phase characteristics of loop gain, as extracted for the parameter values involved in Example 2.32, are displayed in Fig. 2.51(d). Extraction is done with the following input file `dt-loopgain.sp`:

```
filename: dt-loopgain.sp
***************ota***************
*ninv: noninverting input
*inv: inverting input
*out: output
.param vdir=5 gm0=0.5m ro=7meg
.subckt ota ninv inv out
gm 0 out cur='gm0*vdir*tanh((v(ninv)-v(inv))/vdir)'
rout out 0 ro
bb 0 out i=0.9/ro
.ends ota
*************amplifier************
.subckt amplifier 3 4
x1 1 2 3 ota
c1 1 2 2p
c2 2 4 1p
ci 4 0 0.5p
vb 1 0 0.9
e1 6 1 4 5 1
vo 5 0 0.9
l1 6 2 1g
.ends amplifier
**subckt for voltage return ratio**
xvolt 10 11 amplifier
vx 10 11 dc 0 ac 1
**subckt for current return ratio**
xcurr 20 21 amplifier
ix 0 22 dc 0 ac 1
vs1 20 22 0
vs2 22 21 0
*******analysis and plotting*******
.control
ac dec 10 1 1g
let rv=-v(10)/v(11) ; voltage return ratio
let ri=-i(vs1)/i(vs2) ; current return ratio
plot db((rv*ri-1)/(2+rv+ri)) ph((rv*ri-1)/(2+rv+ri))
.endc
.end
```

Notice the gigahenry inductance `l1` used for setting the operation point of the floating node at the inverting input. This surreal inductance, coupled with the picofarad capacitances of the external network, affects the magnitude and phase responses only at frequencies in between 1 and 10 Hz, whereas the unity-gain frequency at which the phase margin is read out equals is no less than 20 MHz. It is therefore obvious that the artifacts arising from the presence of the inductance have no effect on the accuracy of phase-margin extraction. The reader may benefit from comparing the simulated values of the unity-gain frequency and phase margin with theoretical predictions, and repeating the simulation for the sampling phase of this amplifier example.

2.12 Fully-Differential Discrete-Time Amplifiers

Benefits of adopting a fully-differential configuration for implementing a discrete-time amplifier include not only the doubling of output range, immunity against second-order harmonic distortion, and cancelation of coupled noise but also the reduction of the charge-injection and clock-feedthrough effects.

Shown in Fig. 2.52 is the fully-differential version of the discrete-time amplifier of Fig. 2.44, in which, the symmetry of the upper and lower external networks is established by setting $C_{B1} = C_{A1}$ and $C_{B2} = C_{A2}$. Notice the application of the output common-mode voltage V_{OCM} of the OTA also to C_{A1} and C_{B1} during the amplification phase $\varphi_2 = 1$. Also note that, unlike the case of fully-differential continuous-time amplifiers, V_{ICM} can be made different from V_{OCM} without causing any bias-current flow through the external network because the input and output ports are separated capacitively.

2.12.1 DC Analysis with Charge Conservation

The first-order dc response of a fully-differential discrete-time amplifier can be analyzed by applying the principle of charge conservation exactly as described in Subsection 2.11.1 for single-ended amplifiers. Here, we demonstrate its application to the amplifier configuration of Fig. 2.52.

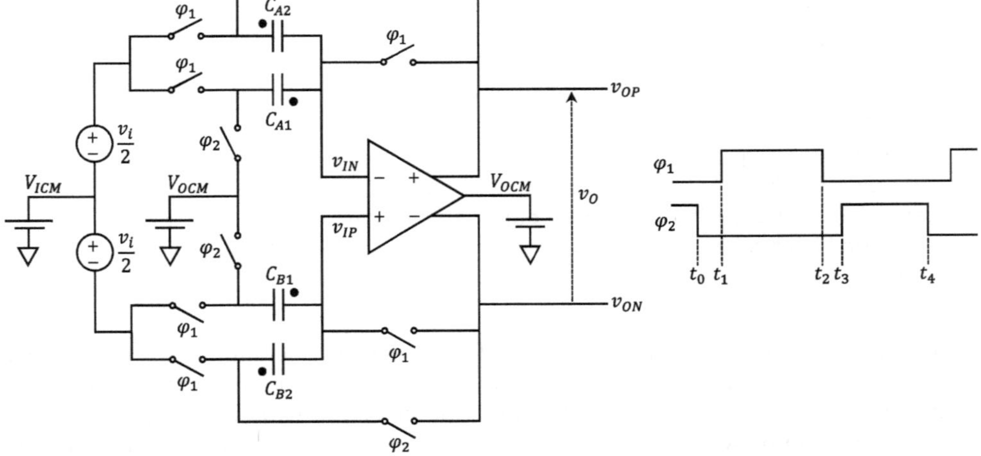

FIGURE 2.52 Fully-differential discrete-time amplifier schematic.

First, consider the instant $t = t_2$ when the sampling phase expires, and note

$$v_{IN}(t_2) = v_{IP}(t_2) = v_{OP}(t_2) = v_{ON}(t_2) = V_{OCM} \tag{2.203}$$

due to (a) the closed feedback switches implying $v_{IN}(t_2) = v_{OP}(t_2)$ and $v_{IP}(t_2) = v_{ON}(t_2)$, (b) virtual short leading to $v_{IN}(t_2) = v_{IP}(t_2)$, and (c) common-mode feedback action maintaining $v_{OP}(t_2) + v_{ON}(t_2) = 2V_{OCM}$. Considering (2.203), the charges accumulated at the reference plates at this instant can be expressed as follows:

$$Q_{A1}(t_2) = C_{A1}\left[V_{OCM} - V_{ICM} - \frac{v_i(t_2)}{2}\right], \tag{2.204}$$

$$Q_{A2}(t_2) = C_{A2}\left[V_{ICM} + \frac{v_i(t_2)}{2} - V_{OCM}\right], \tag{2.205}$$

$$Q_{B1}(t_2) = C_{B1}\left[V_{OCM} - V_{ICM} + \frac{v_i(t_2)}{2}\right], \tag{2.206}$$

$$Q_{B2}(t_2) = C_{B2}\left[V_{ICM} - \frac{v_i(t_2)}{2} - V_{OCM}\right]. \tag{2.207}$$

Next, express the very same charges at the instant $t = t_4$ when the amplification phase expires:

$$Q_{A1}(t_4) = C_{A1}\left[v_{IN}(t_4) - V_{OCM}\right], \tag{2.208}$$

$$Q_{A2}(t_4) = C_{A2}\left[V_{OCM} + \frac{v_o(t_4)}{2} - v_{IN}(t_4)\right], \tag{2.209}$$

$$Q_{B1}(t_4) = C_{B1}\left[v_{IP}(t_4) - V_{OCM}\right], \tag{2.210}$$

$$Q_{B2}(t_4) = C_{B2}\left[V_{OCM} - \frac{v_o(t_4)}{2} - v_{IP}(t_4)\right]. \tag{2.211}$$

Since the input nodes of the OTA become afloat at $t = t_2$, charge conservation implies

$$Q_{A1}(t_4) - Q_{A2}(t_4) = Q_{A1}(t_2) - Q_{A2}(t_2),$$

and

$$Q_{B1}(t_4) - Q_{B2}(t_4) = Q_{B1}(t_2) - Q_{B2}(t_2).$$

Now, reconstruct these equations with (2.204) to (2.211) and take the following steps: (a) invoke the identity of the corresponding capacitances of the two networks, and use the symbol C_1 for $C_{A1} = C_{B1}$ and the symbol C_2 for $C_{A2} = C_{B2}$; (b) replace $v_{IP}(t_4)$ and $v_{IN}(t_4)$ with the common-mode voltage symbol $v_{CM}(t_4)$ as permitted by virtual-short approximation; and (c) solve for $v_o(t_4)$ and $v_{CM}(t_4)$. These steps lead to

$$v_o(t_4) = \left(1 + \frac{C_1}{C_2}\right) v_i(t_2), \tag{2.212}$$

and

$$v_{CM}(t_4) = 2V_{OCM} - V_{ICM}. \tag{2.213}$$

Equation (2.212) is obviously identical to its single-ended counterpart (2.170). As to the common-mode range, it equals V_{OCM} in the sampling phase. According to (2.213), it retains this value in the amplification phase only if $V_{ICM} = V_{OCM}$.

2.12.2 Unified Closed-Loop Model

The unified block diagram shown in Fig. 2.39 also applies to fully-differential discrete-time amplifiers. v_{IN}, v_{IP}, and v_{ID} are still described by (2.135), (2.136), and (2.137), respectively. β is still defined by (2.138), and the extraction procedure described in Subsection 2.10.2 for β_A, α_{Ai}, β_B, and α_{Bi} is still valid. As usual, v_O is related to v_{ID} also by the open-loop transfer function of the OTA.

2.12.3 Accurate Modeling of DC Response

The procedure described in Subsection 2.11.3 for accurately modeling the dc response of single-ended discrete-time amplifiers is applicable also to a fully-differential implementation. In this case, however, the output voltage is described by (2.146) instead of (2.23).

Example 2.33 Analyze the fully-differential amplifier of Fig. 2.52 for accurate dc response and model the total dc error.

Work We begin with the sampling phase. The schematic shown for $\varphi_1 = 1$ in Fig. 2.53(a) indicates the following voltage sources driving the single-ended inputs of the external network:

$$x_A = V_{ICM} + \frac{v_i}{2},$$

$$x_B = V_{ICM} - \frac{v_i}{2}.$$

Input and feedback factors associated with these sources are $\alpha_A = 0$, $\beta_A = 1$, $\alpha_B = 0$, and $\beta_B = 1$. For these factors, (2.146) yields the following expression for the output voltage observed at the end of the sampling phase:

$$v_O(t_2) = -\frac{A_{\text{dif0}}}{1 + A_{\text{dif0}}} V_{os}.$$

According to (2.123) and (2.124), the corresponding single-ended output voltages are described by

$$v_{OP}(t_2) = V_{OCM} - \frac{A_{\text{dif0}}}{1 + A_{\text{dif0}}} \frac{V_{os}}{2},$$

$$v_{ON}(t_2) = V_{OCM} + \frac{A_{\text{dif0}}}{1 + A_{\text{dif0}}} \frac{V_{os}}{2},$$

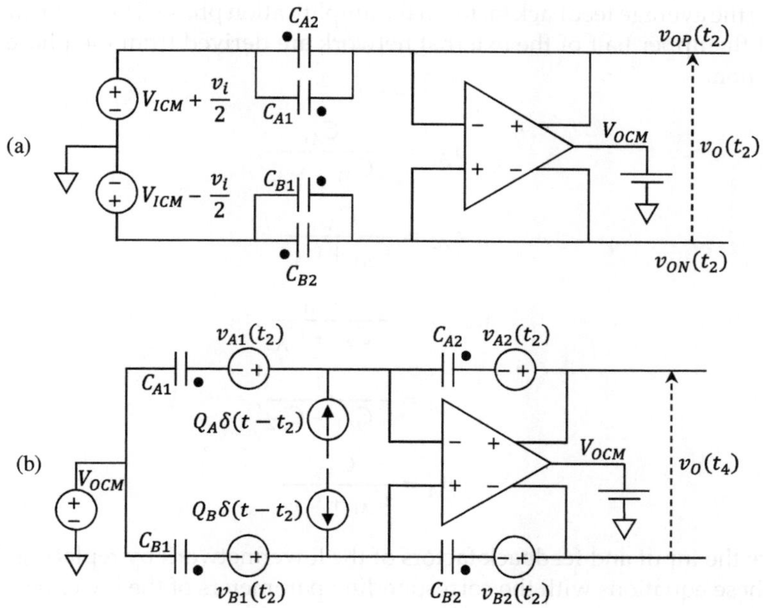

FIGURE 2.53 Phase schematics of the fully-differential amplifier of Example 2.33. (a) Sampling phase. (b) Amplification phase.

which yield the following expressions for the final capacitor voltages to be carried over to the amplification phase:

$$v_{A1}(t_2) = -v_{A2}(t_2) = V_{OCM} - V_{ICM} - \frac{A_{\text{dif0}}}{1+A_{\text{dif0}}}\frac{V_{os}}{2} - \frac{v_i(t_2)}{2},$$

$$v_{B1}(t_2) = -v_{B2}(t_2) = V_{OCM} - V_{ICM} + \frac{A_{\text{dif0}}}{1+A_{\text{dif0}}}\frac{V_{os}}{2} + \frac{v_i(t_2)}{2}.$$

Next, we move to the amplification phase. Phase schematic for $\varphi_2 = 1$ is shown in Fig. 2.53(b). It includes four capacitor initial-voltage sources carried over from the sampling phase, and also contains two impulse-current sources, $Q_A\delta(t-t_2)$ and $Q_B\delta(t-t_2)$, representing the charge-injection and clock-feedthrough effects associated with the opening feedback switches. Note that the charges Q_A and Q_B are nominally identical. Finally, the common-mode voltage source V_{OCM} connecting C_{A1} and C_{B1} is also included. It affects both halves. Considering all of these sources, we obtain from (2.146) the following expression for the output voltage at the end of the amplification phase:

$$v_O(t_4) = \frac{\beta A_{\text{dif0}}}{1+\beta A_{\text{dif0}}}\left[\frac{\alpha_{A1}}{\beta}v_{A1}(t_2) + \frac{\alpha_{A2}}{\beta}v_{A2}(t_2) + \frac{\alpha_{OCMA}}{\beta} + \frac{\alpha_{QA}}{\beta}Q_A - \frac{\alpha_{B1}}{\beta}v_{B1}(t_2)\right.$$
$$\left. - \frac{\alpha_{B2}}{\beta}v_{B2}(t_2) - \frac{\alpha_{OCMB}}{\beta} - \frac{\alpha_{QB}}{\beta}Q_B\right] - \frac{A_{\text{dif0}}}{1+\beta A_{\text{dif0}}}V_{os} + \frac{A_{\text{dif0}}(\beta_B-\beta_A)}{1+\beta A_{\text{dif0}}}V_{OCM}, \quad (2.214)$$

where β is the average feedback factor in the amplification phase. The input and feedback factors of the upper half of the external network are derived from the phase schematic by inspection:

$$\alpha_{A1} = -\frac{C_{A1}}{C_{A1} + C_{A2}},$$

$$\alpha_{A2} = \frac{C_{A2}}{C_{A1} + C_{A2}},$$

$$\alpha_{OCMA} = -\frac{C_{A1}}{C_{A1} + C_{A2}},$$

$$\alpha_{QA} = -\frac{1}{C_{A1} + C_{A2}},$$

$$\beta_A = \frac{C_{A2}}{C_{A1} + C_{A2}}.$$

We derive the input and feedback factors of the lower network by replacing the parameters of these equations with the corresponding parameters of the lower network:

$$\alpha_{B1} = -\frac{C_{B1}}{C_{B1} + C_{B2}},$$

$$\alpha_{B2} = \frac{C_{B2}}{C_{B1} + C_{B2}},$$

$$\alpha_{OCMB} = -\frac{C_{B1}}{C_{B1} + C_{B2}},$$

$$\alpha_{QB} = -\frac{1}{C_{B1} + C_{B2}},$$

$$\beta_B = \frac{C_{B2}}{C_{B1} + C_{B2}}.$$

Substituting these into (2.214) and arranging the outcome lead us to the following expression for the output voltage observed at the end of the amplification phase on a fabricated sample:

$$v'_O(t_4) = \frac{\beta' A_{\text{dif0}}}{1 + \beta' A_{\text{dif0}}} \frac{v_i(t_2)}{\beta'} - \frac{A_{\text{dif0}} V_{os}}{(1 + \beta' A_{\text{dif0}})(1 + A_{\text{dif0}})} - \frac{A_{\text{dif0}}}{1 + \beta' A_{\text{dif0}}} \left(\frac{Q_A}{C_{A1} + C_{A2}} - \frac{Q_B}{C_{B1} + C_{B2}} \right), \quad (2.215)$$

where β' is given by

$$\beta' = \frac{1}{2} \left(\frac{C_{A2}}{C_{A1} + C_{A2}} + \frac{C_{B2}}{C_{B1} + C_{B2}} \right) \quad (2.216)$$

as per (2.138). The target output voltage is described by (2.212). Rewriting it as

$$v_O(t_4) = \frac{v_i(t_2)}{\beta}, \tag{2.217}$$

where

$$\beta \doteq \frac{C_2}{C_1 + C_2} \tag{2.218}$$

is the designed feedback factor of the amplification phase, and subtracting (2.217) from (2.215) yields the the dc-error expression

$$\Delta v_O(t_4) = \underbrace{\left(\frac{1}{\beta'} - \frac{1}{\beta}\right) v_i(t_2)}_{\text{External-network error}} \underbrace{- \frac{1}{1+\beta A_{\text{dif0}}}\frac{v_i}{\beta}}_{\text{Gain error}} \underbrace{- \frac{A_{\text{dif0}} V_{os}}{(1+\beta A_{\text{dif0}})(1+A_{\text{dif0}})}}_{\text{Offset error}}$$

$$\underbrace{- \frac{A_{\text{dif0}}}{1+\beta A_{\text{dif0}}}\left(\frac{Q_A}{C_{A1}+C_{A2}} - \frac{Q_B}{C_{B1}+C_{B2}}\right)}_{\text{Charge error}}.$$

As usual, βA_{dif0} is assumed to be much larger than unity, and β is adopted for β' in all error terms except for the external-network error. Furthermore, we replace v_i/β with $v_O(t_4)$ in the gain-error term and rewrite the charge-error term by replacing $C_{A1} + C_{A2}$ and $C_{B1} + C_{B2}$ with $C_1 + C_2$ and substituting (2.218) for β. The outcome is as follows:

$$\Delta v_O(t_4) = \underbrace{\left(\frac{1}{\beta'} - \frac{1}{\beta}\right) v_i(t_2)}_{\text{External-network error}} \underbrace{- \frac{v_O(t_4)}{\beta A_{\text{dif0}}}}_{\text{Gain error}} \underbrace{- \frac{V_{os}}{\beta A_{\text{dif0}}}}_{\text{Offset error}} \underbrace{- \frac{Q_A - Q_B}{C_2}}_{\text{Charge error}}. \tag{2.219}$$

Comparing this equation with its single-ended counterpart (2.184), and considering the nominally identical Q_A and Q_B, we conclude that the fully-differential operation practically eliminates the charge error. ▲

2.12.4 Transient Analysis and Loop-Gain Extraction

Consider a fully-differential discrete-time amplifier, in which, each half of the OTA features the single-ended parameters r_o, G_m, and V_{DIR}, and each single-ended output drives an effective load capacitance C_L. In the usual case of a sufficiently fast common-mode feedback loop, the transient response of this amplifier will be identical to the transient

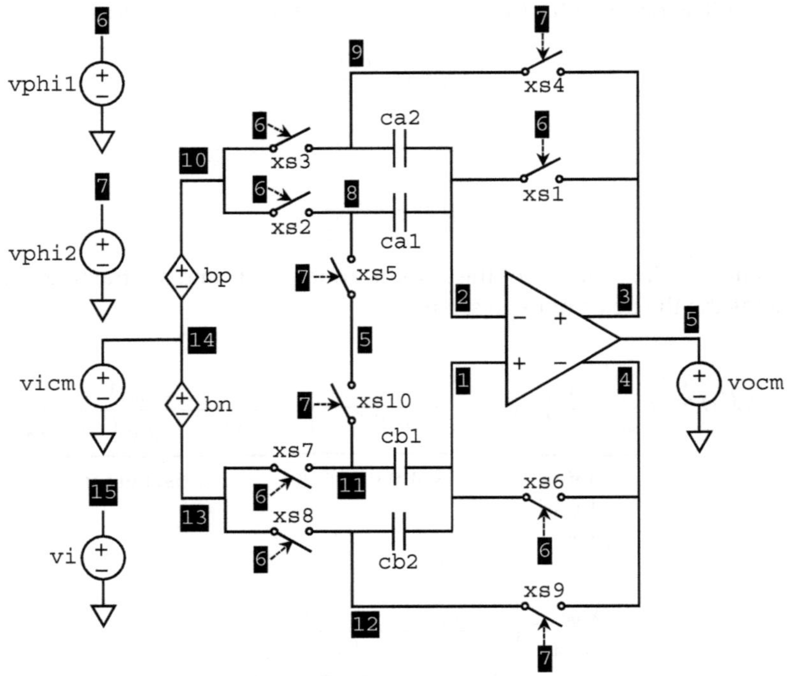

FIGURE 2.54 Simulation schematic of the input file `fdota.sp`.

response of its single-ended counterpart operating with the same V_{DIR}, same C_L, same r_o, twice as large G_m, and $V_B = 0$. For example, the transient-analysis expressions developed in Example 2.32 for the single-ended amplifier of Fig. 2.44(a) are valid also for the fully-differential version shown in Fig. 2.52 provided that we equate V_B to zero, and replace G_m with $2G_m$. The values of V_{ICM} and V_{OCM} have no effect on the outcome. An interested reader may test this claim by simulating the fully-differential version of the amplifier described in Example 2.32. Shown in Fig. 2.54 is the schematic of simulation. Its input file `fdota.sp` is as follows.

```
filename: fdota.sp
.param vdir=5 gm0=0.25m ro=7meg
******ota**************************
*ninv: noninverting input
*inv: inverting input
*outp: noninverting output
*outn: inverting output
*ocm: output common-mode terminal
.subckt ota ninv inv outp outn ocm
gmp ocm outp cur='gm0*vdir*tanh((v(ninv)-v(inv))/vdir)'
routp outp ocm ro
gmn outn ocm cur='gm0*vdir*tanh((v(ninv)-v(inv))/vdir)'
routn outn ocm ro
.ends ota
```

```
******switch**********************
.subckt switch in out cont
s in out cont 0 mod
cin in 0 25f
cout out 0 25f
.model mod sw vt=0.9 ron=1 roff=10g
.ends switch
*****amplifier*******************
xota 1 2 3 4 5 ota
vocm 5 0 0.9
vicm 14 0 0.9
xs1 2 3 6 switch
xs2 8 10 6 switch
xs3 9 10 6 switch
xs4 3 9 7 switch
xs5 5 8 7 switch
xs6 1 4 6 switch
xs7 11 13 6 switch
xs8 12 13 6 switch
xs9 4 12 7 switch
xs10 5 11 7 switch
ca1 2 8 2p
ca2 2 9 1p
cb1 1 11 2p
cb2 1 12 1p
cia 3 0 0.5p
cib 4 0 0.5p
bp 10 14 v=v(15)/2
bn 14 13 v=v(15)/2
vi 15 0 pwl 0 -0.15 74n -0.15 76n 0.15 189n 0.15 191n 0
+  300n 0 302n -0.15 441n -0.15
vphi1 6 0 pulse 1.8 0 47n 2n 2n 58n 112n
vphi2 7 0 pulse 0 1.8 51n 2n 2n 50n 112n
.tran 10p 441n
.options trtol=1
.end
```

Note that the OTA is specified by user-defined parameters V_{DIR}, G_m, and r_o, and is suitable for transient or ac analysis only.[13] Except for G_m, all OTA and external-network parameters match their corresponding values defined in the single-ended input file `ex2-32.sp`. Input and clock waveforms are also the same as those of the single-ended counterpart. G_m, however, is set exactly to one-half of the single-ended version. The reader can use the present file also for simulating slewing transients. Selecting a sufficiently small `vdir` will force the OTA into slewing.

For extracting the differential-mode loop gain of a fully-differential discrete-time amplifier in any phase that does not contain a floating node, we can apply the procedure described in Subsection 2.10.5 and illustrated in Fig. 2.41 for continuous-time amplifiers.

[13] The only dc imperfection this OTA model can handle is gain error.

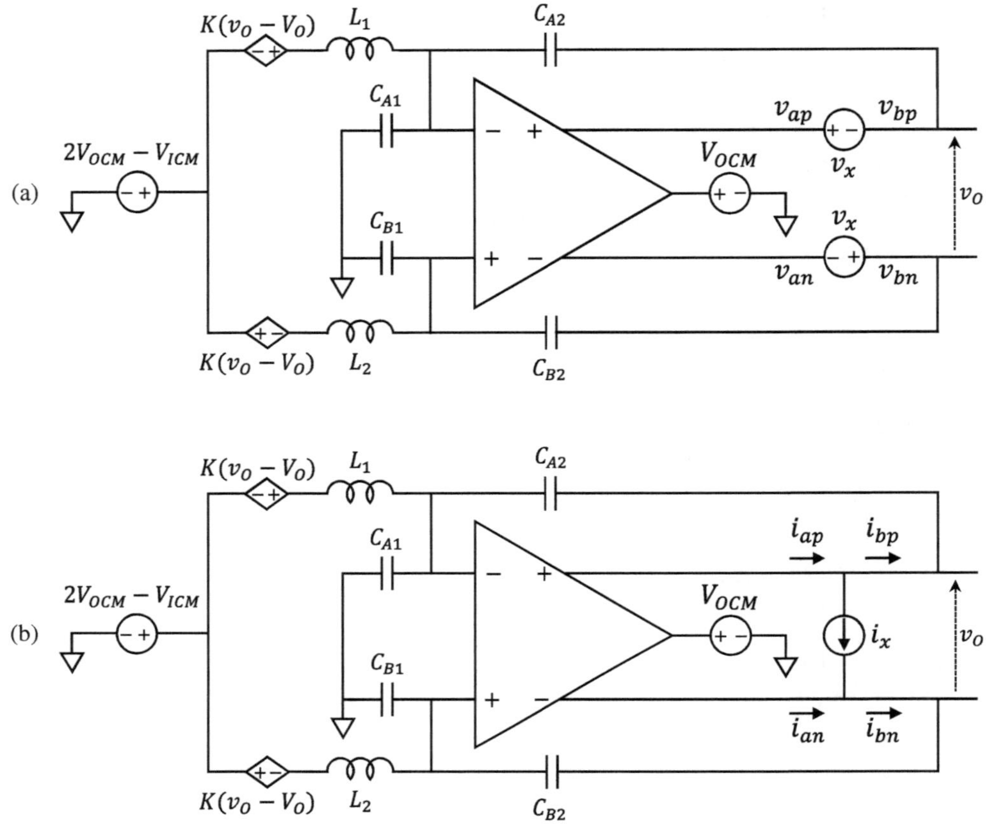

FIGURE 2.55 Loop-gain extraction schematics of the amplifier configuration of Fig. 2.52. (a) Voltage return ratio. (b) Current return ratio.

For extracting the loop gain in the presence of floating nodes, we simply rely on the techniques described in Subsection 2.11.5 for single-ended discrete-time configurations. The amplification phase of the amplifier of Fig. 2.52, for example, can be simulated for loop gain by setting the dc voltages of the floating inverting and noninverting inputs of the OTA with two fictitious dc feedback loops. As depicted in Fig. 2.55, the scheme is just an extension of Figs. 2.51(b) and (c). Notice the inclusion of two voltage-controlled voltage sources amplifying the difference $v_O - V_O$ between the actual and desired values of the differential-output voltage by a voltage gain K. These two sources are connected in series with another voltage source $2V_{OCM} - V_{ICM}$ for setting the common-mode voltage of the opamp as per (2.213). Also notice the large inductors through which the inverting and noninverting inputs of the opamp are driven by the voltage-controlled voltage sources in a complementary fashion. The schematic given in Fig. 2.55(a) is for extracting the voltage return ratio in accordance with (2.158). The current return ratio is extracted as per (2.159) from the schematic given in Fig. 2.55(b). Loop gain is described again by (2.91).

2.13 References

1. S. Ben-Yaakov, "A unified approach to teaching feedback in electronic circuits courses," *IEEE Transactions on Education*, vol. 34, pp. 310–316, 1991.
2. G. Palumbo, and S. Pennisi, "High-frequency harmonic distortion in feedback amplifiers: analysis and applications," *IEEE Transactions on Circuits and Systems–I: Fundamental Theory and Applications*, vol. 50, pp. 328–340, 2003.
3. R. D. Middlebrook, "Measurement of loop gain in feedback systems," *International Journal of Electronics*, vol. 38, pp. 485–512, 1975.

2.14 Exercises

2.1 Assuming virtual short in the application shown in Fig. 2.56, determine the first-order dc transfer function relating the output signal v_o to the input signal v_i in terms of external-network resistances.

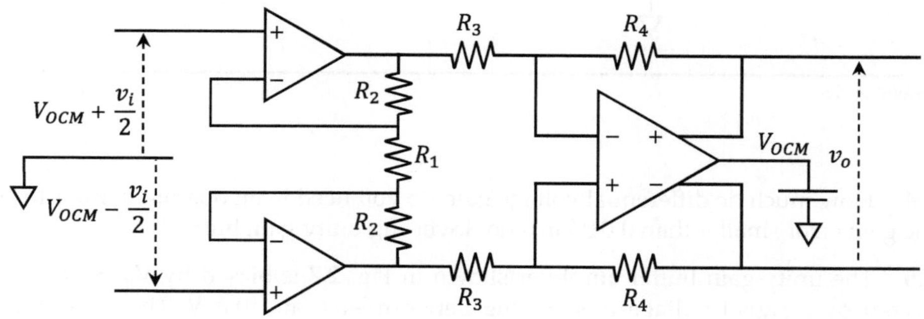

FIGURE 2.56

2.2 The single-supply (V_{DD} = 1.8 V) closed-loop configuration shown in Fig. 2.57 performs a weighted-sum operation on three sinusoidal input signals v_1, v_2, and v_3 around

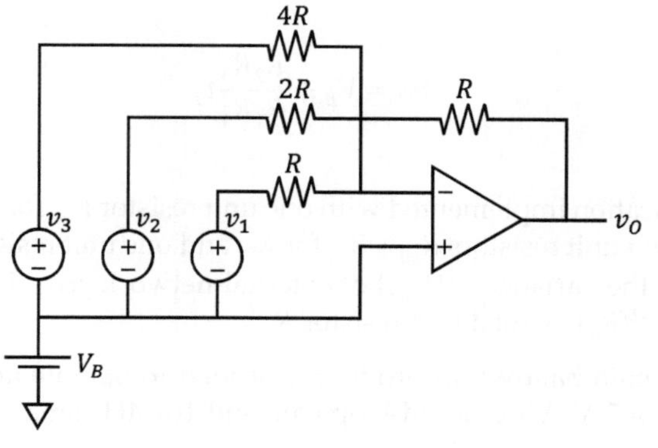

FIGURE 2.57

a bias voltage $V_B = 0.9$ V. The maximum amplitude of each of these signals is 0.4 V. Determine the output-range and common-mode range specifications for the opamp.

2.3 Shown in Fig. 2.58 is the so-called Sallen-Key filter configuration. Write s-domain expressions for β, α_i, and $A_{cl,i}$ in terms of external-network parameters.

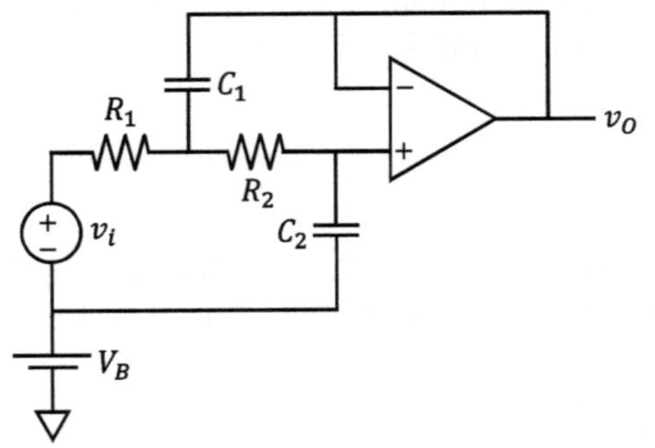

FIGURE 2.58

2.4 How much dc differential voltage gain do you need in an opamp in order to keep the gain error smaller than 0.01% in a noninverting unity-gain buffer?

2.5 The unity-gain buffer amplifier shown in Fig. 2.7 is biased by $V_B = 0.9$ V and driven by a signal voltage v_i swinging between −0.6 and 0.6 V. The opamp is expected to have an offset of $\overline{V_{os}} = 0.2$-mV mean value and $\sigma[V_{os}] = 0.6$-mV standard deviation. How much dc differential gain will you target for the opamp in order to limit the dc error to ±1.5 mV in 95% (±2σ) of the fabricated copies of the buffer?

2.6 The output voltage of the closed-loop configuration shown in Fig. 2.12(b) is described by

$$v_O = V_B + \frac{R_2 R_3}{R_1 R_4} v_i.$$

Consider an application implemented with one unit resistor R_{ua} for R_1, two unit resistors $R_{ub} + R_{uc}$ for R_2, two unit resistors $R_{ud} + R_{ue}$ for R_3, and one unit resistor R_{uf} for R_4. Derive an expression for the variance $\sigma^2[E_{\text{net}}]$ of external-network error in terms of v_i^2 and the relative variance $\sigma^2[R_u]/R_u^2$ of a unit resistor R_u.

2.7 How much gain-bandwidth product is needed to operate an opamp in a noninverting amplifier of 2-V/V dc closed-loop gain and 10-MHz small-signal bandwidth?

2.8 Rework Exercise 2.2 for an inverting amplifier specified for a dc closed-loop gain of −2 V/V and a small-signal bandwidth of 10 MHz.

2.9 An opamp of 0.3-V differential-input range is operated in a noninverting amplifier with a sinusoidal input voltage of 1-V maximum amplitude. If the amplifier is supposed to have a full-power bandwidth of 10 MHz, how much should the small-signal closed-loop bandwidth be?

2.10 Determine the minimum differential-input range needed by an opamp in order to avoid slewing in all four amplifier applications specified in Example 2.13.

2.11 An opamp of $V_{DIR} = 0.1$ V is to be configured as a unity-gain buffer, which is expected to settle in 50 ns for a settling error $\epsilon = 0.01$ in response to a step height $|X_i| =$ 0.62 V. Determine the necessary gain-bandwidth product for this opamp.

2.12 A noninverting amplifier of 1-V/V closed-loop gain is built with an opamp of 0.3-V differential-input range and 10-MHz gain-bandwidth product. How long will the amplifier slew if its input voltage is stepped up or down rapidly by 1 V?

2.13 The inverting amplifier of Fig. 2.8 is built with $R_i = 1$ kΩ, $R_1 = 10$ kΩ, $R_2 = 22$ kΩ, and an opamp of $V_{DIR} = 0.1$ V. Input signal is a step of height $X_i = 0.5$ V. Determine the gain-bandwidth product needed by the opamp for exhibiting 0.1% settling error and 100-ns settling time.

2.14 Rework Example 2.16 for the inverting amplifier of Fig. 2.8 with $R_i = 1$ kΩ, $R_1 =$ 10 kΩ, and $R_2 = 22$ kΩ. Also determine the frequency of the nondominant pole for a phase margin of 65°.

2.15 A noninverting amplifier of 10-V/V closed-loop gain is built with an opamp of 20-MHz gain-bandwidth product. For a minimum phase margin of 65°, how would you set the frequency of the nondominant pole?

2.16 By representing the opamp of Fig. 2.57 with the equivalent circuit identified as "unloaded opamp" in Fig. 2.29, and assuming the presence of a load capacitance C_L at its output port, derive an expression for the differential gain A in terms of Y_m, Z_o, C_L, and all four external-network resistances.

2.17 Assuming $R = 20$ kΩ and $f_{cl} = 5$ MHz, determine the external-network contribution to the mean-square output noise voltage of the closed-loop configuration shown in Fig. 2.57.

2.18 A fully-differential continuous-time voltage amplifier complying with the closed-loop configuration of Fig. 2.37 is specified as follows:

1. Power supply: $V_{DD} = 1.8$ V, $V_{SS} = 0$.
2. $V_{ICM} = V_{OCM} = 0.9$ V.
3. Input signal: Sinewave of amplitude $V_i = 0.25$ V.
4. Closed-loop gain: 5 V/V.
5. Gain error: 2 mV maximum.
6. Slew-free operation with a closed-loop bandwidth $f_{cl} = 4$ MHz.

Determine the following open-loop metrics for the opamp: (a) V_{OH} and V_{OL} assuming $V_M = V_{OCM}$, (b) A_{dif0}, (c) V_{DIR}, and (d) f_{GBW}.

2.19 A discrete-time closed-loop configuration is shown in Fig. 2.59. Relying on the charge-conservation principle, perform a first-order dc analysis for relating the output voltage $v_O(t_4)$ to the input signal $v_i(t_2)$ and bias voltage V_B.

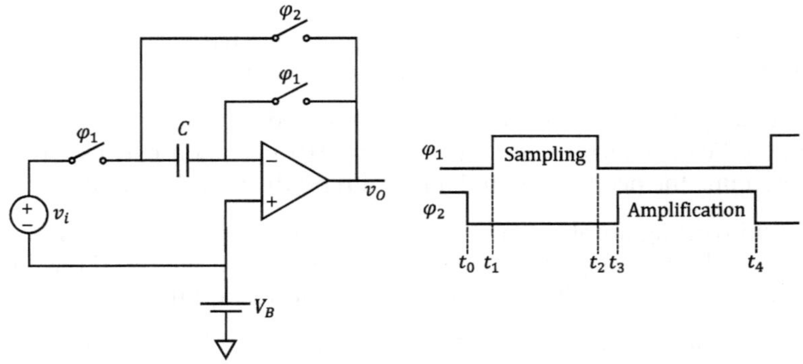

FIGURE 2.59

2.20 Rework Examples 2.29, 2.31, and 2.32 for the discrete-time configuration shown in Fig. 2.59.

2.21 Rework Example 2.33 for the fully-differential version of the discrete-time configuration shown in Fig. 2.59.

CHAPTER 3
Device Layer

3.1 Introduction

The device layer of electrical design is where the bias conditions and physical parameters of all devices are determined from specified values of device performance metrics as depicted in Fig. 1.2. For resistors and capacitors, the relation between these two sets of variables is accurately modeled by simple analytical models. For MOSFET models, however, accuracy and simplicity are mutually exclusive because device operation is much more complicated. First-order analytical MOSFET models are good for gaining an insight into the path connecting the performance metrics to bias conditions and physical parameters but they are grossly inaccurate. Designers utilize them for qualitative purposes only. All quantitative tasks of design are performed with Spice simulations utilizing accurate but complex MOSFET models. However, simulation is a bottom-up tool operating in the opposite direction of device-design flow. It accepts physical parameters and bias conditions as input variables, and puts out electrical variables that represent performance metrics. Inevitably, therefore, design by simulation turns into an iterative process. In each iteration, a simulation is run with some estimates of the physical parameters and bias conditions, and the outcome is compared with the target values specified for performance metrics. This process is repeated in cycles until the differences between the simulated and targeted values of performance metrics are reduced to an acceptable level. Unfortunately, the multidimensional nature of the design space can turn this process into an almost perpetual sequence of trial and error unless (a) the initial values of physical parameters and bias conditions are selected intelligently, and (b) the values of these design variables are updated in the proper direction and amount in subsequent iterations. After briefly reviewing MOSFET basics in Section 3.2, we develop a simulation-based design methodology in Sections 3.3 to 3.6 exactly for this purpose. Based on a few generic design tools and involving only a few and simple calculations at most, it will quickly guide the designer from performance metrics to the optimum set of physical parameters and bias conditions.

All examples presented from this point forward are based on a mature 0.18-μm bulk complementary metal-oxide-semiconductor (CMOS) technology. In Chapter 4 the reader is expected to rework some of the examples and exercises in 90-nm bulk CMOS. We will treat these as twin-well technologies in all examples and exercises except for Example 4.27, where a triple-well technology will be assumed. Although advanced CMOS technologies offer core MOSFETs of a minimum feature size much smaller than these, the maximum voltage rating of such devices happens to be too restrictive for analog and mixed-signal circuits. This is why analog applications are usually implemented with core devices in mature technologies or with optional high-voltage *input/output* (I/O) devices in advanced technologies. The feature size of I/O devices is kept larger than core devices in order to accommodate the higher maximum voltage rating. 0.18-μm is one such option available even in 14-nm CMOS technologies.

As mentioned above, MOSFETs are represented by an accurate model in Spice simulations. The *Berkeley Short-Channel IGFET Model* (BSIM) family is an industry standard in this respect.[1][2] BSIM3v3 and BSIM4 serve pre-0.1-μm and post-0.1-μm CMOS technology nodes, respectively. A third version was released in 2013 for post-40 nm nodes as BSIM6, which was renamed BSIM-BULK in 2017. Each version features hundreds of device parameters, among which only channel dimensions are under the control of an electrical designer. The rest are constants characterized by the manufacturer and made available to the designer as a part of the *process design kit* (PDK). The list of these constants is known as *model deck* in Spice terminology. Actually, a PDK contains multiple model decks describing several different versions of core and I/O devices for design flexibility. These versions differ mainly in the threshold voltage and are known as *threshold options*. Most typically, zero-threshold, low-threshold, and standard-threshold options are offered. Most manufacturers further diversify the model deck of each option by dividing the design space of device dimensions into subspaces, and assigning a slightly different deck to each. This practice of *binning* improves model accuracy. Still further, a deck is replicated for the worst-case and best-case values of parameters representing the extreme cases of variation the selected technology may exhibit among its runs. These are used in so-called *corner simulations* for verifying the robustness of the electrical design. All in all, a large number of model decks are usually present in a PDK. Fortunately, the simulation-based design methodology presented in this chapter is independent of the quantity or complexity of model decks. Based on this fact, we introduce the methodology with only two typical decks describing the standard-threshold NMOS and PMOS devices of a 0.18-μm CMOS technology but will apply it later in Chapter 4 also with low-threshold versions of these decks. The standard-threshold parameter decks are presented in Table 3.1 in a format recognizable as a model declaration by Spice. Saved in a file `bsim.sp`, these decks can be included automatically in any Spice input file containing the line `.incl bsim.sp`. Most of the parameters included in these decks have been characterized on fabricated wafers and made public by the *MOSIS Service*.[3]

Relating open-loop metrics to opamp device metrics is a task undertaken in the circuit layer of electrical design, and will be studied in Chapter 4. However, its proper execution necessitates a firm understanding of the techniques used in MOSFET small-signal analysis. Taking the opportunity provided by the first five sections of the present chapter as a background, we present these techniques in Section 3.7. MOSFET coverage continues in Section 3.8 where we present a model for the inherent noise generated by this device and exemplify its use in determining the noise performance of an analog circuit. Section 3.9 is devoted to the switching performance of MOSFET as an external network component of discrete-time configurations. In Section 3.10, we discuss the design of integrated resistors which find use not only in external networks but also inside the opamp. Finally, in Section 3.11, we move into integrated capacitor structures, which are indispensable not only for constructing the external networks of discrete-time closed-loop configurations but also for stabilizing certain opamp topologies.

```
.MODEL CMOSN NMOS (     LEVEL = 49               VERSION = 3.2           TNOM = 27
+TOX = 4.1E-9           XJ = 1E-7                NCH = 2.3549E17         VTHO = 0.3694303
+K1 = 0.5789116         K2 = 1.110723E-3         K3 = 1E-3               K3B = 0.0297124
+WO = 1E-7              NLX = 2.037748E-7        DVTOW = 0               DVT1W = 0
+DVT2W = 0              DVTO = 1.2953626         DVT1 = 0.3421545        DVT2 = 0.0395588
+U0 = 293.1687573       UA = -1.21942E-9         UB = 2.325738E-18       UC = 7.061289E-11
+VSAT = 1.676164E5      A0 = 2                   AGS = 0.4764546         B0 = 1.617101E-7
+B1 = 5E-6              KETA = -0.0138552        A1 = 1.09168E-3         A2 = 0.3303025
+RDSW = 105.6133217     PRWG = 0.5               PRWB = -0.2             WR = 1
+WINT = 2.885735E-9     LINT = 1.715622E-8       XL = 0                  XW = -1E-8
+DWG = 2.754317E-9      DWB = -3.690793E-9       VOFF = -0.0948017       NFACTOR = 2.1860065
+CIT = 0                CDSC = 2.4E-4            CDSCD = 0               CDSCB = 0
+ETA0 = 2.665034E-3     ETAB = 6.028975E-5       DSUB = 0.0442223        PCLM = 1.746064
+PDIBLC1 = 0.3258185    PDIBLC2 = 2.701992E-3    PDIBLCB = -0.1          DROUT = 0.9787232
+PSCBE1 = 4.494778E10   PSCBE2 = 3.672074E-8     PVAG = 0.0122755        DELTA = 0.01
+RSH = 7                MOBMOD = 1               PRT = 0                 UTE = -1.5
+KT1 = -0.11            KT1L = 0                 KT2 = 0.022             UA1 = 4.31E-9
+UB1 = -7.61E-18        UC1 = -5.6E-11           AT = 3.3E4              WL = 0
+WLN = 1                WW = 0                   WWN = 1                 WWL = 0
+LL = 0                 LLN = 1                  LW = 0                  LWN = 1
+LWL = 0                CAPMOD = 2               XPART = 0.5             CGDO = 8.58E-10
+CGSO = 8.58E-10        CGBO = 1E-12             CJ = 9.471097E-4        PB = 0.8
+MJ = 0.3726161         CJSW = 1.905901E-10      PBSW = 0.8              MJSW = 0.1369758
+CJSWG = 3.3E-10        PBSWG = 0.8              MJSWG = 0.1369758       CF = 0
+PVTH0 = -5.105777E-3   PRDSW = -1.1011726       PK2 = 2.247806E-3       WKETA = -5.071892E-3
+LKETA = 5.324922E-4    PU0 = -4.0206081         PUA = -4.48232E-11      PUB = 5.018589E-24
+PVSAT = 2E3            PETA0 = 1E-4             PKETA = -2.090695E-3    NOIMOD = 2
+NOIA = 1.2E+19         NOIB = 9.58E4            NOIC = 1E-14            EF = 1.02
+EM = 4.1E7             )
*
.MODEL CMOSP PMOS (     LEVEL = 49               VERSION = 3.2           TNOM = 27
+TOX = 4.1E-9           XJ = 1E-7                NCH = 4.1589E17         VTHO = -0.3823437
+K1 = 0.5722049         K2 = 0.0219717           K3 = 0.1576753          K3B = 4.2763642
+WO = 1E-6              NLX = 1.104212E-7        DVTOW = 0               DVT1W = 0
+DVT2W = 0              DVTO = 0.6234839         DVT1 = 0.2479255        DVT2 = 0.1
+U0 = 109.4682454       UA = 1.31646E-9          UB = 1E-21              UC = -1E-10
+VSAT = 1.054892E5      A0 = 1.5796859           AGS = 0.3115024         B0 = 4.729297E-7
+B1 = 1.446715E-6       KETA = 0.0298609         A1 = 0.3886886          A2 = 0.4010376
+RDSW = 199.1594405     PRWG = 0.5               PRWB = -0.4947034       WR = 1
+WINT = 0               LINT = 2.93948E-8        XL = 0                  XW = -1E-8
+DWG = -1.998034E-8     DWB = -2.481453E-9       VOFF = -0.0935653       NFACTOR = 2
+CIT = 0                CDSC = 2.4E-4            CDSCD = 0               CDSCB = 0
+ETA0 = 3.515392E-4     ETAB = -4.804338E-4      DSUB = 1.215087E-5      PCLM = 0.96422
+PDIBLC1 = 3.026627E-3  PDIBLC2 = -1E-5           PDIBLCB = -1E-3         DROUT = 1.117016E-4
+PSCBE1 = 7.999986E10   PSCBE2 = 8.271897E-10    PVAG = 0.0190118        DELTA = 0.01
+RSH = 8.1              MOBMOD = 1               PRT = 0                 UTE = -1.5
+KT1 = -0.11            KT1L = 0                 KT2 = 0.022             UA1 = 4.31E-9
+UB1 = -7.61E-18        UC1 = -5.6E-11           AT = 3.3E4              WL = 0
+WLN = 1                WW = 0                   WWN = 1                 WWL = 0
+LL = 0                 LLN = 1                  LW = 0                  LWN = 1
+LWL = 0                CAPMOD = 2               XPART = 0.5             CGDO = 7.82E-10
+CGSO = 7.82E-10        CGBO = 1E-12             CJ = 1.214428E-3        PB = 0.8461606
+MJ = 0.4192076         CJSW = 2.165642E-10      PBSW = 0.8              MJSW = 0.3202874
+CJSWG = 4.22E-10       PBSWG = 0.8              MJSWG = 0.3202874       CF = 0
+PVTH0 = 5.167913E-4    PRDSW = 9.5068821        PK2 = 1.095907E-3       WKETA = 0.0133232
+LKETA = -3.648003E-3   PU0 = -1.0674346         PUA = -4.30826E-11      PUB = 1E-21
+PVSAT = 50             PETA0 = 1E-4             PKETA = -1.822724E-3    NOIMOD = 2
+NOIA = 2.6E+20         NOIB = 2.82E5            NOIC= 5.4E-12           EF = 1.00
+EM = 1.1E7             )
```

TABLE 3.1 BSIM3v3.2 Model Deck for Standard-Threshold Devices

3.2 MOSFET Basics

In this section, we first present a brief description of CMOS structure, and subsequently identify MOSFET terminals, electrical ports, and port variables. The section concludes with the definition of MOSFET performance metrics and design variables.

3.2.1 Structure and Electrical Ports

CMOS Structure

Cross-sectional and top views of PMOS and NMOS transistor structures in a typical twin-well bulk CMOS technology are shown in Fig. 3.1. At the start of fabrication, the substructure consists of the p-substrate only while no superstructure is yet in place. The structural features first built into the substructure are the silicon dioxide-filled *shallow trench isolation* regions marked STI in the cross-sectional view. These trenches provide lateral isolation between neighboring MOSFETs. The area they collectively occupy on chip surface is called *field*, which complements the so-called *active* area occupied by MOSFETs and contact regions. Next in fabrication, *n-well* regions of PMOS devices and *p-well* regions of NMOS devices are formed by selectively doping the surface of the p-substrate. N-well is needed for providing isolation between the p-type source and drain of a PMOS device.

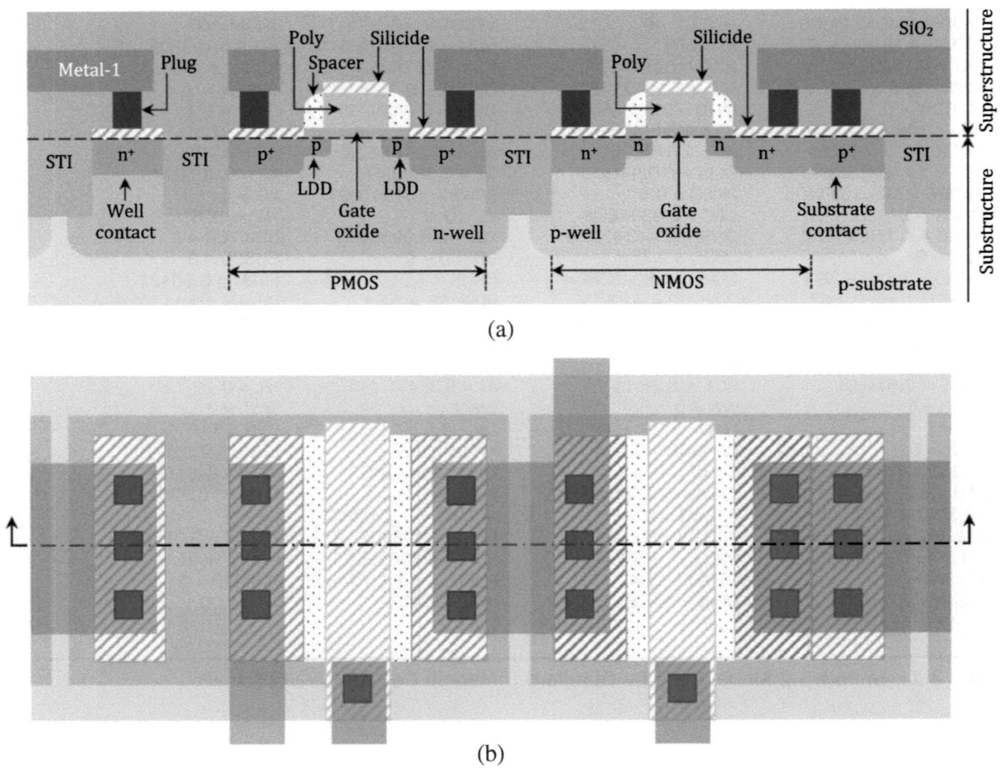

FIGURE 3.1 Twin-well bulk CMOS structure. (a) Cross-sectional view. (b) Top view and line of sectioning.

P-well is not intended for isolation because the n-type source and drain regions of an NMOS device could be isolated by the p-substrate alone, but the presence of a p-well makes it possible to optimize the performance of the NMOS device independently from the PMOS. Next in line of processing is the growth of *gate oxide* as the first structural feature of the superstructure. This is followed by the deposition and patterning of the polycrystalline silicon gate material, which is called *poly* in short. Source and drain regions of both devices are self-aligned with respect to poly gates in a two-step doping process. First is the formation of extensions called lightly doped drain (LDD). These are marked p and n on the cross-sectional view. After masking with a *sidewall spacer*, a second doping creates the relatively deeper and more conductive body of the source and drain regions, which are marked with p^+ and n^+. This doping process is used also for raising the dopant concentration of n-well and p-well regions where metal contacts are to be made at a later stage of fabrication. Without doping enhancement, these contacts would exhibit a rectifying property instead of becoming ohmic. These regions are identified as *well contact* and *substrate contact* on the cross-sectional view. As exemplified on the NMOS structure, a contact region may be abutting a source region provided that both are to operate at the same potential. Otherwise, they are laid out to be disjoint as exemplified in the PMOS structure. The next step of fabrication is the conversion of the surface of the source, drain, and contact regions and poly into *silicide* after capping typically with titanium and heating in an inert environment. The conductance of these layers is significantly increased by this process. Next, a layer of silicon dioxide is deposited onto the surface, and holes are etched through where metal contacts are to be made to the substructure and poly. In order to create a contact, each of these holes is filled with a *plug* made typically of tungsten. This is followed by the deposition and patterning of interconnection metal, typically copper or aluminum. Notice the presence of a single layer of metal, marked *metal-1*, in Fig. 3.1. In actual fact, several layers of metal are available in CMOS technologies. These layers are separated by additional silicon dioxide layers, and are interconnected through holes filled with plugs and called *vias*.

Channel Dimensions

Simplified versions of cross-sectional and top views of NMOS and PMOS devices are presented in Fig. 3.2 for the purpose of defining the reference directions of electrical variables. Also shown are the symbols representing these devices in analog circuit schematics. Note that the current of a MOSFET is controlled inside the *channel*, which is how we name the region in between source and drain. The channel dimension along the direction of current flow between source and drain is called *effective length*, and is denoted by L on the top view of the NMOS device in Fig. 3.2. The channel dimension that is orthogonal to the direction of current flow is called *effective width*, and is denoted by W. As will be explained in Subsection 3.3.1, in all operation modes, MOSFET current is proportional to the so-called *aspect ratio* W/L.

W and L represent the final dimensions of a fabricated channel. They generally differ from the so-called *drawn dimensions* W_{DR} and L_{DR} which are quantified in electrical design and registered on the artwork in physical design. The difference, called *process bias*, is due to dimensional offsetting effects experienced during and after pattern transfer from artwork to silicon in fabrication. The relations between effective and drawn dimensions are described by

$$W = W_{DR} + \mathtt{XW} - 2 \times \mathtt{WINT}, \tag{3.1}$$

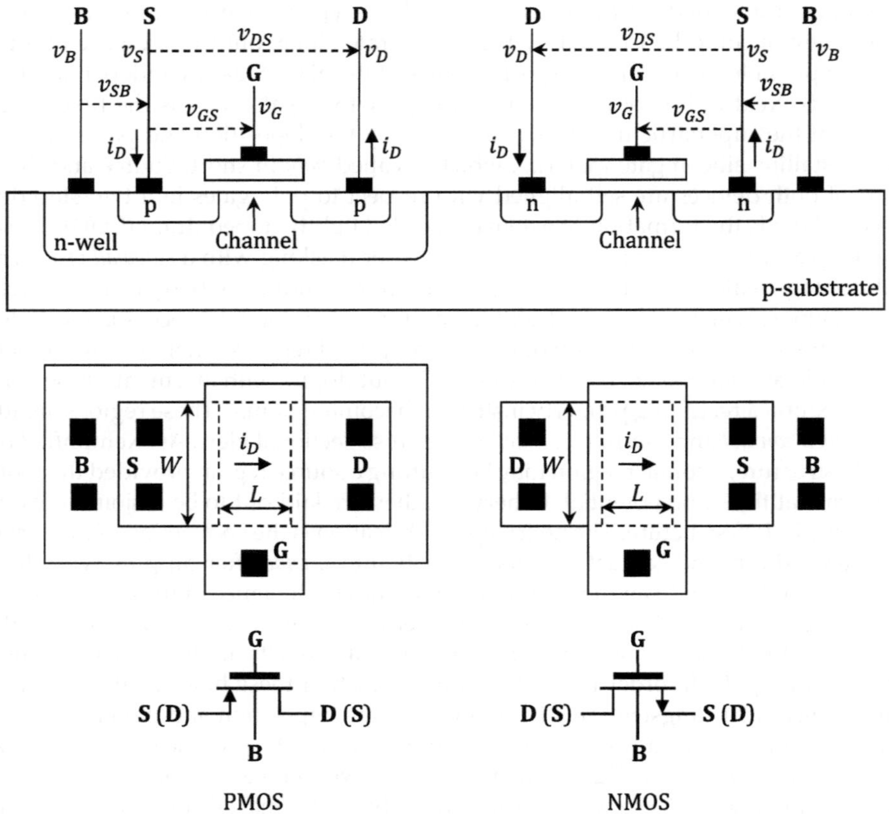

FIGURE 3.2 *Top:* Simplified cross-sectional view and electrical port variables of NMOS and PMOS structures. *Middle:* Top view of the structures. *Bottom:* Symbols representing these devices in circuit schematics.

and

$$L = L_{DR} + \mathtt{XL} - 2 \times \mathtt{LINT}, \tag{3.2}$$

where `XW` (m) and `XL` (m) represent the offsets created during pattern transfer, whereas `WINT` (m) and `LINT` (m) are those created by subsequent processing. All four are BSIM parameters whose values for our exemplar technology can be found in Table 3.1. It is also very important to note and remember that the instance-parameter symbols `W` and `L` used in MOSFET element lines of Spice input files represent not effective but drawn dimensions.

Terminals

A MOSFET has four terminals for electrical access. Namely, drain (D), gate (G), source (S), and bulk (B), whose voltages with respect to ground are denoted with v_D, v_G, v_S, and v_B, respectively. Bulk terminal is the p-substrate in an NMOS and n-well in a PMOS.

These four terminals constitute three independent electrical ports for each device. For modeling purposes, it is most appropriate to define gate source, drain source, and source bulk as independent ports.

Gate-Source Port

The voltage $v_{GS} \doteq v_G - v_S$ of this port is the main determinant of the concentration of current carriers hence conductivity inside the channel. It controls the concentration by imposing a transversal electric field across the gate oxide between gate and channel. In an NMOS, whose carriers are negatively-charged electrons, an increasingly positive v_{GS} enhances the conductivity by attracting a larger concentration of carriers. In a PMOS, positively-charged holes are current carriers. Therefore, an increasingly negative v_{GS} is needed for establishing a larger carrier concentration and thus enhancing the conductivity. The insulating property of the gate oxide prevents this port from conducting any dc steady-state current itself.

Drain-Source Port

The voltage $v_{DS} \doteq v_D - v_S$ of this port is responsible for driving the carriers from source to drain inside the channel, and thus generating the so-called *drain current* i_D. Since carriers are negatively charged in an NMOS, and positively charged in a PMOS, the polarity of the port voltage must satisfy the condition

$$\begin{aligned} v_{DS} &> 0 \text{ in NMOS,} \\ v_{DS} &< 0 \text{ in PMOS} \end{aligned} \tag{3.3}$$

in order to drive them from source to drain. By the same token, the resulting drain current is directed from drain to source in an NMOS, and from source to drain in a PMOS. Also note that $v_{DS} = 0$ implies $i_D = 0$ because carrier motion between source and drain is impossible without a driving force.

The source and drain terminals of a MOSFET are structurally undistinguishable. We use (3.3) as the basis of drain/source designation when we inspect a circuit schematic. Between the two terminals of the drain-source port, the one at the higher potential is designated as drain in NMOS and source in PMOS. If both are at the same potential ($v_{DS} = 0$), it does not matter which way we designate them. Majority of MOSFETs operate with one of these terminals continuously staying at a higher potential than the other. The drain/source designation of such devices never changes during operation. In some MOSFETs, however, the polarity of the potential difference between these two terminals is transposed during operation. An important implication of (3.3) for such a MOSFET is that source/drain designation must also be transposed if and when this happens. For this reason, the arrowhead appearing in analog MOSFET symbols must be interpreted with caution. It must be recognized solely as an indicator of whether the device is an NMOS (outward arrow) or a PMOS (inward arrow), and not necessarily as an indicator of the source terminal.[1]

[1] The arrowhead of the analog MOSFET symbol is inherited from the BJT symbol, where it identifies not only the type of device as *pnp* or *npn* but also the emitter terminal.

Source-Bulk Port

Carrier population and therefore conductivity inside the channel can be controlled also by the source-bulk port voltage $v_{SB} \doteq v_S - v_B$. For a fixed v_{GS}, conductivity decreases as v_{SB} is made more positive in NMOS or more negative in PMOS. However, the efficacy of this so-called *body effect* is about an order of magnitude weaker than that of the gate-source port, which is why source-bulk port is not intentionally utilized for controlling i_D in mainstream applications. Also note that this port may conduct a significant current of its own if the source-bulk junction is forward biased. Such a current is undesirable because it would flow outside the channel beyond the control of the gate terminal. This is why in mainstream applications the polarity of v_{SB} is ensured by design to comply with

$$\begin{aligned} v_{SB} &\geq 0 \text{ in NMOS,} \\ v_{SB} &\leq 0 \text{ in PMOS.} \end{aligned} \tag{3.4}$$

Note that the condition set by (3.4) prevents also the drain-bulk junction from being forward biased because drain potential cannot be lower than source potential in an NMOS or higher than source potential in a PMOS as implied by (3.3).

The only notable exception where source-bulk port is intentionally utilized as a control port is the so-called *bulk-driven MOSFET* application, which occasionally finds use in low-voltage design. This application also violates the condition (3.4).

Bulk Connection

Drain, gate, and source are the primary terminals by which a MOSFET is interconnected with other devices. Bulk is secondary in all mainstream applications but still it must be tied to an appropriate potential satisfying the condition (3.4). For a discussion of the options available in this regard, consider the general case of a split-supply circuit operating between a positive power-supply rail of voltage V_{DD} and a negative power-supply rail of voltage V_{SS}. Four possible ways in which a PMOS device can be interconnected with the rest of the circuit are represented in Fig. 3.3(a) with M1, M2, M3, and M4, all of which have a common bulk tied to V_{DD}. Condition (3.4) is satisfied by all these devices because no source terminal can assume any potential higher than V_{DD}. The source-bulk port of M1 remains unbiased ($v_{SB} = 0$), while those of the other three are reverse biased ($v_{SB} < 0$). This arrangement also yields a very compact layout because we do not need a separate n-well for each device; a single n-well tied to V_{DD} can serve as a common bulk for all.

As explained above, increasing the reverse bias of the source-bulk port leads to a lower channel conductivity through body effect even if v_{GS} is kept constant. In those applications where this is intolerable, the bulk connection of M2, M3, and M4 can be arranged as shown in Fig. 3.3(b). In M2 and M3, upper terminals function permanently as a source because their potential cannot be less than those of the respective lower terminals. By placing each of these two devices in an individual well and short-circuiting the well to the upper terminal, we establish $v_{SB} = 0$ for both. Same can be done for M4 but only if its source/drain designation is invariant during operation. The dashed connection shown in Fig. 3.3(b) exemplifies such a case. It is worth noting that $v_{SB} = 0$ can be maintained in a PMOS even in the case of a changing source/drain designation but it necessitates a specific circuit identifying the source and connecting it to the well in real time.[4] However, the overhead of the additional circuit is justifiable only in very exceptional cases.

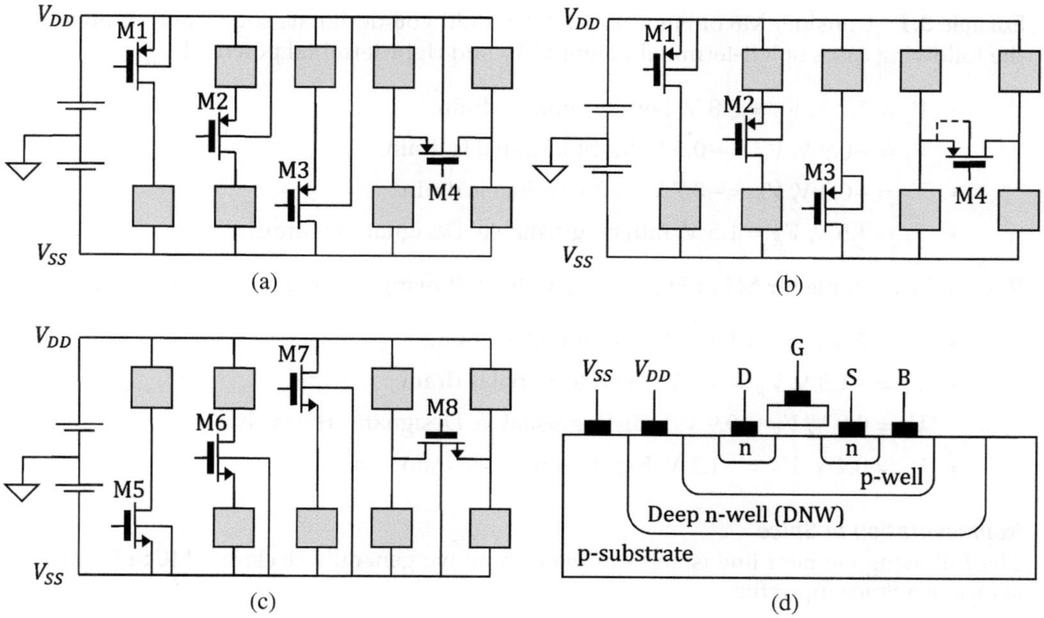

FIGURE 3.3 (a) All PMOS devices sharing a common bulk at V_{DD}. (b) Possibility of individual bulk-source connection in PMOS devices M1, M2, and M3. M4 can have an individual bulk-source connection if source designation does not change during operation. Otherwise, bulk is tied to V_{DD}. (c) All NMOS devices inevitably sharing a common bulk at V_{SS}. (d) A deep n-well NMOS structure in triple-well CMOS technology.

Next, consider Fig. 3.3(c), where all possible ways in which an NMOS device is interconnected with the rest of the circuit. Notice that the source of M5 is tied to the lowest potential on the chip. The only way we can satisfy (3.4) for this device is to connect the substrate also to V_{SS}. With this bulk connection, M5 will operate with $v_{SB} = 0$ but, since substrate is common to all NMOS devices, M6, M7, and M8 will operate with $v_{SB} > 0$. Condition (3.4) is therefore satisfied by all devices. Unlike the case of PMOS devices, however, it is impossible in a twin-well CMOS structure to establish $v_{SB} = 0$ for an NMOS device whose source potential is above V_{SS} as exemplified with M6, M7, or M8. This is because substrate is an inseparable common bulk for all NMOS devices in this technology; it cannot be individualized for a specific NMOS device. This limitation on NMOS devices is removed in the *triple-well* version of the CMOS technology, which offers a second type of NMOS device whose simplified cross-sectional view is shown in Fig. 3.3(d). Notice that the device is built inside a p-well, which, in turn, is built inside a deep n-well (DNW). By connecting DNW to V_{DD}, we keep it reverse biased with respect to the p-substrate, which stays at V_{SS}, and thus prevent forward biasing of the junction between the two. The isolated p-well can now be short-circuited to the source. Since their common potential never exceeds V_{DD}, the junction between p-well and DNW is never forward biased either. $V_{SB} = 0$ is therefore established for the NMOS device regardless of its source potential.

Example 3.1 Consider M8 of Fig. 3.3(c). How would you designate source and drain in the following cases of left-terminal potential V_L and right-terminal potential V_R?

- $V_L = 1.2$ V, $V_R = 0.8$ V: Left terminal is drain.
- $V_L = -0.9$ V, $V_R = -0.5$ V: Right terminal is drain.
- $V_L = -0.3$ V, $V_R = -0.7$ V: Left terminal is drain.
- $V_L = 1.5$ V, $V_R = 1.5$ V: Indistinguishable. Designate arbitrarily.

Repeat the example for M4 of Fig. 3.3(a) and the following cases of terminal potentials:

- $V_L = 0.3$ V, $V_R = 1.1$ V: Left terminal is drain.
- $V_L = -1.3$ V, $V_R = -0.5$ V: Left terminal is drain.
- $V_L = 0.9$ V, $V_R = 0.9$ V: Indistinguishable. Designate arbitrarily.
- $V_L = 0.4$ V, $V_R = -1.2$ V: Right terminal is drain. ▲

Representation in Spice

The following element line is an example of how we generally declare a MOSFET instance in a Spice input file.

```
m18 3 12 7 8 cmosn w=2.5u l=0.7u ad=1.38p pd=6.1u nrd=0.08
+ as=1.38p ps=6.1u nrs=0.08 m=1
```

The line begins with the name "`m18`" of the instance where `m` identifies a MOSFET. The four subsequent characters "`3 12 7 8`" are the names of circuit nodes where the four terminals of the device are connected to. The second and fourth of these nodes are those of the gate and bulk, respectively. The drain and source terminals connect the first and third nodes but we are not concerned with which is which. In each simulation, Spice itself makes the correct assignment by comparing the voltages of these two nodes. What comes next in the element line is the model name of the device. The present example "`cmosn`" defines the NMOS device model of the deck presented in Table 3.1. The rest of the declarations in the element line are optional and not subject to any particular order. However, at least the drawn width "`w`" (m) and drawn length "`l`" (m) of the channel are necessary for any meaningful simulation. If not declared, Spice will assume a default value of 1 m for these parameters. "`ad`" (m^2) and "`pd`" (m) are, respectively, the area and perimeter of the source/drain region connecting the first of the four nodes. "`nrd`" is the number of squares traversed by device current inside this very same region. For understanding how we estimate these three instance parameters, consider the source/drain region tied to node 3 in the example layout of M18 shown in Fig. 3.4, and note

$$\texttt{ad} = W_{DR}L_1 = 2.5(\mu\text{m}) \times 0.55(\mu\text{m}) = 1.38 \times 10^{-12}\text{m}^2,$$

$$\texttt{pd} = 2W_{DR} + 2L_1 = 2 \times 2.5(\mu\text{m}) + 2 \times 0.55(\mu\text{m}) = 6.1 \times 10^{-6}\text{m},$$

and

$$\texttt{nrd} = L_{CG}/W_{DR} = 0.2(\mu\text{m})/2.5(\mu\text{m}) = 0.08.$$

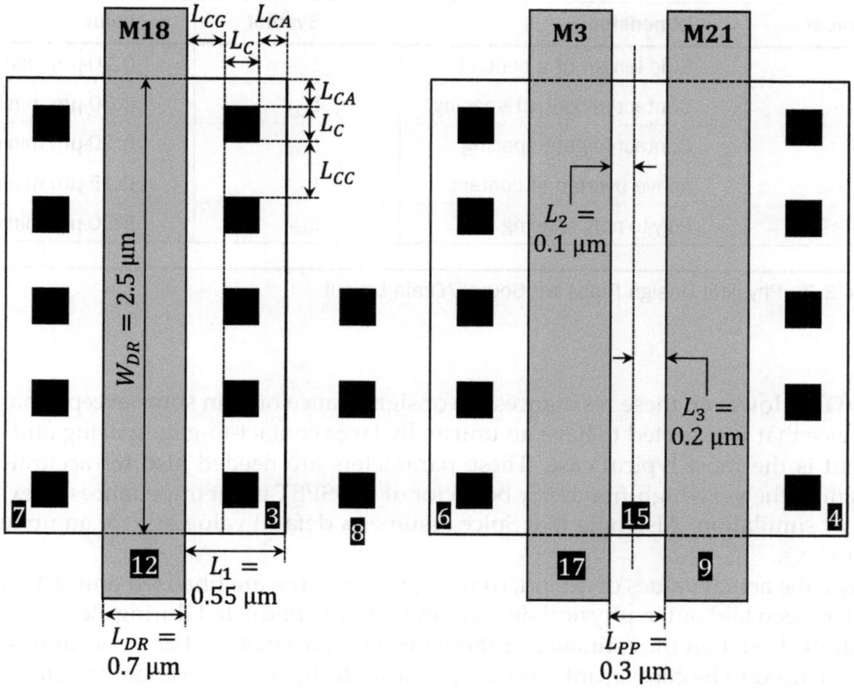

FIGURE 3.4 A layout example of three MOSFETs illustrating the design rules listed in Table 3.2. Gray-shaded features are polysilicon gates.

Notice the calculation of `nrd` by dividing the distance L_{CG} between the gate and inner side of contacts by the width W_{DR} of the region. Had the contacts been placed, say, one W_{DR} away from the gate, device current would traverse just 1 square in this region. In the present example, the division yields much lesser than 1 square because the contacts are pretty close to the gate. Returning to the exemplar element line, also notice the optional parameters `as`, `ps`, and `nrs`. These parameters represent the area, perimeter, and number of squares of the other source/drain region connecting the third of the four nodes. Since exemplar device M18 happens to be symmetrical in source/drain layout, the values of these parameters for node 7 are identical to their counterparts declared for node 3.

Spice needs the optional instance parameters `ad`, `pd`, `as`, and `ps` for taking into account certain parasitic effects associated with the source/drain regions. Junction capacitances, which are of foremost importance in transient, ac, and noise simulations, are among such parasitics. It is therefore essential in these types of simulation to declare the area and perimeter for any source/drain region whose voltage with respect to device bulk is not a constant. If not declared, Spice assumes a zero for each of these four parameters.

The optional instance parameters `nrd` and `nrs` enable Spice to determine the parasitic series resistances of source/drain regions by multiplying each with the model parameter

Number	Dimension	Symbol	Value
1	Side length of a contact	L_C	0.20-μm exact
2	Contact-to-contact spacing	L_{CC}	0.30-μm minimum
3	Contact-to-gate spacing	L_{CG}	0.20-μm minimum
4	Active overlap of contact	L_{CA}	0.15-μm minimum
5	Poly-to-poly spacing	L_{PP}	0.30-μm minimum

TABLE 3.2 Physical Design Rules for Source/Drain Layout

rsh (Ω).[2] However, these resistances are of significance only in some exceptional cases. A device that is expected to have an unusually large contact-to-gate spacing and a large current is the most typical case. These parameters are needed also for accurately determining the very-high-frequency behavior of MOSFET input impedances as extracted with ac simulation. Also note that Spice assumes a default value of 1 for an undeclared nrd or nrs.

Since the actual values of ad, pd, nrd, as, ps, and nrs are finalized only after the device has been laid out in physical design, simulations conducted during electrical design should be based on the estimates of these instance parameters. For an accurate estimation, we need to be conversant with the physical-design rules pertinent to source/drain layout. Listed in Table 3.2 are such rules to be followed throughout this book. The first four of these rules are illustrated in Fig. 3.4 on device M18. Notice that the side length $L_1 = 0.55$ μm used above for estimating ad and pd is the minimum permitted by rules 1, 3, and 4 for a source/drain region containing contacts. The length $L_{CG} = 0.2$ μm of the parasitic resistance used for estimating nrd is also the minimum permitted by rule 3.

Each contact used for accessing a source/drain region has approximately 10 Ω resistance in a typical 0.18-μm CMOS technology. Paralleling as many contacts as possible along the width W_{DR} not only helps minimize the equivalent resistance, but also provides redundancy against possible contact failure. Rule 2 is added to Table 3.2 for enabling the reader to determine the maximum number of contacts that can be accommodated by a source/drain region of a given W_{DR}.

Finally, notice rule 5 for the spacing between two parallel polysilicon lines. For an explanation of the utility of this rule, consider the layout of the two series-connected devices M3 and M21 in Fig. 3.4. These devices share a common source/drain region at node 15 without any metal interconnection, hence, without any contact. Since the polysilicon gates at nodes 17 and 9 can be laid out as close as permitted by rule 5 of Table 3.2, their common source/drain region can be as narrow as $L_{PP} = 0.30$ μm. This helps reduce the parasitic capacitance of this node nearly by half in comparison with a contacted source/drain region of the same W_{DR}, and is therefore practiced in physical design wherever two or more devices of the same conductivity type are connected in series without any interconnection branching out from their common nodes. Cascoded devices, as will be presented in Chapter 4, represent the most common case where this practice is followed in physical design. One outstanding question is how a common source/drain

[2]The reader is referred to Table 3.1 for the NMOS and PMOS values of rsh in our exemplar technology.

region is to be partitioned between two devices in a Spice input file. The answer is that any fictitious line drawn widthwise across the region will do the partitioning. Although the positioning of the line is arbitrary, it is important to avoid assigning all of the region to one of the devices and none to the other because assigning a perimeter lesser than W_{DR} to any of the devices will result in an erroneous calculation of parasitic capacitance. In the example shown in Fig. 3.4, the line is drawn at a distance $L_2 = 0.1$ µm from the gate of M3, and, therefore, at a distance $L_3 = 0.2$ µm from the gate of M21. This leads to the element lines

```
m3 15 17 6 8 cmosn w=2.5u l=0.7u ad=0.25p pd=2.7u nrd=0.04
+ as=1.38p ps=6.1u nrs=0.08
m21 4 9 15 8 cmosn w=2.5u l=0.7u ad=1.38p pd=6.1u nrd=0.08
+ as=0.5p ps=2.9u nrs=0.08
```

where the dimensional parameters of the common-node 15 are estimated from `ad` $= W_{DR}L_2$, `pd` $= W_{DR} + 2L_2$, and `nrd` $= L_2/W_{DR}$ for M3, and from `as` $= W_{DR}L_3$, `ps` $= W_{DR} + 2L_3$, and `nrs` $= L_3/W_{DR}$ for M21. Note that the sum of `ad` of M3 and `as` of M21 equals the actual area of this common region. Similarly, the sum of `pd` of M3 and `ps` of M21 equals the actual perimeter. Same is true also for the sum of `nrd` of M3 and `nrs` of M21.

3.2.2 Performance Metrics and Design Variables

Inside opamps, MOSFETs usually function as transconductors or current sources/sinks. Device metrics associated with these two types of functionality are defined in this section. MOSFETs find use also as external-network switches in discrete-time configurations. The unique performance metrics of this type of application are discussed in Section 3.9. Occasionally, MOSFETs operate also as passive components in the form of a resistor or a capacitor. Performance metrics involved in these two types of operation are presented in Subsections 3.10.3 and 3.11.2, respectively.

A MOSFET operating as a transconductor converts a variation in the gate-source port voltage v_{GS} around a bias value V_{GS} into a variation in the drain-source port current i_D around a bias value I_D. This conversion is characterized with a performance metric called *transconductance* or, more specifically, *gate transconductance*, and defined by

$$g_m \doteq \begin{cases} \left.\dfrac{\partial i_D}{\partial v_{GS}}\right|_{v_{GS}=V_{GS}} & \text{for NMOS,} \\ -\left.\dfrac{\partial i_D}{\partial v_{GS}}\right|_{v_{GS}=V_{GS}} & \text{for PMOS,} \end{cases} \tag{3.5}$$

where the sign convention yields a positive value for both types of MOSFET. Formally, g_m is measured in units of Siemens or A/V but the unit of preference in practice is µA/V or mA/V.

In some transconductor applications, v_{SB} may also vary around a bias value V_{SB}, and cause a variation in i_D around a bias value I_D. The performance metric we use for characterizing this effect is called *bulk transconductance.* It is defined by

$$g_{mb} \doteq \begin{cases} -\left.\dfrac{\partial i_D}{\partial v_{SB}}\right|_{v_{SB}=V_{SB}} & \text{for NMOS,} \\ \left.\dfrac{\partial i_D}{\partial v_{SB}}\right|_{v_{SB}=V_{SB}} & \text{for PMOS.} \end{cases} \tag{3.6}$$

The sign convention of (3.6) makes g_{mb} a positive-valued parameter for both types of MOSFET. g_{mb} is also measured in units of μA/V or mA/V.

A MOSFET operating as a current source or a current sink is expected to keep its drain-source port current i_D constant at a nominal value I_D independently of the port voltage v_{DS}. I_D is the primary performance metric of this application. Its value is set by driving the gate-source port with a dc voltage V_{GS}. The source-bulk port is usually short-circuited but in rare cases it may hold a dc voltage V_{SB}. In practice, i_D cannot be made absolutely independent of v_{DS}. The extent of its dependence is characterized with two performance metrics. For a description of these two, the reader is referred to Fig. 3.5, where typical NMOS and PMOS *output characteristics* relating i_D to v_{DS} are shown. These characteristics indicate a reasonably independent current only for $v_{DS} > V_{DSsat}$ in NMOS and for $v_{DS} < V_{DSsat}$ in PMOS. The so-called *drain-source saturation voltage* V_{DSsat} obviously limits the domain in which a MOSFET can operate as a current source. It is also evident that the current supplied by a MOSFET is not perfectly independent of v_{DS} even if the limit set by V_{DSsat} is not breached. The rate by which it varies with v_{DS} is characterized with a second performance metric called *output conductance*, defined by

$$g_{ds} \doteq \begin{cases} \left.\dfrac{\partial i_D}{\partial v_{DS}}\right|_{v_{DS}=V_{DS}} & \text{for NMOS,} \\ -\left.\dfrac{\partial i_D}{\partial v_{DS}}\right|_{v_{DS}=V_{DS}} & \text{for PMOS,} \end{cases} \tag{3.7}$$

and measured in units of μA/V or mA/V. The nonlinearity of the output characteristics makes g_{ds} a function of v_{DS}. This is why the definitions made in (3.7) are for a specific drain-source voltage V_{DS}. In most design scenarios, this will be the bias voltage across the drain-source port. Also note that the inverse of the output conductance, defined by $r_{ds} \doteq 1/g_{ds}$ (Ω) and called *output resistance* in this book, finds even more widespread use for characterizing the quality of the current source/sink performance.

In summary, the target performance of a MOSFET is specified in the circuit layer of electrical design with a subset of the following metrics: (a) dc drain current I_D, (b) drain-source saturation voltage V_{DSsat}, (c) transconductance g_m, (d) output conductance g_{ds} or its inverse r_{ds}, and (e) bulk transconductance g_{mb}. There exist at most five design variables at the disposal of a designer to implement the specified values of these metrics in

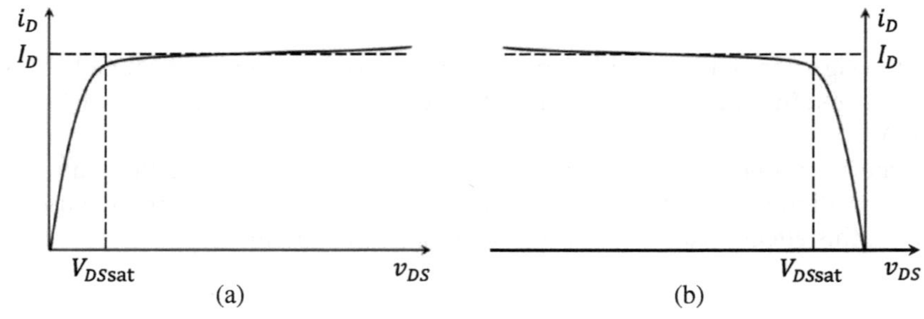

FIGURE 3.5 Typical variation of drain current i_D with drain-source voltage v_{DS} for a constant gate-source voltage V_{GS} in (a) NMOS and (b) PMOS.

the device layer of electrical design: (a) Drawn channel width W_{DR}, (b) drawn channel-length L_{DR}, (c) gate-source bias voltage V_{GS}, (d) drain-source bias voltage V_{DS}, and (e) source-bulk bias voltage V_{SB}. In many cases, V_{DS} and V_{SB} are already fixed by the topological properties of the opamp or constrained by the bias voltages of other devices. In some cases, even V_{GS} is imposed by another device. W_{DR} and L_{DR}, however, are almost always available as independent design variables. Designing a MOSFET therefore turns into a process of sizing the device under constrained bias conditions. The relations connecting device performance metrics to design variables are the tools of this process. A detailed study of these relations and their use in NMOS device design are the subject of the following section. The PMOS version of the relations is presented in Section 3.4. The content of these two sections constitutes the basis of the device-design methodology presented subsequently in Section 3.6.

3.3 NMOS Design Relations and Tools

The design relations presented in this section are derived from the dc relationship between i_D and its variables W, L, v_{GS}, v_{DS}, and v_{SB}. This relationship has a very complex makeup in BSIM models due to the inclusion of numerous secondary physical phenomena observed in devices of near-minimum channel dimensions. Fortunately, channel dimensions needed in analog applications are usually large enough to subdue most of these phenomena. Based on this premise, we present in Subsection 3.3.1 the so-called *long-channel models*, which offer a very simple analytical description of MOSFET operation by ignoring the secondary phenomena. However, these models are good enough only for drawing a qualitative picture of design relations because, as will be explained next in Subsection 3.3.1, certain additional assumptions built into their derivation are not fully justified in analog applications.

While long-channel models provide the designer with a qualitative insight, the quantitative tasks of design have to be based on simulation for accuracy. Starting with Subsection 3.3.2 and ending with Subsection 3.3.6, we will describe the way in which the performance metrics of NMOS devices can be related to design variables by simulation. Subsequently in Section 3.4, we will do the same for PMOS devices.

3.3.1 Long-Channel Models

Long-channel models relate i_D to W, L, v_{GS}, v_{DS}, and v_{SB} for two asymptotic cases depending on the dominant physical mechanism of carrier flow inside the channel. In the case of *strong inversion*, carriers are assumed to flow purely by drifting under the influence of the lateral electric field created by a nonzero v_{DS}. This case prevails if carrier concentration at the source boundary of the channel is much larger than bulk-doping concentration. The second case, called *subthreshold*, is based on the assumption that the carriers move purely by diffusing thermally, which prevails when the carrier concentration at the same boundary is very low. Since these are two asymptotic cases of device operation, long-channel models are least accurate if the device operates in between the two. Since, unfortunately, this is where device performance is usually optimized in analog operation, we can utilize long-channel models only for qualitative purposes.

In this section, we present the analytical expressions of strong-inversion and subthreshold models. Some of the parameters appearing in these expressions are available in the BSIM model deck presented in Table 3.1. Such parameters are printed in the `typewriter` font.

Strong-Inversion Model

The strong-inversion model divides the operation space of a MOSFET into three modes called *cutoff, saturation,* and *nonsaturation.*

Cutoff prevails if gate-source voltage is smaller than a *threshold voltage* parameter V_{TN}, that is,

$$v_{GS} - V_{TN} \leq 0 \text{ (condition of cutoff).} \tag{3.8}$$

Drain current is presumed to vanish in cutoff:

$$i_D = 0 \text{ (in cutoff).} \tag{3.9}$$

Threshold voltage is described by

$$V_{TN} = \texttt{VTH0} + \texttt{K1}(\sqrt{2\phi_{FB} + v_{SB}} - \sqrt{2\phi_{FB}}), \tag{3.10}$$

which contains two BSIM parameters: (a) *Zero-bias threshold voltage* `VTH0` (V) and (b) *body factor* `K1` ($V^{1/2}$). ϕ_{FB} (V) is known as *Fermi potential*. It is described by

$$\phi_{FB} = \frac{kT}{q} \ln \frac{\text{NCH}}{n_i}, \tag{3.11}$$

where $k = 1.38 \times 10^{-23}$, J/K is Boltzmann's constant, T (K) is absolute temperature, $q = 1.6 \times 10^{-19}$ C is the absolute value of electron charge, `NCH` (cm^{-3}) is a BSIM parameter called *substrate-doping concentration* (cm^{-3}), and n_i (cm^{-3}) is *intrinsic carrier concentration* whose value at room temperature ($T = 300$ K) equals 1.1×10^{10} cm^{-3}. Note that kT/q (V) is known as *thermal voltage*; its equals 25.8 mV at room temperature.

As indicated by (3.10), V_{TN} is an increasing function of v_{SB}. This dependence is known as *body effect*. Also note that V_{TN} is reduced to `VTH0` for $v_{SB} = 0$.

The device is in saturation if

$$v_{GS} - V_{TN} \geq 0 \text{ and } v_{DS} \geq V_{DS\text{sat}} \text{ (condition of saturation).} \tag{3.12}$$

Drain-source saturation voltage $V_{DS\text{sat}}$ is related to v_{GS} through

$$V_{DS\text{sat}} \doteq \frac{v_{GS} - V_{TN}}{A_N}. \tag{3.13}$$

Although A_N is generally defined by

$$A_N \doteq 1 + \frac{\mathtt{K1}}{2\sqrt{2\phi_{FB} + v_{SB}}}, \tag{3.14}$$

the following simplified but approximate description is more preferred for the purpose of evaluating V_{DSsat}:

$$A_N \approx 1 + \frac{\mathtt{K1}}{2\sqrt{2\phi_{FB}}}. \tag{3.15}$$

In saturation, drain current is described by

$$i_D = \frac{W}{L}\frac{\mu_N C_{OX}}{2A_N}\left(v_{GS} - V_{TN}\right)^2 \quad \text{(in saturation)}, \tag{3.16}$$

where μ_N (m^2/Vs) represents *electron mobility*, which is a physical parameter relating the velocity of drifting carriers to lateral electric field. Although it is a decreasing function of v_{GS}, we treat it as a constant in hand calculations. A highly simplified form of its BSIM model is described by

$$\mu_N \approx \frac{\mathtt{U0}}{1 + \mathtt{UA}\frac{2\times\mathtt{VTH0}}{\mathtt{TOX}} + \mathtt{UB}\left(\frac{2\times\mathtt{VTH0}}{\mathtt{TOX}}\right)^2}. \tag{3.17}$$

The units used for UA and UB in BSIM model decks are m/V and $(m/V)^2$, respectively. U0, on the other hand, is cited in units of cm^2/Vs, which must be converted into m^2/Vs before being used in (3.17). Also appearing in (3.16) is C_{OX} (F/m^2), which represents the *oxide capacitance per unit gate area*. It is related to *gate-oxide thickness* TOX (m) by

$$C_{OX} \doteq \frac{\epsilon_{OX}}{\mathtt{TOX}}, \tag{3.18}$$

where $\epsilon_{OX} = 3.5 \times 10^{-11}$ F/m is the *dielectric constant* of the gate oxide.

An NMOS device is in nonsaturation if

$$v_{GS} - V_{TN} \geq 0 \text{ and } v_{DS} \leq V_{DSsat} \text{ (condition of nonsaturation)}, \tag{3.19}$$

and the current is described by

$$i_D = \frac{W}{L}\mu_N C_{OX}\left(v_{GS} - V_{TN} - \frac{A_N}{2}v_{DS}\right)v_{DS} \quad \text{(in nonsaturation)}. \tag{3.20}$$

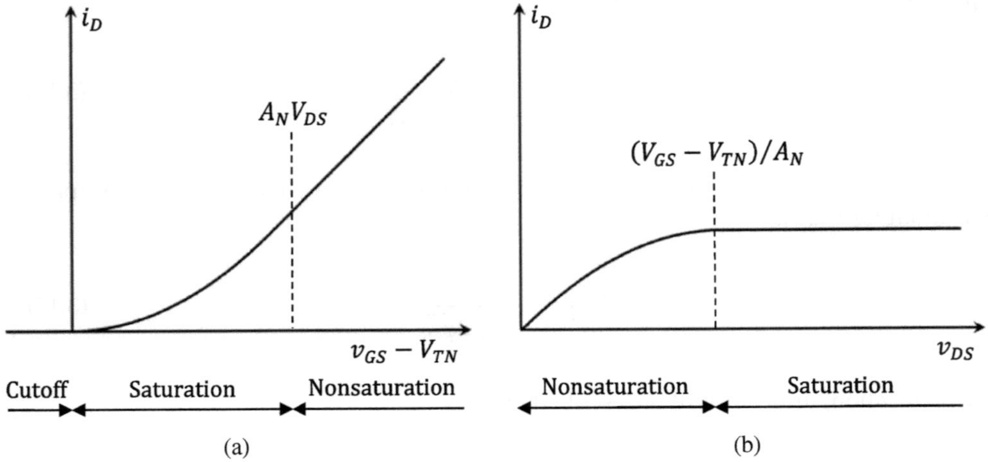

FIGURE 3.6 Strong-inversion model plots of NMOS drain-current versus (a) overdrive voltage, and (b) drain-source voltage.

Notice from (3.9), (3.16), and (3.20) the continuity of current-voltage characteristics at the boundaries $v_{GS} = V_{TN}$ and $v_{DS} = V_{DSsat}$ of the three operation modes. Also notice that v_{GS} never appears in any of the modeling equations without being subtracted by V_{TN}. The difference $v_{GS} - V_{TN}$ is indeed the most influential variable of MOSFET operation and deserves to have a specific name; we call it *overdrive voltage*.

Predictions of the long-channel model for strong inversion are illustrated in Fig. 3.6(a) with a plot of i_D versus overdrive voltage $v_{GS} - V_{TN}$ for a fixed drain-source voltage V_{DS} and a fixed source-bulk voltage V_{SB}. We observe (a) an absence of current in cutoff for negative values of the overdrive voltage $v_{GS} - V_{TN}$, (b) a quadratically increasing current in saturation for positive values of $v_{GS} - V_{TN}$ smaller than $A_N V_{DS}$, and (c) a linearly increasing current for positive values of $v_{GS} - V_{TN}$ greater than $A_N V_{DS}$. An alternative illustration of the same long-channel model predictions is presented in Fig. 3.6(b) with an output characteristic for a fixed positive overdrive voltage $V_{GS} - V_{TN}$ precluding cutoff and a fixed source-bulk voltage V_{SB}. Notice that i_D is predicted to be independent of v_{DS} only in the saturation mode prevailing for $v_{DS} \geq V_{DSsat} = (V_{GS} - V_{TN})/A_N$. In actual fact, it somewhat depends on v_{DS} even in saturation as explained in Subsection 3.2.2, but, compared with nonsaturation, the rate of dependence is much weaker. This is why saturation is essential in current-source and transconductor applications.

Subthreshold Model

Strong-inversion model is based on the assumption that drain current can flow only if the carrier concentration at the source boundary of the channel exceeds the doping concentration, which corresponds to the condition $v_{GS} - V_{TN} \geq 0$ in terms of NMOS overdrive voltage. This is why strong-inversion model predicts $i_D = 0$ unless this condition is met. In actual fact, any finite carrier concentration, however small it may be, yields a drain current as long as $v_{DS} \neq 0$ but, of course, the smaller the concentration, the weaker the current. Therefore, a weak but finite current flows even under the cutoff condition of the strong-inversion model but the model ignores it. This is why model accuracy rapidly deteriorates as the overdrive voltage is lowered toward zero. The small but finite drain current flowing in cutoff is called subthreshold current, and is described approximately by

$$i_D = \frac{W}{L}\mu_N C_{OX}(A_N - 1)\left(\frac{kT}{q}\right)^2 \exp\left(\frac{q}{kT}\frac{v_{GS} - V_{TN}}{A_N}\right)\left[1 - \exp\left(-\frac{q}{kT}v_{DS}\right)\right], \tag{3.21}$$

without specifically distinguishing between saturation and nonsaturation modes. However, considering the current-source and transconductor applications of a MOSFET, we still need to identify the operation mode in which i_D can be considered reasonably independent of v_{DS}. Inspecting (3.21), we observe that such a mode prevails as long as v_{DS} is greater than about $2kT/q$. Therefore, we can define a saturation mode for subthreshold operation with

$$v_{GS} - V_{TN} < 0 \text{ and } v_{DS} \geq V_{DS\text{sat}} \text{ (condition of saturation in subthreshold)}, \tag{3.22}$$

where $V_{DS\text{sat}}$ is given by

$$V_{DS\text{sat}} \approx 2\frac{kT}{q}. \tag{3.23}$$

In saturation, (3.21) reduces to

$$i_D = \frac{W}{L}\mu_N C_{OX}(A_N - 1)\left(\frac{kT}{q}\right)^2 \exp\left(\frac{q}{kT}\frac{v_{GS} - V_{TN}}{A_N}\right) \quad \text{(in saturation)}. \tag{3.24}$$

According to (3.21) or (3.24), i_D is an exponentially increasing function of the overdrive voltage $v_{GS} - V_{TN}$. However, this observation is an accurate representation of the actual subthreshold behavior only for $v_{GS} - V_{TN} \ll 0$ because (3.21) is derived under the assumption of a very small carrier population. Obviously, neither the strong-inversion model nor the subthreshold model can provide us with an accurate description of MOSFET behavior when the overdrive voltage is close to zero. This is unfortunate, because, as stated previously, the overall performance of a MOSFET in current-source and transconductor applications is maximized for small values of the overdrive voltage. We overcome this limitation by resorting to simulation for extracting i_D and other performance metrics wherever we need accuracy in device design. Incidentally, BSIM3 and BSIM4 also model strong inversion and subthreshold separately, and then, bridge the gap around zero-overdrive with mathematical extrapolation. Only in BSIM-BULK, a physically unified model is offered.

3.3.2 Threshold Voltage

The relation between the threshold voltage and design variables is one of the tools of device design. Recall from Subsection 3.2.2 that design variables include the drawn dimensions W_{DR}, L_{DR}, and bias voltages V_{GS}, V_{DS}, and V_{SB}. According to (3.10) and (3.11), the sole design variable that affects V_{TN} among this set is V_{SB}. The rest of the entries of these equations are either physical constants or BSIM parameters, which, as explained in Section 3.1, depend only on temperature for a given threshold option and process corner. In practice, V_{TN} varies also with W_{DR}, L_{DR}, and V_{DS} but among these only the variation with L_{DR} is significant in analog applications. Therefore, a family of plots relating V_{TN} to V_{SB} for the typical range of channel length of analog applications can serve as a common design tool for all devices of the same threshold option and process corner. This design tool can be extracted accurately with a single Spice dc-sweep analysis. The only critical point in simulation is to set the width of the channel around what is typically encountered in practice. A good choice is $W_{DR} = 20W_{DR(\text{min})}$, where $W_{DR(\text{min})}$ is the *minimum drawn width* permitted by physical design rules. Selection of V_{GS} or V_{DS} is not critical; setting each to, say, one half of the maximum permissible port voltage V_{MAX} will do. Note $V_{MAX} = 1.8$ V for our exemplar technology.

Shown in Fig. 3.7 is the variation of V_{TN} with V_{SB} for eight different values of L_{DR} in our exemplar technology, in which, the minimum permissible dimensions are $W_{DR(\text{min})} = 0.3$ μm and $L_{DR(\text{min})} = 0.2$ μm. The following input file extracts and directly plots these curves:

```
filename: nmos1.sp (nmos vth)
.incl bsim.sp
m 2 1 3 0 cmosn w=6u l=0.2u
vds 2 3 0.9
```

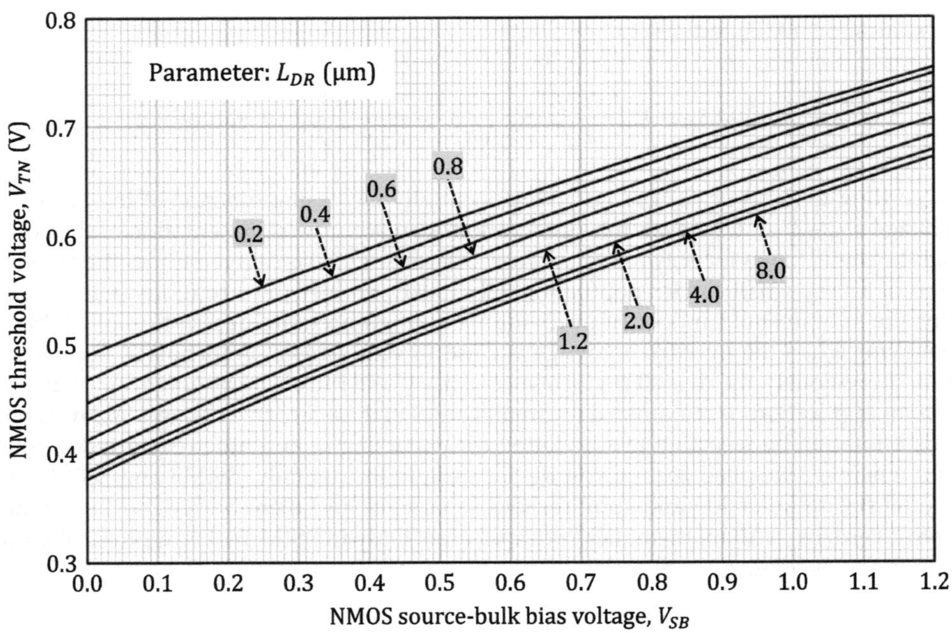

FIGURE 3.7 NMOS threshold voltage V_{TN} as a function of source-bulk bias voltage V_{SB} and drawn channel-length L_{DR}.

```
vgs 1 3 0.9
vsb 3 0 0
.save @m[vth]
.control
destroy all
foreach length 0.2u 0.4u 0.6u 0.8u 1.2u 2u 4u 8u
alter @m[l]=$length
dc vsb 0 1.2 20m
end
plot dc1.@m[vth] dc2.@m[vth] dc3.@m[vth] dc4.@m[vth]
+ dc5.@m[vth] dc6.@m[vth] dc7.@m[vth] dc8.@m[vth]
.endc
.end
```

Notice that V_{TN} is an increasing function of V_{SB} as predicted by (3.10) and a decreasing function of L_{DR}. In some technologies, however, it may be maximized at an L_{DR} larger than $L_{DR(\min)}$. Also note that its value for $V_{SB} = 0$, known as *zero-bias threshold voltage*, is generally greater than the BSIM parameter `VTH0`= 0.37 V which is what the long-channel model equation (3.10) predicts for $V_{SB} = 0$. As expected, however, the difference vanishes for large values of L_{DR} as the MOSFET becomes better qualified as a long-channel device.

3.3.3 Drain-Source Saturation Voltage

Drain-source saturation voltage $V_{DS\text{sat}}$ is among the device performance metrics defined in Subsection 3.2.2. Assuming the simplification offered by (3.15), the only design variable indicated by (3.13) for this metric in strong inversion is the overdrive voltage $V_{GS} - V_{TN}$. In subthreshold, (3.23) indicates a constant $V_{DS\text{sat}}$. Actually, $V_{DS\text{sat}}$ also varies with L_{DR} and V_{SB} but these variations are secondary for the typical values of channel length and overdrive voltage involved in analog applications. A single plot of $V_{DS\text{sat}}$ versus $V_{GS} - V_{TN}$ can therefore be used as a common design tool for all devices of a given set of threshold option, process corner, and temperature. We can extract the plot with a single simulation, and use it as a reasonably accurate design tool for any MOSFET of the same threshold option, same process corner, and same temperature just as we did in Subsection 3.3.2 for V_{TN}.

The simulation is run by setting $V_{SB} = 0$, and sweeping V_{GS}. As long as $V_{DS\text{sat}}$ is plotted as a function of $V_{GS} - V_{TN}$, the plot will also be valid to a good approximation for those cases in which $V_{SB} \neq 0$. Since, in most design scenarios, overdrive voltage turns out to be less than one half of the maximum port voltage V_{MAX}, it usually suffices to set the upper limit of sweep to $V_{GS(\max)} = V_{TN} + (V_{\text{MAX}}/2)$. The lower limit of sweep can be as low as $V_{GS(\min)} = V_{TN}$ or even lower if a subthreshold operation is anticipated. Selection of V_{DS} is not critical; $V_{\text{MAX}}/2$ is a reasonable choice. As to drawn dimensions, $L_{DR} = 4L_{DR(\min)}$, $W_{DR} = 20W_{DR(\min)}$ are reasonable values. The input file `nmos2.sp` used for extracting $V_{DS\text{sat}}$ for our exemplar MOSFET option is as follows:

```
filename: nmos2.sp (nmos vdsat, id, gm, gmbs)
.incl bsim.sp
m 2 1 3 0 cmosn w=6u l=0.8u
vds 2 3 0.9
vgs 1 3
```

```
vsb 3 0 0
.dc vgs 0.2 1.3 10m
.save @m[vth] @m[vdsat] @m[id] @m[gm] @m[gmbs] v(1,3)
.end
```

Note that this file is also used in the following three subsections for developing design tools associated with metrics I_D, g_m, and g_{mb}. This is why these three metrics are also saved in addition to drain-source saturation voltage `@m[vdsat]` and gate-source voltage `v(1,3)`. The command

```
plot @m[vdsat] vs v(1,3)-@m[vth]
```

will generate the plot shown in Fig. 3.8. Qualitatively, the variation is consistent with the prediction of (3.13) for large values of overdrive implying strong inversion and that of (3.23) for small values implying subthreshold. The plot smoothly connects the two in weak inversion where each model is most inaccurate.

As explained in Subsection 3.2.2, V_{DSsat} represents the lowest drain-source voltage for which an NMOS device can remain in saturation as required in current-source and transconductor applications. Therefore, a wide range of V_{DS} is possible in these two types of applications only if V_{DSsat} is small. We now know that this calls for a small overdrive voltage.

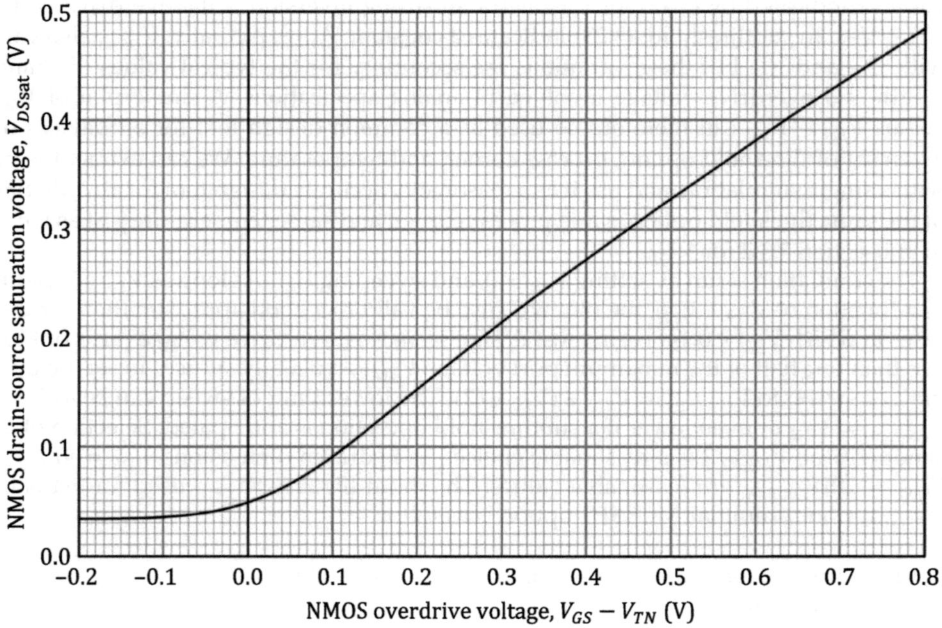

FIGURE 3.8 NMOS drain-source saturation voltage V_{DSsat} as a function of overdrive voltage $V_{GS} - V_{TN}$.

3.3.4 Sheet Current

The dc current I_D is yet another metric of device performance. Considering the saturation mode of operation needed in typical applications, the long-channel model prescribes (3.16) or (3.24) for I_D. According to these expressions, drain current is controlled by three design variables, namely, channel width, channel length, and overdrive voltage. But if we normalize the current to aspect ratio W/L, the outcome

$$I_{\square} \doteq \frac{I_D}{W/L} \tag{3.25}$$

will contain only the overdrive voltage as a design variable. Although $I_{\square}$ also depends weakly on the drain-source voltage, the consequences of this dependence are handled by the output resistance parameter r_{ds} as will be explained in Subsection 3.3.6. In this book, we refer to $I_{\square}$ as *sheet current*, and quantify it in units of A/sq (amperes per square) because it represents the current of one square channel segment of dimensions $L \times L$ as illustrated in Fig. 3.9 on an exemplar MOSFET channel of $W/L = 6.5$. Since there are altogether 6.5 squares along the width, the drain current of this example is given by $I_D = 6.5 \times I_{\square}$.

A single plot of $I_{\square}$ versus $V_{GS} - V_{TN}$ can be used as a common design tool for all devices of a given set of threshold option, process corner, and temperature. This tool is extracted accurately with a single dc-sweep analysis run with the input file `nmos2.sp` described in Subsection 3.3.3. In addition to selecting typical channel dimensions, setting $V_{SB} = 0$, and sweeping V_{GS} in accordance with the comments made in the same section, we need to make sure that the device remains in saturation for the entire range of sweep. Since V_{DSsat} is already available to us as a function of overdrive voltage from Fig. 3.8, we can read its maximum value within the range of sweep, and set V_{DS} at least several hundred millivolts larger than this maximum.

Note that the effective aspect ratio W/L appearing in (3.25) is generally different from the *drawn aspect ratio* W_{DR}/L_{DR}. Although the difference is likely to be small for the

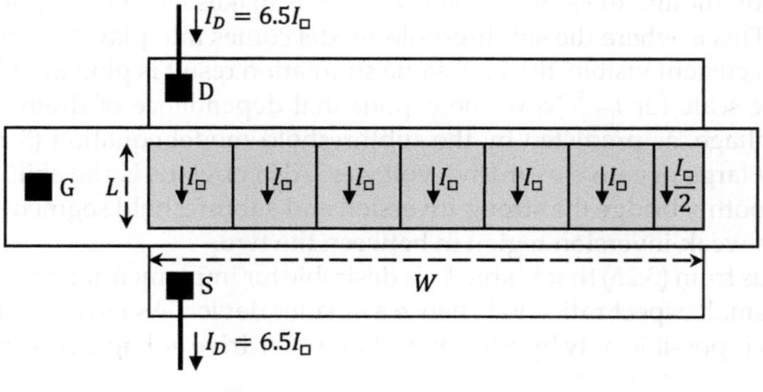

FIGURE 3.9 Sheet current $I_{\square}$ is defined as the fraction of I_D flowing in one square channel segment of dimensions $L \times L$. The channel, shaded in gray, comprises 6.5 squares in this example.

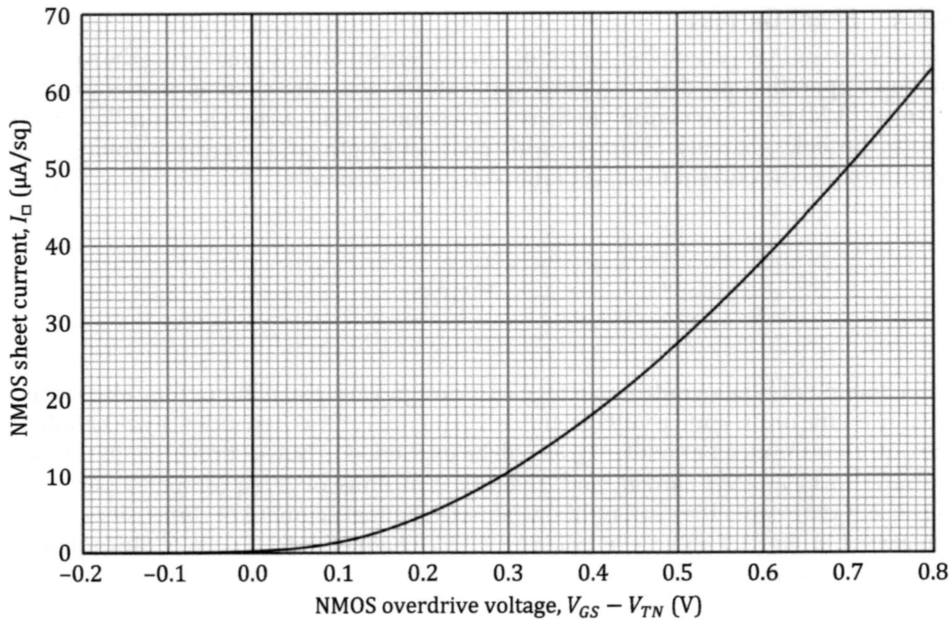

FIGURE 3.10 NMOS sheet current $I_{\square}$ as a function of overdrive voltage $V_{GS} - V_{TN}$.

relatively long and wide device used in simulation, it is prudent to calculate W/L from (3.1) and (3.2), and use the result for normalization. Typing

```
plot @m[id]/7.81 vs v(1,3)-@m[vth]
```

will generate the data shown in Fig. 3.10. The quadratic dependence of $I_{\square}$ on the overdrive voltage, as predicted by the strong-inversion model (3.16), is qualitatively evident from the plot for relatively larger values of the overdrive voltage. However, the prediction that the current should vanish for $V_{GS} - V_{TN} \leq 0$ is not fully corroborated by the simulated plot, though the scale of the vertical axis makes it difficult to detect the residual current. This is where the subthreshold model comes into play. In order to make the subthreshold current visible, the very same simulation result is plotted in Fig. 3.11 using a logarithmic scale for $I_{\square}$.[3] Now, the exponential dependence of drain current on the overdrive voltage, as predicted by the subthreshold model equation (3.24), is obvious for relatively large negative overdrive voltages. Also obvious is the ability of the BSIM model to smoothly bridge the strong-inversion and subthreshold segments of the plot to represent the weak-inversion region in between the two.

It is obvious from (3.25) that a large $I_{\square}$ is desirable for implementing the specified value of I_D with a small aspect ratio W/L, hence a smaller device. As indicated by Fig. 3.10 or Fig. 3.11, this is possible only by adopting a large overdrive voltage. However, it is worth

[3] For a logarithmic $I_{\square}$-axis, plot with the command `plot @m[id]/7.81 vs v(1,3)-@m[vth] ylog`.

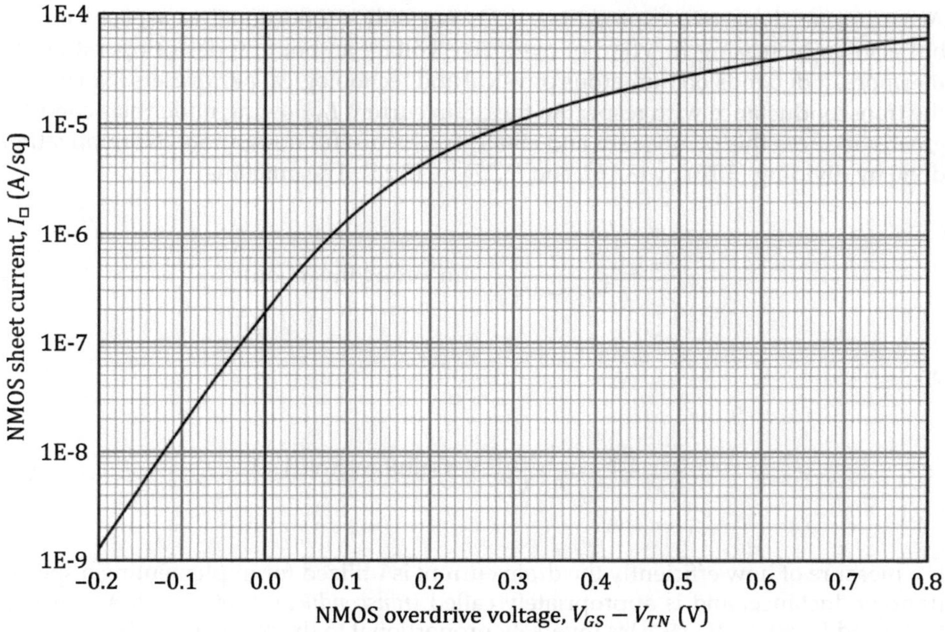

FIGURE 3.11 NMOS sheet current $I_{\square}$ as a function of overdrive voltage $V_{GS} - V_{TN}$. Notice the logarithmic scaling of the vertical axis.

remembering from Subsection 3.3.3 that a large overdrive voltage is counterproductive in terms of V_{DSsat}.

3.3.5 Transconductance Efficiency

Gate Transconductance

As explained in Subsection 3.2.2, g_m is the main performance metric in transconductor applications. Considering the definition (3.5), predictions of the long-channel model for this metric can be formulated by differentiating the saturation-mode drain-current expression (3.16) or (3.24) with respect to v_{GS}. The result is

$$g_m = \frac{W}{L}\frac{\mu_N C_{OX}}{A_N}\left(V_{GS} - V_{TN}\right) \quad \text{(in strong inversion)} \tag{3.26}$$

and

$$g_m = \frac{W}{L}\mu_N C_{OX}\frac{A_N - 1}{A_N}\frac{kT}{q}\exp\left(\frac{q}{kT}\frac{V_{GS} - V_{TN}}{A_N}\right) \quad \text{(in subthreshold).} \tag{3.27}$$

According to these two expressions, g_m is not only proportional to aspect ratio W/L but also an increasing function of overdrive voltage in both modes of operation. Furthermore, it depends on numerous physical and technological parameters. Normalizing the transconductance with respect to the drain current I_D flowing under bias conditions greatly reduces this complexity and creates a very useful design tool. Normalization is done by dividing (3.26) by (3.16), and (3.27) by (3.24). The outcome

$$\frac{g_m}{I_D} = \frac{2}{V_{GS} - V_{TN}} \quad \text{(in strong inversion)} \tag{3.28}$$

or

$$\frac{g_m}{I_D} = \frac{1}{A_N}\frac{q}{kT} \quad \text{(in subthreshold)} \tag{3.29}$$

is a measure of how efficiently the drain current is utilized for implementing a specified transconductance, and is appropriately called *transconductance efficiency*. According to (3.28) and (3.29), g_m/I_D (V^{-1}) is inversely proportional to the overdrive voltage in strong inversion and a constant in subthreshold. In fact, it is also a function of channel dimensions, V_{DS} and V_{SB} but for the typical dimensions and overdrive voltage involved in analog applications, these dependencies are secondary. A single plot of g_m/I_D versus $V_{GS} - V_{TN}$ can therefore be used as a common design tool for all devices of a given set of threshold option, process corner, and temperature. This design tool can be extracted accurately with a single dc-sweep analysis run with the input file `nmos2.sp` described in Subsection 3.3.3. Typing

```
plot @m[gm]/@m[id] vs v(1,3)-@m[vth]
```

will generate the plot shown in Fig. 3.12 for our exemplar technology.

The prediction of (3.28) that g_m/I_D is inversely proportional to overdrive in strong inversion is verified by the plot. Also obvious is the saturation tendency of g_m/I_D as predicted by (3.29) for subthreshold operation. Full constancy can be observed only in deep-subthreshold operation beyond the overdrive range displayed in Fig. 3.12.

A large g_m/I_D enables us to realize a given transconductance with less current, hence less power. Obviously, lowering the overdrive is the only way in which this can be accomplished.

Bulk Transconductance and Total Transconductance

Bulk transconductance g_{mb}, as defined by (3.6), is a metric of the sensitivity of drain current to source-bulk voltage. Since the variation of i_D with v_{SB} is due mainly to the dependence of V_{TN} on v_{SB}, (3.6) for NMOS can be rewritten as

$$g_{mb} = -\frac{\partial i_D}{\partial V_{TN}} \frac{dV_{TN}}{dv_{SB}}\bigg|_{v_{SB}=V_{SB}} . \tag{3.30}$$

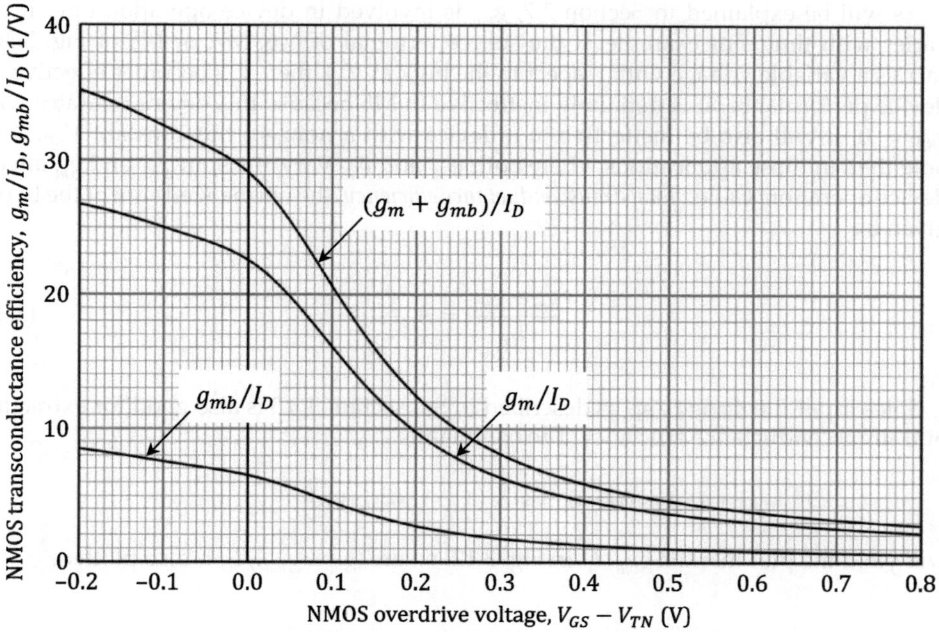

FIGURE 3.12 NMOS transconductance efficiency g_m/I_D and bulk-transconductance efficiency g_{mb}/I_D as functions of overdrive voltage $V_{GS} - V_{TN}$. Also shown is the plot of the sum $(g_m + g_{mb})/I_D$.

Inspecting (3.16) and (3.24), and considering (3.5), we reach the conclusion $\partial i_D/\partial V_{TN} = -\partial i_D/\partial v_{GS} = -g_m$, which turns (3.30) into

$$g_{mb} = g_m \left. \frac{\mathrm{d}V_{TN}}{\mathrm{d}v_{SB}} \right|_{v_{SB}=V_{SB}}. \tag{3.31}$$

Differentiating (3.10) with respect to v_{SB}, and substituting the outcome into (3.31), and comparing the outcome with (3.14), we finally arrive at

$$g_{mb} = (A_N - 1)g_m. \tag{3.32}$$

Although we used the more accurate expression (3.14) of A_N to arrive at (3.32), we will continue treating A_N in (3.32) as a constant described by (3.15). In our exemplar technology, it equals 1.31, and therefore indicates $g_{mb} = 0.31 \times g_m$ from (3.32). Obviously, g_{mb} is a scaled-down copy of g_m, which is why *bulk-transconductance efficiency* is also a fixed fraction of the transconductance efficiency as described by

$$\frac{g_{mb}}{I_D} = (A_N - 1)\,\frac{g_m}{I_D}. \tag{3.33}$$

As will be explained in Section 3.7, g_{mb} is involved in device operation only if v_{SB} varies with time. For example, if the source terminal of M6, M7, or M8 of Fig. 3.3(c) carries a variable voltage with respect to the bulk at V_{SS}, then, g_{mb} becomes effective on device performance. It is definitely ineffective in M5 because this device operates with $v_{SB} = 0$. It will also be ineffective in a device with a nonzero but constant V_{SB}. Also note that, in those applications where both v_{GS} and v_{SB} vary, the sum g_{mb} and g_m affects device operation. Called *total-transconductance efficiency*, the normalized sum of the two is given by

$$\frac{g_m + g_{mb}}{I_D} = A_N \frac{g_m}{I_D}. \tag{3.34}$$

Spice input file `nmos2.sp` as described in Subsection 3.3.3 is also good for extracting bulk-transconductance efficiency. The command

```
plot @m[gmbs]/@m[id] vs v(1,3)-@m[vth]
```

will yield the plot marked g_{mb}/I_D in Fig. 3.12, where $(g_m + g_{mb})/I_D$ is also plotted.

3.3.6 Output Resistance and Early Voltage

The long-channel models presented in Subsection 3.3.1 give us no clue about the variables of the output resistance r_{ds} for a device in saturation. Both (3.16) and (3.24) predict a drain current independent of v_{DS}, hence $r_{ds} = \infty$. In practice, r_{ds} is a very complex function of W_{DR}, L_{DR}, $V_{DS} - V_{DSsat}$, and $V_{GS} - V_{TN}$. Since all these variables have comparable influence, we cannot develop a general design tool of a single variable for r_{ds} unlike what we have already done for V_{DSsat}, $I_{\square}$, and transconductance efficiencies. Any device whose specifications include r_{ds} must therefore be designed by iterative simulation without relying on a general tool. One simplifying opportunity, however, is offered by the fact that the so-called *Early voltage* defined by

$$V_A \doteq r_{ds} I_D \tag{3.35}$$

is independent of W_{DR}.[4] By specifying V_A instead of r_{ds}, the number of variables involved in design is thus reduced to three. We now discuss the way in which V_A varies with each of the three.

$V_{DS} - V_{DSsat}$ is a measure of how deep a device is in saturation. Shown in Fig. 3.13 is the variation of V_A with $V_{DS} - V_{DSsat}$ for eight different values of $V_{GS} - V_{TN}$ ranging from −0.2 V in subthreshold to 0.5 V in strong inversion. These plots are extracted with simulation for $L_{DR} = 0.8$ μm in our exemplar technology. Before starting to review them, note that there are typically three secondary effects that prevent r_{ds} and therefore V_A from being infinitely high in saturation: (a) Channel-length modulation (CLM), which

[4] The term "Early voltage" is also a BJT inheritance. It is named after James M. Early who was first to describe the physical cause of the finite output resistance of BJT in 1952.

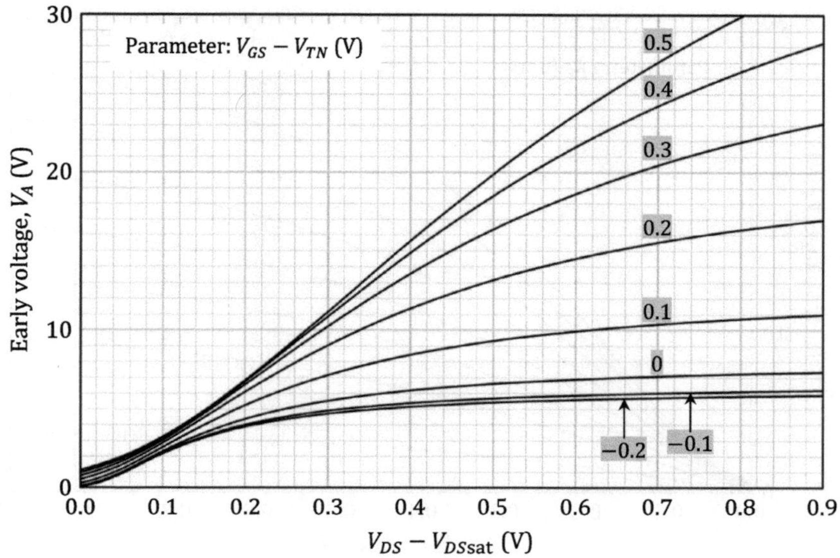

FIGURE 3.13 Early voltage V_A as a function of depth of saturation $V_{DS} - V_{DSsat}$ and overdrive voltage $V_{GS} - V_{TN}$ in an NMOS device of $L_{DR} = 0.8$ µm.

is dominant for small values of $V_{DS} - V_{DSsat}$; (b) drain-induced barrier lowering (DIBL), which is dominant for small values of $V_{GS} - V_{TN}$; and (c) substrate-current-induced body bias (SCBE), which is dominant for large values of $V_{DS} - V_{DSsat}$. Now, let us return to Fig. 3.13, where we make the following three observations:

- V_A is small and relatively independent of the overdrive voltage for small values of $V_{DS} - V_{DSsat}$ lesser than about 150 mV. This is where CLM is dominant.
- For higher values of $V_{DS} - V_{DSsat}$, the device gradually escapes from CLM, and therefore V_A improves. However, the extent of improvement depends strongly on the overdrive voltage due to the dominance of DIBL. Obviously the higher the overdrive, the larger the Early voltage.
- SCBE is nowhere near effective within the range of $V_{DS} - V_{DSsat}$ in Fig. 3.13. Had it been effective, the plots would decline at the high end of $V_{DS} - V_{DSsat}$.

According to these observations, raising $V_{DS} - V_{DSsat}$ to escape from CLM and adopting a large $V_{GS} - V_{TN}$ to suppress DIBL are beneficial for a large V_A. Since V_{DS} is not an influential variable of any of the performance metrics discussed previously, we indeed manipulate it for raising V_A as long as it is not restricted by topological considerations. Treating $V_{GS} - V_{TN}$ as a design variable for V_A, however, is hardly practical because overdrive is the sole variable appearing in all design tools we have developed so far for other device metrics except for the threshold voltage. Only in rare cases, we can manipulate it for achieving a large V_A without adversely affecting one or more of these metrics. Even in such cases, we need to be aware of the fact that a large overdrive voltage leads to a

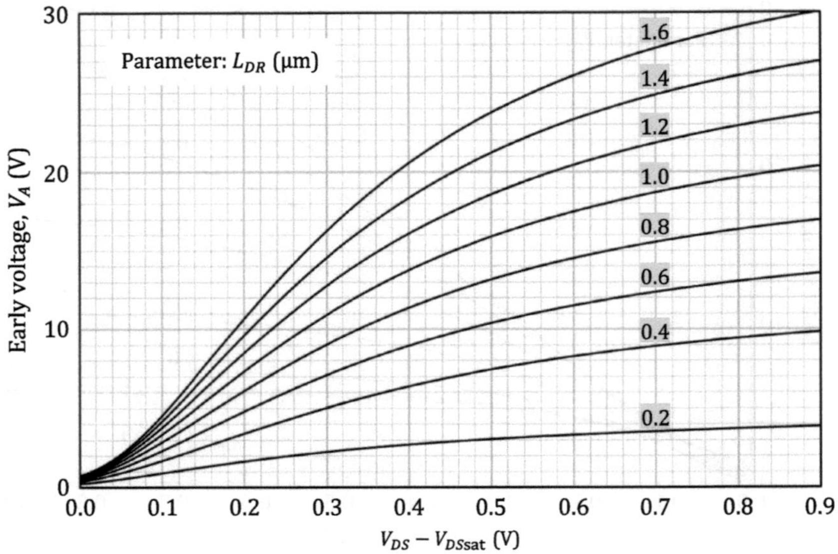

FIGURE 3.14 Early voltage V_A as a function of $V_{DS} - V_{DSsat}$ and L_{DR} in an NMOS device of $V_{GS} - V_{TN} = 0.2$ V.

large V_{DSsat}, hence a small $V_{DS} - V_{DSsat}$ for a given V_{DS}. Unless we compensate for this loss by raising V_{DS}, V_A may actually decrease rather than increase.

Finally, we move to the effect of channel length on V_A. Shown in Fig. 3.14 is the simulated variation of V_A with $V_{DS} - V_{DSsat}$ in our exemplar technology for eight different values of L_{DR} starting with the minimum permissible $L_{DR(min)} = 0.2$ μm. Overdrive voltage is fixed at 0.2 V. Obviously, V_A is a strongly increasing function of the channel length, which not only explains why minimum-geometry MOSFETs are rarely used in analog circuits but also indicates L_{DR} as the prime design variable of V_A.

Example 3.2 An NMOS transistor is specified by $W_{DR} = 20$ μm and $L_{DR} = 0.9$ μm. Determine its drain current for the bias condition $V_{GS} = 0.55$ V, $V_{DS} = 0.9$ V, and $V_{SB} = 0$ by using (a) long-channel model equations, (b) design tools if applicable, and (c) Spice operation-point analysis.

Work First, calculate the technological constants appearing in long-channel model equations:

- (3.18) $\Rightarrow C_{OX} = \frac{3.5\times10^{-11}}{4.1\times10^{-9}} = 8.54 \times 10^{-3}$ F/m^2.
- (3.11) $\Rightarrow \phi_{FB} = 0.0258 \times \ln \frac{2.35\times10^{17}}{1.1\times10^{10}} = 0.435$ V.
- (3.15) $\Rightarrow A_N = 1 + \frac{0.578}{2\sqrt{2\times0.435}} = 1.31$.

Next, calculate effective dimensions:

- (3.1) $\Rightarrow W = 20 \times 10^{-6} - 1 \times 10^{-8} - 2 \times 2.88 \times 10^{-9} = 2 \times 10^{-5}$ m.
- (3.2) $\Rightarrow L = 0.9 \times 10^{-6} + 0 - 2 \times 1.71 \times 10^{-8} = 8.66 \times 10^{-7}$ m.

Now determine I_D:

- From long-channel model equations:
  - (3.10) $\Rightarrow V_{TN} = \texttt{VTH0} = 0.37$ V because $V_{SB} = 0$. $V_{GS} - V_{TN} = 0.18$ V. According to (3.8) device is not in cutoff.
  - (3.13) $\Rightarrow V_{DSsat} = \frac{0.55-0.37}{1.31} = 0.137$ V, which is less than $V_{DS} = 0.9$ V. Therefore (3.12) implies saturation.
  - (3.17) $\Rightarrow \mu_N = \dfrac{293\times10^{-4}}{1-1.22\times10^{-9}\frac{2\times0.37}{4.1\times10^{-9}}+2.32\times10^{-18}\left(\frac{2\times0.37}{4.1\times10^{-9}}\right)^2} = 3.42 \times 10^{-2}$ m^2/Vs.
  - (3.16) $\Rightarrow I_D = \frac{2\times10^{-5}}{8.66\times10^{-7}} \frac{3.42\times10^{-2}\times8.54\times10^{-3}}{2\times1.31} 0.18^2 = 8.34 \times 10^{-5}$ A (83.4 μA).
- From design tools:
  - Figure 3.7 $\Rightarrow V_{TN} = 0.425$ V $\Rightarrow V_{GS} - V_{TN} = 0.125$ V.
  - Figure 3.8 $\Rightarrow V_{DSsat} = 0.105$ V, which is less than $V_{DS} = 0.9$ V. Therefore (3.12) implies saturation.
  - Figures 3.10 and 3.11 are applicable because device is in saturation. They indicate $I_\square = 2 \times 10^{-6}$ A/sq.
  - (3.25) $\Rightarrow I_D = 2 \times 10^{-6} \frac{2\times10^{-5}}{8.66\times10^{-7}} = 4.62 \times 10^{-5}$ A (46.2 μA).
- Spice operation-point analysis: Use the following input file.

```
filename: ex3-2.sp
.incl bsim.sp
m 2 1 0 0 cmosn w=20u l=0.9u
vds 2 0 0.9
vgs 1 0 0.55
.op
.end
```

  The command `show` yields a list of all variables and parameters including the drain current $I_D = 44.8$ μA.

The reader is invited to take this opportunity to compare graphical design tools with long-channel model in (a) the volume of calculation involved, and (b) the accuracy of prediction offered. ▲

3.4 PMOS Design Relations and Tools

The PMOS counterparts of long-channel model equations, design relations, and tools are presented in this section. Sign convention will be consistent with the references set up in Fig. 3.2 for PMOS devices.

3.4.1 Strong-Inversion Model

Cutoff prevails for

$$v_{GS} - V_{TP} \geq 0 \text{ (condition of cutoff)}, \tag{3.36}$$

and implies

$$i_D = 0 \text{ (in cutoff)}. \tag{3.37}$$

PMOS threshold voltage is given by

$$V_{TP} = \texttt{VTH0} - \texttt{K1}(\sqrt{2\phi_{FB} - v_{SB}} - \sqrt{2\phi_{FB}}), \tag{3.38}$$

where

$$\phi_{FB} = \frac{kT}{q} \ln \frac{\text{NCH}}{n_i}. \tag{3.39}$$

According to (3.38), V_{TP} equals VTH0 for $v_{SB} = 0$. As listed in Table 3.1, VTH0 equals −0.38 V in our exemplar technology. Also note from (3.38) that V_{TP} varies in the negative direction due to body effect as v_{SB} is made increasingly negative.

Saturation prevails for

$$v_{GS} - V_{TP} \leq 0 \text{ and } v_{DS} \leq V_{DS\text{sat}} \text{ (condition of saturation)}, \tag{3.40}$$

where $V_{DS\text{sat}}$ is given by

$$V_{DS\text{sat}} = \frac{v_{GS} - V_{TP}}{A_P}, \tag{3.41}$$

and A_P is defined by

$$A_P \doteq 1 + \frac{\texttt{K1}}{2\sqrt{2\phi_{FB} - v_{SB}}}. \tag{3.42}$$

To keep the model simple, we usually ignore the dependence of A_P on v_{SB}, and simplify (3.42) into

$$A_P = 1 + \frac{\texttt{K1}}{2\sqrt{2\phi_{FB}}}. \tag{3.43}$$

In saturation, drain current is described by

$$i_D = \frac{W}{L}\frac{\mu_P C_{OX}}{2A_P}(v_{GS} - V_{TP})^2 \quad \text{(in saturation)}, \tag{3.44}$$

where hole mobility μ_P (m^2/Vs) is given approximately by

$$\mu_P \approx \frac{\text{U0}}{1 + \text{UA}\frac{|2\times\text{VTH0}|}{\text{TOX}} + \text{UB}\left(\frac{|2\times\text{VTH0}|}{\text{TOX}}\right)^2}, \tag{3.45}$$

and C_{OX} is described by

$$C_{OX} \doteq \frac{\epsilon_{ox}}{\text{TOX}}. \tag{3.46}$$

Nonsaturation prevails for

$$v_{GS} - V_{TP} \le 0 \text{ and } v_{DS} \ge V_{DS\text{sat}} \text{ (condition of nonsaturation)}. \tag{3.47}$$

Drain current in nonsaturation is described by

$$i_D = \frac{W}{L}\mu_P C_{OX}\left(v_{GS} - V_{TP} - \frac{A_P}{2}v_{DS}\right)v_{DS} \quad \text{(in nonsaturation)}. \tag{3.48}$$

Drain current, as predicted by the strong-inversion model, is plotted in Fig. 3.15(a) as a function of overdrive voltage $v_{GS} - V_{TP}$ for a fixed drain-source voltage V_{DS}. Its variation with v_{DS} for a fixed negative overdrive voltage $V_{GS} - V_{TP}$ is shown in Fig. 3.15(b).

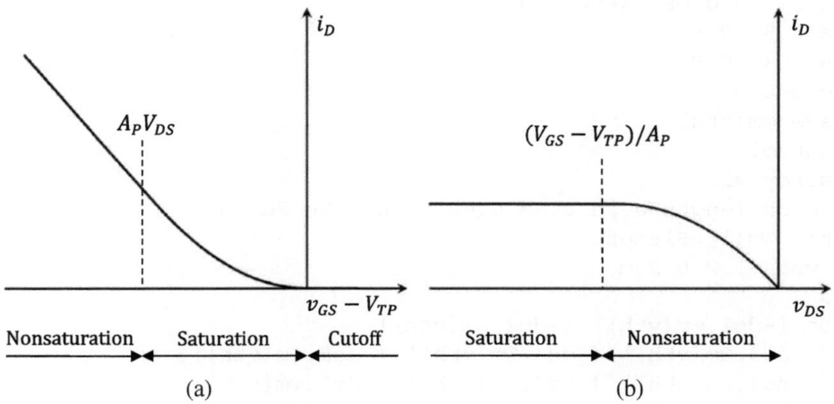

FIGURE 3.15 Strong-inversion model plots of PMOS drain-current versus (a) overdrive voltage and (b) drain-source voltage.

3.4.2 Subthreshold Model

The subthreshold drain current of a PMOS device is modeled by

$$i_D = \frac{W}{L}\mu_P C_{OX}(A_P - 1)\left(\frac{kT}{q}\right)^2 \exp\left(-\frac{q}{kT}\frac{v_{GS} - V_{TP}}{A_P}\right)\left[1 - \exp\left(\frac{q}{kT}v_{DS}\right)\right]. \tag{3.49}$$

Saturation in subthreshold is defined by

$$v_{GS} - V_{TP} \geq 0 \text{ and } v_{DS} \leq V_{DSsat} \text{ (condition of saturation)}, \tag{3.50}$$

where V_{DSsat} is given by

$$V_{DSsat} \approx -2\frac{kT}{q}. \tag{3.51}$$

In saturation, (3.49) is simplified into

$$i_D = \frac{W}{L}\mu_P C_{OX}(A_P - 1)\left(\frac{kT}{q}\right)^2 \exp\left(-\frac{q}{kT}\frac{v_{GS} - V_{TP}}{A_P}\right) \quad \text{(in saturation)}. \tag{3.52}$$

3.4.3 Threshold Voltage

The following Spice input file can be used for extracting and plotting the design tool relating V_{TP} to V_{SB} and L_{DR}.

```
filename: pmos1.sp (pmos vth)
.incl bsim.sp
m 2 1 3 0 cmosp w=6u l=0.2u
vds 2 3 -0.9
vgs 1 3 -0.9
vsb 3 0 0
.save @m[vth]
.control
destroy all
foreach length 0.2u 0.4u 0.6u 0.8u 1.2u 2u 4u 8u
alter @m[l]=$length
dc vsb -1.2 0 20m
end
plot (-dc1.@m[vth]) (-dc2.@m[vth])
+   (-dc3.@m[vth]) (-dc4.@m[vth]) (-dc5.@m[vth])
+   (-dc6.@m[vth]) (-dc7.@m[vth]) (-dc8.@m[vth])
.endc
.end
```

The resulting plots are shown in Fig. 3.16.

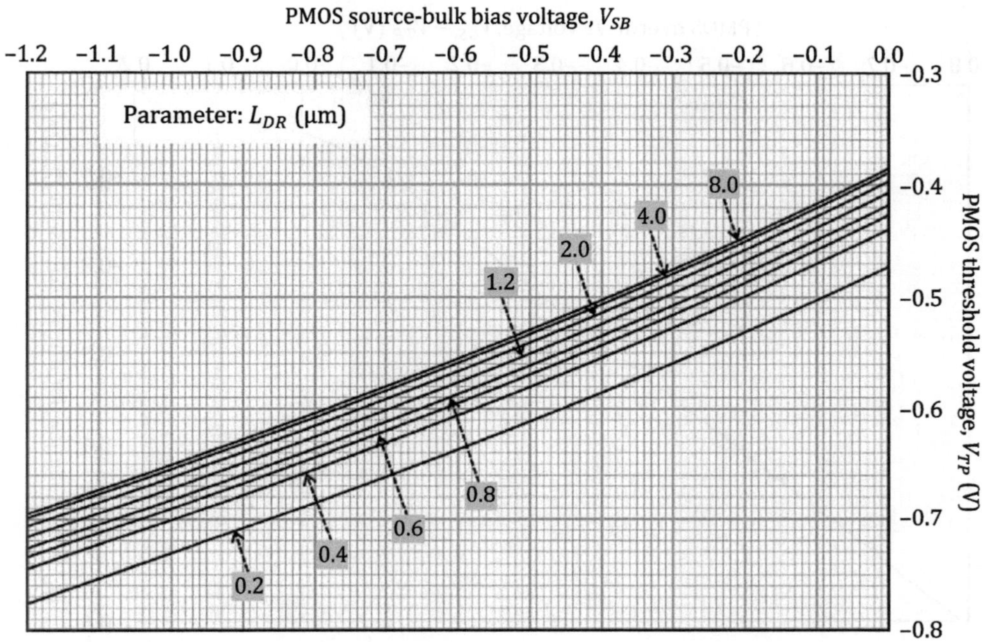

FIGURE 3.16 PMOS threshold voltage V_{TP} as a function of source-bulk bias voltage V_{SB} and drawn channel length L_{DR}.

3.4.4 Drain-Source Saturation Voltage, Sheet Current, and Transconductance Efficiency

The following Spice input file is appropriate for extracting the design tools relating V_{DSsat}, $I_{\square}$, g_m/I_D, and g_{mb}/I_D to overdrive voltage $V_{GS} - V_{TP}$ for a PMOS device.

```
filename: pmos2.sp (pmos vdsat, id, gm, gmbs)
.incl bsim.sp
m 2 1 3 0 cmosp w=6u l=0.8u
vds 2 3 -0.9
vgs 1 3
vsb 3 0 0
.dc vgs -1.3 -0.2 10m
.save @m[vth] @m[vdsat] @m[id] @m[gm] @m[gmbs] v(1,3)
.end
```

The resulting plots are given in Figs. 3.17 to 3.20. For plotting the variables in accordance with our sign convention, we must take into account the fact that V_{DSsat} and V_{TP} are saved in absolute values in Spice. Proper commands are as follows:

V_{DSsat}: `plot -@m[vdsat] vs v(1,3)+@m[vth]`

$I_{\square}$: `plot @m[id]/8.08 vs v(1,3)+@m[vth]`

g_m/I_D: `plot @m[gm]/@m[id] vs v(1,3)+@m[vth]`

g_{mb}/I_D: `plot @m[gmbs]/@m[id] vs v(1,3)+@m[vth]`

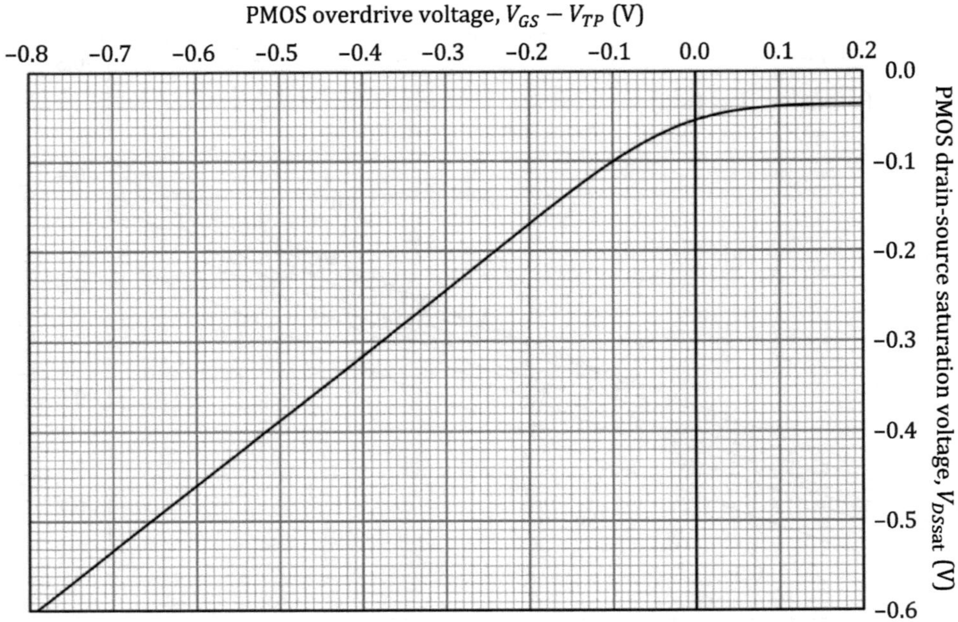

FIGURE 3.17 PMOS drain-source saturation voltage V_{DSsat} as a function of overdrive voltage $V_{GS} - V_{TP}$.

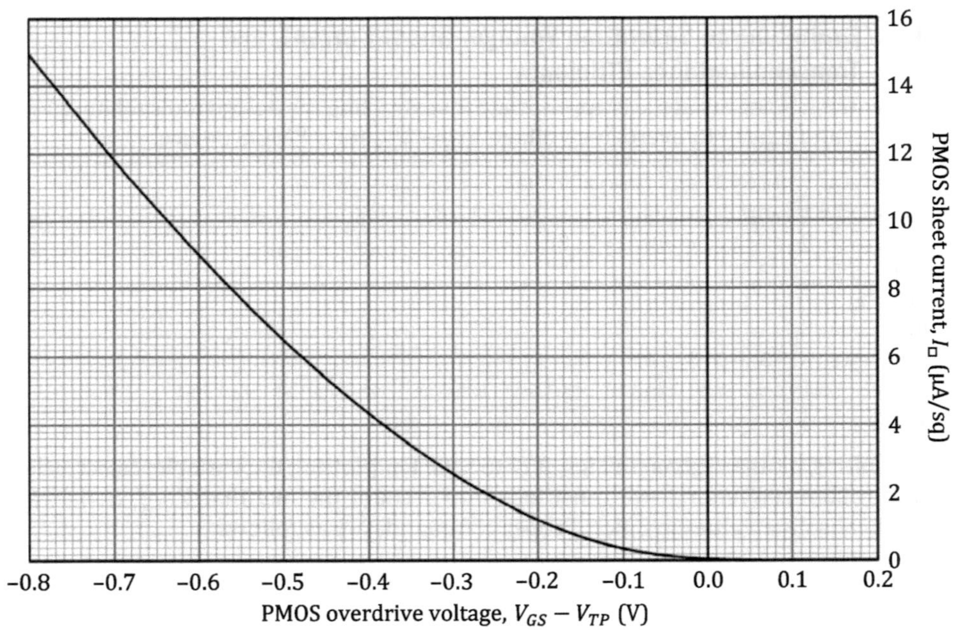

FIGURE 3.18 PMOS sheet current $I_{\square}$ as a function of overdrive voltage $V_{GS} - V_{TP}$.

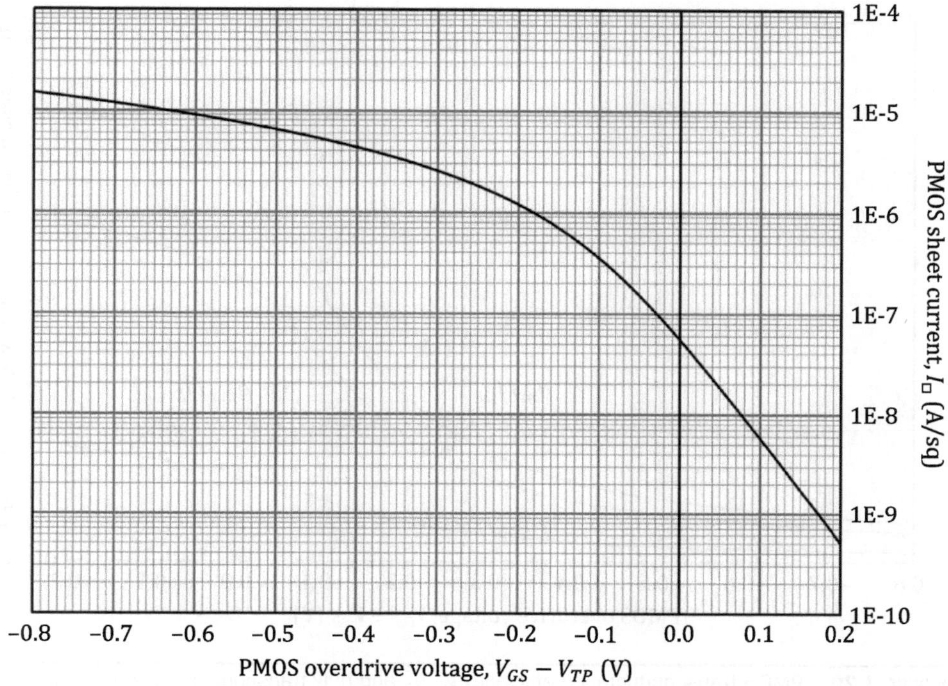

FIGURE 3.19 PMOS sheet current $I_\square$ as a function of overdrive voltage $V_{GS} - V_{TP}$. Notice the logarithmic calibration of the vertical axis.

3.4.5 Output Resistance and Early Voltage

All conclusions reached in Subsection 3.3.6 for the output resistance r_{ds} and the Early voltage V_A are also valid for PMOS devices. In summary, V_A is more convenient to handle than r_{ds} due to its independence of W but an iterative simulation is still needed for designing a PMOS device specified for a particular value of V_A.

Example 3.3 A PMOS transistor is specified by $W_{DR} = 10$ μm and $L_{DR} = 1$ μm. Its dc terminal voltages with respect to circuit ground are as follows: Bulk $V_B = 1.8$ V, source $V_S = 1.4$ V, drain $V_D = 0.9$ V, gate $V_G = 0.6$ V. Determine its drain current by using (a) long-channel model equations, (b) design tools if applicable, and (c) Spice operation-point simulation.

Work First, calculate the technological constants appearing in long-channel model equations:

- (3.46) $\Rightarrow C_{OX} = \frac{3.5\text{x}10^{-11}}{4.1\text{x}10^{-9}} = 8.54 \times 10^{-3}$ F/m^2.
- (3.39) $\Rightarrow \phi_{FB} = 0.0258 \times \ln \frac{4.16\text{x}10^{17}}{1.1\text{x}10^{10}} = 0.450$ V.
- (3.43) $\Rightarrow A_P = 1 + \frac{0.572}{2\sqrt{2\text{x}0.450}} = 1.30$.

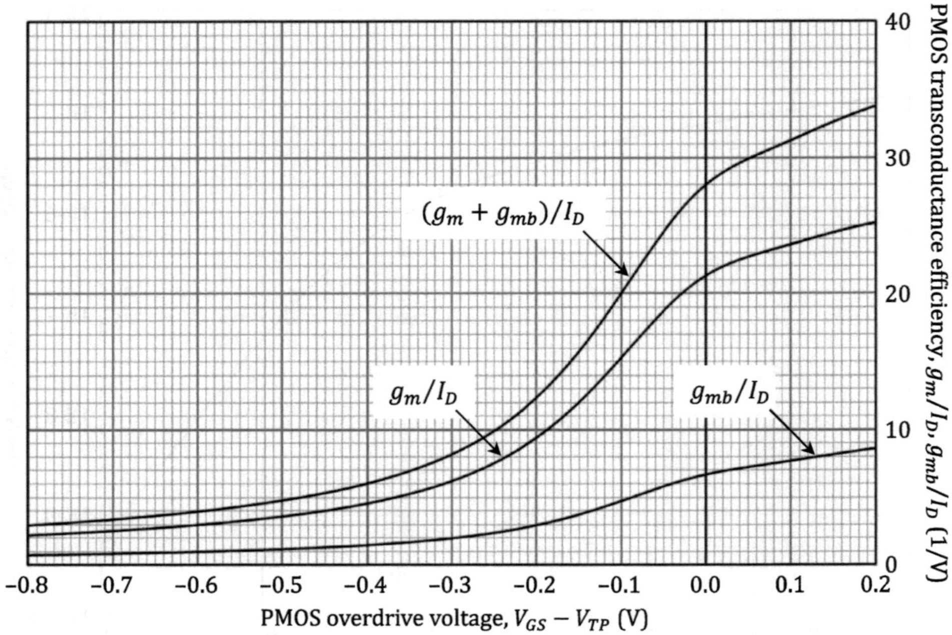

FIGURE 3.20 PMOS transconductance efficiency g_m/I_D and bulk-transconductance efficiency g_{mb}/I_D as functions of overdrive voltage $V_{GS} - V_{TP}$. Also shown is the plot of the sum $(g_m + g_{mb})/I_D$.

Next, calculate effective dimensions and port voltages:

- (3.1) $\Rightarrow W = 10 \times 10^{-6} - 1 \times 10^{-8} - 2 \times 0 = 9.99 \times 10^{-6}$ m.
- (3.2) $\Rightarrow L = 1 \times 10^{-6} - 0 - 2 \times 2.94 \times 10^{-8} = 9.41 \times 10^{-7}$ m.
- $V_{GS} = 0.6 - 1.4 = -0.8$ V, $V_{DS} = 0.9 - 1.4 = -0.5$ V, $V_{SB} = 1.4 - 1.8 = -0.4$ V.

Now determine I_D:

- From long-channel model equations:
  - (3.38) $\Rightarrow V_{TP} = -0.38 - 0.57\left(\sqrt{2 \times 0.45 - (-0.4)} - \sqrt{2 \times 0.45}\right) = -0.489$ V.
  - $V_{GS} - V_{TP} = -0.8 - (-0.489) = -0.311$ V, which rules out cutoff.
  - (3.41) $\Rightarrow V_{DSsat} = \frac{-0.311}{1.30} = -0.239$ V, which is less negative than $V_{DS} = -0.5$ V. Therefore (3.40) implies saturation.
  - (3.45) $\Rightarrow \mu_P = \frac{109\times10^{-4}}{1+1.31\times10^{-9}\frac{|2\times(-0.38)|}{4.1\times10^{-9}}+1\times10^{-21}\left(\frac{|2\times(-0.38)|}{4.1\times10^{-9}}\right)^2} = 8.77 \times 10^{-3}\ \mathrm{m^2/Vs}$.
  - (3.44) $\Rightarrow I_D = \frac{9.99\times10^{-5}}{9.41\times10^{-7}} \frac{8.77\times10^{-3}\times8.54\times10^{-3}}{2\times1.30}(-0.311)^2 = 2.96 \times 10^{-5}$ A (29.6 μA).

- From design tools:
  - Figure 3.16 ⇒ $V_{TP} = -0.535$ V for $V_{SB} = -0.4$ V. Therefore $V_{GS} - V_{TP} = -0.8 - (-0.535) = -0.265$ V.
  - Figure 3.17 ⇒ $V_{DSsat} = -0.22$ V, which is less negative than $V_{DS} = -0.5$ V. Therefore (3.40) implies saturation.
  - Figures 3.18 and 3.19 are applicable because device is in saturation. They indicate $I_{\square} = 2\ \mu\text{A/sq}$.
  - (3.25) ⇒ $I_D = \frac{9.99\times10^{-6}}{9.41\times10^{-7}} 2\times10^{-6} = 21.2\ \mu\text{A}$.
- Spice operation-point analysis: Use the following input file.

```
filename: ex3-3.sp
.incl bsim.sp
m 2 1 3 4 cmosp w=10u l=1u
vb 4 0 1.8
vs 3 0 1.4
vd 2 0 0.9
vg 1 0 0.6
.op
.end
```

The command show will display $I_D = 21.6\ \mu\text{A}$. ▲

Example 3.4 A MOSFET whose gate is interconnected with one of the current-conducting terminals as shown in Fig. 3.21 finds extensive use in analog circuits. It is called *diode-connected MOSFET*. Using the strong-inversion model, show that the device operates unconditionally in saturation when a positive current i_D is imposed on it.

Work First consider the diode-connected NMOS device. A positive i_D implies a positive v, and thus makes the lower terminal source in accordance with (3.3). The gate terminal being short-circuited to the drain, the device therefore operates with $v_{GS} = v_{DS} = v$. Since a current is imposed, the device cannot be in cutoff. This implies a positive overdrive $v - V_{TN} > 0$, which satisfies the first part of the saturation condition (3.12). As to the second part of (3.12), compare $v_{DS} = v$ with $V_{DSsat} = (v - V_{TN})/A_N$. Since V_{TN} is positive, and A_N is positive and greater than 1, V_{DSsat} is definitely smaller than v_{DS}, hence unconditional saturation.

Now consider the diode-connected PMOS device. A positive i_D implies a positive v, and thus makes the upper terminal source in accordance with (3.3). Gate being shorted to

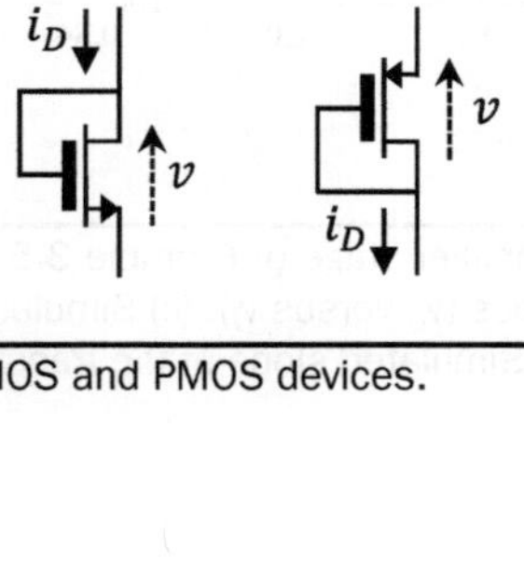

FIGURE 3.21 Diode-connected NMOS and PMOS devices.

the drain, the device therefore operates with $v_{GS} = v_{DS} = -v$. Since current is nonzero, the device cannot be in cutoff. This implies a negative overdrive $-v - V_{TP} < 0$, which satisfies the first part of the saturation condition (3.40). As to the second part of (3.40), compare $v_{DS} = -v$ with $V_{DSsat} = (-v - V_{TP})/A_P$. Since V_{TP} is negative, and A_P is positive and greater than 1, V_{DSsat} is definitely less negative than v_{DS}, hence unconditional saturation.

Actually, a diode-connected MOSFET operates in saturation even under subthreshold conditions as long as v is more positive than a few kT/q in NMOS, or more negative than a few $-kT/q$ in PMOS. See (3.23) and (3.51) for a proof. ▲

Example 3.5 Shown in Fig. 3.22(a) is the schematic of a *common-source* voltage-amplifier stage. M1 operates as a transconductor with its source and bulk terminals tied to circuit ground. M2 operates as a current source with its source and bulk terminals tied to a positive power supply $V_{DD} = 1.8$ V. Dimensions are $W_{DR1} = 6$ μm and $L_{DR1} = 0.8$ μm in M1, and $W_{DR2} = 24$ μm and $L_{DR2} = 0.8$ μm in M2. The gate of M2 is tied to a bias voltage $V_B = 1.25$ V. The gate of M1 receives the input voltage v_I of the amplifier. v_O represents the output voltage. All these voltages are defined with respect to ground. Based on the strong-inversion model, interpret the simulated plot of the *transfer characteristic* (v_O versus v_I) shown in Fig. 3.22(b).

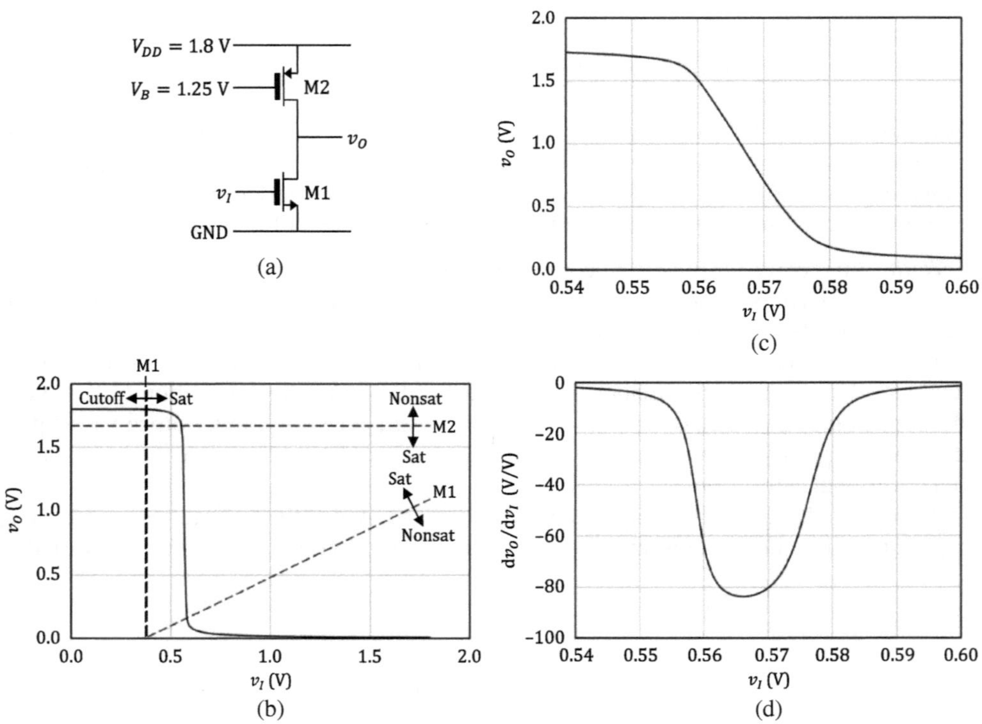

FIGURE 3.22 Common-source amplifier stage of Example 3.5. (a) Electrical schematic. (b) Simulated transfer characteristics (v_O versus v_I). (c) Simulated transfer characteristic zoomed in around the high-gain segment. (d) Simulated slope of the transfer characteristic (dv_O/dv_I versus v_I).

Work In order to interpret the transfer characteristics properly, we need to determine the operation modes of the two devices.

Consider M2 first. This device operates with $v_{SB} = 0$ implying $V_{TP} = \mathtt{VTH0} = -0.38$ V from Table 3.1. Its overdrive voltage is fixed at $V_B - V_{DD} - V_{TP} = 1.25 - 1.8 - (-0.38) = -0.17$ V, which, being negative, implies inversion. Whether this device operates in nonsaturation or saturation should be determined by comparing its drain-source voltage $v_{DS2} = v_O - V_{DD} = v_O - 1.8$ with V_{DSsat2}. For the overdrive calculated above, (3.41) yields $V_{DSsat2} = -0.131$ V ($A_P = 1.30$ is borrowed from Example 3.3). Carrying these into (3.40) and (3.47), we conclude that nonsaturation prevails for $v_O > 1.669$ V, and saturation prevails otherwise. The boundary between the two modes of operation is shown with a dashed line in Fig. 3.22(b).

Next, consider the operation modes of M1. This device also operates with $v_{SB} = 0$; hence, a zero-bias threshold voltage $V_{TN} = \mathtt{VTH0} = 0.369$ V from Table 3.1. According to (3.8), $v_I < 0.369$ V implies cutoff. Otherwise, the device is in inversion, where saturation and nonsaturation modes are demarcated by the line $v_O = V_{DSsat1} = (v_I - V_{TN})/A_N = (v_I - 0.369)/1.31$. This line is also plotted in Fig. 3.22(b). As indicated by (3.12) and (3.19), saturation prevails above the line, and nonsaturation prevails below the line.

Now we can interpret the characteristic. First, consider the part of the characteristic for $v_I \leq V_{TN} = 0.369$ V where M1 is in cutoff. Since M1 does not conduct a current, the current source M2 cannot conduct a current either. However, since M2 operates with a negative overdrive -0.170 V, its current must somehow vanish outside cutoff! Since a finite overdrive in a saturated device would definitely give rise to a finite current as implied by (3.44), this device shouldn't be in saturation. This leaves nonsaturation as the only possible mode of operation. Inspecting (3.48) of nonsaturation, we indeed observe that a vanishing drain-source voltage rules out current flow. This conclusion implies $v_O = V_{DD} = 1.8$ V, which is indeed what the simulated transfer characteristic shown in Fig. 3.22(b) exhibits for $v_I < 0.369$ V.

Moving v_I above 0.369 V forces M1 into saturation. The current of this device now increases with v_I in accordance with (3.16). Since M2 must conduct the same amount of current while remaining in nonsaturation with a fixed overdrive, its drain-source voltage must grow in the negative direction. As a result, v_O starts declining as observed in Fig. 3.22(b).

As the declining v_O drops below 1.669 V, M2 also enters into saturation. This occurs around $v_I = 0.57$ V, and causes the transfer characteristic to drop off abruptly before M1 goes into nonsaturation. Shortly, we will examine this very important segment in detail. Presently, let us continue with the rest of the transfer characteristics, where M1 moves into nonsaturation with an increasing overdrive voltage $v_I - V_{TN}$ while M2 remains in saturation with a fixed overdrive voltage $V_B - V_{DD} - V_{TP}$ hence a fixed current. The only way in which the nonsaturated M1 can conduct the fixed current of M2 with an increasing overdrive is to reduce its own drain-source voltage v_O. This is why v_O continues to decline as v_I increases beyond 0.57 V.

The near-vertical segment is where this circuit is operated as an amplifier stage. Its slope corresponds to the voltage gain available from the stage. Now, let us attempt to characterize this segment analytically by equating the long-channel model equations of saturation (3.16) and (3.44). The result predicts a perfectly vertical segment located at

$$V_I = \sqrt{\frac{\mu_P}{\mu_N}\frac{A_N}{A_P}\frac{W_2}{L_2}\frac{L_1}{W_1}\left(V_B - V_{DD} - V_{TP}\right)^2} + V_{TN},$$

hence an infinitely large voltage gain at this particular input voltage! In actual fact, the slope of the segment is finite as observed in the close-up shown in Fig. 3.22(c). The derivative of the close-up shown in Fig. 3.22(d) clearly shows that the gain is no larger than −83.6 V/V. The reason for this discrepancy between the predictions of long-channel model and simulation is that the variation of i_D with v_{DS} in saturation is ignored in (3.16) and (3.44). As explained in Subsection 3.3.6, this variation is characterized with the output-resistance parameter r_{ds}. Indeed, the significance of r_{ds} in analog design cannot be overemphasized! ▲

3.5 Thermal Effects

MOSFET characteristics are generally temperature dependent. In strong inversion, where drift is the underlying carrier motion, this dependence arises mainly from the temperature sensitivity of the threshold voltage and mobility. In subthreshold operation, in which current flows predominantly by diffusion, additional thermal sensitivities arise. In this section, we discuss the effect of temperature on the design tools we have developed for threshold voltage, drain-source saturation voltage, sheet current, and transconductance efficiency. The plots presented originally in Figs. 3.7 through 3.12 for an NMOS device are extracted for room temperature T = 27°C. They are redrawn in Figs. 3.23(a) to (e) for two additional temperatures T = −40°C and T = 125°C.[5] The reader would benefit from extracting the PMOS counterparts of these plots with input file `pmos2.sp`.

As observed in Fig. 3.23(a), V_{TN} is a decreasing function of temperature. Plots of V_{DSsat}, shown in Fig. 3.23(b), indicate a negligible temperature dependence except for large values of overdrive. In contrast, the variation of sheet current with temperature is quite significant as indicated by Fig. 3.23(c) on linear axis and by Fig. 3.23(d) on logarithmic axis. What is interesting about these plots is that they indicate a positive temperature sensitivity in deep subthreshold but a negative one in strong inversion. The latter is mainly due to the negative temperature sensitivity of the mobility μ_N appearing in (3.16). Although mobility also appears in the subthreshold drain-current expression (3.24), the quadratic dependence on the thermal voltage kT/q overrides its effect in subthreshold, and gives rise to the positive sensitivity observed in these plots. The plots shown in Fig. 3.23(e) indicate no significant temperature sensitivity for transconductance efficiency and bulk-transconductance efficiency in strong inversion. In subthreshold, however, these two become strongly decreasing functions of temperature as predicted by (3.29).

As an interested reader can verify with simulation, the effect of rising temperature on PMOS performance metrics is as follows: V_{TP} becomes less negative, V_{DSsat} becomes more negative in deep subthreshold but less negative in strong inversion, $I_{\square}$ becomes higher in deep subthreshold but lower in strong inversion, g_m/I_D and g_{mb}/I_D become smaller in deep subthreshold and slightly higher in strong inversion.

[5] For a Spice simulation at any temperature other than T = 27°C, just add a line `.options temp=X` to the input file, where X is the desired simulation temperature in degree centigrade.

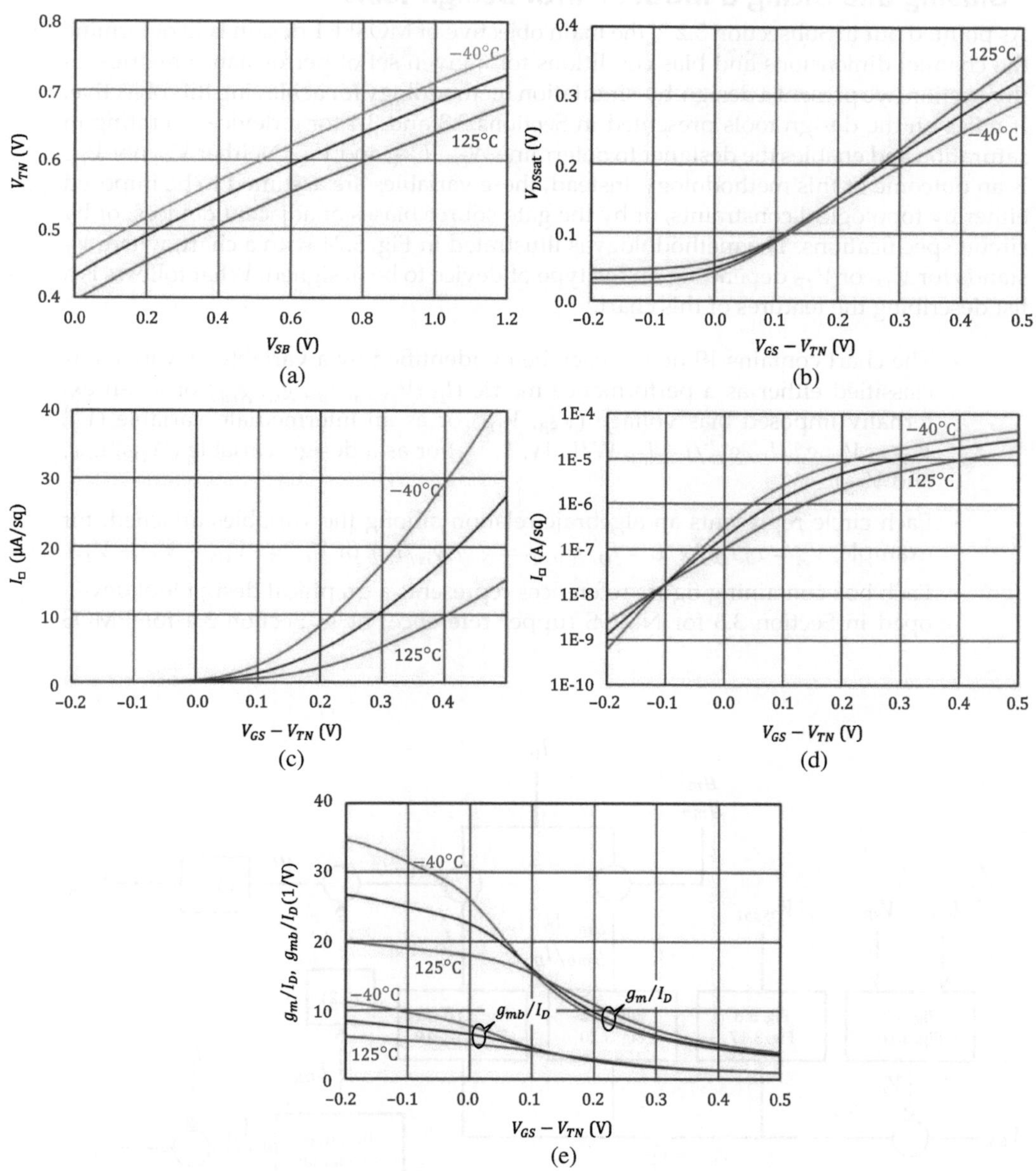

FIGURE 3.23 NMOS design-tool plots for $T = -40°C$, $T = 27°C$, and $T = 125°C$. (a) Threshold voltage for $L_{DR} = 0.8$ μm. (b) Drain-source saturation voltage. (c) Sheet current on a linear axis. (d) Sheet current on a logarithmic axis. (e) Transconductance efficiencies.

3.6 Biasing and Sizing a MOSFET with Design Tools

As pointed out in Subsection 3.2.2, the main objective of MOSFET design is to determine the channel dimensions and bias conditions for a given set of performance metrics. In this section, we present a design-by-simulation methodology for achieving this objective. It relies on the design tools presented in Sections 3.3 and 3.4 for a device operating in saturation and enables the designer to determine W_{DR}, L_{DR}, and V_{GS}. Neither V_{SB} nor V_{DS} is an outcome of this methodology. Instead, these variables are assumed to be imposed either by topological constraints, or by the gate-source biases of adjacent devices, or by circuit specifications. The methodology is illustrated in Fig. 3.24 with a chart, where V_T stands for V_{TN} or V_{TP} depending on the type of device to be designed. What follows is a list describing the features of this chart.

- The chart contains 19 nodes, each being identified by a variable. A variable is classified either as a performance metric (I_D, V_{DSsat}, r_{ds}, g_m, g_{mb}) or as an externally imposed bias voltage (V_{SB}, V_{DS}) or as an intermediate variable (V_T, $V_{GS} - V_T$, g_m/I_D, g_{mb}/I_D, $I_\square$, W/L, W, L, V_A) or as a design variable (W_{DR}, L_{DR}, and V_{GS}).
- Each circle represents an algebraic relation among the variables attached, for example, $V_A = r_{ds}I_D$, $W/L = I_D/I_\square$, $I_D = g_m/(g_m/I_D)$, or $V_{GS} = (V_{GS} - V_T) + V_T$.
- Each box containing figure references represents a graphical design tool developed in Section 3.3 for NMOS (upper reference) or in Section 3.4 for PMOS

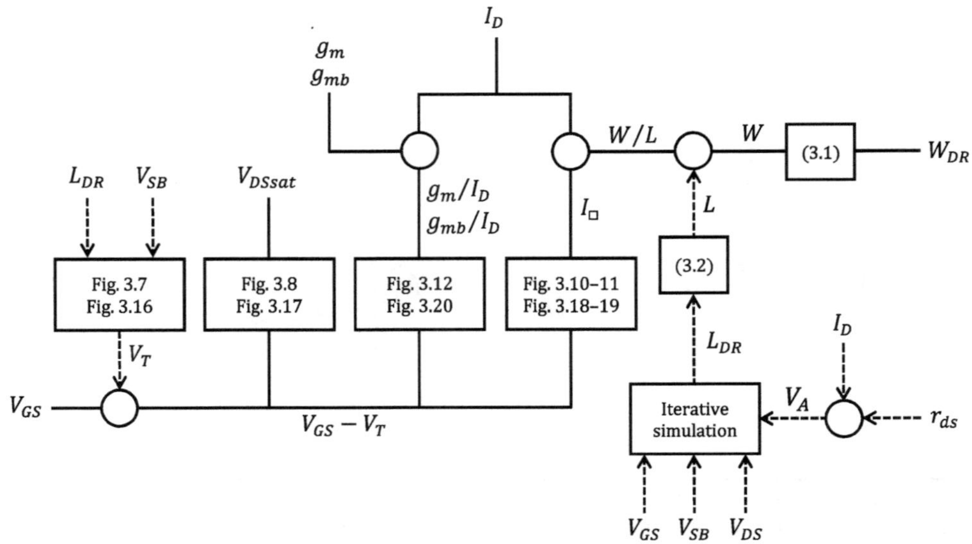

FIGURE 3.24 Flowchart of relations in a saturated MOSFET among device performance metrics (I_D, V_{DSsat}, r_{ds}, g_m, g_{mb}), externally imposed bias voltages (V_{SB}, V_{DS}), intermediate variables (V_T, $V_{GS} - V_T$, g_m/I_D, g_{mb}/I_D, $I_\square$, W/L, W, L, V_A), and design variables (W_{DR}, L_{DR}, and V_{GS}). Circles represent algebraic relations and boxes represent graphical design-tools and equations.

(lower reference). As stated previously, these tools are common to all devices of the same model deck.

- The box marked "iterative simulation" represents the relation between V_A and its variables. As discussed in Subsection 3.3.6, this relation involves multiple variables and cannot be generalized to all devices of the same model deck. An individual simulation is needed for each device whose biasing and sizing involves V_A.
- Boxes marked (3.1) and (3.2) represent the equations numbered accordingly.
- A solid line connecting a node to a circle or to a box defines a bidirectional path. The variable associated with the node may be moved into or out of the circle or box along such a line. For example, you may feed g_m/I_D into the box of Fig. 3.12 or Fig. 3.20, and thus determine the corresponding $V_{GS} - V_T$, or may do the opposite.
- A dashed line connecting a node to a circle or a box is unidirectional. The variable associated with the node should be moved only in the direction indicated by the arrowhead. For example, we may feed L_{DR} and V_{SB} into the box of Fig. 3.7 or Fig. 3.16 for the purpose of determining V_T, but should not attempt to determine V_{SB} or L_{DR} from V_T.

Let us now use the chart to determine the viable paths and steps leading from performance metrics to design variables W_{DR}, L_{DR}, and V_{GS}. As pointed out above, we assume prior knowledge of V_{DS} and V_{SB}. The chart indicates three well-constrained cases in which all three outcomes can be uniquely determined from specified performance metrics. These cases and their design flows are explained next.

1. V_{DSsat}, r_{ds}, and either g_m or g_{mb} are specified. Determine $V_{GS} - V_T$ from the design tool of Fig. 3.8 or Fig. 3.17, and translate the result into g_m/I_D (or g_{mb}/I_D) through the design tool of Fig. 3.12 or Fig. 3.20. Since g_m (or g_{mb}) is available as a specification item, calculate I_D. Next, determine $I_\square$ from the design tool of Figs. 3.10 to 3.11 or Figs. 3.18 to 3.19 and calculate W/L from I_D and $I_\square$. Also calculate V_A from I_D and r_{ds}. It is now possible to determine L_{DR} and V_{GS} by simulation. Set the simulation environment with (a) the specified values of V_{SB} and V_{DS}, (b) approximately $20W_{DR(\min)}$ for W_{DR}, and (c) an arbitrary value for L_{DR} and the corresponding value for V_{GS}. The latter is calculated from $V_{GS} - V_T$ and V_T, where V_T is determined from the design tool of Fig. 3.7 or Fig. 3.16 for the specified value of V_{SB} and the arbitrarily selected value of L_{DR}. Run an operation-point analysis and read the value of V_A. If the latter turns out to be smaller than its target value calculated previously, then, increment L_{DR}, update the corresponding value of V_{GS}, and rerun the simulation. Otherwise, decrement L_{DR}, update V_{GS}, and rerun the simulation. You don't need to readjust W_{DR} in these iterations until the target value of V_A has been reached. Once the final values of L_{DR} and V_{GS} have thus been determined, calculate L from (3.2), and use the outcome together with W/L for determining W. Finally, calculate W_{DR} from (3.1).
2. I_D and r_{ds} and either g_m or g_{mb} are specified. Calculate g_m/I_D (or g_{mb}/I_D) from g_m (g_{mb}) and I_D. Determine firstly $V_{GS} - V_T$ from the design tool of Fig. 3.12 or Fig. 3.20, and subsequently $I_\square$ from the design tool of Figs. 3.10 to 3.11 or

Figs. 3.18 to 3.19. Now, calculate W/L from I_D and $I_{\square}$, and V_A from I_D and r_{ds}. Refer to the first case for iterative simulation and subsequent calculations.

3. I_D and r_{ds} and V_{DSsat} are specified. Determine $V_{GS} - V_T$ from the design tool of Fig. 3.8 or Fig. 3.17, and $I_{\square}$ from the design tool of Figs. 3.10 to 3.11 or Figs. 3.18 to 3.19. Next, calculate W/L from I_D and $I_{\square}$. Refer to the first case for iterative simulation and subsequent calculations.

Also note that specifying all three of I_D, V_{DSsat}, and g_m (or g_{mb}) simultaneously would result in an overconstrained case in which no unique value can be found for $V_{GS} - V_T$. At most two out of these three metrics can be specified without overconstraining the design. There may be underconstrained cases as well. Any such case provides the designer with an opportunity to impose his/her own specification(s) in accordance with good-design principles. Also note that there are many design scenarios in which one or more of W_{DR}, L_{DR}, and V_{GS} are topologically imposed or directly specified rather than being left as free-design variables. You can still use the chart of Fig. 3.24 in all these cases as long as you stick to the rules of use as itemized above. Several examples highlighting the universality of the chart follow next.

Example 3.6 An NMOS device is specified by $V_{SB} = 0$, $V_{DSsat} = 0.2$ V, $g_m = 2.1$ mA/V, and $r_{ds} = 47$ kΩ at $V_{DS} = 0.7$ V. Design and verify.

Work This is an example of the first well-constrained case described above. Design flow is as follows.

- $V_{DSsat} = 0.2 \text{ V} \rightarrow$ Fig. 3.8 $\Rightarrow V_{GS} - V_{TN} = 0.275$ V.
- $V_{GS} - V_{TN} = 0.275 \text{ V} \rightarrow$ Fig. 3.12 $\Rightarrow g_m/I_D = 7 \text{ V}^{-1}$.
- $g_m/I_D = 7 \text{ V}^{-1}$ and $g_m = 2.1 \text{ mA/V} \Rightarrow I_D = 300$ μA.
- $V_{GS} - V_{TN} = 0.275 \text{ V} \rightarrow$ Fig. 3.10 $\Rightarrow I_{\square} = 9$ μA/sq.
- $I_D = 300$ μA and $I_{\square} = 9$ μA/sq $\Rightarrow W/L = 33.3$.
- $I_D = 300$ μA and $r_{ds} = 47$ kΩ $\Rightarrow V_A = 14.1$ V.
- Simulation: Input file is as follows:

```
filename: ex3-6.sp
.incl bsim.sp
m 2 1 0 0 cmosn w=6u l=0.4u
vgs 1 0 0.745
vds 2 0 0.7
.op
.end
```

Note that we arbitrarily set the drawn width to $W_{DR} = 20W_{DR(\text{min})} = 6$ μm in all iterations. The initial value of the channel length is also arbitrarily set to 0.4 μm, for which $V_{TN} = 0.47$ V is indicated by Fig. 3.7. Since $V_{GS} - V_{TN}$ equals 0.275 V, V_{GS} is set to 0.745 V in this initial simulation. After each iteration, V_A is printed with the command

```
print @m[id]/@m[gds]
```

Channel length (μm)	0.4	0.6	0.7	0.8
Gate-source voltage (V)	0.745	0.722	0.715	0.705
Early voltage (V)	9.95	12.99	14.38	15.70

TABLE 3.3 Early Voltage versus Channel Length in Example 3.6

The result is shown in Table 3.3.

Obviously, $L_{DR} = 0.7$ μm satisfies the specified V_A.

- $L_{DR} = 0.7\ \mu\text{m} \rightarrow (3.2) \Rightarrow L = 0.66\ \mu\text{m}$.
- $L = 0.66$ μm and $W/L = 33.3 \Rightarrow W = 22.2$ μm.
- $W = 22.2\ \mu\text{m} \rightarrow (3.1) \Rightarrow W_{DR} = 22.2$ μm. Actually, this step is redundant because it was obvious from the large effective width found in the preceding step that the relative difference between W and W_{DR} would be negligible.
- As always done in sizing a device, we need to verify the outcome of the design. We now rerun the input file after redefining the MOSFET and vgs element lines as follows:

```
m 2 1 0 0 cmosn w=22.2u l=0.7u
vgs 1 0 0.715
```

Verification results are as follows: $V_{DSsat} = 0.197$ V, $g_m = 2.06$ mA/V, $r_{ds} = 48.6$ kΩ, and $I_D = 294$ μA. ▲

Example 3.7 The circuit schematic shown in Fig. 3.25 belongs to a current sink, which is how we name an independent-current source if it directs the current into V_{SS} or ground. It remains functional as long as the MOSFET operates in saturation. The minimum output voltage that still satisfies this condition is called compliance limit, and is described by

$$V_{O(\min)} = V_{DSsat} + R_S I_D.$$

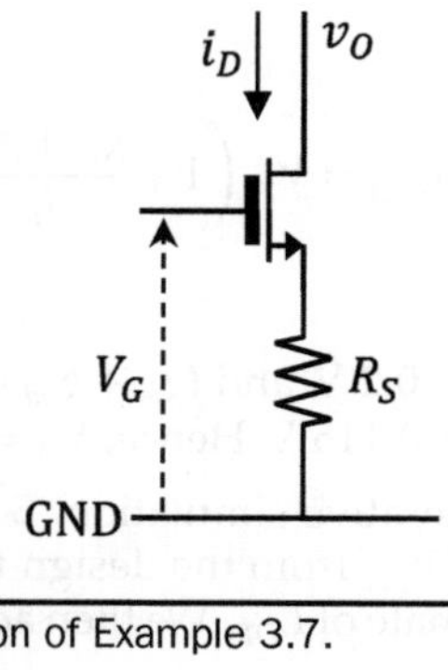

FIGURE 3.25 Current-sink application of Example 3.7.

Output resistance is yet another performance metric of a current sink. For the present sink, it is described by

$$r_o \doteq \left(\frac{\partial i_D}{\partial v_O}\right)^{-1},$$

which is related to MOSFET and resistor parameters through

$$r_o = r_{ds} + (g_m + g_{mb})r_{ds}R_S$$

as the reader will appreciate after studying Section 3.7.

Assume $R_S = 20$ kΩ, and consider an application specified by a compliance-limit $V_{O(\mathrm{min})} = 0.3$ V in addition to a nominal current $I_D = 10$ μA and an output resistance $r_o = 5$ MΩ at $V_O = 0.9$ V. Size the device and determine the necessary gate bias V_G with respect to ground.

Work Design flow is as follows.

- For the specified values of $V_{O(\mathrm{min})}$, R_S, and I_D, the expression presented above for the compliance-limit yields $V_{DSsat} = 100$ mV, for which, Fig. 3.8 indicates $V_{GS} - V_{TN} = 0.115$ V.
- $V_{GS} - V_{TN} = 0.115$ V → Fig. 3.10 ⇒ $I_\square = 1.75$ μA/sq.
- $I_D = 10$ μA and $I_\square = 1.75$ μA/sq ⇒ $W/L = 5.71$.
- Multiply both sides of the output-resistance expression by I_D:

$$r_o I_D = r_{ds} I_D + \frac{g_m + g_{mb}}{I_D} r_{ds} I_D \times R_S I_D.$$

Since $r_{ds}I_D$ equals the Early voltage V_A, and $R_S I_D$ equals the source-bulk voltage V_{SB}, this equation can be rewritten as follows:

$$r_o I_D = V_A\left(1 + \frac{g_m + g_{mb}}{I_D} V_{SB}\right),$$

where $r_o I_D = 50$ V, $V_{SB} = 0.2$ V, and $(g_m + g_{mb})/I_D = 19$ V^{-1} as determined from Fig. 3.12 for $V_{GS} - V_{TN} = 0.115$ V. Hence, $V_A = 10.4$ V.

- We are now ready for iterative simulation. Before running a simulation, however, we first determine V_{TN} from the design tool of Fig. 3.7 for (a) $V_{SB} = 0.2$ V, and (b) an arbitrary estimate of L_{DR}. We then add this V_{TN} to $V_{GS} - V_{TN} = 0.115$ V to obtain V_{GS}. Finally, adding $V_{SB} = 0.2$ V to V_{GS} will yield the voltage V_G to be applied between gate and ground. Input file is as follows:

```
filename: ex3-7a.sp
.incl bsim.sp
m 2 1 3 0 cmosn w=6u l=0.4u
vd 2 0 0.9
vg 1 0 0.838
vs 3 0 0.2
.op
.end
```

Iterative simulation eventually indicates an Early voltage of $V_A = 10.65$ V for $L_{DR} = 0.8$ µm.

- $L_{DR} = 0.8$ µm → (3.2) ⇒ $L = 0.76$ µm.
- $L = 0.76$ µm and $W/L = 5.71$ ⇒ $W = 4.37$ µm ⇒ $W_{DR} = 4.4$ µm.
- Now that design is complete, we need to verify the specifications by simulating the entire circuit. Input file is as follows:

```
filename: ex3-7b.sp
.incl bsim.sp
m 2 1 3 0 cmosn w=4.4u l=0.8u
rs 3 0 20k
vo 2 0 0.9
vg 1 0 0.805
.tf v(2) vo
.end
```

Note that the verification of r_o is performed with a transfer-function analysis where `vo` is the input source, and `v(2)` is the output variable. This type of analysis yields the following three pieces of information: (a) Small-signal transfer function from input source to output variable, which amounts to 1 in our case because `vo` and `v(2)` are identical; (b) small-signal resistance loading the output port (called `output_impedance` in Spice terminology), which amounts to zero in our case because the independent voltage source `vo` shunts the output port of node 2 and ground; and (c) the small-signal resistance seen by the input source `vo` (called `input_impedance` in Spice terminology), which is exactly what the output resistance r_o of our current-source circuit is. For our present purposes, only this final piece of information of the transfer-function analysis is meaningful. Entering the command

```
print vo#input_impedance
```

yields $r_o = 5.02$ MΩ. Since Spice automatically runs an operation-point analysis prior to the transfer-function analysis, we also read $I_D = 10.1$ µA and $V_{DSsat} =$ 103 mV simply by entering the command `show`.

The reader may have noticed that this example belongs to the third well-constrained case of design discussed prior to Example 3.6. ▲

Example 3.8 A PMOS device is specified by $V_{SB} = 0$ and $g_m = 160$ µA/V in addition to $I_D = 16$ µA and $r_{ds} = 2$ MΩ at $V_{DS} = -0.9$ V. Design and verify.

Work This is an example of the second well-constrained case discussed prior to Example 3.6. Design flow is as follows.

- $I_D = 16$ µA and $g_m = 160$ µA/V ⇒ $g_m/I_D = 10\ \text{V}^{-1}$.
- $g_m/I_D = 10\ \text{V}^{-1}$ → Fig. 3.20 ⇒ $V_{GS} - V_{TP} = -0.185$ V.
- $V_{GS} - V_{TP} = -0.185$ V → Fig. 3.18 ⇒ $I_{\square} = 1$ µA/sq.
- $I_D = 16$ µA and $I_{\square} = 1$ µA/sq ⇒ $W/L = 16$.
- $I_D = 16$ µA and $r_{ds} = 2$ MΩ ⇒ $V_A = 32$ V.
- Simulation: Input file is as follows:

```
filename: ex3-8.sp
.incl bsim.sp
m 2 1 0 0 cmosp w=6u l=5.5u
vds 2 0 -0.9
vgs 1 0 -0.575
.op
.end
```

  $L_{DR} = 5.5$ µm satisfies the specification with $V_A = 32.39$ V.
- $L_{DR} = 5.5$ µm → (3.2) ⇒ $L = 5.44$ µm.
- $L = 5.44$ µm and $W/L = 16$ ⇒ $W = 87.1$ µm.
- $W = 87.1$ µm → (3.1) ⇒ $W_{DR} = 87.1$ µm.

Verification will indicate $g_m = 160$ µA/V, $I_D = 16.23$ µA, and $r_{ds} = 2.00$ MΩ. ▲

Example 3.9 An NMOS device is specified by $I_D = 150$ µA and $g_m = 1.1$ mA/V. Gate area WL should not exceed 65 µm^2. Design for the largest possible r_{ds}.

Work Design flow is as follows:

- $I_D = 150$ µA and $g_m = 1.1$ mA/V ⇒ $g_m/I_D = 7.33\ \text{V}^{-1}$.
- $g_m/I_D = 7.33\ \text{V}^{-1}$ → Fig. 3.12 ⇒ $V_{GS} - V_{TN} = 0.26$ V.
- $V_{GS} - V_{TN} = 0.26$ V → Fig. 3.10 ⇒ $I_{\square} = 7$ µA/sq.
- $I_D = 150$ µA and $I_{\square} = 7$ µA/sq ⇒ $W/L = 21.4$.
- Since the largest possible r_{ds} is desired, L must be maximized. Therefore adopt the largest permissible gate area. Solve $WL = 65$ µm^2 together with $W/L = 21.4$. ⇒ $W = 37.3$ µm, $L = 1.74$ µm. ▲

3.7 Small-Signal Modeling and Circuit Analysis

In Section 3.6, we developed a design-by-simulation technique for sizing and biasing a MOSFET for a given set of device performance metrics. Three of these performance metrics, g_m, g_{mb}, and r_{ds} are defined in terms of variations in port variables. The relationship between these three and open-loop performance metrics are derived with a *small-signal analysis* of the opamp. MOSFET modeling for dc small-signal analysis is the first subject of this section. Based on this model, we will develop a technique by which we can determine the dc small-signal behavior of an analog circuit by visual inspection of the electrical schematic only. Finally, we will discuss the parasitic capacitances of a MOSFET, and develop a high-frequency small-signal device model based on these parasitics.

3.7.1 MOSFET DC Small-Signal Model

MOSFET drain current i_D depends nonlinearly on port voltages v_{GS}, v_{DS}, and v_{SB}. We can represent the relation generally with

$$i_D = f(v_{GS}, v_{DS}, v_{SB}). \tag{3.53}$$

Now, suppose that these three port voltages vary around their respective bias voltages V_{GS}, V_{DS}, and V_{SB}, and variations are represented by v_{gs}, v_{ds}, and v_{sb}. We can expand (3.53) into Taylor series around the bias voltages and find the corresponding variation i_d in the drain current i_D around its bias value I_D. If we adopt small-signal approximation, expansion will involve only the first-order terms. Hence,

$$i_d \approx \left.\frac{\partial i_D}{\partial v_{GS}}\right|_{v_{GS}=V_{GS}} v_{gs} + \left.\frac{\partial i_D}{\partial v_{DS}}\right|_{v_{DS}=V_{DS}} v_{ds} + \left.\frac{\partial i_D}{\partial v_{SB}}\right|_{v_{SB}=V_{SB}} v_{sb}. \tag{3.54}$$

According to the definitions (3.5), (3.6), and (3.7), we can rewrite (3.54) as

$$i_d = g_m v_{gs} + \frac{1}{r_{ds}} v_{ds} - g_{mb} v_{sb} \tag{3.55}$$

for an NMOS device, and $i_d = -g_m v_{gs} - (1/r_{ds}) v_{ds} + g_{mb} v_{sb}$ for a PMOS. Recall that these equations comply with the reference convention set for the MOSFET variables in Fig. 3.2; that is, i_D is positive if it flows into the drain in NMOS, or out of drain in PMOS. Now on, we transpose the reference direction of i_d for PMOS, so that it will be positive when it flows into the drain exactly like the case of NMOS. This enables us to use (3.55) to describe the small-signal drain current of both types of devices. Do not forget that we transpose the reference direction in PMOS only for i_d and not for i_D. In other words, we adopt $i_D = I_D - i_d$ for PMOS while retaining $i_D = I_D + i_d$ for NMOS. Another fact to bear in mind is that g_m, r_{ds}, and g_{mb} are characterized for the particular bias voltages around which the device operates.

Equation (3.55) relates the dc small-signal drain current to dc small-signal port voltages. It is accurate only for small dc signals. In dc small-signal analysis of a circuit, we can represent a MOSFET with a *small-signal equivalent circuit* based on this equation. The equivalent circuit is shown in Fig. 3.26.

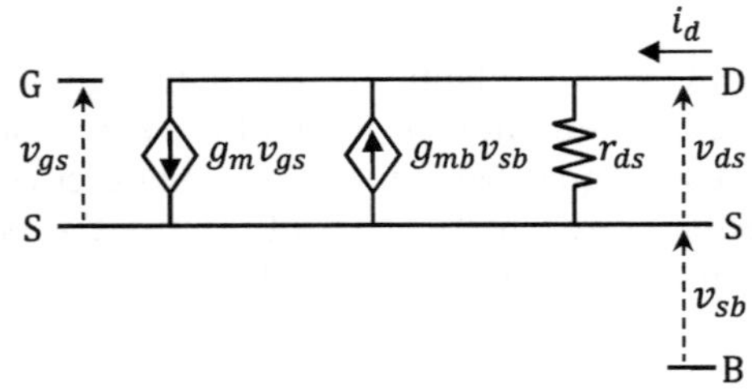

FIGURE 3.26 MOSFET dc small-signal equivalent circuit.

The reason why we qualify the equivalent circuit of Fig. 3.26 for dc excitation only is that the capacitive effects associated with MOSFET operation are not taken into account in developing (3.55). Nevertheless, we rely on this equivalent circuit also in ac analyses as long as the frequency of operation is sufficiently low to make capacitive currents negligible. It is therefore more descriptive to name it as *dc and low-frequency equivalent circuit*.

3.7.2 DC Small-Signal Circuit Analysis

Formally, the dc small-signal analysis of an analog circuit is performed on an equivalent circuit obtained by applying the following procedure:

Step 1 Replace each capacitance with an open circuit since a capacitive impedance is infinite for dc.

Step 2 Replace each inductance with a short circuit since an inductive impedance is zero for dc.

Step 3 Replace each dc bias or supply source with an open circuit in the case of a current source, and by a short circuit in the case of a voltage source.

Step 4 Replace each MOSFET with the equivalent circuit given in Fig. 3.26.

Let us now illustrate the application of this formal procedure with two simple examples.

Example 3.10 Formulate the dc voltage gain of the amplifier of Example 3.5.

Work The amplifier schematic given in Fig. 3.22(a) is translated into a dc small-signal equivalent circuit in Fig. 3.27(a) where gray-shaded boxes represent the equivalent circuits of M1 and M2. Referring also to Fig. 3.26, notice the following features of the equivalent circuit:

- Both MOSFETs operate with a source/bulk short, which implies $v_{sb} = 0$ for both devices. The dependent current-source $g_{mb}v_{sb}$ thus becoming an open circuit is removed from the equivalent circuits of both MOSFETs.
- Short-circuiting the dc voltage sources V_{DD} and V_B implies grounded source and gate for M2. Since v_{gs} is therefore zero for M2, the dependent current-source $g_m v_{gs}$ is also replaced by an open circuit in M2.

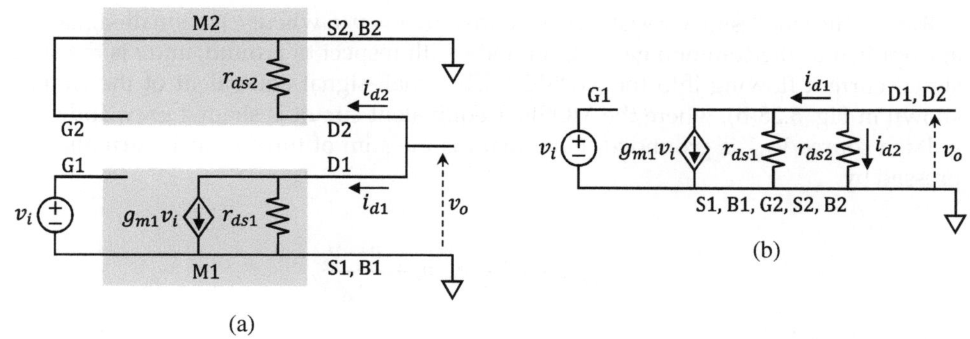

FIGURE 3.27 (a) Detailed dc small-signal equivalent circuit of the common-source amplifier stage of Example 3.10. (b) Usual depiction of the equivalent circuit.

- Input signal v_i is applied between the gate and source of M1. Therefore, the dependent current-source $g_m v_{gs}$ equals $g_{m1} v_i$ in M1.
- Output signal v_o is defined between the common drain of both devices and ground.
- The reference direction of drain signal currents i_{d1} and i_{d2} is directed from drain toward the inside of both devices. Notice $i_{d2} = -i_{d1}$.

Figure 3.27(b) depicts a more usual form of the equivalent circuit, which indicates

$$v_o = -g_{m1}\left(r_{ds1} \parallel r_{ds2}\right) v_i.$$

Therefore, voltage gain is formulated as

$$A_v \doteq \frac{v_o}{v_i} = -g_{m1}\left(r_{ds1} \parallel r_{ds2}\right) = -g_{m1}\frac{r_{ds1} r_{ds2}}{r_{ds1} + r_{ds2}}. \quad \blacktriangle \qquad (3.56)$$

Example 3.11 Shown in Fig. 3.28(a) is a diode-connected PMOS device operating with a resistor R_S between its source and power-supply rail V_{DD}. Determine the small-signal resistance r seen between the common gate/drain node and signal ground.

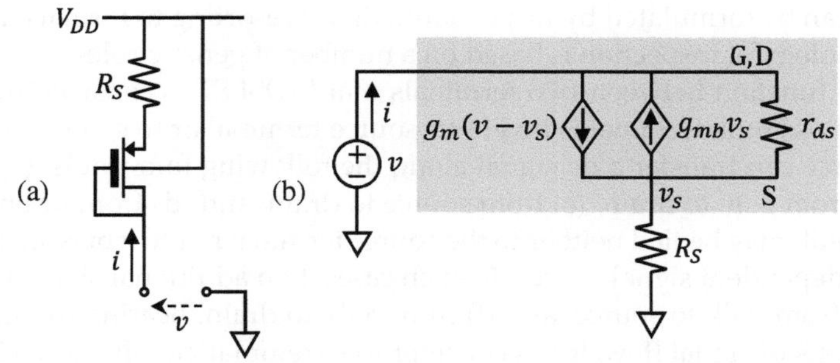

FIGURE 3.28 (a) A diode-connected PMOS with a resistor at its source. (b) Small-signal equivalent circuit for determining the resistance seen into the common gate/drain node.

Work The small-signal resistance is defined by $r \doteq v/i$, where v is a small-signal voltage applied to the common gate/drain node with respect to ground, and i is the small-signal current flowing into the MOSFET. The small-signal equivalent of the circuit is shown in Fig. 3.28(b), where the MOSFET equivalent circuit is shaded gray, and the dc voltage source V_{DD} is grounded. Note that i is the sum of three branch currents as expressed by

$$i = g_m(v - v_s) - g_{mb}v_s + \frac{v - v_s}{r_{ds}},$$

and v_s is related to i by $v_s = iR_s$. Solving for r, we obtain

$$r = \frac{r_{ds}}{1 + r_{ds}g_m}\left(1 + g_{mb}R_S\right) + R_S.$$

Diode-connected MOSFET finds widespread use in analog circuits. In most cases, it does not contain a source resistor; instead, its source is tied directly to a power-supply rail or ground. In such cases, the expression of small-signal resistance is reduced to

$$r = \frac{r_{ds}}{1 + r_{ds}g_m}.$$

$r_{ds}g_m$ is usually much larger than unity, which enables us to simplify this equation into

$$r = \frac{1}{g_m}. \quad \blacktriangle$$

In Examples 3.10 and 3.11, we determined the required expressions rapidly and painlessly because the circuit topologies involved were very simple. In more complex topologies, the formal method of circuit analysis with MOSFET equivalent circuits may prove highly time consuming and error prone. Fortunately, the dc small-signal behavior of a circuit can be formulated by inspection without resorting to formal circuit analysis. The methodology of inspection is based on a number of generic rules by which the voltage transfer function between two terminals of a MOSFET is constructed. Note that a MOSFET whose bulk terminal is tied to its source terminal or to ground or to a dc bias-voltage source can transfer a dc signal along the following four paths: (a) from gate to source, (b) from gate to drain, (c) from source to drain, and (d) from drain to source. In rare cases, bulk may be tied neither to the source terminal nor to a bias source or ground but to an independent signal source. In such cases, two additional signal flow paths are created: (e) from bulk to source, and (f) from bulk to drain. Bearing in mind these possible directions of signal flow, let us consider the exemplar circuit shown in Fig. 3.29(a) for a first taste of analysis by inspection. This is a split-supply *differential amplifier* operating between a positive rail V_{DD} and a negative rail V_{SS}, and receiving its input signal v_i at the gate of M1 with respect to ground. All NMOS devices have their bulks at V_{SS},

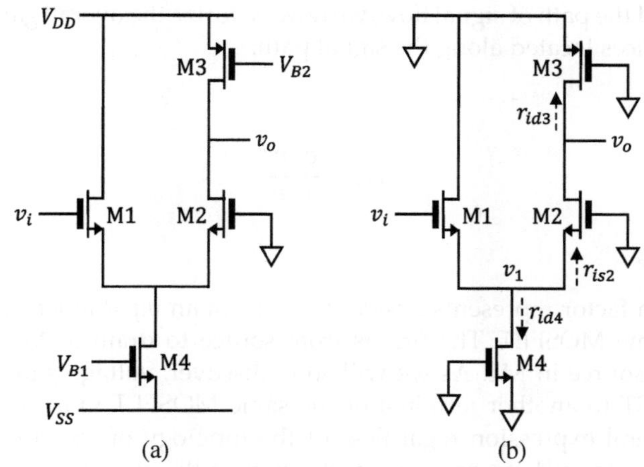

FIGURE 3.29 (a) A differential amplifier stage. (b) Equivalent circuit for analysis by inspection.

whereas the bulk of M3 is at V_{DD}. Output signal v_o is defined between the drain of M2 and ground. Suppose that our objective is to formulate the voltage gain

$$A_v \doteq \frac{v_o}{v_i}.$$

Formally, we would sketch an equivalent circuit applying all four steps of the procedure described at the beginning of this subsection. In analysis by inspection, we apply the first three only. We do not replace devices with equivalent circuits; instead, we retain their symbols in the equivalent circuit. In the present example, the first two steps are redundant anyway because no capacitance or inductance is present in the exemplar circuit. Step 3 is applicable to supply sources V_{DD} and V_{SS} as well as bias sources V_{B1} and V_{B2}. After replacing these with short circuits, we arrive at the equivalent circuit shown in Fig. 3.29(b). Next, starting at the gate of M1, we identify the signal-flow path from input to output on this equivalent circuit. Signal generally can flow from gate to drain and source but, in the case of M1, it cannot reach the output through the drain because the latter is grounded. This is why we consider only its flow to the source, where its voltage is denoted by v_1. M2 receives this signal at its own source terminal, and passes it to its drain as the output signal v_o. Notice in the present example that the signal flows from input to output through a single forward path without looping back. In practice we may come across circuits containing multiple forward paths and/or feedback loops. In such cases, we apply superposition and/or loop-cutting techniques to transform the equivalent circuit into multiple circuits each comprising a single forward path. How we do these transformations will be explained and exemplified later. Let us continue with our present example.

Having traced the path of signal flow, we now factorize the overall gain into individual gains of the devices located along the signal path:

$$A_v = \frac{v_o}{v_1}\frac{v_1}{v_i}.$$

Notice that each factor represents a *voltage gain* from an *input terminal* to an *output terminal* of the same MOSFET. The first is from source to drain in M2, and the second is from gate to source in M1. As we will soon discover, voltage gain from one terminal of a MOSFET to another terminal of the same MOSFET in a single forward path has a very general expression regardless of the topology of the circuit in which this MOSFET resides. In such an expression, the part of the circuit connecting the output terminal of the MOSFET is represented by an equivalent resistance between this terminal and circuit ground. As an example, consider M1 whose output terminal is the source. You can visualize the effective resistance connecting its source by inspecting Fig. 3.29(b). Obviously it is the parallel equivalent of two resistances: (a) r_{id4}, which is the drain input resistance of M4, and (b) r_{is2}, which is the source input resistance of M2. Moving to M2, whose output terminal on the signal path is drain, we observe the drain input resistance r_{id3} of M3 as the resistance loading its output. Fortunately, input resistances seen into the drain and source terminals of a MOSFET are also described by two very general expressions as long as the device resides on a single forward path.

We now derive generic voltage-gain and input-resistance expressions of analysis-by-inspection for four possible cases of signal entry through source, drain, gate, or bulk. Although each case will be illustrated with an NMOS device, the results will be perfectly valid also for a PMOS device thanks to the unified small-signal equivalent circuit presented in Subsection 3.7.1.

Source Drive

First consider a MOSFET whose bulk is tied to a supply or bias voltage or to ground, and suppose that it receives a dc or low-frequency small-signal voltage v_s at its source terminal with respect to circuit ground as shown in the small-signal equivalent circuit of Fig. 3.30(a). Note that the bulk terminal is shown grounded regardless of whether it is physically grounded or tied to a supply or bias source in the actual circuit. This is because all dc voltage sources are short-circuited in a small-signal analysis. Also notice the grounding of the gate terminal. Although gate is physically connected to some node of the circuit in which this MOSFET operates, it does not support a signal because (a) in a circuit of single forward signal path, only one of the MOSFET terminals can be an input terminal, which is assumed to be the source terminal in the present case, and (b) as explained previously, a dc or low-frequency signal applied to source cannot flow to gate. Finally, notice that the external circuit loading the drain terminal is represented by an equivalent resistance R_D placed between drain and ground. The output signal at the drain terminal, represented by v_d with respect to ground, is therefore the signal voltage across R_D.

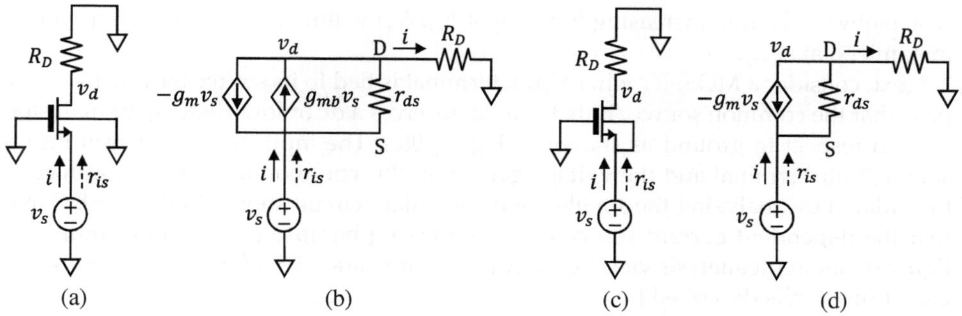

FIGURE 3.30 (a) A MOSFET operating without source-bulk short and receiving a small-signal input voltage at its source terminal. (b) Small-signal equivalent of the circuit shown in (a). (c) A MOSFET operating with source-bulk short and receiving a small-signal input signal at its source terminal. (d) Small-signal equivalent of the circuit shown in (c).

By replacing the MOSFET with its own equivalent circuit, we turn the circuit of Fig. 3.30(a) into what is shown in Fig. 3.30(b), where

$$i = (g_m + g_{mb})v_s + \frac{1}{r_{ds}}(v_s - v_d),$$

and

$$v_d = iR_D.$$

Using these two equations, the MOSFET input resistance r_{is} seen by the signal source between the source terminal and ground is formulated as

$$r_{is} \doteq \frac{v_s}{i} = \frac{r_{ds} + R_D}{A_{max}}, \tag{3.57}$$

where

$$A_{max} = 1 + r_{ds}(g_m + g_{mb}) \approx r_{ds}(g_m + g_{mb}) \tag{3.58}$$

is the maximum voltage gain offered by a MOSFET, as will be proven soon. The approximation made in (3.58) is due to the fact that the product $r_{ds}(g_m + g_{mb})$ is typically around 100.

In order to formulate the gain from source to drain, we replace i in $r_{is} \doteq v_s/i$ with $i = v_d/R_D$, and therefore obtain

$$\frac{v_d}{v_s} = \frac{R_D}{r_{is}}. \tag{3.59}$$

We can now prove that A_{max} is the maximum gain of a MOSFET. Just substitute (3.57) into (3.59) to obtain

$$\frac{v_d}{v_s} = \frac{R_D}{r_{ds} + R_D} A_{max},$$

which obviously is an increasing function of R_D. At the limit $R_D \to \infty$, the gain is indeed maximized at $A_{\max}$.

Next, consider a MOSFET whose bulk terminal is tied to its source terminal, and suppose that the common source/bulk terminal receives a dc or low-frequency small signal v_s with respect to ground as shown in Fig. 3.30(c). The input resistance seen into the source/bulk terminal and the voltage gain from this common terminal to drain can be formulated by analyzing the small-signal equivalent circuit shown in Fig. 3.30(d). Note that the dependent current source $g_{mb}v_{sb}$ is missing because of the source-bulk short. Equivalent-circuit analysis yields exactly (3.57) for r_{is} and (3.59) for v_d/v_s. The maximum gain, however, is described by

$$A_{\max} = 1 + r_{ds}g_m \approx r_{ds}g_m. \tag{3.60}$$

Drain Drive

Now we apply a dc small-signal voltage v_d to the drain of a MOSFET whose bulk is either grounded or tied to a supply or bias source. As shown in Fig. 3.31(a), the signal can flow only to source, which is assumed to be loaded by an external circuit of an equivalent resistance R_S. The corresponding small-signal equivalent circuit is shown in Fig. 3.22(b). Its analysis yields

$$r_{id} = r_{ds} + A_{\max}R_S \tag{3.61}$$

for drain input resistance. Note that $A_{\max}$ is still described by (3.58) in this case. The gain from drain to source is described by

$$\frac{v_s}{v_d} = \frac{R_S}{r_{id}}. \tag{3.62}$$

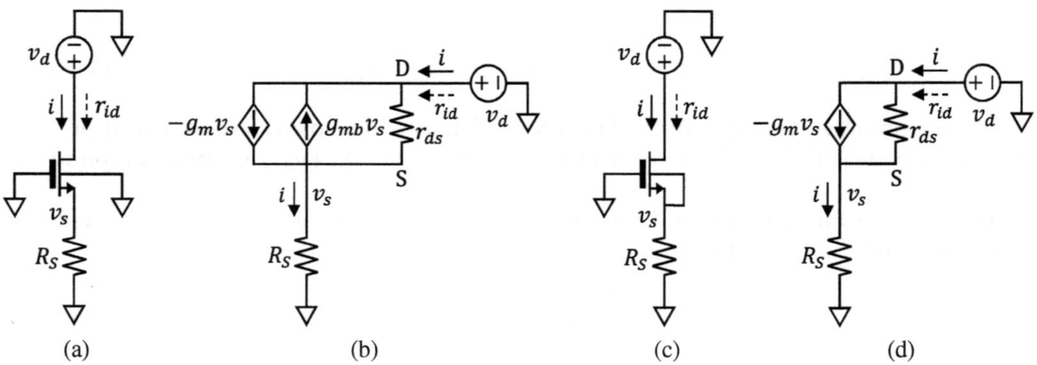

FIGURE 3.31 (a) A MOSFET operating without source-bulk short and receiving a small-signal input voltage at its drain terminal. (b) Small-signal equivalent of the circuit shown in (a). (c) A MOSFET operating with source-bulk short and receiving a small-signal input signal at its drain terminal. (d) Small-signal equivalent of the circuit shown in (c).

If the MOSFET receiving an input signal at the drain terminal operates with a source-bulk short as depicted in Fig. 3.31(c), then, small-signal analysis will be based on the equivalent circuit shown in Fig. 3.31(d). Analysis again yields (3.61) and (3.62) for drain input resistance and gain, respectively, but $A_{\max}$ is described by (3.60) in this case.

Gate Drive

Shown in Fig. 3.32(a) is a MOSFET whose bulk is either gounded or tied to a supply or bias source, and gate receives a small-signal input voltage v_g. Since a signal applied to the gate terminal can flow to drain and source, both of the latter two can be designated as output terminals. Let R_D and R_S represent the equivalent resistance of the external circuit loading the drain and source, respectively. The corresponding equivalent circuit is given in Fig. 3.32(b). Note that the input resistance seen by the signal source is infinite. Analyzing the circuit and doing some algebra, we arrive at the following generic gain expressions for signal flow from gate to drain, and from gate to source:

$$\frac{v_d}{v_g} = -\frac{r_{ds}g_m}{1 + r_{ds}(g_m + g_{mb})}\frac{R_D}{R_S + r_{is}} \approx -\frac{g_m}{g_m + g_{mb}}\frac{R_D}{R_S + r_{is}}, \tag{3.63}$$

$$\frac{v_s}{v_g} = \frac{r_{ds}g_m}{1 + r_{ds}(g_m + g_{mb})}\frac{R_S}{R_S + r_{is}} \approx \frac{g_m}{g_m + g_{mb}}\frac{R_S}{R_S + r_{is}}, \tag{3.64}$$

where approximations are again due to the fact that $r_{ds}(g_m + g_{mb})$ is much larger than unity.

The case of a MOSFET operating with a source-bulk short and receiving an input signal v_g at its gate is shown in Fig. 3.32(c). The corresponding small-signal equivalent circuit is given in Fig. 3.32(d). An analysis of the latter yields the following generic gain expressions:

$$\frac{v_d}{v_g} = -\frac{r_{ds}g_m}{1 + r_{ds}g_m}\frac{R_D}{R_S + r_{is}} \approx -\frac{R_D}{R_S + r_{is}}, \tag{3.65}$$

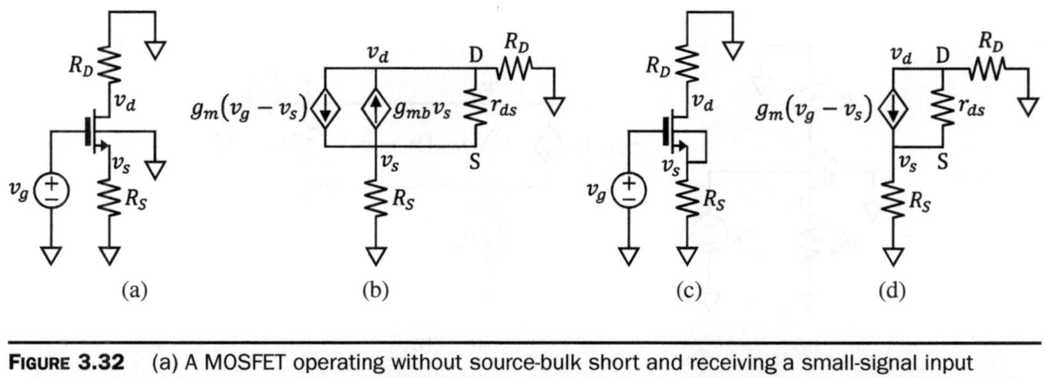

FIGURE 3.32 (a) A MOSFET operating without source-bulk short and receiving a small-signal input voltage at its gate terminal. (b) Small-signal equivalent of the circuit shown in (a). (c) A MOSFET operating with source-bulk short and receiving a small-signal input signal at its gate terminal. (d) Small-signal equivalent of the circuit shown in (c).

and

$$\frac{v_s}{v_g} = \frac{r_{ds}g_m}{1 + r_{ds}g_m} \frac{R_S}{R_S + r_{is}} \approx \frac{R_S}{R_S + r_{is}}. \tag{3.66}$$

Bulk Drive

As mentioned previously, in some rare cases, a MOSFET may receive an input signal only at its bulk terminal as shown in Fig. 3.33(a). This signal can flow to source or drain. The equivalent circuit appropriate for this case is shown in Fig. 3.33(b). Its analysis yields the following gain expressions:

$$\frac{v_d}{v_b} = -\frac{r_{ds}g_{mb}}{1 + r_{ds}(g_m + g_{mb})} \frac{R_D}{R_S + r_{is}} \approx -\frac{g_{mb}}{g_m + g_{mb}} \frac{R_D}{R_S + r_{is}}, \tag{3.67}$$

$$\frac{v_s}{v_b} = \frac{r_{ds}g_{mb}}{1 + r_{ds}(g_m + g_{mb})} \frac{R_S}{R_S + r_{is}} \approx \frac{g_{mb}}{g_m + g_{mb}} \frac{R_S}{R_S + r_{is}}, \tag{3.68}$$

where approximations are again due to the fact that $r_{ds}(g_m + g_{mb})$ is much larger than unity. Notice the similarity between these two expressions and (3.63) and (3.64) of gate drive. The only difference is the transposed roles of g_m and g_{mb}.

Rules of Analysis by Inspection

Analysis by inspection is based on two input-resistance expressions for r_{is} and r_{id}, and six gain expressions for v_d/v_s, v_s/v_d, v_d/v_g, v_s/v_g, v_d/v_b, and v_s/v_b. These expressions can be interpreted in terms of a few common rules which are much easier to remember than the expressions themselves. We now present these rules.

1. *Input Resistances* The four resistances seen when viewed from outside or inside a MOSFET through source or drain terminals are illustrated in Fig. 3.34. The main rule to remember when determining these resistances by inspection is that the source terminal magnifies a resistance by A_{max} when the resistance is viewed from inside the device, and demagnifies it by the same factor when the

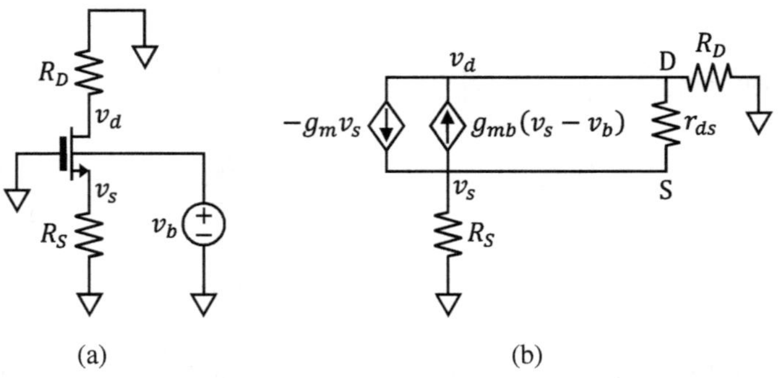

FIGURE 3.33 (a) A MOSFET operating without source-bulk short and receiving a small-signal input voltage at its bulk terminal. (b) Small-signal equivalent of the circuit of (a).

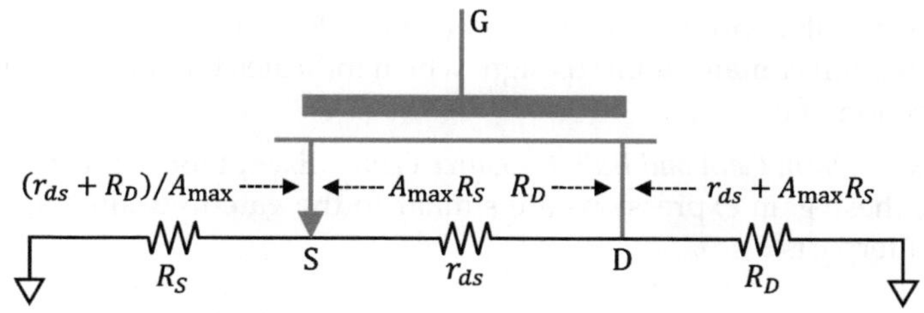

FIGURE 3.34 Resistance magnification and demagnification at the source terminal of a MOSFET.

resistance is viewed from outside. The drain terminal has no such property; it does not magnify or demagnify the value of a resistance regardless of the direction in which you view the resistance. Let us apply these rules firstly for constructing the expression of r_{id}. Looking into the device from outside the drain, we immediately see r_{ds} between drain and source without any magnification or demagnification. There exists also an external resistance R_S lying in series with r_{ds} between the source and ground but since we are viewing it from inside out through the source terminal, what we see is $A_{max}R_S$. Since the two resistances are in series, the equivalent resistance equals their sum, hence (3.61). If the source terminal is grounded ($R_S = 0$), we obviously will see only r_{ds} for r_{id}.

Let us now apply the rule to r_{is}. Since drain does not modify the view of a resistance, the total resistance seen between the inside of the source terminal and ground through drain equals $r_{ds} + R_D$. However, r_{is} is the resistance seen from outside the source. Since source demagnifies by A_{max} when viewed from outside, we indeed end up with (3.57) for r_{is}. Also note that if the drain is held at signal ground ($R_D = 0$), r_{is} will consist only of r_{ds}/A_{max}, which equals $1/(g_m+g_{mb})$ or $1/g_m$ depending on whether the bulk is at signal ground or tied to the source, respectively.

2. *Maximum Gain* Maximum gain A_{max} equals the product $r_{ds}(g_m+g_{mb})$ in the case of a bulk tied to ground or to a dc voltage source. In a device operating with a source-bulk short, maximum gain equals $r_{ds}g_m$ only.

3. *Source-to-Drain Gain and Drain-to-Source Gain* These gain expressions are constructed by dividing the external resistance between the output terminal and ground (R_D for drain output and R_S for source output) by input resistance; hence, (3.59) for source-to-drain gain and (3.62) for drain-to-source gain.

   Considering (3.61) and (3.62) together, we conclude that the gain v_s/v_d from drain to source equals $1/A_{max}$ for $R_S \to \infty$ and vanishes for $R_S = 0$. This is why we never use this signal path when we intend to obtain voltage gain from a MOSFET.

4. *Gate-to-Drain Gain and Gate-to-Source Gain* As indicated by (3.63) and (3.64), each of these two gain expressions is generally a product of two factors. One of these factors is the ratio of the external resistance between the output terminal and ground (R_D for drain output and R_S for source output) to the total resistance $R_S + r_{is}$ associated with the source terminal. The second factor is the ratio $g_m/(g_m+g_{mb})$, which equals unity in the case of a device operating with a

source-bulk short as in (3.65) and (3.66). Also note that the gate-to-drain gain expression contains a minus sign, which indicates that v_d varies in the opposite direction of v_g.

5. *Bulk-to-Drain Gain and Bulk-to-Source Gain* Except for the transposal of g_m and g_{mb}, these gain expressions are similar to the gate-to-drain and gate-to-source counterparts.

Example 3.12 Determine by inspection the small-signal voltage gain v_o/v_i of the common-source amplifier stage shown in Fig. 3.22(a), and compare the result with (3.56).

Work Gain v_o/v_i is from gate to drain in M1. Also note that none of the devices has an external source resistance R_S, while both operate with source-bulk shorts. After grounding V_{DD} and V_B, we obtain the equivalent circuit shown in Fig. 3.35.

Adopting gate-drive expressions, we analyze the circuit as follows.

- (3.65) $\Rightarrow \frac{v_o}{v_i} = -\frac{R_{D1}}{r_{is1}}$.
- Figure 3.35 $\Rightarrow R_{D1} = r_{id2}$, and (3.61) $\Rightarrow r_{id2} = r_{ds2}$.
- (3.57) $\Rightarrow r_{is1} = \frac{r_{ds1}+r_{ds2}}{A_{\max 1}}$.
- (3.60) $\Rightarrow A_{\max 1} = r_{ds1} g_{m1}$.
- Combining above equations, we arrive at

$$\frac{v_o}{v_i} = -g_{m1}\frac{r_{ds1}r_{ds2}}{r_{ds1}+r_{ds2}},$$

which is identical to (3.56). ▲

Example 3.13 Determine by inspection the small-signal voltage gain v_o/v_i of the differential amplifier shown in Fig. 3.29. Assume (a) a twin-well CMOS technology where chip substrate is the common bulk of all NMOS devices, and (b) identical M1 and M2 whose gate and bulk transconductances are denoted by g_m and g_{mb}.

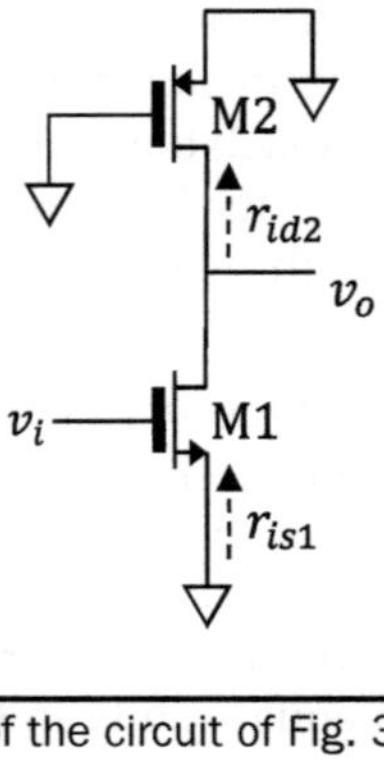

FIGURE 3.35 Small-signal equivalent of the circuit of Fig. 3.22(a).

Work Since the substrate has to be tied to V_{SS}, all NMOS devices operate with a common bulk voltage at V_{SS}. This implies a source/bulk short only for M4 among the three NMOS devices. The PMOS device M3 also operates with a source/bulk short because its source is already at the highest available voltage V_{DD}. Relying on the small-signal equivalent circuit given in Fig. 3.29(b), we identify a signal path from the gate of M1 to the source of M1, and from the source of M2 to the drain of M2. Therefore, the circuit has a single forward path from input to output. We can directly apply the appropriate gain and input-resistance expressions derived in this section, and arrive at

$$\frac{v_o}{v_i} = \underbrace{\frac{g_m}{g_m + g_{mb}} \frac{r_{ds4} \parallel r_{is2}}{(r_{ds4} \parallel r_{is2}) + r_{is1}}}_{v_1/v_i} \underbrace{\frac{r_{ds3}}{r_{is2}}}_{v_o/v_1}, \tag{3.69}$$

where

$$r_{is1} = \frac{1}{g_m + g_{mb}},$$

and

$$r_{is2} = \frac{r_{ds3} + r_{ds2}}{r_{ds2}(g_m + g_{mb})}.$$

It is obvious that the final gain expression obtained from these three equations will be pretty complicated. Fortunately, we can simplify it with an acceptable approximation based on the fact that, in most cases of small-signal analysis, we may assume $1/(g_m + g_{mb})$ to be negligibly small in comparison with an r_{ds} even if the two belong to two different devices. To that end, first, consider the expression of r_{is2}. Since we expect r_{ds2} and r_{ds3} to be on the same order of magnitude, r_{is2} should be comparable to $1/(g_m + g_{mb})$. Since the latter is expected to be much smaller than r_{ds4}, the parallel equivalent $r_{ds4} \parallel r_{is2}$ appearing in (3.69) equals r_{is2} to a good approximation. Also considering the expressions of r_{is1} and r_{is2}, we simplify (3.69) into

$$\frac{v_o}{v_i} \approx g_m \frac{r_{ds3} r_{ds2}}{r_{ds3} + 2r_{ds2}}. \quad \blacktriangle$$

Example 3.14 Shown in Fig. 3.36(a) is the electrical schematic of the split-supply version of a very popular single-stage operational transconductance amplifier called *symmetrical OTA*. I_B is a dc bias-current source assumed to be ideal in this example. M1 and M2 have identical dimensions, and their gates are biased at ground. The diode-connected M3 and M4 also have identical dimensions. Furthermore, M5 and M7 have the same channel length but not necessarily the same channel width. Same is valid for M6 and M8. Needless to say, the wells of all four PMOS devices are tied to V_{DD}. We first analyze the bias condition of the circuit in some detail, and then perform a small-signal analysis for the purpose of determining the voltage gain v_o/v_i.

Work First, notice from Fig. 3.36(a) that M1 and M2 operate with the same gate-source bias voltage V_{GS} because (a) their gates are commonly tied to ground, and (b) their source terminals are physically interconnected. Having the same V_{SB} and the same L_{DR}, they are identical also in V_{TN}. Therefore, both operate with the same overdrive $V_{GS} - V_{TN}$, hence the same $I_{\square}$ (see Fig. 3.24), which, in turn, implies identical bias currents ($I_B/2$

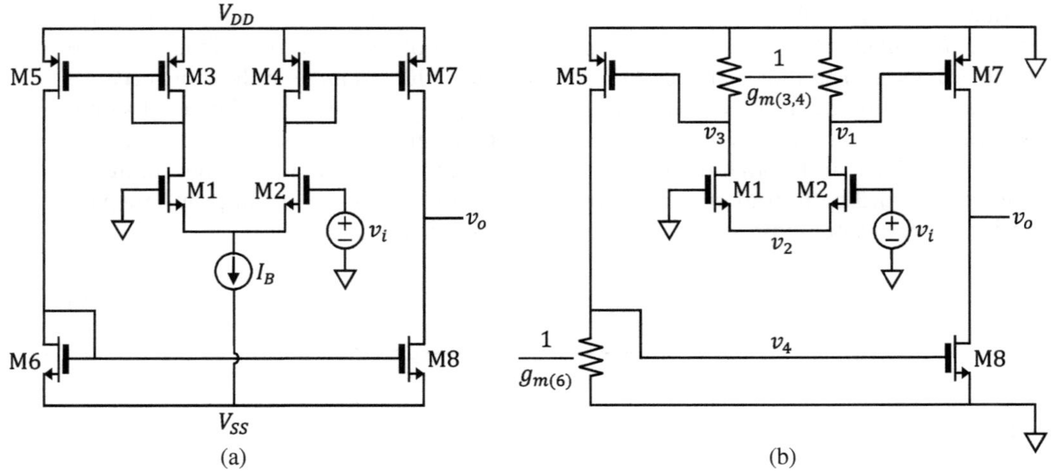

FIGURE 3.36 (a) Electrical schematic of symmetrical OTA. (b) Small-signal equivalent circuit.

each) because their dimensions are identical. Having identical bias currents, and identical overdrive voltages, their transconductances must also be identical (once more, refer to Fig. 3.24). We denote this transconductance with $g_{m(1,2)}$.

Next, consider M3 and M4. Each of these operates with a bias current $I_B/2$ imposed by M1 or M2. Also being dimensionally identical, their $I_{\square}$, hence overdrive voltages, are also identical. This implies identical transconductances as well. We denote this transconductance with $g_{m(3,4)}$.

The transconductances of M5, M6, M7, and M8 are also interrelated. To see why, first, notice that M5 and M7 operate with identical overdrive voltages, hence identical transconductance efficiencies because (a) gates of M3 and M5 are common, (b) gates of M4 and M7 are common, (c) all four share the same source, (d) M3 and M4 have identical gate bias voltages, and (e) the threshold voltages of all four are identical due to the common $V_{SB} = 0$ and identical L_{DR}. Therefore,

$$\frac{g_{m5}}{I_{D5}} = \frac{g_{m7}}{I_{D7}}.$$

Next, consider M6 and M8. These two have a common V_{GS} and identical threshold voltages and therefore identical transconductance efficiencies. But also notice that M6 conducts the bias current I_{D5} of M5, and M8 conducts the bias current I_{D7} as M7. Hence,

$$\frac{g_{m6}}{I_{D5}} = \frac{g_{m8}}{I_{D7}}.$$

Comparing this equation with the previous one, we conclude

$$\frac{g_{m5}}{g_{m6}} = \frac{g_{m7}}{g_{m8}}.$$

We now move to small-signal analysis. The equivalent circuit appropriate for this purpose is shown in Fig. 3.36(b). Notice the absence of the ideal dc current source I_B, and grounded power-supply rails V_{DD} and V_{SS}. Also note that the diode-connected devices are replaced by their equivalent resistances $1/g_{m(3,4)}$ and $1/g_{m6}$, as we concluded in Example 3.11.

This equivalent circuit contains two forward signal paths, both of which begin at the gate of M2. One path moves from gate to drain through M2 and gate to drain through M7. The other moves from gate to source through M2, source to drain through M1, gate to drain through M5, and gate to drain through M8. If only the former path did exist, the output signal would be expressed as

$$v_o' = \underbrace{\frac{v_o}{v_1}\frac{v_1}{v_i}}_{A_v'} v_i.$$

Similarly, the second path would yield

$$v_o'' = \underbrace{\frac{v_o}{v_4}\frac{v_4}{v_3}\frac{v_3}{v_2}\frac{v_2}{v_i}}_{A_v''} v_i.$$

Since small-signal equivalent circuits are linear by definition, we can apply superposition to write $v_o = v_o' + v_o''$, hence

$$A_v \doteq \frac{v_o}{v_i} = A_v' + A_v''.$$

We first determine by inspection each of the two factors of A_v'. We assume any r_{ds} to be much larger than any $1/g_m$:

$$A_v' = \underbrace{-\left(r_{ds7} \parallel r_{ds8}\right) g_{m7}}_{v_o/v_1} \times \underbrace{\frac{-g_{m(1,2)}}{g_{m(1,2)} + g_{mb(1,2)}} \frac{1/g_{m(3,4)}}{r_{is1} + r_{is2}}}_{v_1/v_i},$$

where

$$r_{is1} = \frac{r_{ds1} + \frac{1}{g_{m(3,4)}}}{r_{ds1}\left(g_{m(1,2)} + g_{mb(1,2)}\right)} \cong \frac{1}{g_{m(1,2)} + g_{mb(1,2)}},$$

and

$$r_{is2} = \frac{r_{ds2} + \frac{1}{g_{m(3,4)}}}{r_{ds2}(g_{m(1,2)} + g_{mb(1,2)})} \cong \frac{1}{g_{m(1,2)} + g_{mb(1,2)}}.$$

Hence,

$$A'_v = (r_{ds7} \parallel r_{ds8}) \frac{g_{m7}g_{m(1,2)}}{2g_{m(3,4)}}.$$

We now repeat the analysis for A''_v:

$$A''_v = \underbrace{-(r_{ds7} \parallel r_{ds8})\, g_{m8}}_{v_o/v_4} \times \underbrace{\frac{-1}{g_{m6}r_{is5}}}_{v_4/v_3} \times \underbrace{\frac{1}{g_{m(3,4)}r_{is1}}}_{v_3/v_2} \times \underbrace{\frac{g_{m(1,2)}}{g_{m(1,2)} + g_{mb(1,2)}} \frac{r_{is1}}{r_{is1} + r_{is2}}}_{v_2/v_i},$$

where

$$r_{is5} = \frac{r_{ds5} + \frac{1}{g_{m6}}}{r_{ds5}g_{m5}} \cong \frac{1}{g_{m5}}.$$

Hence

$$A''_v = (r_{ds7} \parallel r_{ds8}) \frac{g_{m8}g_{m(1,2)}g_{m5}}{2g_{m(3,4)}g_{m6}}.$$

Considering $g_{m5}/g_{m6} = g_{m7}/g_{m8}$ as derived earlier in the bias analysis, this equation can be rewritten as

$$A''_v = (r_{ds7} \parallel r_{ds8}) \frac{g_{m7}g_{m(1,2)}}{2g_{m(3,4)}}.$$

Finally, summing up A'_v and A''_v, we obtain

$$A_v = (r_{ds7} \parallel r_{ds8}) \frac{g_{m7}g_{m(1,2)}}{g_{m(3,4)}}. \quad \blacktriangle \tag{3.70}$$

Example 3.15 Shown in Fig. 3.37(a) is a small-signal equivalent circuit with a negative feedback loop. Assuming any r_{ds} to be much larger than any $1/g_m$, determine the voltage gain v_o/v_i.

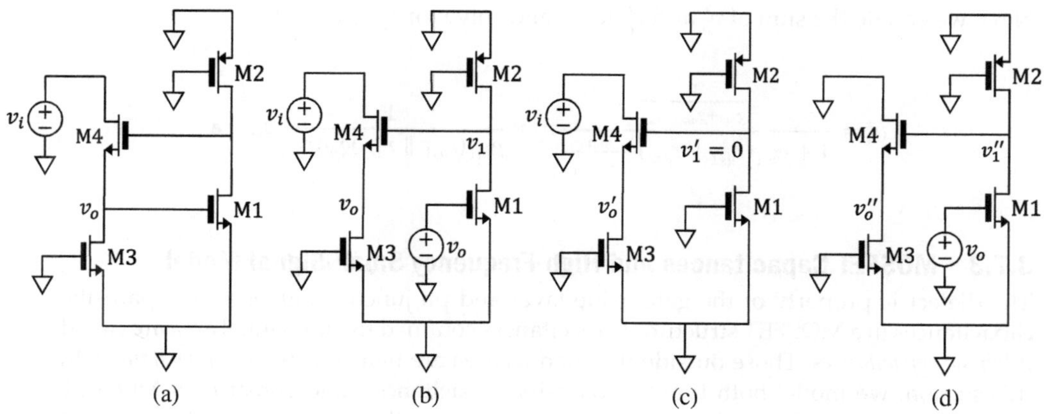

FIGURE 3.37 (a) A small-signal equivalent circuit with feedback. (b) Cutting the feedback loop at the gate of M1. (c) Equivalent circuit for determining v_o' and v_1' when only v_i is present. (d) Equivalent circuit for determining v_o'' and v_1'' when only v_o is present.

Work As always, we begin by tracing the signal path. The signal, applied to the drain terminal of M4 as v_i, flows to the source of the same device as v_o but it does not stop there. It is fed back to the gate of M4 through the gate-to-drain path of M1, and then, ends up again at the output through the gate-to-source path of M4. In order to be able to analyze the circuit by inspection, we must break this feedback loop and thus turn the circuit into an equivalent circuit containing forward paths only. Breaking at any node along the feedback path will do but, for the sake of analytical simplicity, it is most convenient to break the loop where it connects the output, that is, the gate of M1. Shown in Fig. 3.37(b) is the equivalent circuit obtained after breaking the loop. Notice the addition of a voltage source v_o to the gate of M1 for driving the broken feedback path with its original voltage. Generally, we also need to attach a resistance to the broken node of the forward path for representing the original loading of the feedback loop on this node. No such resistance is added in Fig. 3.37(b) because the load resistance seen into the gate of M1 in the original circuit of Fig. 3.37(a) is infinite.

The topology depicted in Fig. 3.37(b) has two voltage sources, v_i and v_o, but no feedback path. We now apply superposition to two forward paths, one from voltage-source v_i to output, and another from voltage-source v_o to output. First, we replace the source v_o with a short but retain the source v_i as shown in Fig. 3.37(c), and determine the corresponding output signal v_o' as a function of v_i. Inspection yields

$$v_o' = \frac{r_{ds3}}{r_{ds4} + r_{ds3}(g_{m4} + g_{mb4})r_{ds4}} v_i \approx \frac{1}{(g_{m4} + g_{mb4})r_{ds4}} v_i.$$

Next, we replace the source v_i with ground, and revive the source v_o as shown in Fig. 3.37(d), and determine the corresponding output-node signal v_o'' as a function of v_o. Inspection yields

$$v_o'' = -g_{m1}(r_{ds1} \parallel r_{ds2}) \times \frac{g_{m4}}{g_{m4} + g_{mb4}} \frac{r_{ds3}}{r_{ds3} + \frac{1}{g_{m4}+g_{mb4}}} v_o \approx -g_{m1}(r_{ds1} \parallel r_{ds2}) \times \frac{g_{m4}}{g_{m4} + g_{mb4}} v_o.$$

Next, we equate the sum of v_o' and v_o'' to v_o, and solve for v_o to obtain

$$v_o = \frac{\frac{1}{(g_{m4}+g_{mb4})r_{ds4}}}{1+g_{m1}(r_{ds1} \parallel r_{ds2})\frac{g_{m4}}{g_{m4}+g_{mb4}}} v_i \approx \frac{1}{g_{m1}(r_{ds1} \parallel r_{ds2})g_{m4}r_{ds4}} v_i. \quad ▲$$

3.7.3 MOSFET Capacitances and High-Frequency Small-Signal Model

The dielectric property of the gate-oxide layer and pn junctions gives rise to parasitic capacitances in a MOSFET structure. Capacitances confined to the channel area are called *intrinsic capacitances*. Those outside the channel area are named *extrinsic capacitances*. In this section, we model both types of parasitic capacitances, and construct a MOSFET high-frequency small-signal model by adding them to the dc small-signal equivalent circuit developed in Subsection 3.7.1.

Intrinsic Capacitances

Intrinsic capacitances are confined to the shaded channel/gate region shown on a cross-sectional view in Fig. 3.38(a). This region generally supports a total charge Q_C (C) inside the channel, and a total charge Q_G (C) inside the gate. Since electric field is assumed to vanish at the boundaries of this region, Gauss's law implies $Q_G = -Q_C$. Note that in general Q_C and therefore Q_G are functions of all four terminal voltages v_G, v_D, v_S, and v_B with respect to circuit ground. Now consider a dc steady state in which none of these terminal voltages changes. In this state, channel receives a dc current I_D across the drain boundary, and sends out the same I_D across the source boundary. Since no current flows across the bulk or gate-oxide boundaries either, no net charge is added to or extracted from the channel. This is why Q_C remains time invariant in a dc steady state. Q_G also remains time invariant because no current flows into or out of the gate.

Next, suppose that one or more of the terminal voltages are changed. In response, Q_C and therefore Q_G should also change. Consider the change in Q_C first. If it is in

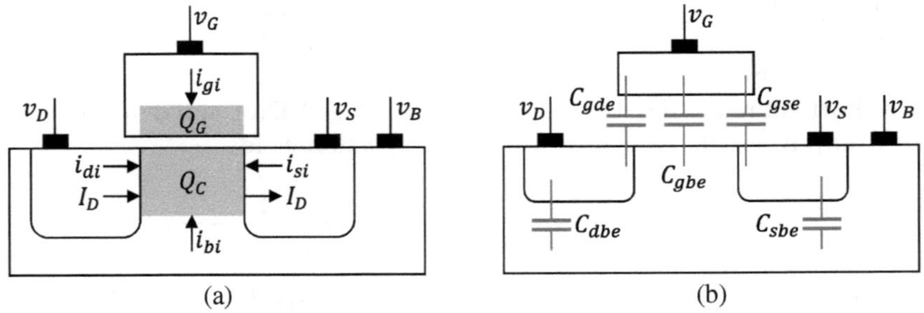

FIGURE 3.38 (a) MOSFET intrinsic capacitances are confined to the shaded channel/gate region. i_{gi}, i_{si}, i_{di}, and i_{bi} represent intrinsic charging currents flowing at the boundaries of this region. (b) Extrinsic capacitances C_{gde}, C_{gbe}, C_{gse}, C_{dbe}, and C_{sbe}.

the positive direction, a current should temporarily flow into the channel in order to bring in the additional positive charge. If the change is in the negative direction, the temporal current will flow out of the channel in order to remove the excess positive charge. In both cases, this current can flow only through the source, drain, or bulk boundaries of the channel but not through the gate-oxide boundary since the insulating property of the latter prevents charge transport. These three temporal current components flowing under ac or transient conditions are represented with i_{di}, i_{bi}, and i_{si} in Fig. 3.38(a). The channel-charge components altered by these currents are denoted by Q_D, Q_B, and Q_S, respectively. Therefore, $Q_C = Q_D + Q_B + Q_S$. Now, consider the variation in Q_G. Obviously, only a current i_{gi} temporarily flowing into or out of the gate can alter Q_G because no net charge exchange can occur through the gate-oxide boundary.

We refer to i_{gi}, i_{di}, i_{bi}, and i_{si} as *intrinsic charging current components*. For the reference directions identified in Fig. 3.38(a), they are related to the temporal derivatives of the associated charge components by the following four equations:

$$i_{gi} = \frac{dQ_G}{dt} = \underbrace{\frac{\partial Q_G}{\partial v_G}}_{C_{ggi}}\frac{dv_G}{dt} + \underbrace{\frac{\partial Q_G}{\partial v_D}}_{C_{gdi}}\frac{dv_D}{dt} + \underbrace{\frac{\partial Q_G}{\partial v_B}}_{C_{gbi}}\frac{dv_B}{dt} + \underbrace{\frac{\partial Q_G}{\partial v_S}}_{C_{gsi}}\frac{dv_S}{dt}, \tag{3.71}$$

$$i_{di} = \frac{dQ_D}{dt} = \underbrace{\frac{\partial Q_D}{\partial v_G}}_{C_{dgi}}\frac{dv_G}{dt} + \underbrace{\frac{\partial Q_D}{\partial v_D}}_{C_{ddi}}\frac{dv_D}{dt} + \underbrace{\frac{\partial Q_D}{\partial v_B}}_{C_{dbi}}\frac{dv_B}{dt} + \underbrace{\frac{\partial Q_D}{\partial v_S}}_{C_{dsi}}\frac{dv_S}{dt}, \tag{3.72}$$

$$i_{bi} = \frac{dQ_B}{dt} = \underbrace{\frac{\partial Q_B}{\partial v_G}}_{C_{bgi}}\frac{dv_G}{dt} + \underbrace{\frac{\partial Q_B}{\partial v_D}}_{C_{bdi}}\frac{dv_D}{dt} + \underbrace{\frac{\partial Q_B}{\partial v_B}}_{C_{bbi}}\frac{dv_B}{dt} + \underbrace{\frac{\partial Q_B}{\partial v_S}}_{C_{bsi}}\frac{dv_S}{dt}, \tag{3.73}$$

$$i_{si} = \frac{dQ_S}{dt} = \underbrace{\frac{\partial Q_S}{\partial v_G}}_{C_{sgi}}\frac{dv_G}{dt} + \underbrace{\frac{\partial Q_S}{\partial v_D}}_{C_{sdi}}\frac{dv_D}{dt} + \underbrace{\frac{\partial Q_S}{\partial v_B}}_{C_{sbi}}\frac{dv_B}{dt} + \underbrace{\frac{\partial Q_S}{\partial v_S}}_{C_{ssi}}\frac{dv_S}{dt}. \tag{3.74}$$

These four modeling equations contain sixteen capacitances. However, only nine out of these sixteen are independent. To understand why, first, consider a specific case in which all terminal voltages change with time in an identical fashion; that is,

$$\frac{dv_G}{dt} = \frac{dv_D}{dt} = \frac{dv_B}{dt} = \frac{dv_S}{dt}, \tag{3.75}$$

which implies time-invariant port voltages V_{GS}, V_{DS}, and V_{SB}, hence time-invariant terminal currents. In other words, the time-dependent charging currents i_{gi}, i_{di}, i_{bi}, and i_{si} must vanish in this case. Replacing the left-hand sides of (3.71) to (3.74) with zero, and applying (3.75) to their right-hand sides, we obtain

$$C_{ggi} = -C_{gdi} - C_{gbi} - C_{gsi}, \tag{3.76}$$

$$C_{ddi} = -C_{dgi} - C_{dbi} - C_{dsi}, \tag{3.77}$$

$$C_{bbi} = -C_{bgi} - C_{bdi} - C_{bsi}, \tag{3.78}$$

$$C_{ssi} = -C_{sgi} - C_{sdi} - C_{sbi}. \tag{3.79}$$

Another set of relations among these capacitances is obtained by applying Kirchhoff's current law to the intrinsic volume. Irrespective of how the four terminal voltages change with time, the sum of the four charging currents must be zero all the time. Now, suppose that only v_G varies with time while v_D, v_B, and v_S are time invariant. Summing up (3.71) to (3.74) after deleting the second, third, and fourth terms on the right-hand sides, and equating the sum to zero, we find

$$C_{sgi} = -C_{ggi} - C_{dgi} - C_{bgi}. \tag{3.80}$$

Assuming only v_D to be time variant, and applying Kirchhoff's law, we obtain a second equation:

$$C_{sdi} = -C_{gdi} - C_{ddi} - C_{bdi}. \tag{3.81}$$

Repeating the procedure for the case of v_B being the sole time-variant voltage, we obtain a third equation:

$$C_{sbi} = -C_{gbi} - C_{dbi} - C_{bbi}. \tag{3.82}$$

We could repeat the procedure one more time by assuming v_S to be time-variant while the remaining terminal voltages to be constant but this would not result in an independent equation, unlike (3.76) to (3.82). These seven equations render nine of the sixteen capacitances independent. As long as any nine are available, you can calculate the rest using (3.76) to (3.82).

Equations (3.71) to (3.74) represent the intrinsic capacitive behavior in terms of terminal voltages defined with respect to circuit ground. In order to comply with our port convention set earlier in Fig. 3.2, we now convert the voltage variables of these equations into v_{GS}, v_{DS}, and v_{SB}. Actually, we need only three of these equations for constructing an equivalent circuit because Kirchhoff's current law constrains them. We select (3.71) to

(3.73) for conversion. Substituting $v_G = v_{GS} + v_S$, $v_D = v_{DS} + v_S$, $v_B = -v_{SB} + v_S$ into these three equations, and considering (3.76) to (3.78), we arrive at

$$i_{gi} = C_{ggi}\frac{dv_{GS}}{dt} + C_{gdi}\frac{dv_{DS}}{dt} - C_{gbi}\frac{dv_{SB}}{dt}, \tag{3.83}$$

$$i_{di} = C_{dgi}\frac{dv_{GS}}{dt} + C_{ddi}\frac{dv_{DS}}{dt} - C_{dbi}\frac{dv_{SB}}{dt}, \tag{3.84}$$

$$i_{bi} = C_{bgi}\frac{dv_{GS}}{dt} + C_{bdi}\frac{dv_{DS}}{dt} - C_{bbi}\frac{dv_{SB}}{dt}. \tag{3.85}$$

Finally, we rewrite these three expressions in s domain for later use in the high-frequency equivalent circuit of a MOSFET:

$$i_{gi} = sC_{ggi}v_{gs} + sC_{gdi}v_{ds} - sC_{gbi}v_{sb}, \tag{3.86}$$

$$i_{di} = sC_{dgi}v_{gs} + sC_{ddi}v_{ds} - sC_{dbi}v_{sb}, \tag{3.87}$$

$$i_{bi} = sC_{bgi}v_{gs} + sC_{bdi}v_{ds} - sC_{bbi}v_{sb}. \tag{3.88}$$

Intrinsic capacitances are bias dependent. In saturation, they vary mainly with the overdrive voltage. They are also proportional to gate area WL. Their areal density defined by

$$\overline{C_{xyi}} \doteq \frac{C_{xyi}}{WL} \tag{3.89}$$

is, therefore, independent of device size but dependent on the overdrive voltage. We can characterize the areal densities in a device-independent manner by running a dc analysis by sweeping V_{GS}, and use the outcome for estimating the actual capacitances for any device of the same model deck. Shown in Figs. 3.39 and 3.40 are such areal-density plots (in units of fF/μm^2) of the capacitances appearing in (3.86) to (3.88). These plots belong to the NMOS and PMOS model decks of Table 3.1, and have been extracted with Spice input files `nmos2.sp` and `pmos2.sp` (see Subsections 3.3.3 and 3.4.4) after changing the `.save` lines into

```
.save @m[cgg] @m[cgd] @m[cgs] @m[cdg] @m[cdd] @m[cds]
+  @m[cbg] @m[cbd] @m[cbs] @m[vth] v(1,3)
```

Except for $\overline{C_{gbi}}$, $\overline{C_{dbi}}$, and $\overline{C_{bbi}}$, areal densities in units of fF/μm^2 are plotted with the command

```
plot @m[cxy]/4.6e-15 vs v(1,2)-@m[vth]
```

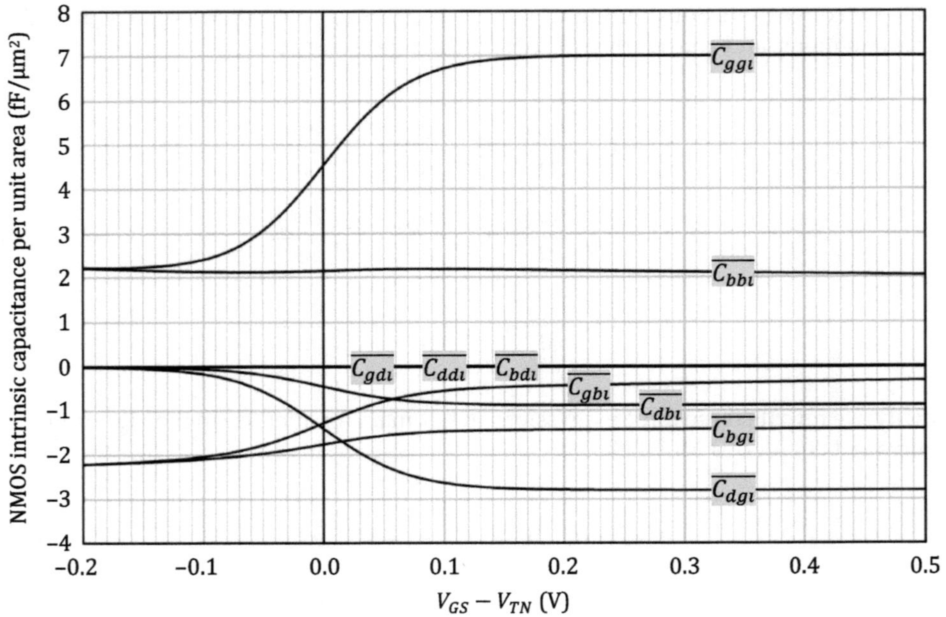

FIGURE 3.39 Variation of the density (per unit channel area) of intrinsic capacitances with overdrive voltage in an NMOS in saturation.

for NMOS whose effective channel area equals 4.60 μm^2, and with the command

```
plot @m[cxy]/4.44e-15 vs v(1,3)-@m[vth]
```

for PMOS whose effective channel area equals 4.44 μm^2. Since C_{gbi}, C_{dbi}, and C_{bbi} are not directly accessible in the particular version of Ngspice used in these simulations, their areal densities have been plotted for NMOS with

```
plot (-@m[cgg]-@m[cgd]-@m[cgs])/4.6e-15 vs v(1,3)-@m[vth]
plot (-@m[cdd]-@m[cdg]-@m[cds])/4.6e-15 vs v(1,3)-@m[vth]
plot (-@m[cbg]-@m[cbd]-@m[cbs])/4.6e-15 vs v(1,3)-@m[vth]
```

as implied by (3.76) to (3.78), respectively. Same entries have been used for PMOS counterparts after changing the factor of normalization from 4.6×10^{-15} to 4.44×10^{-15}.

Notice the following features of the areal density plots observed in Figs. 3.39 and 3.40 for a device operating in saturation:

- Some of the capacitances have negative values. This is a result of the reference directions implicitly assumed in defining intrinsic capacitances in (3.71) to (3.74). As an example, consider C_{gbi}. It is defined as $C_{gbi} \doteq \partial Q_G/\partial v_B$ in (3.71). Physically, an increasing v_B causes the gate charge Q_G to decrease. Therefore, the partial derivative is negative.
- $\overline{C_{gdi}}$, $\overline{C_{ddi}}$, and $\overline{C_{bdi}}$ are negligibly small.

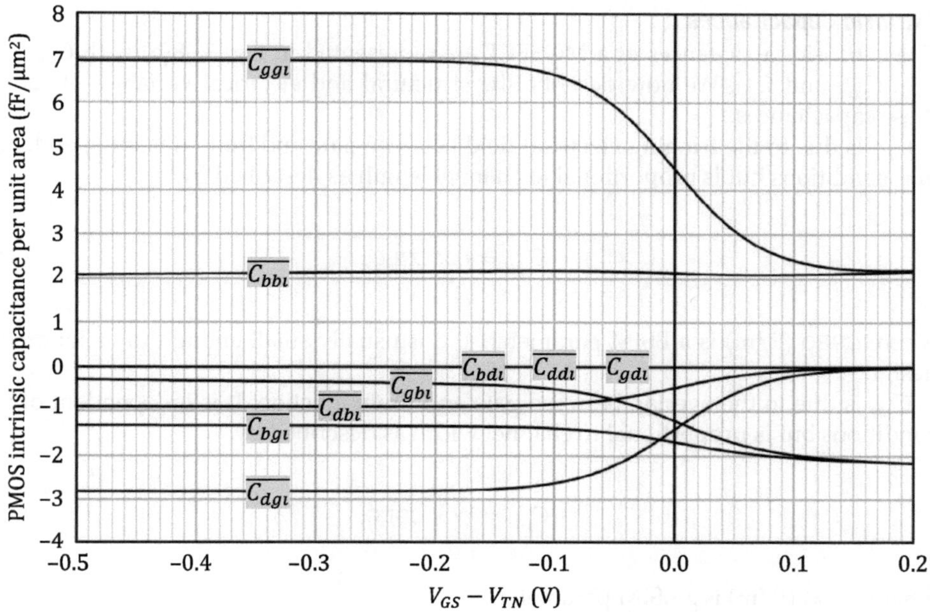

FIGURE 3.40 Variation of the intrinsic capacitances per unit gate area with overdrive voltage in a PMOS in saturation.

- $\overline{C_{bbi}}$ is virtually constant regardless of the strength of inversion.
- $\overline{C_{ggi}}$ is virtually independent of the overdrive voltage in strong inversion for approximately $V_{GS} - V_{TN} > 0.1$ V in NMOS and $V_{GS} - V_{TP} < -0.1$ V in PMOS. It decreases as the device moves into weak inversion and subthreshold.
- $\overline{C_{gbi}}$ and $\overline{C_{bgi}}$ are virtually independent of the overdrive voltage in strong inversion for approximately $V_{GS} - V_{TN} > 0.1$ V in NMOS and $V_{GS} - V_{TP} < -0.1$ V in PMOS. Their magnitude increases as the device moves into weak inversion and subthreshold.
- $\overline{C_{dbi}}$ and $\overline{C_{dgi}}$ are virtually independent of the overdrive voltage in strong inversion for approximately $V_{GS} - V_{TN} > 0.1$ V in NMOS and $V_{GS} - V_{TP} < -0.1$ V in PMOS. Their magnitude decreases as the device moves into weak inversion and subthreshold.

As a final word, note that C_{ggi} happens to be the most influential intrinsic capacitance in ac or transient behavior of a MOSFET. It is worth remembering that its areal density is reasonably well approximated by

$$\overline{C_{ggi}} \approx \frac{2}{3} C_{ox} \tag{3.90}$$

under strong-inversion conditions in saturation.

Extrinsic Capacitances

The extrinsic capacitances of a MOSFET are represented in Fig. 3.38(b) with C_{gbe}, C_{gde}, C_{gse}, C_{dbe}, and C_{sbe}. We now describe the structural and electrical variables of each of these capacitances.

C_{gbe} is due to the fringing electrical field between gate and bulk. It is independent of bias conditions but is proportional to channel length as described by

$$C_{gbe} = L \times \texttt{CGBO}, \tag{3.91}$$

where CGBO (F/m) is a BSIM parameter. C_{gbe} is usually small enough to be ignored in design.

C_{gde} is due to the overlap between gate and drain junction. It is independent of bias conditions but is proportional to channel width as described by

$$C_{gde} = W \times \texttt{CGDO}, \tag{3.92}$$

where CGDO (F/m) is a BSIM parameter.

C_{gse} is due to the overlap between gate and source junction. It is also independent of bias conditions and proportional to channel width, and is described by

$$C_{gse} = W \times \texttt{CGSO}, \tag{3.93}$$

where CGSO (F/m) is a BSIM parameter.

C_{dbe} is the junction capacitance associated with drain-bulk junction. It has three components, all of which depend on drain-bulk bias V_{DB} as follows:

$$C_{dbe} = \frac{A_D \times \texttt{CJ}}{\left(1 + \frac{|V_{DB}|}{\texttt{PB}}\right)^{\texttt{MJ}}} + \frac{(P_D - W) \times \texttt{CJSW}}{\left(1 + \frac{|V_{DB}|}{\texttt{PBSW}}\right)^{\texttt{MJSW}}} + \frac{W \times \texttt{CJSWG}}{\left(1 + \frac{|V_{DB}|}{\texttt{PBSWG}}\right)^{\texttt{MJSWG}}}, \tag{3.94}$$

where A_D (m^2) is the area of the drain-bulk junction, and P_D (m) is the total perimeter of the same junction. In a Spice input file, these two parameters are specified as instance parameters in the element line of the MOSFET they belong to. Among the BSIM3 parameters appearing in (3.94), MJ, MJSW, and MJSWG are dimensionless. The rest are included in the model deck with the following units: CJ in F/m^2; CJSW and CJSWG in F/m; PB, PBSW, and PBSWG in V.

C_{sbe} is the junction capacitance associated with the source-bulk junction. It has three components, all of which depend on the source-bulk bias V_{SB} as follows:

$$C_{sbe} = \frac{A_S \times \texttt{CJ}}{\left(1 + \frac{|V_{SB}|}{\texttt{PB}}\right)^{\texttt{MJ}}} + \frac{(P_S - W) \times \texttt{CJSW}}{\left(1 + \frac{|V_{SB}|}{\texttt{PBSW}}\right)^{\texttt{MJSW}}} + \frac{W \times \texttt{CJSWG}}{\left(1 + \frac{|V_{SB}|}{\texttt{PBSWG}}\right)^{\texttt{MJSWG}}}, \tag{3.95}$$

where A_S (m^2) is junction area and P_S (m) is total junction perimeter. In a Spice input file, these two parameters are also specified as instance parameters in MOSFET element line.

In (3.94) and (3.95), the first term on the right-hand side represents the capacitance of the bottom surface of the junction, whereas the second and third terms are the junction sidewall capacitances outside the channel and inside the channel, respectively.

Having modeled each of the five extrinsic capacitances, we now express the *extrinsic charging currents* i_{ge}, i_{de}, and i_{be} flowing into the device at the gate, drain, and bulk terminals due to these capacitances:

$$i_{ge} = C_{gde}\frac{dv_{GD}}{dt} + C_{gse}\frac{dv_{GS}}{dt} + C_{gbe}\frac{dv_{GB}}{dt},$$

$$i_{de} = -C_{gde}\frac{dv_{GD}}{dt} + C_{dbe}\frac{dv_{DB}}{dt},$$

$$i_{be} = -C_{gbe}\frac{dv_{GB}}{dt} - C_{sbe}\frac{dv_{SB}}{dt} - C_{dbe}\frac{dv_{DB}}{dt}.$$

Substituting $v_{GS} - v_{DS}$ for v_{GD}, $v_{GS} + v_{SB}$ for v_{GB}, and $v_{DS} + v_{SB}$ for v_{DB}, we rewrite these three expressions in terms of our conventional port voltages v_{GS}, v_{DS}, and v_{SB}:

$$i_{ge} = C_{gge}\frac{dv_{GS}}{dt} - C_{gde}\frac{dv_{DS}}{dt} + C_{gbe}\frac{dv_{SB}}{dt}, \tag{3.96}$$

$$i_{de} = -C_{gde}\frac{dv_{GS}}{dt} + C_{dde}\frac{dv_{DS}}{dt} + C_{dbe}\frac{dv_{SB}}{dt}, \tag{3.97}$$

$$i_{be} = -C_{gbe}\frac{dv_{GS}}{dt} - C_{dbe}\frac{dv_{DS}}{dt} - C_{bbe}\frac{dv_{SB}}{dt}, \tag{3.98}$$

where

$$C_{gge} \doteq C_{gse} + C_{gde} + C_{gbe}, \tag{3.99}$$

$$C_{dde} \doteq C_{gde} + C_{dbe}, \tag{3.100}$$

$$C_{bbe} \doteq C_{gbe} + C_{dbe} + C_{sbe}. \tag{3.101}$$

Finally, we convert (3.96) to (3.98) into s-domain expressions:

$$i_{ge} = sC_{gge}v_{gs} - sC_{gde}v_{ds} + sC_{gbe}v_{sb}, \tag{3.102}$$

$$i_{de} = -sC_{gde}v_{gs} + sC_{dde}v_{ds} + sC_{dbe}v_{sb}, \tag{3.103}$$

$$i_{be} = -sC_{gbe}v_{gs} - sC_{dbe}v_{ds} - sC_{bbe}v_{sb}. \tag{3.104}$$

High-Frequency Equivalent Circuit

The total capacitive charging current flowing into the device at a given terminal is the sum of intrinsic and extrinsic components. Using (3.86) to (3.88) and (3.102) to (3.104) for these components, we can express the total charging currents as follows

$$i_g = i_{gi} + i_{ge} = s\,\underbrace{(C_{gge} + C_{ggi})}_{C_{gg}}\,v_{gs} - s\,\underbrace{(C_{gde} - C_{gdi})}_{C_{gd}}\,v_{ds} + s\,\underbrace{(C_{gbe} - C_{gbi})}_{C_{gb}}\,v_{sb},$$

$$i_d = i_{di} + i_{de} = -s\,\underbrace{(C_{gde} - C_{dgi})}_{C_{dg}}\,v_{gs} + s\,\underbrace{(C_{dde} + C_{ddi})}_{C_{dd}}\,v_{ds} + s\,\underbrace{(C_{dbe} - C_{dbi})}_{C_{db}}\,v_{sb},$$

$$i_b = i_{bi} + i_{be} = -s\,\underbrace{(C_{gbe} - C_{bgi})}_{C_{bg}}\,v_{gs} - s\,\underbrace{(C_{dbe} - C_{bdi})}_{C_{bd}}\,v_{ds} - s\,\underbrace{(C_{bbe} - C_{bbi})}_{C_{bb}}\,v_{sb},$$

where we define

$$C_{gg} \doteq C_{gge} + C_{ggi}, \quad (3.105)$$

$$C_{gd} \doteq C_{gde} - C_{gdi}, \quad (3.106)$$

$$C_{gb} \doteq C_{gbe} - C_{gbi}, \quad (3.107)$$

$$C_{dd} \doteq C_{dde} + C_{ddi}, \quad (3.108)$$

$$C_{dg} \doteq C_{gde} - C_{dgi}, \quad (3.109)$$

$$C_{db} \doteq C_{dbe} - C_{dbi}, \quad (3.110)$$

$$C_{bb} \doteq C_{bbe} + C_{bbi}, \quad (3.111)$$

$$C_{bg} \doteq C_{gbe} - C_{bgi}, \quad (3.112)$$

$$C_{bd} \doteq C_{dbe} - C_{bdi}. \quad (3.113)$$

Finally, supplementing the dc equivalent circuit of Fig. 3.26 with these capacitive components leads us to the high-frequency small-signal equivalent circuit shown in Fig. 3.41.

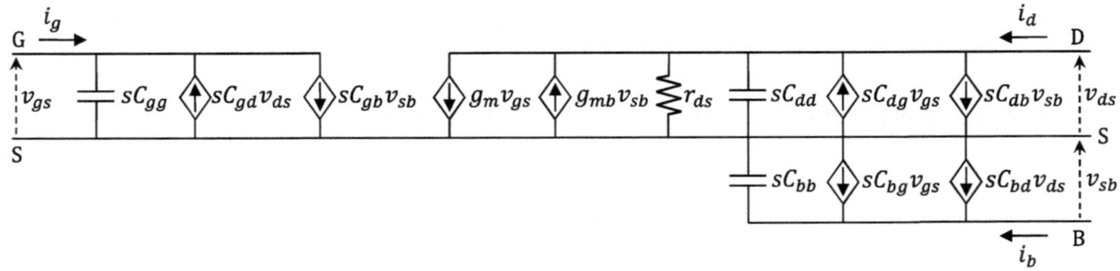

FIGURE 3.41 High-frequency small-signal equivalent circuit of a MOSFET.

Note that among nine capacitive components only C_{gg}, C_{dd}, and C_{bb} are represented by a capacitance symbol. Each of the remaining six capacitances relates the capacitive current of one port to the variation in the voltage of another, and therefore, does not fit into the proper definition of a capacitance. This is why such components are referred to as *trans-capacitance*. They function as a voltage-controlled current source, and this is how they are represented in the equivalent circuit.

Example 3.16 This is our fourth visit to the common-source amplifier stage of Fig. 3.22 (after Examples 3.5, 3.10, and 3.12). This time we load its output with a capacitance $C_L = 150$ fF as shown in Fig. 3.42(a), and analyze its frequency response in terms of the voltage gain $A_v \equiv v_o/v_i$ and input admittance $Y_i \doteq i_i/v_i$ around an output bias level $V_O = 0.9$ V. For both devices, we assume a rectangular drain-bulk junction geometry of side lengths W_{DR} and 0.55 μm. As usual, we verify the results with a Spice ac analysis.

Work For the purpose of small-signal analysis, the source and bulk terminals of M1 and the source, bulk, and gate terminals of M2 are at signal ground. Using the MOSFET equivalent circuit of Fig. 3.41 for both devices after grounding these terminals, we represent the amplifier stage with the equivalent circuit shown in Fig. 3.42(b). Notice that only r_{ds2} and C_{dd2} are included from the equivalent circuit of M2. This is a consequence of both v_{sb2} and v_{gs2} being zero. Therefore, M2 acts merely as a parallel equivalent of resistance and a capacitance. Straightforward circuit analysis yields

$$A_v \doteq \frac{v_o}{v_i} = -g_{m1}r\frac{1 - s\frac{C_{dg1}}{g_{m1}}}{1 + sC_L'r}, \tag{3.114}$$

for the transfer function of voltage gain, and

$$Y_i \doteq \frac{i_i}{v_i} = s\left(C_{gg1} + g_{m1}r\frac{1 - s\frac{C_{dg1}}{g_{m1}}}{1 + sC_L'r}C_{gd1}\right) = s\left(C_{gg1} - A_vC_{gd1}\right), \tag{3.115}$$

for input admittance. Note the definitions $C_L' \doteq C_L + C_{dd1} + C_{dd2}$ and $r \doteq r_{ds1} \parallel r_{ds2}$.

For proper interpretation of (3.114) and (3.115), we need to extract the numerical values of the device parameters appearing in these expressions. This calls for a Spice operation-point analysis. But, first, we perform a dc-sweep for accurately determining the input bias voltage V_I for which the specified $V_O = 0.9$ V is realized. Accuracy is essential because even a slight deviation from the necessary input bias will shift V_O far away from the specified 0.9 V due to the large gain of the stage. We use the `.dc` line of the following input file, and determine the necessary input bias voltage as $V_I = 567$ mV.[6]

[6] First, plot V_O versus V_I by typing `plot v(2)`. Then, point $V_O = 0.9$ V on the transfer characteristic, and left click once. Spice puts out the corresponding V_I as `x0 = 0.567619`.

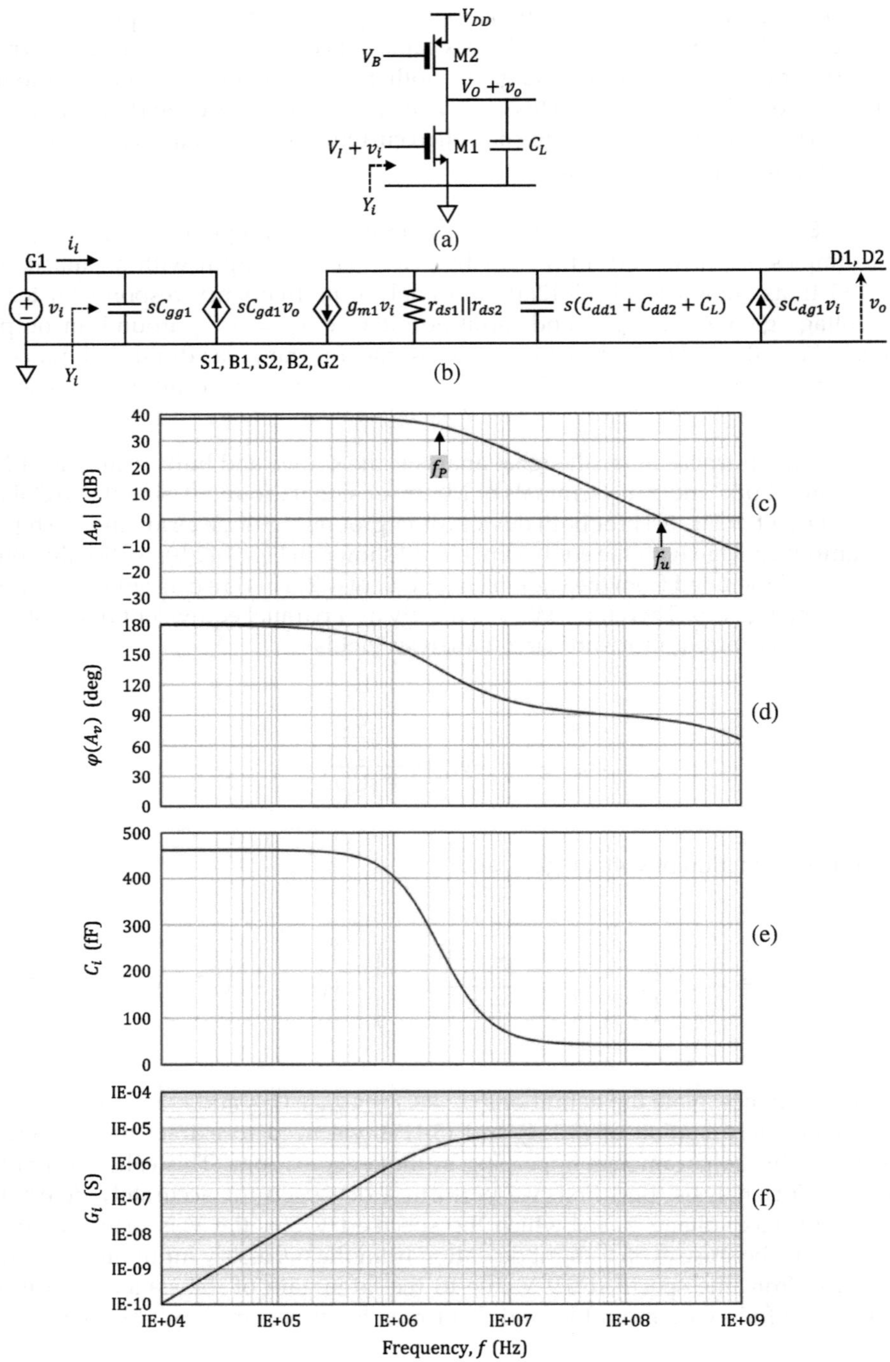

FIGURE 3.42 (a) Common-source amplifier stage loaded by a capacitance C_L. (b) High-frequency small-signal equivalent circuit for analyzing the voltage gain A_v and input admittance Y_i. (c) Magnitude response of the gain. (d) Phase response of the gain. (e) Input capacitance. (f) Input conductance.

```
filename: ex3-16.sp
.incl bsim.sp
m1 2 1 0 0 cmosn w=6u l=0.8u ad=3.3p pd=13.1u
+  nrd=3.33e-2 nrs=3.33e-2
m2 2 3 4 4 cmosp w=24u l=0.8u ad=13.2p pd=49.1u
+  nrd=8.33e-3 nrs=8.33e-3
vb 3 0 1.25
vdd 4 0 1.8
cl 2 0 0.15p
vi 1 0 dc 567m ac 1
.dc vi 0.5 0.6 1m
*.op
*.tf v(2) vi
*.ac dec 100 10k 1g
*.noise v(2) vi dec 100 10k 1g ; used in Example 3.17
.end
```

Before proceeding to the operation-point analysis, notice the declaration of the area (ad) and perimeter (pd) of the drain-bulk junctions of both MOSFETs as instance parameters. As pointed out in Subsection 3.2.1, these are necessary for the simulator to properly calculate the extrinsic drain-bulk capacitances. No area or perimeter is declared for source-bulk junctions because these junctions do not support any signal in the present circuit. Also notice the declaration of nrd and nrs as estimated by assuming a minimum contact-to-gate spacing L_{CG} = 0.2 μm as per Table 3.2. This is recommended in any ac simulation where input admittance is to be evaluated because otherwise the resistances placed by the simulator in series with the driver device would be unrealistically large. This would lead to an inaccurate evaluation of the input admittance at high frequencies. Although declaring these two parameters only for the driver device will suffice, it is prudent to follow the practice for all MOSFETs as a precaution against possible oversight in identifying the devices that are influential in this respect.

Now, we run an operation-point simulation, and read the following parameter values for M1 with the command show: g_{m1} = 244 μA/V, r_{ds1} = 693 kΩ, C_{ggi1} = 31.79 fF, C_{gdi1} = 69 aF, C_{ddi1} = −38 aF, C_{dgi1} = −12.59 fF, and C_{dbe1} = 5.32 fF for M1.[7] For M2, we read r_{ds2} = 667 kΩ, C_{ddi2} = −42 aF, and C_{dbe2} = 24.3 fF.

Next, we calculate the extrinsic capacitance C_{gse1} = 5.15 fF from (3.93), C_{gde1} = 5.15 fF and C_{gde2} = 18.8 fF from (3.92), and C_{gbe1} = 0.8 aF from (3.91). These numbers lead to C_{gge1} = 10.3 fF from (3.99), and C_{dde1} = 10.5 fF and C_{dde2} = 43.1 fF from (3.100).

Using these parameter values in (3.105), (3.106), (3.108), and (3.109), we obtain the following values for the capacitances appearing in (3.114) and (3.115): C_{gg1} = 42.1 fF, C_{gd1} = 5.08 fF, C_{dd1} = 10.5 fF, C_{dd2} = 43.1 fF, and C_{dg1} = 17.7 fF. Calculation also yields C'_L = 204 fF and r = 340 kΩ.

Now that all the necessary numerical data are available to us, we can start interpreting (3.114) and (3.115). According to (3.114), voltage gain A_v has a dc value

$$A_{v0} = -g_{m1}r = -83 \text{ V/V}$$

[7] Drain-bulk extrinsic capacitance is represented with parameter capbd in Ngspice. Source-bulk counterpart is capbs.

corresponding to 38.4 dB, which can be verified by running the transfer-function analysis included in the input file, and reading `transfer_function` after typing `print all`. A_v also has a right-half-plane zero at the frequency

$$f_Z = \frac{g_{m1}}{2\pi C_{dg1}} = 2.19 \text{ GHz},$$

and a pole at the frequency

$$f_P = \frac{1}{2\pi r C'_L} = 2.29 \text{ MHz}. \tag{3.116}$$

The magnitude and phase responses of A_v are shown in Figs. 3.42(c) and (d), respectively. These are obtained with the ac analysis included in the input file. Simply, run the simulation, set the unit of phase to degrees with

```
set units=degrees
```

and subsequently plot with

```
plot vdb(2)-vdb(1) vp(2)-vp(1)
```

Also notice a unity-gain frequency f_u around 206 MHz. Since the zero is located at a much higher frequency, the amplifier exhibits a single-pole characteristic all the way up to f_u. The phase equals 180° at low frequencies but drops to 135° at $f = f_P$ as expected from the left half-plane pole. Had there been no zero, the phase would settle at 90° at high frequencies. However, it decreases further at higher frequencies under the influence of the right half-plane zero, and reaches 75° at $f = f_u$.

The input admittance described by (3.115) has a capacitive component C_i and a conductive component G_i. Both vary with frequency as shown in Figs. 3.42(e) and (f) with simulated plots. These plots are generated by typing

```
plot imag(-i(vi)/v(1))/(2*pi*frequency)
```

for C_i, and

```
plot real(-i(vi)/v(1)) loglog
```

for G_i after running the ac analysis. A straightforward piecewise analysis of (3.115) in frequency domain yields the following asymptotic values for the two components:

$$C_i = \begin{cases} C_{gg1} + g_{m1} r C_{gd1} & \text{for } f \ll f_P \ll \sqrt{f_P f_Z} \\ C_{gg1} + g_{m1} r C_{gd1} \left(\frac{f_P}{f}\right)^2 & \text{for } f_P \ll f \ll \sqrt{f_P f_Z} \\ C_{gg1} - \frac{C_{dg1}}{C'_L} C_{gd1} & \text{for } f_P \ll \sqrt{f_P f_Z} \ll f \end{cases} \tag{3.117}$$

$$G_i = \begin{cases} g_{m1} \frac{C_{gd1}}{C'_L} \left(\frac{f}{f_P}\right)^2 & \text{for } f \ll f_P \\ g_{m1} \frac{C_{gd1}}{C'_L} & \text{for } f_P \ll f \end{cases}.$$

For $f \ll f_P$, (3.117) indicates a frequency-independent C_i comprising not only C_{gg1} (42.1 fF) but also C_{gd1} (5.08 fF) multiplied by the dc gain $g_{m1}r$ (83 V/V). This dominating contribution of C_{gd1} to input capacitance is known as *Miller effect*, which is represented in the equivalent circuit of Fig. 3.42(a) by the voltage-dependent current source $sC_{gd1}v_o$ attached to the input port. Since v_o equals $A_v v_i$, the current of this source is essentially controlled by v_i, and therefore the source itself is equivalent to an admittance $-sA_vC_{gd1}$ shunting the input port. Its low-frequency form $sg_{m1}rC_{gd1}$ based on $A_v = A_{v0} = -g_{m1}r$ is indeed capacitive. However, as f rises above f_P, and therefore the gain becomes $A_v \approx -g_{m1}/sC'_L$ as approximated from (3.114), the admittance due to Miller effect turns into $g_{m1}C_{gd1}/C'_L$, which, obviously, is conductive. ▲

3.8 MOSFET Noise Model

Two distinct types of inherent noise are dominant in MOSFET devices belonging to the technology nodes covered by BSIM3v3. The random motion of thermally agitated carriers inside the channel is the underlying mechanism of one of these two. It gives rise to thermal noise, which has a frequency-independent flat spectrum exactly like its counterpart observed in resistors. The other type arises from the random fluctuation of carrier mobility and/or the random capture and release of carriers by traps located at or near the channel-oxide interface. This type of noise is called *flicker noise*, which is also known as *1/f noise* because its power spectral density (PSD) is inversely proportional to the frequency f. The effect of each is represented by a current source placed in parallel with the drain-source port as shown in Fig. 3.43(a).

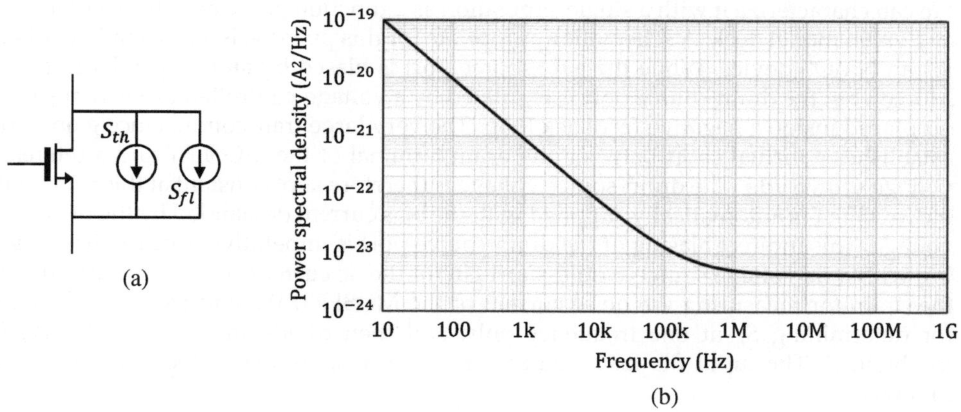

FIGURE 3.43 (a) Representing noise with independent current sources in a MOSFET, where S_{th} and S_{fl} are the power spectral densities of the thermal and flicker noise components. (b) A typical variation of $S_{th} + S_{fl}$ with frequency.

Thermal Noise

BSIM3v3 offers two PSD models for thermal noise current. The simpler model, known as *Spice2 model* and identified by `noimod=1` or `noimod=3` in the model deck, is described by

$$S_{th} = \frac{8}{3}E_{th}\left(g_m + g_{mb} + g_{ds}\right), \tag{3.118}$$

where E_{th} is the thermal energy defined right after (2.108). This is also how we model S_{th} for manual analysis and design purposes throughout this book. Note that g_{mb} has to be retained in this expression even if the device operates with a source-bulk short but g_{ds} may be ignored in saturation. The more complex model, known as *charge-based model* and identified by `noimod=2` or `noimod=4`, is not amenable to manual analysis or design. The reader is referred to [5] for its details. In BSIM4 and BSIM-BULK the Spice2 model is abandoned. The charge-based model is retained and identified by `tnoimod=0`. A new *holistic model* is also introduced in these later BSIM versions.[6][7]

Flicker Noise

BSIM3v3 also offers two options for the PSD of flicker noise current. Model identifiers are `noimod=1` or `noimod=4` for the simple Spice2 option, and `noimod=2` or `noimod=3` for the complex charge-based option. These two models are retained in BSIM4 with identifiers `fnoimod=0` and `fnoimod=1`, respectively.[6] BSIM-BULK offers only the complex model with the same `fnoimod=1` identifier.[7] Since the complex model is what model decks usually feature, we capture its description of the PSD S_{fl} in terms of design variables with

$$S_{fl} = K\frac{W}{fL^3}, \tag{3.119}$$

where K is a coefficient depending only on the overdrive voltage for a given model deck. This dependence resembles that of $I_\square$ in strong inversion and $I_\square^2$ in deep subthreshold. We can characterize it with a single simulation as a function of the overdrive voltage for any given model deck. A schematic appropriate for this purpose is shown in Fig. 3.44(a) for an NMOS device. While the gate-source port is biased by an independent voltage source `vgs`, the drain-source port is regulated by a voltage-controlled dc current source `gd` placed inside a negative feedback loop. The very large transconductance gain of `gd` establishes a virtual short between the drain terminal of the MOSFET and a reference voltage source `vds`. The drain-source voltage is therefore kept constant at this `vds` while the MOSFET is automatically supplied with the bias current demanded by the overdrive voltage. This indirect biasing of the drain-source port is imperative because direct biasing with a dc voltage source would short-circuit the ac current source `ino` attached to the drain for measuring the noise current of the MOSFET. We will use this schematic for determining S_{fl} at one frequency only, and then divide the outcome by W/fL^3 to obtain K. The input file `nmos-flicker.sp` corresponding to the schematic is as follows:

```
filename: nmos-flicker.sp
.incl bsim.sp
m 2 1 0 0 cmosn w=6u l=0.8u nrd=0 nrs=0
vgs 1 0 dc 0.6
```

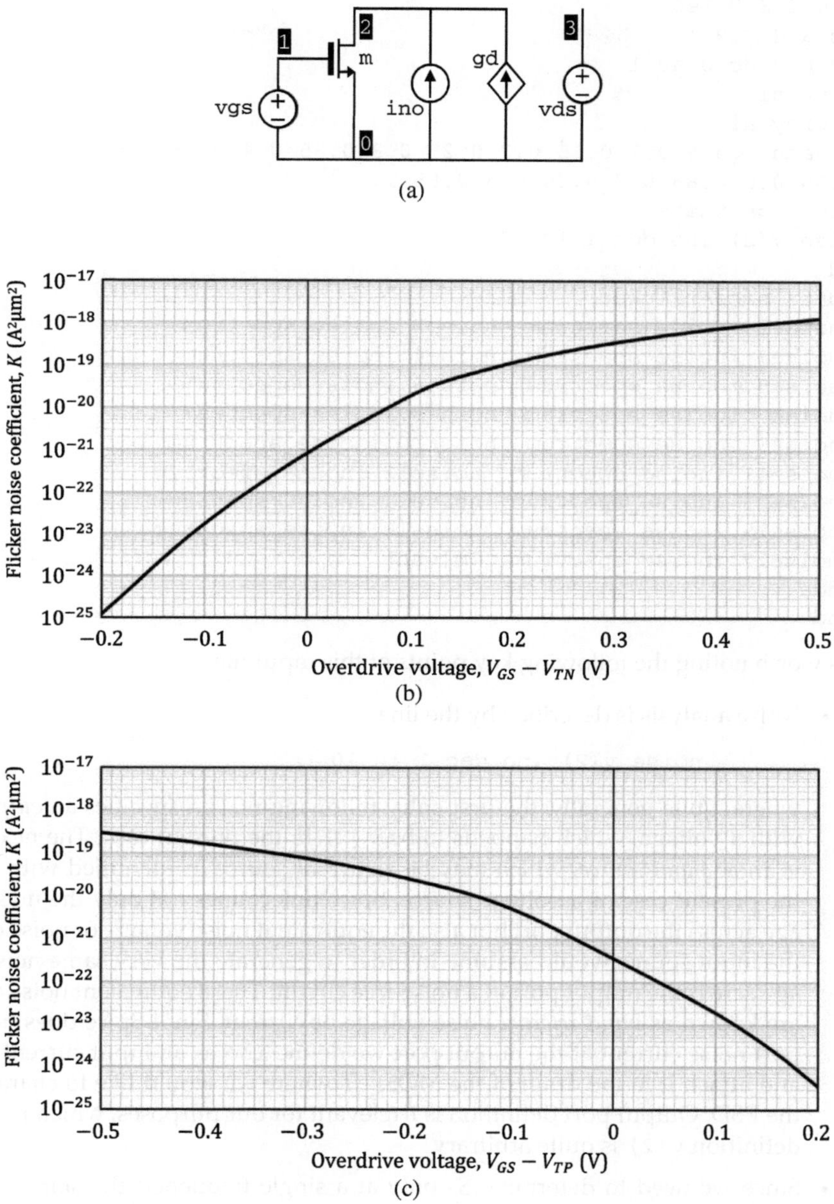

FIGURE 3.44 (a) Schematic for extracting the flicker noise coefficient K for NMOS devices. (b) K as a function of overdrive voltage in NMOS devices. (c) K as a function of overdrive voltage in PMOS devices.

```
gd 0 2 3 2 1e6
vds 3 0 0.9
ino 0 2 dc 0 ac 1
.control
destroy all
foreach  gate 0.1 0.15 0.2 0.25 0.3 0.35 0.4 0.45 0.5
+0.55 0.6 0.65 0.7 0.75 0.8 0.85 1.0
alter vgs=$gate
noise v(2) ino dec 1 10 10
end
wrdata nmos-flicker
+ noise1.inoise_spectrum  noise2.inoise_spectrum
+ noise3.inoise_spectrum  noise4.inoise_spectrum
+ noise5.inoise_spectrum  noise6.inoise_spectrum
+ noise7.inoise_spectrum  noise8.inoise_spectrum
+ noise9.inoise_spectrum  noise10.inoise_spectrum
+ noise11.inoise_spectrum  noise12.inoise_spectrum
+ noise13.inoise_spectrum  noise14.inoise_spectrum
+ noise15.inoise_spectrum  noise16.inoise_spectrum
+ noise17.inoise_spectrum  @m[vth]
.endc
.end
```

It is worth noting the following key points of this input file:

- Noise analysis is described by the line

  ```
  noise v(2) ino dec 1 10 10
  ```

  where, quite generally, the first entry following `noise` specifies an *output port* with a voltage, which happens to be `v(2)` in the present case. The next entry is an ac *input source*, which may be a current source as identified with `ino` in the present case or a voltage source. Spice determines not only the noise voltage across the output port but also the equivalent noise voltage or noise current the input source would assume in order to generate the very same noise voltage across the output port in a noise-free circuit. These equivalent noise entities are known as *input-referred* noise voltage or current. Since Spice does not support noise current at the output port we define `ino` as an input current source and attach it to the drain of the MOSFET where we would like to characterize the PSD. Output port definition is irrelevant for our purposes, which is why its definition `v(2)` is quite arbitrary.

- Since we need to determine S_{fl} only at a single frequency, the start and stop frequencies of the noise analysis are set to the same 10 Hz. The reason why we select $f = 10$ Hz is that it is low enough to ensure $S_{fl} \gg S_{th}$ so that the PSD determined by simulation can be attributed almost entirely to flicker noise.

- `nrs` and `nrd` are set to zero in order to not involve the noise due to the parasitic resistances of the source and drain terminals of the MOSFET.

- Noise analysis is repeated for 17 different values of the gate-source voltage as defined inside the `foreach–end` loop. These analyses are numbered in ascending order starting with 1 for $V_{GS} = 0.1$ V and ending with 17 for $V_{GS} = 1.0$ V.

The square root $S_{fl}^{1/2}$ of the PSD of noise current flowing in `ino` and therefore in the MOSFET is identified by

```
noiseX.inoise_spectrum
```

where X = [1, … , 17] is the order of the analysis in which it is determined. These values of $S_{fl}^{1/2}$ are written into the data file `nmos-flicker` as instructed with the `wrdata` line. Each such value, preceded by an entry of the frequency 10 Hz at which the simulation has been run, will appear in a data string inside this data file. The threshold voltage `@m[vth]` will appear as the last entry of the string. We process this data by (a) copying the entire string to a spreadsheet software, (b) deleting all 10-Hz entries, (c) squaring the PSD data to obtain S_{fl}, and dividing the outcome by

$$\frac{W}{fL^3} = \frac{W_{DR} + \mathtt{XW} - 2 \times \mathtt{WINT}}{f\left(L_{DR} + \mathtt{XL} - 2 \times \mathtt{LINT}\right)^3} = 1.33 \frac{1}{\mathrm{Hz} \cdot \mu\mathrm{m}^2},$$

as stated just before the input file, and (d) pairing each PSD value with the corresponding overdrive voltage calculated from the corresponding `vgs` − `@m[vth]`. Plotting the outcome yields K in units of $\mathrm{A}^2\mu\mathrm{m}^2$ as a function of $V_{GS} - V_{TN}$. The plot, presented in Fig. 3.44(b), is valid for any NMOS device of the model deck `bsim.sp`.

The reader is now invited to construct a circuit schematic and the corresponding Spice input file for extracting K for PMOS devices. For our exemplar model deck `bsim.sp`, simulation should yield the plot shown in Fig. 3.44(c).

Power Spectral Density of Total Noise Current

The schematic given in Fig. 3.44(a) and its PMOS counterpart are also good for characterizing the PSD $S_{th}+S_{fl}$ of the total noise current of a MOSFET as a function of frequency for a given bias condition. The following input file `nmos-noise.sp` is an example serving this purpose for an NMOS device:

```
filename: nmos-noise.sp
.incl bsim.sp
m 2 1 0 0 cmosn w=6u l=0.8u nrd=0 nrs=0
vgs 1 0 dc 0.6
gd 0 2 3 2 1e6
vds 3 0 0.9
ino 0 2 dc 0 ac 1
.control
destroy all
noise v(2) ino dec 100 10 1g
setplot noise1
plot inoise_spectrum^2 loglog
.endc
.end
```

The outcome of the simulation run with this file is plotted in Fig. 3.43(b). Notice the dominance of the flicker noise at low frequencies and of the thermal noise at high frequencies.

Example 3.17 Determine the root mean square (rms) output noise voltage of the common-source amplifier stage of Example 3.16 for $f > 10$ Hz at room temperature. Verify the result with a Spice simulation.

Work Shown in Fig. 3.45(a) is the circuit schematic of the amplifier, where $i_{n1}(t)$ and $i_{n2}(t)$ represent the noise current sources of the two devices. We first need to derive the transfer function from each of these sources to the output noise voltage v_n. To that end, we construct the high-frequency equivalent circuit corresponding to this circuit schematic. We do this by referring back to Fig. 3.42(b) of Example 3.16, and (a) replacing the input signal source v_i with a short circuit (assuming zero-source resistance), and (b) representing the parallel equivalent of C_{dd1}, C_{dd2}, and C_L with C_L'. The result, shown in Fig. 3.45(b), indicates a common transfer function for both noise sources.

Straightforward circuit analysis yields

$$A(f) = \frac{v_n}{i_n} = r_{ds1} \parallel r_{ds2} \parallel \frac{1}{j2\pi f C_L'} = \frac{r}{1 + j2\pi f r C_L'},$$

for the transfer function, in which $r \doteq r_{ds1} \parallel r_{ds2}$. The magnitude squared of this transfer function is given by

$$|A(f)|^2 = \frac{r^2}{1 + \left(\frac{f}{f_o}\right)^2}, \tag{3.120}$$

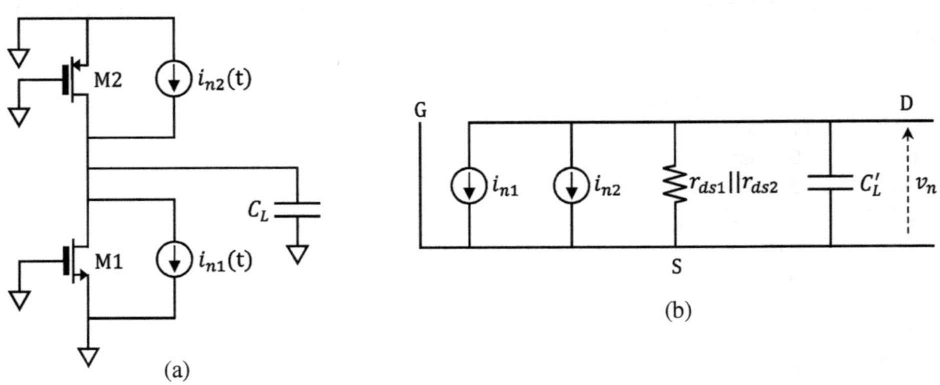

FIGURE 3.45 (a) Noise sources in the current-source-loaded amplifier stage of Example 3.17. (b) Small-signal equivalent circuit for determining the transfer function from noise current to output voltage.

where $f_o \doteq 1/2\pi r C_L'$. Next, we adopt (3.118) and (3.119) for the power spectral densities of thermal and flicker noise components:

$$S_{th1} = \frac{8}{3}E_{th}(g_{m1} + g_{mb1} + g_{ds1}),$$

$$S_{th2} = \frac{8}{3}E_{th}(g_{m2} + g_{mb2} + g_{ds2}),$$

$$S_{fl1} = K_1\frac{W_1}{fL_1^3},$$

$$S_{fl2} = K_2\frac{W_2}{fL_2^3}.$$

Now, we determine the output power spectral density $S_o(f)$ by substituting these four equations together with (3.120) into (2.107), and integrate the latter over f between $f = f_1$ and $f = \infty$ to obtain the following expression for the ms output noise voltage:

$$V_{no}^2 = r^2 f_o(S_{th1} + S_{th2})\left(\frac{\pi}{2} - \arctan\frac{f_1}{f_o}\right) + r^2\left(K_1\frac{W_1}{L_1^3} + K_2\frac{W_2}{L_2^3}\right)\ln\sqrt{1 + \left(\frac{f_o}{f_1}\right)^2}.$$

Assuming $f_o \gg f_1 = 10$ Hz, this equation is well approximated by

$$V_{no}^2 = \frac{\pi}{2}r^2 f_o(S_{th1} + S_{th2}) + r^2\left(K_1\frac{W_1}{L_1^3} + K_2\frac{W_2}{L_2^3}\right)\ln\left(\frac{f_o}{f_1}\right).$$

For numerical calculation, we adopt $C_L' = 204$ fF, $g_{m1} = 244$ μA/V, and $r = 340$ kΩ directly from the solution of Example 3.16. We also rerun the operation-point analysis of the input file `ex3-16.sp` to obtain $g_{m2} = 240$ μA/V, $g_{mb1} = 68.3$ μA/V, $g_{mb2} = 74.6$ μA/V in addition to the overdrive voltages $V_{GS1} - V_{TN} = 0.167$ V, $V_{GS2} - V_{TP} = -0.134$ V for which Figs. 3.44(b) and (c) indicate $K_1 = 4.4 \times 10^{-20}$ A²μm² and $K_2 = 9.9 \times 10^{-21}$ A²μm². Effective dimensions, calculated from (3.1) and (3.2), are $W_1 = 5.98$ μm, $L_1 = 0.76$ μm, $W_2 = 24$ μm, and $L_2 = 0.74$ μm. These numbers lead to $f_o = 2.29$ MHz. Considering also $f_1 = 10$ Hz, we arrive at $V_{no} = 2.05$ mV.

Spice verification can be performed with the very same input file `ex3-16.sp` of Example 3.16 after being supplemented with a noise-analysis line

```
.noise v(2) vi dec 100 10 1g
```

V_n, printed with `print onoise_total`, turns out to be 2.06 mV. ▲

3.9 MOSFET as a Switch

As discussed in Sections 2.11 and 2.12, external networks of discrete-time amplifiers contain switches. A switch is implemented with either a single n-channel or p-channel MOSFET, or a parallel combination of the two. The latter type of switch is known as *transmission gate*. In this section, we begin with the first-order properties of these two types of switches, and subsequently move to a description of the parasitic *charge-injection* and *clock-feedthrough* effects associated with switching MOSFETs. The reader will recall from the contents of Subsection 2.11.3 that these two effects are responsible for the charge error observed in the dc response of discrete-time amplifier configurations.

3.9.1 Single-Device and Transmission-Gate Switch Properties

An NMOS-only switch is depicted in Fig. 3.46(a), where v_I is the input voltage, v_O is the output voltage, and v_G is the gate voltage. All of these voltages are defined with respect to the ground. Also notice the termination of the output port by an equivalent load capacitance C_L representing the switched capacitors and other parasitic capacitances attached to this port.

The state of the switch is controlled primarily by the binary v_G. When the latter is at the negative rail in a split-supply configuration or at the ground in a single-supply

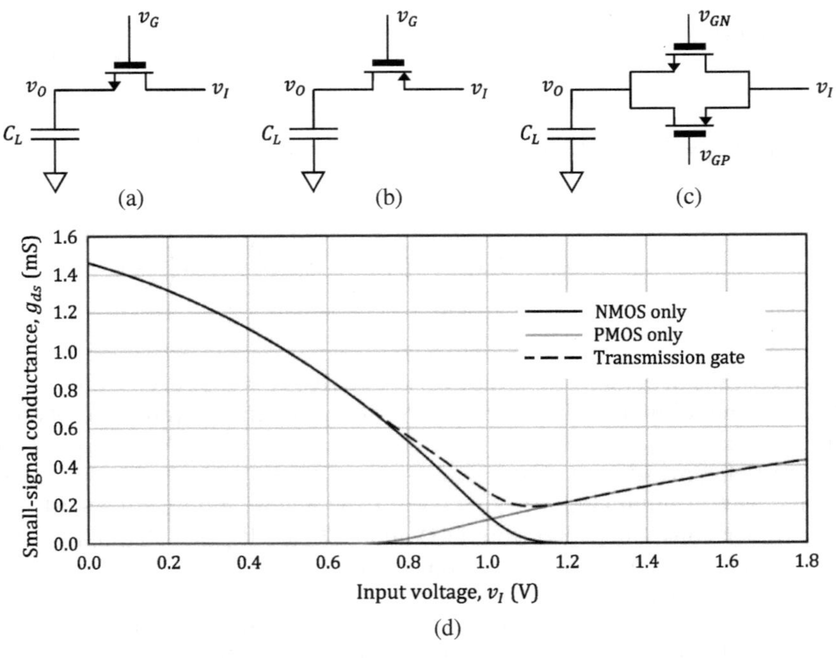

FIGURE 3.46 (a) NMOS-only switch. (b) PMOS-only switch. (c) Transmission-gate switch. (d) Variation of small-signal conductance with input voltage in NMOS, PMOS, and transmission-gate switches. Device dimensions are $W_{DR} = 1$ µm and $L_{DR} = 0.2$ µm in all cases.

configuration, the NMOS device remains in cutoff regardless of the input and output voltages. Since conductance virtually vanishes in cutoff, the device acts like an open switch. When v_G is raised to the positive rail V_{DD}, the device exits cutoff and acts like a closed switch with a finite conductance provided that the smaller between v_I and v_O is also smaller than $V_{DD} - V_{TN}$. Otherwise it continues staying in cutoff as implied by (3.8).

For $v_G = V_{DD}$, the typical variation of the dc small-signal conductance g_{ds} of an NMOS switch with the input voltage v_I is displayed in Fig. 3.46(d) by the plot labeled "NMOS only." This plot has been extracted with a dc-sweep simulation on a standard-threshold device of dimensions $W_{DR} = 1$ µm and $L_{DR} = 0.2$ µm. It clearly shows how the conductance continuously decreases with an increasing v_I and vanishes beyond $v_I = 1.1$ V. This observation indicates a threshold voltage around $V_{TN} = 1.8 - 1.1 = 0.7$ V, which is indeed close to what Fig. 3.7 indicates for $L_{DR} = 0.2$ µm and $V_{SB} = 1.1$ V. Note that a high conductance is desirable for a switch because it lowers the time constant of settling. This is why an NMOS switch can offer an acceptable level of dynamic performance only if v_I is confined to the lower half of the rail-to-rail supply-voltage range.

A PMOS switch, shown in Fig. 3.46(b), acts as an open switch for $v_G = V_{DD}$, and is conditionally closed when v_G is lowered to V_{SS} in a split-supply configuration or to the ground in a single-supply configuration. The condition of closure is that the larger between v_I and v_O should be larger than $V_{SS} - V_{TP}$ in a split-supply configuration, or larger than $-V_{TP}$ in a single-supply configuration.

The conductance g_{ds} of a PMOS switch increases with v_I as indicated by the plot extracted with a dc-sweep simulation on a PMOS device of $W_{DR} = 1$ µm and $L_{DR} = 0.2$ µm and labeled "PMOS only" in Fig. 3.46(d). Obviously, a reasonable dynamic performance can be displayed by a closed PMOS switch only if v_I happens to be in the upper half of the rail-to-rail supply-voltage range where the conductance is high. Also note that the conductance of a PMOS switch is considerably lower than the conductance of a comparably sized NMOS switch. This is due to the fact that the mobility of holes is approximately one-third of the mobility of electrons. However, this deficiency can be compensated for by widening the PMOS switch.

Considering the fact that NMOS and PMOS switches perform well in complementary ranges of v_I, a full rail-to-rail performance can be achieved by paralleling the two devices as depicted in Fig. 3.46(c). This combination is what we call a *transmission gate*. Obviously, a transmission gate is open when v_{GP} is high and v_{GN} is low. Transposing these states closes the switch. The conductance of a transmission gate is expected to be the sum of NMOS and PMOS conductances, which is indeed what the simulated plot marked "transmission gate" in Fig. 3.46(d) indicates.

3.9.2 Charge Injection and Clock Feedthrough

For device-level modeling of charge-injection and clock-feedthrough effects, consider an NMOS switch first, and suppose that it opens at an instant $t = t_1$. Its closed state just before this instant is detailed in Fig. 3.47(a), where C_L represents the capacitive load at the output and C_{ov} represents the extrinsic overlap capacitance modeled in Subsection 3.7.3 by (3.92) or (3.93). Notice the instantaneous value $v_I(t_1^-) = V$ of the input voltage and $v_G(t_1^-) = V_{DD}$ of the closed state. If settling is already complete by t_1, the instantaneous

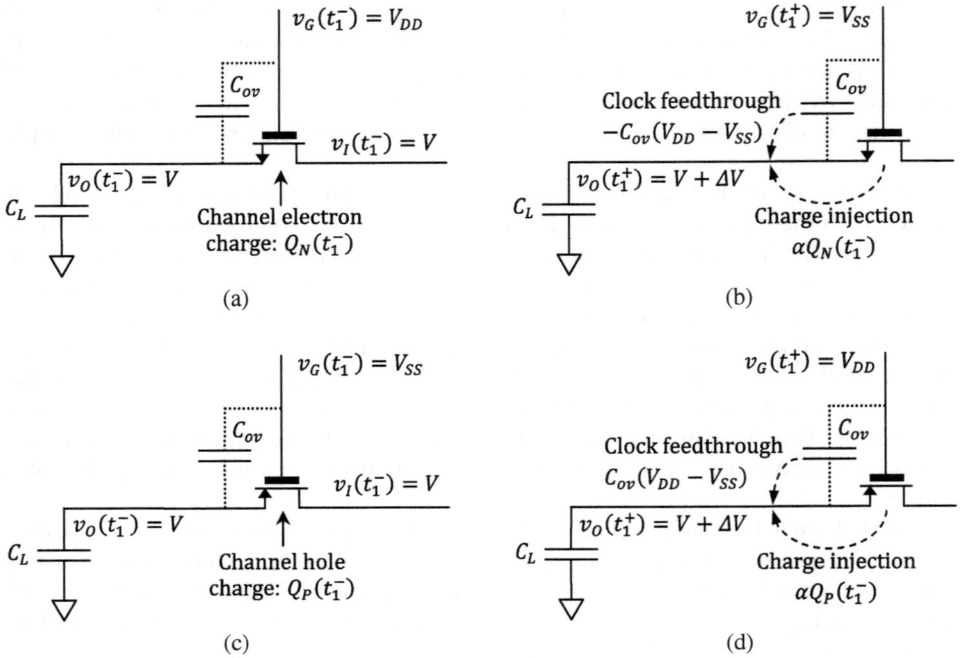

FIGURE 3.47 (a) Closed NMOS switch. (b) Open NMOS switch. (c) Closed PMOS switch. (d) Open PMOS switch.

value of the output voltage equals V too. According to the strong-inversion model, total electron charge associated with the inversion layer of the channel under these conditions is given by

$$Q_N(t_1^-) = -WLC_{ox}(V_{DD} - V - V_{TN}).$$

At $t = t_1$, v_G is lowered to V_{SS} in a split-supply configuration or to ground in a single-supply configuration, and therefore the device is forced into cutoff. As a result of cutoff, $Q_N(t_1^-)$ is transferred to input and output nodes because the inversion layer can no longer be sustained. The part $\alpha Q_N(t_1^-)$ injected to the now floating output node as depicted in Fig. 3.47(b) for $t = t_1^+$ is responsible for what we call the charge-injection effect. It is one of the two components of the charge error Q defined in Subsection 2.11.3. The fraction α depends on various factors including the capacitive load C_L, the transition rate of v_G, and the resistance of the input source, but its value is typically in between 0.25 and 0.75.

Now consider the overlap capacitance C_{ov}. At $t = t_1$, the voltage of the upper plate of this capacitance is lowered from V_{DD} to V_{SS} or to the ground depending on the power-supply configuration. Assuming the general case of a split-supply configuration, and a negligible output voltage perturbation ΔV in comparison with $V_{DD} - V_{SS}$, this change reduces the charge of the upper plate by $C_{ov}(V_{DD} - V_{SS})$. Since an equal amount of charge leaves the lower plate at the output node, the total charge of the output node will change by $-C_{ov}(V_{DD} - V_{SS})$. This is the so-called clock-feedthrough component of the charge error Q.

Combining the contributions of charge injection and clock feedthrough, we can describe the total charge delivered to the output node with

$$Q = \underbrace{-\alpha WLC_{ox}(V_{DD} - V - V_{TN})}_{\text{Charge injection}} \underbrace{-C_{ov}(V_{DD} - V_{SS})}_{\text{Clock feedthrough}} \cdot \tag{3.121}$$

Note that, since both components of Q are negative in an NMOS switch, the resulting perturbation ΔV in the output voltage is in the negative direction. Also notice the proportionality of the charge-injection component to channel length L and width W. Clock feedthrough is also proportional to channel width as indicated by (3.92) and (3.93). This is why a small device size is essential for minimizing Q.

A PMOS switch is closed for $V_G = V_{SS}$ and open for $V_G = V_{DD}$. The hole charge residing in its channel prior to opening is represented by $Q_P(t_1^-)$ in Fig. 3.47(c), and is modeled by

$$Q_P(t_1^-) = -WLC_{ox}(V_{SS} - V - V_{TP}).$$

This charge is positive because the overdrive voltage $V_{SS} - V - V_{TP}$ is negative prior to $t = t_1$. Opening the switch injects a fraction $\alpha Q_P(t_1^-)$ of this charge into the output node. The charge contributed to the same node by clock feedthrough amounts to $C_{ov}(V_{DD} - V_{SS})$. Therefore, the total charge received by the output node due to switch opening is described by

$$Q = \underbrace{-\alpha WLC_{ox}(V_{SS} - V - V_{TP})}_{\text{Charge injection}} \underbrace{+C_{ov}(V_{DD} - V_{SS})}_{\text{Clock feedthrough}} , \tag{3.122}$$

which, being definitely positive, perturbs the output node in the positive direction. Except for the polarity, however, all of its features are qualitatively similar to those of an NMOS switch.

As pointed out above, charge error is an increasing function of MOSFET channel dimensions. Adopting the minimum permissible dimensions therefore helps minimize this error. If the charge error thus minimized is still unacceptably large, we may prefer supplementing the output node with a *dummy switch* or using a transmission gate instead of a single MOSFET. We now discuss these two techniques of charge-error reduction.

The dummy-switch technique is illustrated in Fig. 3.48 on an NMOS switch. Notice the placement of short-circuited source and drain terminals of a dummy NMOS device MD to the output node of the main switch MM. The clock driving the gate terminal of MD is complementary to the clock driving the gate of MM. Channel length is minimized in both devices but the width of MD is exactly one half of the width of MM. Each of the gate-to-source and gate-to-drain overlap capacitances of MD is therefore exactly one half of the gate-to-output overlap capacitance C_{ov} of MM. However, since both source and drain terminals of MD are connecting the output node, the total gate-to-output overlap

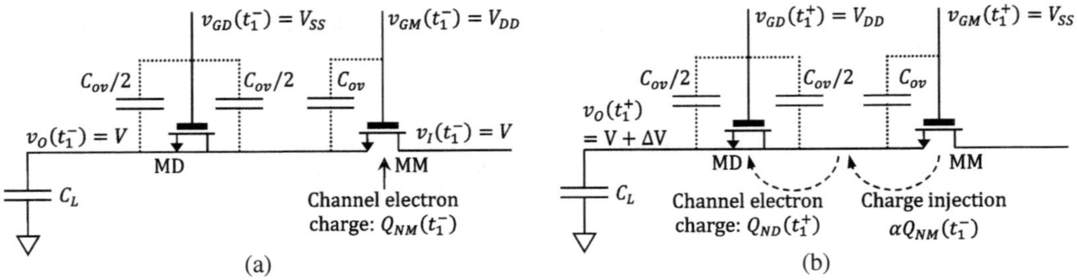

FIGURE 3.48 Dummy switch. (a) Before opening. (b) After opening.

capacitance of this device also equals C_{ov}. Thanks to equal overlap capacitances and complementary clocks, the clock-feedthrough effects of MM and MD cancel out.

Now, consider the state of the MOSFETs at $t = t_1^-$ just prior to switch opening as illustrated in Fig. 3.48(a). Electron charge in the inversion layer of MM is described by

$$Q_{NM}(t_1^-) = -WLC_{ox}(V_{DD} - V - V_{TN}),$$

whereas the cutoff state precludes inversion in MD. Next, consider the state of these devices at $t = t_1^+$ just after switch opening as shown in Fig. 3.48(b). On the one hand, MM enters into cutoff and releases a fraction $\alpha Q_{NM}(t_1^-)$ of its inversion charge to the output node. On the other hand, MD is driven into inversion, and therefore removes a charge

$$Q_{ND}(t_1^+) = -\frac{1}{2}WLC_{ox}(V_{DD} - V - V_{TN})$$

from the same node in order to establish its own inversion layer.[8] Therefore, the net charge injected into the output node is given by

$$Q = \left(\frac{1}{2} - \alpha\right) WLC_{ox}(V_{DD} - V - V_{TN}). \tag{3.123}$$

Comparing this expression with (3.121), we conclude that the dummy-switch technique not only eliminates clock feedthrough but also reduces charge injection because, as pointed out above, α is usually within the range 0.25 to 0.75.

A transmission gate built with identically sized devices has a first-order immunity against clock feedthrough because devices are controlled by complementary clocks. Its charge-injection performance is similar to a single-NMOS switch for small values of the input voltage and to a single-PMOS switch for large values of the input voltage. However, it performs better than single-device switches for midrange input voltages. To see why, consider a closed transmission gate. For $V < V_{SS} - V_{TP}$, its PMOS is in cutoff with a negligible hole charge in its channel. NMOS, on the other hand, is strongly inverted

[8] This equation is valid as long as the perturbation caused by charge injection is negligible in comparison with V.

with a negative electron charge. Therefore, switch opening causes only the NMOS to inject charge to the output node. For $V > V_{DD} - V_{TN}$, the roles of NMOS and PMOS are transposed. Now, only the PMOS device contains charge in its channel when the switch is closed. Opening the switch causes part of this positive charge to be injected. If, however, V is in between $V_{SS} - V_{TP}$ and $V_{DD} - V_{TN}$, both devices contain a mobile charge of opposite polarity in their respective channels when the switch is closed. Opening the switch results in positive charge injection from PMOS, and negative charge injection from NMOS. The net charge injected is therefore less than what would be injected by a single device. As a matter of fact, the net charge injected vanishes if V happens to be midway between V_{SS} and V_{DD}. For an analytical proof of this fact, consider a transmission gate with identically sized NMOS and PMOS devices. Also assuming an identical injection factor α for both devices, and summing up the charge-injection components of (3.121) and (3.122), the net charge injected is formulated as

$$Q = -\alpha WLC_{ox}(V_{DD} + V_{SS} - 2V - V_{TN} - V_{TP}).$$

Obviously the charge vanishes for

$$V = \frac{V_{DD} - V_{TN} + V_{SS} - V_{TP}}{2}.$$

Since $V_{TN} + V_{TP}$ is usually very small, the voltage indicated by this equation is indeed close to $(V_{DD} + V_{SS})/2$.

Example 3.18 In this example, we simulate the transient response of the discrete-time amplifier of Example 2.32 with transmission-gate switches.

Work Input file is as follows:

```
filename: ex3-18.sp
.incl bsim.sp
.param vol=0.15 voh=1.65 vdir=5 gm0=1m ro=1meg
.param width=4u
*****ota*****************************************
.subckt ota ninv inv out
gm 0 out cur='gm0*vdir*tanh((v(ninv)-v(inv))/vdir)'
rout out 0 ro
bb 0 out i=0.9/ro
.ends ota
*****transmission gate************************
.subckt tg in out ngate pgate
mn in ngate out 0 cmosn w=width l=0.2u ad={width*0.55e-6}
+  pd={1.1e-6+2*width} as={width*0.55e-6} ps={1.1e-6+2*width}
mp in pgate out 1 cmosp w=width l=0.2u ad={width*0.55e-6}
+  pd={1.1e-6+2*width} as={width*0.55e-6} ps={1.1e-6+2*width}
vbulk 1 0 1.8
.ends
******amplifier********************************
xota 1 2 3 ota
xs1 2 3 7 9 tg
xs2 6 4 7 9 tg
```

```
xs3 6 5 7 9 tg
xs4 4 3 8 10 tg
xs5 5 1 8 10 tg
c1 5 2 2p
c2 4 2 1p
ci 3 0 0.5p
vb 1 0 0.9
vi 6 1 pwl 0 -0.15 74n -0.15 76n 0.15 189n 0.15
+  191n 0 300n 0 302n -0.15 441n -0.15
vphi1 7 0 pulse 1.8 0 47n 2n 2n 58n 112n
vphi1c 9 0 pulse 0 1.8 47n 2n 2n 58n 112n
vphi2 8 0 pulse 0 1.8 51n 2n 2n 50n 112n
vphi2c 10 0 pulse 1.8 0 51n 2n 2n 50n 112n
.tran 10p 441n
.options trtol=1
.end
```

Notice two differences with respect to the original file `ex2-32.sp`. First is the replacement of the ideal-switch subcircuit with a transmission-gate subcircuit, where the channel width of NMOS and PMOS devices is represented by the parameter `width`. Junction dimensions appearing in element lines are also expressed in terms of this parameter for automatic calculation by the simulator. The second difference is the inclusion of the complementary clocks `vphi1c` and `vphi2c` for driving the PMOS devices.

The reader is invited to examine the differences between (a) the ideal output-voltage waveform obtained with the original file, and (b) the one obtained with the present file for the following values of channel width in micrometers: 0.5, 1, 2, 4, 8, 16, 32, 64, 128, and 256. Which channel width offers the best performance? Try to discover the root causes of inferior performance observed for narrower or wider channels. ▲

3.10 Resistor Design

With the exception of unity-gain buffer, all continuous-time amplifier configurations we studied in Chapter 2 contain resistors in their external network. As will be discussed in Chapter 4, resistors also find use inside opamps, most notably within supplementary circuits that generate a current or voltage reference for biasing purposes. In this section, we elaborate on integrated resistor design in CMOS technology.

3.10.1 Resistor Structures and Resistance Modeling

A resistor can be built in any of the standard semiconductor layers of the CMOS structure or in a dedicated layer if offered as a process option. Inspecting Fig. 3.1, we identify n-well, p^+, n^+, and doped poly as standard layers suitable for resistor implementation in a twin-well CMOS technology. Cross-section and top views of basic strip-geometry resistors built in these layers are shown in Figs. 3.49(a) to (d). The resistor structures shown in Figs. 3.49(e) and (f) are built-in dedicated layers as will be described later. Note that in all options, v_A and v_B represent terminal voltages defined with respect to the

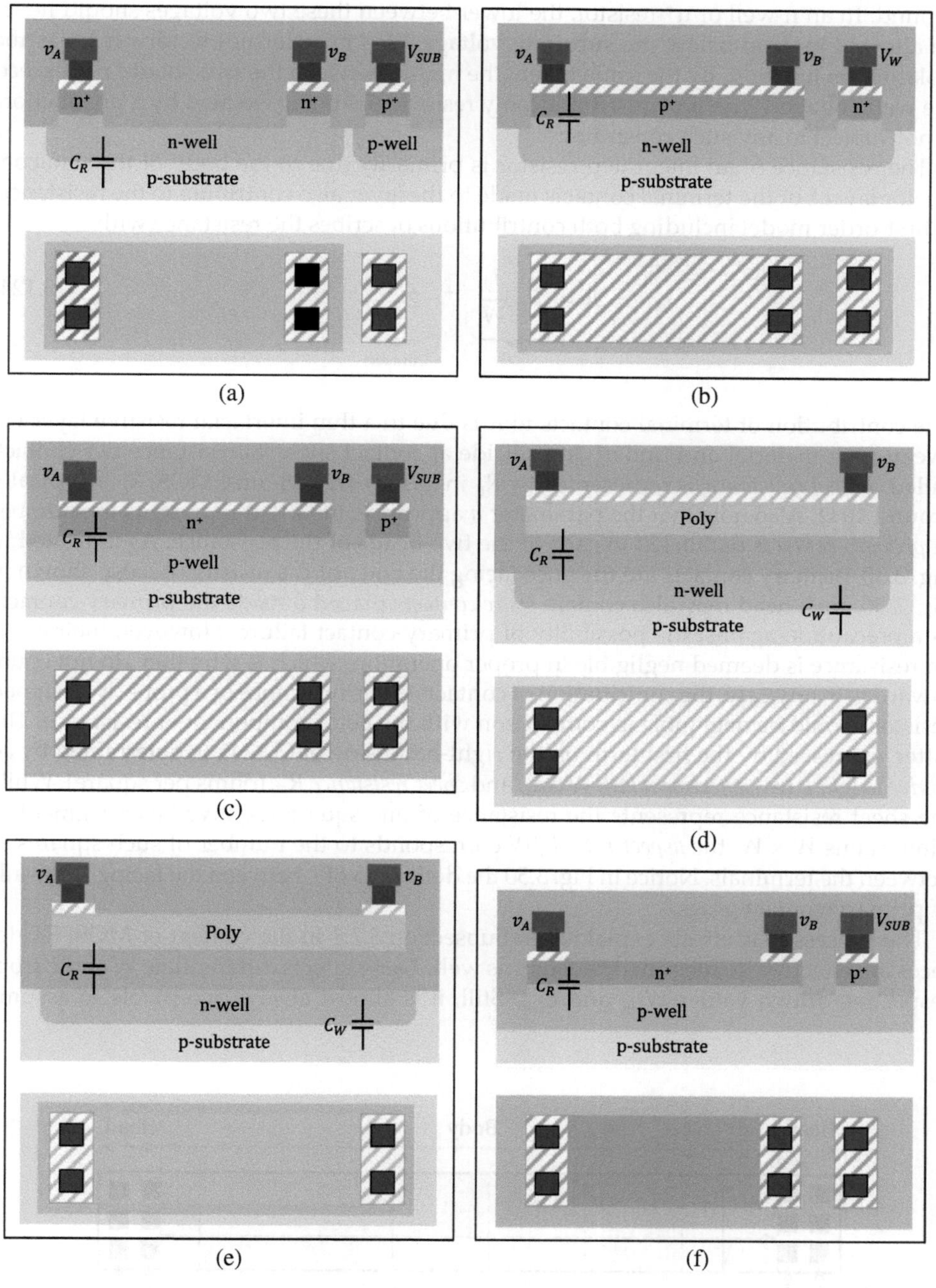

FIGURE 3.49 Resistors built in CMOS semiconductor layers. (a) n-well. (b) p^+. (c) n^+. (d) Poly. (e) Poly with silicide block. (f) n^+ with silicide block.

ground. In an n-well or n^+ resistor, the lower between these two voltages should never be allowed to drop below the substrate voltage V_{SUB} in order not to forward bias the isolating pn junction. By the same token, the higher between the two should not exceed the well voltage V_W in a p^+ resistor. A poly resistor, not being isolated by a pn junction, is not subject to any such constraint.

The resistance of an integrated resistor is primarily due to the body of the semiconductor layer but the terminal contacts made to the layer also contribute to the resistance. A first-order model including both contributions describes the resistance with

$$R = \underbrace{R_{\square}\frac{L}{W}}_{\text{Body}} + \underbrace{2\frac{R_C}{n}}_{\text{Contacts}} . \tag{3.124}$$

The contribution of terminal contacts to R is due to a thin interfacial resistive layer between plug material and underlying silicide at contact sites. Its resistance per contact, called *contact resistance*, is represented by R_C in (3.124). In 0.18-μm CMOS, R_C is typically around 10 Ω. Also note that the parameter n appearing in (3.124) represents the number of *primary contacts* paralleled in each of the two heads of the resistance. As indicated in Fig. 3.50, primary contacts are the ones facing the body of the resistor. As also shown in Fig. 3.50, each head may also contain *spare contacts* placed outside the primary contacts as a precaution against the possibility of primary-contact failure. However, their effect on resistance is deemed negligible in proper operation, which is why they do not count toward n. In any case, the contribution of contacts to the resistance of a properly designed resistor should be negligible in comparison with the body component of resistance. The latter, expressed by the first term on the right-hand side of (3.124), is determined by *effective length* L (m), *effective width* W (m), and *sheet resistance* $R_{\square}$ (ohms per square). While the sheet resistance represents the resistance of one square resistive-layer segment of dimensions $W \times W$, the *aspect ratio* L/W corresponds to the number of such squares in between the terminals. Notice in Fig. 3.50 the definition of L between the facing sidewalls of primary-contact plugs.

The process-bias effects explained in Subsection 3.2.1 in the context of MOSFET devices are effective in integrated resistors as well. These effects differentiate W and L from respective drawn values W_{DR} and L_{DR}. Still, it is almost always acceptable to assume

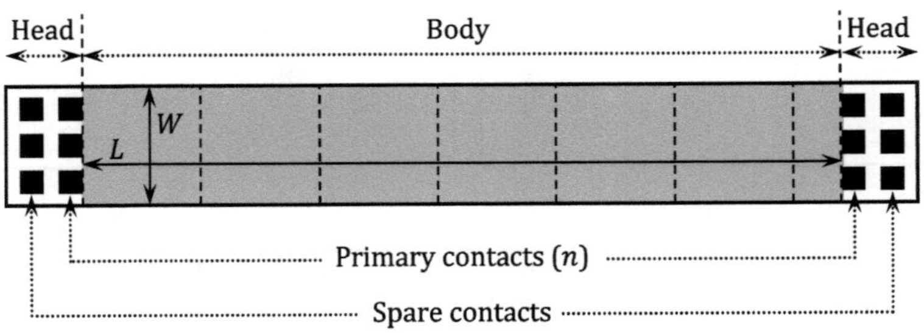

FIGURE 3.50 Length L, width W, and primary and spare contacts in a resistor structure. Gray shading exemplifies a body length of 6.4 squares.

Parameter	Standard layers				Dedicated layers			
	n-well	n^+	p^+	Poly	n^+ with silicide block	p^+ with silicide block	Poly with silicide block	High-resistivity poly
$R_\square$ (Ω/sq)	950	7	8	8	60	120	300	1100
ΔW (μm)	0.2	−0.08	−0.08	−0.06	0	0	0.08	0.08
$W_{DR(\min)}$ (μm)	1.2	0.3	0.3	0.2	0.3	0.3	0.5	0.4

TABLE 3.4 Resistor Layer Data

$L \approx L_{DR}$ because L_{DR} is usually much longer than the process bias associated with the length. This is why we consider only the width bias

$$\Delta W \doteq W_{DR} - W \tag{3.125}$$

in resistance calculations.

Typical values of $R_\square$, ΔW, and minimum drawn width $W_{DR(\min)}$ in a 0.18-μm CMOS technology are presented in Table 3.4 for all layers of interest. Notice that the sheet-resistance values associated with the standard n^+, p^+, and poly layers are relatively low, which is due to the silicidation process applied intentionally for minimizing the parasitic resistances of MOSFET terminals. Implementing, say, a 10-kΩ resistor with any of these three layers would necessitate an aspect ratio in excess of a thousand as one can ascertain from (3.124). Since the minimum of W is technologically limited, such a large aspect ratio implies a very long resistor geometry, hence a large footprint and a large parasitic capacitance. This is why these layers are suitable only for implementing very low resistances. Among the standard layers, only n-well has a sufficiently high sheet resistance for implementing a high-resistance resistor in a relatively small area but its nonlinearity is also high. Obviously, none of the resistor structures shown in Figs. 3.49(a) to (d) is close to ideal.

Two options in most CMOS technologies offer dedicated resistor structures of higher utility. One of the options, called *silicide block*, enables us to turn n^+, p^+, or doped poly into a dedicated layer by locally blocking the silicide formation. As listed in Table 3.4, the sheet resistance of an unsilicided layer is an order of magnitude higher than its silicided counterpart. Thanks to this property, a high-resistance resistor can be implemented in a much smaller area. The cross-sectional and top views of such poly and n^+ resistors are shown in Figs. 3.49(e) and (f). Notice the retention of silicide under the contacts in order to avoid rectification due to Schottky-barrier formation. The second dedicated-layer option is known as *high-resistivity poly*. It relies on the undoped body of a second layer of poly introduced specifically for resistor and capacitor integration. Its geometry is similar to the doped poly resistor shown in Fig. 3.49(e) but, as cited in Table 3.4, it offers a much higher sheet resistance.

The resistance of an integrated resistor is not perfectly linear because the sheet resistance somewhat depends on the applied voltage. This dependence has several different root causes. One, observed in junction-isolated resistors, is the modulation of the effective thickness of the resistor by the voltage-dependent junction depletion region. This effect is more pronounced in layers of relatively lighter doping, which is why n-well resistors are the most nonlinear among all.

Another effect that causes nonlinearity by making the sheet resistance an increasing function of voltage in all types of resistors is the degradation of carrier mobility by the longitudinal electric field. Since this electric field is well approximated by $|v_A - v_B|/L$, the effect is obviously most pronounced in short resistors operating with a high voltage.

The unique grainy morphology of polysilicon is yet another cause for nonlinearity. Being field dependent, this type of nonlinearity is also significant in short poly resistors operating with a high voltage.

Finally, the variation of sheet resistance with temperature is an indirect cause for nonlinearity because the temperature depends on power dissipation hence voltage. This type of nonlinearity is most pronounced in poly resistors whose oxide isolation has a very high-thermal resistance. Still, however, poly resistors are usually the best in overall linearity.

The temperature sensitivity of resistance in junction-isolated resistors is due to mobility degradation and has a positive value. In poly resistors a negative temperature sensitivity may be observed due to additional conduction mechanisms involved at grain boundaries. The effect of voltage and temperature on resistance is conventionally described by generalizing the body component of (3.124) into the polynomial form

$$R = R_{\square}\frac{L}{W}(1 + \mathrm{VCR1} \times v + \mathrm{VCR2} \times v^2)\left[1 + \mathrm{TCR1} \times (T - T_0) + \mathrm{TCR2} \times (T - T_0)^2\right], \tag{3.126}$$

where VCR1 are VCR2 voltage coefficients, TCR1 and TCR2 are temperature coefficients, $v \doteq |v_A - v_B|$ is the absolute value of the voltage across the resistor, and T_0 is the temperature at which resistance parameters are characterized. Most technology providers adopt $T_0 = 25°\mathrm{C}$. The model described by (3.126) is reasonably accurate for temperature sensitivity but not quite so for nonlinearity. A curious reader may find more accurate variants in most PDKs and related literature. Also note that TCR1 is around 3000 ppm/C in standard layers and 1000 ppm/C in dedicated layers.[9] In some technologies, poly layers may exhibit a negative TCR1. TCR2 is typically around 0.5 to 1 ppm/C^2 in most layers except for high-resistivity poly which usually exhibits a slightly higher value.

Dielectric isolation by means of a junction-depletion region or by an oxide film gives rise to a parasitic capacitance between a resistor and the underlying semiconductor region as represented by C_R in Fig. 3.49. The simplest modeling expression of this capacitance is

$$C_R = C_A A + C_F P, \tag{3.127}$$

where A and P are the area and perimeter of the resistor, and C_A and C_F are two coefficients. Note that A and P should include not only the body of the resistor but also the heads. The first term on the right-hand side of (3.127) represents the areal component. The second term is the component associated with the fringing field along the perimeter. Some PDKs may utilize more sophisticated variants of (3.127) involving voltage-dependent descriptions for C_A and C_P or a distributed capacitance.

---

[9] "ppm" stands for *parts per million*, that is, 10^{-6}.

Poly resistors usually exhibit the smallest C_A and C_P thanks to the thick oxide used for their isolation. Parasitic capacitance can be further reduced in poly resistors by placing a floating well on the underlying silicon surface as depicted in Figs. 3.49(d) and (e). The series combination of C_R and the well-to-substrate capacitance C_W reduces the effective capacitance to $C_R C_W/(C_R + C_W)$. Finally, note $C_A = 100$ aF/μm² and $C_F = 40$ aF/μm as the typical coefficient values for dedicated poly resistors in 0.18-μm CMOS technologies.

Example 3.19 Ignoring the contact resistances and relying on the data of Table 3.4, determine the necessary length of a 20-kΩ resistor of minimum width for the following layers: n^+, n^+ with silicide block, p^+ with silicide block, poly with silicide block, and high-resistivity poly.

Work

- n^+: $W = 0.38$ μm from (3.125), and $L = 1085$ μm from (3.124).
- n^+ with silicide block: $W_{DR} = 0.3$ μm from (3.125), and $L_{DR} = 100$ μm from (3.124).
- p^+ with silicide block: $W_{DR} = 0.3$ μm from (3.125), and $L_{DR} = 50$ μm from (3.124).
- Poly with silicide block: $W_{DR} = 0.42$ μm from (3.125), and $L_{DR} = 28$ μm from (3.124).
- High-resistivity poly: $W_{DR} = 0.32$ μm from (3.125), and $L_{DR} = 5.82$ μm from (3.124). ▲

3.10.2 Design Techniques for Accuracy and Precision

Electrical design of an integrated resistor is completed by determining W_{DR} and L_{DR} after selecting a semiconductor layer that can deliver the specified resistance with acceptable nonlinearity, temperature sensitivity, and parasitic capacitance. Design is based on the mean values of $R_{\square}$, ΔW, and R_C as characterized by the technology vendor over the past runs of the target technology and declared in the PDK as typical values. However, none of the fabricated copies of the resistor will exhibit the target resistance exactly. Their resistance will be distributed randomly around a mean value which itself will be offset from what has been targeted in design. As explained in Subsection 2.4.2, these imperfections will contribute to the imprecision and inaccuracy of any closed-loop performance metric that depends on the resistance of this resistor. Fortunately, the resulting uncertainty can be drastically reduced if we base the metric on the ratio of the resistances of two properly laid-out resistors. A ratio-based metric will also be highly insensitive to temperature. The closed-loop gain of a continuous-time inverting or noninverting amplifier is probably the best-known beneficiary of this practice. Any two resistors intended for accurately and precisely implementing a specific resistance ratio are called *ratioed resistors*. *Ratioing* is actually a very general design practice applied to capacitors and MOSFETs as well. These additional applications will be covered in Subsections 3.11.1 and 4.2.2.

The accuracy of the ratio of two resistances is measured by the difference between the mean and target values of the ratio. Of course, the lesser the difference, the better the accuracy. The mean value deviates from the target value mainly due to (a) imperfect repeatability of fabrication processes; (b) on-chip gradients in resistance parameters, temperature, and stress; and (c) proximity effects in patterning. Number rounding in design also contributes to inaccuracy. The specific physical-design techniques applied for neutralizing these root causes of inaccuracy in ratioed resistors are *segmentation*, *common-centroid placement*, and *dummy termination*. How these techniques work is described in this subsection.

The precision of the ratio of two resistances is measured by the extent of the distribution of the ratio among fabricated copies. Obviously, the tighter the distribution, the higher the precision. The distribution is mainly due local random microscopic fluctuations observed in such morphological variables as doping concentration, polysilicon grain size, and edge roughness. As will be discussed in this subsection, the effect of such fluctuations can be reduced only by enlarging the size of ratioed resistors.

Segmentation into Matching Units

Before describing the technique of segmentation, let us go through a hypothetical ratioed-resistor design effort without applying any particular technique for reducing the inaccuracy. Consider the resistor pair whose layout is shown Fig. 3.51(a), and suppose that we intend to implement a ratio $R_A/R_B = 1.5$ with unsilicided poly resistors of $R_A = 15$ kΩ and $R_B = 10$ kΩ. Aiming for a regular and compact layout, we

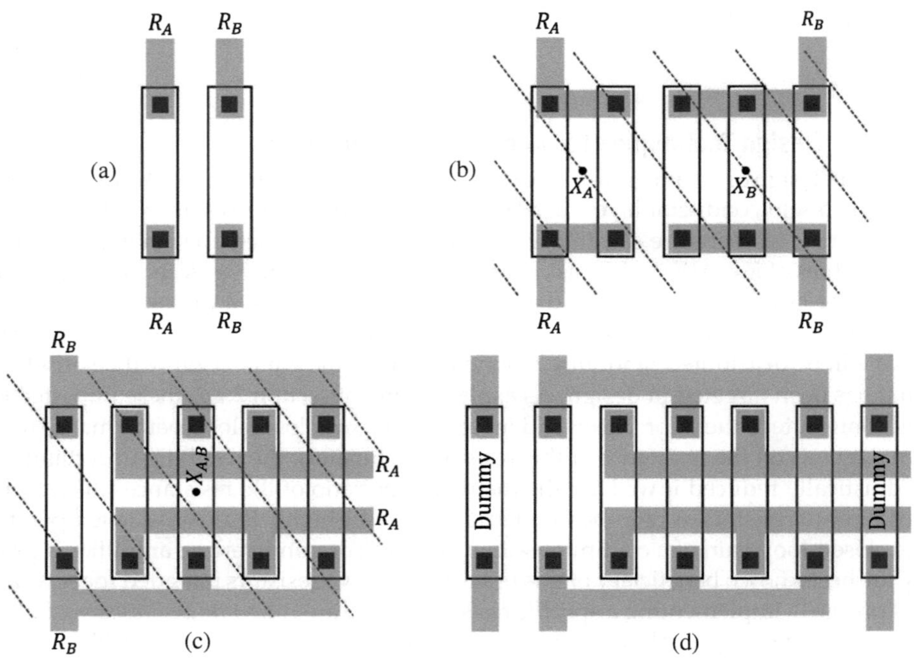

FIGURE 3.51 Resistor ratioing. (a) Ratioing by width adjustment. (b) Ratioing by segmentation into matching unit resistors, where dashed lines are iso-resistivity contours, X_A is the centroid of R_A, X_B is the centroid of X_B. (c) Matching by common-centroid interdigitation where $X_{A,B}$ represents the common centroid. (d) Dummy termination.

decide to adopt equal lengths for the two resistors and the minimum possible drawn-width $W_{DR(A)} = 0.50$ μm for R_A. These numbers, together with $R_C = 10\ \Omega$, $n = 1$, $R_\square = 300$ ohms/sq, and $\Delta W = 0.08$ μm, as cited in Table 3.4, yield $L_A = 20.97$ μm from (3.124) and (3.125) for R_A. Adopting the same length for R_B, we calculate the drawn width of this resistor from (3.124) and (3.125) as $W_{DR(B)} = 0.71$ μm. Notice that all dimensions involved in these calculations are rounded to the nearest hundredth of a micrometer because this is the minimum grid size mandated by the PDK for physical design. If we now calculate the resistance ratio with these rounded numbers, we will end up with $R_A/R_B = 1.4990$, hence a −0.067% systematic deviation with respect to the target value. Therefore, the ratioed resistor layout depicted in Fig. 3.51(a) suffers from a small inaccuracy even before being taped out for fabrication!

Let us now consider the inaccuracies inflicted on the resistance ratio of the layout of Fig. 3.51(a) by the imperfect repeatability of fabrication processes. First, imagine a fabrication run in which the mean value $\overline{R_\square}$ of sheet resistance happens to be 15% smaller than its typical value 300 Ω/sq, while all other process parameters remain at their respective typical values cited in Table 3.4. This results in approximately −15% deviation in the mean values of the two resistances but the mean value of their ratio shifts only by −0.078% including the deviation due to number rounding. The accuracy of the resistance ratio is obviously far better than what each resistance individually offers.

Next, consider a fabrication run in which the mean value $\overline{\Delta W}$ of process bias in fabricated resistors happens to be 0.05 μm instead of the typical 0.08 μm considered in design. Assuming typical values for all other process parameters, and recalculating the deviation from target value in each of the two resistances, we arrive at −6.66% for $\overline{R_A}$, −4.48% for $\overline{R_B}$, and −2.28% for the ratio of the two. Although the ratio of two resistances turns out to be still more accurate than an individual resistance, the extent of improvement is nowhere near what we have observed in the previous case of sheet-resistance deviation. Ratioing alone is obviously good for accuracy but it is not perfect.

Fortunately, the inaccuracy emanating from number rounding or imperfect repeatability of fabrication processes can be eliminated almost entirely in ratioed design by segmenting both resistors into physically *matching* unit resistors connected in series or in parallel. A segmented version of the resistors of Fig. 3.51(a) is shown in part-b of the same figure. Notice the use of two parallel unit resistors for R_A, and three parallel unit resistors for R_B. The mean value of the resistance ratio is therefore related to the mean value $\overline{R_U}$ of a unit resistor by

$$\overline{R_A/R_B} = \frac{(1/2)\times\overline{R_U}}{(1/3)\times\overline{R_U}},$$

which equals 1.5 regardless of how $\overline{R_U}$ is affected by number rounding in design or by imperfections of process repeatability.

Interdigitated Common-Centroid Placement

Segmentation is effective on those inaccuracies arising from the imperfect repeatability of fabrication processes but it is not a cure for inaccuracies that are caused by on-chip nonrandom spatial variations in resistance parameters. Such nonrandom variations are characterized by a common gradient sustained along the same direction in all fabricated copies of the chip. They may be due to spatial nonuniformities in fabrication processes, in operation temperature, or in mechanical stress.[10]

Now, suppose that a resistance parameter, say $R_{\square}$, varies across the segmented pair of resistors R_A and R_B as depicted in Fig. 3.51(b) with dashed iso-resistivity contours. As long as the spatial variation is linear, the resistance of each resistor is still described by (3.124) provided that the value of sheet resistance at the centroid is adopted for $R_{\square}$.[11] This leads us to the conclusion that if the centroids X_A and X_B of the two resistors are placed on the same constant-resistivity contour, the target ratio $R_A/R_B = 1.5$ will still be accurately achieved despite the spatial variation of sheet resistance. However, it is virtually impossible to predict the pattern of on-chip spatial variations. A radical solution to this problem is to lay out the ratioed resistors with coincident centroids. This technique, called *common-centroid placement*, is exemplified in Fig. 3.51(c) where unit resistors are interdigitated in a pattern B A B A B. Another possible common-centroid pattern is A B B B A but this one is less preferable because interdigitation is not as dispersed as it is in the previous one. Dispersion enables common-centroid placement to function as intended even in the case of mildly nonlinear spatial variations. Finally, note that any type of interdigitation is useful for countering the effect of systematic spatial variations of resistance parameters even if a common centroid cannot be established. Common-centroid placement is the ultimately effective form of interdigitation.

Dummy Termination

The process-bias parameter ΔW is affected by the nonuniformity of the etching processes applied in fabrication. Unfortunately, these processes are somewhat sensitive to the proximity of the structural features neighboring a resistor. Consider, for example, the common-centroid poly resistors depicted in Fig. 3.51(c). During fabrication, the outside facing edges of the leftmost and rightmost B units are usually etched faster than all other edges of both units. This leads to a larger effective ΔW hence a narrower W in the leftmost and rightmost units. The resulting resistance ratio will deviate from the target 1.5 despite the perfectly interdigitated common-centroid placement. This problem can be solved by placing dummy features of the same type of resistive layer alongside the outermost resistor segments as shown in Fig. 3.51(d). Being connected to ground or to a dc voltage, these dummies equalize the effect of proximity on etching for all segments without electrically interfering with the operation of the two ratioed resistors.

Sizing for Precision

Distribution of the resistance ratio among the fabricated copies of ratioed resistors leads to imprecision. The main cause of imprecision is the random spatial fluctuation of morphological variables of resistance in microscopic scale. For example, sidewall irregularities lead to a fluctuation in ΔW, whereas irregularities in layer thickness or in doping

[10] Semiconductor resistors are sensitive to mechanical stress due to their piezoelectric property. Chips are permanently stressed in a nonuniform fashion during packaging.

[11] Centroid is defined as the center of mass of an object.

concentration lead to a fluctuation in $R_\square$. The resulting distribution is usually Gaussian with a relative standard deviation described by

$$\frac{\sigma[R]}{R} = \frac{K_R}{\sqrt{LW}}, \tag{3.128}$$

where the so-called Pelgrom coefficient K_R is typically in between 0.04 and 0.15 μm for poly resistors in a typical 0.18-μm CMOS technology.[8] According to this expression, distribution can be tightened by elongating and/or widening the resistor.

Example 3.20 Size the 30-kΩ unsilicided-poly unit resistor used in the ratioed-resistor configuration of Fig. 3.51(d) so that the resistance ratio R_A/R_B does not deviate more than 1% of the target value 1.5 in at least 99.9997% of the fabricated copies. Assume $K_R = 0.04$ μm.

Work Since inaccuracy is practically eliminated by the dummy-terminated common-centroid layout, the specified uncertainty can be attributed solely to imprecision. Since 99.9997% corresponds to 4.5σ, the specification implies

$$\frac{\sigma[R_A/R_B]}{R_A/R_B} \le \frac{0.01}{4.5} = 2.22 \times 10^{-3},$$

which, considering R_A and R_B as two uncorrelated random variables in (2.28), can be expressed as

$$\frac{\sigma[R_A/R_B]}{R_A/R_B} = \sqrt{\frac{\sigma^2[R_A]}{R_A^2} + \frac{\sigma^2[R_B]}{R_B^2}} \le 2.22 \times 10^{-3}. \tag{3.129}$$

Since R_A has two parallel unit resistors and R_B has three, the following expressions hold:

$$R_A = \frac{1}{\frac{1}{R_{U1}} + \frac{1}{R_{U2}}},$$

and

$$R_B = \frac{1}{\frac{1}{R_{U3}} + \frac{1}{R_{U4}} + \frac{1}{R_{U5}}},$$

where R_{Ui}, $i = 1, \ldots, 5$, are five independent random variables representing unit resistances of nominal value R_U and standard deviation $\sigma[R_U]$. Applying (2.28) once more, we arrive at

$$\frac{\sigma^2[R_A]}{R_A^2} = \frac{1}{2}\frac{\sigma^2[R_U]}{R_U^2},$$

and

$$\frac{\sigma^2[R_B]}{R_B^2} = \frac{1}{3}\frac{\sigma^2[R_U]}{R_U^2}.$$

Substitution of these two into (3.129) yields

$$\frac{\sigma[R_A/R_B]}{R_A/R_B} = 0.913\frac{\sigma[R_U]}{R_U} \le 2.22 \times 10^{-3},$$

hence

$$\frac{\sigma[R_U]}{R_U} \le 2.43 \times 10^{-3}.$$

Using this result together with $K_R = 0.04$ μm in (3.128), we obtain

$$WL \ge 271\ \mu\text{m}^2 \tag{3.130}$$

for the effective area of a unit resistor. W and L are interrelated also through (3.124). Ignoring the resistance of the contacts, and adopting $R_U = 30$ kΩ from design specifications and $R_\square = 300$ Ω/sq from Table 3.4, this equation yields $L/W = 100$, which, together with (3.130), yield $W = 1.65$ μm and $L = 165$ μm. According to (3.125) and Table 3.4, drawn width should be set to $W_{DR} = 1.73$ μm. ▲

3.10.3 MOSFET as a Resistor

The drain-source port of a MOSFET operating in nonsaturation can be utilized as a resistor. Considering the strong-inversion characteristic equation (3.20) of the drain-source port of an NMOS device, we can express the small-signal port resistance $r_{ds} \doteq 1/(\partial i_D/\partial v_{DS})$ with the equation

$$r_{ds} = \underbrace{\frac{1}{\mu_N C_{OX}(V_{GS} - V_{TN} - A_N V_{DS})}}_{R_\square}\frac{L}{W} \quad \text{(n-channel in nonsaturation).} \tag{3.131}$$

For a PMOS device, port characteristic in nonsaturation is described by (3.48), and small-signal resistance is defined by $r_{ds} \doteq -1/(\partial i_D/\partial v_{DS})$. These two yield

$$r_{ds} = \underbrace{\frac{1}{\mu_P C_{OX}[A_P V_{DS} - (V_{GS} - V_{TP})]}\frac{L}{W}}_{R_\square} \quad \text{(p-channel in nonsaturation).} \tag{3.132}$$

Notice that these resistance expressions are in the same form as the body resistance component of (3.124). However, $R_\square$ is not a constant but a function of the over-drive and drain-source voltages at the operation point. Dependence on the over-drive voltage makes the resistance electrically controllable, which is a useful property exploited widely in tunable continuous-time *MOSFET-C* filters. Variation with V_{DS}, on the other hand, is undesirable because it implies a significant nonlinearity. Although this problem can be alleviated by deploying fully-differential closed-loop configurations and certain composite MOSFET techniques, it is still virtually impossible for a MOSFET resistor to match the level of linearity available from passive resistors.

Example 3.21 Shown in Fig. 3.52(a) is a low-pass filter in the form of leaky integrator. Its small-signal transfer function as determined by following the extraction technique described in Subsection 2.10.2 is given by

$$A_{cl} \doteq \frac{v_o}{v_i} = -\frac{R_2}{R_1}\frac{1}{1 + sCR_2}, \tag{3.133}$$

where R_1 represents R_{A1} and R_{B1}, R_2 represents R_{A2} and R_{B2}, and C represents C_A and C_B. Notice the dc gain $A_{cl0} = -R_2/R_1$ and the closed-loop bandwidth $f_{cl} = 1/2\pi CR_2$. A MOSFET-C implementation is to be designed in accordance with the following specification set:

1. Input-signal swing: $v_{i(\text{min})} = -0.2$ V, $v_{i(\text{max})} = 0.2$ V.
2. Closed-loop dc gain: $A_{cl0} = 2$ V/V.
3. Closed-loop bandwidth f_{cl} at the operation point: Tunable in between 6 kHz and 12 kHz by means of a dc voltage in between 0.1 and 1.7 V. Exact limits of the control-voltage range are to be determined.
4. Capacitors: 25 pF each.
5. Opamp: `opmacro2.sp` specified by $V_{DD} = 1.8$ V, $V_{SS} = 0$, $V_{OH} = 1.65$ V, $V_{OL} = 0.15$ V, $A_{dif0} = 10000$ V/V, $V_{DIR} = 0.5$ V, $f_{GBW} = 12$ MHz, $f_{nd} = 10$ GHz.
6. Common-mode voltages $V_{ICM} = V_{OCM}$: Value to be determined.

Work Simulation schematic for an NMOS implementation is shown in Fig. 3.52(b), where all MOSFETs have identical geometries and are controlled by the same gate

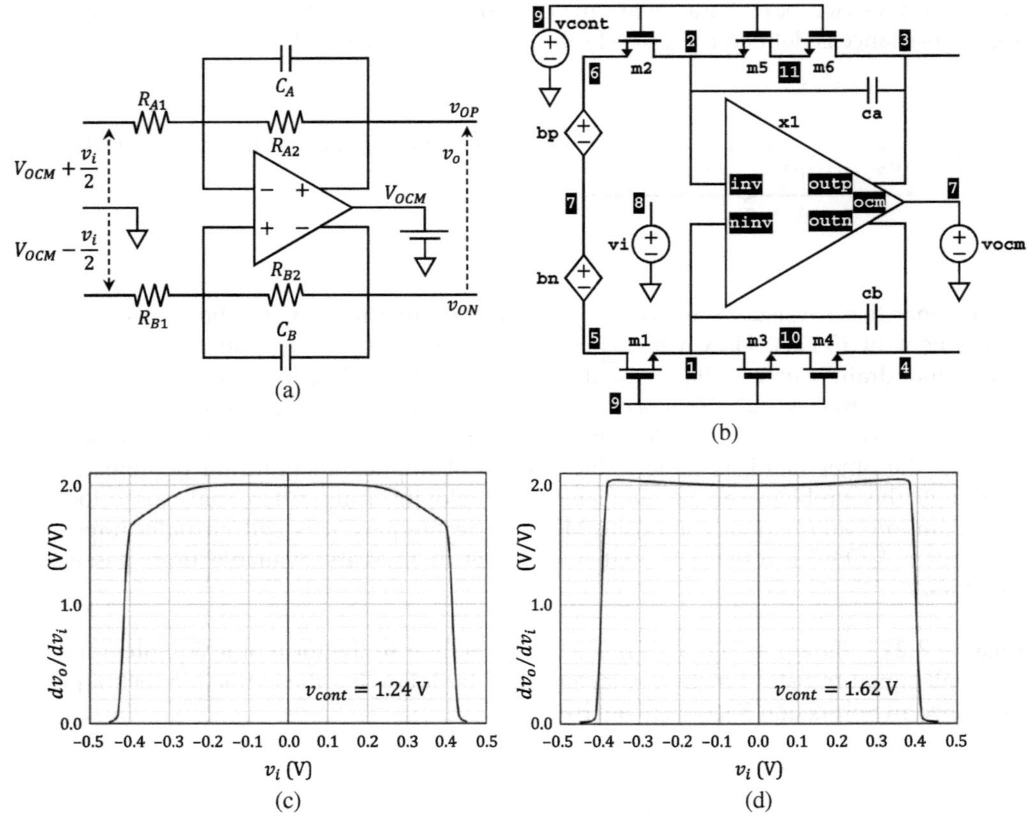

FIGURE 3.52 Low-pass filter of Example 3.21. (a) Circuit schematic. (b) Simulation schematic. (c) DC gain as a function of input signal for $V_{\text{cont}} = 1.24$ V. (d) DC gain as a function of input signal for $V_{\text{cont}} = 1.62$ V.

voltage V_{cont}. R_2 is implemented by the series combination of unit devices M3 and M4 in the lower half, and M5 and M6 in the upper half. As represented by M1 and M2, one unit device is used for implementing R_1 in each half. The resulting setting $R_2/R_1 = 2$ automatically satisfies item 2 of the specification set.

Items 1 and 2 of the specification set imply $V_{O(\text{min})} = V_{OCM} - 0.2$ and $V_{O(\text{max})} = V_{OCM} + 0.2$ for the limits of single-ended output swing. Considering the necessary conditions $V_{O(\text{min})} > V_{OL}$ and $V_{O(\text{max})} < V_{OH}$ in addition to the single-ended output range limits $V_{OL} = 0.15$ V and $V_{OH} = 1.65$ V of the opamp, we arrive at the following constraint on V_{OCM}:

$$0.35\text{ V} \leq V_{OCM} \leq 1.45\text{ V}. \tag{3.134}$$

Another design constraint which is implicit in any MOSFET-resistor application is imposed by the condition of nonsaturation. This condition must be satisfied throughout the entire range of signal swing in order to ensure a reasonable linearity. For the NMOS devices used in this design, it necessitates $V_{DS(\text{max})} < V_{DS\text{sat}(\text{min})}$, where $V_{DS(\text{max})}$ is the largest drain-source voltage and $V_{DSsat(min)}$ is the smallest drain-source saturation voltage experienced by a unit device. In order to identify the unit device that is most unfavorably positioned in this respect, consider the extreme case of $v_i = -0.2$ V. In the upper half of the schematic in Fig. 3.52(b), the node voltages are expected to be `V(6)` $= V_{OCM} - 0.1$, `V(2)` $\approx V_{OCM}$, `V(3)` $\approx V_{OCM} + 0.2$, and `V(11)` $\approx V_{OCM} + 0.1$. In the lower half, we expect `V(5)` $= V_{OCM} + 0.1$, `V(1)` $\approx V_{OCM}$, `V(4)` $\approx V_{OCM} - 0.2$, and `V(10)` $\approx V_{OCM} - 0.1$. Obviously, all of the devices operate with approximately the same maximum drain-source voltage $V_{DS(max)} = 0.1$ V. As to $V_{DSsat(min)}$, we need to identify the device that operates with the lowest overdrive voltage and therefore the highest source voltage. M6 happens to be the one with a source voltage $V_{OCM} + 0.1$. Since the overdrive voltage of this device is absolutely minimized when its gate voltage is lowered to $V_{cont(min)}$, the nonsaturation condition finally turns into

$$\underbrace{\frac{V_{cont(min)} - (V_{OCM} + 0.1) - V_{TN}\big|_{v_S = V_{OCM}+0.1}}{A_N}}_{V_{DSsat(min)}} \geq \underbrace{0.1}_{V_{DS(max)}} \quad . \tag{3.135}$$

The factor-of-two controllability of the bandwidth as required in item 3 of the specification set imposes yet another design constraint. Note that the bandwidth is inversely proportional to R_2, and is specified for the operation point. Since R_2 is described generally by (3.131), in which, the operation point condition $v_i = 0$ of the configuration implies $V_{DS} = 0$ and $V_S = V_{OCM}$, this constraint can be expressed as

$$\frac{V_{cont(max)} - V_{OCM} - V_{TN}\big|_{v_S = V_{OCM}}}{V_{cont(min)} - V_{OCM} - V_{TN}\big|_{v_S = V_{OCM}}} = 2. \tag{3.136}$$

It is worth noting that this expression is not very accurate because it does not take into account the variation of mobility with gate voltage, which may have a significant effect at large overdrive levels.

The design constraints defined so far enable us to determine V_{OCM}, $V_{cont(min)}$, and $V_{cont(max)}$. Assuming V_{OCM} as an independent design variable, we first calculate $V_{cont(min)}$ from (3.135) after adding 0.1 V to the right-hand side as a safety margin. Next, we determine $V_{cont(max)}$ from (3.136). $V_{TN}\big|_{v_S = V_{OCM}+0.1}$ appearing in (3.135) and $V_{TN}\big|_{v_S = V_{CM}}$ appearing in (3.136) are determined from Fig. 3.7. The outcome of these calculations are presented in Table 3.5. Note that the minimum entry for V_{OCM} is 0.35 V as limited by (3.134). In contrast, the 0.5-V maximum entry for the same design variable is much lower than what (3.134) indicates as the upper limit. The reason is that, for this value of V_{OCM}, the upper limit 1.7 V imposed on $V_{cont(max)}$ in item 3 of the specification set is already reached. Obviously, the viable range of V_{OCM} is very narrow. The midpoint $V_{OCM} = 0.40$ V of this range is a reasonable choice. The corresponding limits of the control voltage are $V_{cont(min)} = 1.24$ V and $V_{cont(max)} = 1.59$ V.

The final step of design is device sizing in accordance with the bandwidth specification cited in item 3 of the specification set. Since item 4 prescribes 25 pF for the capacitance,

V_{CM} (V)	$V_{TN}\|_{V_S=V_{CM}+0.1}$ (V) from Fig. 3.7	$V_{cont(min)}$ (V) from (3.135) + 0.1	$V_{TN}\|_{V_S=V_{CM}}$ (V) from Fig. 3.7	$V_{cont(max)}$ (V) from (3.136)
0.35	0.50	1.18	0.48	1.53
0.40	0.51	1.24	0.49	1.59
0.45	0.53	1.31	0.50	1.67
0.50	0.54	1.37	0.51	1.73

TABLE 3.5 Calculated Data of Example 3.21

R_2 should be around 1.06 MΩ in order to realize the specified 6-kHz lower limit of the bandwidth. This implies r_{ds} = 530 kΩ for each unit device operating with a gate voltage $V_G = V_{\text{cont(min)}}$ = 1.24 V. Considering also $v_S = V_{OCM}$ = 0.4 V for the source voltage at the operation point, the aspect ratio W/L of the MOSFET can be calculated from (3.131) after determining the mobility from (3.17). However, sizing by iterative simulation instead of calculation will be more efficient because an input file has to be written for verification anyway. Shown below is the input file `ex3-21.sp` corresponding to the schematic shown in Fig. 3.52(b).

```
filename: ex3-21.sp
.param vdd=1.8 vss=0 voh=1.65 vol=0.15 adif0=10000 vdir=0.25
+  fgbw=10meg fnd=10g
.incl opmacro2.sp
.incl bsim.sp
x1 1 2 3 4 7 fdopamp
ca 2 3 25p
cb 1 4 25p
m1 5 9 1 0 cmosn w=0.5u l=26u
m2 6 9 2 0 cmosn w=0.5u l=26u
m3 1 9 10 0 cmosn w=0.5u l=26u
m4 10 9 4 0 cmosn w=0.5u l=26u
m5 2 9 11 0 cmosn w=0.5u l=26u
m6 11 9 3 0 cmosn w=0.5u l=26u
vcm 7 0 0.40
bp 6 7 v=v(8)/2
bn 7 5 v=v(8)/2
vcont 9 0 1.24 ; 1.62
vi 8 0 dc 0 ac 200m
.ac dec 100 1 100k
.measure ac fcl when vm(4)=141.42m
*.dc vi -450m 450m 10m
.end
```

Considering the rather large target value of r_{ds}, the aspect ratio W/L is likely to be less than unity. It is therefore reasonable to set a narrow width and to iterate an ac analysis by varying the channel length until the corner frequency meets the specification. Notice the selection of W_{DR} = 0.5 μm and the inclusion of a measurement line "`.measure`" in the input file. Since the ac magnitude of the differential input signal source `vi` is set to 0.2 V, and the low-frequency gain from this source to each single-ended output equals 1 V/V, the ac magnitude of a single-ended output signal at the corner frequency must equal $0.2(\text{V}) \times 1(\text{V/V})/\sqrt{2} = 0.14142$ V. This is why the measurement line instructs the simulator to measure the frequency at which the ac magnitude `vm(4)` of one of the single-ended outputs is exactly this much. Simulator puts out the result as `fcl` after running the ac analysis. Iterating with a fixed 1.24-V control voltage provided by `vcont` eventually yields a 26-μm drawn channel length. Incidentally, it is advisable to also double check the outcome of each iteration on the closed-loop magnitude response after plotting with

```
plot vdb(4,3)-vdb(8)
```

Once the proper channel length has thus been determined, repeat the ac simulation after resetting the control voltage to 1.59 V as determined in design for shifting the corner frequency to 12 kHz. Surely, the corner frequency indicated by the simulation will be somewhat different from the target. Iterate by readjusting the control voltage until the 12-kHz target is achieved, and adopt the corresponding control voltage as true $V_{\text{cont(max)}}$. In the present example, it turns out to be 1.62 V.

The dc-analysis line appearing in the input file is for verifying the range performance of the filter. Plotting the derivative `deriv(v(4,3))` of the differential output voltage displays the variation of the fully-differential dc gain with the input signal `vi`. The outcome for $V_{\text{cont}} = 1.24$ V and $V_{\text{cont}} = 1.62$ V are shown in Figs. 3.52(c) and (d), respectively. Both plots indicate a reasonably constant 2-V/V dc gain over the specified input-signal range $-0.2(\text{V}) \leq v_i \leq 0.2(\text{V})$. The reader may quantify the nonlinearity within this range by running a transient simulation with a sinusoidal input signal of say 100-Hz frequency and reading out the total harmonic distortion. ▲

3.11 Capacitor Design

As discussed in Sections 2.11 and 2.12, discrete-time amplifiers are implemented with capacitive external networks. Capacitors also find use as external-network components in continuous-time filtering applications, and as internal components in opamp stabilization by frequency compensation.

Advanced CMOS technologies offer a parallel-plate capacitor structure in the form of two metal layers separated by an insulator. Called MIM, this type of capacitor provides a highly linear capacitance. Another dedicated capacitor, constructed by sandwiching an insulator film in between two polysilicon layers, and called PIP, is offered in 0.25 μm and earlier CMOS technologies. However, the quality of PIP is inferior to MIM in linearity and capacitive/resistive parasitics. In those applications where nonlinearity can be tolerated, a MOSFET can be utilized as a capacitor of significantly larger capacitance per unit area than MIM or PIP. Decoupling is the most common of such applications. Stripped of its source and drain, a MOSFET can also be used as a voltage-controlled capacitor, that is, a varactor, which finds widespread use in frequency-tuning applications.

In this section, the modeling and ratioing aspects of MIM capacitors will be discussed first. Subsequently, the MOSFET will be explored as a decoupling capacitor and as a varactor.

3.11.1 MIM Capacitor

The cross-sectional view of an MIM capacitor is given in Fig. 3.53(a), where metal layers T and $T-1$ represent the two topmost standard metal layers available in CMOS technologies. The former of the two is reserved for accessing the upper and lower plates of the capacitor. The lower plate is built in metal $T-1$, whereas a dedicated metal layer (cap metal) is used for building the upper plate. The plates are laid out typically in a square geometry with slightly shorter sides on the upper plate. The dimensions of the latter therefore defines the effective area and perimeter of the capacitor.

The capacitance of an MIM capacitor, represented by C in Fig. 3.53(a), is usually modeled by

$$C = \left(C_A A + C_F P\right)\left(1 + \mathrm{VCC} \times |v_A - v_B|\right)\left[1 + \mathrm{TCC1} \times (T - T_0) + \mathrm{TCC2} \times (T - T_0)^2\right], \tag{3.137}$$

where A and P are respectively the area and perimeter of the upper plate, $|v_A - v_B|$ is the magnitude of the voltage across the capacitor, C_A and C_F are the coefficients of the

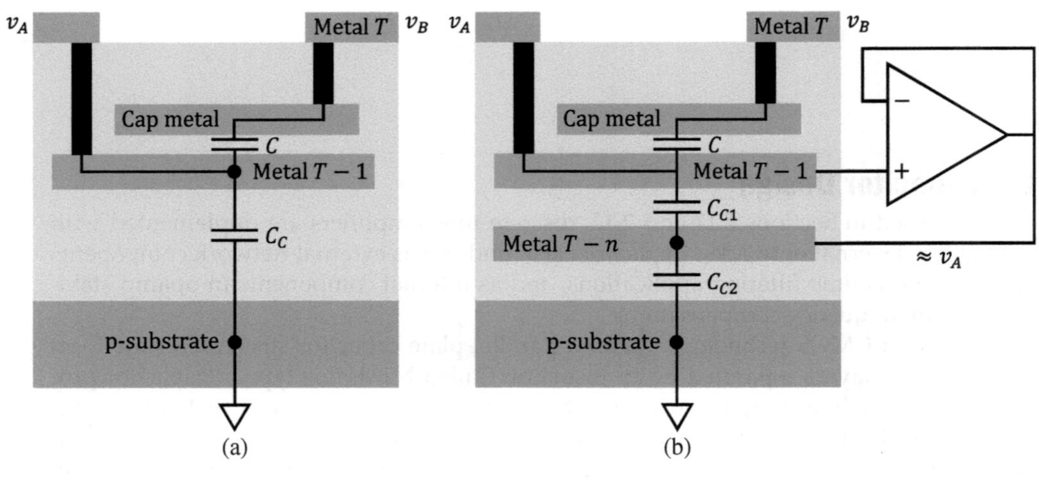

FIGURE 3.53 (a) Cross-sectional view of an MIM capacitor. Top metal layer (*metal T*) is used for accessing the two plates of the capacitor. Lower plate of the capacitor is built in *Metal T* – 1, which is the standard metal layer just below the top metal. Upper plate of the capacitor is built in a dedicated metal layer (*cap metal*). C denotes the main capacitance. C_C is the parasitic capacitance between the lower plate and substrate. (b) Screening with a lower metal layer (*metal T* – *n*) for minimizing the parasitic capacitance of the lower plate.

areal and fringe components of the capacitance, VCC is the voltage coefficient, TCC1 and TCC2 are temperature coefficients, T is the operation temperature, and T_0 is the temperature at which the capacitance parameters are characterized by the manufacturer. Numerical examples of MIM-capacitor parameters in a typical 0.18-μm CMOS technology are as follows: $C_A = 1\ \text{fF}/\mu\text{m}^2$, $C_F = 0.2\ \text{fF}/\mu\text{m}$, VCC $= 2\times10^{-5}\ \text{V}^{-1}$, TCC1 $= 5\times10^{-3}\ \text{C}^{-1}$, TCC2 $= 1\times10^{-5}\ \text{C}^{-2}$, and $T_0 = 25°\text{C}$.

The combination of inaccuracy and imprecision may be as large as 15% of the designed value of an MIM capacitance. The ratio of two MIM capacitances, however, exhibits negligible inaccuracy provided that common-centroid interdigitation and dummy termination are practiced in physical design. As for the imprecision of an MIM capacitance, the simplest model prescribes

$$\frac{\sigma[C]}{C} = \frac{K_C}{\sqrt{A}} \tag{3.138}$$

for the relative standard deviation, where A (μm²) and K_C (μm) represent the area and Pelgrom coefficient, respectively. In 0.18-μm CMOS technologies, K_C is typically around 0.01 μm.

As shown in Fig. 3.53(a), the lower plate of an MIM capacitor is coupled to the substrate via a parasitic capacitance C_C, which can be as large as 5% of the main capacitance. This parasitic capacitance adversely affects the applications in which the lower plate is on the signal path. It may inject substrate noise onto the signal path or alter the closed-loop performance of *stray-sensitive* configurations. Fortunately, these effects can be suppressed by introducing a screening plate in between the bottom plate and the substrate. Any metal layer below $T-1$ can be used for this purpose. This arrangement is depicted in Fig. 3.53(b) with a metal layer labeled $T-n$. For noise suppression, it suffices to connect the screen to a clean ground node but this won't eliminate the effect on the closed-loop properties of a stray-sensitive configuration. As a matter of fact, the bottom plate will be more strongly coupled to ground because its parasitic capacitance C_{C1} to the screen is larger than the original parasitic capacitance C_C to the substrate. A radical technique by which the capacitive coupling of the bottom plate can be virtually eliminated is to drive the screen with the voltage v_A of the bottom plate. The OTA-based unity-gain amplifier shown in Fig. 3.53(b) accomplishes this task almost ideally. Considering the gain error of the amplifier, the voltage across C_{C1} is given by v_A/A_0, where A_0 is the dc-differential gain of the OTA. The parasitic capacitance is therefore effectively reduced to C_{C1}/A_0. Although the screen itself is capacitively coupled to the substrate, the capacitance C_{C2} associated with this coupling is driven by the buffer outside the signal path. This screening technique is of course expensive due to the power consumption and footprint of the amplifier. In less critical applications, the parasitic capacitance between the lower plate and substrate can be reduced simply by placing a floating n-well beneath the MIM capacitor. As described in Subsection 3.10.1 for minimizing the parasitic capacitance of poly resistors, this technique works by replacing C_C with $C_C C_W/(C_C + C_W)$ where C_W is the junction capacitance of the well.

Example 3.22 The external network of the fully-differential discrete-time amplifier of Fig. 2.52 is to be built with MIM capacitors of $C_A = 1\ \text{fF}/\mu\text{m}^2$, $C_F = 0.2\ \text{fF}/\mu\text{m}$, and

$K_C = 0.01$ µm. Determine the size and capacitance of all capacitors in accordance with the following specifications:

1. Target closed-loop gain: $A_{cl} = 7$ V/V.
2. 4.5-σ random external-network error: Equal to or smaller than 0.1% of the output signal.

Work Closed-loop gain A_{cl} is described by (2.212) as $1 + (C_1/C_2)$, where C_1 is the target value of C_{A1} and C_{B1}, and C_2 is the target value of C_{A2} and C_{B2}. Item 1 of specifications therefore implies

$$\frac{C_1}{C_2} = 6. \tag{3.139}$$

Considering (2.219), we can represent the 4.5σ random external-network error with $4.5\sigma[1/\beta']v_i(t_2)$, where β' is described by (2.216). On the other hand, the output signal is given by $A_{cl}v_i(t_2)$. Considering $A_{cl} = 7$ V/V, item 2 of specifications can therefore be expressed as

$$4.5 \times \sigma\left[1/\beta'\right] \times v_i(t_2) \leq 0.001 \times 7 \times v_i(t_2),$$

hence

$$\sigma\left[1/\beta'\right] = 1.56 \times 10^{-3}. \tag{3.140}$$

The variance of $1/\beta'$ can be related to the variances of C_{A1}, C_{A2}, C_{B1}, and C_{B2} through (2.216), (2.28), and (3.139) as follows:

$$\sigma^2\left[1/\beta'\right] = 9\left(\frac{\sigma^2\left[C_{A1}\right]}{C_{A1}^2} + \frac{\sigma^2\left[C_{A2}\right]}{C_{A2}^2} + \frac{\sigma^2\left[C_{B1}\right]}{C_{B1}^2} + \frac{\sigma^2\left[C_{B2}\right]}{C_{B2}^2}\right). \tag{3.141}$$

Assuming that (3.139) is fulfilled by assigning one unit capacitor C_U to each of C_{A2} and C_{B2} and six parallel unit capacitors to each of C_{A1} and C_{B1}, the relative variances appearing on the right-hand side of (3.139) can be expressed as

$$\frac{\sigma^2\left[C_{A1}\right]}{C_{A1}^2} = \frac{\sigma^2\left[C_{B1}\right]}{C_{B1}^2} = \frac{1}{6}\frac{\sigma^2\left[C_U\right]}{C_U^2},$$

$$\frac{\sigma^2\left[C_{A2}\right]}{C_{A2}^2} = \frac{\sigma^2\left[C_{B2}\right]}{C_{B2}^2} = \frac{\sigma^2\left[C_U\right]}{C_U^2},$$

which turn (3.141) into

$$\sigma^2\left[1/\beta'\right] = 21\frac{\sigma^2\left[C_U\right]}{C_U^2}.$$

Combining this equation with (3.138) and (3.140) and adopting $K_C = 0.01$ µm as specified, we finally obtain $A = 864$ µm^2 for the area of a unit capacitor. For a square capacitor layout, this leads to a perimeter $P = 118$ µm. The nominal value of unit capacitance, as calculated from (3.137) for zero voltage and $T = T_0$, is therefore $C_U = 888$ fF. ▲

3.11.2 MOSFET as a Capacitor

A capacitor built out of a MOSFET is called MOSCAP. The capacitance of a MOSCAP is typically an order-of-magnitude higher than that of a MIM capacitor of comparable size. However, it is hardly linear, which is why its utility is mostly in decoupling applications where linearity is secondary to the density of capacitance per unit area.

A MOSFET is configured into a MOSCAP by combining its source and drain into a common terminal called *source*. The consequence $v_{DS} = 0$ of this combination turns the general high-frequency small-signal equivalent circuit of Fig. 3.41 into the three-terminal form depicted in Fig. 3.54(a). Another implication of $v_{DS} = 0$ is $C_{gb} = C_{bg}$, which further transforms the equivalent circuit into what is shown in Fig. 3.54(b). Note that this final form contains three physical interterminal capacitances defined by $C_{GB} \doteq C_{bg}$, $C_{GS} \doteq C_{gg} - C_{bg}$, and $C_{SB} \doteq C_{bb} - C_{bg}$. Recalling (3.112) for C_{bg}, (3.105) for C_{gg}, and (3.111) for C_{bb}, and subsequently using (3.99) for C_{gge}, (3.76) for C_{ggi}, (3.101) for C_{bbe}, and (3.78) for C_{bbi}, we arrive at the following expressions for the capacitances of a MOSCAP:

$$C_{GB} = C_{gbe} - C_{bgi}, \tag{3.142}$$

$$C_{GS} = C_{gse} + C_{gde} - (C_{gsi} + C_{gdi}), \tag{3.143}$$

$$C_{SB} = C_{sbe} + C_{dbe} - (C_{bsi} + C_{bdi}). \tag{3.144}$$

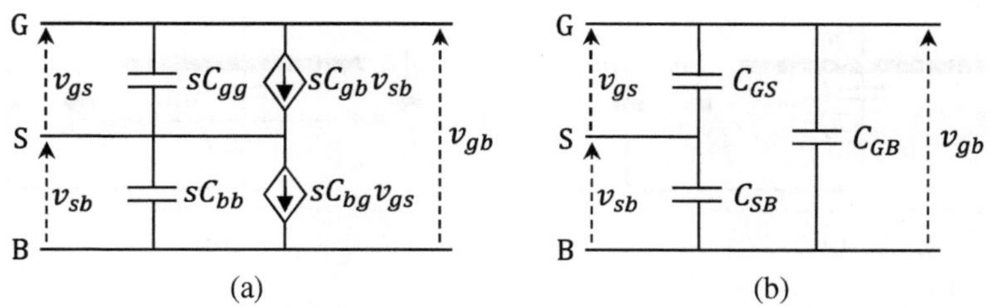

FIGURE 3.54 (a) MOSCAP equivalent circuit as derived from Fig. 3.41 for $v_{ds} = 0$. (b) MOSCAP equivalent circuit represented by physical interterminal capacitances only.

Extrinsic components are related to device dimensions through (3.91) to (3.95). Intrinsic components, on the other hand, are proportional to gate area with an areal density described by (3.89), and are strongly dependent on terminal voltages as observed in Figs. 3.55(a) to (c) on simulated density plots. For an explanation of the physical roots of this strong nonlinearity, consider Fig. 3.55(d) first. It shows a MOSCAP channel operating in subthreshold and therefore supporting only a depletion layer at the silicon surface as highlighted in light gray. This layer, having much lower carrier concentrations in comparison with the source and substrate regions, acts as a dielectric and forms a *depletion capacitance* C_{si} between silicon surface and bulk terminal. The gate-oxide layer overlying the channel is also a dielectric. It forms an *oxide capacitance* C_{oxide} in between gate and silicon surface. These two capacitances quite generally account for all the intrinsic capacitances appearing in (3.142) to (3.144). However, due to the diminishing electron population of the channel in subthreshold, none of these two capacitances is connected to the n-type source terminal. This is why the intrinsic components plotted in Figs. 3.55(a) and (b) for the source-related capacitances C_{GS} and C_{SB} are vanishingly small under the subthreshold condition $V_{GB} < V_{SB} + V_{TN}$. In contrast, Fig. 3.55(c) indicates a maximized intrinsic component for C_{GB} under the same condition. This is because the depletion and

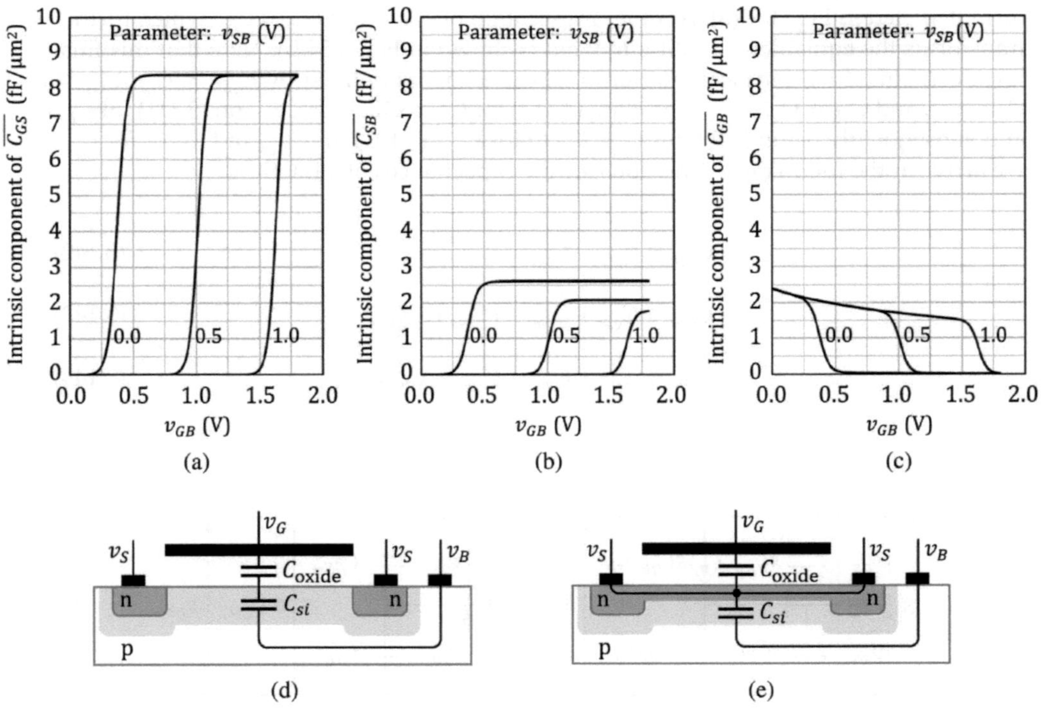

FIGURE 3.55 (a) Gate-source intrinsic capacitance $-(\overline{C_{gsi}} + \overline{C_{gdi}})$. (b) Source-bulk intrinsic capacitance $-(\overline{C_{bsi}} + \overline{C_{bdi}})$. (c) Gate-bulk intrinsic capacitance $-\overline{C_{bgi}}$. (d) Network of capacitances in subthreshold. (e) Network of capacitances in strong inversion.

oxide capacitances are acting in series between the gate and bulk terminals as depicted in Fig. 3.55(d).

Now consider Fig. 3.55(e), where a strongly inverted channel is depicted. The dark gray sheet at the surface of the channel represents the inversion layer of electrons, which connects the common node of the oxide and depletion capacitances to the source terminal. Due to this connection, the oxide capacitance now appears in between gate and source terminals, and thus maximizes the intrinsic component of C_{GS} as observed in Fig. 3.55(a) for $V_{GB} > V_{SB} + V_{TN}$. Since the gate terminal is no more coupled to the bulk, the intrinsic component of C_{GB} vanishes in strong inversion as observed in Fig. 3.55(c). Finally, notice in Fig. 3.55(b) the maximized intrinsic component of C_{SB}. This is due to the fact that the depletion capacitance is now in between the source and bulk terminals.

In decoupling applications, the node to be decoupled is tied to the ground via a large capacitance. The most popular MOSCAP configuration serving this purpose is shown in Fig. 3.56(a). Notice the grounding of source and bulk terminals. Gate is tied to the node to be decoupled. The effective capacitance C_G between gate and ground equals $C_{GS} + C_{GB}$. As illustrated in Fig. 3.56(b), the density of the intrinsic component of this capacitance is maximized in strong inversion to the density C_{OX} of the oxide capacitance. The decoupled voltage v_G should therefore be kept above the threshold voltage in order to maintain this maximum value.

A capacitor serving as a voltage-controlled capacitance is known as *varactor*. The most popular varactor structure is shown in Fig. 3.57(a). Notice the absence of a source terminal. Gate and n-type bulk are the two terminals of its capacitance C_{GB}, which equals the series equivalent of the gate-oxide capacitance C_{oxide} and the depletion capacitance

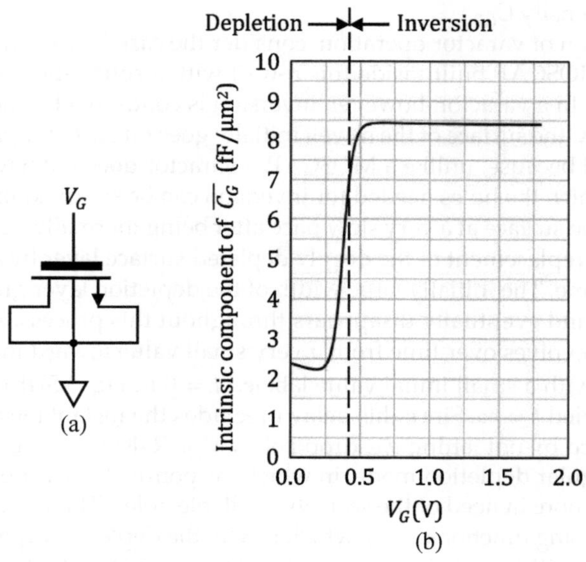

FIGURE 3.56 (a) A decoupling MOSCAP. (b) Areal density of the intrinsic components of the capacitance between gate and ground.

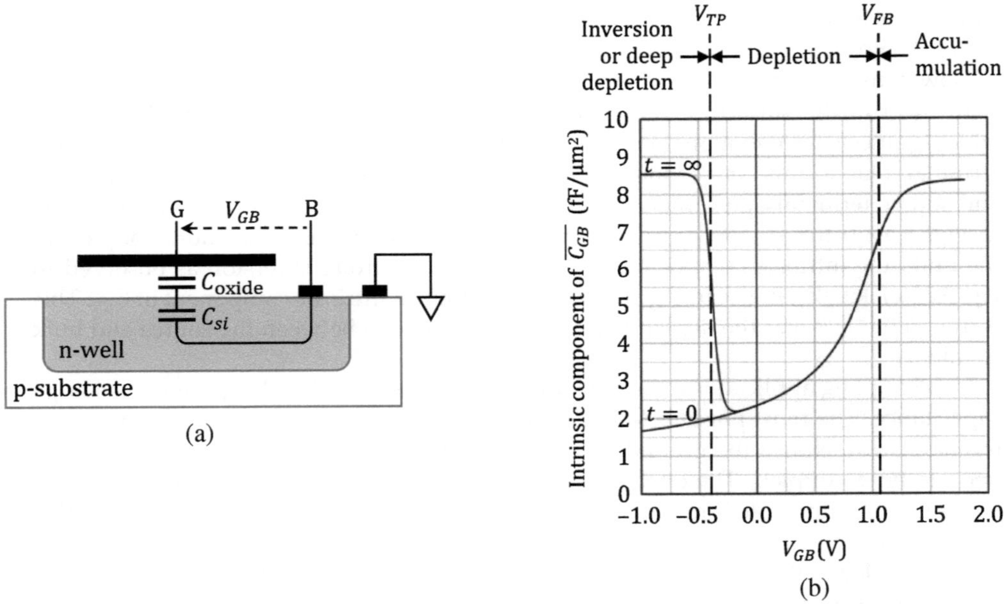

FIGURE 3.57 (a) Varactor structure. (b) The variation of the areal density of gate-bulk capacitance with gate-bulk voltage.

C_{si}. C_{GB} is controlled by the gate-bulk voltage v_{GB} as shown in Fig. 3.57(b) by a simulated plot of its areal density $\overline{C_{GB}}$.

For a description of varactor operation, consider the case $V_{GB} < V_{TP}$, for which, a regular p-channel MOSCAP built inside this n-well with a source-bulk short would be in strong inversion. In a varactor, however, inversion is conditional. When V_{GB} is lowered below V_{TP} rapidly, the surface of the n-well initially goes into a *deep depletion* state instead of being inverted because, unlike a MOSCAP, a varactor does not have a p-type source terminal from which the holes needed for inversion can be supplied immediately. Holes are supplied to the surface at a very slow pace after being thermally generated inside the n-type bulk. The replacement of the deeply depleted surface layer by an inversion layer therefore takes time. The initially large width of the depletion layer gradually shrinks in a transient state and eventually disappears throughout this process. As a result, depletion capacitance evolves over time from a very small value toward infinity. This is why $\overline{C_{GB}}$ also begins with a small initial value labeled $t = 0$ in Fig. 3.57(b), and increases toward C_{OX} as labeled $t = \infty$. Since this delay precludes the instantaneous control of C_{GB}, it must be avoided by not letting v_{GB} drop below V_{TP}. Raising v_{GB} above V_{TP} brings the surface into a regular depletion mode, in which, response to v_{GB} is not delayed because the surface is no more in need of the scarcely available holes. The width of this depletion region is a decreasing function of v_{GB}, which is why the depletion capacitance and therefore C_{GB} increase with v_{GB}. This trend comes to an end as the depletion layer vanishes for $v_{GB} = V_{FB}$, where V_{FB} is called *flatband voltage*. For $v_{GB} > V_{FB}$, an *accumulation* layer of electrons prevails at the surface. These electrons effectively connect the lower node of the oxide capacitance to the bulk terminal, and thus maximize $\overline{C_{GB}}$ at C_{OX}. Since n-well can instantaneously supply the electrons needed by the accumulation layer, response to v_{GB} in this mode is also free of delay.

It is obvious from the foregoing description that the range of v_{GB}, in which, C_{GB} can be varied is limited approximately by V_{TP} at the low end and by V_{FB} at the high end. As observed in Fig. 3.57(b), this range is no less than a volt wide in a typical 0.18-μm CMOS technology. Incidentally, if V_{FB} is not quantified in a BSIM parameter deck, it can still be estimated from

$$V_{FB} \approx \texttt{VTH0} + 2\phi_{FB} + \texttt{K1}\sqrt{2\phi_{FB}}. \tag{3.145}$$

3.12 References

1. Y. S. Chauhan, S. Venugopalan, M. A. Karim, S. Khandelwal, N. Paydavosi, P. Thakur, A. M. Niknejad, et al., "BSIM—Industry standard compact MOSFET models," in *Proceedings of European Solid-State Device Research Conference*, pp. 46–49, 2012.
2. BSIM Group. bsim.berkeley.edu/ (accessed November 1, 2017).
3. The MOSIS Service. www.mosis.com/Technical/Testdata/tsmc-018-prm.html (accessed March 15, 2010).
4. J. Shin, I.-Y. Chung, Y. J. Park, and H. S. Min, "A new charge pump without degradation in threshold voltage due to body effect," *IEEE Journal of Solid-State Circuits*, vol. 35, pp. 1227–1230, 2000.
5. BSIM Group. http://bsim.berkeley.edu/models/bsim3/ (accessed February 12, 2018).
6. BSIM Group. http://bsim.berkeley.edu/models/bsim4/ (accessed February 12, 2018).
7. BSIM Group. http://bsim.berkeley.edu/models/bsimbulk/ (accessed February 12, 2018).
8. R. Thewes, R. Brederlow, C. Dahl, U. Kollmer, C. G. Linnenbank, B. Holzapfl, J. Becker, et al., "Explanation and quantitative model for the matching behaviour of poly-silicon resistors," *Technical Digest, International Electron Devices Meeting*, pp. 771–774, 1998.

3.13 Exercises

3.1 A model deck for standard-threshold devices of 90-nm bulk CMOS technology is provided in `bsim90.sp`. It is a slightly modified version of a model deck prepared by the Nanoscale Integration and Modeling (NIMO) Group at Arizona State University and made public at the following URL of Predictive Technology Model:

```
http://ptm.asu.edu/modelcard/90nm_bulk.txt
```

For this technology, assume $L_{DR(\min)} = 0.1$ µm and $W_{DR(\min)} = 0.3$ µm for minimum permissible dimensions, and one half of the numerical values cited in Table 3.2 for physical design rules. Also assuming a maximum voltage of 1.2 V instead of 1.8 V, extract the design tools appropriate for this technology. When rewriting `nmos1.sp`, `nmos2.sp`, `pmos1.sp`, and `pmos2.sp` for this purpose, it is appropriate to scale down the dimensional variables by a factor of 2. Scaling down the port voltages by a factor around 1.5 is also meaningful since their upper limit is reduced from 1.8 to 1.2 V. Make a record of the extracted design tools because they will be utilized in some of Chapter 4 exercises. It is worth mentioning that the threshold-voltage tools will not exhibit a monotonic relationship between threshold voltages and channel length. As observed in some short-channel bulk CMOS technologies, V_{TN} and $|V_{TP}|$ exhibit a maximum for a channel length close to $L_{DR(\min)}$.

3.2 An NMOS transistor is to operate as a current sink with $V_{SB} = 0$. Its performance is specified by $V_{DSsat} = 200$ mV, $I_D = 100$ µA, and $r_{ds} = 200$ kΩ at $V_{DS} = 0.9$ V. Set up and apply a design plan for determining V_{GS}, L_{DR}, and W_{DR}.

3.3 A PMOS transistor is to operate as a transconductor with $V_{SB} = -300$ mV. Its performance is specified by $V_{DSsat} = -200$ mV, $g_m = 600$ µA/V, and $r_{ds} = 200$ kΩ at $V_{DS} = -0.9$ V. Set up and apply a design plan for determining V_{GS}, L_{DR}, and W_{DR}.

3.4 A MOSFET is specified by V_{GS}, V_{SB}, V_{DS}, I_D, and r_{ds}. Which steps do you take for determining L_{DR} and W_{DR}?

3.5 The PMOS transistor shown in Fig. 3.58 is to operate in saturation with a source-bulk short. Determine V_B as well as W_{DR}, L_{DR} in accordance with the following specifications: $I = 150$ µA, $g_m = 1.5$ mA/V, $r_{ds} = 100$ kΩ for $V_O = 0.9$ V.

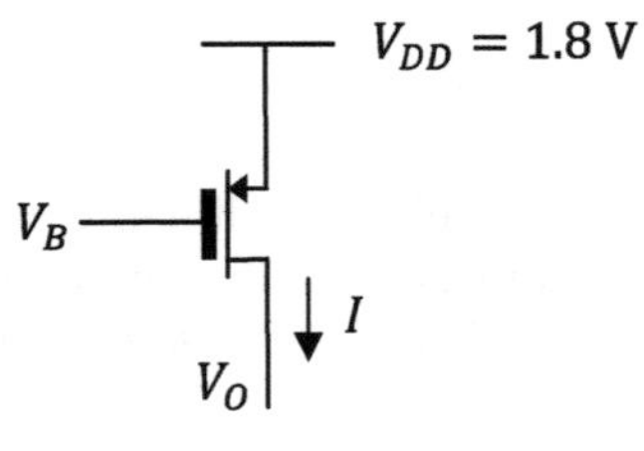

FIGURE 3.58

3.6 The NMOS transistors shown in Fig. 3.59 have identical channel lengths and identical channel widths. They are biased by dc gate voltages V_{B1} and V_{B2}, and conduct a dc current $I = 1$ mA in saturation. While M1 operates with a source-bulk short, M2 doesn't. The drain-source saturation voltage of M1 happens to be 125 mV.

Determine (a) the drain-source saturation voltage of M2, (b) overdrive voltage of M1, (c) overdrive voltage of M2, (d) sheet current of M1, (e) sheet current of M2, (f) aspect ratio of M1, and (g) aspect ratio of M2. Also derive by inspection the relationship between dc small-signal output resistance $r_o \doteq v_o/i$ and dc small-signal device parameters.

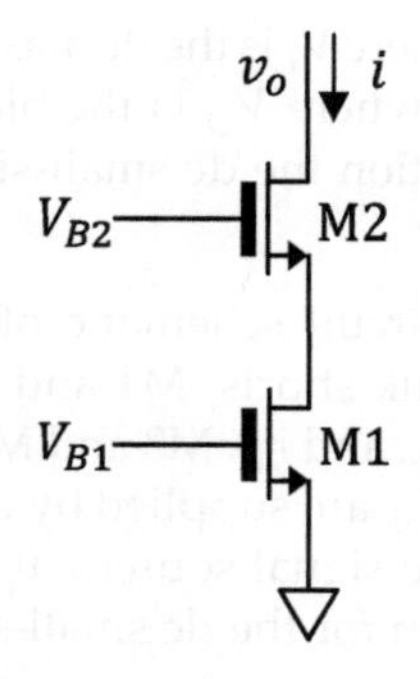

FIGURE 3.59

3.7 V_{B1} and V_{DD} shown in Fig. 3.60 are dc voltage sources. All devices operate with source-bulk shorts. Derive by inspection the relationship between dc small-signal voltage gain v_o/v_i and dc small-signal device parameters.

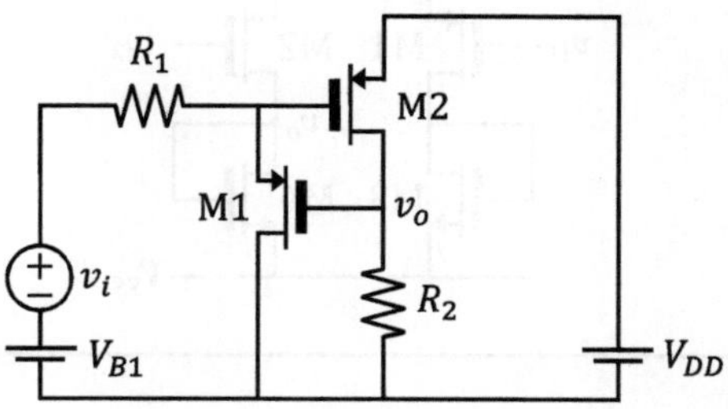

FIGURE 3.60

3.8 Shown in Fig. 3.61 is the circuit schematic of a source-follower stage. The MOSFET operates without a source-bulk short. It is specified by small-signal parameters g_m, g_{mb}, and r_{ds}. V_{DD}, V_{SS}, and V_I are supplied by dc voltage sources. I_B is a dc current source.

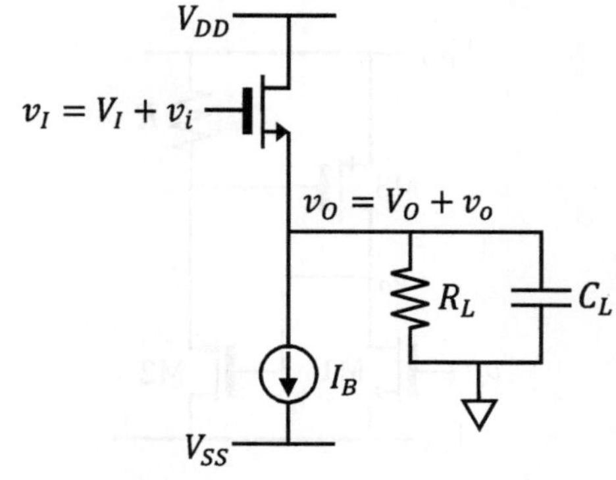

FIGURE 3.61

$v_I = V_I + v_i$ is the input voltage, where V_I is the dc bias, and v_i is the signal. Output voltage is represented by $v_O = V_O + v_o$, where V_O is the bias component, and v_o is the signal component. Determine by inspection the dc small-signal voltage gain v_o/v_i in terms of device parameters.

3.9 Shown in Fig. 3.62 is the circuit schematic of a differential amplifier stage. All MOSFETs operate *with* source-bulk shorts. M1 and M2 are identical devices specified by small-signal parameters g_m, g_{mb}, and r_{ds}. M3 and M4 are also identical. They are specified by g'_m, g'_{mb}, and r'_{ds}. V_{DD} and V_{SS} are supplied by dc voltage sources. I_B is a dc current source. v_{i1} and v_{i2} are two voltage signal sources. v_o is the output signal. Construct by inspection a parametric expression for the dc small-signal v_o as a function of v_{i1} and v_{i2} in terms of device parameters.

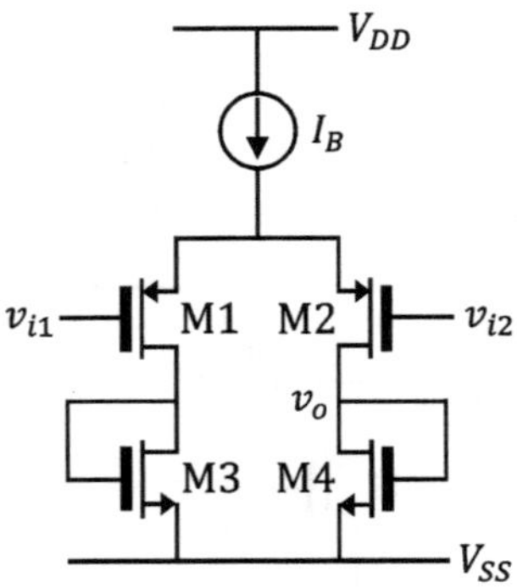

FIGURE 3.62

3.10 Shown in Fig. 3.63 is the circuit schematic of an amplifier stage with a feedback loop. All MOSFETs operate with source-bulk shorts. Small-signal parameters are specified as follows: g_{m1}, g_{mb1}, and r_{ds1} for M1; g_{m2}, g_{mb2}, and r_{ds2} for M2; g_{m3}, g_{mb3}, and r_{ds3} for M3. V_{DD} and V_{SS} are supplied by dc voltage sources. v_i and v_o represent input and output signals, respectively. Determine by inspection the dc small-signal voltage gain v_o/v_i in terms of device parameters.

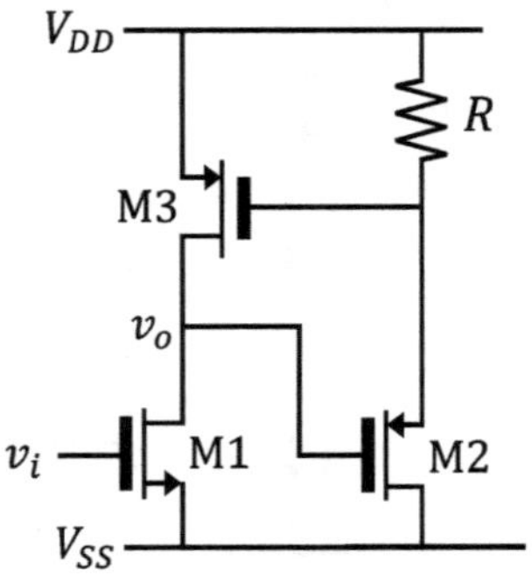

FIGURE 3.63

3.11 An amplifier stage used occasionally as a buffer is shown in Fig. 3.64. V_{SS} and V_{DD} are dc power-supply voltages, V_I is supplied by a dc bias source, and V_O is the dc bias voltage at the output. Prove that the following dc small-signal voltage-gain expression is valid irrespective of device dimensions:

$$\frac{v_o}{v_i} = \frac{V_{GS2} - V_{TN}}{V_{GS1} - V_{TN}}.$$

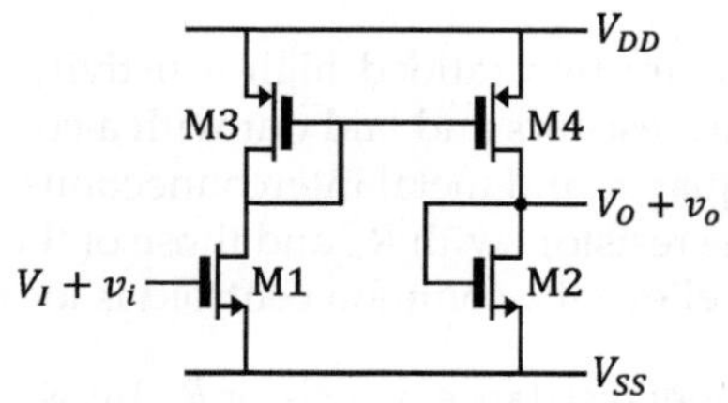

FIGURE 3.64

3.12 A gain stage used in class-AB opamps is shown in Fig. 3.65. It contains a driver device M1, a bias device M4, and two diode-connected devices M2 and M3. Derive by inspection an expression for the dc small-signal voltage gain v_o/v_i in terms of device transconductances and output resistances. Simplify the expression by assuming any g_m to be much larger than any $1/r_{ds}$.

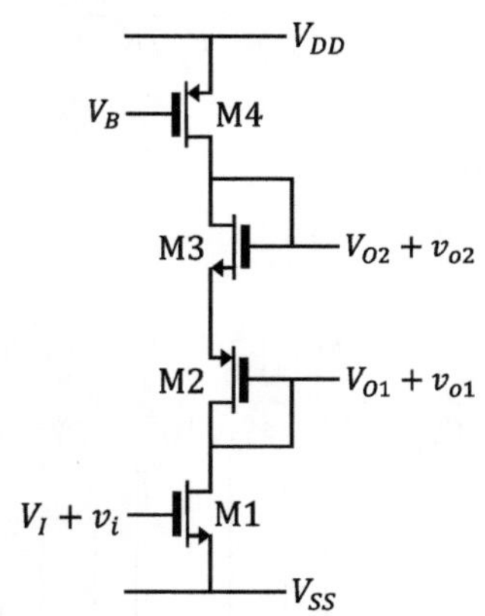

FIGURE 3.65

3.13 Shown in Fig. 3.66 is the circuit schematic of an amplifier stage built with a self-biased inverter. Both MOSFETs operate with source-bulk shorts. Draw the high-frequency small-signal equivalent circuit of this stage without showing ineffective components.

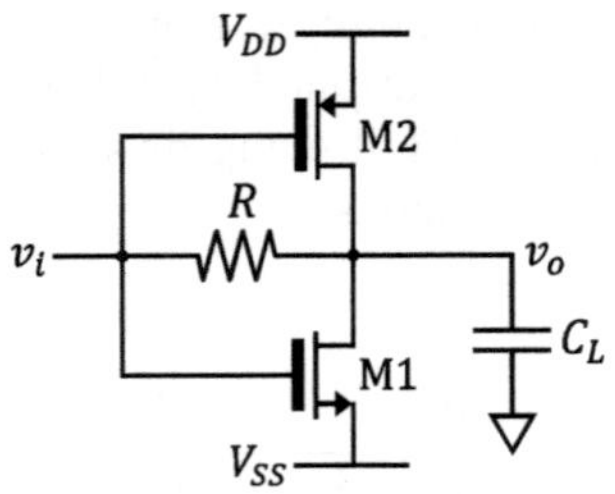

FIGURE 3.66

3.14 Shown in Fig. 3.67 are two ratioed high-resistivity poly resistors that are segmented into matching unit resistors and laid out with a common centroid. Contacts are represented with black squares, and metal interconnections are drawn in gray. Mark the two terminals of one of the resistors with R_A and those of the other resistor with R_B. Also place a cross where you believe the common centroid is located.

1. Determine the target resistance of resistor R_A by assuming $R_C = 10\ \Omega$ for contact resistance and referring to Table 3.4 for other resistance parameters.
2. Repeat part 1 for R_B.
3. Determine R_A/R_B.
4. Assuming $K_R = 0.04\ \mu\text{m}$ for the Pelgrom coefficient, evaluate the relative standard deviation

$$\frac{\sigma[R_A/R_B]}{R_A/R_B}$$

of the resistance ratio.

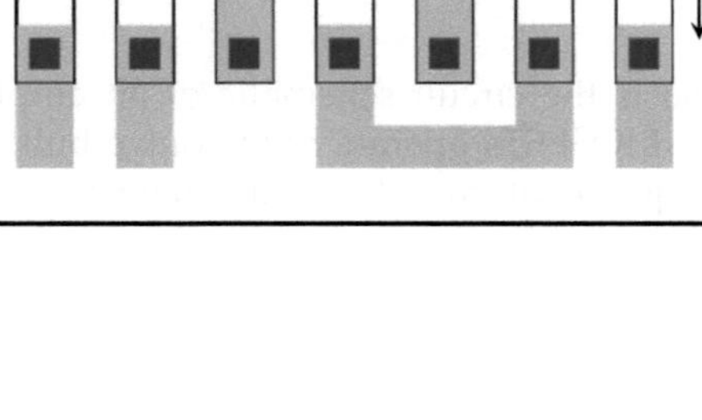

FIGURE 3.67

3.15 As explained in Subsection 3.11.1 and modeled by (3.137), the capacitance of an MIM capacitor has an areal component proportional to A and a fringe component proportional to P. If you wish to accurately establish a certain capacitance ratio $C_2/C_1 = S$ between two MIM capacitors without resorting to segmentation, you need to make sure that both A_2 and P_2 are scaled with respect to A_1 and S_1 by the same S. Now suppose C_1 has a square layout of 10-μm side length, and determine the dimensions of a rectangular C_2 for $S = 1.26$.

CHAPTER 4

Circuit Layer

4.1 Introduction

As explained in Chapter 1, and illustrated in Fig. 1.2, the circuit layer of electrical design includes (a) the selection of an OTA or opamp topology suitable for the application in hand, and (b) the conversion of open-loop performance metrics into device metrics for this selection. There exists a vast collection of OTA and opamp topologies in the literature of analog integrated circuits. In this chapter, the most popular ones are presented and analyzed with a view to interrelate their open-loop metrics and device metrics. In Sections 4.2 to 4.4, we study the basic linear building blocks such as current sources, current and voltage references, and amplifier stages. Sections 4.5 to 4.8 are where the single-ended topologies of basic OTA, symmetrical OTA, folded-cascode OTA, and Miller OTA are introduced. Converting any of these OTAs into an opamp by buffering is the subject of Sections 4.9 and 4.10. Finally, in Section 4.11 the fully-differential versions of these OTAs and opamps are presented.

Since the application and device layers of design have already been covered, the reader should be able to practice the full sequence of electrical design after studying any of the topologies introduced in this chapter. Chapter examples are designed to meet this expectation. Virtually all of them involve the device layer, while many also include the application layer.

4.2 Current Sources, Sinks, and Mirrors

Current sources and current sinks are essential for biasing the amplifier stages of opamps. Equally indispensable for opamps are current mirrors, which copy an exact or scaled replica of a current from one branch to another. Sources, sinks, and mirrors find widespread use also in nonlinear-analog or mixed-signal applications. Their general features and performance metrics are presented in Subsection 4.2.1. Accuracy and precision issues are addressed in Subsection 4.2.2. The remaining four subsections involve high-performance source/sink and mirror topologies utilizing the technique of *cascoding*.

4.2.1 Fundamental Concepts and Performance Metrics

Current sources and sinks share the common property of being independent current sources. What distinguishes one from the other is the topological position with respect to the load circuit they drive. Source is located between the positive power-supply line

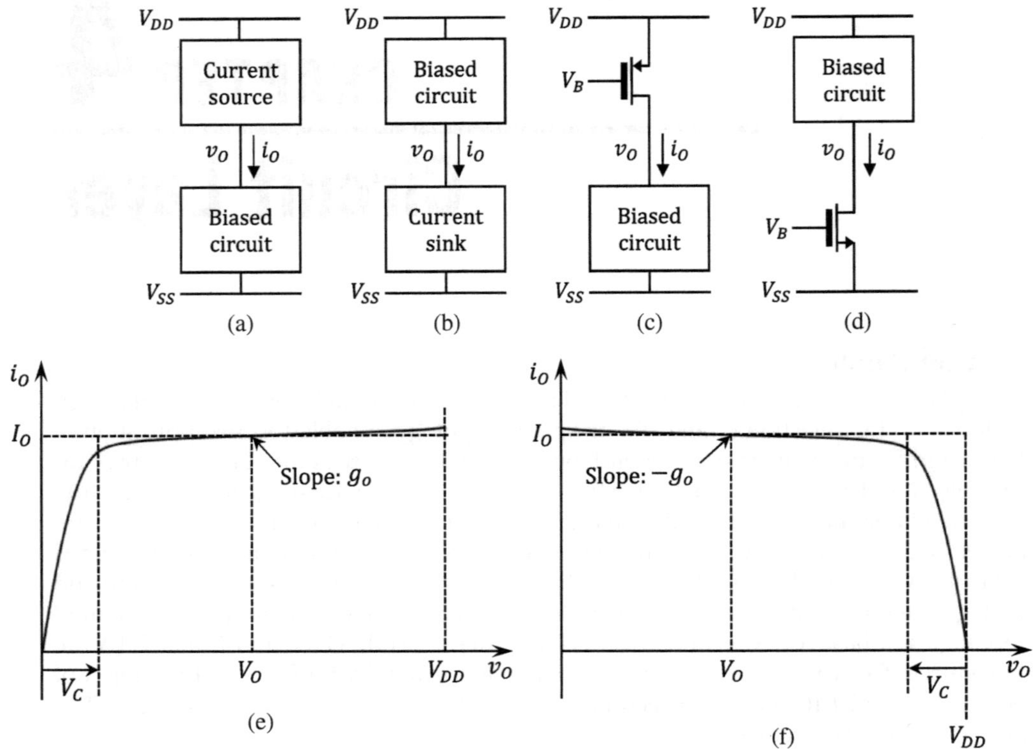

FIGURE 4.1 (a) Block diagram of a current source. (b) Block diagram of a current sink. (c) Single-PMOS current source. (d) Single-NMOS current sink. (e) Conceptual current-voltage characteristic of a single-supply (V_{DD} = 1.8 V) current sink. (f) Conceptual current-voltage characteristic of a single-supply (V_{DD} = 1.8 V) current source.

and the load as depicted in Fig. 4.1(a). Sink is located in between the load and the negative power-supply line (in a split-supply configuration) or ground (in a single-supply configuration) as depicted in Fig. 4.1(b). In both cases i_O denotes the current supplied and v_O denotes the voltage with respect to ground at the output terminal of the source or sink. v_O will be referred to as the output voltage.

The most basic current-source topology is shown in Fig. 4.1(c). It is a single-PMOS device operating in saturation with a source terminal tied to V_{DD} and a gate terminal biased by a dc voltage V_B. The reader may recall the PMOS device M2 of the common-source voltage amplifier stage of Fig. 3.22(a) as a typical example.

As shown in Fig. 4.1(d), a basic current sink is implemented with a single-NMOS device operating in saturation with its gate terminal biased by a dc voltage V_B and the source terminal tied to V_{SS} in a split-supply configuration or to ground in a single-supply configuration.

Source and sink performances are characterized in terms of *nominal output current, compliance limit,* and *output conductance* or *resistance*. These three metrics are described next with reference to the examples of source and sink output characteristics presented in Figs. 4.1(e) and (f) for a single-supply configuration.

Nominal Current Denoted by I_O in this book, nominal current is the constant output current expected to flow independently of the output voltage v_O. In practice, the output current i_O is not constant; it varies with v_O and equals I_O only at one particular value V_O of v_O. Most commonly, V_O is defined as the bias voltage at the output terminal.

Compliance Limit As the voltage across a sink or a source vanishes, so does the current i_O as illustrated in Figs. 4.1(e) and (f) for a single-supply example. The smallest value of this voltage, for which, the sink or source still supplies an acceptable amount of current is called *compliance limit* in this book. It is represented by V_C in Figs. 4.1(e) and (f). Since the basic current source and sink shown in Figs. 4.1(c) and (d) retain proper operation as long as the MOSFET operates in saturation, their compliance limit is described by

$$V_C = V_{DSsat}. \tag{4.1}$$

Output Conductance This parameter characterizes the sensitivity of i_O to v_O around $v_O = V_O$. It is defined by

$$g_o \doteq \begin{cases} \left.\dfrac{\partial i_O}{\partial v_O}\right|_{v_O=V_O} & \text{for a current sink,} \\ -\left.\dfrac{\partial i_O}{\partial v_O}\right|_{v_O=V_O} & \text{for a current source,} \end{cases} \tag{4.2}$$

which corresponds to the absolute value of the slope exhibited by the current-voltage characteristics around $v_O = V_O$, as illustrated in Figs. 4.1(e) and (f). The reciprocal $r_o \doteq 1/g_o$ of the output conductance, called *output resistance*, is also widely used for characterizing the sensitivity of i_O to v_O.

If V_O is defined as the bias voltage at the output of a current source or a sink, the signal component of the output voltage will be represented by $v_o \doteq v_O - V_O$. In the case of a sufficiently small v_o, the signal component $i_o \doteq i_O - I_O$ of the output current can be related to v_o by assuming an approximately linear current-voltage characteristic with a slope $\mp g_o$ around $v_O = V_O$. This leads us to the small-signal sink or source model

$$i_o = \mp g_o v_o = \mp \frac{v_o}{r_o}, \tag{4.3}$$

where the plus sign applies to a sink and the minus sign applies to a source.

It is also worth noting $g_o = g_{ds}$ and $r_o = r_{ds}$ in a single-device source or sink. The reader is referred to (3.7) for g_{ds}, and to Subsection 3.3.6 for r_{ds}.

A typical opamp needs multiple current sources and sinks for fulfilling all biasing requirements. In most cases, these sources and sinks need to be desensitized against power-supply fluctuations, and in many cases, their temperature sensitivity should comply with certain specifications. These additional constraints add to circuit complexity,

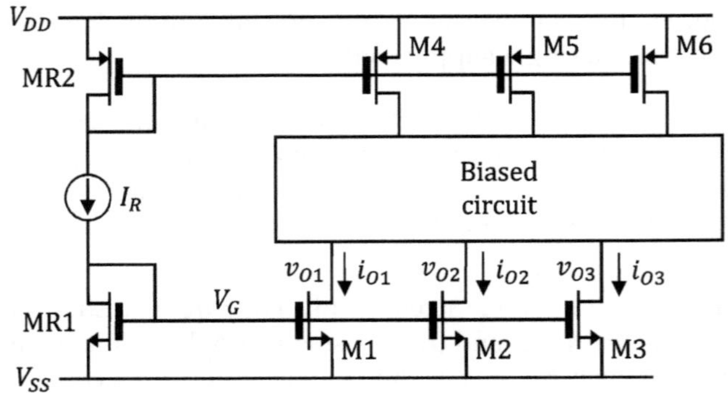

FIGURE 4.2 Simple current mirroring with single-device sources and sinks.

and therefore increase the size and power consumption. Instead of replicating such a complex and expensive source or sink circuit at all points of use throughout the opamp, we prefer to build just one circuit that generates a dc reference current fulfilling the specified performance, and then, to copy its current to individual sources or sinks by appropriate scaling. This central circuit is called *current reference*. For copying, we utilize *current mirrors*.

The principle of mirroring is illustrated in Fig. 4.2 as applied to basic sourcing and sinking. Since the details of current-reference topologies will be presented later, the reference current of this mirroring network is represented conceptually with an independent current source I_R. The bank of NMOS devices M1, M2, and M3 sharing a common-gate rail and a common-source rail exemplifies a network of current sinks biasing three separate ports of a circuit. The similarly configured PMOS bank of M4, M5, and M6 exemplifies a current-source network biasing a different set of three ports.

For an explanation of how the nominal current I_O is set in a mirror, consider the NMOS sink network, and notice the diode-connected and therefore saturated MR1 conducting a reference current I_R and imposing its own overdrive voltage $V_G - V_{SS} - V_{TN}$ onto all sink devices through the common-gate and common-source rails. As explained in Subsection 3.3.4, the sheet current of a MOSFET is determined primarily by the overdrive voltage, and secondarily by the drain-source voltage. Ignoring this secondary dependence as permitted by the definition of nominal current, we conclude that, ideally, the common overdrive voltage equates the sheet current $I_R/(W_R/L_R)$ of the reference device MR1 to the sheet current $I_O/(W/L)$ of a sink device. This equation implies

$$I_O = SI_R, \tag{4.4}$$

where the so-called *scaling factor* S of mirroring is defined by

$$S \doteq \frac{W}{L}\frac{L_R}{W_R}. \tag{4.5}$$

According to (4.4) and (4.5), the nominal current of a sink device is a scaled copy of the reference current, where the scale factor is determined by the ratio of the aspect ratios of the sink and reference devices.

As shown in Fig. 4.2, mirroring for source devices M4, M5, and M6 is implemented with a diode-connected PMOS device MR2 conducting the very same reference current I_R of MR1. The nominal current and scaling factor of a source device are still described by (4.4) and (4.5), respectively. In this case, however, the dimensions appearing in (4.5) belong to the PMOS source and reference devices.

4.2.2 Accuracy and Precision in Current Mirroring

As long as the output voltage of a mirroring sink or source device does not fluctuate, the output current will remain as a perfect dc. Still, this dc current won't be a perfect match to the nominal current I_O targeted in design because the accuracy and precision of mirroring are not perfect. The resulting error is usually of little concern in opamp-biasing applications but may prove to be significant particularly in mixed-signal applications. In this subsection, we identify the causes of inaccuracy and imprecision in mirroring, and determine the techniques by which we can control them.

Accuracy

One systematic contribution to inaccuracy comes from the difference between the drain-source bias voltage V_{DS} of the sink or source device and the drain-source voltage V_{GS} of the diode-connected reference device.[1] The effect of this difference on the sheet currents of the devices was ignored in the derivation of (4.4) and (4.5). If taken into account, it will modify the nominal-current expression (4.4) approximately into

$$I_O = \begin{cases} \left(1 + \frac{V_{DS}-V_{GS}}{V_A}\right) SI_R \text{ for a current sink,} \\ \left(1 - \frac{V_{DS}-V_{GS}}{V_A}\right) SI_R \text{ for a current source,} \end{cases} \tag{4.6}$$

where V_A denotes the average value of the Early voltages of the reference and source/sink devices, and S is the scaling factor described by (4.5). However, the actual dc current I'_O measured in a fabricated copy of the source or sink will be different even from the nominal current predicted by this more accurate expression. The difference $\Delta I_O \doteq I'_O - I_O$ arises mainly from parametric variations experienced in fabrication, and exhibits a distribution around a mean value. The relative mean value $\overline{\Delta I_O / I_O}$ of the distribution is the metric by which we characterize the inaccuracy associated with parametric variations. We now develop a model for this inaccuracy assuming sink mirroring without loss of generality. To this end, we recall from Section 3.3 that the current of an n-channel MOSFET is proportional to the aspect ratio W/L, and is a strongly increasing function of the overdrive voltage $V_{GS} - V_{TN}$. Although the current also weakly depends

[1] For an output bias voltage V_O, the drain-source bias voltage is given by $V_{DS} = V_O - V_{DD}$ in a current source and $V_{DS} = V_O - V_{SS}$ in a current sink.

on the drain-source voltage, we ignore this dependence because its effect on accuracy is separately modeled by (4.6). We therefore represent the sink current with

$$I_O = K\frac{W}{L}\,\mathrm{f}(V_{GS} - V_{TN}), \tag{4.7}$$

where K is a coefficient combining all structural and physical parameters other than the aspect ratio and overdrive voltage. In strong inversion, for example, K is given by $K \doteq \mu_N C_{OX}/2A_N$ as one can ascertain from (3.16). Similarly, we represent the reference current I_R with

$$I_R = K_R\frac{W_R}{L_R}\,\mathrm{f}(V_{GS} - V_{TNR}), \tag{4.8}$$

where the unique parameters or variables of the reference device are identified with a subscript R. Notice V_{GS} as the sole common variable between (4.7) and (4.8).

We now suppose that each parameter or variable X appearing in (4.7) or (4.8) deviates on the average by a $\overline{\Delta X}$ from what is targeted in design. Also assuming sufficiently small deviations to justify a first-order series expansion of (4.7) and (4.8), we can express $\overline{\Delta I_O}$ and $\overline{\Delta I_R}$ with

$$\overline{\Delta I_O} = \frac{I_O}{K}\overline{\Delta K} + \frac{I_O}{W}\overline{\Delta W} - \frac{I_O}{L}\overline{\Delta L} + \frac{\partial I_O}{\partial V_{GS}}\overline{\Delta V_{GS}} + \frac{\partial I_O}{\partial V_{TN}}\overline{\Delta V_{TN}}, \tag{4.9}$$

and

$$\overline{\Delta I_R} = \frac{I_R}{K_R}\overline{\Delta K_R} + \frac{I_R}{W_R}\overline{\Delta W_R} - \frac{I_R}{L_R}\overline{\Delta L_R} + \frac{\partial I_R}{\partial V_{GS}}\overline{\Delta V_{GS}} + \frac{\partial I_R}{\partial V_{TNR}}\overline{\Delta V_{TNR}}, \tag{4.10}$$

respectively. By adopting (3.5) for the derivatives with respect to the gate-source voltage in both (4.9) and (4.10), and also invoking the identity $\partial i_D/\partial V_{TN} = -\partial i_D/\partial v_{GS}$ stated prior to (3.31), we turn these equations into

$$\frac{\overline{\Delta I_O}}{I_O} = \frac{\overline{\Delta K}}{K} + \frac{\overline{\Delta W}}{W} - \frac{\overline{\Delta L}}{L} + \frac{g_m}{I_O}\overline{\Delta V_{GS}} - \frac{g_m}{I_O}\overline{\Delta V_{TN}}, \tag{4.11}$$

and

$$\frac{\overline{\Delta I_R}}{I_R} = \frac{\overline{\Delta K_R}}{K_R} + \frac{\overline{\Delta W_R}}{W_R} - \frac{\overline{\Delta L_R}}{L_R} + \frac{g_{mR}}{I_R}\overline{\Delta V_{GS}} - \frac{g_{mR}}{I_R}\overline{\Delta V_{TNR}}. \tag{4.12}$$

Finally, we solve (4.12) for $\overline{\Delta V_{GS}}$, using the outcome in (4.11), and assume $g_{mR}/I_R \approx g_m/I_O$ as justified by the fact that these two quantities are determined primarily by the nominally identical $V_{GS} - V_{TNR}$ and $V_{GS} - V_{TN}$. These steps lead us to

$$\frac{\overline{\Delta I_O}}{I_O} = \underbrace{\frac{\overline{\Delta I_R}}{I_R}}_{1} + \underbrace{\frac{\overline{\Delta W}}{W} - \frac{\overline{\Delta W_R}}{W_R} - \frac{\overline{\Delta L}}{L} + \frac{\overline{\Delta L_R}}{L_R}}_{2} + \underbrace{\frac{\overline{\Delta K}}{K} - \frac{\overline{\Delta K_R}}{K_R} - \frac{g_m}{I_O}\left(\overline{\Delta V_{TN}} - \overline{\Delta V_{TNR}}\right)}_{3}, \tag{4.13}$$

where the components are numbered for convenience in referencing.

It is obvious from the first component on the right-hand side of (4.13) that the inaccuracy of the reference current I_R is directly copied to the sink current. This component of inaccuracy can therefore be mitigated only by prudent reference design. The remaining components of inaccuracy appearing in (4.13) arise generally from the very same causes we identified in Subsection 3.10.2 for resistors, that is, (a) imperfect repeatability of fabrication processes; (b) on-chip gradients of device parameters, temperature, and stress; and (c) proximity effects in patterning. We now discuss the way in which each of these causes affects the inaccuracy, and identify the design practices by which we can counter their effects.

Consider an abstract case, in which, the imperfect repeatability of fabrication processes is the only cause of inaccuracy. In this case, channel dimensions and device-independent structural and physical parameters will shift with respect to their values set or assumed in design but, since the amount of shift in a dimension or a parameter affects all devices equally, we can assume $\overline{\Delta K} = \overline{\Delta K_R}$ and $\overline{\Delta V_{TN}} = \overline{\Delta V_{TNR}}$. As a result, the third component in (4.13) vanishes. The second component also vanishes but only if $W = W_R$ and $L = L_R$; that is, if the reference and sink devices are designed to be dimensionally identical. This, of course, implies a scale factor $S = 1$. In an application necessitating $S \neq 1$, we can still eliminate the second component by resorting to segmentation into matching unit devices as we did for resistors in Subsection 3.10.2. As depicted in Fig. 4.3, a mirroring network comprising m_R parallel copies of a unit device on the reference side sets the current of each unit to I_R/m_R. Paralleling m copies of the same unit device on the sink side

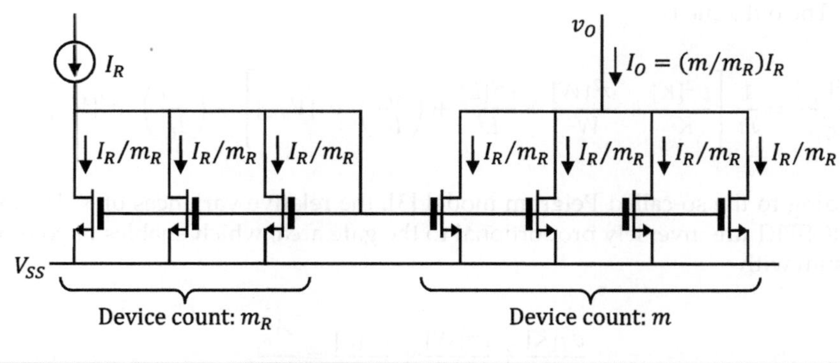

FIGURE 4.3 Mirroring with matching unit devices.

yields $I_O = mI_R/m_R$ for the sink current. This is how a scaling factor $S = m/m_R$ is accurately implemented without being affected by the imperfect repeatability of fabrication processes.

A current mirror, even if segmented, will suffer from inaccuracy due to on-chip gradients in temperature, stress, and device-independent structural and physical parameters causing $\overline{\Delta K} \neq \overline{\Delta K_R}$ and $\overline{\Delta V_{TN}} \neq \overline{\Delta V_{TNR}}$. Gradients may also be present in the channel length and width of the unit devices laid out in the reference and sink sides of the mirror. Interdigitation of these devices around a common centroid will virtually eliminate this type of inaccuracy as long as the gradients remain reasonably constant over the combined footprint of both sides.

Proximity effects in patterning may also lead to $\overline{\Delta W} \neq \overline{\Delta W_R}$ and $\overline{\Delta K} \neq \overline{\Delta K_R}$.[1][2] As described for resistors in Subsection 3.10.2, dummy termination is a radical solution for this type of inaccuracy.

Although we derived (4.13) for sink mirroring, it is applicable also to source mirroring provided that the minus sign of the threshold mismatch term on the right-hand side is replaced with a plus sign.

As a final word on the accuracy of mirroring, it is worth reiterating the role of the disparate drain-source voltages as modeled in (4.6). A large Early voltage V_A is obviously the only way of reducing this type of inaccuracy in a basic current mirror. A more radical solution is to cascode the current mirror. As explained in the remaining subsections, cascoding works by effectively equating the drain-source voltages of the reference and mirror devices.

Precision

Imprecision in mirroring is characterized by the relative standard deviation $\sigma[I_O]/I_O$ of the distribution of I_O among the fabricated copies of the sink/source device. For modeling this metric, we consider the segmented sink mirror depicted in Fig. 4.3, and assume a deterministic I_R. Adopting (4.7) for the nominal current conducted by each unit MOSFET, we can describe the total sink current as

$$I_O = \sum_{j=1}^{m} K_j \frac{W_j}{L_j} \mathrm{f}(V_{GS} - V_{TNj}), \tag{4.14}$$

where j is the index of unit MOSFETs. The variance of I_O is determined by applying (2.28). The outcome is

$$\frac{\sigma^2[I_O]}{I_O^2} = \frac{1}{m}\left[\frac{\sigma^2[K]}{K^2} + \frac{\sigma^2[W]}{W^2} + \frac{\sigma^2[L]}{L^2} + \left(\frac{g_m}{I_U}\right)^2 \sigma^2[V_{TN}]\right] + \left(\frac{g_m}{I_U}\right)^2 \sigma^2[V_{GS}]. \tag{4.15}$$

According to the so-called Pelgrom model [3], the relative variances of K, W, and L in any MOSFET are inversely proportional to the gate area, which enables us to represent their sum with

$$\frac{\sigma^2[K]}{K^2} + \frac{\sigma^2[W]}{W^2} + \frac{\sigma^2[L]}{L^2} = \frac{A_\beta^2}{WL},$$

where A_β is a coefficient whose value in 0.18-μm CMOS technologies is typically around 0.01 μm.[4] The variance of V_{TN} is also inversely proportional to WL, that is, $\sigma^2[V_{TN}] = A_{VT}^2/WL$, where the coefficient A_{VT} is typically around 0.004 V · μm in 0.18-μm CMOS technologies.[4] By applying these modeling expressions we finally turn (4.15) into

$$\frac{\sigma^2[I_O]}{I_O^2} = \frac{1}{mWL}\left[A_\beta^2 + \left(\frac{g_m}{I_U}\right)^2 A_{VT}^2\right] + \left(\frac{g_m}{I_U}\right)^2 \sigma^2[V_{GS}]. \tag{4.16}$$

Note that the variance $\sigma^2[V_{GS}]$ appearing in this expression depends solely on the reference path of the mirror. It can be related to its own random variables only after the topological details of the reference path are specified. The rest of the parameters appearing in (4.16) are associated with the source path. They indicate the following two design measures for improving the precision of mirroring: (a) enlarging the total channel area mWL of all unit devices combined in that path, and (b) lowering g_m/I_U by adopting a large overdrive voltage $V_{GS} - V_{TN}$. The downside of these measures is that they respectively lead to a large footprint and a large V_{DSsat} and therefore a large compliance limit.

Finally, note that (4.16) can also be used for evaluating the imprecision of a current-source built with PMOS unit devices.

Example 4.1 Referring to Figs. 4.2 and 4.3, consider a sink network of $V_{SS} = 0$ and $I_R =$ 20 μA built with (a) a reference device MR1 of 4 unit MOSFETs ($m_R = 4$), (b) a sink device M1 of 28 unit MOSFETs ($m_1 = 28$), and (c) another sink device M2 of 44 unit MOSFETs ($m_2 = 44$). Determine the dimensions of the unit MOSFET for a compliance limit $V_C = 0.15$ V and an output resistance $r_o \geq 100$ kΩ for $V_O = 0.9$ V.

Work The steps we take for determining the dimensions of a unit device in accordance with the flowchart of Fig. 3.24 are itemized below.

1. *Current* $I_D = I_R/m_R = 20(\mu\text{A})/4 = 5$ μA.
2. *Drain-Source Saturation Voltage* $V_{DSsat} = V_C = 0.15$ V from (4.1).
3. *Overdrive Voltage* $V_{GS} - V_{TN} = 0.195$ V from $V_{DSsat} = 0.15$ V and design tool of Fig. 3.8.
4. *Sheet Current* $I_\square = 4.5$ μA/sq from $V_{GS} - V_{TN} = 0.195$ V and design tool of Fig. 3.10 or Fig. 3.11.
5. *Aspect Ratio* $W/L = I_D/I_\square = 1.12$.
6. *Early Voltage* Let us represent the output resistance of a unit MOSFET with r_{ds}. The output resistance of M1 can be expressed as $r_{o1} = r_{ds}/m_1$ because a total of m_1 unit-MOSFETs are paralleled in this device. Similarly, the output resistance of M2 is given by $r_{o2} = r_{ds}/m_2$. Note that r_{o2} is smaller than r_{o1} because $m_2 > m_1$. Therefore, if the output-resistance specification $r_o \geq 100$ kΩ is satisfied by M2, it will be automatically satisfied by M1. This leads to $r_{ds} \geq m_2 r_o = 44 \times 100(\text{k}\Omega)=$ 4.4 MΩ, which implies $V_A \geq r_{ds} I_D = 4.4(\text{M}\Omega) \times 5(\mu\text{A}) = 22$ V for the target value of the Early voltage.
7. *Channel Length* To be determined by iterative simulation conducted with a constant drain-source voltage $V_{DS} = V_O = 0.9$ V as specified for the output

resistance. The overdrive voltage $V_{GS} - V_{TN} = 0.195$ V determined above will also be kept constant. Channel width is not a critical parameter of the simulation because it is not a factor of the Early voltage. We set it arbitrarily to $W_{DR} = 1$ μm. We start iteration with an initial channel-length estimate of $L_{DR} = 0.8$ μm. According to the design tool of Fig. 3.7, the corresponding value of the threshold voltage is approximately 0.43 V. Therefore, the gate-source voltage is set to $V_{GS} = 0.195 + 0.43 = 0.625$ V in this initial run. Simulation is performed with the following input file `ex4-1a.sp`.

```
filename: ex4-1a.sp
.incl bsim.sp
mu 2 1 0 0 cmosn w=1u l=0.8u
vg 1 0 0.625
vd 2 0 0.9
.control
op
print @mu[id]/@mu[gds]
.endc
.end
```

Note that this is an operation-point analysis that directly prints out the Early voltage of the simulated device with the embedded command

```
print @mu[id]/@mu[gds]
```

The initial run yields $V_A = 15.69$ V, which is significantly short of the target. We raise the channel length to $L_{DR} = 1.2$ μm, and determine the corresponding threshold voltage from Fig. 3.7 as $V_{TN} \approx 0.41$ V. Considering $V_{GS} - V_{TN} = 0.195$ V, we reset the gate-source voltage to $V_{GS} = 0.605$ V, and rerun the simulation. In this second run, we get $V_A = 20.61$ V, which is closer to the target but not close enough. We run one more simulation after raising the channel length to $L_{DR} = 1.4$ μm without altering V_{GS} because Fig. 3.7 does not indicate a significant change in the threshold voltage. This run yields $V_A = 23.49$ V, which is slightly higher and therefore better than the target value. We adopt 1.4 μm for the channel length, and calculate the corresponding effective channel length from (3.2) as $L = 1.37$ μm.

8. *Channel Width* $W = 1.53$ μm as determined from $W/L = 1.12$ and $L = 1.37$ μm. The corresponding drawn width is calculated from (3.1): $W_{DR} = 1.55$ μm.
9. *Verification* Performed with the following input file:

```
filename: ex4-1b.sp
.incl bsim.sp
mr1 1 1 0 0 cmosn w=1.55u l=1.4u m=4
m1 2 1 0 0 cmosn w=1.55u l=1.4u m=28
m2 3 1 0 0 cmosn w=1.55u l=1.4u m=44
ir 0 1 20u
vd1 2 0 0.9
vd2 3 0 0.9
.op
.end
```

After the completion of the simulation, entering the command `show` at the prompt displays all device parameters. The reciprocal of `gds` indicates r_o = 164 kΩ for M1 and r_o = 104 kΩ for M2. The very same simulation also indicates 141.9 μA for the nominal current of M1, and 223.1 μA for the nominal current of M2. In a perfect sink, these numbers would be $(m_1/m_R)I_R = (28/4)20(\mu\text{A})$ = 140 μA and $(m_2/m_R)I_R = (44/4)20(\mu\text{A})$ = 220 μA, respectively. As the reader can estimate from (4.6), the differences between the actual and target values of these two currents are due to the mismatching drain-source voltages of the reference and sink devices. Also notice V_{DSsat} = 0.152 V for the simulated value of the drain-source saturation voltages of M1 and M2. It almost exactly matches the target value. ▲

Example 4.2 Repeat Example 4.1 after changing the output-resistance specification to $r_o \geq 200$ kΩ for $V_O = 0.9$ V while retaining other specifications.

Work All design steps prior to the determination of the channel length are identical to their counterparts in Example 4.1. For the channel length, we need to target $V_A = 44$ V because the specified value of r_o is doubled. By running iterative simulations with file `ex4-1a.sp`, we finally arrive at L_{DR} = 4 μm for this much Early voltage. The corresponding width is calculated as W_{DR} = 4.46 μm.

Note that, in order to increase the output resistance from 100 to 200 kΩ, we have to expand the gate area $W_{DR}L_{DR}$ of a unit device more than eightfold from 1.55 × 1.4 = 2.17 μm² to 4.46 × 4 = 17.8 μm². Indeed, the size of a single-device current sink or source increases exponentially with the output resistance targeted in design. As will be explained in the remaining subsections of the present section, this deficiency of single-device solutions is overcome by applying the technique of cascoding. ▲

Example 4.3 As will be described in Subsection 4.3.1, the reference current I_R in a voltage-divider current-sink topology is supplied by a resistor placed between V_{DD} and the common gate/drain terminal of the NMOS unit devices of the reference path. The variance of the gate-source voltage observed in such a reference is described approximately by

$$\sigma^2[V_{GS}] = \frac{A_{VT}^2}{m_R WL} + \left(\frac{I_U}{g_m}\right)^2 \left[\frac{A_\beta^2}{m_R WL} + \frac{K_{\text{RES}}^2}{L_{\text{RES}} W_{\text{RES}}}\right], \tag{4.17}$$

where the last term on the right-hand side represents the relative variance of the resistance as per (3.128). Assuming A_β = 0.01 μm, A_{VT} = 0.004 V · μm, K_{RES} = 0.1 μm, L_{RES} = 85 μm, and W_{RES} = 1.55 μm, repeat Example 4.1 after adding $\sigma[I_O]/I_O \leq 1.5\%$ to its existing design specifications.

Work We first rewrite (4.16) after substituting (4.17) for $\sigma^2[V_{GS}]$:

$$\frac{\sigma^2[I_O]}{I_O^2} = \left(\frac{1}{m} + \frac{1}{m_R}\right) \frac{A_\beta^2 + \left(\frac{g_m}{I_U}\right)^2 A_{VT}^2}{WL} + \frac{K_{\text{RES}}^2}{L_{\text{RES}} W_{\text{RES}}}, \tag{4.18}$$

where m is to be replaced by m_1 or m_2. Notice from (4.18) that $m = m_1$ is appropriate because it implies a larger $\sigma[I_O]/I_O$ due to $m_1 < m_2$. With this substitution, (4.18) leads to

$$WL = \left(\frac{1}{m_1} + \frac{1}{m_R}\right) \frac{A_\beta^2 + \left(\frac{g_m}{I_U}\right)^2 A_{VT}^2}{\frac{\sigma^2[I_{O1}]}{I_{O1}^2} - \frac{K_{RES}^2}{L_{RES} W_{RES}}} \tag{4.19}$$

for the minimum-effective gate area of a unit MOSFET. Transconductance efficiency is the only unknown in this expression. We quantify it as $g_m/I_U = 10\ \text{V}^{-1}$ by referring to the design tool of Fig. 3.12 for the overdrive voltage $V_{GS} - V_{TN} = 0.195$ V determined earlier in Example 4.1. Equation (4.19) therefore yields $WL = 3.24\ \mu\text{m}^2$. Considering also $W/L = 1.12$ of Example 4.1, we arrive at $W = 1.90\ \mu$m and $L = 1.70\ \mu$m, which are sufficiently large to be considered as drawn dimensions. Since the present channel length is larger than the one found in Example 4.1, it will readily satisfy the output-resistance specification. We therefore set channel dimensions to these new values. If the channel length satisfying the specified upper limit of imprecision had turned out to be smaller than the one found in Example 4.1, then, we would have retained the dimensions found in that previous exercise. ▲

Example 4.4 Referring to Fig. 4.2, consider a source network of $V_{DD} = 1.8$ V built with (a) a current reference of $I_R = 20\ \mu$A, (b) a reference device MR2 of 4 unit MOSFETs ($m_R = 4$), (c) a source device M4 of 28 unit MOSFETs ($m_4 = 28$), and (d) another source device M5 of 44 unit MOSFETs ($m_5 = 44$). Determine the dimensions of the unit MOSFET for a compliance limit $V_C = -0.15$ V and an output resistance $r_o \geq 100$ kΩ at $V_O = 0.9$ V.

Work Design flow is similar to the one we followed in Example 4.1:

1. *Current* $I_D = I_R/m_R = 20(\mu\text{A})/4 = 5\ \mu\text{A}$.
2. *Drain-Source Saturation Voltage* $V_{DSsat} = V_C = -0.15$ V from (4.1).
3. *Overdrive Voltage* $V_{GS} - V_{TP} = -0.17$ V from $V_{DSsat} = -0.15$ V and design tool of Fig. 3.17.
4. *Sheet Current* $I_\square = 0.89\ \mu\text{A/sq}$ from $V_{GS} - V_{TP} = -0.17$ V and design tool of Fig. 3.18 or Fig. 3.19.
5. *Aspect Ratio* $W/L = I_D/I_\square = 5.62$.
6. *Early Voltage* $r_{ds} \geq m_5 r_o = 44 \times 100(\text{k}\Omega) = 4.4\ \text{M}\Omega$. Therefore, $V_A = r_{ds} I_D = 4.4(\text{M}\Omega) \times 5(\mu\text{A}) = 22$ V.
7. *Channel Length* To be determined by iterative simulation. Drain-source voltage is set to $V_{DS} = V_O - V_{DD} = -0.9$ V as specified for the output resistance. Gate-source voltage is calculated from (a) $V_{GS} - V_{TP} = -0.17$ V as determined above, and (b) the threshold voltage V_{TP} as estimated from the design tool of Fig. 3.16 for $V_{SB} = 0$ and selected channel length. Simulation is run with the following input file:

```
filename: ex4-4a.sp
.incl bsim.sp
```

```
mu 2 1 0 0 cmosp w=1u l=2.75u
vg 1 0 -0.57
vd 2 0 -0.9
.control
op
print @mu[id]/@mu[gds]
.endc
.end
```

The target set in part 6 is reached for $L_{DR} = 2.75\ \mu m$. The corresponding effective channel length is calculated from (3.2) as $L = 2.69\ \mu m$.

8. *Channel Width* $W = 15.1\ \mu m$ as determined from $W/L = 5.62$ and $L = 2.69\ \mu m$. W is large enough to ignore the process bias. Hence, $W_{DR} = W = 15.1\ \mu m$.
9. *Verification* Performed with the following input file:

```
filename: ex4-4b.sp
.incl bsim.sp
mr2 1 1 4 4 cmosp w=15.1u l=2.75u m=4
m4 2 1 4 4 cmosp w=15.1u l=2.75u m=28
m5 3 1 4 4 cmosp w=15.1u l=2.75u m=44
ir 1 0 20u
vd4 2 0 0.9
vd5 3 0 0.9
vdd 4 0 1.8
.op
.end
```

Simulation yields (a) $r_o = 160\ k\Omega$ and $I_O = 142.1\ \mu A$ for M4, (b) $r_o = 102\ k\Omega$ and $I_O = 223.3\ \mu A$ for M5, and (c) $V_{DSsat} = -0.150$ V for M4 and M5. ▲

4.2.3 Basic Cascoding

As we concluded initially in Example 3.10 and later in Example 3.12, the voltage gain of a simple current-source loaded amplifier stage is proportional to the parallel equivalent of the output resistances of the driver and current-source devices. The very same fact surfaced in the differential amplifier stages involved in Examples 3.13 and 3.14. A large output resistance is indeed necessary for a large voltage gain in all conventional amplifier stages.

For a given nominal current I_O, the output resistance of a basic single-device current source or sink can be increased only by elongating the device or by raising its overdrive voltage as we concluded in Subsection 3.3.6. If we select to elongate the device, we must also widen it proportionately in order to maintain the specified I_O. As a result, the device gets increasingly larger, and therefore becomes more expensive in terms of its footprint. We paid this price in Example 4.2, where raising r_o from 100 to 200 kΩ resulted in an eightfold increase in the active channel area of a unit MOSFET. If, instead, we resort to raising the overdrive voltage for a larger output resistance, a higher V_{DSsat} will follow. This, in turn, will enlarge the compliance limit, and therefore shrink the headroom available for output-voltage swing. Obviously, our options for achieving a large output resistance in a single-device current sink or current source are rather limited. This limitation is overcome by applying a topological technique called *cascoding*. In this section, we present cascoding in its most fundamental form. Three more forms of cascoding, each involving a unique level of trade-off between output-resistance and compliance-limit, will be presented in subsequent subsections.

The technique of cascoding is illustrated in Fig. 4.4(a) on a sink application. The output current i_O is determined primarily by the sink device M1 which operates in saturation with (a) a constant gate-source voltage $V_{GS1} = V_{B1} - V_{SS}$ applied externally, and (b) an approximately constant drain-source voltage $v_{DS1} = v_{D1} - V_{SS}$ imposed by the cascode device M2. For an explanation of how the drain-source voltage of M1 is kept almost constant by M2, first, note that the variable component v_{d1} of the drain voltage v_{D1} is due to the flow of the signal component v_o of the output voltage v_O through the drain-to-source path of M2. Applying (3.61) and (3.62) to M2, we determine the gain along this path as

$$\frac{v_{d1}}{v_o} = \frac{r_{ds1}}{r_o}, \tag{4.20}$$

where the output resistance r_o seen into the drain of M2 is given by

$$r_o = r_{ds2} + A_{\max 2} r_{ds1}. \tag{4.21}$$

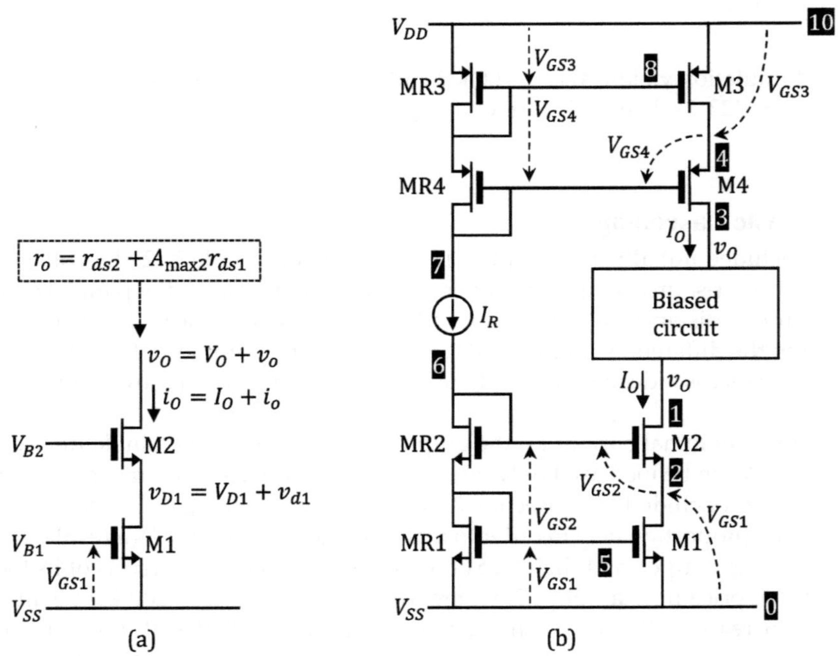

FIGURE 4.4 (a) The principle of cascoding, where r_o denotes the output resistance. (b) Basic-cascode mirror, where node numbers are used in the input file of Example 4.5.

These equations indicate that the output signal v_o is attenuated by a factor no less than A_{max2} as it flows to the drain of M1 as v_{d1}. The reader will recall that A_{max2} is the maximum gain of M2 as described by (3.58) for a regular twin-well device, and by (3.60) for a deep n-well (DNW) device with an individual well tied to its source. In any case, its value is on the order of hundred volts per volt. It is obvious that the drain-source voltage and therefore the current of M1 are nearly constant even in the presence of a large output signal v_o. This is why the output resistance r_o of a cascoded sink surpasses that of a single-device sink which operates with a drain exposed directly to the output signal. Recall that r_o equals r_{ds1} in the case of a single-device current sink. According to (4.21), cascoding multiplies it by A_{max2}, and adds r_{ds2} to the product. Since r_{ds2} is comparable to r_{ds1} while A_{max2} is typically on the order of hundred volts per volt, r_o happens to be about two orders of magnitude larger than what is offered by a single-device sink. For a given v_o, such an enhancement implies two orders of magnitude reduction in the variable part i_o of the output current.

In order for cascoding to function as intended, both M1 and M2 should be kept in saturation. The saturation condition of M1 is satisfied by properly setting the gate voltage V_{B2} of M2. To see how, ignore the very small variable components v_{d1} and i_o, and consider M2 conducting a dc current I_O. In order to conduct this much current, M2 must establish for itself a constant overdrive voltage $V_{B2} - V_{D1} - V_{TN2}$, and thus relate V_{D1} to V_{B2}. By properly adjusting V_{B2}, we can set the drain-source voltage $V_{D1} - V_{SS}$ of M1 above V_{DSsat1}, and thus ensure saturation for M1. As to the saturation condition of M2, we satisfy it by not letting the output voltage v_O drop below $V_{D1} + V_{DSsat2}$.

The most fundamental circuit topology by which a reference current can be mirrored to a cascode sink or source is shown in Fig. 4.4(b). It is called *basic cascode*. Consider the sink network of NMOS devices for an explanation of how this circuit works. M1 is the actual sink device cascoded by M2. MR1 is the reference device from which M1 mirrors the reference current I_R with a scaling factor S to establish a nominal current $I_O = SI_R$. MR2 is used for biasing M2. Most typically, the aspect ratio of M2 is scaled with respect to MR2 by the very same factor S by which their respective currents I_O and I_R are scaled. This leads to the following implications for biasing:

- Since the currents and aspect ratios are scaled by the same factor, the sheet currents of MR2 and M2 are identical (see the flowchart of Fig. 3.24).
- Identical sheet currents imply identical overdrive voltages for MR2 and M2 (see the flowchart of Fig. 3.24).
- Identical overdrive voltages and interconnected gates of MR2 and M2 imply identical sums of source-bulk and threshold voltages for these devices. Since the threshold voltage depends on the source-bulk voltage, this identity, in turn, implies identical source-bulk voltages, and therefore identical gate-source voltages for these two devices. This common gate-source voltage is marked V_{GS2} in Fig. 4.4(b).
- The identical source-bulk voltages of MR2 and M2 imply identical drain-source voltages for M1 and MR1. This common drain-source voltage obviously equals the gate-source voltage V_{GS1} of the diode-connected MR1.
- M1, thus having the same V_{GS1} across its gate-source and drain-source ports, operates unconditionally in saturation as a *virtually* diode-connected device without having a physical short circuit between its gate and drain terminals.

The compliance limit of a basic-cascode sink is given by

$$V_C = V_{GS1} + V_{DSsat2}. \tag{4.22}$$

Comparing this with the compliance limit described by (4.1) for a single-device sink, we observe V_{GS1} as an additional component. For a given I_R and I_O, we can lower this component by widening MR1 and M1 but, except for subthreshold applications, it will be no less than the threshold voltage, which, for the standard-threshold devices of our exemplar technology, is around 0.4 V. Considering also the contribution of V_{DSsat2}, the compliance limit will still be at least half a volt. This much loss in headroom is hardly acceptable in a 1.8-V technology. One way to reduce it is to build the basic cascode with zero-threshold or low-threshold devices if such threshold options are available in the technology selected for fabrication. As will be exemplified in Example 4.5, such devices offer the same overdrive voltage with a smaller V_{GS} and therefore yield a lower V_C.

It is worth mentioning that the threshold voltage of the cascoding pair MR2/M2 of Fig. 4.4(b) is inflated due to the body effect caused by the nonzero source-bulk voltage V_{GS1} of these devices. This leads to a large V_{GS2}. If necessary, this can be avoided by deploying DNW devices, and connecting the p-wells of these devices to their respective source terminals. Incidentally, the threshold voltage of the pair MR1/M1 is free from body effect due to the interconnected source and bulk terminals.

As shown in the upper half of Fig. 4.4(b), mirroring in a basic-cascode current source is implemented with PMOS devices. Although its operation is qualitatively similar to that of a cascode sink, the reader will still benefit from reviewing the contents of the present subsection in terms of a cascode current source before starting to work on Example 4.5.

As a final remark, note that the contents of Subsection 4.2.2 on precision and accuracy in single-device mirroring are also applicable to basic-cascode mirroring. MR1 and M1 of the sink section and MR3 and M3 of the source section are the devices to be considered when applying the conclusions of that subsection. It should also be obvious to the reader that the mirroring inaccuracy caused by disparate drain-source voltages and modeled by (4.6) is practically eliminated in cascode mirroring.

Example 4.5 In this example, we design the basic-cascode current source and current sink shown in Fig. 4.4(b) with low-threshold devices for single-supply operation with $V_{DD} = 1.8$ V in accordance with the following specifications:

1. *Reference Current* $I_R = 5\ \mu$A.
2. *Number of Unit Devices* MR1 and MR2, one NMOS unit device each; M1 and M2, five NMOS unit devices each; MR3 and MR4, one PMOS unit device each; M3 and M4, three PMOS unit devices each.
3. *Each of MR4 and M4 will have an individual n-well tied to source.*
4. *Current-Source Output Resistance* $r_o = 35\ \text{M}\Omega$ for $V_O = 0.9$ V.
5. *Current-Source Compliance Limit* $V_C = -0.3$ V.
6. *Current-Sink Output Resistance* $r_o = 35\ \text{M}\Omega$ for $V_O = 0.9$ V.
7. *Current-Sink Compliance Limit* $V_C = 0.3$ V.

As explained in Section 3.1, most CMOS manufacturers offer a variety of threshold options. The BSIM3v3 parameter deck presented in Table 3.1 and used so far in examples belongs to the standard-threshold option. In this example we introduce the deck `bsimlvt.sp` representing a low-threshold option. It is a fictitious deck obtained by replacing the zero-bias threshold parameters of the original deck `bsim.sp` of Table 3.1 with `vth0=0.05` for NMOS and `vth0=-0.05` for PMOS. Model names of these low-threshold devices are `nmoslvt` and `pmoslvt`. In a real low-threshold model deck, the values of all parameters would be different from their standard-threshold counterparts but the qualitative features of the design tools described in Sections 3.3 and 3.4 would remain the same.

The threshold-voltage design tool of the low-threshold NMOS option is extracted with the following input file `nmoslvt1.sp`:

```
filename: nmoslvt1.sp (low-threshold nmos vth)
.incl bsimlvt.sp
m 2 1 3 0 cmosnlvt w=6u l=0.2u
vds 2 3 0.9
vgs 1 3 0.9
vsb 3 0 0
.save @m[vth]
.control
destroy all
foreach length 0.2u 0.4u 0.6u 0.8u 1.2u 2u 4u 8u
alter @m[l]=$length
dc vsb 0 1.2 20m
end
plot dc1.@m[vth] dc2.@m[vth] dc3.@m[vth] dc4.@m[vth]
+ dc5.@m[vth] dc6.@m[vth] dc7.@m[vth] dc8.@m[vth]
.endc
.end
```

For the same low-threshold NMOS option the following input file `nmoslvt2.sp` is used for extracting V_{DSsat}, $I_{\square}$, g_m, and g_{mb}:

```
filename: nmoslvt2.sp (low-threshold nmos vdsat, id, gm, gmbs)
.incl bsimlvt.sp
m 2 1 3 0 cmosnlvt w=6u l=0.8u
vds 2 3 0.9
vgs 1 3
vsb 3 0 0
.dc vgs -0.1 1 10m
.save @m[vth] @m[vdsat] @m[id] @m[gm] @m[gmbs] v(1,3)
.end
```

NMOS design tools extracted with these files are presented in Figs. 4.5 to 4.9.

The following input file `pmoslvt1.sp` is used for extracting the threshold-voltage design tool of the low-voltage PMOS option:

```
filename: pmoslvt1.sp (low-threshold pmos vth)
.incl bsimlvt.sp
m 2 1 3 0 cmosplvt w=6u l=0.2u
```

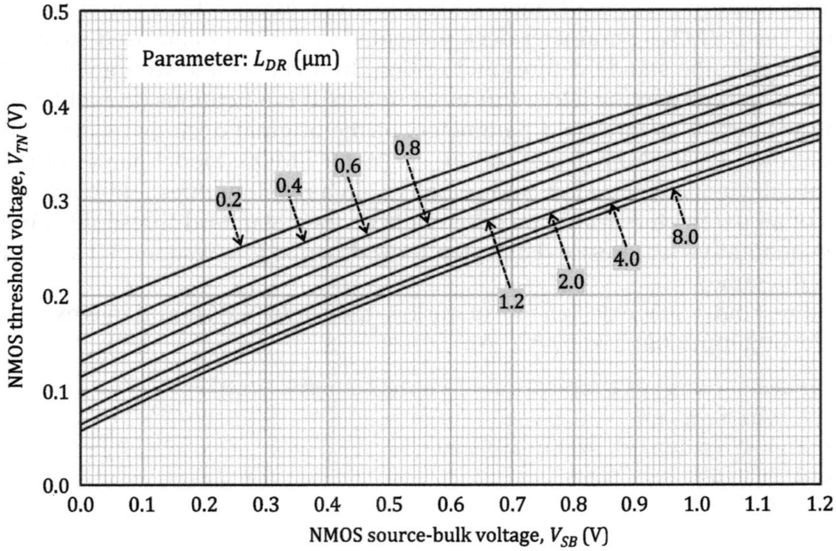

FIGURE 4.5 Low-threshold option NMOS threshold voltage V_{TN} as a function of source-bulk bias voltage V_{SB} and drawn channel length L_{DR}.

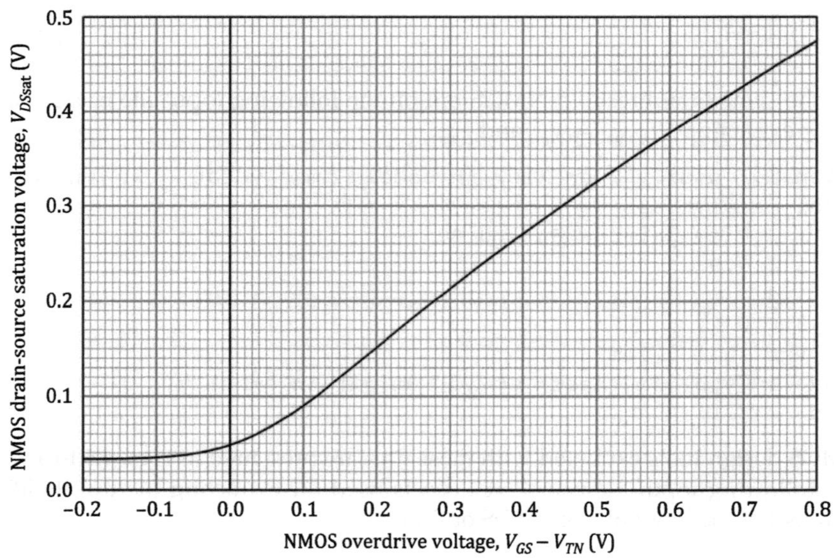

FIGURE 4.6 Low-threshold option NMOS drain-source saturation voltage V_{DSsat} as a function of overdrive voltage $V_{GS} - V_{TN}$.

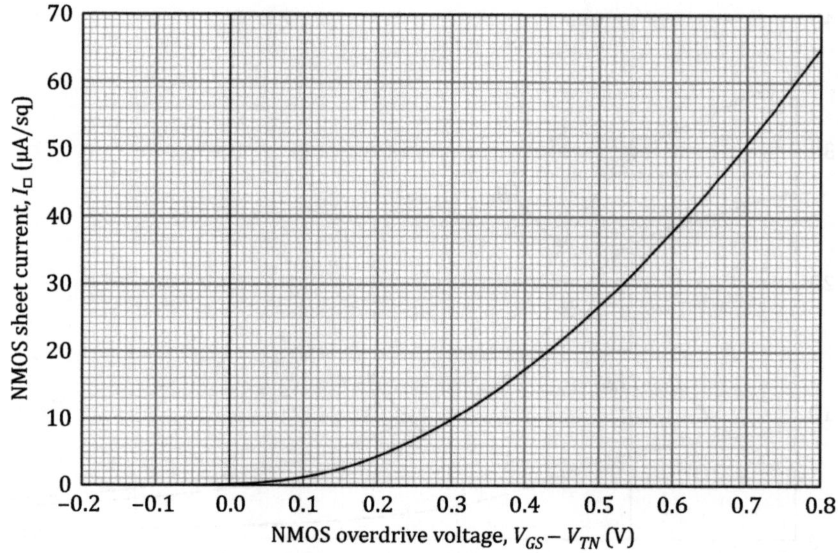

FIGURE 4.7 Low-threshold option NMOS sheet current $I_\square$ as a function of overdrive voltage $V_{GS} - V_{TN}$.

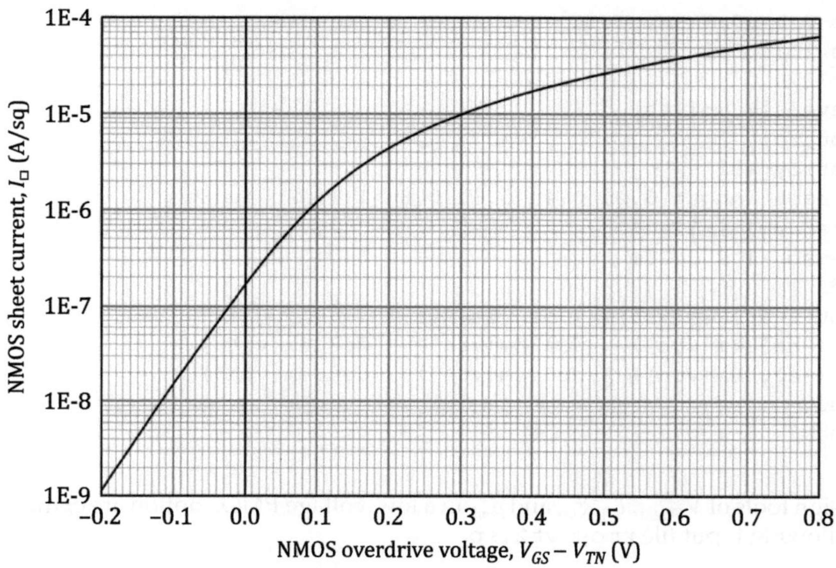

FIGURE 4.8 Low-threshold option NMOS sheet current $I_\square$ as a function of overdrive voltage $V_{GS} - V_{TN}$. Notice the logarithmic scaling of the vertical axis.

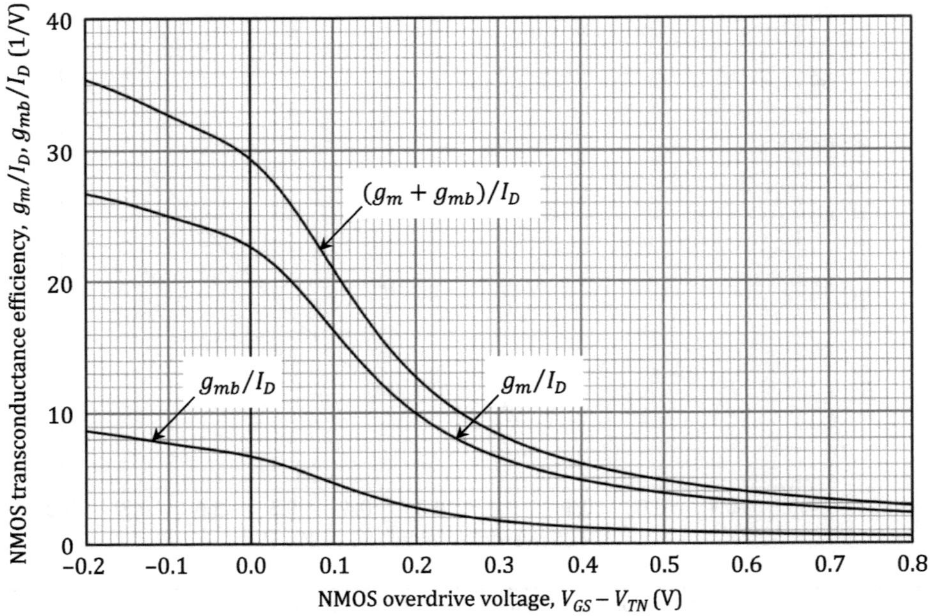

FIGURE 4.9 Low-threshold option NMOS transconductance efficiency g_m/I_D and bulk-transconductance efficiency g_{mb}/I_D as functions of overdrive voltage $V_{GS} - V_{TN}$. Also shown is the plot of the sum $(g_m + g_{mb})/I_D$.

```
vds 2 3 -0.9
vgs 1 3 -0.9
vsb 3 0 0
.save all @m[vth]
.control
destroy all
foreach length 0.2u 0.4u 0.6u 0.8u 1.2u 2u 4u 8u
alter @m[l]=$length
dc vsb -1.2 0 20m
end
plot (-dc1.@m[vth]) (-dc2.@m[vth])
+  (-dc3.@m[vth]) (-dc4.@m[vth]) (-dc5.@m[vth])
+  (-dc6.@m[vth]) (-dc7.@m[vth]) (-dc8.@m[vth])
.endc
.end
```

Design tools of V_{DSsat}, $I_{\square}$, g_m, and g_{mb} in a low-voltage PMOS option are extracted with the following input file pmoslvt2.sp:

```
filename: pmoslvt2.sp (low-threshold pmos vdsat, id, gm, gmbs)
.incl bsimlvt.sp
m 2 1 3 0 cmosplvt w=6u l=0.8u
```

```
vds 2 3 -0.9
vgs 1 3
vsb 3 0 0
.dc vgs -0.9 0.2 10m
.save @m[vth] @m[vdsat] @m[id] @m[gm] @m[gmbs] v(1,3)
.end
```

The PMOS design tools extracted with these files are presented in Figs. 4.10 to 4.14.

Work Consider the sink section first. One of its specified metrics is the compliance limit described by (4.22). Although it may not be obvious at first sight, the right-hand-side terms V_{GS1} and V_{DSsat2} of this equation are interrelated. For an explanation of how, consider the fact that M1 and M2 have identical aspect ratios because they are to be built with the same number of unit devices. Since they also conduct the same current, their sheet currents should therefore be identical. This identity implies identical overdrive voltages, which, in turn, implies identical drain-source saturation voltages. This observation enables us to replace V_{DSsat2} with V_{DSsat1} in (4.22). Also subtracting V_{TN1} from both sides, we turn the equation into

$$V_C - V_{TN1} = V_{GS1} - V_{TN1} + V_{DSsat1}, \tag{4.23}$$

which now exhibits on its right-hand side the overdrive voltage $V_{GS1} - V_{TN1}$ and drain-source saturation voltage V_{DSsat1} of the very same M1. These two voltages are indeed

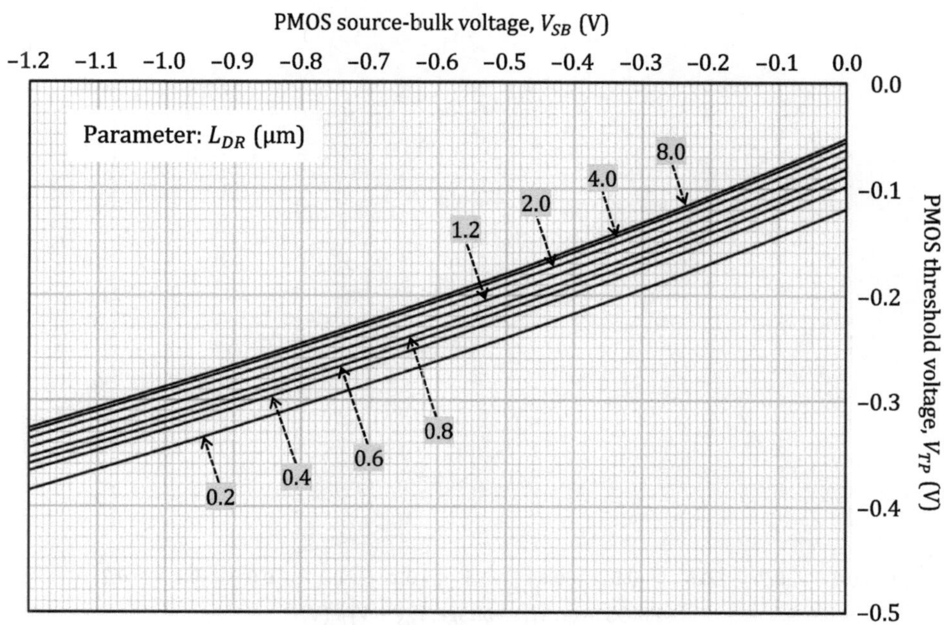

FIGURE 4.10 Low-threshold option PMOS threshold voltage V_{TP} as a function of source-bulk bias voltage V_{SB}.

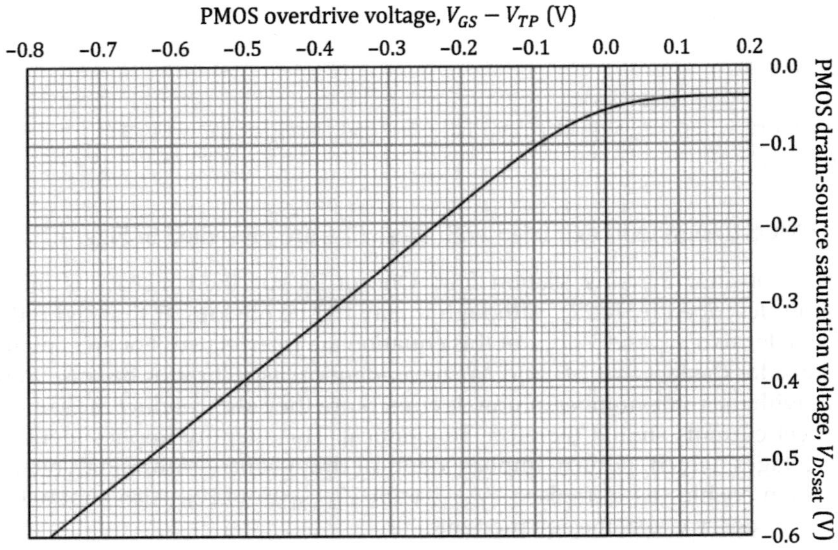

FIGURE 4.11 Low-threshold option PMOS drain-source saturation voltage V_{DSsat} as a function of overdrive voltage $V_{GS} - V_{TP}$.

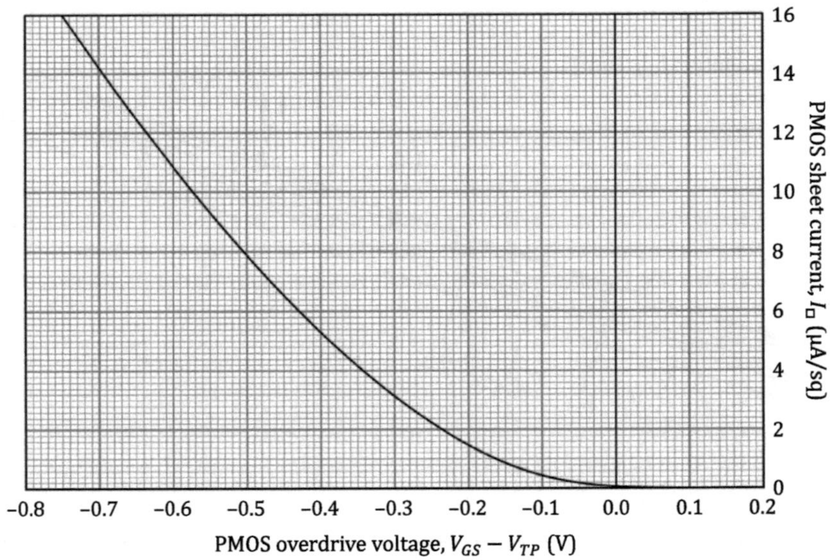

FIGURE 4.12 Low-threshold option PMOS sheet current $I_{\square}$ as a function of overdrive voltage $V_{GS} - V_{TP}$.

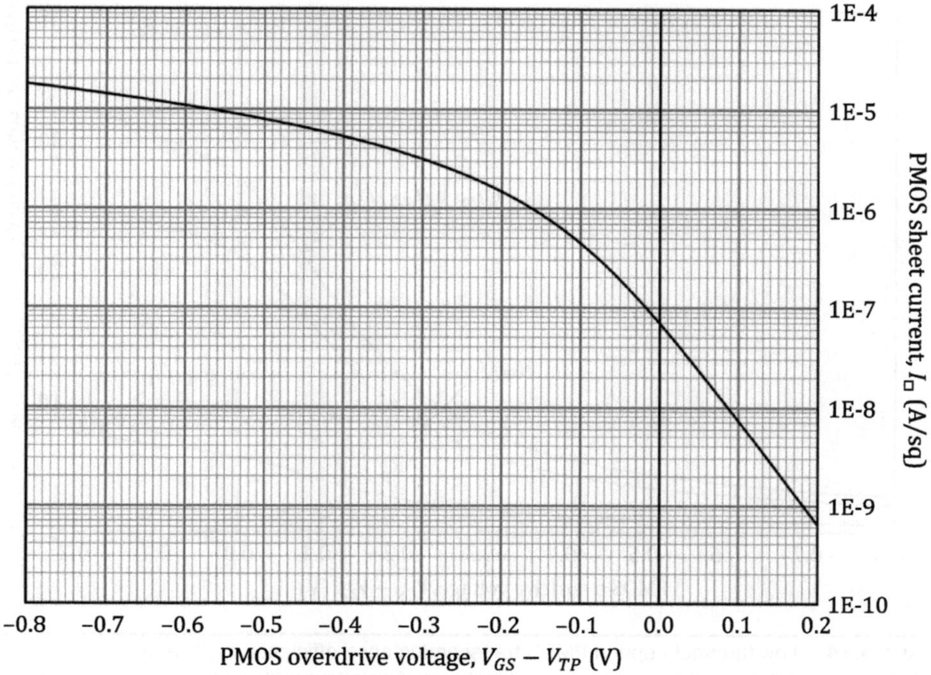

FIGURE 4.13 Low-threshold option PMOS sheet current $I_\square$ as a function of overdrive voltage $V_{GS} - V_{TP}$. Notice the logarithmic calibration of the vertical axis.

intimately related through the design tool given in Fig. 4.6. As to the terms on the left-hand side of (4.23), the target value of V_C is given as a specification item but the value of V_{TN1} is uncertain because, although the source-bulk voltage of this device is known to be zero, the channel length is still unknown. Nevertheless, we assume a reasonable initial value for the channel length, say, $L_{DR} = 4L_{DR(\min)} = 0.8\ \mu\text{m}$, and read the corresponding threshold voltage from Fig. 4.5 as $V_{TN1} \approx 0.115$ V. Considering also the specified $V_C = 0.3$ V, we turn (4.23) into

$$V_{GS1} - V_{TN1} + V_{DSsat1} = 0.185\,\text{V}.$$

For determining the overdrive voltage that satisfies this constraint we refer to Fig. 4.6, and arrive at $V_{GS1} - V_{TN1} \approx 0.095$ V. Considering $V_{TN1} \approx 0.115$ V, the gate-source voltage of M1 is estimated as $V_{GS1} = 0.210$ V. As to M2, the threshold voltage is inflated by body effect due to $V_{SB2} = V_{GS1} = 0.210$ V. Also considering $L_{DR} = 0.8\ \mu\text{m}$, the threshold voltage is determined from Fig. 4.5 as $V_{TN2} = 0.180$ V. Since the overdrive voltage of this device equals that of M1 hence 0.095 V, its gate-source voltage is calculated as $V_{GS2} = 0.095 + 0.180 = 0.275$ V, which implies a gate voltage $V_{G2} = V_{GS1} + V_{GS2} = 0.210 + 0.275 = 0.485$ V. All numbers calculated so far are based on an arbitrarily assigned value of L_{DR}. These numbers constitute the initial condition of the iterative simulation to be conducted next for determining the actual value of the channel length with a view to satisfying the output-resistance specification.

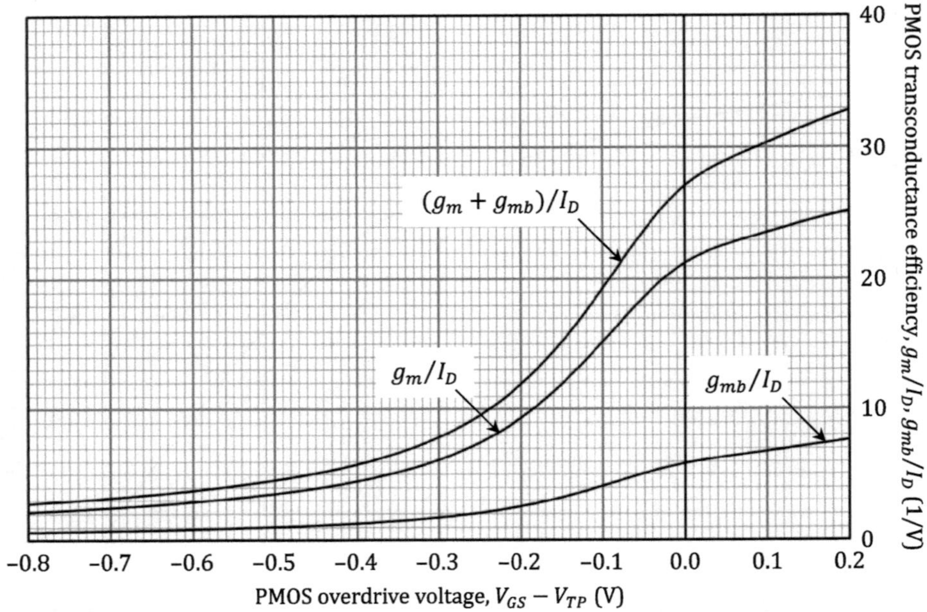

FIGURE 4.14 Low-threshold option PMOS transconductance efficiency g_m/I_D and bulk-transconductance efficiency g_{mb}/I_D as functions of overdrive voltage $V_{GS} - V_{TP}$. Also shown is the plot of the sum $(g_m + g_{mb})/I_D$.

Output resistance is described by (4.21). After substituting the exact form of (3.58) for A_{max}, and multiplying both sides of the equation by the nominal current I_O, we turn this equation into

$$r_o I_O = r_{ds2} I_O + \left(1 + r_{ds2} I_O \frac{g_{m2} + g_{mb2}}{I_O}\right) r_{ds1} I_O. \tag{4.24}$$

Since neither the total transconductance efficiency $(g_{m2} + g_{mb2})/I_O$ nor the Early voltages $r_{ds1}I_O$ and $r_{ds2}I_O$ are dependent on the width of M1 or M2, $r_o I_O$ is also independent of the width hence the number of unit devices paralleled. Targeting $r_o I_O$ instead of r_o alone will therefore enable us to run the simulation on one unit device for each of M1 and M2. The width of this device is set arbitrarily and kept constant throughout the iteration process. Considering the specified values $r_o = 35$ MΩ and $I_O = 5 \times 5$ μA $= 25$ μA, (4.24) yields $r_o I_O = 875$ V for the target. Simulation is performed with the following input file `ex4-5a.sp`, whose node numbers correspond to those marked in Fig. 4.4(b).

```
filename: ex4-5a.sp
.incl bsimlvt.sp
m1 2 5 0 0 cmosnlvt l=0.8u w=10u
vg1 5 0 0.21
m2 1 6 2 0 cmosnlvt l=0.8u w=10u
vg2 6 0 0.485
vo 1 0 0.9
.dc vo 0.8 1.0 0.1m
.end
```

Notice the following features: (a) Channel width is set arbitrarily to 10 μm; (b) channel length is at its initial value 0.8 μm; (c) gate voltages, as calculated above for this channel length, are applied by vg1 and vg2; and (d) output voltage is applied by vo, and is swept around 0.9 V in accordance with the output-resistance specification.

After running the simulation, and plotting the Early voltage with the command

```
plot i(vo)/deriv(i(vo))
```

we read $r_o I_O = 739$ V for $V_O = 0.9$ V. Since this falls short of the target, we increase the channel length, say, to $L_{DR} = 1$ μm for the next iteration. For this new channel length, we determine $V_{TN1} \approx 0.105$ V from Fig. 4.5, and turn (4.23) into

$$V_{GS1} - V_{TN1} + V_{DSsat1} = 0.195\,\text{V},$$

which yields $V_{GS1} - V_{TN1} \approx 0.100$ V from Fig. 4.6. We calculate $V_{GS1} = 0.205$ V from this overdrive voltage, and determine $V_{TN2} = 0.165$ V from Fig. 4.5. This leads to $V_{G2} = 0.470$ V. Rerunning the simulation with these numbers, we obtain $r_o I_O = 913$ V for $V_O = 0.9$ V, which obviously satisfies the specification.

Now that the channel length of the unit devices of the sink section has been determined, we move to the calculation of their channel width. The identical overdrive voltage 0.100 V of these devices implies a sheet current $I_{\square} \approx 1.3$ μA/sq from Fig. 4.7 or Fig. 4.8. For the specified 5-μA unit-device current, this leads to $W/L = 5/1.3 = 3.85$. Considering the effective channel length $L = 0.966$ μm obtained from (3.2), we arrive at $W = 3.72$ μm for the corresponding effective channel. Since the process bias on channel width is negligible, we adopt this value of W as the drawn width.

In qualitative terms, current-source design is not much different from the foregoing current-sink design. As a matter of fact, it is a little less complex thanks to the fact that M3 and M4 are identical not only in their overdrive voltages but also in their threshold voltages because both operate with source-bulk shorts. We begin with the compliance-limit expression

$$V_C = V_{GS3} + V_{DSsat4}, \tag{4.25}$$

where we can omit device identifiers in subscripts due to the fact that M3 and M4 are completely identical. Subtracting V_{TP} from both sides leads us to

$$V_C - V_{TP} = V_{GS} - V_{TP} + V_{DSsat}. \tag{4.26}$$

Next, we set an initial channel length, say, $L_{DR} = 1$ μm, and read the corresponding zero-bias threshold voltage $V_{TP} \approx -0.075$ V from Fig. 4.10. For this much V_{TP} and $V_C = -0.3$ V, (4.26) turns into

$$V_{GS} - V_{TP} + V_{DSsat} = -0.225\,\text{V},$$

which yields $V_{GS} - V_{TP} \approx -0.115$ V from Fig. 4.11. Since this implies $V_{GS} = -0.115 + (-0.075) = -0.190$ V for both devices, the gates are to be held at $V_{G3} = 1.8 - 0.190 = 1.61$ V and $V_{G4} = 1.8 - 0.190 - 0.190 = 1.42$ V in the initial run of iterative simulation. Input file `ex4-5b.sp` is as follows:

```
filename: ex4-5b.sp
.incl bsimlvt.sp
m3 4 8 10 10 cmosplvt l=1u w=10u
vg3 8 0 1.61
m4 3 7 4 4 cmosplvt l=1u w=10u
vg4 7 0 1.42
vo 3 0 0.9
vdd 10 0 1.8
.dc vo 0.8 1.0 0.1m
.end
```

Considering $r_o = 35$ MΩ and $I_O = 3 \times 5 = 15$ µA, we target $r_o I_O = 525$ V. The outcome of the simulation, as obtained for $V_O = 0.9$ V, is $r_o I_O = 557$ V. Our initial channel-length selection is therefore acceptable. Next, we determine the channel width of the unit device. For $V_{GS} - V_{TP} = -0.115$ V, Fig. 4.12 or Fig. 4.13 yields $I_{\square} \approx 0.53$ µA/sq. For the specified nominal current of 5 µA, the aspect ratio of a unit device should therefore be $W/L = 5/0.53 = 9.5$, which, considering the effective channel length $L = 0.941$ µm, implies $W = 8.94$ µm. The corresponding drawn width is $W_{DR} = 9$ µm for each unit device.

Verification is performed with the following input file `ex4-5.sp`.

```
filename: ex4-5.sp
.incl bsimlvt.sp
mr1 5 5 0 0 cmosnlvt w=3.73u l=1u m=1
mr2 6 6 5 0 cmosnlvt w=3.73u l=1u m=1
m1 2 5 0 0 cmosnlvt w=3.73u l=1u m=5
m2 1 6 2 0 cmosnlvt w=3.73u l=1u m=5
mr3 8 8 10 10 cmosplvt w=9u l=1u m=1
mr4 7 7 8 8 cmosplvt w=9u l=1u m=1
m3 4 8 10 10 cmosplvt w=9u l=1u m=3
m4 3 7 4 4 cmosplvt w=9u l=1u m=3
ir 7 6 5u
vosink 1 0 0.9
vosource 3 0 0.9
vdd 10 0 1.8
*.dc vosink 0 1 0.1m
.dc vosource 0.8 1.8 0.1m
.end
```

Notice two separate dc-sweep analyses for sink and source sections. The reader is invited to run these simulations, and plot source and sink currents as well as output resistances as functions of the output voltage. ▲

4.2.4 Low-Voltage Cascoding

As argued in the preceding subsection, the compliance limit of a basic-cascode current sink or source is prohibitively large for implementing a low-voltage application with standard-threshold devices. What makes it large is the virtual diode-connection imposed on M1 and M3 in Fig. 4.4(b) for ensuring saturation. Although saturation can be established with a drain-source voltage as low as V_{DSsat}, the diode connection equates the drain-source voltage to V_{GS}, which, of course, is considerably higher than V_{DSsat} in a standard-threshold device. Low-voltage cascoding is a technique where the saturation of the source or sink device is ensured by keeping the drain-source voltage close to V_{DSsat} instead of equating it to V_{GS}. It enables us to deploy standard-threshold devices for implementing cascoding in low-voltage applications.

Low-voltage cascoding is illustrated on a sink application in Fig. 4.15(a). Notice MR1 and M1 functioning exactly as their counterparts in the basic-cascode topology of Fig. 4.4(b). M1 sinks a scaled copy I_O of the reference current I_R by mirroring from MR1. The cascoding device M2, being forced to conduct this constant I_O, develops a constant V_{GS2}. Since its gate is tied to an external dc voltage V_B, its source stays at $V_{S2} = V_B - V_{GS2}$ with respect to ground. By properly adjusting V_B we therefore can set V_{S2} hence the drain-source voltage of M1 to any desired value. Considering the inevitability of inaccuracy and imprecision we ensure saturation by setting V_{DS1} above V_{DSsat1} by an adequate margin ΔV_{DS1}. Bias voltage and compliance limit are therefore generally described by

$$V_B = V_{SS} + V_{DSsat1} + \Delta V_{DS1} + V_{GS2}, \tag{4.27}$$

and

$$V_C = V_{DSsat1} + \Delta V_{DS1} + V_{DSsat2}, \tag{4.28}$$

respectively.

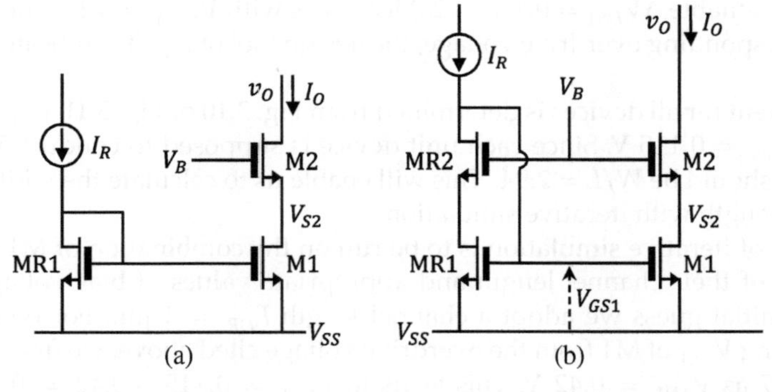

FIGURE 4.15 Low-voltage cascode configuration. (a) Fundamental circuit. (b) Inclusion of MR2 for equalizing the drain-source voltages of MR1 and M1.

One particular deficiency of the cascoding technique shown in Fig. 4.15(a) is the inaccuracy introduced to the scaling factor by the disparate drain-source voltages of MR1 and M1. This problem does not exist in the basic cascode shown in Fig. 4.4(b) because MR2 and M2 equalize the drain-source voltages of MR1 and M1. Here, we can do the same by inserting an additional device between the reference current source and MR1 as represented with MR2 in Fig. 4.15(b). This additional device shares the same gate voltage V_B with M2. Also note that the scale factor from MR2 to M2 is made identical to the scale factor of mirroring from MR1 to M1.

One point of concern in the low-voltage cascode topology of Fig. 4.15(b) is the saturation condition of MR2. The source voltage of this device matches V_{S2} and therefore equals $V_{DSsat1} + \Delta V_{DS1}$. Considering also the fact that its drain voltage is identical to V_{GS1}, we can express the saturation condition $V_{DSR2} > V_{DSsat2}$ as

$$V_{GS1} \geq \Delta V_{DS1} + V_{DSsat1} + V_{DSsat2}, \tag{4.29}$$

whose right-hand side equals V_C as per (4.28). In the case of implementation with MOSFETs of low-threshold option, this condition proves to be very restrictive and in some cases even prohibitive because the gate-source voltage of such MOSFETs is essentially small. Standard-threshold devices are therefore more suitable for implementing a low-voltage cascode mirror. It is worth mentioning that no such restriction is involved in designing the basic-cascode mirror of Fig. 4.4(b) because the condition of saturation is unconditionally satisfied by the diode connection of MR2.

The topology shown in Fig. 4.15(b) is also known as *wide-swing cascode*. Although we introduced and analyzed it on a sink configuration, the reader can easily extend the conclusions to a source configuration as well. As a final word, note that the treatment of accuracy and precision presented in Subsection 4.2.2 is also valid for low-voltage cascode mirroring.

Example 4.6 Using standard-threshold devices, design a low-voltage cascode sink in accordance with the specifications stated in Example 4.5.

Work We begin with the compliance-limit specification $V_C = 0.3$ V, and note $V_{DSsat2} = V_{DSsat1}$ due to the identical overdrive voltages of M1 and M2 as concluded in Example 4.5. For a reasonable $\Delta V_{DS1} = 0.1$ V, (4.28) leaves us with $V_{DSsat} = 0.1$ V for both devices. For the corresponding overdrive voltage, the design tool of Fig. 3.8 indicates $V_{GS} - V_{TN} = 0.115$ V.

Sheet current for all devices is determined from Fig. 3.10 or Fig. 3.11 as $I_\square \approx 1.7\ \mu A/sq$ for $V_{GS} - V_{TN} = 0.115$ V. Since each unit device is supposed to conduct 5 μA, the unit aspect ratio should be $W/L = 2.94$. This will enable us to calculate the width after determining the length with iterative simulation.

Each step of iterative simulation is to be run on the combination of M1 and M2 with an estimate of their channel length and appropriate values of bias voltages V_{GS1} and V_B. As an initial guess we adopt a channel length $L_{DR} = 1\ \mu m$. For determining the corresponding V_{GS1} of M1 from the overdrive voltage cited above, we first estimate V_{TN1} from Fig. 3.7 as $V_{TN1} = 0.42$ V. This leads to $V_{GS1} = 0.115 + 0.42 = 0.535$ V, which obviously satisfies the saturation condition (4.29). For V_B, we move to M2, and firstly determine its source-bulk voltage as $V_{S2} = V_{DSsat1} + \Delta V_{DS1} = 0.1 + 0.1 = 0.2$ V. The corresponding threshold voltage for $L_{DR} = 1$ μm is estimated from the design tool of Fig. 3.7 as $V_{TN2} = 0.48$ V, which leads to $V_{GS2} = 0.115 + 0.48 = 0.595$ V for the overdrive voltage determined earlier. Therefore, the bias voltage should be set to $V_B = V_{S2} + V_{GS2} = 0.2 + 0.595 = 0.795$ V when running the simulation for $L_{DR} = 1$ μm.

Iterative simulation targets the product $r_o I_O = 875$ V of the specified output-resistance and nominal current as calculated in Example 4.5. It is run with the following input file ex4-6a.sp.

```
filename: ex4-6a.sp
.incl bsim.sp
m1 2 3 0 0 cmosn l=1u w=10u
vg1 3 0 0.535
m2 1 5 2 0 cmosn l=1u w=10u
vg2 5 0 0.795
vo 1 0 0.9
.dc vo 0.8 1.0 0.1m
.end
```

The initial value $L_{DR} = 1$ μm yields $r_o I_O = 801$ V, which is slightly short of the target. We raise the channel length to $L_{DR} = 1.1$ μm, and rerun the simulation without altering the gate voltage of M1 or M2 because the correction indicated by Fig. 3.7 for the two threshold voltages is insignificant. This iteration yields $r_o I_O = 884$ V, and thus meets the target. Effective channel length is calculated from (3.2) as $L = 1.066$ μm, for which, $W/L = 2.94$ implies $W = 3.13$ μm. Drawn width is calculated from (3.1) as $W_{DR} = 3.15$ μm.

The following input file is used in verification:

```
filename: ex4-6.sp
.incl bsim.sp
mr1 4 3 0 0 cmosn w=3.15u l=1.1u
mr2 3 5 4 0 cmosn w=3.15u l=1.1u
m1 2 3 0 0 cmosn w=3.15u l=1.1u m=5
m2 1 5 2 0 cmosn w=3.15u l=1.1u m=5
ir 0 3 5u
vb 5 0 0.795
vo 1 0 0.9
.dc vo 0 1.0 0.1m
.end
```

Simulation indicates $r_o = 34.5$ MΩ, which is slightly below the target $r_o = 35$ MΩ. Raising V_B by just 5 mV will move it to 36 MΩ as the reader can verify with an additional simulation run. What is also recommended is to rerun the simulation after replacing the dc-sweep analysis with an operation-point analysis, and reading out V_{DSsat} in M1 and M2 for a comparison with the value targeted in design. ▲

4.2.5 Regulated Cascoding

Occasionally, we may need a current sink or source with much higher output resistance than what is available from basic or low-voltage cascoding. The *regulated cascode* topology

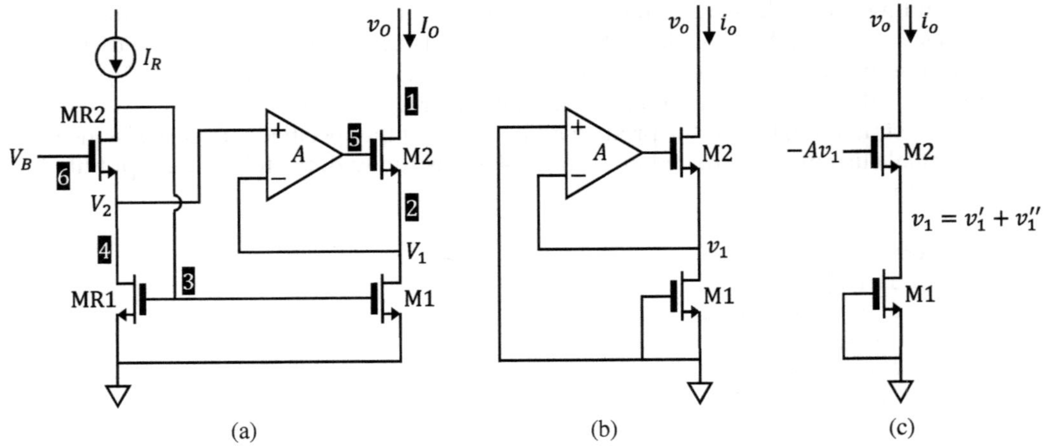

FIGURE 4.16 Regulated low-voltage cascode configuration. (a) Electrical schematic with node numbers used in Example 4.7. (b) Simplified electrical schematic for small-signal analysis by inspection. (c) Small-signal schematic obtained after breaking the feedback loop.

presented in Fig. 4.16(a) as a single-supply sink example caters for this need while keeping the compliance limit on par with what the low-voltage cascode offers. Note that the bias voltage V_B is now applied only to the gate of MR2, whereas the gate of M2 is driven by an OTA of differential gain A. The noninverting and inverting inputs of the OTA receive the drain voltages V_2 and V_1 of MR1 and M1, respectively. V_2 is fixed by V_B, and V_1 is equated to the fixed V_2 by the virtual short existing between OTA input terminals. M1, thus furnished with a virtually constant drain-source voltage, generates a nearly constant output current I_O without being significantly affected by the output voltage v_O. This is how a very high output resistance is achieved.

For modeling the output resistance r_o, we now apply a small-signal voltage v_o to the output, and determine the corresponding small-signal component i_o of the output current. The ratio v_o/i_o will yield the output resistance. Since a dc small-signal analysis is needed for this purpose, we will rely on the equivalent circuit shown in Fig. 4.16(b), where the gate of M1 and the noninverting input of the OTA are grounded because none of these nodes carries a signal in the original schematic of Fig. 4.16(a). As to M2, we observe signal entry not only through the drain due to the test source v_o but also through the gate due to the feedback signal $-Av_1$, where v_1 is the small-signal voltage at the source of M2. We need to break the feedback loop, and apply superposition, as we did previously in Example 3.15 of Subsection 3.7.2. By breaking the loop at the gate of M2, we finally arrive at the equivalent circuit shown in Fig. 4.16(c).

In the first part of the small-signal analysis, we short out the voltage source $-Av_1$, and determine the source voltage v_1' of M2 due to v_o only. According to (3.61) and (3.62), v_1' is given by

$$v_1' = \frac{r_{ds1}}{r_{ds2} + A_{\max 2} r_{ds1}} v_o. \tag{4.30}$$

Next, we disable v_o, apply $-Av_1$, and determine the voltage v_1'' at the source of M2. According to (3.64), v_1'' is given by

$$v_1'' = \frac{\alpha_2 r_{ds1}}{r_{ds1} + \frac{r_{ds2}}{A_{\max 2}}} \left(-Av_1\right), \tag{4.31}$$

in which, α_2 equals either 1 or $g_{m2}/(g_{m2} + g_{mb2})$ depending on whether M2 operates with a source-bulk short or not.

Superposition $v_1 = v_1' + v_1''$ with (4.30) and (4.31) yields

$$v_1 = \frac{r_{ds1}}{r_{ds2} + (1 + \alpha_2 A)A_{\max 2} r_{ds1}} v_o.$$

Finally, by replacing v_1 with $r_{ds1} i_o$, and rearranging the equation, we arrive at

$$r_o \doteq \frac{v_o}{i_o} = r_{ds2} + (1 + \alpha_2 A)A_{\max 2} r_{ds1}. \tag{4.32}$$

The output resistance indicated by this equation is approximately a factor of $\alpha_2 A$ larger than what (4.21) indicates for the basic cascode or low-voltage cascode. Also note that (4.28) of the low-voltage cascode sink is also valid for the compliance limit of a regulated cascode sink. Needless to say, regulation is also possible for a current-source application. The reader will benefit from reviewing this subsection in terms of such an application.

As a final word, note that the contents of Subsection 4.2.2 are valid for the accuracy and precision of mirroring with a regulated cascode.

Example 4.7 How much an output resistance r_o does the low-voltage cascode design of Example 4.6 offer if it is regulated by an OTA of A = 1000-V/V differential gain? Calculate and verify with a transfer-function simulation.

Work The output resistance of the regulated version is expected to be $r_o \approx \alpha_2 A \times 35\ \text{M}\Omega$. Since the overdrive voltage of M2 equals 0.115 V, Fig. 3.12 yields $g_{m2}/I_O = 15\ \text{V}^{-1}$ and $(g_{m2} + g_{mb2})/I_O = 19\ \text{V}^{-1}$, and therefore $\alpha_2 \doteq g_{m2}/(g_{m2} + g_{mb2}) = 0.789$. Since A is specified as 1000 V/V, the regulated cascode is expected to offer $r_o = 0.789 \times 1000 \times 35(\text{M}\Omega) = 27.6\ \text{G}\Omega$. Verification is performed with the following file `ex4-7.sp`, which includes the OTA subcircuit presented in Example 2.32 of Section 2.11 and complies with the node numbers marked in Fig. 4.16:[2]

---

[2]Notice the options line resetting the value of the simulation parameter `gmin`, which is the minimum conductance imposed by Spice across nonlinear devices for convergence purposes. Since its default value 1×10^{-12} A/V is not negligible in comparison with the very small output conductance expected in the present simulation, its use without resetting to a sufficiently small value would impair the accuracy of the simulation. The reader may check its effect on the accuracy by repeating the simulation after closing the options line.

```
filename: ex4-7.sp
.incl bsim.sp
.param vdir=5 gm0=1m ro=1meg
********************ota*************************
*ninv: noninverting input
*inv: inverting input
*out: output
.subckt ota ninv inv out
gm 0 out cur='gm0*vdir*tanh((v(ninv)-v(inv))/vdir)'
rout out 0 ro
bb 0 out i=0.9/ro
.ends ota
************************************************
xota 4 2 5 ota
mr1 4 3 0 0 cmosn w=3.15u l=1.1u
mr2 3 6 4 0 cmosn w=3.15u l=1.1u
m1 2 3 0 0 cmosn w=3.15u l=1.1u m=5
m2 1 5 2 0 cmosn w=3.15u l=1.1u m=5
ir 0 3 5u
vb 6 0 0.795
vo 1 0 0.9
.tf v(1) vo
.options gmin=1e-18
.end
```

Simulation yields $r_o = 26.4$ GΩ. ▲

4.2.6 Self-Cascoding

The reader may have noticed the selection of identical aspect ratios for M1 and M2 in Examples 4.5 to 4.7. Although this is a very common practice, it is by no means a necessity. As a matter of fact, it has to be abandoned in one particular cascoding technique called *self-cascoding*, which is practiced whenever the compliance limit needs to be very low. As shown in Fig. 4.17(a) in a sink application of this technique, the gates of the current-sinking M1 and the cascoding M2 are interconnected and tied to similarly configured gates of the reference devices MR1 and MR2. As in all previously discussed cascoding techniques, the factor of scaling from MR2 to M2 is identical to the factor of scaling from MR1 to M1. For a large output resistance, however, M1 and MR1 need to be narrower than M2 and MR2. For an explanation of the underlying reason, notice that the nonzero drain-source voltage V_{DS1} of M1 implies $V_{GS1} > V_{GS2}$. Furthermore, in the usual case of conventional twin-well devices, the very same V_{DS1} also raises the threshold voltage of M2 above that of M1. As a result of these differences, the overdrive voltage of M1 exceeds the overdrive voltage of M2. Unless M1 is built with a smaller aspect ratio, the only way in which it can equate its current to that of M2 despite the large overdrive is to operate in nonsaturation with a small V_{DS1}. This would lead to a very small r_{ds1}, and therefore to a very low output resistance r_o as the reader can ascertain from (4.21).

One interesting fact about self-cascoding is that the saturation condition of M1 implies subthreshold operation for M2. This is because the drain boundary of the channel

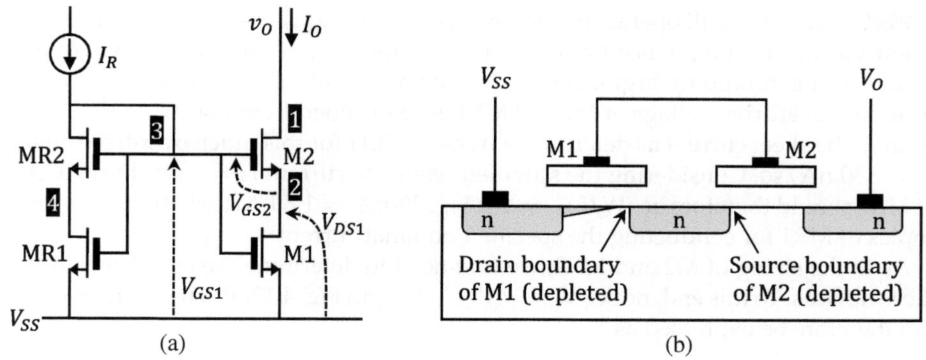

FIGURE 4.17 (a) Electrical schematic of a self-cascode sink topology. (b) Cross-sectional view of M1 and M2, where electron-rich regions are highlighted in gray.

in M1 and the source boundary of the channel in M2 operate under the same physical condition due to both being controlled by the same junction voltage and the same gate voltage as illustrated in Fig. 4.17(b). Since the carrier concentration at the drain boundary of M1 is lower than the doping concentration due to the saturated operation of this device, the carrier concentration at the source boundary of M2 is also below the doping concentration. As explained in Subsection 3.3.1, a MOSFET with such a carrier-depleted source boundary operates in subthreshold.

The compliance limit of a self-cascoded current sink is given by

$$V_C = V_{DS1} + V_{DSsat2}. \tag{4.33}$$

Since M2 operates in subthreshold, V_{DSsat2} is no larger than 50 mV as indicated by Fig. 3.8. V_{DS1}, on the other hand, can be made as low as V_{DSsat1} without breaching the saturation condition of M1. For a near-threshold operation, this implies a V_{DS1} close to 50 mV too. It is therefore possible to reduce the compliance limit of a self-cascode sink down to 100 mV. As to the output resistance, (4.21) continues to be valid.

On a final note, a self-cascoded current source is constructed with two series connected PMOS devices of a common gate terminal. Its output resistance and compliance limit are described by (4.21) and (4.33), respectively.

Example 4.8 The single-supply version (V_{DD} = 1.8 V) of the self-cascoded sink of Fig. 4.17 is to be designed in accordance with the following specifications:

1. *Reference Current* $I_R = 1$ μA.
2. *Nominal Output Current* $I_O = 8$ μA.
3. *Output Resistance* $r_o \geq 35$ MΩ for $V_O = 0.9$ V.
4. *Compliance Limit* $V_C = 120$ mV.

Work Since M2 will operate in subthreshold with $V_{DSsat2} \approx 40$ mV (see Fig. 3.8), what is left out of the compliance-limit specification for V_{DS1} is around 80 mV. We reserve 15 mV as a saturation margin, and assign the remaining 65 mV to V_{DSsat1}. The corresponding overdrive voltage of M1 and MR1 is determined from Fig. 3.8 as $V_{GS1} - V_{TN1} = 50$ mV. The sheet current as determined from Fig. 3.11 for this much overdrive voltage is $I_{\square} \approx 530$ nA/sq. Considering the specified reference current $I_R = 1$ μA, the aspect ratio of MR1 should therefore be $(W/L)_{R1} = 1(\mu\text{A})/530(\text{nA}) = 1.89$. M1 will be built with eight copies of MR1 for conducting the specified nominal current.

As to the sizing of M2 and MR2, we first need to determine the overdrive voltage of these devices. To this end, notice $V_{GS2} = V_{GS1} - V_{DS1}$ in Fig. 4.17(a). The overdrive voltage can therefore be expressed as

$$V_{GS2} - V_{TN2} = V_{GS1} - V_{DS1} - V_{TN2},$$

and further rearranged into

$$V_{GS2} - V_{TN2} = V_{GS1} - V_{TN1} - V_{DS1} + V_{TN1} - V_{TN2}, \tag{4.34}$$

in which, $V_{GS1} - V_{TN1} = 50$ mV and $V_{DS1} = 80$ mV, as already determined. For $V_{TN1} - V_{TN2}$, we refer to Fig. 3.7. Although the exact values of the threshold voltages are still unknown at this stage, their difference for $V_{SB2} = V_{DS1} = 80$ mV and $V_{SB1} = 0$ is fairly constant at $V_{TN1} - V_{TN2} \approx -20$ mV regardless of the channel length. Therefore, (4.34) yields $V_{GS2} - V_{TN2} = -50$ mV, which indeed confirms the subthreshold operation of M2 and MR2. Next, we determine $I_{\square} \approx 58$ nA/sq from Fig. 3.11 for the sheet current corresponding to this much overdrive voltage. Considering the specified reference current $I_R = 1$ μA, the aspect ratio of MR2 is calculated as $(W/L)_{R2} = 1(\mu\text{A})/58(\text{nA}) = 17.2$. This is a factor of 9.1 larger than the aspect ratio calculated for MR1. Rounding this ratio to an integer will enable us to construct all four devices with copies of the same unit MOSFET. Rounding up to 10 is better than rounding down to 9 as it will lead to a higher V_{DS1} hence a higher saturation margin for M1. We therefore construct MR1 with one unit device, and use 10 parallel copies for MR2. Since I_O is to be a factor of 8 higher than I_R, M2 will be built with 80 copies. As determined previously, M1 will contain eight copies.

Next step involves iterative simulation for determining the channel length corresponding to the product $r_o I_O = 280$ V specified for the output port. The following input file `ex4-8a.sp` is used for this purpose.

```
filename: ex4-8a.sp
.incl bsim.sp
m1 2 3 0 0 cmosn w=1u l=1u m=8
m2 1 3 2 0 cmosn w=1u l=1u m=80
vo 1 0 dc 0.9
vg 3 0 0.475
.dc vo 0.8 1.0 1m
.end
```

The file excludes the reference branch and contains M1 and M2 in addition to (a) a dc voltage source `vg` driving the common gate of these two devices, and (b) another dc source `vo` sweeping the output-port voltage around 0.9 V. For node numbers, the reader is referred to Fig. 4.17(a). Before running each iteration, we determine the common gate voltage vg by reading from Fig. 3.7 the zero-bias threshold voltage V_{TN1} corresponding to the selected channel length, and adding it to $V_{GS1} - V_{TN1} = 50$ mV. For $L = 1$ μm, for example, Fig. 3.7 indicates $V_{TN1} = 0.425$ V hence $V_{GS1} = 0.475$ mV. As usual, we keep the channel width of the unit device fixed at an arbitrarily selected value. $r_o I_O$ is plotted with the command

```
plot i(vo)/deriv(i(vo))
```

and compared with the specified 280 V for $V_O = 0.9$ V. Iterative simulation eventually leads to $r_o I_O = 289$ V for $L_{DR} = 4.5$ μm. Verification is done with the following input file ex4-8b.sp, which also incorporates the reference branch.

```
filename: ex4-8b.sp
.incl bsim.sp
m1 2 3 0 0 cmosn w=10u l=4.5u m=8
m2 1 3 2 0 cmosn w=10u l=4.5u m=80
mr1 4 3 0 0 cmosn w=10u l=4.5u m=1
mr2 3 3 4 0 cmosn w=10u l=4.5u m=10
ir 0 3 1u
vo 1 0 dc 0.9
.tf v(1) vo
*.dc vo 0 1.8 0.1m
.end
```

The reader is invited to run the transfer-function analysis, and enter the command print all to read r_o. Subsequent entry of show will enable the reader to verify V_{DSsat1}, V_{DSsat2}, and I_O (as the drain current of M1 or M2). The dc-analysis line is intended for obtaining the current-voltage characteristic of the output port. ▲

4.3 Current and Voltage References

All current-source and current-sink topologies described so far generate their current by mirroring from a reference source. The task of generating the reference current itself is addressed in this section. Also covered in this section is the so-called *bandgap reference* (BGR) technique used for generating a reference voltage of low sensitivity to power-supply voltage and temperature.

4.3.1 Voltage-Divider Current Reference

Simple current references can be built with voltage-divider circuits. Shown in Fig. 4.18(a) is an example suitable for the single-device source/sink network depicted in Fig. 4.2. The reference current I_R is set by the voltage across a resistor whose resistance is determined from

$$R = \frac{V_{DD} + V_{GS2} - (V_{SS} + V_{GS1})}{I_R}, \tag{4.35}$$

where the target values of I_R, V_{GS1}, and V_{GS2} are imposed by sink/source specifications.

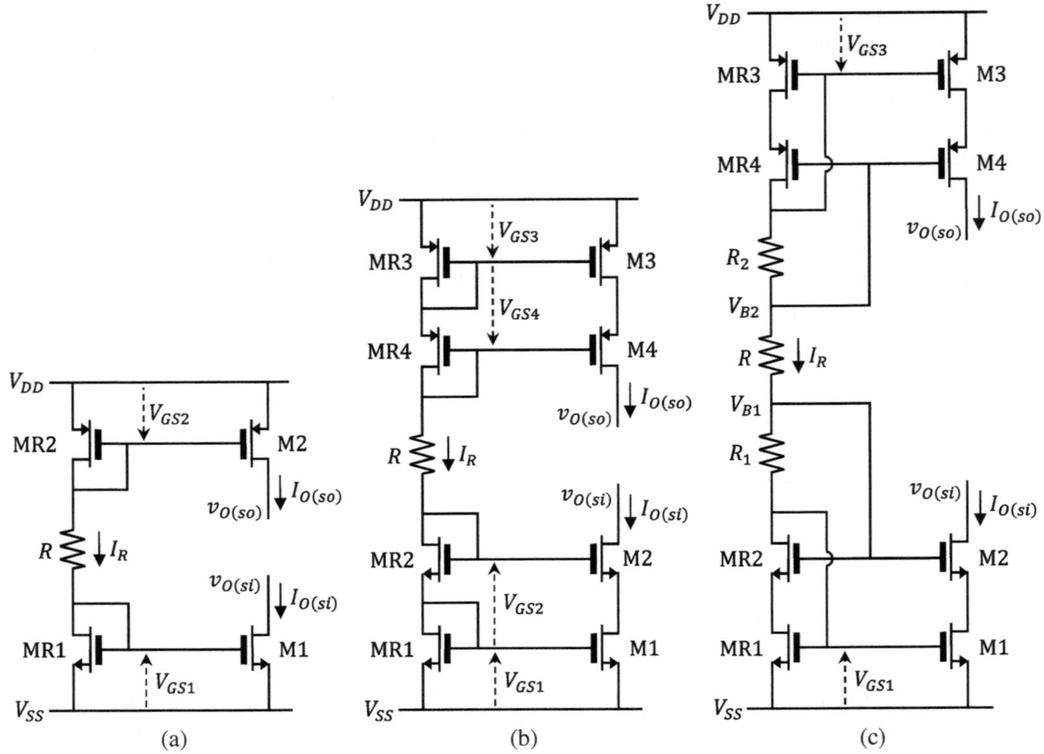

FIGURE 4.18 Voltage-divider current-reference topologies. (a) Single-device reference. (b) Basic-cascode reference. (c) Low-voltage cascode reference.

As depicted in Fig. 4.18(b), the voltage-divider concept can be extended to basic-cascode mirroring. In this case, R is determined from

$$R = \frac{V_{DD} + V_{GS3} + V_{GS4} - \left(V_{SS} + V_{GS1} + V_{GS2}\right)}{I_R}. \tag{4.36}$$

As explained in Subsection 4.2.3, a basic-cascode source or a sink can offer a reasonably low-compliance limit only if built with devices of low-threshold option. Such an option is actually imperative in basic-cascode reference design because otherwise the stacking of four gate-source voltages along the reference path would leave no voltage headroom for the resistor. Standard-threshold devices will work if the reference is needed only for sinking or for sourcing. In each of these two cases, we can reduce the number of stacked devices to two by omitting the upper or lower pair of MOSFETs and interconnecting the appropriate terminal of the resistor directly to the power-supply rail.

The schematic of a voltage-divider current reference suitable for low-voltage cascoding is shown in Fig. 4.18(c). Notice the presence of two additional resistors R_1 and R_2

for generating the cascode bias voltages V_{B1} and V_{B2}. The resistances of this circuit are calculated from

$$R_1 = \frac{V_{B1} - V_{SS} - V_{GS1}}{I_R}, \tag{4.37}$$

$$R = \frac{V_{B2} - V_{B1}}{I_R}, \tag{4.38}$$

$$R_2 = \frac{V_{DD} + V_{GS3} - V_{B2}}{I_R}. \tag{4.39}$$

Finally, note that voltage-divider current-reference circuits are very sensitive to supply-voltage fluctuations as will be demonstrated next in Example 4.9. Their use is limited to those cases in which such a sensitivity is not a concern.

Example 4.9 A basic-cascode current-reference circuit is shown in Fig. 4.19(a). Assuming a source-bulk short for each MOSFET, and adopting the current-source specifications of Example 4.5, determine the resistance R and the relative dc sensitivity of the output current I_O to the supply voltage V_{DD}.

Work R is determined from

$$R = \frac{V_{DD} + V_{GS3} + V_{GS4}}{I_R},$$

where $V_{DD} = 1.8$ V, $I_R = 5$ μA, $V_{GS3} = V_{GS4} = -0.190$ V. Therefore, $R = 284$ kΩ.

The dc absolute sensitivity of I_O to V_{DD} is given by the transfer function i_o/v_{dd} of dc small-signal components. Relative sensitivity is defined by $i_o/v_{dd}I_O$, which equals the relative sensitivity $i_r/v_{dd}I_R$ of the reference current because I_O is just a scaled replica of I_R. This observation enables us to consider only the reference section of the circuit for the purpose of determining the relative sensitivity. Recognizing MR3 and MR4 as

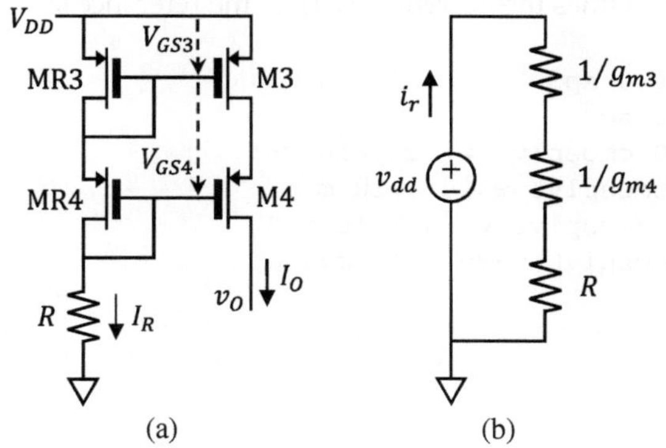

FIGURE 4.19 (a) Basic-cascode current-reference circuit of Example 4.7. (b) Small-signal equivalent circuit.

diode-connected devices operating with individual source-bulk shorts, we directly represent each with a small-signal resistance $1/g_m$ as we concluded in Example 3.11 in Subsection 3.7.2, and thus arrive at the small-signal equivalent circuit shown in Fig. 4.19(b) for the reference section. Since this circuit indicates

$$v_{dd} = i_r\left(\frac{1}{g_{m3}} + \frac{1}{g_{m4}} + R\right),$$

the relative sensitivity is given by

$$\frac{i_r}{v_{dd}I_R} = \frac{1}{\frac{I_R}{g_{m3}} + \frac{I_R}{g_{m4}} + RI_R}. \tag{4.40}$$

We recall from Example 4.5 the overdrive voltage −0.115 V for MR3 and MR4, and therefore estimate from Fig. 4.14 the transconductance efficiencies $g_{m3}/I_R = g_{m4}/I_R \approx 14\ \text{V}^{-1}$ for these two devices. Also taking into account $RI_R = 284(\text{k}\Omega) \times 5(\mu\text{A}) = 1.42$ V, we calculate $i_r/v_{dd}I_R = 64\%$ per volt from (4.40).

The shortest way of verifying this result is to run a Spice transfer-function (`.tf`) analysis. The input file `ex4-9.sp` presented below defines two such analyses: one for the sink section and another for the reference section. The reader is invited to run each separately, read the absolute sensitivity by entering

```
print transfer_function
```

Dividing the outcome by $I_R = 5\ \mu\text{A}$ yields the relative sensitivity of the reference section. For the source section divide the outcome by $I_O = 15\ \mu\text{A}$.

The file also includes an operation-point analysis for verifying the currents in both sections. It is advisable to run it before the transfer-function analysis, and trim R for setting I_R to the specified 5 μA if necessary. Notice the inclusion of an ammeter

```
vs 9 0 0
```

for this purpose. It defines the current `i(vs)` of the reference section.

```
filename: ex4-9.sp
.incl bsimlvt.sp
mr3 8 8 10 10 cmosplvt w=9u l=1u m=1
mr4 7 7 8 8 cmosplvt w=9u l=1u m=1
m3 4 8 10 10 cmosplvt w=9u l=1u m=3
m4 3 7 4 4 cmosplvt w=9u l=1u m=3
R 7 9 284k
vs 9 0 0
vos 3 0 0.9
vdd 10 0 1.8
.op
*.tf i(vs) vdd ; absolute sensitivity of reference section
*.tf i(vos) vdd ; absolute sensitivity of source section
.end
```
▲

4.3.2 Beta-Multiplier Current Reference

As we observed in Example 4.9, voltage-divider current references are highly sensitive to power-supply voltage. This problem can be alleviated if the reference circuit is made self-biased instead of relying on supply-voltage division. The so-called *beta multiplier* is probably the most popular of such self-biased current references. Its single-device and cascode topologies are presented in this subsection.

Single-Device Beta Multiplier

The circuit schematic of a single-device beta multiplier is given in Fig. 4.20. Recognizing M1 and M2 as examples of sink and source devices, and ignoring the *start-up* section for the time being, we concentrate on the beta-multiplier core comprising MR1, MR2, MR3, MR4, and R_S. Among these MOSFETs, MR4 and the diode-connected MR2 are matching pairs. MR4 therefore mirrors onto MR1 the common current of MR2 and MR3. The currents of MR1 and MR3 are thus equalized as denoted by I_R on both paths of the core but, because MR3 is built with a larger aspect ratio in comparison with MR1, V_{GS3} turns out to be smaller than V_{GS1}. For strong inversion, these two gate-source voltages are described by (3.16) as

$$V_{GS1} = \sqrt{\frac{2A_N I_R}{\mu_N C_{ox}(W/L)_{R1}}} + V_{TN},$$

$$V_{GS3} = \sqrt{\frac{2A_N I_R}{\mu_N C_{ox}(W/L)_{R3}}} + V_{TN},$$

where identical threshold voltages are assumed for the sake of analytical simplicity. Considering the presence of R_S at the source of MR3, these gate-source voltages are also

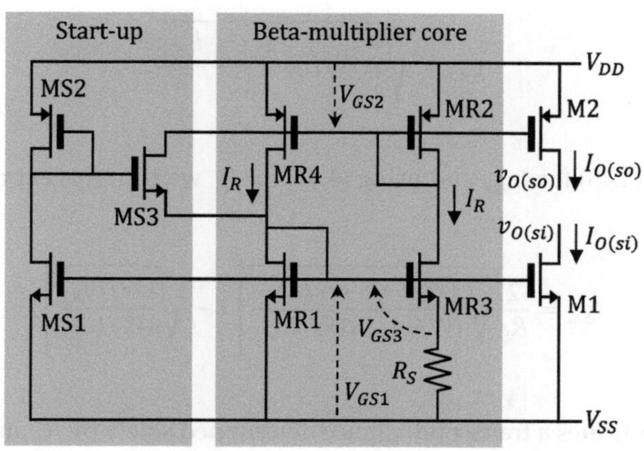

FIGURE 4.20 Beta-multiplier current reference for single-device sourcing and sinking.

interrelated by $V_{GS1} = V_{GS3} + I_R R_S$. Solving these three equations for I_R, we obtain

$$I_R = \frac{2A_N}{\mu_N C_{ox} R_S^2}\left(\frac{1}{\sqrt{(W/L)_{R1}}} - \frac{1}{\sqrt{(W/L)_{R3}}}\right)^2, \tag{4.41}$$

which doesn't show any dependence of the reference current I_R on V_{DD} or V_{SS}. In actual fact, I_R exhibits some degree of power-supply dependence due to the secondary effects not represented in (3.16) but this dependence is much weaker than what is observed in voltage-divider-based references.

It is obvious from Fig. 4.20 that the nominal source and sink currents are related to I_R by

$$I_{O(so)} = \frac{(W/L)_2}{(W/L)_{R2}} I_R, \tag{4.42}$$

and

$$I_{O(si)} = \frac{(W/L)_1}{(W/L)_{R1}} I_R. \tag{4.43}$$

One interesting property of the beta multiplier is that the transconductance of a MOSFET biased by $I_{O(so)}$ or $I_{O(si)}$ or by some fraction of one of these currents is determined solely by R_S and device aspect ratios. For an explanation of this so-called *constant-g_m* biasing property, suppose that an NMOS device MX of aspect ratio $(W/L)_X$ is biased by a fraction $I_{O(si)}/\alpha$ of the current sunk by M1. Assuming strong inversion, the transconductance of MX can be expressed from (3.26) and (3.28) as

$$g_{mx} = \sqrt{2(W/L)_X \frac{\mu_N C_{ox}}{A_N} \frac{I_{O(si)}}{\alpha}}.$$

By substituting (4.43) for $I_{O(si)}$ and using (4.41) for I_R, we turn this expression into

$$g_m = \frac{2}{R_S}\sqrt{\frac{(W/L)_1}{(W/L)_{R1}} \frac{(W/L)_X}{\alpha(W/L)_{R1}}}\left[1 - \sqrt{\frac{(W/L)_{R1}}{(W/L)_{R3}}}\right], \tag{4.44}$$

which indeed indicates a transconductance determined solely by R_S and ratios of device aspect ratios. Provided that proper physical design practices are followed in laying out the MOSFETs, the accuracy, precision, and temperature sensitivity of g_m will therefore be determined by R_S alone. Constant-g_m biasing is applied most typically with $(W/L)_{R3} = 4(W/L)_{R1}$. In this special case, (4.44) turns into

$$g_m = \frac{1}{R_S}\sqrt{\frac{(W/L)_1}{(W/L)_{R1}}\frac{(W/L)_X}{\alpha(W/L)_{R1}}}. \tag{4.45}$$

Although our explanation of the constant-g_m property is based on strong inversion, the conclusion that the transconductance is determined solely by R_S and ratios of device aspect ratios is valid irrespective of the strength of inversion. An interested reader may repeat the foregoing analysis for subthreshold operation. The counterpart of (4.44) will be found as follows.

$$g_m = \frac{1}{R_S}\frac{(W/L)_1}{\alpha(W/L)_{R1}}\ln\frac{(W/L)_{R3}}{(W/L)_{R1}}. \tag{4.46}$$

Notice in Fig. 4.20 that the core of the beta multiplier forms a feedback loop, in which, signal flows from gate to drain in each of MR3 and MR4. Since the polarity of a gate-to-drain transfer is negative, two such transfers along the loop result in a positive loop gain. However, the magnitude of this gain is lesser than unity even if the MOSFETs operate in subthreshold, which is why a beta multiplier is unconditionally stable despite the positive feedback.

Although the beta-multiplier core has a single stable state, it will need an unacceptably long time to settle into this state if its MOSFETs are initially in a deep subthreshold state. This is usually what happens when the power supply is turned on. For an explanation of the root cause, continue ignoring the start-up section, and suppose that the power-supply voltage is applied but core devices are temporarily forced to have $V_{GS1} = 0$ and $V_{GS2} = 0$. This will place all of these devices into deep subthreshold, in which, MR1 and MR2 do not conduct any current since their drain-source voltages are zero, but MR3 and MR4 conduct very small subthreshold currents due to their nonzero drain-source voltages. The current flowing in MR3 starts discharging the capacitance of the common gate nodes of MR2 and MR4, and thus moves V_{GS2} in the negative direction toward the stable state. The current of MR4 also drives the core toward the stable state by charging the capacitance of the common gate nodes of MR1 and MR3 and thus moving V_{GS1} in the positive direction. However, the initial time rates of change in V_{GS1} and V_{GS2} are extremely low due to the vanishingly small values of these currents. Although these rates increase exponentially over time as the overdrive voltages of core devices grow progressively, the overall duration of this transient state is usually too long to tolerate. The start-up circuit shown in Fig. 4.20 helps shorten it to an acceptable level.

For an explanation of how the start-up circuit works, assume $V_{GS1} \approx 0$ just after power up. MS1, being in a deep subthreshold state due to this initial condition, does not conduct any significant current. The diode-connected MS2, thus starving for current, also remains in a deep subthreshold state with a gate voltage close to V_{DD}. MS3 operates with this high voltage at its gate and a low voltage near V_{SS} at its source, and therefore responds by drawing a large amount of current from the capacitance associated with the common gate nodes of MR2 and MR4. As a result, V_{GS2} rapidly falls toward its stable-state level.

Since the large current drawn by MS3 is sourced onto the capacitance associated with the gate nodes of MR1 and MR3, V_{GS1} also rises rapidly toward its stable-state level. As the core thus escapes from its initial current-starving state, the rising V_{GS1} causes MS1 to conduct current. MS2, conducting the same current, lowers the gate voltage of MS3 below the source voltage $V_{SS} + V_{GS1}$. Since the current of MS3 is thus cut off, the start-up circuit is electrically separated from the beta-multiplier core. The core thereafter operates on its own in its stable state.

Beta-multiplier design commences after unit-device sizing has been completed for the sink and source networks as done in Examples 4.1 and 4.4. Since unit-device counts for the reference devices MR1 and MR2 are also known at this stage, and MR4 is just a replica of MR2, the only device left to be sized for the core is MR3. Typically, it is sized as a fourfold replica of MR1. Next in design, we determine R_S. Although we can calculate R_S from (4.41), the inaccuracy of this equation would necessitate trimming with an iterative simulation in final verification anyway. It is therefore meaningful to skip the calculation and rely directly on iterative simulation for determining this resistance. Finally, we design a start-up circuit. Its design is subject to two primary requirements. One is to force a sufficiently high current through MS3 to start up the core rapidly from the worst-case initial condition of $V_{GS1} = 0$ and $V_{GS2} = 0$. The other is to drive the same device into cutoff with a reasonably small current flowing in MS1 and MS2 after start-up.

Cascode Versions of Beta Multiplier

Basic and low-voltage cascode versions of the beta multiplier are shown in Fig. 4.21 for mirroring to a sink pair M1/M2 and a source pair M3/M4. In both circuits, the sizing

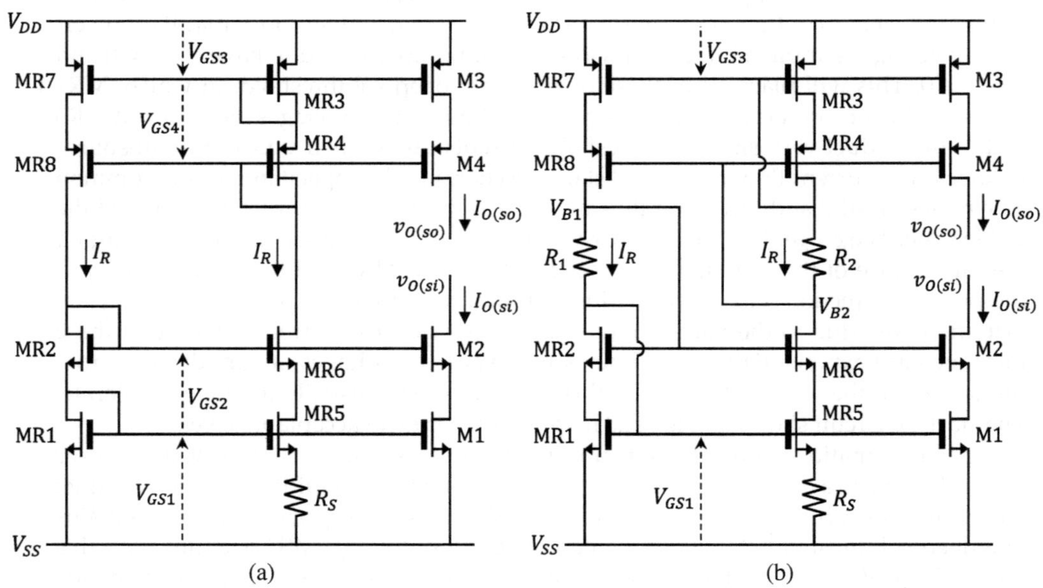

FIGURE 4.21 Cascode beta-multiplier current references. (a) Basic version. (b) Low-voltage version.

of the diode-connected devices MR1 and MR2 is a part of sink design as exemplified in Examples 4.5 and 4.6. For MR6, we deploy a replica of MR2. As to MR5, we usually size it four times wider than MR1. Source specifications dictate the sizing of the pair MR3/MR4. MR7 and MR8 are just replicas of these devices. R_1 and R_2 of the low-voltage cascode beta-multiplier are determined from

$$R_1 = \frac{V_{B1} - (V_{SS} + V_{GS1})}{I_R}, \tag{4.47}$$

and

$$R_2 = \frac{(V_{DD} + V_{GS3}) - V_{B2}}{I_R}, \tag{4.48}$$

where V_{B1}, V_{B2}, V_{GS1}, and V_{GS3} are available from sink/source design. R_S is determined with iterative simulation once the MOSFETs have been sized, and R_1 and R_2 have been determined from (4.47) and (4.48).

As in all self-biased circuit topologies, a start-up circuit is also needed in the cascoded versions of the beta multiplier. The topology shown in Fig. 4.20 is also suitable for this purpose. MS3 should connect the common gate nodes of MR3/MR7 to the common gate-nodes of MR1/MR5.

Example 4.10 A beta-multiplier current reference is to be designed for handling the single-device mirroring applications specified in Examples 4.1 and 4.4. A complete schematic including the sink/source paths and start-up circuit is shown in Fig. 4.22. Proceed through the following steps:

1. Determine R_S for $(W/L)_{R3} = 4(W/L)_{R1}$.
2. Determine the relative power-supply sensitivity of sink currents flowing in M1 and M2 of Example 4.1 and source currents flowing in M4 and M5 of

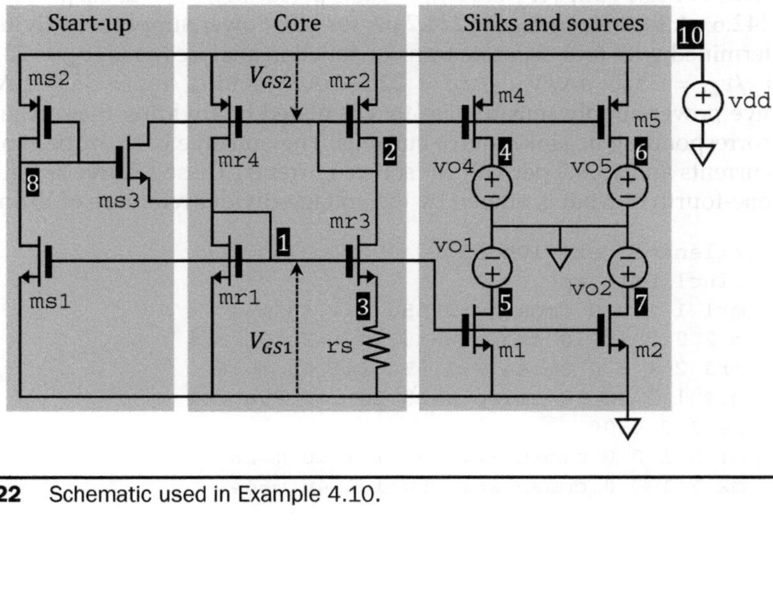

FIGURE 4.22 Schematic used in Example 4.10.

Example 4.4. How does the outcome compare with the relative power-supply sensitivity of the basic-cascode current source designed in Example 4.9?

3. Design a start-up circuit that conducts (a) approximately 20 μA through MS3 in the initial state of $V_{GS1} = 0$ and $V_{GS2} = 0$, and (b) approximately 2 μA through MS1/MS2 in steady state after start-up.

Work

1. R_S is determined with an iterative operation-point analysis of the core section only. As observed in the following input file `ex4-10a.sp`, device sizes are copied directly from Examples 4.1 and 4.4. A print line reads out the currents flowing in MR1 and MR2 for comparison with the target reference current $I_R = 20$ μA.

```
filename: ex4-10a.sp
.incl bsim.sp
mr1 1 1 0 0 cmosn w=1.55u l=1.4u m=4
mr2 2 2 10 10 cmosp w=15.1u l=2.69u m=4
mr3 2 1 3 0 cmosn w=1.55u l=1.4u m=16
mr4 1 2 10 10 cmosp w=15.1u l=2.69u m=4
rs 3 0 4300
vdd 10 0 1.8
.control
op
print @mr1[id] @mr2[id]
.endc
.end
```

$R_S = 4300\ \Omega$ satisfies the specification by setting the currents of MR1 and MR2 to 20.67 and 20.09 μA, respectively.

2. Sink and source currents and their power-supply sensitivities are characterized with the following input file `ex4-10b.sp`, which complements the core with sink/source paths. An operation-point analysis is run for extracting the dc currents of these paths. It yields $I_{O1} = 146.5$ μA for M1, $I_{O2} = 230.2$ μA for M2, $I_{O4} = 142.6$ μA for M4, and $I_{O5} = 224.2$ μA for M5. Power-supply sensitivities are determined with four separate transfer-function analyses as $i_{o1}/v_{dd} = 22.78$ μA/V, $i_{o2}/v_{dd} = 35.79$ μA/V, $i_{o4}/v_{dd} = 22.64$ μA/V, and $i_{o5}/v_{dd} = 35.59$ μA/V. Relative power-supply sensitivities are calculated by dividing these results by the corresponding dc sink/source currents. The outcome is 15.5% per volt for sink currents and 15.9% per volt for source currents. These relative sensitivities are one-fourth of what is offered by the voltage-divider reference of Example 4.9.

```
filename: ex4-10b.sp
.incl bsim.sp
mr1 1 1 0 0 cmosn w=1.55u l=1.4u m=4
mr2 2 2 10 10 cmosp w=15.1u l=2.75u m=4
mr3 2 1 3 0 cmosn w=1.55u l=1.4u m=16
mr4 1 2 10 10 cmosp w=15.1u l=2.75u m=4
rs 3 0 4300
m1 5 1 0 0 cmosn w=1.55u l=1.4u m=28
m2 7 1 0 0 cmosn w=1.55u l=1.4u m=44
```

```
m4 4 2 10 10 cmosp w=15.1u l=2.75u m=28
m5 6 2 10 10 cmosp w=15.1u l=2.75u m=44
vo1 5 0 0.9
vo2 7 0 0.9
vo4 4 0 0.9
vo5 6 0 0.9
vdd 10 0 1.8
.op
*.tf i(vo1) vdd
*.tf i(vo2) vdd
*.tf i(vo4) vdd
*.tf i(vo5) vdd
.end
```

3. Dimensions of MS1 are determined first. Once the core has started up, this device will operate in a steady state with $V_{GS1} - V_{TN} = 0.195$ V as determined for sink devices in Example 4.1. According to the design tool displayed in Fig. 3.10, its sheet current should therefore be 4.5 μA/sq. For the specified 2-μA current, this implies an aspect ratio of 0.444. Since there is no specific constraint on either the length or the width, we can select one of these dimensions arbitrarily, and then determine the other from this aspect ratio. Size and reliability concerns may guide us in this selection. Selecting the minimum width that can accommodate, say, two contacts at each of the source and drain terminals will be a reasonable trade-off between these two concerns. This leads us to $W_{S1} = 1$ μm, and therefore to $L_{S1} = 2.25$ μm. Note that we are not concerned about the differences between effective and drawn dimensions because the current is specified approximately anyway.

   Now, we run an iterative operation-point analysis with `ex4-10b.sp` after including MS1 and MS2 but excluding MS3. In this analysis, MS1 is specified with the dimensions just determined. As to the dimensions of MS2, we may select $W_{S2} = 1$ μm as justified in the preceding paragraph for MS1, and iterate L_{S2} until the gate voltages of MS2 and MS1 are equalized and thus MS3 is cut off with a zero gate-source voltage. Needless to say, the current of MS1 should also be looked into in each iteration to make sure that it remains around 2 μA. This step of design yields $L_{S2} = 7$ μm.

   After finalizing the sizing of MS2, we add MS3 to the input file, and imitate the initial condition of start-up by short-circuiting node 1 to ground and node 2 to V_{DD}. We complete the design by selecting $W_{S3} = 1$ μm, and iterating L_{S3} until the current of MS3 reaches the specified 20 μA. $L_{S3} = 9$ μm turns out to be an appropriate channel length. ▲

4.3.3 Bandgap Voltage Reference

A reference voltage that is insensitive to power-supply voltage, device parameters, and temperature is indispensable in certain types of analog or mixed-signal processing. For example, all linear regulators and most analog-to-digital converters need one such voltage. This need is served most commonly by a *bandgap reference* (BGR) circuit, which relies on self-biasing for supply insensitivity, matching for device-parameter insensitivity and

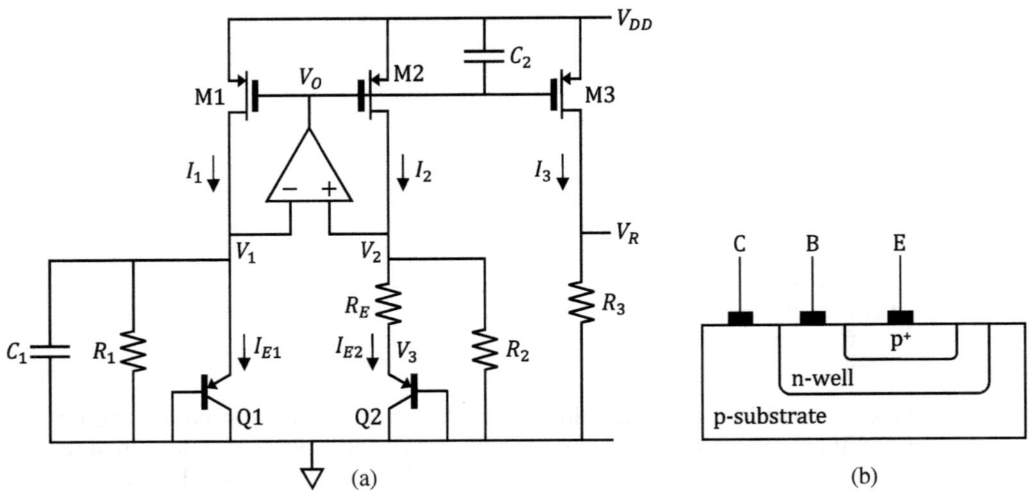

FIGURE 4.23 (a) A bandgap reference circuit. (b) Building a pnp BJT out of a p-channel MOSFET structure.

the bandgap of silicon for temperature insensitivity. The term "bandgap" refers to the energy needed by a valence electron for escaping the silicon atom and becoming a free electron inside a semiconductor crystal.

A BGR topology suitable for low-voltage applications is shown in Fig. 4.23(a) in a single-supply configuration, where the reference voltage is denoted by V_R.[5] It is built around a self-biasing negative feedback loop, in which, the matching pair M1 and M2 equalizes I_1 and I_2 whereas an OTA equalizes V_1 and V_2. Q1 and Q2 are two pnp BJT devices each emulating a diode. Q2 comprises N parallel replicas of Q1. In a pure CMOS technology, these are usually built with the junctions of a PMOS device. As depicted in Fig. 4.23(b), the p-type substrate acts as a collector, an n-well acts as a base, and a p-active region inside the n-well acts as an emitter. The interconnection between the n-well and p-type substrate forces Q1 and Q2 to operate at the borderline between saturation and active modes of operation, for which, the emitter currents are described by

$$I_{E1} = \frac{I_S}{\alpha_F} \exp\left(\frac{q}{kT} V_1\right), \tag{4.49}$$

and

$$I_{E2} = \frac{NI_S}{\alpha_F} \exp\left(\frac{q}{kT} V_3\right), \tag{4.50}$$

where I_S and α_F represent the *intercept-current* and *forward alpha* parameters, respectively. Due to $V_1 = V_2$, the matching resistors R_1 and R_2 steal identical amounts of current from

I_1 and I_2. Since the latter two are also identical, Q1 and Q2 are thus forced to maintain $I_{E1} = I_{E2}$ despite the disparity between their sizes. Equating the right-hand sides of (4.49) and (4.50) leads us to $V_1 - V_3 = (kT/q)\ln N$. Since $V_1 = V_2$, this equation also describes the voltage $V_2 - V_3$ across R_E. Dividing this voltage by R_E yields

$$I_{E1} = I_{E2} = \frac{kT}{q}\frac{1}{R_E}\ln N \tag{4.51}$$

for the identical emitter currents. Next, we add V_1/R_1 to the right-hand side of this equation to arrive at

$$I_1 = I_2 = \frac{V_1}{R_1} + \frac{kT}{q}\frac{1}{R_E}\ln N \tag{4.52}$$

for the currents sourced by M1 and M2. Assuming M3 to be scaled by a factor S with respect to M1 and M2, we express the current flowing in R_3 as $I_3 = SI_1$, where I_1 is given by (4.52). The reference voltage $V_R = R_3 I_3$ is therefore described by

$$V_R = SR_3\left(\frac{V_1}{R_1} + \frac{kT}{q}\frac{\ln N}{R_E}\right). \tag{4.53}$$

Among the parameters appearing in this expression, neither the MOSFET scaling factor S nor the BJT scaling factor N is temperature dependent. Although resistances depend on temperature, this dependence is almost perfectly canceled out in their ratios provided that they are properly laid out. This leaves only the emitter-base voltage V_1 and the thermal voltage kT/q in (4.53) as temperature-dependent variables. Since temperature sensitivity is negative in V_1 but positive in kT/q, we can nullify the temperature sensitivity of V_R by properly setting the resistance ratios and N. By differentiating (4.53) with respect to T, and equating the outcome to zero, we determine the necessary condition of zero thermal-sensitivity as

$$\frac{kT}{q}\frac{R_1}{R_E}\ln N = -\frac{\mathrm{d}V_1}{\mathrm{d}T}T. \tag{4.54}$$

If this condition is met by design, then, (4.53) leads to the following equation for the temperature insensitive V_R:

$$V_R = S\frac{R_3}{R_1}\left(V_1 - \frac{\mathrm{d}V_1}{\mathrm{d}T}T\right). \tag{4.55}$$

On the other hand, a detailed modeling of the relative thermal-sensitivity of the emitter-base voltage yields

$$\frac{dV_1}{dT}T = V_1 - V_{BG} - 3\frac{kT}{q}, \tag{4.56}$$

where V_{BG} is the bandgap voltage of silicon, which is a very weak function of the temperature. Its value at room temperature ($T = 300$ K) is 1.12 V. We finally substitute (4.56) into (4.54) and (4.55) to obtain

$$\frac{R_1}{R_E}\ln N = \frac{q}{kT}\left(V_{BG} - V_1\right) + 3, \tag{4.57}$$

for the condition of temperature insensitivity, and

$$V_R = S\frac{R_3}{R_1}\left(V_{BG} + 3\frac{kT}{q}\right) \tag{4.58}$$

for the reference voltage. Since V_{BG} is much larger than $3kT/q$, (4.58) indicates a reference voltage proportional to V_{BG}, which is why the circuit of Fig. 4.23(a) is qualified as a BGR. It is also obvious from (4.58) that the coefficient of proportionality is defined by the ratio of two resistances and the scaling factor between two MOSFETs, and therefore, can be implemented with great accuracy. However, it should be reiterated that V_R is temperature insensitive only around the particular temperature for which the condition (4.57) is met by design. Sensitivity can be reduced at additional temperatures by applying one of the many *curvature-correction* techniques available for this purpose but the details of such techniques are beyond the scope of this book.[6][7]

The sensitivity of V_R to power-supply voltage is characterized with a performance metric called *power-supply gain* (*PSG*), which is defined by

$$PSG \doteq 20\log\left|\frac{v_r}{v_{dd}}\right|, \tag{4.59}$$

where v_r is the variation of the reference voltage in response to a variation v_{dd} in the power-supply voltage. PSG is a function of frequency and is expressed in units of decibel.

An exact small-signal analysis performed on the BGR schematic of Fig. 4.23(a) yields the following expression for the dc value of v_r/v_{dd}:

$$\frac{v_r}{v_{dd}} = \frac{1 + \left(g_{m3}r_{ds3}\right)\left(1 - A_{dd}\right) + A_0\frac{r_{ds(1,2)}\left(r_{d2}-r_{d1}\right)\left(g_{m(1,2)}r_{ds(1,2)}-g_{m3}r_{ds3}\right)}{\left(r_{ds(1,2)}+r_{d2}\right)\left(r_{ds(1,2)}+r_{d1}\right)}}{\left(1 + \frac{r_{ds3}}{R_3}\right)\left[1 + A_0\frac{g_{m(1,2)}r_{ds(1,2)}^2\left(r_{d2}-r_{d1}\right)}{\left(r_{ds(1,2)}+r_{d2}\right)\left(r_{ds(1,2)}+r_{d1}\right)}\right]}, \tag{4.60}$$

where the symbols have the following meaning:

- $g_{m(1,2)}$ and $r_{ds(1,2)}$: Transconductance and output resistance of M1 and M2.
- g_{m3} and r_{ds3}: Transconductance and output resistance of M3.
- A_0: DC differential gain of the OTA.
- A_{dd}: PSG of the OTA as defined by the ratio v_o/v_{dd}, in which, v_{dd} is a dc small-signal voltage applied to the power pin of the OTA, and v_o is the corresponding small-signal voltage observed at the output while noninverting and inverting inputs are kept at signal ground.
- r_{d1}: External small-signal resistance between the drain terminal of M1 and ground.
- r_{d2}: External small-signal resistance between the drain terminal of M2 and ground.

Note that the resistances r_{d1} and r_{d2} are described by

$$r_{d1} = R_{1,2} \parallel r_e \approx r_e, \tag{4.61}$$

and

$$r_{d2} = R_{1,2} \parallel (r_e + R_E) \approx r_e + R_E, \tag{4.62}$$

in which, $R_{1,2}$ stands for the matching resistances R_1 and R_2, and r_e denotes the small-signal emitter resistance of Q1 and Q2.[3]

It is obvious from (4.61) and (4.62) that the difference $r_{d2} - r_{d1}$ appearing in (4.60) can be replaced by R_E. Furthermore, assuming $r_{d1} \ll r_{ds(1,2)}$, $r_{d2} \ll r_{ds(1,2)}$, $R_3 \ll r_{ds3}$, and $A_0 g_{m(1,2)} R_E \gg 1$, we can simplify (4.60) into

$$\frac{v_r}{v_{dd}} = \frac{g_{m3}R_3}{A_0 g_{m(1,2)} R_E}\left[1 + \frac{1}{g_{m3}r_{ds3}} - A_{dd} + A_0 g_{m(1,2)} R_E \left(\frac{1}{g_{m3}r_{ds3}} - \frac{1}{g_{m(1,2)}r_{ds(1,2)}}\right)\right].$$

[3]For justifying the removal of $R_{1,2}$ in the approximate forms of (4.61) and (4.62), note that r_e is described by $r_e = kT/qI_E$, where I_E represents the identical emitter currents I_{E1} and I_{E2}. Considering this description together with (4.51) and (4.57), we can relate $R_{1,2}$ to r_e with

$$R_{1,2} = \left[\frac{q}{kT}\left(V_{BG} - V_1\right) + 3\right] r_e.$$

For the typical $V_1 \approx 0.6$ V, this expression indicates $R_{1,2} \approx 20r_e$ at room temperature. According to (4.57), $R_{1,2}$ should also be larger than R_E by a factor of 5 to 10 for the usual range $10 < N < 100$ of the BJT scaling factor.

We reorganize this expression by (a) considering $g_{m3} = Sg_{m(1,2)}$ as implied by $I_3 = SI_1$ and the common overdrive voltage of M1 and M3, and (b) using (4.57) and (4.58) for R_E and R_3, respectively. The result is

$$\frac{v_r}{v_{dd}} = \frac{\left(V_{BG} - V_1 + 3\frac{kT}{q}\right) V_R}{\left(V_{BG} + 3\frac{kT}{q}\right) \frac{kT}{q} A_0 \ln N} \underbrace{\left[1 + \frac{1}{g_{m3} r_{ds3}} - A_{dd} + A_0 R_E \left(\frac{1}{S r_{ds3}} - \frac{1}{r_{ds(1,2)}}\right)\right]}_{\eta}. \tag{4.63}$$

Ignoring for a moment the factor η, this expression indicates A_0 and N as the primary variables of design for setting the PSG for a given V_R because (a) V_{BG} and kT/q are not design variables, and (b) V_1 is almost constant around 0.6 V for any reasonable value of the emitter current. To be specific, PSG decreases and therefore improves as these two design variables are increased. However, raising N is a costly approach because it improves PSG only logarithmically while adding linearly to the footprint. Raising A_0 is definitely a more effective approach in this respect.

Returning to the parameters of the factor η in (4.63), note that A_{dd} is slightly smaller than 1 in most OTA designs, whereas $(g_{m3} r_{ds3})^{-1}$ is on the order of 10^{-2}. We also expect $Sr_{ds3} = r_{ds(1,2)}$ but only if $V_R = V_1$. Otherwise, the disparate drain-source voltages of M3 and the pair M1/M2 usually lead to $Sr_{ds3} > r_{ds(1,2)}$ for $V_R < V_1$ and to $Sr_{ds3} < r_{ds(1,2)}$ for $V_R > V_1$. All in all, η can be negative or positive but in both cases it is meaningful to keep it as small as possible. This is accomplished by targeting an A_{dd} close to 1, and setting $V_R \approx V_1$ if possible. The latter of these two design measures works even if the desired reference voltage happens to be smaller than V_1 because we can always generate a fraction of V_R with resistive division on R_3.

The reader may wonder whether it is permissible to transpose the inverting- and noninverting-input terminals of the OTA in the BGR circuit of Fig. 4.23(a). The answer is not affirmative because it would create a positive feedback loop. To see why, let us extract the feedback factor β of the closed-loop configuration of the BGR. To this end, we disconnect the output of the OTA from the schematic, force a small-signal voltage v_o to the interconnected gates of M1 and M2, and relate the small-signal voltages v_{in} and v_{ip} of the inverting- and noninverting-input voltages to v_o through

$$v_{in} = -g_{m1} \left(r_{ds1} \parallel R_1 \parallel r_e\right) v_o, \tag{4.64}$$

and

$$v_{ip} = -g_{m2} \left[r_{ds2} \parallel R_2 \parallel (r_e + R_E)\right] v_o. \tag{4.65}$$

Since $g_{m2} = g_{m1}$, $r_{ds2} = r_{ds1}$, and $R_2 = R_1$, (4.64) and (4.65) indicate a positive feedback factor $\beta \doteq (v_{in} - v_{ip})/v_o$, hence a negative feedback loop. Transposing the input terminals will transpose the left-hand sides of (4.64) and (4.65), and thus lead to a negative feedback factor, hence a positive feedback loop. It is therefore obvious that the noninverting terminal of the OTA has to be connected to the branch where R_E is placed.

A BGR circuit, being self-biased, needs a start-up unit or a power-on reset unit for ensuring operation in its intended stable state. Start-up circuits appropriate for the BGR circuit of Fig. 4.23(a) are presented in [8]. In Example 4.11, we will avoid this problem by utilizing a power-on reset technique as originally proposed in [5].

One final note on the schematic shown in Fig. 4.23(a) is that the capacitors marked C_1 and C_2 may be needed for keeping the BGR stable throughout the start-up process.

Example 4.11 The BGR circuit of Fig. 4.23(a) is redrawn in Fig. 4.24 together with a power-on reset (POR) unit. This unit is added for avoiding improper start-up. It comprises a digital inverter built with M5 and M6, and a pull-down device M4 driven by the inverter. For an explanation of how it works, consider a start-up event, during which, V_{DD} gradually rises from zero to its nominal value 1.8 V. During this event V_{POR} is kept at ground, which forces cutoff on M6. M5 also remains in cutoff until V_{DD} crosses over $-V_{TP}$. The voltage of node 47 is therefore undefined during this initial period of start-up. M5 turns on and connects node 47 to the power rail as V_{DD} crosses over $-V_{TP}$. M4 now

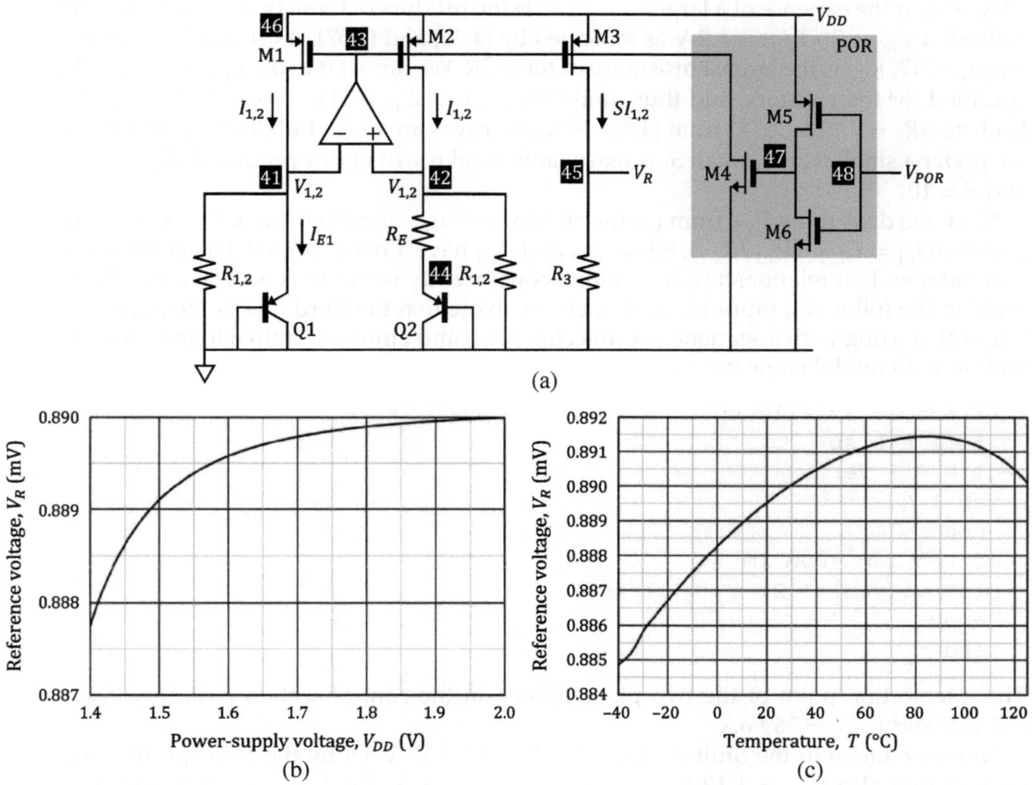

FIGURE 4.24 (a) The BGR schematic of Example 4.11, where M4, M5, and M6 belong to the power-on reset unit. (b) Simulated reference voltage versus supply voltage. (b) Simulated reference voltage versus temperature.

with V_{DD} at its gate also turns on for $V_{DD} > V_{TN}$ and starts stealing current from the OTA. The output voltage of the OTA is thus prevented from catching up with the rising V_{DD}. The resulting negative gate-source voltage enables M1 and M2 to source current onto nodes 41 and 42 and thus raise the voltage of these nodes into the common-mode range of the OTA. This is how the OTA is pulled out of a possible open-loop state while start-up is in progress. Start-up is completed by decoupling the power-on reset unit from the BGR. This is done by raising V_{POR} to V_{DD}, and thus cutting off M4.

The BGR circuit shown in Fig. 4.24 is expected to generate $V_R = 900$ mV with a *PSG* better than −55 dB at $V_{DD} = 1.8$ V, and remain functional within $1.4\,\text{V} \le V_{DD} \le 2.0\,\text{V}$ and $-40°\,\text{C} \le T \le 125°\,\text{C}$. Design should also target low power and small footprint, and avoid resistances in excess of 1 MΩ. The temperature coefficients of resistances are specified as TCR1 = 3000 ppm/C and TCR2 = 1 ppm/C^2 for all resistors. The OTA available for this design is netlisted as a subcircuit in the file named `ota.sp`. It offers $A_0 = 4000$ V/V and $A_{dd} = 0.941$ V/V at $V_O = 0.9$ V and $T = 27$°C. The unit BJT available for this design is specified by the model deck `vert` presented in the file named `bjt.sp`.

Work Excluding the OTA, the power consumed by the BGR equals $(2 + S)I_{1,2}V_{DD}$. Obviously, $I_{1,2}$ has to be minimized for a low-power design. However, a small $I_{1,2}$ is possible only at the expense of a large $R_{1,2}$ because the product of these two design variables is fixed at $V_{BG} + 3(kT/q) \approx 1.2$ V as indicated by (4.52) and (4.57). As stated in footnote 3 on page 317, $R_{1,2}$ is the largest resistance of the BGR. We can set it to the upper limit 1 MΩ specified for the resistors, and thus minimize $I_{1,2}$ to 1.2 μA. This choice about $R_{1,2}$ also leads to $SR_3 = 7.51 \times 10^5$ Ω from (4.58). Since resistors are much bulkier than MOSFETs, we prefer a small R_3 to a small S. A reasonable breakdown of this product is $R_3 = 75$ kΩ and $S = 10$.

Next, we determine $V_{1,2}$ from (a) the current-voltage characteristic of Q1, and (b) the equation $I_{E1} = I_{1,2} - (V_{1,2}/R_{1,2})$, where $I_{1,2}$ and $R_{1,2}$ have already been determined. These two independent relationships are plotted concurrently with the dc-sweep analysis defined in the following input file `ex4-11a.sp`. Note that the third line of the netlist defines Q1 starting with a sequence of collector, base, and emitter terminal identifiers, and ending with model name `vert`.

```
filename: ex4-11a.sp
.incl bjt.sp
q1 0 0 1 vert
veb 1 0
.control
dc veb 10m 700m 1m
plot -i(veb) 1.2e-6-(v(1)/1e6)
.endc
.end
```

The intersection point of the two plots defined in the control section indicates $V_{1,2} =$ 631 mV and $I_{E(1,2)} = 569$ nA.

Next, we move to the limit $(v_r/v_{dd}) \le 1.78 \times 10^{-3}$ V/V set by the *PSG* specification. Considering also $V_{BG} = 1.12$ V, $V_1 = 0.631$ V, $kT/q = 0.0258$ V, $V_R = 0.9$ V, and $A_0 =$ 4000 V/V, we obtain from (4.63) the constraint

$$\ln N \ge 2.31\eta,$$

which can be satisfied by $N = 10$ provided that η can be kept lower than 1. This is a very favorable scaling factor as far as the footprint of Q2 is concerned. Also note from (4.57) the implication $R_E = 105$ kΩ of this selection.

For determining the ways in which η can be lowered below 1, we refer to (4.63). Since $1 - A_{dd}$ and $1/g_{m3}r_{ds3}$ are much smaller than 1 anyway, we obviously need to manipulate the difference $1/Sr_{ds3} - 1/r_{ds(1,2)}$ for this purpose. Although the disparate drain-source voltages of M3 and M1/M2 prevent us from reducing this difference to zero, we can minimize it by raising the output resistances. This calls for long-channel MOSFETs.

Turning our attention to the MOSFETs, we determine their least negative drain-source voltages as $V_{DS(1,2)} = V_{1,2} - V_{DD(\min)} = 0.631 - 1.4 = -0.769$ V and $V_{DS3} = V_R - V_{DD(\min)} = 0.9 - 1.4 = -0.5$ V. Saturation for all three devices is therefore ensured for $V_{DSsat} \geq -0.5$ V. Assuming standard-threshold devices, the corresponding limit of overdrive voltage is determined from Fig. 3.17 as −0.65 V. We target $V_{GS(1,2,3)} - V_{TP} = -0.45$ V for two reasons. First of all, it will allow for a reasonable saturation margin even for the worst-case drain-source voltage. Secondly, considering $V_{TP} \approx -0.45$ V as indicated by Fig. 3.16, it will bias the OTA around $V_O \approx 0.9$ V, for which, A_0 is known to be maximized. This much overdrive implies $I_\square \approx 5.35$ μA/sq from Fig. 3.18, which, for $I_{1,2} = 1.2$ μA, indicates $(W/L)_{1,2} = 0.224$ for the aspect ratio of M1 and M2. We break this into $W = 1.5$ μm and $L = 6.7$ μm to have a long and reasonably wide channel. As decided upon earlier, M3 is to comprise 10 parallel replicas of this device.

Device sizing for the POR unit is not very critical. Inverter devices M5 and M6 can be implemented with minimum length and width. A relatively long channel is desirable for M4 in order to limit the current it draws from the OTA during start-up.

Verification is performed with the input file `ex4-11b.sp` presented below.

```
filename: ex4-11b.sp
.incl bsimlvt.sp
.incl bsim.sp
.incl bjt.sp
.incl ota.sp
x1 42 41 43 46 ota
m1 41 43 46 46 cmosp w=1.5u l=6.7u ad=0.83p pd=4.1u
m2 42 43 46 46 cmosp w=1.5u l=6.7u ad=0.83p pd=4.1u
m3 45 43 46 46 cmosp w=1.5u l=6.7u ad=0.83p pd=4.1u m=10
q1 0 0 41 vert
q2 0 0 44 vert m=10
r1 41 0 1meg tc1=3000e-6 tc2=1e-6
r2 42 0 1meg tc1=3000e-6 tc2=1e-6
r3 45 0 75k tc1=3000e-6 tc2=1e-6
re 42 44 105k tc1=3000e-6 tc2=1e-6
m4 43 47 0 0 cmosn w=1u l=8u ad=0.55p pd=3.1u
m5 47 48 46 46 cmosp w=1u l=0.2u ad=0.55p pd=3.1u
m6 47 48 0 0 cmosn w=1u l=0.2u ad=0.55p pd=3.1u
*******************power-supply sweep*********************
vpor 48 0 1.8
vdd 46 0
.save all @m3[gm] @m3[gds] @m1[gds]
.control
dc vdd 1.4 2 0.1m
plot v(45) ; vr
```

```
plot deriv(v(45)) ; psg
meas dc gm3 find @m3[gm] when v(46)=1.8
meas dc gds3 find @m3[gds] when v(46)=1.8
meas dc gds1 find @m1[gds] when v(46)=1.8
print 1.79e-3*(5.9e-2+gds3/gm3+4.2e8*(gds3/10-gds1)) ; (4.62)
.endc
********************temperature sweep**********************
*vpor 48 0 1.8
*vdd 46 0 1.8
*.control
*dc temp -40 125 5
*plot v(45)
*.endc
********************power-supply gain**********************
*vpor 48 0 1.8
*vdd 46 0 dc 1.8 ac 1
*.control
*ac dec 100 10 100meg
*plot vdb(45)-vdb(46)
*.endc
************************transient*************************
*vpor 48 0 pulse 0 1.8 110m
*vdd 46 0 pwl 0 0 100m 1.8
*.control
*tran 20u 170m
*plot v(46) v(45) v(48)
*.endc
**********************************************************
.end
```

The reader need not be concerned with the details of the OTA defined in ota.sp but it is worth noting that the power-supply pin of this OTA is connected to the very same V_{DD} that powers the rest of the BGR in the main input file. Simulated PSG analysis will thus take into account the nonzero A_{dd} associated with the OTA. Also notice the inclusion of the parameter deck bsimlvt.sp of the low-threshold MOSFET option. This is due to the use of this option for some of the OTA devices netlisted in ota.sp.

Four distinct simulations are defined in separate control sections. First is a dc power-supply sweep, in which, the power-on reset circuit is disabled by applying $V_{POR} = 1.8$ V, and V_{DD} is swept over the specified power-supply range at room temperature. It generates two plots, one for V_R and another for the derivative v_r/v_{dd}. The first plot enables us (a) to compare the actual reference voltage with its value specified for $V_{DD} = 1.8$ V, and (b) to evaluate the extent of the variation of V_R within the specified range of V_{DD}. As to the plot of the derivative, the reader is invited to zoom in and read its value at $V_{DD} = 1.8$ V, and compare the outcome with what the specified value of PSG corresponds to. The very same simulation also prints out the prediction of (4.63) for v_r/v_{dd}. This equation is evaluated after measuring g_{m3}, g_{ds3}, and g_{ds1} for $V_{DD} = 1.8$ V, and then, combining the outcome with $V_{BG} = 1.12$ V, $V_1 = 0.631$ V, $kT/q = 0.0258$ V, $V_R = 0.9$ V, $A_0 = 4000$ V/V, $N = 10$, $A_{dd} = 0.941$ V/V, $R_E = 105$ kΩ, and $S = 10$. A comparison with the directly simulated value of v_r/v_{dd} will enable the reader to assess the reliability of (4.63).

The second simulation sweeps the temperature within the specified range while keeping V_{DD} constant at 1.8 V. Its outcome will give a fair idea as to how immune a BGR is to temperature variations.

The third simulation is an ac analysis performed for extracting the variation of PSG with the frequency of power-supply variation. The dc value of PSG, which is close to −62 dB, is retained up to the open-loop bandwidth of the OTA. At higher frequencies, PSG rises as the differential gain of the OTA rolls off. This will be observed beyond 10 kHz on the transfer characteristic generated by this simulation.

The final simulation involves a transient analysis of a start-up event, in which, V_{DD} is ramped from 0 to 1.8 V in 100 ms, and V_{POR} is stepped from 0 to 1.8 V after V_{DD} has reached its final value. Its purpose is to verify the reliability of the start-up process. ▲

4.4 Basic Amplifier Stages

The input stage of all opamps is in the form of what we call a *differential pair*. If additional gain is needed, a *common-source* stage supplements the input stage. For buffering, either a common-source stage or a *source-follower* stage is used. These three basic building blocks of opamps are the subject of this section. The first two subsections cover the common-source and source-follower stages. The third is devoted to the basic-differential pair. Two high-performance versions of the differential pair are studied in the final two subsections.

4.4.1 Common-Source Stage

We introduced the topology of the basic common-source stage in Fig. 3.22(a) of Example 3.5, and studied it further in Examples 3.10, 3.12, 3.16, and 3.17. What is probably worth adding to what we already know about this stage is that the roles of its two MOSFETs can be interchanged so that the PMOS operates as a driver while the NMOS supplies the bias current. Gain expression (3.56) is valid also for this case provided that g_{m1} is replaced by the transconductance of the PMOS driver. Both versions are good for achieving a moderate voltage gain on the order of 100 V/V across a purely capacitive load. None of the two, however, can maintain this much gain in the presence of a resistive load unless the latter is much higher than the output resistances of the driver and current-source devices. A comparison between (3.56) and the gain expression

$$A_{v0} = -g_{m1}\left(r_{ds1} \parallel r_{ds2} \parallel R_L\right)$$

derived for the case of loading by a resistance R_L will affirm this statement.

If necessary, the gain of a common-source stage can be increased by cascoding the driver and current-source devices as exemplified in Fig. 4.25 on a stage with an NMOS driver. Input signal v_i is biased by V_I and applied to the gate of the driver device M1, which is cascoded by M2. M3, cascoded by M4 and acting as a current source, loads the driver. The bias voltages V_{B1}, V_{B2}, and V_{B3} are generated by an appropriate cascode current-reference circuit as depicted in Figs. 4.18(b) and (c) or in Fig. 4.21.

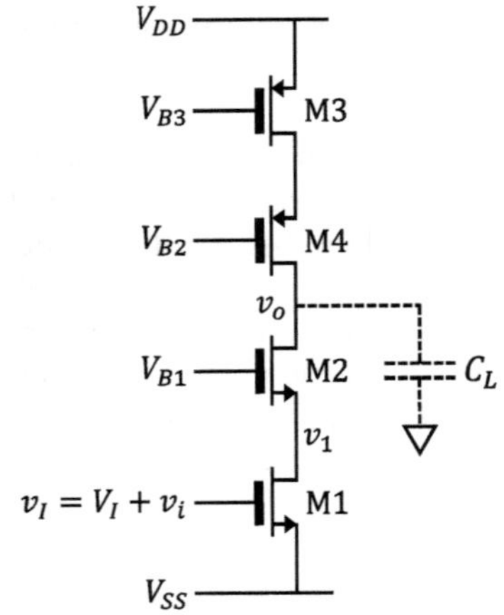

FIGURE 4.25 A cascode common-source amplifier with NMOS driver. C_L represents the load capacitance.

The dc gain of the cascode stage is described by $A_{v0} \doteq v_o/v_i = (v_o/v_1)(v_1/v_i)$. Analysis by inspection yields

$$\frac{v_o}{v_1} = \frac{r_{id4}}{r_{is2}},$$

and

$$\frac{v_1}{v_i} = -\frac{r_{is2}}{r_{is1}},$$

which lead to

$$A_{v0} = -\frac{r_{id4}}{r_{is1}}.$$

Since r_{is1} is described by

$$r_{is1} = \frac{r_{ds1} + r_{is2}}{g_{m1} r_{ds1}} = \frac{r_{ds1} + \frac{r_{ds2} + r_{id4}}{(g_{m2}+g_{mb2})r_{ds2}}}{g_{m1} r_{ds1}} = \frac{\overbrace{(g_{m2} + g_{mb2}) r_{ds2} r_{ds1} + r_{ds2}}^{r_{id2}} + r_{id4}}{\underbrace{(g_{m2} + g_{mb2}) r_{ds2} r_{ds1}}_{\approx r_{id2}} g_{m1}},$$

the gain expression turns into

$$A_{v0} = -(r_{id2} \parallel r_{id4}) g_{m1}, \tag{4.66}$$

in which, the following approximate forms are appropriate for r_{id4} and r_{id2}:

$$r_{id4} = (g_{m4} + g_{mb4})r_{ds4}r_{ds3}, \tag{4.67}$$

and

$$r_{id2} = (g_{m2} + g_{mb2})r_{ds2}r_{ds1}. \tag{4.68}$$

Equation (4.66) is similar to (3.56) of the basic common-source stage of Fig. 3.22(a) in that the gain equals the product of output resistance and driver transconductance but of course the output resistance of a cascoded stage is much higher than what a basic common-source stage offers.

The transfer function (3.114) derived for the basic common-source stage is also valid to a good approximation for a cascoded common-source stage provided that the output resistance is interpreted as $r \doteq r_{id2} \parallel r_{id4}$, and the effective load resistance is interpreted as $C_L' = C_L + C_{dd2} + C_{dd4}$. For identical sizing, biasing, and loading, the basic and cascoded versions of a common-source stage will exhibit the same right half-plane zero frequency. The pole frequency, however, will be lower in the cascoded version due to the higher output resistance.

4.4.2 Source Follower

Source followers are used for shifting the bias level of a signal or for transferring a signal from a high-resistance source to a low-resistance load without significant attenuation. These two functions are named *level shifting* and *buffering*, respectively.

Topology

As shown in Fig. 4.26 for the general case of a split-supply configuration, the source follower comprises a driver MOSFET biased by a current source or a current sink. An input signal v_i biased around a dc voltage V_I is fed to the gate terminal of the driver, and

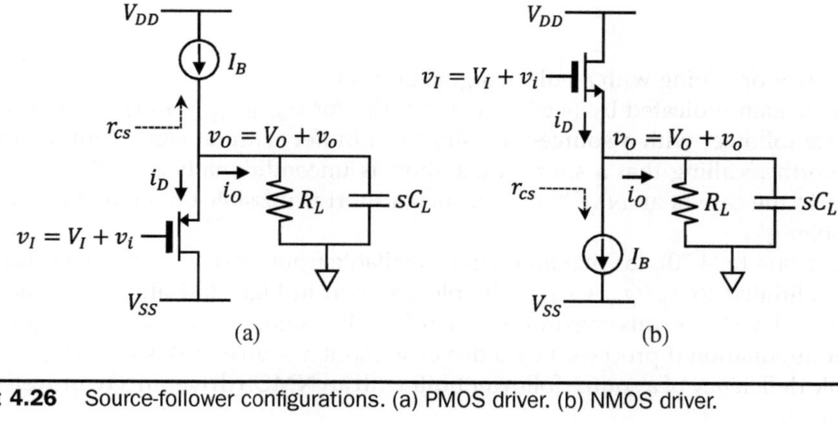

FIGURE 4.26 Source-follower configurations. (a) PMOS driver. (b) NMOS driver.

the corresponding output signal v_o is delivered to a capacitive and/or resistive load at the source terminal. The driver operates unconditionally in saturation due to its drain terminal being connected directly to V_{SS} in the case of PMOS implementation or to V_{DD} in the case of NMOS. It translates the input bias V_I to an output bias $V_O = V_I - V_{GS}$, where the dc gate-source voltage V_{GS} represents the amount of level shifting. Obviously, the direction of level shifting is positive in the case of a PMOS driver, and negative in the case of an NMOS driver.

The reader can visualize a single-supply source follower by replacing V_{SS} with ground, and connecting the lower terminal of the load resistance R_L to a bias voltage V_B in between V_{DD} and ground.

DC Gain

Since signal flow in a source follower is from gate to source, the associated voltage gain is less than unity as indicated by (3.64) or (3.66). By proper design, however, the gain can be set arbitrarily close to unity even when driving a load of low resistance. This property, combined with the infinitely high input resistance seen at the gate terminal of the driver device, enables a source follower to act as a buffer transferring the signal of a high-resistance source to a low-resistance load without much attenuation. The reader will recall our first mention of buffering in Section 2.8.

For modeling the dc gain of a source follower, we represent the small-signal resistance of the current source or sink with r_{cs}, and use (a) the exact form of the gate-to-source gain expression (3.64) or (3.66), (b) the source input-resistance expression (3.57), and (c) the exact form of the maximum-gain expression (3.58) or (3.60). The outcome is

$$A_{v0} \doteq \frac{v_o}{v_i} = \frac{(r_{ds} \parallel r_{cs} \parallel R_L)g_m}{1 + (r_{ds} \parallel r_{cs} \parallel R_L)g_m}, \tag{4.69}$$

for a driver operating with source-bulk short, and

$$A_{v0} \doteq \frac{v_o}{v_i} = \frac{(r_{ds} \parallel r_{cs} \parallel R_L)g_m}{1 + (r_{ds} \parallel r_{cs} \parallel R_L)(g_m + g_{mb})}, \tag{4.70}$$

for a driver operating with a bulk at signal ground.

The dc gain indicated by (4.69) tends to unity for $(r_{ds} \parallel r_{cs} \parallel R_L)g_m \gg 1$. Therefore, a source follower with a source-bulk short can buffer without significant attenuation. It is worth recalling that a source-bulk short is unconditionally available to a PMOS driver. In the case of an NMOS driver, such a short is possible only if a DNW device is available.

According to (4.70), the maximum gain available from a driver without a source-bulk short is limited to $g_m/(g_m + g_{mb})$. The plots shown in Figs. 3.12 and 3.20 typically indicate 0.75 V/V for this maximum. Therefore, the signal is subject to a significantly higher attenuation if processed by a driver without a source-bulk short. This is one inevitable deficiency of a source follower built with an NMOS driver in a twin-well CMOS technology.

Output Range

The driver device of a source follower remains unconditionally in saturation. Compliance limit is therefore a concern only for the current source or sink used for biasing the driver. It sets an upper limit V_{OH} on the output range in the case of a PMOS source follower and a lower limit V_{OL} on the output range in the case of an NMOS source follower. If the load is purely capacitive, the output voltage v_O can swing as far as this limit without any hindrance. In the case of a resistive load, however, one potential restriction must be eliminated by design in order to extend the output range that far. For an explanation of this restriction, consider without loss of generality the PMOS source follower of Fig. 4.26(a). As v_O rises toward the upper limit of the output range, the current i_D of the driver decreases due to the diversion of an increasingly larger fraction of I_B to the load resistance in accordance with $i_D = I_B - (v_O/R_L)$. Unless I_B is made larger than V_{OH}/R_L, where V_{OH} is the upper limit of the output range as defined by the compliance limit of the current source, i_D and therefore g_m will obviously vanish at an output voltage lesser than V_{OH}. Since the gain described by (4.69) or (4.70) collapses as a result, v_O will be top limited by this current-starving condition of the driver rather than V_{OH} of the current source. This limitation can be prevented only by making I_B larger than the output current i_O that is expected to flow in R_L for $v_O = V_{OH}$. Note that the larger the difference $I_B - i_O$, the lesser the gain deterioration at $v_O = V_{OH}$. We will further elaborate on this issue in Example 4.12. Although the foregoing discussion is based on a source follower with a PMOS driver, the conclusions are also valid for the lower limit V_{OL} of the output range in the case of an NMOS driver.

Although the driver device of a source follower is unconditionally saturated, this doesn't necessarily allow for an unrestricted output swing toward V_{SS} in the PMOS version or toward V_{DD} in the NMOS version. Even when the input voltage v_I swings all the way to the power-supply rail, v_O will remain one gate-source voltage away at $V_{SS} - V_{GS}$ in PMOS and at $V_{DD} - V_{GS}$ in NMOS. In a driver implemented with a standard-threshold device the restriction caused by the negative V_{GS} of the PMOS or by the positive V_{GS} of the NMOS can be sizeable even if a large aspect ratio is adopted for the driver. A low-threshold or zero-threshold device is definitely much more preferable in this respect.

Frequency Response

The frequency response of a source follower can be derived from the small-signal equivalent circuit shown in Fig. 4.27, where the bulk is assumed to be at signal ground. A

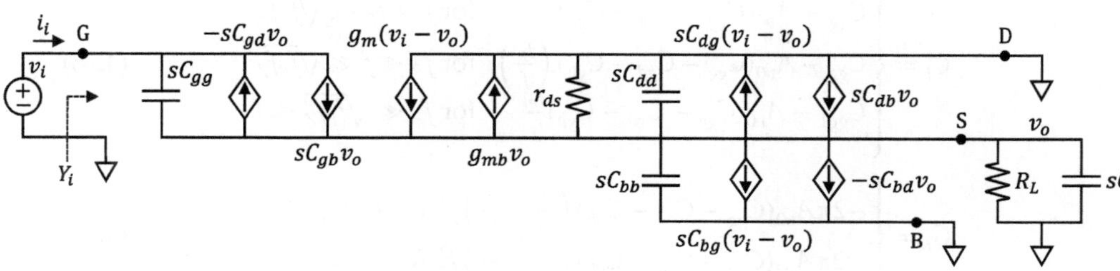

FIGURE 4.27 Small-signal equivalent circuit of a source follower whose driver operates with a bulk at signal ground.

straightforward analysis of this circuit yields

$$A_v \doteq \frac{v_o}{v_i} = A_{v0}\frac{1+\frac{s}{2\pi f_Z}}{1+\frac{s}{2\pi f_P}}, \tag{4.71}$$

where A_{v0} is given by (4.70), and zero and pole frequencies are described by

$$f_Z = \frac{g_m}{2\pi\left(C_{gg} - C_{dg} - C_{bg}\right)}, \tag{4.72}$$

and

$$f_P = \frac{g_m + g_{mb} + (1/R_L')}{2\pi\left(C_L + C_{gg} + C_{dd} + C_{bb} - C_{dg} - C_{bg} - C_{db} - C_{gb} - C_{bd} - C_{gd}\right)}, \tag{4.73}$$

respectively, where

$$R_L' \doteq r_{ds} \parallel r_{cs} \parallel R_L. \tag{4.74}$$

Note that f_P is lower than f_Z. The bandwidth defined by this pole is much wider than the bandwidth of a common-source stage of comparable driver dimensions, comparable bias current, and comparable capacitive load. This is due to the fact that the numerator of (4.73) is much larger than the equivalent conductance $1/r$ that determines the pole frequency of the common-source gain described by (3.114).

The equivalent circuit shown in Fig. 4.27 also yields

$$Y_i = s\left[C_{gg} - (C_{gg} - C_{gd} - C_{gb})A_v\right] \tag{4.75}$$

for the input admittance of a source follower whose driver bulk is at signal ground. Considering (4.71) for A_v, we can show that this admittance has a capacitive component C_i and a conductive component G_i. A piecewise analysis of (4.75) yields

$$C_i = \begin{cases} C_{gg} - A_{v0}(C_{gg} - C_{gd} - C_{gb}) & \text{for } f \ll f_P \ll \sqrt{f_P f_Z} \\ C_{gg} - A_{v0}(C_{gg} - C_{gd} - C_{gb})\left(\frac{f_P}{f}\right)^2 & \text{for } f_P \ll f \ll \sqrt{f_P f_Z} \\ C_{gg1} - A_{v0}(C_{gg} - C_{gd} - C_{gb})\frac{f_P}{f_Z} & \text{for } f_P \ll \sqrt{f_P f_Z} \ll f \end{cases} \tag{4.76}$$

$$G_i = \begin{cases} -2\pi A_{v0}(C_{gg} - C_{gd} - C_{gb})\left(\frac{1}{f_P} - \frac{1}{f_Z}\right)f^2 & \text{for } f \ll f_P \\ -2\pi A_{v0}(C_{gg} - C_{gd} - C_{gb})\left(\frac{1}{f_P} - \frac{1}{f_Z}\right)f_P^2 & \text{for } f_P \ll f \end{cases} \tag{4.77}$$

where f denotes the frequency.

As a final note, (4.71) to (4.77) are also valid for the case of source-bulk short provided that (a) (4.69) is used for A_{v0}, and (b) a zero is substituted for each of g_{mb}, C_{gb}, C_{db}, C_{bb}, C_{bg}, and C_{bd}.

Example 4.12 We can construct an opamp by buffering the output of an OTA. Between the two approaches explained in Section 2.8 for buffering, the one implemented with a source follower works by lowering the output resistance r_o below R_L.

In this example we design a buffer for converting an OTA into an opamp, and use the outcome in a unity-gain amplifier driving a load resistance $R_L = 1$ kΩ with respect to a bias level $V_B = 0.9$ V. Its closed-loop schematic is shown in Fig. 4.28. The single-supply OTA, marked `xota`, is buffered by a source follower comprising (a) a PMOS driver device M1 operating with a source-bulk short, and (b) low-voltage cascode current-source devices M2 and M3 generating a bias current I_B by mirroring from reference devices M4 and M5. Note that cascoding is used not for raising the gain of the source follower by providing a large r_{cs}. After all, even if a single-device current source were deployed, r_{cs} would already be far greater than the specified 1-kΩ load and therefore be ineffective on the gain. The purpose of cascoding is to desensitize I_B against v_O in order to prevent I_B and therefore power consumption from going up as the output voltage swings down to its minimum. The opamp thus constructed operates in a single-supply unity-gain amplifier configuration with a supply voltage $V_{DD} = 1.8$ V and an input signal $v_i = \pm 0.6$ V with respect to a bias voltage $V_I = 0.9$ V. Targeting a worst-case gain $A_{v0} = 0.5$ V/V for the source follower, we are expected to size all devices of the source follower, and determine the reference current. Verification will be performed with the OTA macro described in Example 2.32 of Subsection 2.11.4.

Work The gain of the source-follower stage is given generally by (4.69) but, since r_{ds} of M1 and r_{cs} of the current source will certainly turn out to be much larger than the load

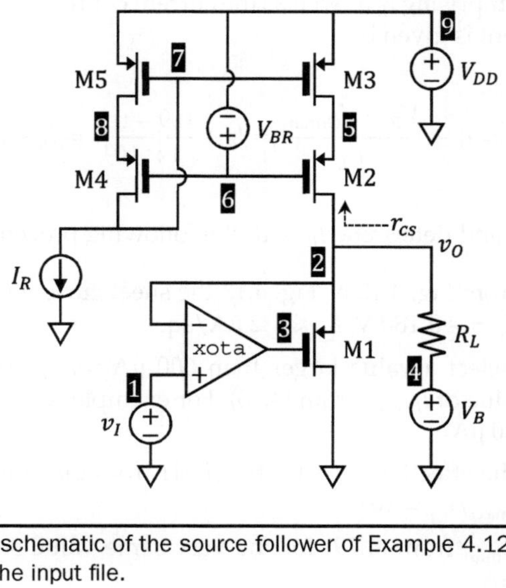

FIGURE 4.28 Electrical schematic of the source follower of Example 4.12. Node numbers correspond to those of the input file.

resistance specified as $R_L = 1\ \text{k}\Omega$, we can simplify that equation into

$$A_{v0} \approx \frac{g_{m1} R_L}{1 + g_{m1} R_L}. \tag{4.78}$$

For the worst-case gain specified as 0.5 V/V, the minimum acceptable transconductance is calculated from (4.78) as $g_{m1(\text{min})} = 1$ mA/V. M1 is to be designed to exhibit this much transconductance when v_O swings up to the upper limit V_{OH} of the output range. This is because the current $i_D = I_B - i_O$ hence g_{m1} are minimized at this limit due to the maximization of i_O. Considering the specified unity gain of the application alongside $V_I = 0.9$ V and $v_i = 0.6$ V, we determine the upper limit as $V_{OH} = 0.9 + 0.6 = 1.5$ V. The minimized current conducted by M1 at this limit is therefore described by

$$i_{D(\text{min})} = I_B - \frac{V_{OH} - V_B}{R_L} = I_B - \frac{1.5 - 0.9}{1 \times 10^3} = I_B - 6 \times 10^{-4}. \tag{4.79}$$

It is obvious from this equation that I_B has to be higher than 600 μA in order to prevent current starvation in M1.

Let us now consider the lower limit $V_{O(\text{min})}$ of output swing. For $V_I = 0.9$ V and $v_i = -0.6$ V, this limit is calculated as $V_{O(\text{min})} = 0.9 - 0.6 = 0.3$ V. Assuming that the OTA can lower its own output almost to ground, the margin left for accommodating the gate-source voltage of M1 is therefore a mere −0.3 V at this extreme, which precludes the standard-threshold option for this device. Fortunately, the low-threshold PMOS option can comply with this restriction as attested by Fig. 4.10. Since r_{ds} is not critical, we select the minimum permissible channel length $L_{DR} = 0.2$ μm, and read $V_{TP} = -0.12$ V for $V_{SB} = 0$. The overdrive voltage for the case of $v_O = V_{O(\text{min})} = 0.3$ V will therefore be $V_{GS} - V_{TP} = -0.3 - (-0.12) = -0.18$ V. With this much overdrive voltage, M1 will conduct its maximum current comprising I_B as well as the current contributed by the load resistance. This maximum current is given by

$$i_{D(\text{max})} = I_B + \frac{V_B - V_{O(\text{min})}}{R_L} = I_B + \frac{0.9 - 0.3}{1 \times 10^3} = I_B + 6 \times 10^{-4}. \tag{4.80}$$

We can now size M1 and determine I_B with the following procedure:

1. Determine from Fig. 4.12 or Fig. 4.13 the sheet current corresponding to $i_{D(\text{max})}$ for $V_{GS} - V_{TP} = -0.180$ V: $I_\square \approx 1.2$ μA/sq.
2. Arbitrarily select a value larger than 600 μA for I_B, and calculate the corresponding value of $i_{D(\text{max})}$ from (4.80). For example, selecting $I_B = 620$ μA yields $i_{D(\text{max})} = 1220$ μA.
3. Determine the effective aspect ratio of M1 from the outcomes of steps 1 and 2: $W_1/L_1 = I_{D(\text{max})}/I_\square = 1017$.
4. Calculate $i_{D(\text{min})}$ from (4.79) for the very same value selected for I_B in step 2: $i_{D(\text{min})} = 20$ μA.

5. Calculate the sheet current of $i_{D(\min)}$ from the aspect ratio found in step 3 and $i_{D(\min)}$ found in step 4: $I_{\square} = 20 \times 10^{-6}/1017 = 1.97 \times 10^{-8}$ A/sq.
6. Referring to Fig. 4.13, determine the overdrive voltage corresponding to the sheet current calculated in step 5: $V_{GS} - V_{TP} \approx 0.060$ V.
7. Referring to Fig. 4.14, determine the transconductance efficiency corresponding to the overdrive voltage calculated in step 6: $g_{m1(\min)}/i_{D(\min)} \approx 22.9\ \text{V}^{-1}$.
8. Determine $g_{m1(\min)}$ from the outcomes of steps 4 and 7. This step yields $g_{m1(\min)} =$ 458 μA/V, which is significantly lesser than the target 1 mA/V obtained from (4.78). Therefore, start all over again from step 2 with a higher value for I_B, and iterate until an adequate match is found for $g_{m1(\min)}$.

Iteration with this procedure leads to $I_B = 650$ μA, for which we expect $i_{D(\max)} = 1250$ μA, $W_1/L_1 = 1042$, $i_{D(\min)} = 50$ μA, and $g_{m(\min)} = 1090$ μA/V. Since the effective channel length corresponding to $L_{DR} = 0.2$ μm is $L = 0.141$ μm, the drawn width of M1 turns out to be $W_{DR} = 147$ μm.

Next, we size the devices of the current source. We assume source-bulk short, and break down the compliance limit −0.3 V into (a) $V_{DSsat} = -0.1$ V for each of M2 and M3, and (b) a saturation margin $\Delta V_{DS3} = -0.1$ V for M3. Figure 3.17 yields $V_{GS} - V_{TP} = -0.1$ V, which, in turn, indicates $I_{\square} = 350$ nA/sq. Since these devices are expected to conduct 650 μA, their aspect ratio should therefore be $W_{2,3}/L_{2,3} = 1860$. We decide to build each of these devices with 10 unit devices of aspect ratio 186 each, so that their reference counterparts can be constructed with one unit device conducting $I_B/10 = 65$ μA. Considering the insignificance of their output resistance, we select a drawn channel length $L_{DR} = 0.4$ μm close to the minimum. The corresponding drawn channel width of a unit device is calculated as $W_{DR} = 64$ μm. Finally, we estimate $V_{BR} - V_{DD} = -0.74$ V from the sum of V_{DSsat}, ΔV_{DS3}, $V_{GS} - V_{TP}$, and V_{TP}.

Design can be verified with the following input file `ex4-12.sp`.

```
filename: ex4-12.sp
.incl bsim.sp
.incl bsimlvt.sp
.param vdir=5 gm0=1m ro=1meg
*********************ota************************
*ninv: noninverting input
*inv: inverting input
*out: output
.subckt ota ninv inv out
gm 0 out cur='gm0*vdir*tanh((v(ninv)-v(inv))/vdir)'
rout out 0 ro
bb 0 out i=0.9/ro
.ends ota
*******************buffer**********************
xota 1 2 3 ota
m1 0 3 2 2 cmosplvt w=147u l=0.2u
m2 2 6 5 5 cmosp w=64u l=0.4u m=10
m3 5 7 9 9 cmosp w=64u l=0.4u m=10
m4 7 6 8 8 cmosp w=64u l=0.4u
m5 8 7 9 9 cmosp w=64u l=0.4u
rl 2 4 1k
```

```
vb 4 0 0.9
ir 7 0 65u
vbr 6 9 -0.74
vdd 9 0 1.8
vi 1 0 dc
.save v(2) v(3) @m1[gm] @m1[id] @m2[id]
.control
dc vi 0.3 1.5 1m
plot v(2)
plot deriv(v(2))/deriv(v(3))
plot @m1[id] @m2[id]
plot @m1[gm]
.endc
.end
```

Verification is based on dc-sweeping the input voltage of the unity-gain configuration, and plotting the corresponding output voltage v_O as `v(2)`, source follower gain A_{v0} as `deriv(v(2))/deriv(v(3))`, driver current I_D as `@m1[id]`, bias current I_B as `@m2[id]`, and driver transconductance g_m as `@m1[gm]`.

This example provides us with an opportunity to determine the parameters of the effective transconductance of an opamp buffered by a source follower. To this end, note that the overall dc differential gain of the opamp equals the product of the OTA gain $A_{0(\mathrm{ota})}$ and source-follower gain A_{v0}; that is, $A_0 = A_{0(\mathrm{ota})}A_{v0}$. Substituting into this equation the right-hand sides of (2.96) and (4.78) for A_0 and A_{v0}, respectively, and considering $r_o = 1/g_{m1}$ for the output resistance of the opamp as defined by M1, we arrive at $G_m = A_{0(\mathrm{ota})}g_{m1}$ for the effective transconductance. ▲

4.4.3 Basic Differential Pair

Input stages of opamps and OTAs are built around a *differential pair*, which acts as a differential transconductor by putting out two complementary currents controlled by a differential-input voltage. Differential pairs also find use in stand-alone voltage-to-current conversion applications without being embedded into an opamp or OTA. In this subsection, we study this indispensable building block in its basic form.

Topology and Operation

The NMOS and PMOS variants of the basic differential-pair topology are depicted in Figs. 4.29(a) and (b), respectively. Notice the presence of a pair of matching driver devices M1 and M2, whose source terminals are interconnected and tied to a current sink or current source for receiving a common-bias current I_B. Drains are terminated inside an external load block whose topological details depend on the overall application as will be exemplified in Section 4.5 and thereafter. Presently, it suffices to note that the drain voltages v_{D1} and v_{D2} imposed by this block ensure saturation for the drivers in proper operation. Each of the two gates constitutes an input terminal, whose voltage with respect to ground is denoted by v_{I1} or v_{I2}. In an opamp or an OTA, these two voltages correspond to the single-ended input voltages v_{IP} and v_{IN} as defined in Subsection 2.2.1 but which of the two is inverting or noninverting cannot be ascertained unless the complete circuit schematic of the opamp or OTA is available.

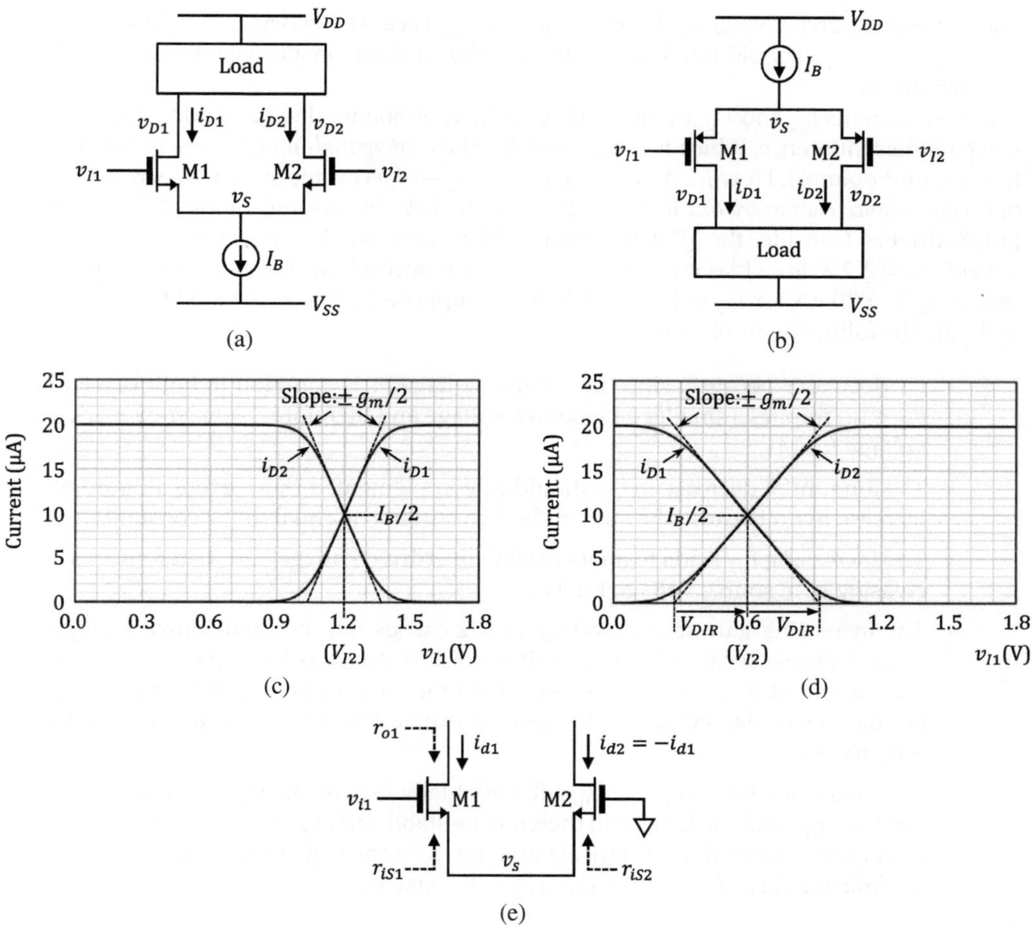

FIGURE 4.29 Basic differential pair. (a) NMOS version. (b) PMOS version. (c) Transconductance characteristics of an NMOS differential pair of $I_B = 20$ μA and $V_{I2} = 1.2$ V. (d) Transconductance characteristics of a PMOS differential pair of $I_B = 20$ μA and $V_{I2} = 0.6$ V. (e) Small-signal equivalent circuit for determining the effective transconductance G_m.

For explaining the properties of an *ideal* differential pair, we assume infinitely large output resistances for the bias-current source/sink and driver devices. This idealization has two implications: (a) I_B becomes perfectly independent of the common-source voltage v_S, and (b) i_{D1} and i_{D2} become perfectly independent of v_{D1} and v_{D2}. Now, suppose that the input voltages v_{I1} and v_{I2} are equalized and held at a common-mode voltage v_{CM}. In this case, we expect $i_{D1} = i_{D2} = I_B/2$ because the drivers have (a) identical aspect ratios due to matching, and (b) identical overdrive voltages due to their interconnected source terminals and equalized gate voltages. Note that the driver currents remain identical and equal to $I_B/2$ regardless of how large or small the common-mode voltage is. Ideally, therefore, these currents are not sensitive to the common-mode input.

The common-source voltage v_S, however, tracks v_{CM} because the driver overdrive voltage $v_{CM} - v_S - V_{TN}$ should remain constant in order to maintain a constant current $I_B/2$ for each driver.

Driver currents i_{D1} and i_{D2} are insensitive to the common mode of v_{I1} and v_{I2} but sensitive to their difference, which is what we defined as *differential-input voltage* in Subsection 2.2.1 for opamps. How i_{D1} and i_{D2} vary with $v_{I1} - v_{I2}$ is exemplified with simulated *transconductance characteristics* in Fig. 4.29(c) for NMOS drivers and in Fig. 4.29(d) for PMOS drivers. Consider the NMOS version, which is extracted by sweeping v_{I1} around a fixed $V_{I2} = 1.2$ V for a bias current $I_B = 20$ µA, and notice how the currents are equalized at $I_B/2 = 10$ µA for $v_{I1} = V_{I2} = 1.2$ V. Next, suppose that v_{I1} is lowered below V_{I2}, and note the following implications:

- i_{D1} decreases because, since the source voltage v_S is common to both drivers, $v_{I1} < V_{I2}$ implies a smaller gate-source voltage and therefore a smaller overdrive voltage for M1.
- The amount of decrease in i_{D1} should be complemented by an equal amount of increase in i_{D2} because the sum of the two currents is equal to the constant I_B.
- The increasing i_{D2} leads to an increasing overdrive voltage and therefore an increasing gate-source voltage for M2.
- The increasing gate-source voltage of M2 causes the common-source voltage v_S to decrease because the gate voltage of this driver is kept fixed. However, the amount of decrease in v_S is lesser than the amount by which v_{I1} is lowered because, as explained above, the gate-source voltage of M1 has to decrease in the process.
- If we continue lowering v_{I1}, i_{D1} will eventually become negligible in comparison with i_{D2}, and the latter will therefore be stabilized at I_B. Lowering v_{I1} further won't change any of the electrical variables. As a consequence, the transconductor functionality of the driver devices will collapse.

Raising v_{I1} above V_{I2} transposes the roles of M1 and M2. It causes the gate-source voltage of M1 and therefore i_{D1} to increase. The gate-source voltage of M2 and i_{D2} decrease in response. v_S also increases but not as much as the amount by which v_{I1} is raised because the gate-source voltage of M1 has to increase. If we continue raising v_{I1}, eventually, i_{D2} will become negligible in comparison with i_{D1}, and the latter will be equalized to I_B. Since electrical variables will become insensitive to the differential-input voltage beyond that point, the drivers will cease to function as transconductors.

As observed in Fig. 4.29(d), the variation of i_{D1} and i_{D2} with differential voltage in the PMOS version of the differential pair is qualitatively similar to the NMOS counterpart with the exception that the direction of variation is opposite to what is observed in Fig. 4.29(c). This contrast is due to the fact that, although v_S increases with v_{I1} as also observed in the NMOS version, here it leads to a less negative overdrive for M1 and a more negative overdrive for M2.

The foregoing discussion of differential-pair operation is based on the assumption of infinitely high output resistance for the bias-current source/sink. The fact that the output impedance of a real source/sink is finite makes I_B a function of v_S. Since v_S tracks v_{CM} as explained above, I_B and therefore i_{D1} and i_{D2} are actually sensitive to the common-mode voltage. For dc and low-frequency common-mode signals, this sensitivity can be reduced to any desired level by raising the output resistance of the source/sink sufficiently above the small-signal resistances r_{is1} and r_{is2} seen into the source terminals of driver devices. At high frequencies, however, the high output resistance of the current source/sink is shunted by the decreasing impedance of the parasitic capacitance associated with the common-source node. This leads to a higher common-mode sensitivity for the currents.

Transconductance and Output Resistance

The sensitivities of i_{D1} and i_{D2} to the differential-input voltage $v_{I1} - v_{I2}$ are represented by the slopes of the transconductance characteristics belonging to M1 and M2, respectively. As observed in Figs. 4.29(c) and (d), the magnitudes of these slopes are equalized and maximized at the operation point where $v_{I1} = v_{I2}$. The magnitude of this maximized slope is the effective transconductance of the differential pair. As usual, we denote it with G_m, and relate it to device parameters by performing a small-signal analysis for i_{d1}/v_{i1} assuming $v_{i2} = 0$, or for i_{d2}/v_{i2} assuming $v_{i1} = 0$. The small-signal equivalent circuit appropriate for the former choice is shown in Fig. 4.29(e) for the NMOS version. We assume an ideal bias-current source, and therefore replace it with an open circuit. We also continue assuming infinite output resistances r_{ds1} and r_{ds2} for the drivers. Observing

$$i_{d1} = -i_{d2} = \frac{v_s}{r_{is2}}$$

from the equivalent circuit, where r_{is2} is the small-signal resistance seen into the source terminal of M2, and writing

$$v_s = \frac{g_m}{g_m + g_{mb}} \frac{r_{is2}}{r_{is2} + r_{is1}} v_{i1} \tag{4.81}$$

from (3.64), where r_{is1} is the small-signal resistance seen into the source terminal of M1, we arrive at

$$G_m \doteq \frac{i_{d1}}{v_{i1}} = \frac{g_m}{g_m + g_{mb}} \frac{1}{r_{is2} + r_{is1}}$$

for the effective transconductance. Since (3.57) and (3.58) yield $r_{is2} = r_{is1} = 1/(g_m + g_{mb})$ for $r_{ds} \to \infty$, this equation turns into

$$G_m = \frac{g_m}{2}. \tag{4.82}$$

One interesting aspect of (4.82) is the absence of g_{mb}. Indeed, this equation is valid irrespective of whether or not the drivers operate with a source-bulk short.

Yet another differential-pair performance metric is the small-signal output resistance seen into the drain terminal of a driver. Denoted by r_{o1} for the drain of M1 in the small-signal equivalent circuit of Fig. 4.29(e), this resistance is related to device parameters by

$$r_{o1} = 2r_{ds1} + \frac{r_{ds1}}{r_{ds2}} R_{D2}, \tag{4.83}$$

where R_{D2} is the equivalent external resistance between the drain of M2 and ground, and r_{ds1} and r_{ds2} represent the output resistances of M1 and M2, respectively. Since r_{o1} is defined for a common-mode operation point, M1 and M2 are assumed to have identical transconductances and identical bulk transconductances in the derivation of (4.83). No such identity is assumed for output resistances because the drain-source voltages of the two drivers may be disparate. Equation (4.83) can also be used for determining the resistance r_{o2} seen into the drain terminal of M2 provided that device indexes 1 and 2 appearing in the subscripts are transposed.

Although the foregoing small-signal derivation is based on an NMOS differential pair, the outcome represented by (4.82) and (4.83) is also perfectly valid for a PMOS differential pair.

Differential-Input Range

It is apparent from Figs. 4.29(c) and (d) that the slope of the nonlinear transconductance characteristics gradually decreases from its maximum value $\pm g_m/2$ as the differential-input voltage moves away from the common-mode operation point $v_{I1} - v_{I2} = 0$. By drawing a line tangent to the characteristic at the operation point, and extrapolating it to the voltage axis, we can define a differential-input range within which the transconductance functionality of the differential pair is retained to a reasonable extent. The range thus defined and displayed in Fig. 4.29(d) corresponds to what we introduced in Subsection 2.5.2 as V_{DIR}. Considering the slope $g_m/2$ of the extrapolated line and the current $I_B/2$ flowing at the operation point, we can describe the differential-input range of a basic differential pair with

$$V_{DIR} = \frac{I_B/2}{g_m/2} \tag{4.84}$$

but since $g_m/(I_B/2)$ is the transconductance efficiency of a driver device around the operation point, this equation can also be expressed as

$$V_{DIR} = \frac{2}{\text{Driver transconductance efficiency}}. \tag{4.85}$$

Since transconductance efficiency depends almost solely on the overdrive voltage, V_{DIR} of a basic differential pair can be manipulated only by varying the overdrive voltage of driver devices at the operation point $v_{I1} = v_{I2}$. For NMOS drivers, the asymptotic values of V_{DIR} in strong inversion and in deep subthreshold are obtained from (3.28) and (3.29) as

$$V_{DIR} = V_{GS} - V_{TN} \quad \text{(in strong inversion)} \tag{4.86}$$

and

$$V_{DIR} = 2A_N\frac{kT}{q} \quad \text{(in deep subthreshold).} \tag{4.87}$$

For PMOS drivers, $|V_{GS} - V_{TP}|$ replaces $V_{GS} - V_{TN}$ in (4.86), and A_P replaces A_N in (4.87). It is worth reemphasizing that the value of the gate-source voltage to be used in (4.86) is the one associated with the common-mode operation point.

Input Admittance Network

As stated at the beginning of this subsection, OTAs and opamps invariably contain a differential pair in their input stages. The admittance seen into the input terminals of such a differential pair may significantly affect the feedback and input factors of the closed-loop application unless it is much lower than the admittances associated with the external network. For modeling this admittance, we consider the equivalent circuit shown in Fig. 4.30(a), where a capacitance C_S is included to represent the capacitive loading of the bias-current source/sink. Also note that we continue assuming negligible load resistances and an infinite source/sink resistance. v_{i1} and v_{i2} are two independent voltage sources added for driving the single-ended inputs. The small-signal currents flowing into these input terminals are represented with i_1 and i_2. The corresponding high-frequency small-signal equivalent circuit is shown in Fig. 4.30(b). Its analysis yields

$$i_1 = Y_d(v_{i1} - v_{i2}) + Y_{se}v_{i1} \tag{4.88}$$

$$i_2 = -Y_d(v_{i1} - v_{i2}) + Y_{se}v_{i2}, \tag{4.89}$$

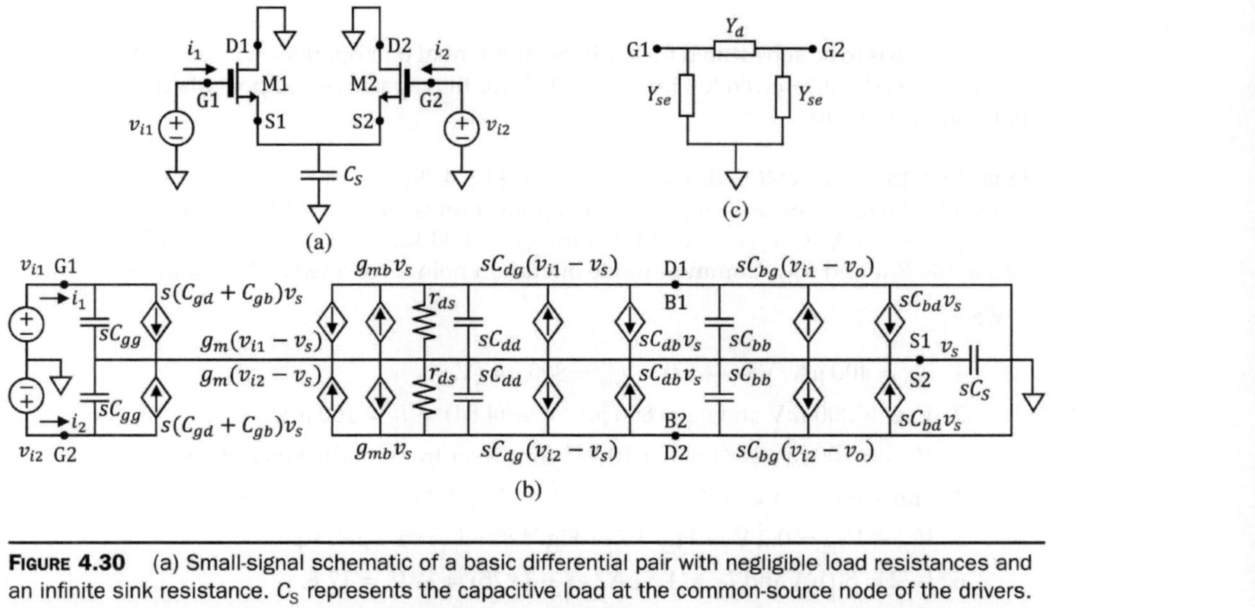

FIGURE 4.30 (a) Small-signal schematic of a basic differential pair with negligible load resistances and an infinite sink resistance. C_S represents the capacitive load at the common-source node of the drivers. (b) High-frequency equivalent circuit of the schematic presented in part a. (c) The admittance network exhibited by the differential pair. Refer to (4.90) or (4.92) for the differential admittance Y_d, and (4.91) or (4.93) for the single-ended admittance Y_{se}.

where

$$Y_d \doteq \frac{\left(1-\frac{C_{gd}+C_{gb}}{C_{gg}}\right)\left[g_m+s\left(C_{gg}-C_{dg}-C_{bg}\right)\right]}{g_m+g_{mb}+s\left(C_{gg}-C_{dg}-C_{bg}+C_{dd}+C_{bb}-C_{gd}-C_{gb}-C_{db}+C_S\right)}\frac{sC_{gg}}{2}, \tag{4.90}$$

$$Y_{se} \doteq \frac{\frac{C_{gd}+C_{gb}}{C_{gg}}g_m+g_{mb}+s\left[C_{dd}+C_{bb}-C_{db}-\frac{C_{gd}+C_{gb}}{C_{gg}}\left(C_{dg}+C_{bg}\right)+C_S\right]}{g_m+g_{mb}+s\left(C_{gg}-C_{dg}-C_{bg}+C_{dd}+C_{bb}-C_{gd}-C_{gb}-C_{db}+C_S\right)}sC_{gg}. \tag{4.91}$$

The admittance network corresponding to (4.88) and (4.89) is shown in Fig. 4.30(c). According to (4.90) and (4.91), the low-frequency values of the differential admittance Y_d and the single-ended admittance Y_{se} are purely capacitive as described by

$$Y_d = \left(1-\frac{C_{gd}+C_{gb}}{C_{gg}}\right)\frac{g_m}{g_m+g_{mb}}\frac{sC_{gg}}{2}, \tag{4.92}$$

$$Y_{se} = \left(\frac{C_{gd}+C_{gb}}{C_{gg}}\frac{g_m}{g_m+g_{mb}}+\frac{g_{mb}}{g_m+g_{mb}}\right)sC_{gg}, \tag{4.93}$$

where a zero is to be substituted for g_{mb} if the differential pair operates with a source-bulk short. The reader is referred to (3.105) to (3.107) for the capacitive components appearing in these expressions.

Example 4.13 The NMOS differential pair of Fig. 4.29(a) is to be designed with low-threshold devices. This is a single-supply application with $V_{DD} = 1.8$ V. Specified items are $G_m = 400$ μA/V, $V_{DIR} = 200$ mV, and $r_{o1} = 600$ kΩ for $V_{D1} = V_{D2} = 1.5$ V and negligible R_{D1} and R_{D2}. Common-mode operation point is defined as $V_{CM} = 0.9$ V.

Work

1. $G_m = 400$ μA/V → (4.82) ⇒ $g_m = 800$ μA/V.
2. $V_{DIR} = 200$ mV and $g_m = 800$ μA/V → (4.84) ⇒ $I_B = 160$ μA.
3. $V_{DIR} = 200$ mV → (4.85) ⇒ 10 V^{-1} for driver transconductance efficiency.
4. Transconductance efficiency 10 V^{-1} → Fig. 4.9 ⇒ $V_{GS} - V_{TN} = 0.2$ V.
5. $V_{GS} - V_{TN} = 0.2$ V → Fig. 4.7 or Fig. 4.8 ⇒ $I_\square = 4.5$ μA/sq.
6. $I_B/2 = 80$ μA and $I_\square = 4.5$ μA/sq → (3.25) ⇒ $W/L = 17.8$.
7. Determining the drawn channel length of driver devices with iterative simulation: For the identical V_{D1} and V_{D2} specified for r_{o1}, we expect $r_{ds1} = r_{ds2}$. Since $R_{D2} \approx 0$, (4.83) turns into $r_{o1} = 2r_{ds}$, where r_{ds} denotes the output resistance of

each driver device. The specified $r_{o1} = 600\ \text{k}\Omega$ therefore implies $r_{ds} = 300\ \text{k}\Omega$. Considering $I_B/2 = 80\ \mu\text{A}$, the target value of the Early voltage of a driver is calculated as $V_A = 24$ V. Iterative simulation is performed on a single device through the following steps:

a. Arbitrarily select a drawn channel width and a drawn channel length. For example, $W_{DR} = 10\ \mu\text{m}$ and $L_{DR} = 0.8\ \mu\text{m}$.
b. Since simulation is to be conducted at the operation point, the gate voltage of the device equals the specified common-mode voltage: $V_G = V_{CM} = 0.9$ V.
c. Since the overdrive voltage $V_G - V_S - V_{TN}$ at the operation point has already been determined as 0.2 V, and V_G is to be 0.9 V, the sum $V_S + V_{TN}$ is constrained by

$$V_S + V_{TN} = 0.9 - 0.2 = 0.7\,\text{V}.$$

While complying with this constraint, determine V_S from Fig. 4.5 for the value of L_{DR} selected above: $V_S \approx 0.45$ V. This step will also yield $V_{TN} \approx 0.25$ V but we won't need this outcome anywhere in the simulation. Incidentally, if reading $V_S + V_{TN}$ by summing coordinates on the plots of Fig. 4.5 proves challenging, refer directly to the plots of $V_S + V_{TN}$ versus V_S as obtained by running the following input file `nmoslvt3.sp`.

```
filename: nmoslvt3.sp (low-threshold nmos vth+vsb)
.incl bsimlvt.sp
m 2 1 3 0 cmosnlvt w=6u l=0.2u
vds 2 3 0.9
vgs 1 3 0.9
vsb 3 0 0
.save @m[vth] v(3)
.control
destroy all
foreach length 0.2u 0.4u 0.6u 0.8u 1.2u 2u 4u 8u
alter @m[l]=$length
dc vsb 0 1.2 20m
end
plot dc1.@m[vth]+v(3) dc2.@m[vth]+v(3)
+ dc3.@m[vth]+v(3) dc4.@m[vth]+v(3) dc5.@m[vth]+v(3)
+ dc6.@m[vth]+v(3) dc7.@m[vth]+v(3) dc8.@m[vth]+v(3)
.endc
.end
```

d. Run the following input file `ex4-13a.sp` for performing an operation-point analysis on a device biased by $V_G = 0.9$ V, $V_D = 1.5$ V, and $V_S = 0.45$ V.

```
filename: ex4-13a.sp
.incl bsimlvt.sp
m 2 1 3 0 cmosnlvt w=10u l=0.8u
vd 2 0 1.5
vg 1 0 0.9
vs 3 0 0.45
.control
```

```
op
print -i(vd)/@m[gds]
.endc
.end
```

Simulation directly prints out an Early voltage $V_A = 17.57$ V, which falls short of the target.

e. Reset L_{DR} to a larger value without changing W_{DR}, and repeat step 7 until the target value of V_A has been reached. This leads to $L_{DR} = 1.5$ µm, which meets the target with $V_A = 25.37$ V.

8. Since $L_{DR} = 1.5$ µm is long enough to ignore the process bias, we assume $L = 1.5$ µm, and calculate $W = 26.7$ µm from $W/L = 17.78$. Since this is also large enough to ignore the process bias, W_{DR} is set to this value.

Verification can be performed with the following input file `ex4-13b.sp`:

```
filename: ex4-13b.sp
.incl bsimlvt.sp
m1 4 1 3 0 cmosnlvt w=26.7u l=1.5u
m2 5 2 3 0 cmosnlvt w=26.7u l=1.5u
vd1 4 0 1.5
vd2 5 0 1.5
vi1 1 0 0.9
vi2 2 0 0.9
ib 3 0 160u
.tf v(4) vd1
*.dc vi1 0 1.5 1m
.end
```

The transfer-function analysis yields $r_{o1} = 619.2$ kΩ in response to the command

```
print all
```

The dc-sweep analysis is run for verifying G_m and V_{DIR}. The former of these two metrics is obtained as $G_m = 409$ µA/V by plotting the derivative of i_{D1} with respect to v_{I1} with the command

```
plot -deriv(i(vd1))
```

and reading the outcome for $V_{I1} = 0.9$ V. Additionally, entering the following command will enable the reader to verify V_{DIR} by plotting (a) the transconductance characteristics i_{D1} and i_{D2} versus v_{I1}, and (b) the line tangent to the transconductance characteristic of i_{D1} at the common-mode operation point:

```
plot -i(vd1) (-i(vd2)) 409e-6*(v(1)-0.9)+80e-6 ylimit 0 160u
```

▲

Example 4.14 Extract with simulation the differential and single-ended input capacitances C_d and C_{se} of the differential pair designed in Example 4.13, and compare the outcome with the predictions of (4.92) and (4.93).

Work For extraction with simulation, we modify the input file `ex4-13b.sp` of Example 4.13 by defining an ac component for v_{I1} and replacing the lines of its transfer-function and dc-sweep analyses with an ac-analysis line. The resulting input file `ex4-14.sp` is as follows.

```
filename: ex4-14.sp
.incl bsimlvt.sp
m1 4 1 3 0 cmosnlvt w=26.7u l=1.5u as=14.7p ps=54.5u
+  nrs=0.0075 nrd=0.0075
m2 5 2 3 0 cmosnlvt w=26.7u l=1.5u as=14.7p ps=54.5u
+  nrs=0.0075 nrd=0.0075
vd1 4 0 1.5
vd2 5 0 1.5
vi1 1 0 dc 0.9 ac 1
vi2 2 0 0.9
ib 3 0 160u
.ac dec 100 10k 1g
.end
```

Notice the inclusion of (a) `as` and `ps` for properly representing the contribution of the source-bulk junctions to C_S, and (b) `nrs` and `nrd` for properly representing the parasitic resistances of the source and drain junctions as it is prudent to do so in any high-frequency ac simulation. Also note the absence of `ad` and `pd`, which is justified by the fact that drain-bulk junction capacitances are ineffective due to the signal ground at drain terminals.

Since this file also keeps the gate terminal G2 of M2 at signal ground, the admittance i_1/v_{i1} seen by the signal source v_{i1} equals $Y_d + Y_{se}$ as the reader can ascertain from Figs. 4.30(a) and (c). Dividing the imaginary part of this admittance by $2\pi f$ yields the sum $C_d + C_{se}$ of the corresponding capacitances. Y_d itself is given by the transadmittance $-i_2/v_{i1}$, where $-i_2$ is the ac current flowing into the dc voltage source V_{I2}. Dividing its imaginary part by $2\pi f$ yields C_d. Therefore, the plots of $C_d + C_{se}$ and C_d are generated as functions of frequency by entering

```
plot imag(-i(vi1)/v(1))/(2*pi*frequency)
plot imag(i(vi2)/v(1))/(2*pi*frequency)
```

The plots thus obtained are displayed in Fig. 4.31. For frequencies lower than about 100 MHz they indicate $C_d + C_{se} = 202$ fF and $C_d = 110.8$ fF and therefore $C_{se} = 91.2$ fF.

For determining the predictions of (4.92) and (4.93), we first extract device parameters with a `show` command. What are immediately available are $g_m = 822.8\ \mu\text{A/V}$, $g_{mb} = 192.5\ \mu\text{A/V}$, $C_{ggi} = 266.9$ fF, and $C_{gdi} = 0.26$ fF. C_{gbi} is unavailable but considering (3.76), it can be expressed as

$$C_{gbi} = -C_{ggi} - C_{gdi} - C_{gsi},$$

where $C_{gsi} = -251.4$ fF is also available. Therefore,

$$C_{gbi} = -266.9 - 0.26 + 251.4 = -15.7 \text{ fF}.$$

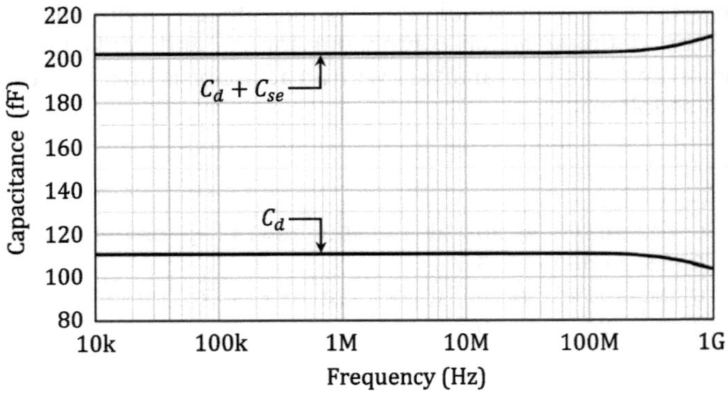

FIGURE 4.31 Simulated plots of the differential-input capacitance C_d and the sum $C_d + C_{se}$ of differential and single-ended input capacitances as obtained in Example 4.14.

Next, we calculate the extrinsic capacitances $C_{gbe} = 0.001$ fF, $C_{gde} = 22.9$ fF, and $C_{gge} = 45.8$ fF from (3.91), (3.92), and (3.99), respectively, and combine the intrinsic and extrinsic components in accordance with (3.105), (3.106), and (3.107) to arrive at $C_{gg} = 312.7$ fF, $C_{gd} = 22.6$ fF, and $C_{gb} = 15.7$ fF. Finally, we substitute into (4.92) and (4.93) these numbers together with the simulated values of g_m and g_{mb}, and obtain $C_d = 111.2$ fF and $C_{se} = 90.3$ fF. ▲

4.4.4 Source-Degenerated Differential Pair

As observed in Figs. 4.29(c) and (d), the transconductance characteristics of a basic differential pair are considerably nonlinear. In a closed-loop opamp or OTA application, the effect of this nonlinearity is greatly suppressed by the negative feedback loop. In open-loop OTA applications, however, we need to rely on an alternative technique if nonlinearity cannot be tolerated. *Source degeneration* is one such technique, which is also occasionally applied to closed-loop applications. As shown in Figs. 4.32(a) and (b) in two alternative topologies with NMOS drivers, it is implemented by resistively interconnecting the source terminals of the two driver devices. By performing a small-signal analysis on any of these two topologies with the assumption that the drivers operate with a bulk at signal ground and exhibit very large output resistances, we obtain

$$G_m = \frac{g_m}{2 + (g_m + g_{mb})R}$$

for the effective transconductance. Considering (3.32) and its PMOS counterpart, we can rewrite this expression as

$$G_m = \frac{g_m}{2 + A_{NP} g_m R}, \tag{4.94}$$

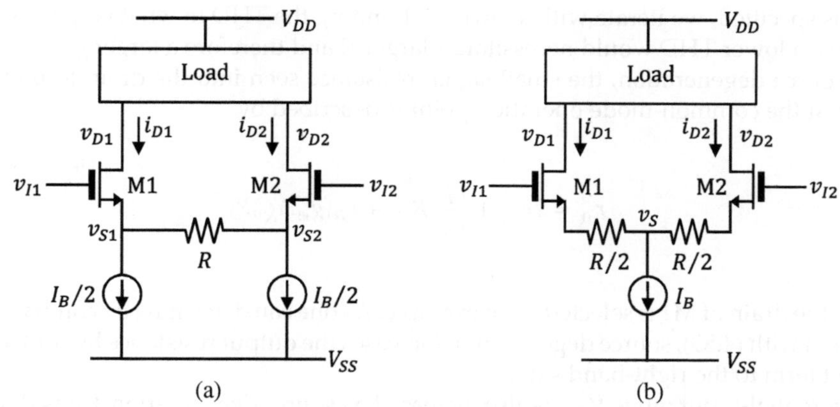

FIGURE 4.32 Source-degenerated differential pair topologies. (a) Double-sink topology. (b) Single-sink topology.

where A_{NP} stands for A_N or A_P depending on whether the drivers are NMOS or PMOS, respectively. In the case of operation with a source-bulk short, A_{NP} has to be replaced by unity.

It is obvious from (4.94) that R, instead of g_m, will be the dominant variable of G_m if the product g_mR is made increasingly larger than $2/A_{NP}$. Since R is much more linear than g_m, the linearity of the transconductance characteristic improves as a consequence.

It is important to recognize from (4.94) that, for a given G_m, g_m is not independent of R. To be specific, it is related to R by

$$g_m = \frac{2G_m}{1 - A_{NP}G_mR}, \tag{4.95}$$

which, for any meaningful value of g_m, imposes the following upper limit on R:

$$R_{\max} = \frac{1}{A_{NP}G_m}. \tag{4.96}$$

What is also obvious from (4.95) is that g_m grows very rapidly as R comes close to $R_{\max}$. It is therefore meaningful not to set R higher than what is absolutely needed.

Source-degenerated differential-pair design is based typically on the specified values of G_m and total harmonic distortion (THD) observed on the output current i_{D1} or i_{D2}. THD cannot be expressed in any simple way in terms of the interrelated design variables R and g_m. Instead, we resort to iterative simulation. We begin by calculating $R_{\max}$ from (4.96). Next, selecting an arbitrary value for R lesser than $R_{\max}$, we calculate from (4.95) the corresponding value of g_m, and use it in driver sizing. If, after completing the design with the rest of the constraints, the THD performance falls short of the target, then, we reset the value of R closer to $R_{\max}$ and repeat the cycle. If THD turns out to be lower than what is specified, we iterate with a lower R. Limiting the THD to what is specified pays off since a lower THD would necessitate a larger R and therefore a larger g_m.

In source degeneration, the small-signal resistance seen into the drain terminal of a driver at the common-mode operation point is described by

$$r_{o1} = 2r_{ds1} + \frac{r_{ds1}}{r_{ds2}} R_{D2} + r_{ds1} g_m A_{NP} R, \tag{4.97}$$

where the drain of M1 is selected as an example. As one can determine by comparing this equation with (4.83), source degeneration increases the output resistance by contributing the last term to the right-hand side.

Differential-input range V_{DIR} is also increased by source degeneration. Considering its definition $V_{DIR} \doteq (I_B/2)/G_m$, and referring to (4.94) for G_m, we can describe its degenerated version with

$$V_{DIR} = \frac{I_B/2}{g_m/2} + A_{NP} R \frac{I_B}{2} = \frac{2}{\text{Driver transconductance efficiency}} + A_{NP} R \frac{I_B}{2}, \tag{4.98}$$

where again the last term on the far right represents the contribution of source degeneration.

Example 4.15 The source-degenerated NMOS differential pair of Fig. 4.32(a) is to be designed in a single-supply configuration (V_{DD} = 1.8 V) with low-threshold devices. Specified items include (a) G_m = 400 μA/V, (b) r_{o1} = 600 kΩ for a negligible R_{D2}, and (c) THD = 0.2% on i_{D1} for 1-kHz sinewave input signal of 100-mV amplitude. The common-mode operation point is specified by V_{CM} = 0.9 V and V_{D1} = V_{D2} = 1.5 V with an overdrive voltage $V_{GS} - V_{TN}$ = 0.2 V. Note that G_m and r_{o1} are identical to their counterparts specified in Example 4.13. The specified overdrive voltage is also the same as the one derived in Example 4.13. Present example differs from Example 4.13 in that THD is specified instead of V_{DIR}.

Work

1. Considering the specified G_m = 400 μA/V, and $A_N \approx 1.3$, (4.96) yields R_{max} = 1923 Ω. Select R = 1500 Ω as a first guess.
2. The corresponding target value of driver transconductance is calculated from (4.95) as g_m = 3.64 mA/V. The reader may have already noticed that this is a factor of 4.5 larger than the driver transconductance of the basic differential pair of Example 4.13.
3. $V_{GS} - V_{TN}$ = 0.2 V → Fig. 4.9 ⇒ 10 V^{-1} for driver transconductance efficiency.
4. $g_m/(I_B/2)$ = 10 V^{-1} and g_m = 3.64 mA/V ⇒ $I_B/2$ = 364 μA. Again a 4.5-fold increase in comparison with Example 4.13.
5. $V_{GS} - V_{TN}$ = 0.2 V → Fig. 4.7 or Fig. 4.8 ⇒ $I_\square$ = 4.5 μA/sq.
6. $I_B/2$ = 364 μA and $I_\square$ = 4.5 μA/sq → (3.25) ⇒ W/L = 81.

7. Determining the drawn channel length of driver devices with iterative simulation: $V_{D1} = V_{D2}$ implies $r_{ds1} = r_{ds2}$. Denoting these resistances with r_{ds}, and considering the specified $R_{D2} \approx 0$ and $r_{o1} = 600\ \text{k}\Omega$, we obtain $r_{ds} = 60\ \text{k}\Omega$ from (4.97). Since $I_B/2 = 364\ \mu\text{A}$, the target Early voltage of the drivers is calculated as $V_A = 22$ V. Iterative simulation is performed with `ex4-13a.sp` of Example 4.13. It yields $L_{DR} = 1.5\ \mu\text{m}$.
8. $L_{DR} = 1.5\ \mu\text{m}$ and $W/L = 81 \Rightarrow W_{DR} = 119\ \mu\text{m}$.
9. Verification is performed with the following file

```
filename: ex4-15.sp
.incl bsimlvt.sp
m1 3 1 5 0 cmosnlvt w=133u l=1.5u nrd=0.0015 nrs=0.0015
m2 4 2 6 0 cmosnlvt w=133u l=1.5u nrd=0.0015 nrs=0.0015
r 5 6 1538
vd1 3 0 1.5
vd2 4 0 1.5
vi1 1 0 sin 0.9 0.1 1000
vi2 2 0 0.9
ib1 5 0 364u
ib2 6 0 364u
.tf v(3) vd1
*.dc vi1 0 1.5 1m
*.tran 2u 20m
.end
```

The transfer-function analysis is used for verifying r_{o1}. It yields 605.2 kΩ. The dc sweep is for plotting the transconductance characteristics. The reader is invited to compare the outcome of this analysis with its counterpart in Example 4.13. Finally, the transient analysis is used for verifying THD. We read its outcome with command

```
fourier 1000 i(vd1)
```

It yields THD = 0.112%, which, being nearly one half of the specified 0.2%, is far better than what is needed. Considering the price paid in terms of driver size and bias current, it is appropriate to bring it closer to the specified value. It is now up to the reader to do so by repeating the design after lowering R. It is also worth running the file `ex4-13b.sp` of Example 4.13 for a transient analysis similar to the present one, and comparing the THD performance of the basic differential pair with what is achieved here with source degeneration. ▲

4.4.5 Super-GM Differential Pair

The linearity of a source-degenerated differential pair can be further improved by imposing the differential-input voltage directly across the degeneration resistance by means of two feedback loops. This so-called *super-gm* technique is illustrated in Fig. 4.33 on an

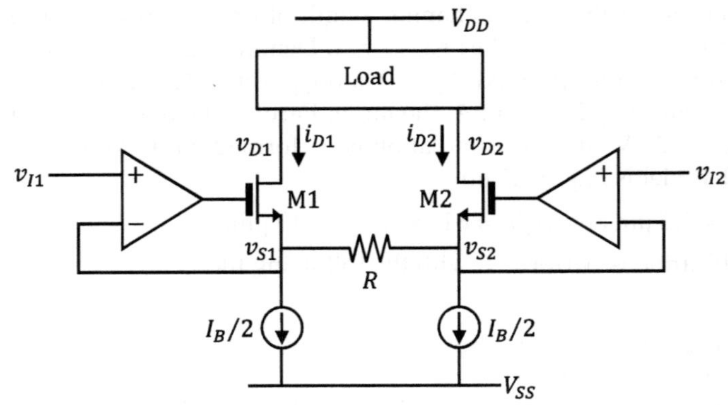

FIGURE 4.33 Super-gm differential-pair topology.

NMOS differential pair. Assuming a high differential gain and therefore virtual short for the OTAs, the source voltages are related to the corresponding input voltages by $v_{S1} = v_{I1}$ and $v_{S2} = v_{I2}$, which imply

$$i_{D1} = \frac{I_B}{2} + \frac{v_{I1} - v_{I2}}{R}$$

and

$$i_{D2} = \frac{I_B}{2} - \frac{v_{I1} - v_{I2}}{R}$$

for driver currents. These equations indicate

$$G_m \doteq \frac{\partial i_{D1}}{\partial(v_{I1} - v_{I2})} = \frac{1}{R} \tag{4.99}$$

for the effective transconductance, and

$$V_{DIR} \doteq |v_{I1} - v_{I2}|_{i_{D2}=0} = R\frac{I_B}{2} \tag{4.100}$$

for the differential input range. Since G_m is determined solely by R, the linearity of a super-gm differential pair is as good as the linearity of the resistor used for degeneration. This, of course, is true if the differential gain of the OTAs is indeed high.

It is left to the reader to construct the PMOS version of the super-gm schematic shown in Fig. 4.33, and to revise the expressions of i_{D1} and i_{D2} accordingly. As to G_m and V_{DIR}, (4.99) and (4.100) are also valid for the PMOS version.

The output resistance seen into the drain terminal of a super-gm driver far exceeds what a regular source-degenerated pair offers. Assuming the same r_{ds} for the drivers, it is described to a very good approximation by

$$r_{id} = A_0 g_m r_{ds} R, \tag{4.101}$$

where A_0 is the dc differential gain of each OTA.

Example 4.16 This design involves a single-supply ($V_{DD} = 1.8$ V) NMOS super-gm differential pair to be used as a stand-alone voltage-to-current converter in open loop. Specifications are as follows:

1. v_{I2} will be fixed at the common-mode operation point V_{CM}. v_{I1} will swing in both directions around V_{CM} with a peak magnitude $V_{id} = 0.3$ V.
2. *Common-Mode Operation Point* V_{CM} As low as possible.
3. *Effective Transconductance* $G_m = 400$ μA/V.
4. *Fixed Drain Voltages* $V_{D1} = V_{D2} = 1.4$ V but make sure that the drivers operate in saturation with a 0.1-V margin at a minimum.
5. *Current Sinks* The low-voltage cascode current reference and sink designed and verified in Example 4.6 of Subsection 4.2.4 will be used. Recall from that example the compliance limit $V_C = 0.3$ V and the reference current $I_R = 5$ μA. Operate the sink with at least a 0.1-V margin above the compliance limit.
6. *OTAs* DC differential gain $A_0 = 1000$ V/V. Lower limit of the output range $V_{OL} = 0.2$ V. Upper limit of the output range $V_{OH} = 1.6$ V. Other properties are immaterial. Use `opmacro1.sp` in verification.

Work $G_m = 400$ μA/V of item 3 implies $R = 2500\ \Omega$ from (4.99) while $V_{id} = 0.3$ V of item 1 defines a lower limit for V_{DIR}. Since a V_{DIR} larger than this limit would unnecessarily elevate the bias current as indicated by (4.100), we decide to target $V_{DIR} = 0.3$ V, which leads to $I_B/2 = 120$ μA.

Next, we consider item 2. The lowest acceptable value of V_{CM} is the sum of (a) the compliance limit $V_C = 0.3$ V defined in item 5 for the sinks, (b) the 0.1-V margin specified in the same item, and (c) the peak swing $V_{id} = 0.3$ V specified in item 1 for v_{I1} hence for v_{S1}. We therefore target $V_{CM} = 0.3 + 0.1 + 0.3 = 0.7$ V.

According to item 1, the peak input voltage $v_{I1(\max)}$ hence the peak source voltage $v_{S1(\max)}$ of M1 will be

$$v_{I1(\max)} = v_{S1(\max)} = V_{CM} + V_{id} = 0.7 + 0.3 = 1\text{ V}.$$

At this peak M1 conducts its maximum current, and therefore operates with its maximum gate voltage. Since the gate voltage cannot be larger than $V_{OH} = 1.6$ V as specified in item 6, the upper limit of the maximum gate-source voltage of M1 is therefore determined as $V_{GS1(\max)} = 1.6 - 1 = 0.6$ V. Adopting this upper limit for $V_{GS1(\max)}$ is meaningful as it will minimize driver dimensions. However, we need to make sure that this setting is compatible with the saturation condition of these devices. We can test this condition after determining the maximum overdrive voltage $V_{GS1(\max)} - V_{TN}$. Inspecting

Fig. 3.7, we observe a standard-threshold voltage around 0.7 V for $v_{S1(\max)} = 1$ V, which is too high to avoid subthreshold operation. This leads us to the low-threshold option for driver devices. Since none of the design specifications is pertinent to channel length, we select a short but not minimum $L_{DR} = 0.4$ μm, and read $V_{TN} \approx 0.4$ V from Fig. 4.5 for $v_{S1(\max)} = 1$ V. The corresponding overdrive voltage is determined as $V_{GS1(\max)} - V_{TN} = 0.6 - 0.4 = 0.2$ V.

We can now test the saturation condition of M1 when receiving the peak input. First of all, Fig. 4.6 indicates $V_{DSsat1(\max)} = 0.15$ V. Next, referring to item 4, we calculate the drain-source voltage as $V_{DS1(\min)} = V_{D1} - v_{S1(\max)} = 1.4 - 1 = 0.4$ V. It is obvious that the saturation condition is satisfied with a better-than-specified 0.25-V margin. We therefore move to the following final steps of design:

- $V_{GS1(\max)} - V_{TN} = 0.2\text{ V} \rightarrow$ Fig. 4.7 or Fig. 4.8 $\Rightarrow I_{\square 1(\max)} = 4.4\ \mu\text{A/sq}$.
- $I_{\square 1(\max)} = 4.4\ \mu\text{A/sq}$ and $i_{D1(\max)} = I_B = 240\ \mu\text{A} \rightarrow W/L = I_{D1(\max)}/I_{\square 1(\max)} \Rightarrow W/L = 55$.
- $L_{DR} = 0.4\ \mu\text{m} \rightarrow (3.2) \Rightarrow L = 0.366\ \mu\text{m}$.
- $L = 0.366\ \mu\text{m}$ and $W/L = 55 \Rightarrow W = 20\ \mu\text{m}$, which is large enough to assume $W_{DR} = 20\ \mu\text{m}$.

Verification is performed with the following input file `ex4-16.sp`, whose netlist corresponds to the schematic given in Fig. 4.34.

```
filename: ex4-16.sp
.incl opmacro1.sp
.incl bsim.sp
.incl bsimlvt.sp
.param vdd=1.8 vss=0 voh=1.6 vol=0.2 a0=1000 fgbw=10meg
+vdir=0.5 fnd=30meg
****************drivers+resistor+OTAs****************
```

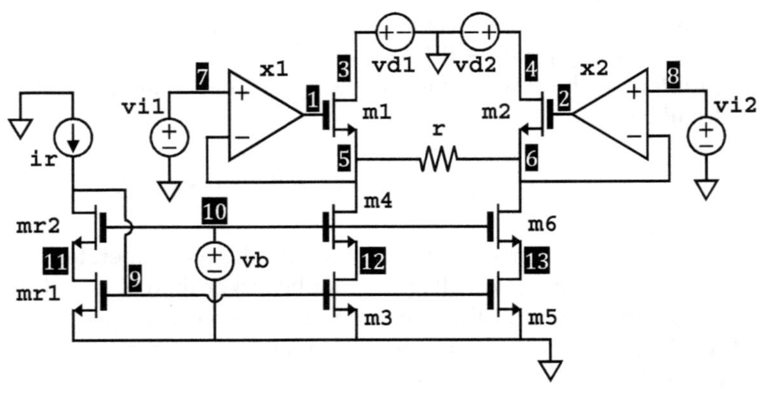

FIGURE 4.34 Electrical schematic of the super-gm differential pair designed in Example 4.16.

```
m1 3 1 5 0 cmosnlvt w=20u l=0.4u nrd=0.01 nrs=0.01
m2 4 2 6 0 cmosnlvt w=20u l=0.4u nrd=0.01 nrs=0.01
r 5 6 2500
x1 7 5 1 opamp
x2 8 6 2 opamp
vd1 3 0 1.4
vd2 4 0 1.4
vi2 8 0 0.7
*************current reference and sinks*************
mr1 11 9 0 0 cmosn w=3.15u l=1.1u
mr2 9 10 11 0 cmosn w=3.15u l=1.1u
m3 12 9 0 0 cmosn w=3.15u l=1.1u m=24
m4 5 10 12 0 cmosn w=3.15u l=1.1u m=24
m5 13 9 0 0 cmosn w=3.15u l=1.1u m=24
m6 6 10 13 0 cmosn w=3.15u l=1.1u m=24
ir 0 9 5u
vb 10 0 0.795
**********************dc sweep**********************
vi1 7 0 0.7
.dc vi1 0.3 1.1 0.1m
*********************transient**********************
*vi1 7 0 sin 0.7 0.3 1000
*.tran 2u 20m
******************transfer function*****************
*vi1 7 0 0.7
*.tf v(3) vd1
****************************************************
.end
```

The dc analysis sweeps v_{I1} by ± 400 mV around the operation point established with $V_{I2} = 0.7$ V. Transconductance characteristics are plotted with

```
plot -i(vd1) (-i(vd2))
```

for a visual assessment of linearity and V_{DIR}. The derivative plotted with

```
plot deriv(-i(vd1))
```

is used for verifying V_{DIR} and G_m quantitatively.

The transient analysis is intended for determining THD. It is run with a sinusoidal voltage source defined for v_{I1}. The reader is invited to compare the outcome with the THD obtained in Example 4.15.

The transfer-function analysis is run for determining the output resistance R_{id1}. Again, a comparison with the corresponding outcome of Example 4.15 might be of interest to the reader, who may also wish to assess the accuracy of (4.101) by comparing the output resistance obtained from the transfer-function analysis with what this equation predicts. The values of g_m and r_{ds} needed in this comparison can be obtained with a `show` command after running the transfer-function analysis. ▲

4.5 Basic OTA

Shown in Figs. 4.35(a) and (b) are the NMOS- and PMOS-driver versions of the basic OTA topology consisting of a differential pair M1/M2 biased by M5 and a current mirror load M3/M4. Output node is defined at the interconnected drain terminals of M1 and M3. Notice the capacitive termination of the output. As explained in Section 2.8, an OTA cannot drive a resistive load due to the lack of a buffering output stage, which is why none of the continuous-time closed-loop voltage amplifier configurations other than the unity-gain amplifier can be constructed with an OTA. In discrete-time closed-loop configurations, however, OTAs find unlimited utility thanks to the purely capacitive external network of such configurations.

4.5.1 DC Transfer Characteristic as a Voltage Amplifier

For an explanation of the dc operation of the basic OTA as a voltage amplifier, the reader is referred to the NMOS-driver version shown in Fig. 4.35(a). The output voltage v_O and branch currents i_{D1} and i_{D2} of this basic OTA are plotted in Figs. 4.35(c) and (d) against the

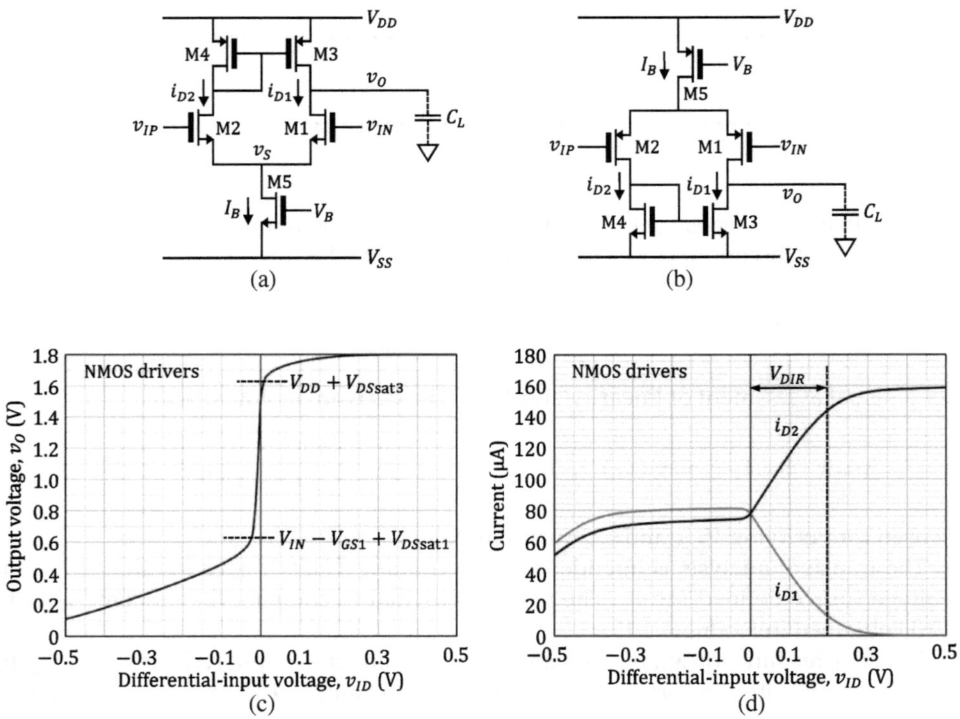

FIGURE 4.35 (a) Basic OTA with NMOS drivers. (b) Basic OTA with PMOS drivers. (c) DC transfer characteristic of a single-supply basic OTA with NMOS drivers. (d) Transconductance characteristics of the basic OTA whose dc transfer characteristic is presented in part c.

differential-input voltage $v_{ID} \doteq v_{IP} - v_{IN}$. These are simulated plots obtained by keeping v_{IN} constant at a V_{IN} and sweeping v_{IP} around V_{IN}.

For large positive values of v_{ID}, I_B is routed almost in its entirety to M2 and therefore to M4 as $i_{D2} = I_B$. No current is left for M1 and M3 to conduct, hence $i_{D1} = 0$. What keeps M1 in this cutoff state is a large source voltage v_S lowering the gate-source voltage $V_{IN} - v_S$ below the threshold voltage. M3, on the other hand, starves for current despite having a large negative gate-source voltage as imposed by the fully conducting M4. Since only nonsaturation with a zero drain-source voltage can hold it in this state, we observe $v_O = V_{DD}$ for large v_{ID}.

As v_{ID} is lowered, the differential pair starts raising i_{D1} above zero and lowering i_{D2} below I_B. In order for M3 to be able to conduct the nonzero i_{D1} while still being in nonsaturation, the drain-source voltage of this device has to become increasingly negative. This is why v_O starts decreasing as v_{ID} is lowered.

M3 enters saturation as v_O drops to $V_{DD} + V_{DSsat3}$. This occurs around $v_{ID} = 0$ because, while saturation enables M3 to equalize i_{D1} and i_{D2} by acting as a current mirror, the differential pair permits this equality only for $v_{ID} = 0$.

For $v_{ID} < 0$, the differential pair raises i_{D1} above i_{D2}. Since M3 is in saturation with the same gate-source voltage as M4, it can support a higher current only by making its drain-source voltage more negative than that of M4. But, since the sensitivity of current to drain-source voltage is very low in saturation, it takes a large drop in v_O for making i_{D1} even slightly larger than i_{D2}. As a result, v_O declines very steeply. Indeed, Fig. 4.35(c) indicates a mere −10 mV or so variation in v_{ID} for lowering v_O down to $V_{IN} - V_{GS1} + V_{DSsat1}$ and thus taking M1 out of saturation.

As v_{ID} is lowered further, i_{D1} and i_{D2} retain their small difference because the saturated M3 and M4 cannot conduct much disparate currents. As long as M5 remains in saturation and thus sinks I_B, each of i_{D1} and i_{D2} should therefore remain close to $I_B/2$. This is indeed what the transconductance characteristics exhibit in Fig. 4.35(d) for $-0.3\,\text{V} < v_{ID} < 0$. However, in order for the saturated M2 to conduct a constant current while having a decreasing gate voltage v_{IP}, its source voltage v_S has to decrease also. The nonsaturated M1, on the other hand, needs to lower its drain voltage v_O in response to the decreasing v_S because, having a constant-gate voltage V_{IN}, it cannot sustain a constant current otherwise. This is exactly what is observed in Fig. 4.35(c). As a final observation, notice in Fig. 4.35(d) the declining currents for $v_{ID} < -0.3\,\text{V}$. This is where M5 is taken out of saturation by the decreasing v_S.

4.5.2 Range Limitations

As observed in Fig. 4.35(c), the voltage-transfer characteristic of a basic OTA exhibits a large slope and therefore a large dc differential voltage gain only for the case of saturation in both M1 and M3. The range of v_O in which this condition is satisfied is the output range of the OTA. In the NMOS-driver version, the upper limit V_{OH} of this range is determined by the saturation condition of M3. It is described by

$$V_{OH} = V_{DD} + V_{DSsat3} \quad \text{(NMOS drivers).} \tag{4.102}$$

The lower limit V_{OL} brings M1 to the borderline of saturation. In the preceding subsection we identified this limit as $V_{IN} - V_{GS1} + V_{DSsat1}$ for a specific dc voltage V_{IN} applied to the gate of M1. Considering the validity of $v_{IP} \approx v_{IN}$ for the lower limit of the output range, we can assume a common-mode voltage at the gates of M1 and M2 for the purpose of defining V_{OL}. It is also obvious that the highest expected value $V_{CM(\max)}$ of common-mode swing will impose the most severe restriction on V_{OL}. Hence

$$V_{OL} = V_{CM(\max)} - V_{GS(1,2)} + V_{DS\text{sat}1} \quad \text{(NMOS drivers).} \tag{4.103}$$

Note that unless M1 and M2 are implemented with DNW devices operating with a source-bulk short, the gate-source voltage $V_{GS(1,2)}$ of these devices will also be a function of $V_{CM(\max)}$ due to body effect.

The limits of the output range in the PMOS-driver version are described by

$$V_{OH} = V_{CM(\min)} - V_{GS(1,2)} + V_{DS\text{sat}1} \quad \text{(PMOS drivers),} \tag{4.104}$$

and

$$V_{OL} = V_{SS} + V_{DS\text{sat}3} \quad \text{(PMOS drivers),} \tag{4.105}$$

where $V_{CM(\min)}$ denotes the lowest expected value of common-mode swing. Needless to say, $V_{GS(1,2)}$ appearing in (4.104) will be dependent on $V_{CM(\min)}$ unless M1 and M2 operate with a source-bulk short.

It is obvious from these expressions that a wide output range necessitates a small drain-source saturation voltage and therefore a small overdrive voltage for M3. It also calls for a low $V_{CM(\max)}$ in the NMOS-driver version and a high $V_{CM(\min)}$ in the PMOS-driver version. The dependence of output range on common-mode swing is one undesirable property of the basic OTA. This deficiency is eliminated in advanced OTA topologies at the expense of higher device count and higher power. Such topologies will be presented in Section 4.6 and thereafter.

As the reader may recall from Subsection 2.2.1, dc performance metrics also include the limits of the common-mode range. In the NMOS-driver version, the lower limit V_{IL} is dictated by the saturation condition of M5, and is described by

$$V_{IL} = V_{SS} + V_{DS\text{sat}5} + V_{GS(1,2)} \quad \text{(NMOS drivers).} \tag{4.106}$$

In the case of driver operation without source-bulk short, $V_{GS(1,2)}$ should be treated as the specific value of the driver gate-source voltage at the lower limit of the common-mode range.

The upper limit V_{IH} of the common-mode range of the NMOS-driver version is dictated by the saturation condition of either M2 or M1. The former of the two devices sets the limit if the minimum output voltage exceeds the drain voltage of M2, that is, $V_{O(\min)} \geq V_{DD} + V_{GS(3,4)}$. In this case, the limit is described by

$$V_{IH} = V_{DD} + V_{GS(3,4)} - V_{DS\text{sat}(1,2)} + V_{GS(1,2)} \quad \text{(NMOS drivers).} \tag{4.107}$$

In the case of $V_{O(\min)} < V_{DD} + V_{GS(3,4)}$, the limit is imposed by M1, and $V_{DD} + V_{GS(3,4)}$ should be replaced by $V_{O(\min)}$ in (4.107). In both cases, $V_{GS(1,2)}$ is specific to the upper limit of the common-mode range if drivers operate without source-bulk short.

In the PMOS-driver version, lower limit is described by

$$V_{IL} = V_{SS} + V_{GS(3,4)} - V_{DSsat(1,2)} + V_{GS(1,2)} \quad \text{(PMOS drivers)} \tag{4.108}$$

if M2 has a more restrictive saturation condition than M1, that is, $V_{O(\max)} \leq V_{SS} + V_{GS(3,4)}$. Otherwise, we need to replace $V_{SS} + V_{GS(3,4)}$ with $V_{O(\max)}$ in (4.108). In both cases, $V_{GS(1,2)}$ is specific to the lower limit if drivers operate without source-bulk short.

The higher limit of the common-mode range in the PMOS-driver version is imposed by the saturation condition of M5, and is described by

$$V_{IH} = V_{DD} + V_{DSsat5} + V_{GS(1,2)} \quad \text{(PMOS drivers)}, \tag{4.109}$$

where $V_{GS(1,2)}$ is specific to the upper limit if drivers operate without source-bulk short.

4.5.3 DC Differential Gain and Offset

As defined in Subsection 2.4.1, the dc differential gain A_0 of an opamp or OTA corresponds to the slope of the dc transfer characteristic within output range. We now relate this important open-loop metric to the device parameters of a basic OTA. For the implications of finite A_0 on closed-loop metrics, the reader is referred to Section 2.4.

The relationship between A_0 and its variables can be determined with a small-signal dc analysis performed by inspection on any of the two schematics shown in Figs. 4.35(a) and (b). An equivalent circuit appropriate for such an analysis on the NMOS-driver version is shown in Fig. 4.36(a) assuming an ideal current sink of infinite output resistance $r_{ds5} = \infty$. Notice the presence of one feedforward path and one feedback path for the signal to flow from v_i to v_o. The feedforward path is directed from gate to source in M2 and source to drain in M1. Along the feedback path, signal moves from gate to drain in M2, where it is represented by v_4, and then reaches the output by moving from gate to drain in M3, but it doesn't stop there. It moves back to the drain of M2 through the drain-to-source path of M1 and the source-to-drain path of M2. We cut the feedback path and introduce a signal source v_4 to the gate of M3 as shown in Fig. 4.36(b), and apply superposition with the two schematics presented in Figs. 4.36(c) and (d). The former of the two yields two expressions in the form of $v_o' = f_1(v_i)$ and $v_4' = f_2(v_i)$. The other yields $v_o'' = f_3(v_4)$ and $v_4'' = f_4(v_4)$. For v_4 and v_o, we write

$$v_4 = v_4' + v_4'' = f_2(v_i) + f_4(v_4),$$

and

$$v_o = v_o' + v_o'' = f_1(v_i) + f_3(v_4).$$

By eliminating v_4 between these two, and subsequently, solving for v_o/v_i and assuming $r_{ds2} \gg 1/g_{m4}$, we finally obtain

$$A_0 \doteq \frac{v_o}{v_i} = \underbrace{g_{m(1,2)}}_{G_m} \underbrace{\left(r_{ds1} \parallel r_{ds3}\right)}_{r_o} . \tag{4.110}$$

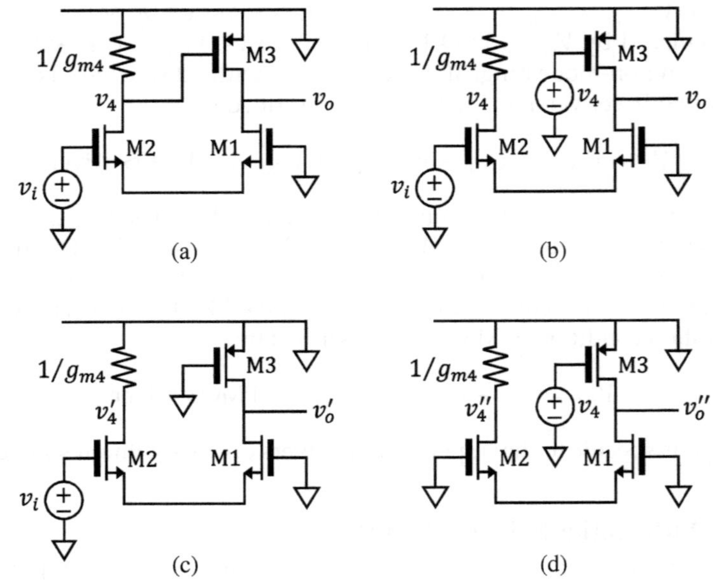

FIGURE 4.36 (a) Equivalent circuit for the dc analysis of the basic OTA with NMOS drivers. (b) Equivalent circuit after replacing the feedback path with two feedforward paths. (c) Driving with v_i. (d) Driving with v_4.

Note that G_m and r_o represent the effective-transconductance and output-resistance parameters defined in Section 2.8 for the generic dc small-signal model of opamps. Considering the nominal bias condition $v_{ID} = 0$, hence $i_{D1} = i_{D2} = I_B/2$, this equation can be transformed into

$$A_0 = \frac{g_{m(1,2)}}{I_B/2}\frac{r_{ds1}(I_B/2) \times r_{ds3}(I_B/2)}{r_{ds1}(I_B/2) + r_{ds3}(I_B/2)} = \frac{g_{m(1,2)}}{I_B/2}\frac{V_{A1}V_{A3}}{V_{A1} + V_{A3}}. \tag{4.111}$$

It is obvious from (4.111) that a large dc differential gain calls for large Early voltages and a large transconductance efficiency.

What is also worth noting is that the strong dependence of the output resistance of a MOSFET on the drain-source voltage makes A_0 sensitive to v_O. This sensitivity can be formulated from (4.110) as

$$\frac{dA_0}{A_0 dv_O} = \frac{r_{ds3}}{r_{ds1} + r_{ds3}}\frac{dr_{ds1}}{r_{ds1}dv_O} + \frac{r_{ds1}}{r_{ds1} + r_{ds3}}\frac{dr_{ds3}}{r_{ds3}dv_O}. \tag{4.112}$$

Since $dr_{ds1}/r_{ds1}dv_O$ and $dr_{ds3}/r_{ds3}dv_O$ are opposite in polarity but usually comparable in magnitude for an operation point around the middle of the output range, selecting $r_{ds1} \approx r_{ds3}$ or equivalently $V_{A1} \approx V_{A2}$ can reduce the sensitivity of A_0 to v_O as the reader can ascertain from (4.112). Although this approach helps improve the linearity of the

OTA, its practice is meaningful as long as it does not require highly disparate channel lengths for M1 and M3. If it does, the total size of these two devices and therefore their parasitic capacitances may turn out to be too large to tolerate.

The dc differential gain provided by a basic OTA is on the order of 100 V/V, which is good enough only in open-loop transconductor applications or in closed-loop applications that can tolerate large amounts of gain error and nonlinearity.

As explained in Subsection 2.4.2, the offset voltage V_{os} exhibits a Gaussian distribution of variance $\sigma^2[V_{os}]$ among the fabricated copies of an opamp or OTA around a mean value $\overline{V_{os}}$. Practicing segmentation with matching unit devices for the drivers and mirrors, and arranging the segments into dummy-terminated common-centroid layouts help minimize the contribution of imperfect repeatability of manufacturing processes, on-chip gradients, and proximity effects to $\overline{V_{os}}$. Provided that these measures are taken, the residual $\overline{V_{os}}$ will be due to the asymmetry caused by the difference between output voltage and the common-drain voltage of M2 and M4. As to the variance of offset, we can reduce it by enlarging MOSFET channels.

4.5.4 Frequency Response and Step Response

Analytical modeling of the frequency response of a basic OTA with exact high-frequency device models is impractical due to the complexity of the topology. Fortunately, an approximate analysis of adequate accuracy can be performed on a dc small-signal equivalent circuit supplemented by lumped parasitic capacitances between each node and ground. The resulting schematic shown in Fig. 4.37 for the NMOS-driver version includes internal capacitances C_1 and C_2 in addition to the capacitance C_L loading the output node. Notice the application of $v_i/2$ to the noninverting input and $-v_i/2$ to the inverting input in order to establish a purely differential-input drive so that the transfer function is not corrupted by a common-mode component.[4] For such an input drive, the common-source node of the drivers is approximately signal free, which renders C_2 practically ineffective on the transfer function. Assuming that all MOSFET output

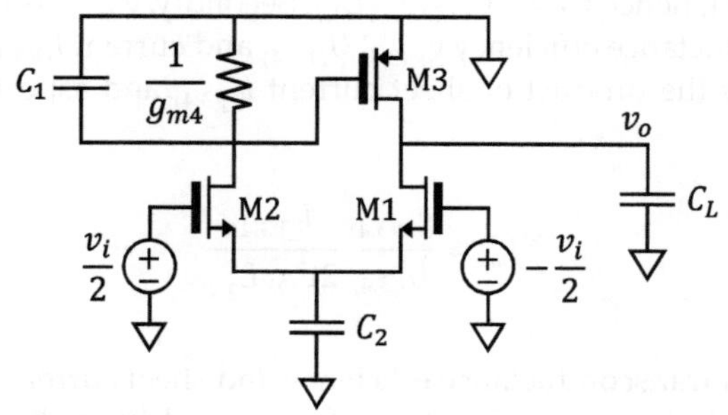

FIGURE 4.37 Simplified high-frequency equivalent circuit of a basic OTA with NMOS drivers. C_1 and C_2 are lumped parasitic capacitances at internal nodes.

[4] As explained previously in Subsection 4.4.3, branch currents of a differential pair become more sensitive to the common-mode signal at high frequencies.

conductances are negligible in comparison with transconductances, an analysis of this equivalent circuit yields the transfer function

$$A = g_{m(1,2)} r_o \frac{1 + s\frac{C_1}{2g_{m(3,4)}}}{\left(1 + sr_o C_L\right)\left(1 + s\frac{C_1}{g_{m(3,4)}}\right)}, \tag{4.113}$$

where $r_o C_L$ is by far the largest time constant defining the dominant-pole frequency

$$f_o = \frac{1}{2\pi r_o C_L}. \tag{4.114}$$

Substituting this equation together with (4.110) into (2.43) yields the gain-bandwidth product expression

$$f_{GBW} = \frac{g_{m(1,2)}}{2\pi C_L}. \tag{4.115}$$

Although these results have been derived on an equivalent circuit with NMOS drivers, they are valid also for a basic OTA with PMOS drivers.

Returning to (4.113), we observe a nondominant pole at the frequency

$$f_{nd} = \frac{g_{m(3,4)}}{2\pi C_1}. \tag{4.116}$$

and a zero at a frequency twice as high. For identifying the available options for setting these frequencies high and therefore maintaining a large phase margin, we reshape (4.116) in three steps. First, we assume C_1 to be dominated by the intrinsic gate capacitances of M3 and M4, hence $C_1 = 2C_{OX}W_{3,4}L_{3,4}$. Secondly, $g_{m(3,4)}$ is replaced by the product of the transconductance efficiency $g_{m(3,4)}/I_{D(3,4)}$ and current $I_{D(3,4)}$. Finally, the current $I_{D(3,4)}$ is replaced by the product of sheet current $I_{\square(3,4)}$ and aspect ratio $W_{3,4}/L_{3,4}$. The outcome is

$$f_{nd} \approx \frac{g_{m(3,4)}}{I_{D(3,4)}} \frac{I_{\square(3,4)}}{2C_{OX}L_{3,4}^2}. \tag{4.117}$$

Since the product of transconductance efficiency and sheet current appearing in this expression is a strongly increasing function of the overdrive voltage as exemplified by Figs. 3.11 and 3.12, setting a large overdrive voltage in M3 and M4 is one obvious option for moving f_{nd} and the coupled zero to high frequencies. Another very effective option indicated by (4.117) is to keep the channels of these devices short. Also note that the PMOS-driver version of the basic OTA is superior to the NMOS-driver version in offering a high f_{nd} because, for the same overdrive voltage, the sheet current of its n-channel M3/M4 is roughly three times higher than the sheet current of the p-channel M3/M4 of the NMOS version.

As a final word, it is worth recalling from Subsections 2.5.2 and 2.6.2 that the frequency response or step response of a closed-loop configuration is subject to slewing if the magnitude of the differential-input voltage exceeds the differential-input range parameter V_{DIR}. In a basic OTA, this open-loop parameter is related solely to the properties of the differential pair. The reader is referred to Subsection 4.4.3 for a basic differential pair, Subsection 4.4.4 for a degenerated differential pair, and Subsection 4.4.5 for a super-gm differential pair.

4.5.5 Noise-Related Properties

As explained in Subsection 2.9.3 and described by (2.116), the contribution of an opamp to the output noise of a closed-loop configuration is characterized by the power spectral density (PSD) $S_{i(\text{opamp})}(f)$ of a fictitious input-referred noise voltage source. This PSD is described by (2.115), where $A_{k(f)}$ is the transfer function from the noise source associated with device-k to the output of the opamp in open loop, $S_k(f)$ is the PSD of the same source, and $A(f)$ is the differential gain of the opamp in open loop. All these descriptions also apply to OTAs.

For a basic OTA, $A_k(f)$ of all five devices can be constructed by inspection from the small-signal equivalent circuits presented in Fig. 4.38. Although these circuits belong to the NMOS-driver version, the transfer functions they yield are also valid for the PMOS-driver version. Notice the following features:

- As in any small-signal analysis, all independent voltage sources of signals, bias, and power supply are short-circuited. This leads to grounded supply rails and input nodes.
- Resistances seen into the drain terminals of MOSFETs are assumed to be infinite.
- Internal node capacitances are ignored, and the frequency dependence of $A_k(f)$ is attributed solely to the equivalent impedance Z_L of the output port.
- Noise current source of each device is represented by i_{nk} where k is device index.

Referring to Fig. 4.38(a) for the noise transfer function of M1, we observe the noise current i_{n1} of this device being split into two nearly equal parts at the common-source node of M1 and M2 due to the nearly equal source input resistances of these devices. The half that flows into M2 is mirrored from M4 to M3. The latter sinks it from output node to ground. The other half of i_{n1} loops back to the output through M1. The net current flowing to the output node from Z_L is therefore equal to i_{n1}, which implies $|A_1(f)| = |v_o/i_{n1}| = |Z_L|$ for the magnitude of the noise transfer function of M1.

The case of i_{n2} is depicted in Fig. 4.38(b). It is also split into two equal parts at the common-source node of M1 and M2. The half flowing into M2 limits the current of M4 to $i_{n2}/2$, which is mirrored to the output node by M3. Since M1 also conducts $i_{n2}/2$, the current flowing to Z_L equals i_{n2}, hence $|A_2(f)| = |Z_L|$ for the magnitude of the noise transfer function of M2.

As depicted in Fig. 4.38(c), i_{n3} flows into Z_L in its entirety because the resistances seen into the drain terminals of M1 and M3 are assumed to be infinite. This leads to $|Z_3(f)| = |Z_L|$ for the magnitude of the noise transfer function of M3.

i_{n4}, shown in Fig. 4.38(d), flows in its entirety into the diode-connected M4 because the resistance seen into the drain of M2 is assumed to be infinite. It is mirrored to M3, and

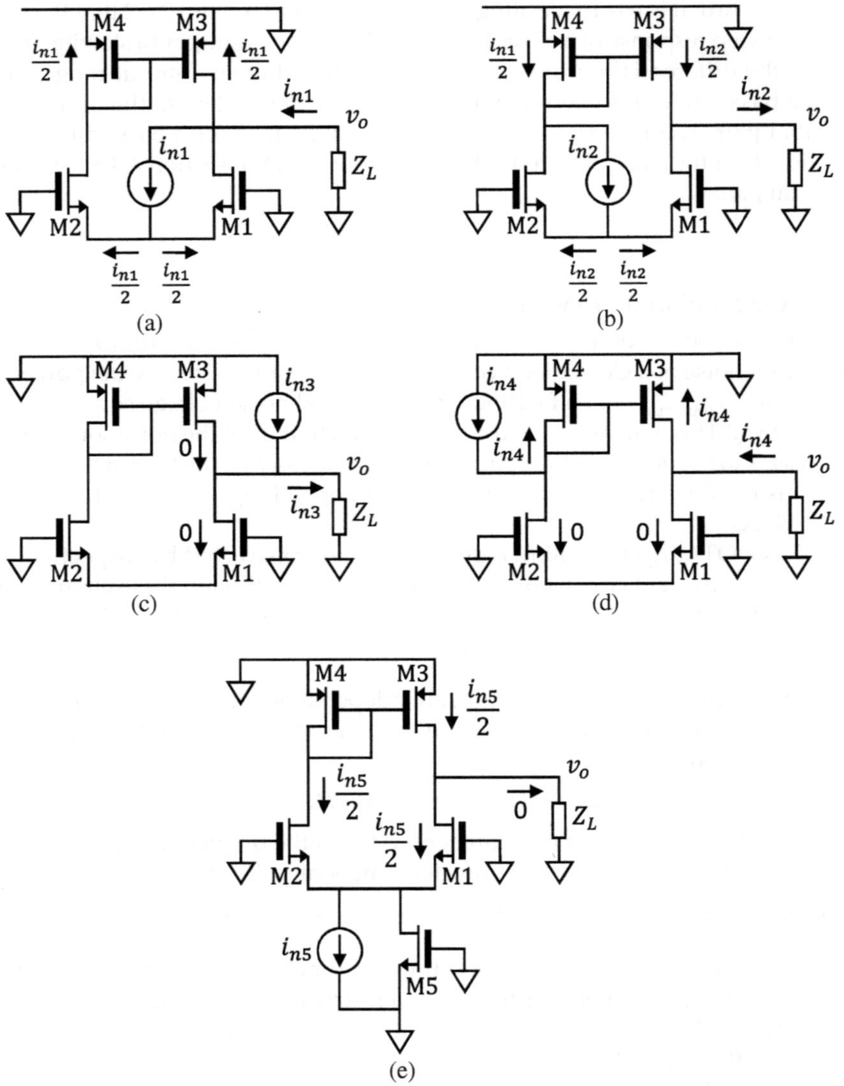

FIGURE 4.38 Equivalent circuits for determining the transfer functions needed in the noise analysis of a basic OTA. (a) v_o/i_{n1}. (b) v_o/i_{n2}. (c) v_o/i_{n3}. (d) v_o/i_{n4}. (e) v_o/i_{n5}.

drawn from Z_L. Therefore, $|A_4(f)| = |Z_L|$ for the magnitude of the noise transfer function of M4.

Referring to Fig. 4.38(e), note that the noise current i_{n5} of M5 is split into two nearly equal parts, one flowing in M1 and the other in M2. M5 itself does not conduct any significant fraction of i_{n5} because the resistance seen into its drain is much higher than the approximately equal resistances seen into the source terminals of M1 and M2. The half of i_{n5} flowing into M2 is mirrored from M4 to M3, which sources it to the output node. Since the other half is sunk from the output node by M1, no current is left to flow into Z_L and to create a noise voltage at the output. Therefore, $|A_5(f)| \doteq |v_o/i_{n5}| = 0$ for the noise transfer function of M5.

In summary, the magnitude $|A_k(f)|$ of the noise transfer function is zero for M5 but $|Z_L|$ for M1, M2, M3, and M4. The magnitude of the differential gain, on the other hand, equals $g_{m(1,2)} \times |Z_L|$ as long as the frequency response is attributed to the equivalent impedance attached to the output port. These two turn (2.115) into

$$S_{i(\text{opamp})}(f) = \frac{1}{g_{m(1,2)}^2} \sum_{k=1}^{4} S_k(f). \tag{4.118}$$

For the thermal component of $S_k(f)$, we ignore g_{ds} in (3.118), and represent $g_m + g_{mb}$ with $A_{NP} g_{mb}$, where A_{NP} represents the approximately equal A_N and A_P, which is around 1.3 as indicated by Figs. 3.12 and 3.20 for standard-threshold devices and by Figs. 4.9 and 4.14 for low-threshold devices. For the MOSFET of index k, these considerations lead to the following expression for the thermal component of $S_k(f)$:

$$S_{th(k)} = \frac{8}{3} E_{th} A_{NP} g_{m(k)}. \tag{4.119}$$

The PSD of the flicker component of $S_k(f)$ is described directly by (3.119) as

$$S_{fl(k)} = K_k \frac{W_k}{f L_k^3}. \tag{4.120}$$

After substituting into (4.118) the sum $S_{th(k)} + S_{fl(k)}$ for $S_k(f)$, and using (4.119) and (4.120) for $S_{th(k)}$ and $S_{fl(k)}$, respectively, we arrive at an expression in the form of (2.119), in which, k_{th} and k_{fl} are described by

$$k_{th} = \frac{8 E_{th} A_{NP}}{3 g_{m(1,2)}^2} \sum_{k=1}^{4} g_{mk}, \tag{4.121}$$

and

$$k_{fl} = \frac{1}{g_{m(1,2)}^2} \sum_{k=1}^{4} K_k \frac{W_k}{L_k^3}. \tag{4.122}$$

The contribution of a basic OTA to the output noise voltage of a continuous-time closed-loop configuration is described quite generally by the following expression obtained by substituting (4.121) and (4.122) into (2.120):

$$V_{no(\text{opamp})}^2 = \underbrace{\frac{4\pi A_{cl0,ni}^2 f_{cl} E_{th} A_{NP}}{3 g_{m(1,2)}^2} \sum_{k=1}^{4} g_{mk}}_{\text{Thermal noise}} + \underbrace{\frac{A_{cl0,ni}^2}{g_{m(1,2)}^2} \ln\left(\frac{f_{cl}}{f_1}\right) \sum_{k=1}^{4} \frac{K_k W_k}{L_k^3}}_{\text{Flicker noise}}. \tag{4.123}$$

With a view to determining the design options available to us for keeping the output noise low, let us consider the flicker component of (4.123) first. Dismissing $A^2_{cl0,ni}$ because it is a closed-loop metric specified by concerns unrelated to noise, and also ignoring the weakly effective logarithmic term, we transform the rest of this component through the following sequence of substitutions:

$$\frac{1}{g^2_{m(1,2)}}\sum_{k=1}^{4}\frac{K_kW_k}{L^3_k} = \frac{(I_B/2)^2}{g^2_{m(1,2)}}\sum_{k=1}^{4}\frac{K_kW_k}{(I_B/2)^2L^3_k}$$
$$= \frac{(I_B/2)^2}{g^2_{m(1,2)}}\sum_{k=1}^{4}\frac{K_kW_k}{I^2_{\square k}(W_k/L_k)^2L^3_k} = \frac{(I_B/2)^2}{g^2_{m(1,2)}}\sum_{k=1}^{4}\frac{K_k}{I^2_{\square k}W_kL_k}.$$

Two design options leading to a small flicker noise are immediately obvious on the far right of this equation. One is to use large channel area W_kL_k in all devices. The other is to keep the overdrive voltage of M1 and M2 low, so that they exhibit a high transconductance efficiency $g_{m(1,2)}/(I_B/2)$. Although not immediately obvious, it is also beneficial to operate M3 and M4 with large overdrive voltages because the ratio $K_k/I^2_{\square k}$ is a decreasing function of the magnitude of the overdrive voltage as the reader can determine from sheet-current plots and Fig. 3.44.

Moving to the thermal-noise component in (4.123), and again dismissing $A_{cl0,ni}$, we derive two design directives for reducing this component: (a) Limiting the closed-loop bandwidth f_{cl} to the absolute minimum needed for the signal-processing task at hand, and (b) targeting a large transconductance for driver devices and a small transconductances for the rest.

For amplifier applications, which are based on a frequency-independent external network, (4.123) can be transformed into a more instructive form by using (2.48) for f_{cl}, (4.115) for f_{GBW}, and $1/\beta$ for $A_{cl0,ni}$. The outcome

$$V^2_{no(\text{opamp})} = \underbrace{\frac{2E_{th}A_{NP}}{3\beta C_L}\left(2+\sum_{k=3}^{4}\frac{g_{mk}}{g_{m(1,2)}}\right)}_{\text{Thermal noise}} + \underbrace{\frac{1}{\beta^2 g^2_{m(1,2)}}\ln\left(\frac{\beta g_{m(1,2)}}{2\pi f_1 C_L}\right)\sum_{k=1}^{4}\frac{K_kW_k}{L^3_k}}_{\text{Flicker noise}} \qquad (4.124)$$

indicates that the only open-loop tool by which we can manipulate the thermal component in this case is the ratio $g_{m(3,4)}/g_{m(1,2)}$. Since all devices conduct the very same bias current, this ratio also equals the ratio of transconductance efficiencies, which, of course, is controlled solely by the overdrive voltages of these devices. Setting the overdrive voltages of the two driver devices low and those of M3 and M4 high is what we need to do for reducing the contribution of thermal noise in closed-loop amplifier applications.

Example 4.17 In this example, we design a basic OTA in accordance with the following specifications:

1. Closed-loop continuous-time unity-gain amplifier application loaded by a 2.4-pF capacitor.
2. Single-supply configuration of $V_{DD} = 1.8$ V.

3. Pulsed input signal stepping between 0.6 and 0.8 V around 0.7-V bias.
4. Slew-free pulse response with 70-MHz closed-loop bandwidth.
5. Capacitance seen into the input port of the unity-gain amplifier: Less than 140 fF within closed-loop bandwidth.
6. PMOS drivers operating with source-bulk short.
7. Minimized power consumption.
8. Minimized systematic offset.

Work also includes the estimation and simulated verification of the output mean-square noise voltage of the closed-loop configuration.

Work Specifications are too many to permit a casual design flow. A systematic approach is definitely needed in this example. We begin by converting application-layer specifications into circuit constraints and expressing these constraints in terms of device performance metrics. The reader is referred to Fig. 4.35(b) for the OTA schematic.

- We add 0.1 pF to the external load to account for the output capacitance of the OTA itself, and thus specify $C_L = 2.5$ pF. Since the unity-gain application implies $f_{GBW} = f_{cl}$, the specified $f_{cl} = 70$ MHz necessitates $g_{m(1,2)} = 1.1$ mA/V in accordance with (4.115).
- The requirement of a slew-free operation together with the specified input-signal step height $V_i = 0.8 - 0.6 = 0.2$ V implies $V_{DIR} \geq 0.2$ V from (2.79). On the other hand, the requirement of minimum power consumption necessitates a minimized bias current I_B, which, in turn, calls for a minimized V_{DIR} as implied by (4.84) for a fixed $g_{m(1,2)}$. We therefore target $V_{DIR} = 0.2$ V, for which, (4.84) yields $I_B = 220$ μA.
- Minimization of systematic offset necessitates equal drain-source bias voltages for M3 and M4. Since the specified bias implies $V_{DS3} = 0.7$ V, we also target 0.7 V for V_{DS4} hence V_{GS4}.
- The limits of the common-mode range are constrained by (2.7) and (4.108) for V_{IL} and by (2.8) and (4.109) for V_{IH}. Considering $V_{CM(\min)} = 0.6$ V, $V_{CM(\max)} = 0.8$ V, $V_{DD} = 1.8$ V, and $V_{GS(3,4)} = 0.7$ V, we turn these expressions into

$$V_{GS(1,2)} < -0.1 + V_{DSsat(1,2)} \tag{4.125}$$

and

$$V_{GS(1,2)} > -1.0 - V_{DSsat5}. \tag{4.126}$$

Note that $V_{GS(1,2)}$ has a constant value regardless of the common-mode level because the source-bulk short of M1 and M2 precludes body effect.

- The lower limit of the output range is constrained by (2.4) and (4.105). These two imply $V_{DSsat(3,4)} < 0.6$ V, which, considering the previously set $V_{GS(3,4)} = 0.7$ V, is redundant because it will definitely be satisfied even if the low-threshold option is exercised for M3 and M4.

The upper limit of the output range is generally constrained by the saturation condition of M1 and is described by (2.5) and (4.104). This condition is automatically satisfied in the present unity-gain amplifier application whose configuration turns M1 into a diode-connected device by interconnecting the output and inverting-input terminals of the OTA.

- The limitation imposed on the capacitance of the input port sets a dimensional constraint on driver devices. With a view to developing an analytical representation for this constraint in the present unity-gain amplifier configuration, we go back to Fig. 4.30(c), and consider G1 and G2 as the inverting and noninverting terminals, respectively. For frequencies much lower than f_{cl}, both of these terminals receive the same signal due to virtual short. Since current flow through Y_d is thus suppressed, the admittance seen into the input port between G2 and ground is Y_{se} only. At frequencies much higher than f_{cl} this admittance rises to $Y_{se} + Y_d$ as a result of the signal ground established at the inverting-input terminal G1. Y_{se} and Y_d are mainly capacitive as described by (4.92) and (4.93). $C_{se} + C_d$ is therefore an upper limit to the input capacitance of a unity-gain amplifier. Although observed for frequencies well above f_{cl}, this sum can be regarded as a conservative estimate for the maximum input capacitance observed within the bandwidth.

For $g_{mb} = 0$, as implied by the source-bulk short specified for driver devices, (4.92) and (4.93) indicate $C_d = (C_{gg} - C_{gd} - C_{gb})/2$ and $C_{se} = C_{gd} + C_{gb}$. We now relate these two capacitances to the dimensions of driver devices. The first step is to substitute (3.105) for C_{gg}, (3.106) for C_{gd}, and (3.107) for C_{gb}. In the outcome, we use (3.99) for C_{gge}, and ignore the essentially small components C_{gdi} and C_{gbe}. Also considering (3.92) for C_{gde}, (3.93) for C_{gse}, and (3.89) for C_{ggi} and C_{gbi}, we finally arrive at

$$C_d = \frac{\mathsf{CGSO} \times W_{1,2} + (\overline{C_{ggi}} + \overline{C_{gbi}})W_{1,2}L_{1,2}}{2}$$

and

$$C_{se} = \mathsf{CGDO} \times W_{1,2} - \overline{C_{gbi}}W_{1,2}L_{1,2}.$$

The upper limit specified for the input capacitance can now be stated with the aid of these two expressions as follows:

$$C_{se} + C_d = \left(\mathsf{CGDO} + \frac{\mathsf{CGSO}}{2}\right) W_{1,2} + \frac{\overline{C_{ggi}} - \overline{C_{gbi}}}{2} W_{1,2}L_{1,2} \le 140 \text{ fF}. \qquad (4.127)$$

Having finished the conversion of specifications into device constraints, we can now construct a design plan for device-level design. A reasonable starting point is the driver transconductance efficiency $g_{m(1,2)}/I_{D(1,2)} = 10\ \text{V}^{-1}$ as calculated from $g_{m(1,2)} = 1.1$ mA/V and $I_{D(1,2)} = I_B/2 = 110$ μA. After tentatively selecting a threshold option for driver devices, we determine the overdrive voltage $V_{GS(1,2)} - V_{TP}$ corresponding to this transconductance efficiency. Next, sheet current $I_{\square(1,2)}$ is determined from the overdrive voltage, and used together with $I_{D(1,2)}$ for calculating $(W/L)_{1,2}$. The outcome, combined with (4.127), yields the largest acceptable values of $L_{1,2}$ and $W_{1,2}$. It is preferable to adopt these largest values unless they turn out to be excessive in terms of footprint because a shorter $L_{1,2}$ would lower the dc differential gain A_0. At this stage, we test the selected threshold option against the constraints (4.125) or (4.126). We do this by determining the threshold voltage of driver devices first, and then calculating $V_{GS(1,2)}$ from the overdrive voltage. $V_{DSsat(1,2)}$ appearing in (4.125) is also determined from the overdrive voltage. As to V_{DSsat5} appearing in (4.126), any value more negative than a hundred millivolts or so is acceptable. If (4.125) or (4.126) is not satisfied, we will switch to an alternative threshold option that can possibly satisfy both. If the alternative does not work either, we will abandon the basic OTA and seek an alternative OTA topology with relaxed range limitations.

Assuming all goes well, we next move to the sizing of M3 and M4, whose large gate-source bias voltage $V_{GS(3,4)} = 0.7$ V can easily accommodate the standard-threshold option. The channel length $L_{3,4}$ of these devices is a parameter of A_0. Setting it with a view to establishing $V_{A3} \approx V_{A1}$ is meaningful for better linearity. Since $L_{1,2}$ is already available at this stage, we determine V_{A1} with a single simulation first. $L_{3,4}$ is determined next with an iterative simulation. After extracting the threshold voltage of these devices from Fig. 3.7, we calculate their overdrive voltage $V_{GS(3,4)} - V_{TN}$, and use the outcome for determining $I_{\square(3,4)}$ and therefore $(W/L)_{3,4}$ and $W_{3,4}$.

Sizing of M5 is based on the specified value of I_B as well as the target value of V_{DSsat5} determined in compliance with (4.126). V_{DSsat5} enables us to fix the overdrive voltage $V_{GS5} - V_{TP}$, and therefore the sheet current $I_{\square 5}$, and eventually, the aspect ratio $(W/L)_5$. Channel length L_5 is not constrained by any of the specified items, but since this device is a current source, it has to be long enough to provide an output resistance r_{ds5} that is at least an order of magnitude larger than the resistance seen into the source terminal of each driver device as approximated by $1/g_{m(1,2)}$. Calculating W_5 for the selected value of L_5 is the final step of device sizing. As usual, we will conclude by verifying the entire set of specifications. Steps of design flow are presented next.

1. *Sizing M1 and M2* We begin with the standard-threshold option, for which, Fig. 3.20 yields $V_{GS(1,2)} - V_{TP} = -0.185$ V for $g_{m(1,2)}/I_{D(1,2)} = 10\ \text{V}^{-1}$. The corresponding sheet current is determined from Fig. 3.19 as $I_{\square(1,2)} = 1\ \mu\text{A/sq}$. Considering $I_{D(1,2)} = I_B/2 = 110\ \mu\text{A}$, this much sheet current implies $(W/L)_{1,2} = 110$. Another relationship between $W_{1,2}$ and $L_{1,2}$ is provided by (4.127). We solve these two equations after substituting $\overline{C_{ggi}} = 7 \times 10^{-3}\ \text{F/m}^2$ and $\overline{C_{gbi}} = -0.3 \times 10^{-3}\ \text{F/m}^2$ from Fig. 3.40 and CGDO = CGSO = 7.82×10^{-10} F/m from Table 3.1. The outcome is $W_{1,2} = 49\ \mu\text{m}$ and $L_{1,2} = 0.44\ \mu\text{m}$. We adopt these two directly as drawn dimensions because the accuracy of (4.127) is far from being perfect anyway.

   Next, we evaluate the range constraints. To this end, we first obtain $V_{TP} = -0.435$ V from Fig. 3.16 for $V_{SB(1,2)} = 0$ and $L_{DR(1,2)} = 0.44\ \mu\text{m}$. Considering the overdrive voltage determined in step 1, this threshold voltage implies $V_{GS(1,2)} = -0.185 - 0.435 = -0.620$ V. The very same overdrive voltage also yields $V_{DSsat(1,2)} = -0.160$ V from Fig. 3.17. These numbers satisfy (4.125) by a large margin, and turn (4.126) into $V_{DSsat5} > -0.380$ V.

2. *Sizing M3 and M4* We first determine $V_{A1} = 9.9$ V by simulating M1 with the following input file `ex4-17a.sp`.

```
filename: ex4-17a.sp
.incl bsim.sp
m1 2 1 0 0 cmosp w=49u l=0.44u
vgs 1 0 -0.620
vds 2 0 -0.620
.control
op
print @m1[id]/@m1[gds]
.endc
.end
```

Note that V_{DS1} is equated to V_{GS1} in the netlist because the unity-gain application implies identical gate and drain voltages.

Next, an iterative simulation is run for determining the channel length of M3 for a target Early voltage 9.9 V in a standard-threshold option. The following input file ex4-17b.sp is used for this purpose.

```
filename: ex4-17b.sp
.incl bsim.sp
m3 2 1 0 0 cmosn w=8u l=0.45u
vgs 1 0 0.7
vds 2 0 0.7
.control
op
print @m3[id]/@m3[gds]
.endc
.end
```

This step yields $L_{DR(3,4)} = 0.45$ μA with $V_{A3} = 10.3$ V. Since Fig. 3.7 indicates $V_{TN} = 0.46$ V for this channel length, the overdrive voltage of M3 and M4 is determined as $V_{GS(3,4)} - V_{TN} = 0.7 - 0.46 = 0.24$ V. This leads to $I_{\square(3,4)} = 6.8$ μA/sq from Fig. 3.10, and therefore to $(W/L)_{3,4} = 16.2$ for $I_{D(3,4)} = 110$ μA. Taking into consideration the process bias, we end up with $W_{DR(3,4)} = 6.8$ μm.

Having already determined $g_{m(1,2)}/I_{D(1,2)}$, V_{A1}, and V_{A3}, we estimate the dc differential gain from (4.111) as $A_0 = 50.4$ V/V.

3. *Sizing M5* We satisfy the constraint $V_{DSsat5} > -0.380$ V obtained in step 1 by targeting $V_{DSsat5} = -0.200$ V plus a saturation margin $\Delta V_{DS5} = -0.180$ V. Threshold option is not critical for this device; we select the standard-threshold option. Figure 3.17 indicates $V_{GS5} - V_{TP} = -0.240$ V, for which, $I_{\square 5} = 1.65$ μA/V is obtained from Fig. 3.19. The corresponding aspect ratio is $(W/L)_5 = 134$ for $I_B = 220$ μA. For determining L_5, we target $r_{ds5} \geq 20 \times (1/g_{m(1,2)}) = 9.09$ kΩ, hence $V_{A5} \doteq r_{ds5} I_B = 4$ V. Iterative simulation is performed with the following input file ex4-17c.sp.

```
filename: ex4-17c.sp
.incl bsim.sp
m5 2 1 0 0 cmosp w=10u l=0.4u
vgs 1 0 -0.680
vds 2 0 -0.480
.control
```

```
op
print @m5[id]/@m5[gds]
.endc
.end
```

Since the limit set for the Early voltage is very low, it can be achieved with a pretty short channel. L_{DR} = 0.4 μm, for example, yields V_{A5} = 7.62 V, which is almost twice as high as the limit targeted in simulation. We adopt this channel length, and determine the corresponding width as W_{DR5} = 46 μm.

4. *Verification* The following input file `ex4-17d.sp` is used for verification. It reflects the OTA schematic shown in Fig. 4.39(a), where `ir` represents the reference current mirrored from `m6` to `m5` for biasing the OTA.

```
filename: ex4-17d.sp
.incl bsim.sp
*******OTA*********
m1 5 2 3 3 cmosp w=49u l=0.44u nrd=0.003 nrs=0.003
+  ad=27p pd=99.1u
m2 4 1 3 3 cmosp w=49u l=0.44u nrd=0.003 nrs=0.003
+  ad=27p pd=99.1u
m3 5 4 0 0 cmosn w=6.8u l=0.45u nrd=0.03 nrs=0.03
+  ad=3.74p pd=14.7u
m4 4 4 0 0 cmosn w=6.8u l=0.45u nrd=0.03 nrs=0.03
+  ad=3.74p pd=14.7u
m5 3 6 7 7 cmosp w=46u l=0.4u nrd=0.003 nrs=0.003
+  ad=25.3p pd=93.1u
m6 6 6 7 7 cmosp w=46u l=0.4u nrd=0.003 nrs=0.003
+  ad=25.3p pd=93.1u
cl 5 0 2.4p
ir 6 0 220u
vdd 7 0 1.8
******open-loop******
vin 2 0 dc 0.7
vip 1 0 dc 0.7 ac 1
.dc vip 0 1.8 0.1m
*.ac dec 100 10k 10g
*****closed-loop*****
*vs 2 5 0
*vip 1 0 dc 0.7 ac 1
*.ac dec 100 10k 10g
*******tran********
*vs 2 5 0
*vip 1 0 pulse 0.6 0.8 5n 0.1n 0.1n 20n
*.tran 0.01n 40n
*******noise*******
*vs 2 5 0
*vip 1 0 dc 0.7 ac 1
*.noise v(5) vip dec 100 10 10g
********************
.end
```

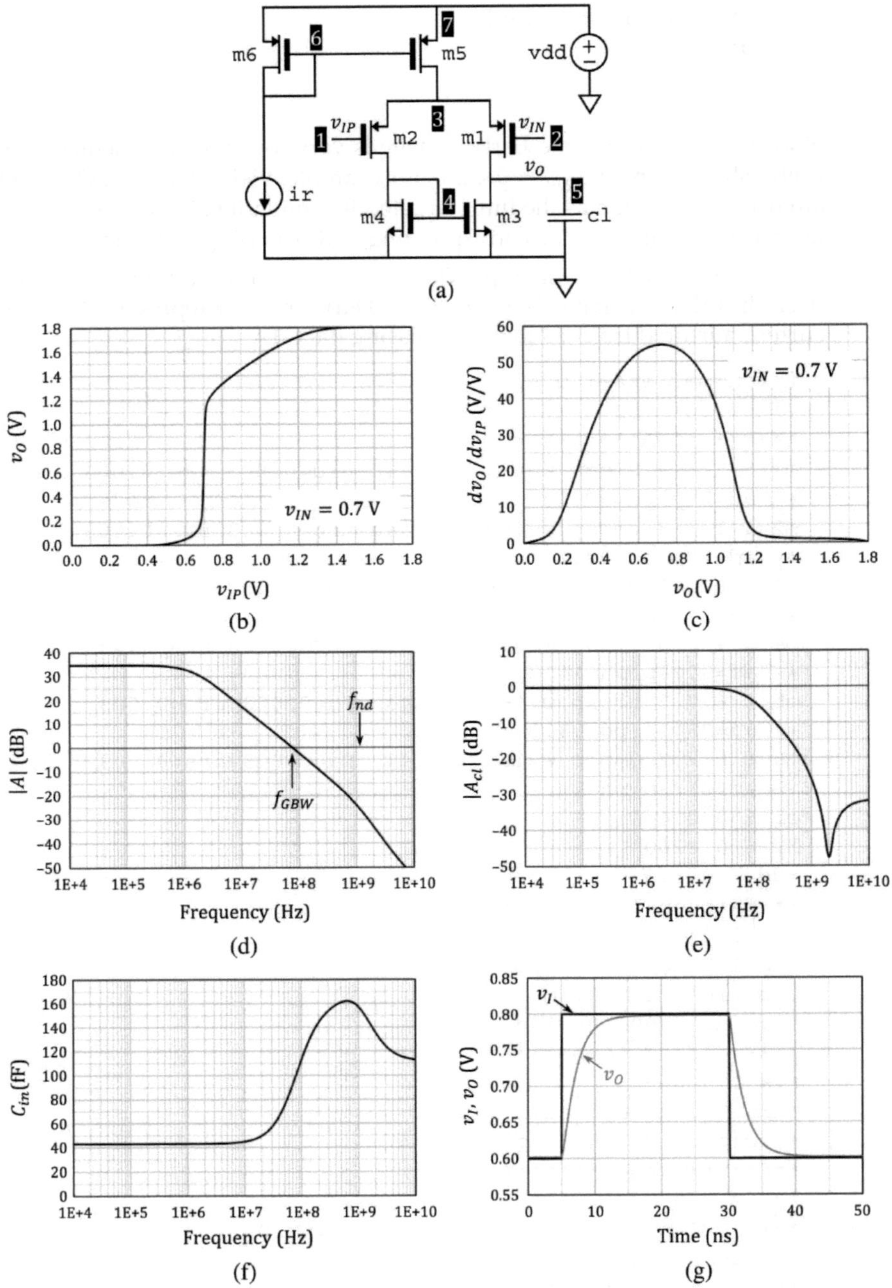

FIGURE 4.39 (a) OTA schematic netlisted in `ex4-17d.sp` of Example 4.17. (b) Open-loop dc transfer characteristic. (c) Open-loop dc voltage gain as a function of output voltage. (d) Open-loop magnitude response. (e) Closed-loop magnitude response. (f) Input capacitance as a function of frequency. (g) Step response.

Two separate open-loop simulations are run. One is a dc sweep performed for evaluating the properties of the dc transfer characteristic. The other is an ac simulation performed for extracting the gain-bandwidth product and input capacitance. The reader is reminded that when running one of the two, the line describing the other and the lines belonging to closed-loop and transient simulations should be closed. Upon completion of the dc sweep, the transfer characteristic is plotted with the command

```
plot v(5)
```

The outcome, shown in Fig. 4.39(b), exhibits an output range much wider than the output swing specified in between 0.6 and 0.8 V. This is not surprising because, as the reader will recall, the constraints associated with the limits of the output range proved to be redundant in design. Offset also appears to be small. It can be measured by zooming in on the gain segment, and reading the deviation of v_I from 0.7 V for v_O = 0.7 V. The outcome is V_{os} = 125 μV.

DC differential gain A_0 is extracted from the derivative of v_O with respect to v_I as plotted against v_O with the command

```
plot deriv(v(5)) vs v(5)
```

As observed in Fig. 4.39(c), A_0 equals 54.8 V/V at v_O = 0.7 V, which compares reasonably well with the predicted 50.4 V/V.

The open-loop ac simulation is run for assessing the frequency response. Plotting with

```
plot vdb(5)-vdb(1)
```

yields the magnitude response shown in Fig. 4.39(d). This plot exhibits a unity-gain frequency around 77 MHz and a nondominant-pole frequency f_{nd} around 1 GHz. The order-of-magnitude difference between the two indicates an almost ideal single-pole frequency response, which implies the equality of the unity-gain frequency and the gain-bandwidth product. Since the application has a unity-gain configuration, we also expect the closed-loop bandwidth to be equal to the gain-bandwidth product. The outcome of the ac simulation is therefore indicative of $f_{cl} = f_{GBW}$ = 77 MHz, which is 10% higher than the target set for closed-loop bandwidth. According to (4.115), this discrepancy may result from a lesser-than-expected OTA contribution to C_L or from a larger-than-targeted $g_{m(1,2)}$. The latter of the two factors is dominant in the present case. As the reader can print with a `show` command, the simulated value of $g_{m(1,2)}$ equals 1.21 mA/V instead of the targeted 1.1 mA/V. Incidentally, a mildly excessive f_{GBW} can be lowered back to the target simply by trimming down I_B hence $g_{m(1,2)}$ without significantly affecting the rest of the performance metrics. This is true regardless of whether C_L or $g_{m(1,2)}$ is the cause of excessive f_{GBW}. The present discrepancy, however, is too small to warrant trimming.

A closed-loop ac simulation is run for verifying f_{cl} directly, and for evaluating the input capacitance. Feedback is established by adding a short-circuit source `vs` to the netlist. For f_{cl}, we plot the magnitude response with the same command used for its open-loop counterpart, and read the frequency at which the magnitude drops by 3 dB with respect to its low-frequency value. The outcome displayed in Fig. 4.39(e) indicates f_{cl} = 80 MHz, which is indeed close to what is estimated in the preceding paragraph. As to the input capacitance, we plot it as a function of frequency with the command

```
plot imag(-i(vip)/v(1))/(2*pi*frequency)
```

The plot shown in Fig. 4.39(f) verifies the specified 140-fF maximum capacitance within closed-loop bandwidth.

A transient simulation is run for visually evaluating the step response in closed loop. The output waveform shown in Fig. 4.39(g) exhibits an overdamped response, which is consistent with the wide separation between f_{nd} and f_{cl} as previously discussed in Subsection 2.7.1 and displayed in Figs. 2.26(a) and (b). Also evident is the purely exponential variation of the rising and falling edges of the output waveform, which is typical of a slew-free response.

Finally, we evaluate the noise performance of the unity-gain amplifier. Since the external network is nothing but a short circuit, output noise is due solely to the OTA itself, and is described by (4.124) for $\beta = 1$. It yields $V_{no} = 76.5\ \mu V$ for the following parameter values: $E_{th} = 4.14 \times 10^{-21}$ J, $A_{NP} = 1.3$, $C_L = 2.5$ pF, $g_{m(3,4)} = 880\ \mu A/V$ as determined from Fig. 3.12 for $V_{GS(3,4)} - V_{TN} = 0.24$ V and $I_{D(3,4)} = 110\ \mu A$, $g_{m(1,2)} = 1.1$ mA/V, $f_{cl} = 70$ MHz, $f_1 = 10$ Hz, $K_{1,2} = 2 \times 10^{-20}\ A^2\mu m^2$ as determined from Fig. 3.44(c) for $V_{GS(1,2)} - V_{TP} = -0.185$ mV, $W_{1,2} = 49\ \mu m$, $L_{1,2} = 0.44\ \mu m$, $K_{3,4} = 2 \times 10^{-19}\ A^2\mu m^2$ as determined from Fig. 3.44(b) for $V_{GS(3,4)} - V_{TN} = 0.24$ V, $W_{3,4} = 6.8\ \mu m$, and $L_{3,4} = 0.45\ \mu m$. Simulated verification is performed with the noise section of `ex4-17d.sp`, and is printed with the command

```
print onoise_total
```

The outcome is $V_{no} = 71.7\ \mu V$. ▲

Example 4.18 OTAs are good for driving capacitive loads, which is why they find unrestricted use in discrete-time closed-loop configurations. Among continuous-time closed-loop configurations only the unity-gain amplifier can be constructed with an OTA because it is the only application that doesn't deploy a resistive external network. It is possible to build OTA-based continuous-time amplifiers or filters with larger-than-unity gain in open loop but only at the expense of inferior linearity. One such open-loop amplifier is shown in Fig. 4.40(a), where V_B and v_i represent the bias and input-signal sources, v_O is the output voltage, and v_o is the output signal. Notice the application of v_i directly to the differential input of OTA-1. This OTA, acting as a transconductor in open loop, generates an output current ideally described by $i_{o1} = G_{m1}v_i$, where G_{m1} is the effective transconductance. OTA-2, on the other hand, receives $-v_o$ as its differential-input signal, and therefore generates an output current $i_{o2} = -G_{m2}v_o$. Since the purely capacitive load implies $i_{o2} = -i_{o1}$ at low frequencies, the dc voltage gain v_o/v_i equals G_{m1}/G_{m2}, which can be set to any desired value by ratioing the effective transconductances. For $v_i = 0$, OTA-1 does not generate any output current, which also implies $i_{o2} = 0$, hence $v_o = 0$ and $v_O = V_B$ at the operation point.

Unless driven by an input signal of small amplitude, the open-loop amplifier of Fig. 4.40(a) exhibits poor linearity due to the fact that the differential-input voltages of the OTAs are as large as v_i and v_o instead of being virtually short-circuited by a negative feedback loop. If necessary, linearity can be improved at the expense of topological

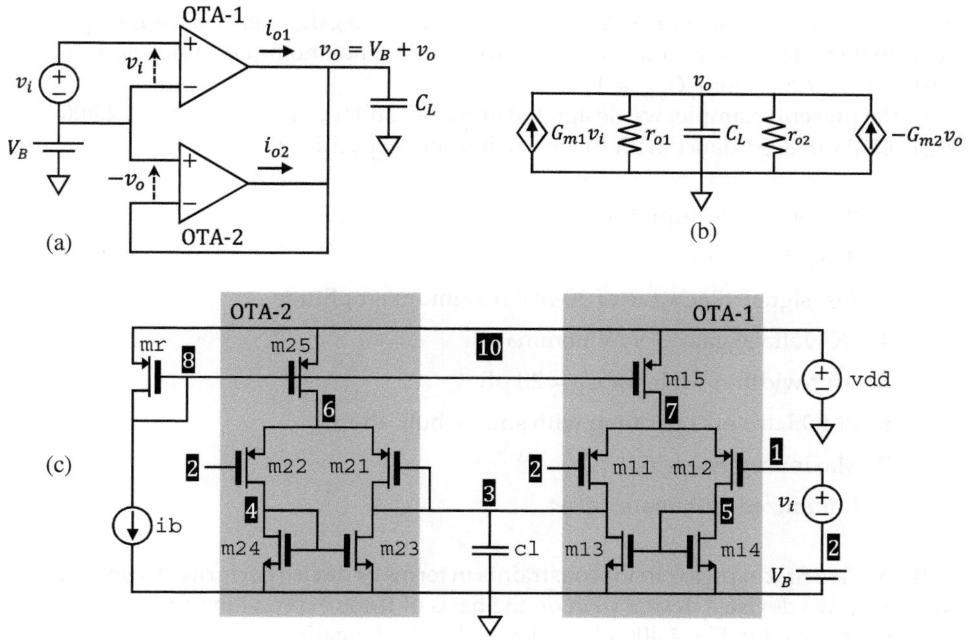

FIGURE 4.40 (a) Open-loop amplifier/filter application schematic of Example 4.18. (b) DC small-signal equivalent circuit. (c) Circuit schematic netlisted in `ex4-18d.sp`.

complexity by applying source degeneration to the differential pair as described in Subsection 4.4.4, or by deploying the super-gm technique presented in Subsection 4.4.5.

For an accurate small-signal analysis of the amplifier, consider the small-signal equivalent circuit given in Fig. 4.40(b), where each OTA is represented by its effective transconductance G_m and output resistance r_o in compliance with the dc version of the generic equivalent circuit presented earlier in Fig. 2.29. The dc representation is justified by the fact that the dominant pole of the amplifier is usually defined by the load capacitance C_L instead of the internal capacitances of the OTAs. A straightforward analysis yields the following transfer function

$$\frac{v_o}{v_i} = \frac{A_0}{1 + \frac{s}{2\pi f_o}}, \tag{4.128}$$

where

$$A_0 = \frac{G_{m2} r_o}{1 + G_{m2} r_o} \frac{G_{m1}}{G_{m2}}, \tag{4.129}$$

$$f_o = \frac{1 + G_{m2} r_o}{G_{m2} r_o} \frac{G_{m2}}{2\pi C_L}, \tag{4.130}$$

and $r_o \doteq r_{o1} \parallel r_{o2}$, where the subscripts 1 and 2 refer to OTA-1 and OTA-2, respectively. Note that the dc voltage gain and bandwidth tend to their nominal values $A_0 = G_{m1}/G_{m2}$ and $f_o = G_{m2}/2\pi C_L$ for $r_o G_{m2} \gg 1$.

In the present example, we design the amplifier/filter application of Fig. 4.40(a) with basic OTAs in accordance with the following set of specifications:

1. Power: Single-supply $V_{DD} = 1.8$ V.
2. Bias: $V_B = 0.9$ V.
3. Test signal: Sinewave of 30-mV maximum amplitude.
4. DC voltage gain: 3 V/V nominal.
5. Bandwidth: 1 MHz for $C_L = 20$ pF.
6. PMOS drivers operating with source-bulk short.
7. Maximized linearity.
8. Minimized systematic offset.

Work We first express circuit constraints in terms of device performance metrics. Subsequently, we devise a design plan on the basis of these expressions and execute it. The reader is referred to Fig. 4.40(c) for a device-level schematic of the amplifier.

- Without source degeneration, we can improve the linearity only by enlarging the differential-input range V_{DIR} of the OTAs as much as possible. According to (4.85), a large V_{DIR} calls for a low transconductance efficiency and therefore a large driver overdrive voltage at the operation point. A large overdrive voltage can be achieved by implementing the drivers with low-threshold devices and operating them with a large gate-source voltage. Since their gate voltage at the operation point is fixed by specification, we need to raise their source voltage for this purpose. What sets a limit on the source voltage in a differential pair is the compliance condition of the bias-current source. This condition is threatened most severely in OTA-2 when the output signal at node 3 rises to the peak $v_{o(\max)} = 3 \times 30\,\text{mV} = 90$ mV and thus lifts the source voltage at node 6 to its highest level. The signal component of the source voltage equals $v_{o(\max)}/2$ as indicated by (4.81) for $g_{mb(21,22)} = 0$ and $r_{is21} = r_{is22}$. Therefore, the maximum source voltage can be expressed as $v_{S(\max)} = V_S + v_{o(\max)}/2 = V_S + 0.045\,\text{V}$, where V_S represents the bias value. The compliance condition of M25 allows for a common-source voltage as high as $V_{DD} + V_{DSsat25} + \Delta V_{DS25}$, where ΔV_{DS25} is the saturation margin of M25. Targeting a comfortable $V_{DSsat25} + \Delta V_{DS25} = -0.25$ V, we thus obtain $v_{S(\max)} = 1.8 - 0.25 = 1.55$ V, and, therefore, $V_S = 1.505$ V.

  Considering $V_B = 0.9$ V for the bias voltage at driver gates, this result leads us to $V_{GS(21,22)} = V_B - V_S = 0.9 - 1.505 = -0.605$ V for driver gate-source voltages at the operation point. Since the threshold voltage indicated by Fig. 4.10 for the low-threshold option is -0.1 V or less for any reasonable channel length, the corresponding overdrive voltage of M21 and M22 is determined as $V_{GS(21,22)} - V_{TP} = -0.5$ V. This is what we will target for maximizing the linearity.

- According to (4.129), the nominal value specified for the dc voltage gain implies

$$\frac{G_{m1}}{G_{m2}} = \frac{g_{m(11,12)}}{g_{m(21,22)}} = 3,$$

where $g_{m(11,12)}$ and $g_{m(21,22)}$ are the driver transconductances of OTA-1 and OTA-2, respectively. It is preferable to set this gain with identical overdrive voltages for OTA-1 and OTA-2. Such an identity will provide the OTAs with identical transconductance efficiencies and identical sheet currents, and therefore enable us to implement the gain by scaling the aspect ratios and bias currents by the same factor as $(W/L)_{11,12} = 3(W/L)_{21,22}$ and $I_{B1} = 3I_{B2}$. Implementing each device in OTA-1 with three parallel copies of the corresponding device in OTA-2 will automatically ensure this scaling.

As indicated by (4.129) and (4.130), the sensitivity of A_0 and f_o to the effective output resistance $r_o \doteq r_{o1} \parallel r_{o2}$ can be reduced only by making the product $G_{m2}r_o$ much larger than 1. This product can be expressed in a more appropriate form by (a) considering $r_{o1} = r_{o2}/3$ due to scaling by 3, and thus replacing r_o with $r_{o2}/4$, (b) replacing G_{m2} with $g_{m(21,22)}$, (c) dividing $g_{m(21,22)}$ by $I_{B2}/2$ and multiplying r_{o2} by $I_{B2}/2$, where I_{B2} is the bias voltage of OTA-2, and (d) replacing $g_{m(21,22)}/(I_{B2}/2)$ with 3.5 V^{-1} as indicated by Fig. 4.14 for $V_{GS(21,22)} - V_{TP} = -0.5$ V. The outcome is

$$G_{m2}r_o = 0.875 \times V_{A2}, \tag{4.131}$$

where $V_{A2} \doteq r_{o2}(I_B/2)$ is the effective Early voltage of OTA-2. Considering $r_{o2} = r_{ds21} \parallel r_{ds23}$, V_{A2} can be expressed as

$$V_{A2} = \frac{V_{A21}V_{A23}}{V_{A21} + V_{A23}}. \tag{4.132}$$

- Minimized systematic offset calls for a 0.9-V gate-source voltage for M13, M14, M23, and M24 because this is how much the output bias is. This gate-source voltage is large enough to justify the standard-threshold option for these devices.
- The bandwidth specification implies $g_{m(21,22)} = 126\ \mu\text{A/V}$ from (4.130) provided that $G_{m2}r_o$ is sufficiently larger than 1.

Our design plan is focused on OTA-2 because OTA-1 is just a scaled-up replica of the former. Since $g_{m(21,22)}/(I_{B2}/2)$ and $g_{m(21,22)}$ have already been determined, we first calculate I_{B2}, and use the outcome together with the specific $I_{\square(21,22)}$ corresponding to $V_{GS(21,22)} - V_{TP} = -0.5$ V for determining $(W/L)_{21,22}$. What follows is an evaluation of $G_{m2}r_o$ from (4.131) as a function of $L_{DR(21,22)}$ and $L_{DR(23,24)}$. We do this by extracting the Early voltages of M21 and M23 as functions of channel length, and calculating the corresponding V_{A2}. After deciding on a reasonable channel-length combination, we calculate $W_{DR(21,22)}$ from $(W/L)_{21,22}$. The overdrive voltage hence sheet current of M23 and M24 are determined next. This is done with (a) the threshold voltage as determined for the channel length selected for these devices, and (b) the gate-source voltage set to 0.9 V for minimizing the offset. Since the bias current of these devices is already available, we determine their aspect ratio $(W/L)_{23,24}$ and therefore the channel width $W_{DR(23,24)}$. The final step of OTA-2 design is the sizing of M25 whose smallest drain-source voltage has been determined as −0.25 V on the basis of compliance considerations. We allocate a reasonable fraction of this voltage to $V_{DSsat25}$, and determine the corresponding overdrive voltage $V_{GS25} - V_{TP}$, which, in turn, enables us to determine the sheet current $I_{\square 25}$ and therefore the aspect ratio of this device. Its channel length is determined with an iterative simulation targeting an r_{ds25} much larger than the resistance seen into the source terminal of M21 or M22. The channel width of M25 is calculated from the aspect ratio and the channel length.

As to the devices of OTA-1, each is sized by paralleling three copies of the corresponding device of OTA-2.

Steps of design flow are presented next.

1. $g_{m(21,22)} = 126\ \mu A/V$ and $g_{m(21,22)}/(I_{B2}/2) = 3.5\ V^{-1} \Rightarrow I_{B2} = 72\ \mu A$.
2. $V_{GS(21,22)} - V_{TP} = -0.5\ V \rightarrow$ Fig. 4.12 $\Rightarrow I_{\square(21,22)} = 7.9\ \mu A/sq$.
3. $(W/L)_{(21,22)} = (I_{B2}/2)/I_{\square(21,22)} = 36/7.9 = 4.56$.
4. Early voltages $V_{A(21,22)}$ of M21/M22 and $V_{A(23,24)}$ of M23/M24 are determined with simulation for six different values of L_{DR} in between 0.4 and 4 μm. This is done with the following input files: `ex4-18a.sp` and `ex4-18b.sp`.

```
filename: ex4-18a.sp
.incl bsimlvt.sp
m21 1 1 0 0 cmosplvt w=10u l=0.2u
vgs 1 0 -0.62
.control
op
print @m21[id]/@m21[gds]
.endc
.end
```

```
filename: ex4-18b.sp
.incl bsim.sp
m23 1 1 0 0 cmosn w=10u l=0.2u
vgs 1 0 0.9
.control
op
print @m23[id]/@m23[gds]
.endc
.end
```

The former of these two files is for extracting $V_{A(21,22)}$. For each value of L_{DR}, V_{TP} is determined from Fig. 4.10 and added to $V_{GS(21,22)} - V_{TP} = -0.5$ V for setting the gate-source voltage `vgs` in this file.

The file `ex4-18b.sp` is used for extracting $V_{A(23,24)}$. Regardless of channel-length selection, `vgs` in this file is kept constant at 0.9 V.

For a reasonable lower limit 10 V/V for $G_{m2}r_o$, (4.131) indicates the necessity of at least 11.4 V for V_{A2}. According to (4.132) and the simulation data presented in Table 4.1, $L_{DR(21,22)} = L_{DR(23,24)} = 2\ \mu m$ leads to $V_{A2} = 11.7$ V, and therefore to $G_{m2}r_o = 10.2$ V.

L_{DR} (μm)	0.4	0.6	0.8	1.2	2.0	4.0
$V_{A(21,22)}$ (V)	7.0	9.1	10.7	13.0	16.0	19.3
$V_{A(23,24)}$ (V)	14.9	19.0	22.7	29.9	43.0	65.0

TABLE 4.1 Early Voltages versus Channel Length in Example 4.18

5. $L_{DR(21,22)} = 2$ μm and $(W/L)_{(21,22)} = 4.56 \Rightarrow W_{DR(21,22)} = 9.2$ μm.
6. $L_{DR(23,24)} = 2$ μm → Fig. 3.7 $\Rightarrow V_{TN(23,24)} = 0.4$ V, and therefore $V_{GS(23,24)} - V_{TN} = 0.9 - 0.4 = 0.5$ V.
7. $V_{GS(23,24)} - V_{TN} = 0.5$ V → Fig. 3.10 $\Rightarrow I_{\square(23,24)} = 27$ μA/sq, and therefore $(W/L)_{23,24} = (I_{B2}/2)/I_{\square(23,24)} = 1.33$.
8. $(W/L)_{23,24} = 1.33$ and $L_{DR(23,24)} = 2$ μm $\Rightarrow W_{DR(23,24)} = 2.7$ μm.
9. Select the standard-threshold option for M25. We partition its worst-case drain-source voltage of −0.25 V into $V_{DSsat25} = -0.150$ V and $\Delta V_{DS25} = -0.1$ V.
10. $V_{DSsat25} = -0.150$ V → Fig. 3.17 $\Rightarrow V_{GS25} - V_{TP} = -0.170$ V.
11. $V_{GS25} - V_{TP} = -0.170$ V → Fig. 3.19 $\Rightarrow I_{\square 25} = 0.84$ μA/sq, and therefore $(W/L)_{25} = I_B/I_{\square 25} = 72/0.84 = 86$.
12. Determine L_{DR25} with iterative simulation. Since $g_{m(21,22)}$ is very low, targeting an r_{ds25} much larger than $15/g_{m(21,22)} \approx 80$ kΩ will probably lead to an impractically long channel. For $I_B = 72$ μA, this much resistance corresponds to $V_{A25} = 8.6$ V. The drain-source bias voltage of M25 is given by

$$V_{DS25} = V_S - V_{DD} = 1.505 - 1.8 = -0.295\,\text{V}.$$

Its gate-source voltage is calculated from $V_{GS25} = (V_{GS25} - V_{TP}) + V_{TP} = -0.17 + V_{TP}$ following the characterization of V_{TP} from Fig. 3.16 for the channel length selected in each iteration. Simulation is run with the following input file `ex4-18c.sp`.

```
filename: ex4-18c.sp
.incl bsim.sp
m25 2 1 0 0 cmosp w=10u l=1.4u
vgs 1 0 -0.575
vds 2 0 -0.295
.control
op
print @m25[id]/@m25[gds]
.endc
.end
```

Iterative simulation yields $L_{DR25} = 1.4$ μm and $V_{A25} = 8.78$ V. The corresponding channel width W_{DR25} is 115 μm.

13. Verification is performed on the schematic shown in Fig. 4.40(c) and netlisted in the following input file `ex4-18d.sp`:

```
filename: ex4-18d.sp
.incl bsimlvt.sp
.incl bsim.sp
*************ota2*************
m21 3 3 6 6 cmosplvt w=9.2u l=2u nrd=0.02 nrs=0.02
+  ad=5.1p pd=20u
m22 4 2 6 6 cmosplvt w=9.2u l=2u nrd=0.02 nrs=0.02
+  ad=5.1p pd=20u
m23 3 4 0 0 cmosn w=2.7u l=2u nrd=0.08 nrs=0.08
+  ad=1.5p pd=6.5u
m24 4 4 0 0 cmosn w=2.7u l=2u nrd=0.08 nrs=0.08
+  ad=1.5p pd=6.5u
m25 6 8 10 10 cmosp w=115u l=1.4u nrd=0.002 nrs=0.002
+  ad=63p pd=231u
*************ota1**************
m11 3 2 7 7 cmosplvt w=9.2u l=2u nrd=0.02 nrs=0.02
+  ad=5.1p pd=20u m=3
m12 5 1 7 7 cmosplvt w=9.2u l=2u nrd=0.02 nrs=0.02
+  ad=5.1p pd=20u m=3
m13 3 5 0 0 cmosn w=2.7u l=2u nrd=0.08 nrs=0.08
+  ad=1.5p pd=6.5u m=3
m14 5 5 0 0 cmosn w=2.7u l=2u nrd=0.08 nrs=0.08
+  ad=1.5p pd=6.5u m=3
m15 7 8 10 10 cmosp w=115u l=1.4u nrd=0.002 nrs=0.002
+  ad=63p pd=231u m=3
*****bias sources and load*****
mr 8 8 10 10 cmosp w=115u l=1.4u nrd=0.002 nrs=0.002
+  ad=63p pd=231u
ib 8 0 72u
cl 3 0 20p
vb 2 0 0.9
vdd 10 0 1.8
**************dc***************
vi 1 2
.control
dc vi -30m 30m 0.1m
plot v(3)
plot deriv(v(3))
.endc
**************ac***************
*vi 1 2 dc 0 ac 1
*.control
*ac dec 100 100 10meg
*plot vdb(3)
*.endc
************tran*************
*vi 1 2 sin 0 30m 1meg
*.control
*tran 1n 15u
*plot v(3)
```

```
*fourier 1meg v(3)
*.endc
******************************
.end
```

The file defines three separate simulations. A dc simulation is performed by sweeping v_i between –30 and 30 mV. The resulting dc transfer characteristic and its derivative with respect to v_i are plotted automatically. The reader is invited to compare the derivative at $v_i = 0$ with the exact estimate of A_0 as obtained from (4.129) for $G_{m2}r_o = 10.2$ V.

An ac simulation is run for verifying the bandwidth, which is extracted from the plot as the frequency at which the magnitude drops by 3 dB. For an exact estimate of the bandwidth, the reader is referred to (4.130) for $G_{m2}r_o = 10.2$ V.

The final simulation is a transient analysis performed for evaluating the linearity of the amplifier in terms of THD. A sinusoidal input signal of 1-MHz frequency and 30-mV amplitude is applied. This simulation plots the output-voltage waveform and prints out the distortion data. ▲

4.6 Symmetrical OTA

As explained in Subsection 4.5.2, the output range of a basic OTA is not independent of the common-mode swing. Symmetrical OTA is the simplest OTA in which this limitation is removed. Its dc features, frequency and step responses, and noise properties are studied in the following five subsections. In its elementary form, the symmetrical OTA provides a moderate dc differential gain not much different from the basic OTA. Its cascoded version, however, enhances the gain to the order of 1000 V/V. This very popular extension is the subject of the final subsection.

4.6.1 Topology and DC Transfer Characteristic

The symmetrical OTA topology in its elementary form is presented in Figs. 4.41(a) and (b) with NMOS and PMOS drivers, respectively. A differential pair built with matching drivers M1 and M2 and bias-current sink/source M9 constitutes the input stage of the OTA. The noninverting- and inverting-input terminals of this stage are identified with voltages v_{IP} and v_{IN}, respectively. The common-drain terminal of M7 and M8 is designated as the output; its voltage is denoted by v_O. M3, M4, M5, and M7 function as current mirrors. So do M6 and M8. Saturation is guaranteed for M3, M4, and M6 by virtue of their diode connection, and is ensured by design for the rest of the devices within the specified common-mode range and output range.

Shown in Fig. 4.41(c) is a typical dc transfer characteristic relating the output voltage v_O to the differential-input voltage $v_{ID} \doteq v_{IP} - v_{IN}$ in a single-supply ($V_{DD} = 1.8$ V) symmetrical OTA of NMOS drivers. Part d of the same figure exhibits the transconductance characteristics relating the driver currents i_{D1} and i_{D2} to v_{ID}, and also shows the variation of the dc current $i_{D(7,8)}$ flowing in the output branch through M7 and M8.

We now interpret these characteristics starting with a negative v_{ID}, for which, the differential pair keeps i_{D2} below i_{D1}. As will be validated shortly, M7 is saturated in this case, which enables M4 to mirror the smaller i_{D2} of the two driver currents onto the

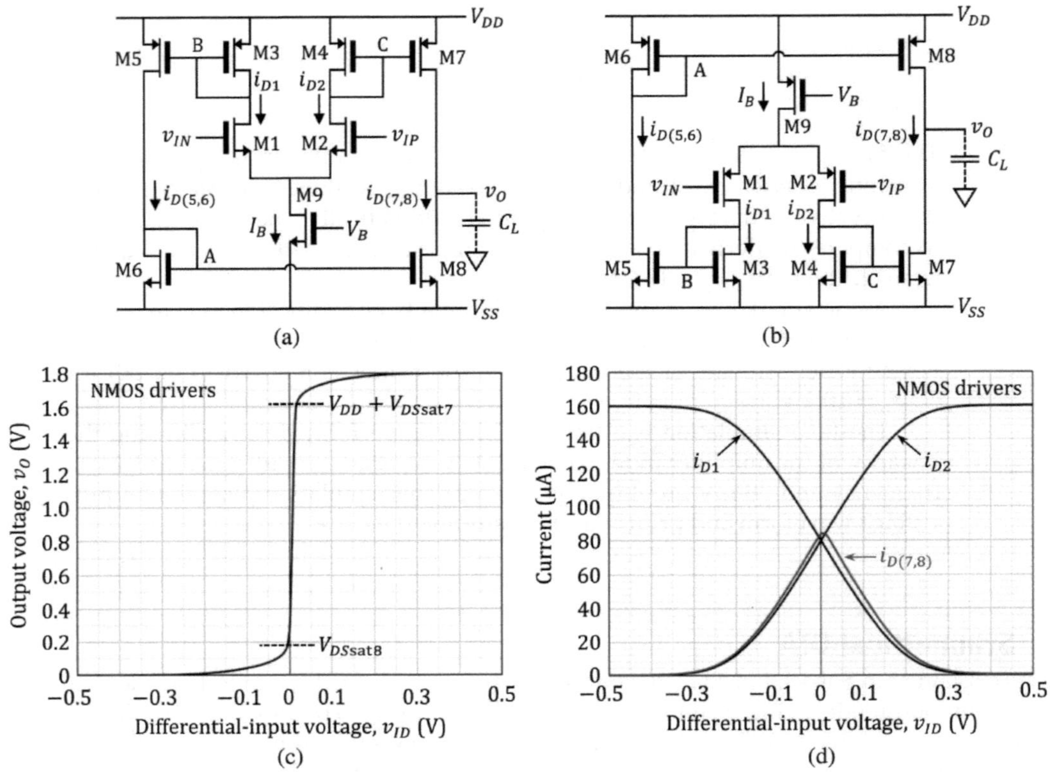

FIGURE 4.41 (a) Symmetrical OTA with NMOS drivers. (b) Symmetrical OTA with PMOS drivers. (c) DC transfer characteristic of a single-supply symmetrical OTA with NMOS drivers. (d) Transconductance characteristics of the symmetrical OTA whose dc transfer characteristic is presented in part c.

output branch as $i_{D(7,8)} = i_{D2}$. On the other side of the differential pair, the unconditionally saturated M3 mirrors the larger current i_{D1} to M5 which is kept in saturation by the low voltage maintained at node A by the diode-connected M6. However, this common current $i_{D(5,6)} = i_{D1}$ of M5 and M6 cannot be further mirrored by M6 to M8 because the latter of these two devices has to conduct the smaller current i_{D2} imposed on it by M7. Mirroring is suppressed by the nonsaturation mode of M8 as secured by a v_O lesser than the drain-source saturation voltage V_{DSsat8} of this device. Such a low v_O is what keeps M7 in saturation as presumed above for $v_{ID} < 0$.

As we push the negative v_{ID} closer to zero, the smaller i_{D2} increases while the larger i_{D1} decreases. The increasing i_{D2} elevates $i_{D(7,8)}$. The decreasing i_{D1}, on the other hand, lowers $i_{D(5,6)}$, which, in turn, lowers the common gate-source voltage of M6 and M8. The nonsaturated M8 is thus forced to conduct a higher current with a lower gate-source voltage, which can be done only by raising the drain-source voltage. This is why v_O increases at a growing rate with v_{ID}.

As v_O reaches V_{DSsat8}, M8 enters saturation and thus starts conducting i_{D1} as mirrored by M6. But, as long as v_O is lower than $V_{DD} + V_{DSsat7}$, M7 also remains in saturation, and continues to conduct i_{D2} as mirrored by M4. Since these two devices have to conduct the same current, this concurrent saturation condition exists only for $i_{D2} = i_{D1}$, hence $v_{ID} = 0$. Making v_{ID} even slightly positive moves i_{D2} above i_{D1}, and therefore forces M7 out of saturation in order to limit its current to the smaller i_{D1} enforced by M8. This is why v_O rises very steeply to $V_{DD} + V_{DSsat7}$ around $v_{ID} = 0$.

The reader can explain the part of the characteristics observed for $v_{ID} > 0$ simply by transposing the roles of M7 and M8 in the foregoing interpretation.

4.6.2 Range Limitations

The output range of a symmetrical OTA is limited by the saturation conditions of M7 and M8. The limits are therefore described by

$$V_{OH} = V_{DD} + V_{DSsat7} \quad \text{(NMOS drivers)}, \tag{4.133}$$

$$V_{OL} = V_{SS} + V_{DSsat8} \quad \text{(NMOS drivers)} \tag{4.134}$$

for the topology shown in Fig. 4.41(a), and by

$$V_{OH} = V_{DD} + V_{DSsat8} \quad \text{(PMOS drivers)}, \tag{4.135}$$

$$V_{OL} = V_{SS} + V_{DSsat7} \quad \text{(PMOS drivers)} \tag{4.136}$$

for the topology shown in Fig. 4.41(b).

Note that V_{OL} or V_{OH} are independent of $V_{CM(\text{min})}$ and $V_{CM(\text{max})}$ in contrast to the case of basic-OTA. An almost rail-to-rail output range can therefore be realized in a symmetrical OTA simply by selecting small drain-source saturation voltages. This calls for small overdrive voltages for M7 and M8.

The lower limit of the common-mode range in a symmetrical OTA with NMOS drivers is described by

$$V_{IL} = V_{SS} + V_{DSsat9} + V_{GS(1,2)} \quad \text{(NMOS drivers)}. \tag{4.137}$$

If M1 and M2 operate without source-bulk shorts, $V_{GS(1,2)}$ will be specific to the lower limit of the common-mode range.

The upper limit of the common-mode range in the case of NMOS drivers is given by

$$V_{IH} = V_{DD} + V_{GS(3,4)} - V_{DSsat(1,2)} + V_{GS(1,2)} \quad \text{(NMOS drivers)}, \tag{4.138}$$

where $V_{GS(1,2)}$ is specific to the upper limit if M1 and M2 operate without source-bulk shorts.

For a symmetrical OTA with PMOS drivers, common-mode range limits are described by

$$V_{IL} = V_{SS} + V_{GS(3,4)} - V_{DSsat(1,2)} + V_{GS(1,2)} \quad \text{(PMOS drivers)}, \tag{4.139}$$

where $V_{GS(1,2)}$ is specific to the lower limit if M1 and M2 operate without source-bulk shorts, and by

$$V_{IH} = V_{DD} + V_{DSssat9} + V_{GS(1,2)} \quad \text{(PMOS drivers)}, \tag{4.140}$$

where $V_{GS(1,2)}$ is specific to the upper limit if M1 and M2 operate without source-bulk shorts.

4.6.3 DC Differential Gain and Offset

The dc differential gain A_0 of the symmetrical OTA topology is described in Example 3.14 of Subsection 3.7.1 by (3.70) in terms of device parameters. Since M7 and M4 have identical transconductance efficiencies due to their common overdrive voltage and conduct identical currents, the ratio $g_{m7}/g_{m(3,4)}$ appearing in that equation equals 1, which turns (3.70) into

$$A_0 = \underbrace{g_{m(1,2)}}_{G_m} \underbrace{r_{ds7} \parallel r_{ds8}}_{r_o} . \tag{4.141}$$

Note that G_m and r_o are the effective-transconductance and output-resistance parameters of the OTA.

A_0 can be expressed also along the lines of (4.111) as

$$A_0 = \frac{g_{m(1,2)}}{I_B/2} \frac{V_{A7} V_{A8}}{V_{A7} + V_{A8}}, \tag{4.142}$$

which prescribes high Early voltages in output devices and a large transconductance efficiency in driver devices for achieving a large A_0. It is also desirable to target comparable V_{A7} and V_{A8} in order to reduce the sensitivity of A_0 to v_O as explained in Subsection 4.5.3. In any case, A_0 is on the order of 100 V/V at most. The OTA topologies of Fig. 4.41 are therefore suitable for use in open-loop transconductor applications and in discrete-time closed-loop amplifier/filter applications that can tolerate large gain error and nonlinearity. Fortunately, the symmetrical OTA has a favorable topology for cascoding, which enhances the gain typically to the order of 1000 V/V. The details of cascoded-symmetrical OTAs are presented in Subsection 4.6.6.

Offset properties of the symmetrical OTA are also very similar to those of the basic OTA. As usual, the variance of offset is minimized by enlarging the dimensions of all matching devices. Provided that the measures described in Subsection 4.5.3 are taken, the mean value of offset will be due to the asymmetry caused by the difference between the drain voltages of M6 and M8 at the operation point. The drain voltage of the diode-connected M6, also being the gate voltage of M8, is the main variable of the output range through V_{DSsat8} as indicated by (4.134) and (4.135). In any application that necessitates a wide output range, it has to be set close to V_{SS} for the NMOS-driver version and to V_{DD} for the PMOS-driver version. The drain voltage V_O of M8, on the other hand, has to be close to midway between V_{SS} and V_{DD} for a wide output range. A significant difference between the two drain voltages is therefore inevitable in most applications. Fortunately, this restriction is eliminated in the high-gain cascoded version of the symmetrical OTA as will be discussed in Subsection 4.6.6.

4.6.4 Frequency Response and Step Response

An equivalent circuit appropriate for analyzing the frequency response of a symmetrical OTA is shown in Fig. 4.42 with NMOS drivers and lumped capacitances. It yields

$$A = g_{m(1,2)} r_o \frac{1 + s\frac{C_2}{2g_{m(6,8)}}}{\left(1 + s r_o C_L\right)\left(1 + s\frac{C_2}{g_{m(6,8)}}\right)\left(1 + s\frac{C_1}{g_{m(3,4,5,7)}}\right)}, \tag{4.143}$$

where $r_o C_L$ is usually by far the largest time constant. The dominant-pole frequency and gain-bandwidth product are therefore described by

$$f_o = \frac{1}{2\pi r_o C_L}, \tag{4.144}$$

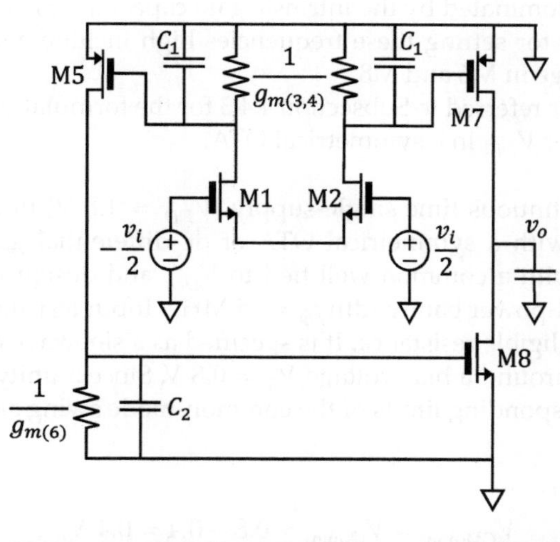

FIGURE 4.42 Simplified high-frequency equivalent circuit of a symmetrical OTA with NMOS drivers. C_1 and C_2 are lumped parasitic capacitances at internal nodes.

and

$$f_{GBW} = \frac{g_{m(1,2)}}{2\pi C_L}, \tag{4.145}$$

which are also valid for a symmetrical OTA with PMOS drivers.

Notice the presence of two more poles and one zero in (4.143). The frequency of one of these poles is described by

$$f_{P2} = \frac{g_{m(3,4,5,7)}}{2\pi C_1}. \tag{4.146}$$

This equation is similar to (4.116) of the nondominant pole of a basic OTA, and, therefore is indicative of similar design options for setting a high frequency for the pole. Based on the conclusions reached in Subsection 4.5.4, these options include a short channel length and a large overdrive voltage in M3, M4, M5, and M7.

The remaining pole and zero are located at the frequencies

$$f_{P3} = \frac{g_{m(6,8)}}{2\pi C_2}, \tag{4.147}$$

and

$$f_Z = \frac{g_{m(6,8)}}{\pi C_2}, \tag{4.148}$$

where C_2 is usually dominated by the intrinsic gate capacitances of M6 and M8. Again, the options available for setting these frequencies high include a short channel and a large overdrive voltage in M6 and M8.

Finally, the reader is referred to Subsection 4.4.3 for the formulation of the differential-input range parameter V_{DIR} in a symmetrical OTA.

Example 4.19 A continuous-time single-supply (V_{DD} = 1.8 V) unity-gain amplifier is to be implemented with a symmetrical OTA of dc differential gain A_0 = 125 V/V. Use PMOS drivers with a common well tied to V_{DD}, and design for an external load C_L = 3.4 pF and a full-power bandwidth f_{fp} = 15 MHz. Input test signal is supplied by a voltage source of negligible resistance. It is specified as a sinewave of maximum amplitude $V_{i(\max)}$ = 0.4 V around a bias voltage V_B = 0.8 V. Since a unity-gain configuration is involved, the corresponding limits of the common-mode swing and the output swing are as follows:

$$V_{CM(\min)} = V_{O(\min)} = 0.8 - 0.4 = 0.4 \text{ V}, \tag{4.149a}$$

$$V_{CM(\max)} = V_{O(\max)} = 0.8 + 0.4 = 1.2 \text{ V}. \tag{4.149b}$$

Work We begin by deriving open-loop constraints and expressing them in terms of device metrics of the symmetrical OTA.

- Equations (2.4) and (4.149a) imply $V_{OL} < 0.4$ V, whose substitution into (4.136) for V_{OL} yields

$$V_{DSsat7} < 0.4 \text{ V}. \tag{4.150}$$

- Equations (2.5) and (4.149b) imply $V_{OH} > 1.2$ V, whose substitution into (4.135) for V_{OH} yields

$$V_{DSsat8} > -1.8 + 1.2 = -0.6 \text{ V}. \tag{4.151}$$

- Equations (2.7) and (4.149a) imply $V_{IL} < 0.4$ V, whose substitution into (4.139) for V_{IL} yields

$$V_{GS(3,4)} - V_{DSsat(1,2)} + V_{GS(1,2)L} < 0.4 \text{ V}, \tag{4.152}$$

where $V_{GS(1,2)L}$ denotes the specific value of the gate-source voltage for the lower limit of the common-mode range. This distinction is necessary because driver operates without a source-bulk short.

- Equations (2.8) and (4.149b) imply $V_{IH} > 1.2$ V, whose substitution into (4.140) for V_{IH} together with 1.8 V for V_{DD} yields

$$V_{DSsat9} + V_{GS(1,2)U} > -0.6 \text{ V}, \tag{4.153}$$

where $V_{GS(1,2)U}$ denotes the specific value of the gate-source voltage for the upper limit of the common-mode range.

- For $\alpha_i = 1$, $V_{i(\text{max})} = 0.4$ V, and $f_{fp} = 15$ MHz, (2.59a) and (2.59b) imply

$$f_{cl} = 15 \times 10^6 \quad \text{for } V_{DIR} > 0.283 \text{ V}, \tag{4.154a}$$

$$f_{cl} = 15 \times 10^6 \times \sqrt{\left(\frac{0.4}{V_{DIR}}\right)^2 - 1} \quad \text{for } V_{DIR} < 0.283 \text{ V}. \tag{4.154b}$$

Next, we express f_{cl} and V_{DIR} in terms of device parameters and variables. Consider f_{cl} first. $\beta = 1$ of the unity-gain configuration implies $f_{cl} = f_{GBW}$ from (2.48). Using (4.145) for f_{GBW}, assuming 3.5 pF for C_L to safely cover the contribution of the OTA output capacitance, and replacing $g_{m(1,2)}$ with the product of the transconductance efficiency and current, we arrive at

$$f_{cl} = \frac{g_{m(1,2)}/(I_B/2)}{2.2 \times 10^{-11}} \frac{I_B}{2}. \tag{4.155}$$

For V_{DIR}, (4.85) indicates

$$V_{DIR} = \frac{2}{g_{m(1,2)}/\left(I_B/2\right)}. \tag{4.156}$$

- Targeting $A_0 = 125$ V/V and $V_{A8} = V_{A7}$, we obtain the following Early voltage expression from (4.142):

$$V_{A(7,8)} = \frac{250}{g_{m(1,2)}/(I_B/2)}. \tag{4.157}$$

Now, we can put together a design plan. An inspection of the constraints for this purpose shows that all of the performance metrics involved in (4.154a) to (4.157) are uniquely determined by the overdrive voltage $V_{GS(1,2)} - V_{TP}$ of M1 and M2 at the operation point. This overdrive voltage directly yields the corresponding transconductance efficiency $g_{m(1,2)}/(I_B/2)$ from the related design tool of the threshold option selected for the drivers. Subsequently, V_{DIR} is determined from (4.156), f_{cl} from (4.154a) or (4.154b), I_B from (4.155), and finally $V_{A(7,8)}$ from (4.157). Overdrive voltage also defines $I_{\square(1,2)}$, and thus enables us to determine $(W/L)_{1,2}$ from I_B. Although the existing specifications do not impose a unique value for the overdrive voltage itself, the constraints imposed by (4.152) and (4.153) obviously impose upper and lower limits on it. It is also obvious that the channel length of driver devices is not constrained in any way by the existing specifications. In conclusion, driver devices will be biased and sized by relying on additional considerations associated with such generic performance metrics as phase margin, power consumption, precision, etc.

As to the matching NMOS devices M3, M4, M5, and M7, note that the bias current $I_B/2$ of these devices are imposed by driver design as already explained. Furthermore, their overdrive voltage is constrained by the more restrictive between (4.150) and (4.152). The specific value of this overdrive is to be determined by considering additional constraints based on generic metrics. Biasing and sizing will then be performed routinely.

The matching NMOS devices M6 and M8 conduct the same bias current as M3, M4, M5, and M7. The overdrive voltage of these two devices is constrained by (4.151). We will determine the specific value of this bias by also relying on generic considerations, and using the outcome for sizing M6 and M8.

Finally, we will design a current source to supply the bias current I_B specified in driver design. It will target an appropriate output resistance for M9 and comply with the constraint imposed on V_{DSsat9} by (4.153).

Steps of design flow are presented next.

1. *Channel Length of M1 and M2* For a given aspect ratio, elongating the channel will lead to a proportional rise in the width and a quadratic rise in the area. Too wide a channel may contribute a significant drain-bulk capacitance to C_1, and thus impair the phase margin by lowering f_{P2}. Also note that a large area implies a large parasitic capacitance for the input terminals of the OTA. Although not an issue in the present application thanks to the negligible source resistance, this capacitance generally has an adverse effect on the frequency response of an OTA as will be elaborated later in Example 4.20. The only notable benefit of a large area is the reduction of the standard deviation of mismatch between the two drivers as discussed in Subsection 4.2.2 in the context of current mirroring. However, considering the relatively large systematic offset arising from the disparate drain voltages of M6 and M8, mismatch should not be a prime concern in the present application. Too short a channel, on the other hand, would lead to a large flicker noise. We therefore opt for a short but not minimum channel $L_{DR(1,2)} = 0.4$ μm.
2. *Overdrive Voltage of M1 and M2* As explained in the design plan, this voltage uniquely determines f_{cl}, I_B, $(W/L)_{1,2}$, and $V_{A(7,8)}$. A large f_{cl} impairs the phase margin as indicated by (2.94). A large I_B leads to a large power consumption. A large $(W/L)_{1,2}$, if excessive, can further impair the phase margin as explained in step 1. Finally, a large $V_{A(7,8)}$ necessitates long output devices, and thus adversely affects the frequency response by lowering f_{P2}, f_{P3}, and f_Z as discussed in Subsection 4.6.4. In determining the overdrive voltage, we need to take into consideration all of these facts as well as the limits imposed by (4.152) and (4.153). What is immediately obvious is that (4.153) can hardly be complied with by devices of standard-threshold option, for which, the threshold voltage alone would be around −0.5 V due to body effect. We therefore adopt low-threshold driver devices.

   Although the trade-offs involved in determining $V_{GS(1,2)} - V_{TP}$ can be visualized with the aid of graphical design tools, a much more accurate approach is to rely on plots of f_{cl}, I_B, $(W/L)_{1,2}$, and $V_{A(7,8)}$ obtained with simulation as functions of the overdrive voltage. A single dc-sweep based on the following input file `ex4-19a.sp` will provide the necessary data:

```
filename: ex4-19a.sp
.incl bsimlvt.sp
m 2 1 0 0 cmosplvt w=6u l=0.4u
vds 2 0 -1
vgs 1 0 dc
.save @m[vth] @m[id] @m[gm] v(1)
.control
dc vgs -0.6 -0.05 1m
*********defining the rhs of (4.154b)*********
let f=real(15e6*sqrt((0.2*@m[gm]/@m[id])^2-1))
***************defining fcl*****************
let fcl='(f>=15e6)*f+(f<=15e6)*15e6'
********plotting fcl against overdrive********
plot fcl vs v(1)+@m[vth]
```

```
****************defining IB*******************
let ib=4.4e-11*fcl*@m[id]/@m[gm]
********plotting IB against overdrive*********
plot ib vs v(1)+@m[vth]
*************defining (W/L)1,2****************
let aspect=8.8*ib/@m[id]
*****plotting (W/L)1,2 against overdrive******
plot aspect vs v(1)+@m[vth] ylog
****************defining VA*******************
let va=250*@m[id]/@m[gm]
********plotting VA against overdrive*********
plot va vs v(1)+@m[vth]
.endc
**********************************************
.end
```

The netlist includes a low-threshold PMOS device of 0.4-μm channel length as determined in step 1 and an arbitrarily selected channel width. Unlike the actual driver devices of the OTA, whose source-bulk voltage is nonzero and dependent on the gate-bulk voltage, this device is simulated with a source-bulk short for the sake of simplicity. This discrepancy does not invalidate the outcome of the simulation as long as the metrics are plotted against the overdrive voltage. Note that the drain-source voltage is kept at an arbitrary but saturating dc level −1 V while the gate-source voltage is swept as observed in the first line of control section. As instructed in the second line of control section, only the threshold voltage, drain current, transconductance, and gate-source voltage are saved because all of the required calculations involve only these four. Also note that the expression of f_{cl} is defined in two steps. First, the right-hand side of (4.154b) is defined as `f` after substituting (4.156) for V_{DIR}. Next, the larger between `f` and 15 MHz is defined as f_{cl}. This is followed by a plot command for f_{cl} as a function of the overdrive voltage `v(1)+@m[vth]`, where the subtraction of the threshold-voltage parameter is implemented with a plus sign because Ngspice reports the absolute value of this parameter. Definition of I_B is based on (4.155), where the reciprocal of $g_{m(1,2)}/(I_B/2)$ is represented by `@m[id]/@m[gm]`. Aspect ratio is determined from

$$\left(\frac{W}{L}\right)_{1,2} = \frac{I_B}{2I_{\square(1,2)}}, \tag{4.158}$$

where, considering the effective aspect ratio 17.6 of the simulated device, $2I_{\square(1,2)}$ is replaced with `@m[id]/8.8`. Finally, the target value of $V_{A(7,8)}$ is determined by calculating (4.157).

All four plots obtained with simulation are displayed in Figs. 4.43(a) to (d) as functions of the overdrive voltage. Judging solely from the plots of f_{cl}, I_B, and $(W/L)_{1,2}$, an overdrive voltage around $V_{GS(1,2)} - V_{TP} = -0.26$ V appears to be the best choice as it offers a reasonably small aspect ratio alongside a minimized bandwidth and a minimized current. Unfortunately, however, the value of $V_{A(7,8)}$ corresponding to this much overdrive voltage is on the high side. The corresponding channel length is most likely to be beyond the range in

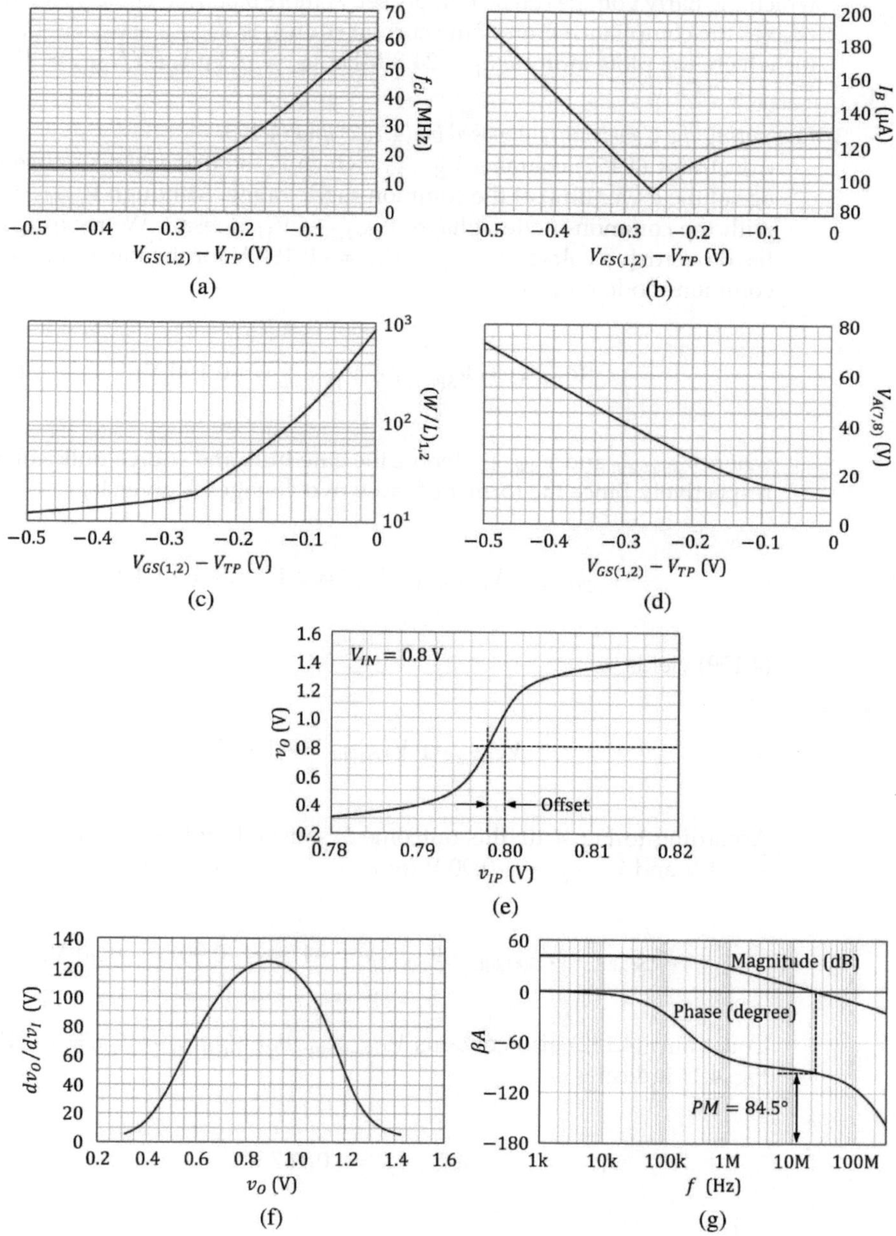

FIGURE 4.43 Plots used in design or verification in Example 4.19. (a) Closed-loop bandwidth versus driver overdrive as obtained from (4.154a) and (4.154b). (b) Bias current versus driver overdrive as obtained from (4.155). (c) Driver aspect ratio versus driver overdrive as obtained from (4.158). (d) Early voltage of output devices versus driver overdrive as obtained from (4.157). (e) Simulated dc transfer characteristic v_O versus v_{IP} for $V_{IN} = 0.8$ V. (f) Simulated slope of the dc transfer characteristic versus output voltage. (g) Simulated magnitude and phase of loop gain.

which the Early voltage can be reliably set. A more balanced choice, which also relaxes the compliance limit of the current source, is $V_{GS(1,2)} - V_{TP} = -0.190$ V, for which, the plots indicate $f_{cl} = 24.8$ MHz, $I_B = 113$ μA, $(W/L)_{1,2} = 40$, and $V_{A(7,8)} = 25.8$ V.

3. *Evaluating the Constraints Imposed by (4.152) and (4.153)*
   a. Constraint (4.152) contains $V_{GS(1,2)L}$, which is the driver gate-source voltage at the lower limit of the common-mode range. Although $V_{GS(1,2)}$ varies with the common-mode voltage, $V_{GS(1,2)} - V_{TP}$ doesn't. We reexpress this fixed overdrive voltage $V_{GS(1,2)} - V_{TP} = -0.190$ V for the lower limit of the common-mode range as

$$V_{GB(1,2)L} - V_{SB(1,2)L} - V_{TP(1,2)L} = -0.190 \text{ V}, \qquad (4.159)$$

where $V_{GB(1,2)L}$ and $V_{SB(1,2)L}$ denote the gate-bulk and source-bulk voltages, respectively. Since the former of these two voltages is given by

$$V_{GB(1,2)L} = V_{CM(\min)} - V_{DD} = 0.4 - 1.8 = -1.4 \text{ V},$$

(4.159) yields

$$V_{SB(1,2)L} + V_{TP(1,2)L} = -1.21 \text{ V}.$$

According to Fig. 4.10, this outcome is satisfied by the sum of $V_{TP(1,2)L} = -0.31$ V and $V_{SB(1,2)L} = -0.90$ V for $L_{DR(1,2)} = 0.4$ μm. This leads to

$$V_{GS(1,2)L} = V_{GB(1,2)L} - V_{SB(1,2)L} = -1.4 - (-0.90) = -0.50 \text{ V}.$$

Constraint (4.152) also contains $V_{DSsat(1,2)}$. For $V_{GS(1,2)} - V_{TP} = -0.190$ V, Fig. 4.11 indicates

$$V_{DSsat(1,2)} = -0.167 \text{ V}.$$

As a consequence, (4.152) turns into

$$V_{GS(3,4)} < 0.733 \text{ V}. \qquad (4.160)$$

We will take this limit into consideration in step 5, where we will start designing M3, M4, M5, and M7.

b. Constraint (4.153) contains $V_{GS(1,2)U}$. We determine this specific limit of driver gate-source voltage by applying the procedure used in the preceding paragraph for calculating $V_{GS(1,2)L}$. First, the fixed overdrive voltage $V_{GS(1,2)} - V_{TP} = -0.190$ V is rewritten as

$$V_{GB(1,2)U} - V_{SB(1,2)U} - V_{TP(1,2)U} = -0.190 \text{ V}. \tag{4.161}$$

Next,

$$V_{GB(1,2)U} = V_{CM(\max)} - V_{DD} = 1.2 - 1.8 = -0.6 \text{ V}$$

is substituted into (4.161) to obtain

$$V_{SB(1,2)U} + V_{TP(1,2)U} = -0.41 \text{ V},$$

for which, the plot specific to $L_{DR(1,2)} = 0.4$ µm in Fig. 4.10 yields $V_{TP(1,2)U} = -0.165$ V and $V_{SB(1,2)U} = -0.245$ V. The latter, together with $V_{GB(1,2)U} = -0.6$ V, lead to

$$V_{GS(1,2)U} = V_{GB(1,2)U} - V_{SB(1,2)U} = -0.6 - (-0.245) = -0.355 \text{ V}.$$

Constraint (4.153) therefore indicates

$$V_{DSsat9} > -0.245 \text{ V}. \tag{4.162}$$

We will take this result into account when we design M9 in step 9.

4. *Channel Width of M1 and M2* $L_{DR(1,2)} = 0.4$ µm and $(W/L)_{1,2} = 40$ imply $W_{DR(1,2)} = 13.6$ µm.
5. *Overdrive Voltage of M3, M4, M5, and M7* A high overdrive voltage is desirable for M3, M4, M5, and M7 as it would positively affect the frequency response by leading to a high f_{P2}. However, (4.150) or (4.160) places an upper limit on this overdrive voltage. Assuming a saturation margin $\Delta V_{DS7} = 50$ mV for M7, the former of these two constraints yields $V_{DSsat(3,4,5,7)} < 0.35$ V, for which, Figs. 3.8 and 4.6 indicate a maximum overdrive voltage $V_{GS(3,4,5,7)} - V_{TN} = 0.54$ V. The standard-threshold option, whose threshold voltage is no less than 0.4 V for any reasonable channel length, is not viable for setting this much overdrive voltage because the corresponding gate-source voltage cannot satisfy (4.160). The low-threshold option offers a threshold voltage no larger than 0.18 V, which necessitates $V_{GS(3,4,5,7)} = 0.54 + 0.18 = 0.72$ V for maximizing the overdrive voltage. This much gate-source voltage is obviously compatible with (4.160). We therefore select the low-threshold option for these devices, and proceed to sizing.

6. *Sizing M3, M4, M5, and M7*
   a. $V_{GS(3,4,5,7)} - V_{TN} = 0.54$ V implies $I_{\square(3,4,5,7)} = 31$ μA/sq from Fig. 4.7. For $I_B/2 = 56.5$ μA this leads to $(W/L)_{3,4,5,7} = 1.83$.
   b. Next, we determine $L_{DR(3,4,5,7)} = 1.4$ μm with an iterative simulation by targeting $V_{A7} = 25.8$ V for $V_{DS7} = 0.8$ V. Considering $(W/L)_{3,4,5,7} = 1.83$, the corresponding width is calculated as $W_{DR(3,4,5,7)} = 2.5$.
7. *Overdrive Voltage of M6 and M8* Selection of threshold option for M6 and M8 is not as critical as it is for other devices. We select the low-threshold option. Considering the fact that a high overdrive voltage will set f_{P3} and f_Z high, and thus positively affect the frequency response, we select a large $V_{DSsat(6,8)} = -0.55$ V close to the limit defined by (4.151). The corresponding overdrive voltage is determined from Fig. 4.11 as $V_{GS(6,8)} - V_{TP} = -0.705$ V.
8. *Sizing M6 and M8*
   a. $V_{GS(6,8)} - V_{TP} = -0.705$ V implies $I_{\square(6,8)} = 14.2$ μA/sq from Fig. 4.12. For $I_B/2 = 56.5$ μA, this leads to $(W/L)_{6,8} = 3.98$.
   b. Next, we determine $L_{DR(6,8)} = 1.2$ μm with an iterative simulation by targeting $V_{A8} = 25.8$ V with $V_{DS8} = -1$ V. For $(W/L)_{6,8} = 3.98$, this leads to $W_{DR(6,8)} = 4.6$ μm.
9. *Sizing M9* This device is expected to conduct $I_B = 113$ μA. Its compliance limit is constrained by (4.162). There is however no specification for its output resistance. As usual, we specify a target much larger than the resistance $1/g_{m(1,2)}$ seen into the source terminal of M1 or M2. For determining $g_{m(1,2)}$, we first refer to Fig. 4.14 and obtain $g_{m(1,2)}/(I_B/2) = 9.9\ \text{V}^{-1}$ for $V_{GS(1,2)} - V_{TP} = -0.190$ V. Considering $I_B/2 = 56.5$ μA, we calculate 1.78 kΩ for $1/g_{m(1,2)}$ and target $r_{ds9} = 10/g_{m(1,2)} = 17.8$ kΩ for the output resistance. The corresponding Early voltage $V_{A9} = r_{ds9}I_B = 2$ V can be easily implemented with a close-to-minimum channel length. Threshold option is not critical for M9; we select the standard-threshold option and target $V_{DSsat9} = -0.150$ V to satisfy the compliance constraint of (4.162) with a saturation margin $\Delta V_{DS9} = -95$ mV. The corresponding overdrive voltage $V_{GS9} - V_{TP} = -0.173$ V is obtained from Fig. 3.17. It yields $I_{\square 9} = 0.9$ μA/sq from Fig. 3.18. This much sheet current implies $(W/L)_9 = 126$ for $I_B = 113$ μA. Next, we determine $L_{DR9} = 0.3$ μm with an iterative simulation by targeting $V_{A9} = 2$ V with $V_{DS9} = -0.245$ V. This leads to $W_9 = 30.2$ μm for the already calculated aspect ratio. In order to keep the power consumption of the reference branch of the current source low, we decide to mirror I_B with a scaling factor $S = 10$. This leads to a unit-device effective channel width $W_u = W_9/10 = 3$ μm, which is large enough to adopt $W_{DRu} = 3$ μm.
10. *Verification* Performed with the following input file `ex4-19b.sp`.

```
filename: ex4-19b.sp
.incl bsimlvt.sp
***********ota subckt***********
.subckt ota 1 2 7
m1 3 1 5 10 cmosplvt w=13.6u l=0.4u nrd=0.015 nrs=0.015
+ ad=7.48p pd=28.3u as=7.48p ps=28.3u
m2 4 2 5 10 cmosplvt w=13.6u l=0.4u nrd=0.015 nrs=0.015
+ ad=7.48p pd=28.3u as=7.48p ps=28.3u
```

```
m3 3 3 0 0 cmosnlvt w=2.5u l=1.4u nrd=0.08 nrs=0.08
+  ad=1.38p pd=6.1u
m4 4 4 0 0 cmosnlvt w=2.5u l=1.4u nrd=0.08 nrs=0.08
+  ad=1.38p pd=6.1u
m5 6 3 0 0 cmosnlvt w=2.5u l=1.4u nrd=0.08 nrs=0.08
+  ad=1.38p pd=6.1u
m6 6 6 10 10 cmosplvt w=4.6u l=1.2u nrd=0.045 nrs=0.045
+  ad=2.5p pd=10.3u
m7 7 4 0 0 cmosnlvt w=2.5u l=1.4u nrd=0.08 nrs=0.08
+  ad=1.38p pd=6.1u
m8 7 6 10 10 cmosplvt w=4.6u l=1.2u nrd=0.045 nrs=0.045
+  ad=2.5p pd=10.3u
m9 5 8 10 10 cmosplvt w=3u l=0.3u nrd=0.07 nrs=0.07
+ ad=1.65p pd=7.1u m=10
m10 8 8 10 10 cmosplvt w=3u l=0.3u nrd=0.07 nrs=0.07
+ ad=1.65p pd=7.1u
ir 8 0 11.3u
vdd 10 0 1.8
.ends
***********open loop************
x1 11 12 13 ota
vin 11 0 dc 0.8
vip1 12 0 dc 0.8
cl1 13 0 3.4p
.dc vip1 0.78 0.82 0.01m
.save v(13) x1.6
.save @m.x1.m3[gm] @m.x1.m6[gm] @m.x1.m3[cgg] @m.x1.m6[cgg]
***********loop gain************
*x2 21 22 23 ota      ; for voltage return ratio
*cl2 21 0 3.4p        ; for voltage return ratio
*vip2 22 0 dc 0.8     ; for voltage return ratio
*vx 21 23 dc 0 ac 1 ; for voltage return ratio
*x3 31 32 33 ota      ; for current return ratio
*cl3 31 0 3.4p        ; for current return ratio
*vip3 32 0 dc 0.8     ; for current return ratio
*ix 0 34 dc 0 ac 1    ; for current return ratio
*vs1 33 34 0          ; for current return ratio
*vs2 34 31 0          ; for current return ratio
*.control
*ac dec 100 1k 1g
*let rv=-v(23)/v(21)
*let ri=-i(vs1)/i(vs2)
*plot db((rv*ri-1)/(2+rv+ri)) ph((rv*ri-1)/(2+rv+ri))
*let magnitude=mag((rv*ri-1)/(2+rv+ri))
*let phasedif=ph((rv*ri-1)/(2+rv+ri))+180
*meas ac fu when magnitude=1
*meas ac pm find phasedif at=fu
*.endc
*************tran*************
*x4 41 42 41 ota
```

```
*vip4 42 0 sin 0.8 0.4 15meg
*cl4 41 0 3.4p
*.control
*tran 0.01n 1u
*plot v(41)
*fourier 15meg v(41)
*.endc
******noise in Example 4.21*****
*x4 41 42 41 ota
*vip4 42 0 dc 0.8 ac 1
*cl4 41 0 3.4p
*.control
*noise v(41) vip4 dec 100 10 1g
*print onoise_total
*.endc
********************************
.end
```

The OTA is defined as a subcircuit whose schematic is given in Fig. 4.44(a).

The open-loop simulation is a dc sweep performed on the top-level schematic shown in Fig. 4.44(b). Its purpose is to evaluate the properties of the dc transfer characteristic. The lines belonging to the loop-gain, transient, and noise simulations must be closed while performing this simulation. The dc transfer characteristic presented in Fig. 4.43(e) is plotted with the command

```
plot v(13)
```

Notice the presence of an offset as large as almost −2 mV. As explained in Subsection 4.6.3, this systematic offset arises from the asymmetry caused by the disparate drain voltages of M6 and M8 at the operation point. The output-voltage specification implies 0.8 V for the drain voltage of M8. The reader can characterize the drain voltage of M6 by plotting it against the output voltage v_O with the command

```
plot x1.6 vs v(13)
```

and reading the outcome for $v_O = 0.8$ V.

Another plot derived from the same open-loop simulation is the dc differential gain versus the output voltage. Plotted with the command

```
plot deriv(v(13)) vs v(13)
```

and shown in Fig. 4.43(f), it indicates $A_0 = 119$ V/V for $V_O = 0.8$ V, which is slightly lower than the target value.

The top-level schematic used for characterizing the loop gain and therefore the phase margin is shown in Figs. 4.44(c) and (d). This self-running simulation implements the technique described in Subsection 2.7.2, and directly plots the magnitude and phase responses of the loop gain as shown in Fig. 4.43(g). It also prints out the unity-gain frequency f_u in units of hertz and phase margin PM in units of degree as follows[5]

[5]See footnote 7 in Subsection 2.7.2 for setting the unit of phase to degree.

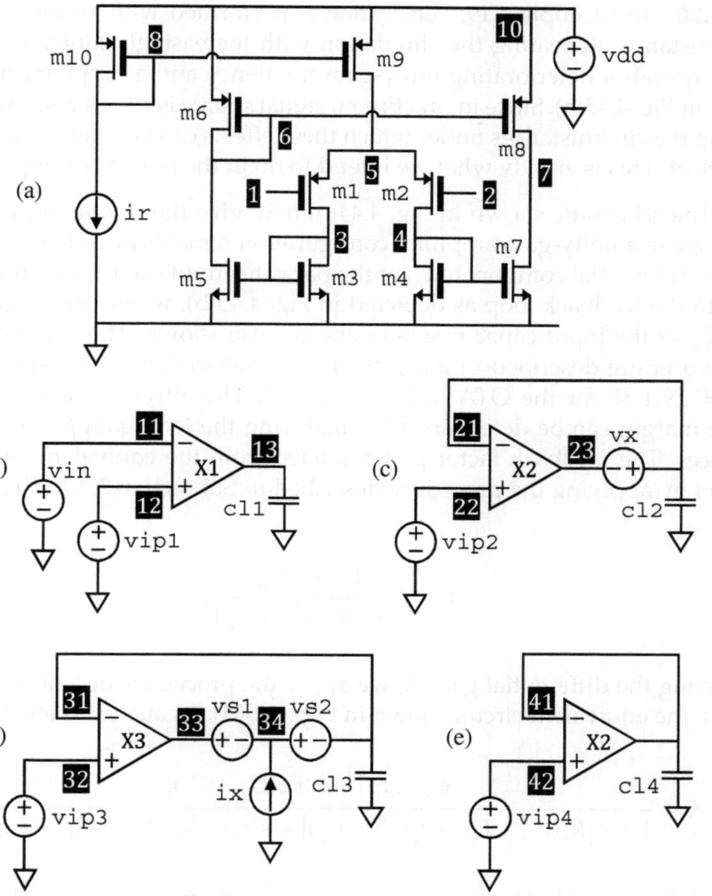

FIGURE 4.44 Schematics netlisted in `ex4-19b.sp`. (a) OTA subcircuit. (b) Open-loop simulation. (c and d) Characterization of phase margin. (e) Closed-loop transient simulation.

```
fu = 2.362354e+07
pm = 8.420860e+01
```

Since an almost perfect single-pole response is indicated, f_{GBW} and therefore f_{cl} are well approximated by the unity-gain frequency f_u = 23.7 MHz.

A closed-loop transient analysis is run on the top-level schematic shown in Fig. 4.44(e). This is also a self-running simulation. It generates the output waveform and evaluates the harmonic distortion in response to an input waveform of maximum specified frequency and amplitude. It indicates THD = 2.33%, which is quite reasonable for an application based on a low-gain OTA.

The noise analysis defined as the last item of the input file is reserved for Example 4.21. ▲

Example 4.20 In Example 4.19, verification is performed with an input-signal source of zero resistance. Repeating the simulation with increasingly larger values of source resistance reveals a deteriorating unity-gain frequency and a dipping phase margin as displayed in Fig. 4.45(a). Since in practice no signal source is free of resistance, it is worth identifying the circumstances under which these effects can be safely ignored as done in Example 4.19. This is exactly what we intend to do in the present example.

Work The schematic shown in Fig. 4.44(e) may give the wrong impression that the signal source in a unity-gain amplifier configuration remains outside the feedback loop. In fact, the differential component C_d of the parasitic input capacitance brings the signal source into the feedback loop as depicted in Fig. 4.45(b), where the single-ended components C_{se} of the input capacitive network are also shown. The simulation-based extraction procedure described in Example 4.14 of Subsection 4.4.3 yields $C_{se} = 20.6$ fF and $C_d = 29.1$ fF for the OTA of Example 4.19. The effect of these capacitances on the phase margin can be determined by analyzing the loop gain βA in terms of these capacitances. The feedback factor β is extracted from the equivalent circuit shown in Fig. 4.45(c) by applying the procedure described in Subsection 2.3.2. The outcome is as follows:

$$\beta = \frac{1 + sR_S C_{se}}{1 + sR_S\left(C_{se} + C_d\right)}. \tag{4.163}$$

For extracting the differential gain A, we apply the procedure described in Section 2.8, and obtain the equivalent circuit shown in Fig. 4.45(d). Its analysis yields

$$A = \frac{g_{m(1,2)} r_o\left[1 + sR_S(C_{se} + C_d)\right]}{1 + s\left[R_S(C_{se} + C_d) + r_o(C_L' + C_d)\right] + s^2 r_o R_S\left[C_L'(C_{se} + C_d) + C_{se}C_d\right]},$$

where $C_L' \doteq C_L + C_{se}$. However, considering the conditions $C_L \gg C_{se}$ and $C_L \gg C_d$ of the present application, this equation is well approximated by

$$A = \frac{g_{m(1,2)} r_o}{1 + sr_o C_L},$$

which, together with (4.163), yield the following loop-gain expression:

$$\beta A = \frac{g_{m(1,2)} r_o\left(1 + sR_S C_{se}\right)}{\left(1 + sr_o C_L\right)\left[1 + sR_S(C_{se} + C_d)\right]}. \tag{4.164}$$

According to (4.164), R_S introduces a pole at the frequency $1/2\pi R_S(C_{se} + C_d)$ and a zero at the higher frequency $1/2\pi R_S C_{se}$. The ratio of the two frequencies is 2.4 in the present application. For $R_S = 50$ kΩ, for example, the frequencies are 64 and 154 MHz. Since these are lower than the frequencies of the internal poles and zero, phase margin is determined primarily by this additional pole/zero pair. As illustrated in Fig. 4.45(e), the pole located at 64 MHz obviously enhances the roll-off rate of the magnitude toward −40 dB/dec while rotating the phase further toward −180°. Although the zero

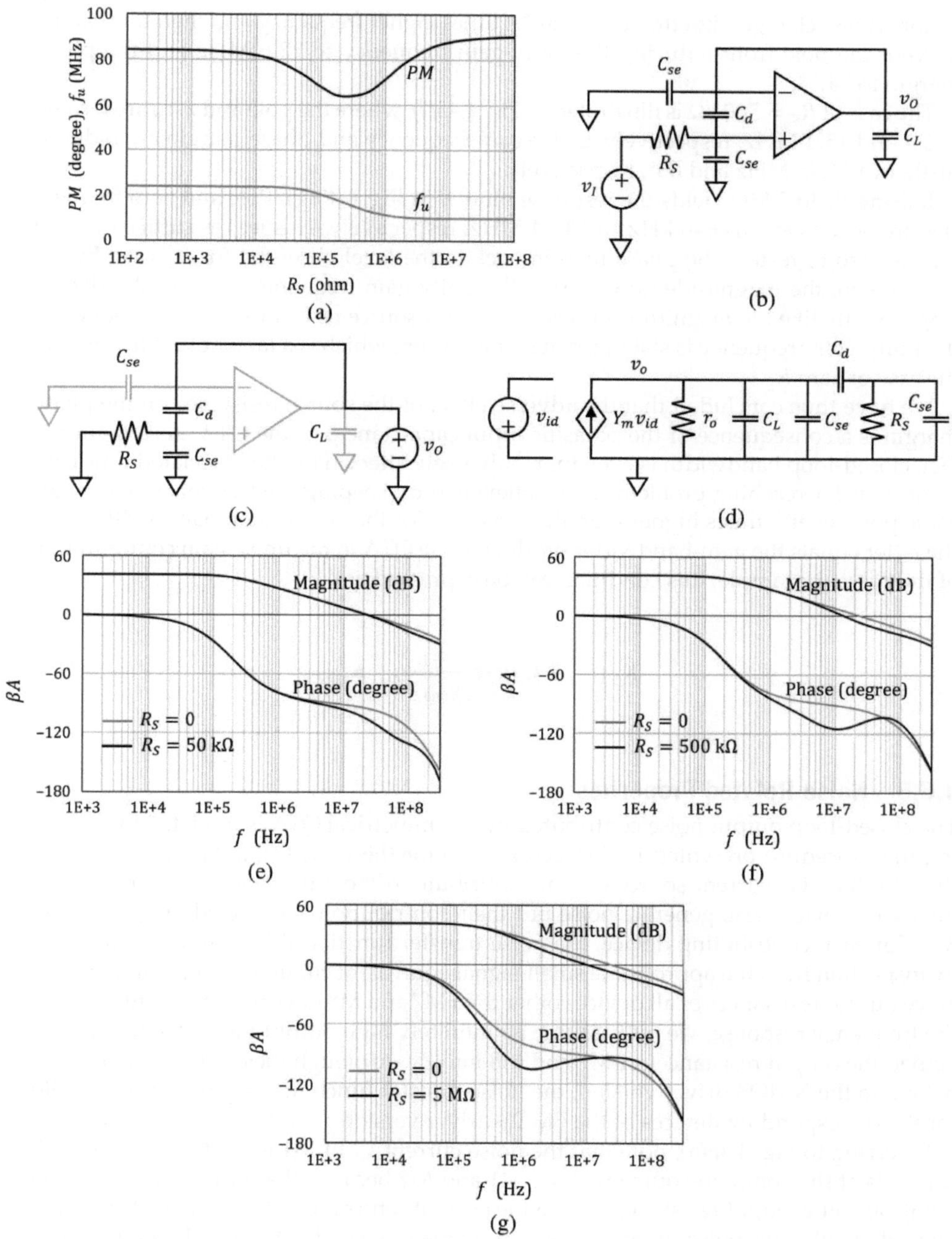

FIGURE 4.45 (a) Variation of phase margin PM and unity-gain frequency f_u with signal-source resistance R_S in the symmetrical OTA designed in Example 4.19. (b) Unity-gain amplifier configuration including R_S and the parasitic input capacitances C_{se} and C_d of the OTA. (c) Equivalent circuit for extracting the feedback factor β. (d) Equivalent circuit for extracting the differential gain A. (e) to (g) Magnitude and phase response of loop gain βA for $R_S = 50$ kΩ, $R_S = 500$ kΩ, and $R_S = 5$ MΩ.

restores these changes, its effect comes at 2.4 times higher frequency, and therefore cannot prevent the pole from reducing the unity-gain frequency to 22.6 MHz and the phase margin to 74.5°.

The case of $R_S = 500$ kΩ is illustrated in Fig. 4.45(f), where the pole and zero are located at 6.4 and 15.4 MHz, respectively. Unity-gain frequency and phase margin are reduced further to 12.95 MHz and 66°, respectively.

Raising R_S to 5 MΩ yields the response plotted in Fig. 4.45(g). Pole and zero frequencies are now as low as 640 kHz and 1.54 MHz, respectively. This leaves sufficient room for the zero to restore the phase margin back to the level observed for $R_S = 0$. By the same token, the magnitude now crosses the unity-gain frequency with a −20 dB/dec slope exactly like the magnitude observed for zero-source resistance. As a consequence, the unity-gain frequency is stabilized at its minimum, which is a factor of 2.4 lower than the case of zero R_S.

We have thus concluded that the adverse effect of the source resistance on the phase margin is a consequence of the parasitic input capacitances of the OTA. As a matter of fact, closed-loop bandwidth is even more adversely affected by the associated time constant. This far-reaching problem can be alleviated by keeping the frequency of the parasitic pole several times higher than the target set for the closed-loop bandwidth. Since the latter equals the gain-bandwidth product of the OTA in the unity-gain configuration of the present example, this condition can be expressed as

$$R_S\left(C_{se} + C_d\right) \ll \frac{C_L}{g_{m(1,2)}}. \quad \blacktriangle$$

4.6.5 Noise-Related Properties

The closed-loop output noise contributed by a symmetrical OTA is modeled by following the procedure presented in Subsection 4.5.5 for the basic OTA. All devices except those of the bias-current source or sink contribute to the output noise. Source or sink device or devices also generate noise but their contribution is canceled out in M7 and M8. For each contributing device, the noise transfer function $A_k(f)$ can be constructed by inspection from the appropriate small-signal equivalent circuit presented in Fig. 4.46. Since output resistances of all devices other than M7 and M8 have no significant effect on the frequency response, we will assume an infinitely high output resistance for all, and embed the output resistances of M7 and M8 into Z_L. Although these equivalent circuits belong to the NMOS-driver version, the noise transfer functions they indicate are valid for the corresponding devices of the PMOS-driver version as well.

Referring to Fig. 4.46(a), note that the noise current i_{n1} of M1 is split into two identical parts at the common-source node of M1 and M2 because these two devices exhibit identical source input resistances. Since the current returning to the drain of M1 is only one half of what the noise source removes from the same node, the other half must come from M3. M5 also conducts $i_{n1}/2$ because it mirrors what M3 conducts. The very same half is mirrored by M6 to M8. The other half of i_{n1} flowing originally into M2 is mirrored by M4 to M7, which, together with M8, sink a total of i_{n1} from the load impedance Z_L.

The case of i_{n2} is depicted in Fig. 4.46(b). Its analysis is not different from that of i_{n1} with the exception that the roles of M1 and M2 are transposed. Again, the noise current in its entirety flows into the load impedance.

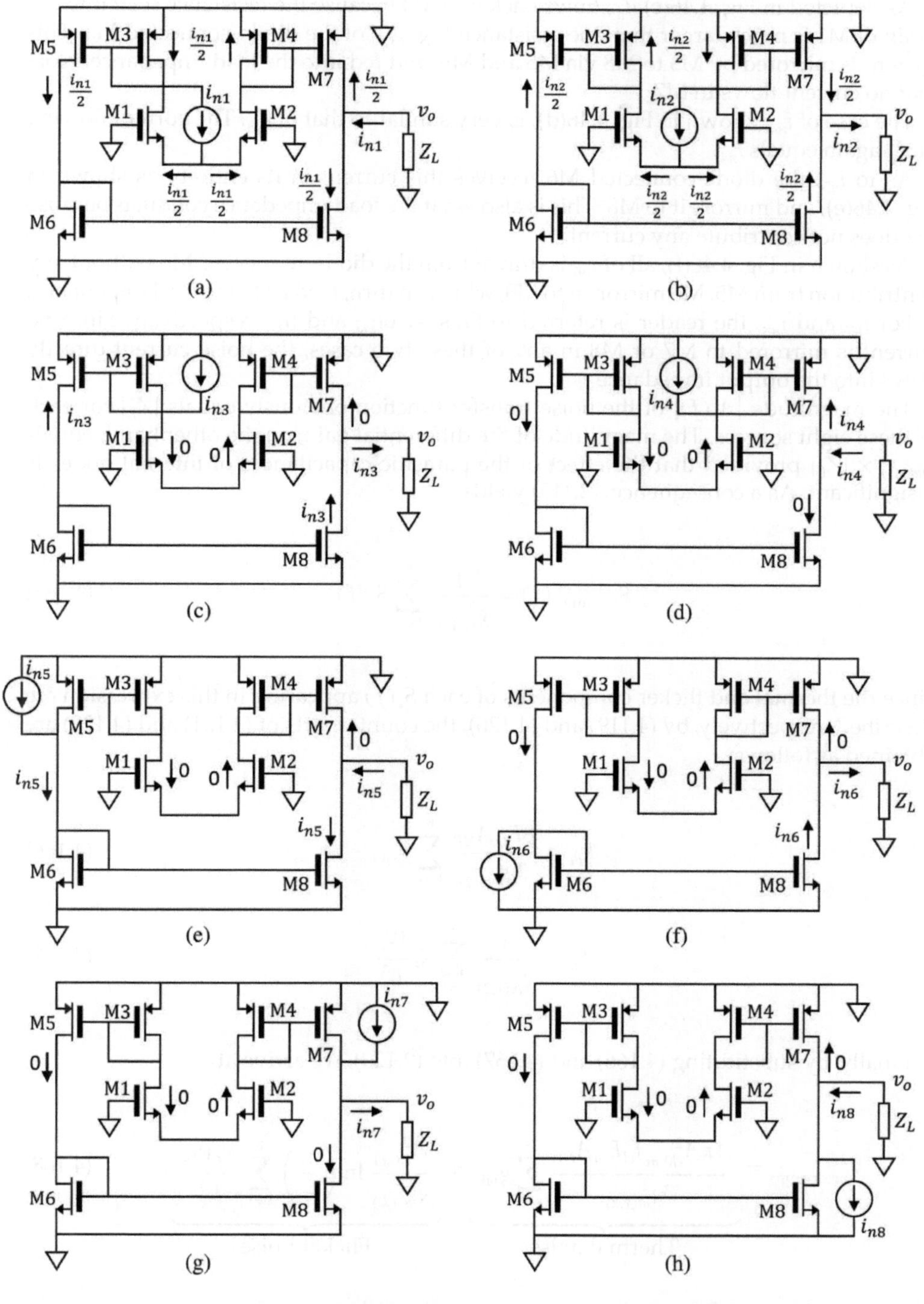

Figure 4.46 Equivalent circuits for determining the transfer functions needed in noise analysis. (a) v_o/i_{n1}. (b) v_o/i_{n2}. (c) v_o/i_{n3}. (d) v_o/i_{n4}. (e) v_o/i_{n5}. (f) v_o/i_{n6}. (g) v_o/i_{n7}. (h) v_o/i_{n8}.

As depicted in Fig. 4.46(c), i_{n3} flows back into M3 because the resistance seen into the drain of M1 is much larger than the resistance $1/g_{m(3,4)}$ of the diode-connected M3. This current is mirrored by M3 to M8 via M5 and M6, and fed into the load impedance. Note that no current flows in M7.

The case of i_{n4}, shown in Fig. 4.46(d), is very similar to that of i_{n3}. The current flowing in Z_L again equals i_{n4}.

As to i_{n5}, the diode-connected M6 receives this current in its entirety as shown in Fig. 4.46(e), and mirrors it to M8. This is also what the load impedance conducts because M7 does not contribute any current.

As shown in Fig. 4.46(f), all of i_{n6} is drawn from the diode-connected M6 without any contribution from M5. M6 mirrors it to M8, which, in turn, feeds it to the load impedance.

For i_{n7} and i_{n8}, the reader is referred to Figs. 4.46(g) and (h), respectively. Since no current is mirrored to M7 or M8 in any of these two cases, the noise current directly flows into the output impedance.

The magnitude $|A_k(f)|$ of the noise transfer function obviously equals $|Z_L|$ for each of these eight sources. The magnitude of the differential gain, on the other hand, equals $g_{m(1,2)} \times |Z_L|$ provided that the effect of the parasitic capacitances of internal nodes is insignificant. As a consequence, (2.115) yields

$$S_{i(\text{opamp})}(f) = \frac{1}{g_{m(1,2)}^2} \sum_{k=1}^{8} S_k(f). \tag{4.165}$$

Since the thermal and flicker components of each $S_k(f)$ appearing in this expression are described, respectively, by (4.119) and (4.120), the counterparts of (4.121) and (4.122) are obtained as follows:

$$k_{th} = \frac{8E_{th}A_{NP}}{3g_{m(1,2)}^2} \sum_{k=1}^{8} g_{mk}, \tag{4.166}$$

$$k_{fl} = \frac{1}{g_{m(1,2)}^2} \sum_{k=1}^{8} K_k \frac{W_k}{L_k^3}. \tag{4.167}$$

Finally, by substituting (4.166) and (4.167) into (2.120), we arrive at

$$V_{no(\text{opamp})}^2 = \underbrace{\frac{4\pi A_{cl0,ni}^2 f_{cl} E_{th} A_{NP}}{3g_{m(1,2)}^2} \sum_{k=1}^{8} g_{mk}}_{\text{Thermal noise}} + \underbrace{\frac{A_{cl0,ni}^2}{g_{m(1,2)}^2} \ln\left(\frac{f_{cl}}{f_1}\right) \sum_{k=1}^{8} \frac{K_k W_k}{L_k^3}}_{\text{Flicker noise}}, \tag{4.168}$$

for the contribution of a symmetrical-OTA to the output noise voltage of a continuous-time closed-loop configuration. This equation differs from (4.123) only in the upper limit of device index k. Design directives explained on the basis of (4.123) are also therefore perfectly valid for a symmetrical OTA.

For the case of an amplifier application utilizing a frequency-independent external network, (4.168) is simplified into

$$V^2_{no(\text{opamp})} = \underbrace{\frac{2E_{th}A_{NP}}{3\beta C_L}\left(2+\sum_{k=3}^{8}\frac{g_{mk}}{g_{m(1,2)}}\right)}_{\text{Thermal noise}} + \underbrace{\frac{1}{\beta^2 g^2_{m(1,2)}}\ln\left(\frac{\beta g_{m(1,2)}}{2\pi f_1 C_L}\right)\sum_{k=1}^{8}\frac{K_k W_k}{L_k^3}}_{\text{Flicker noise}}, \quad (4.169)$$

which indicates the very same design directive extracted earlier from (4.124). That is, the sole open-loop tool by which we can manipulate the thermal component is the ratio $g_{mk}/g_{m(1,2)}$ for $k = 3, \ldots, 8$. We can reduce this ratio, hence thermal noise, by targeting low overdrive voltages for driver devices and high overdrive voltages for the rest.

Example 4.21 Estimate the room-temperature rms output noise voltage of the unity-gain amplifier designed in Example 4.19, and verify with a noise simulation for $f \geq 10$ Hz. For flicker noise coefficients of low-threshold devices, refer to the plots presented in Fig. 4.47. These plots are extracted with the model deck `bsimlvt.sp` by applying the technique described in Section 3.8. Incidentally, the kink observed in the PMOS plot is an artifact of the method used in BSIM3v3 for integrating the subthreshold and strong-inversion models of flicker noise. A smooth transition is ensured in BSIM4 and BSIM-BULK.

Work Since the closed-loop configuration of a unity-gain amplifier does not contain any noise-generating component other than the OTA, its output noise voltage is described directly by (4.169), where all parameters except $g_{m(3,4,5,7)}$ and $g_{m(6,8)}$ are already quantified in Example 4.19.

The ratio $g_{mk}/g_{m(1,2)}$ appearing in the thermal noise component equals the ratio of the corresponding transconductance efficiencies because all devices conduct the same current. The transconductance efficiencies, in turn, are determined from Fig. 4.9 or Fig. 4.14 as follows:

- $V_{GS(1,2)} - V_{TP} = -0.190$ V → Fig. 4.14 ⇒ $g_{m(1,2)}/(I_B/2) = 9.9\ \text{V}^{-1}$,
- $V_{GS(3,4,5,7)} - V_{TN} = 0.540$ V → Fig. 4.9 ⇒ $g_{m(3,4,5,7)}/(I_B/2) = 3.6\ \text{V}^{-1}$,
- $V_{GS(6,8)} - V_{TP} = -0.725$ V → Fig. 4.14 ⇒ $g_{m(6,8)}/(I_B/2) = 2.4\ \text{V}^{-1}$.

These numbers yield $g_{m(3,4,5,7)}/g_{m(1,2)} = 0.363$ and $g_{m(6,8)}/g_{m(1,2)} = 0.242$. The rest of the parameters have the following values: $g_{m(1,2)} = 560$ μA/V, $A_{cl0,ni} = 1$, $\beta = 1$, $E_{th} = 4.14 \times 10^{-21}$ J, $A_{NP} = 1.3$, $C_L = 3.5$ pF, $f_1 = 10$ Hz, $K_{1,2} = 3.4 \times 10^{-20}\ \text{A}^2\mu\text{m}^2$, $W_{1,2} = 13.6$ μm, $L_{1,2} = 0.341$ μm, $K_{3,4,5,7} = 1.5 \times 10^{-18}\ \text{A}^2\mu\text{m}^2$, $W_{3,4,5,7} = 2.5$ μm, $L_{3,4,5,7} = 1.366$ μm, $K_{6,8} = 1.2 \times 10^{-18}\ \text{A}^2\mu\text{m}^2$, $W_{6,8} = 4.6$ μm, $L_{6,8} = 1.14$ μm. We therefore obtain

$$V^2_{no} = \underbrace{4.03 \times 10^{-9}}_{\text{Thermal noise}} + \underbrace{1.72 \times 10^{-9}}_{\text{Flicker noise}} = 5.75 \times 10^{-9}\ \text{V}^2,$$

hence $V_{no} = 75.9$ μV.

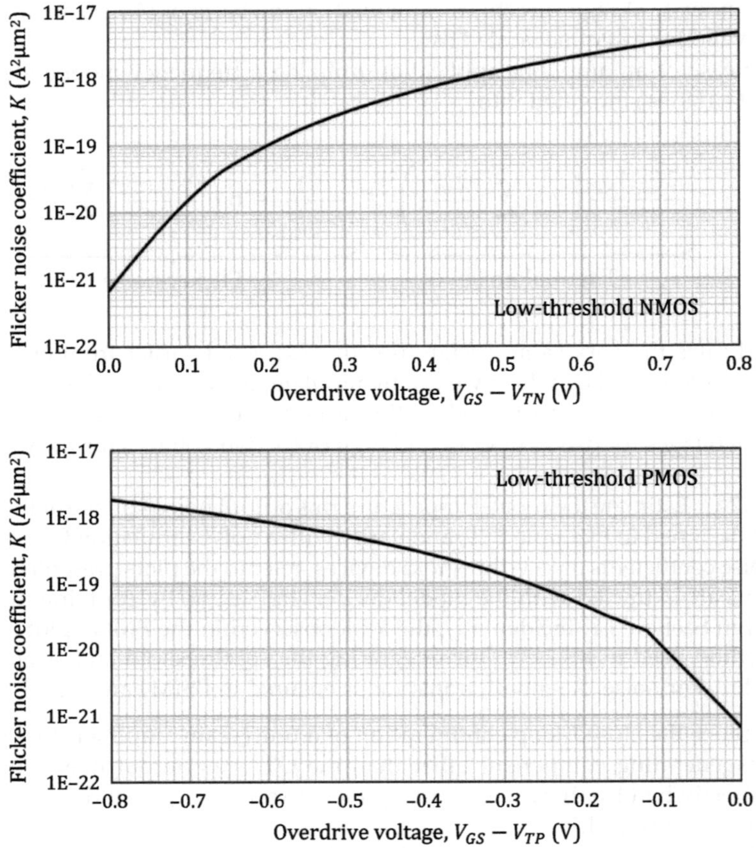

FIGURE 4.47 Variation of the flicker noise coefficient with overdrive voltage in low-threshold NMOS and PMOS devices of model deck `bsimlvt.sp`.

Verification is performed with the noise analysis defined as the last item in the input file `ex4-19b.sp`. It is based on the schematic depicted in Fig. 4.44(e). Simulation yields $V_{no} = 78.1\ \mu V$. ▲

4.6.6 Cascoded-Symmetrical OTA

As explained in Subsection 4.6.3 and observed in Example 4.19, the elementary version of the symmetrical OTA offers a low dc differential gain, and may exhibit a large systematic offset. Both of these deficiencies can be alleviated by cascoding M5, M6, M7, and M8 as depicted in Fig. 4.48. M7c and M8c increase the output resistance hence the dc gain by roughly two orders of magnitude. M5c and M7c improve the symmetry by equalizing the drain-source voltages of M5 and M7. Same is done by M6c and M8c for M6 and M8. Offset is reduced as a result.

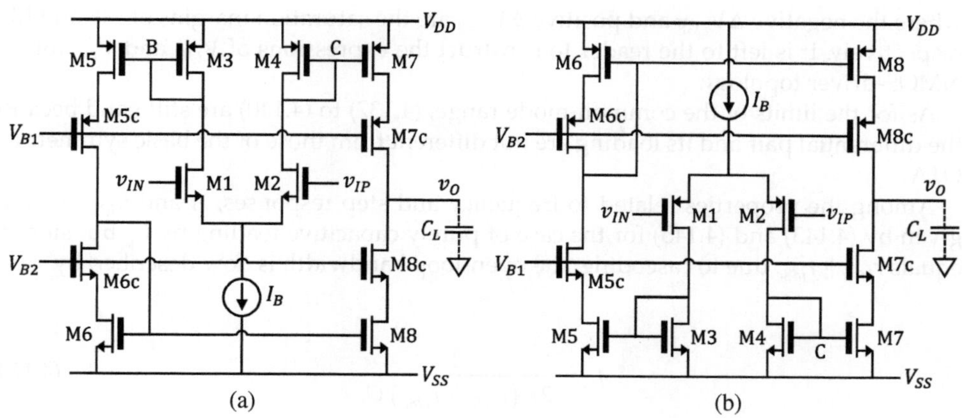

FIGURE 4.48 Cascoded-symmetrical OTA. (a) NMOS drivers. (b) PMOS drivers.

The enhanced dc differential gain is described by

$$A_0 = \underbrace{g_{m(1,2)}}_{G_m} \underbrace{r_{id7c} \parallel r_{id8c}}_{r_o}, \tag{4.170}$$

where the input resistances r_{id7c} and r_{id8c} seen into the drain terminals of M7c and M8c are generally given by

$$r_{id7c} = \underbrace{(g_{m7c} + g_{mb7c})\, r_{ds7c}}_{A_{\max 7c}} r_{ds7}, \tag{4.171a}$$

$$r_{id8c} = \underbrace{(g_{m8c} + g_{mb8c})\, r_{ds8c}}_{A_{\max 8c}} r_{ds8}. \tag{4.171b}$$

As usual, the bulk transconductance parameters appearing in these equations are to be ignored if the cascoding devices operate with source-bulk shorts.

The output range limits in the PMOS-driver topology are described by

$$V_{OH} = V_{DD} + V_{DS\text{sat}8} + \Delta V_{DS8} + V_{DS\text{sat}8c}, \tag{4.172}$$

and

$$V_{OL} = V_{SS} + V_{DS\text{sat}7} + \Delta V_{DS7} + V_{DS\text{sat}7c}, \tag{4.173}$$

where the negative ΔV_{DS8} and positive ΔV_{DS7} are the saturation margins of M8 and M7, respectively. It is left to the reader to construct the expressions of V_{OH} and V_{OL} for the NMOS-driver topology.

As for the limits of the common-mode range, (4.137) to (4.140) are still valid because the differential pair and its loading are not different from those of the basic symmetrical OTA.

Among the properties related to frequency and step responses, A and f_{GBW} are still given by (4.143) and (4.145) for the case of purely capacitive loading by C_L but since r_o equals $r_{id7c} \parallel r_{id8c}$ due to cascoding, the open-loop bandwidth is now described by

$$f_o = \frac{1}{2\pi \left(r_{id7c} \parallel r_{id8c}\right) C_L}. \tag{4.174}$$

Substituting this expression together with (4.170) into (2.43), we indeed obtain the very same (4.145) for the gain-bandwidth product. Also note that the options cited in Subsection 4.6.4 for setting the frequencies of internal poles and zero are also valid for the cascoded version.

Also applicable are the contents of Subsection 4.6.5 for noise-related properties because the cascoding devices M5c, M6c, M7c, and M8 do not significantly contribute to the output noise. For verifying this claim, consider a noise-current source across any of these cascoding devices, for example, M5c. The current it injects into the common node of M5 and M5c flows back into M5c almost in its entirety because the impedance seen into a source terminal is much lower than the one seen into a drain terminal. After exiting the drain of M5c, this current is collected by the current source again. In other words, the noise current of a cascoding device circulates inside the device itself rather than finding a way to the output. The magnitude of its noise transfer function is therefore negligible.

Example 4.22 The discrete-time amplifier of Fig. 2.44 is to be implemented with a cascoded-symmetrical OTA of PMOS drivers in accordance with the following specifications:

1. Power-supply configuration: Single supply $V_{DD} = 1.8$ V, $V_{SS} = 0$.
2. All PMOS devices in a common n-well tied to V_{DD}.
3. Closed-loop gain: $A_{cl} = 3$ V/V.
4. Input signal: $v_i = \pm 0.15$ V around $V_B = 0.9$ V. Waveform to be used in verification is similar to the one displayed in Fig. 2.50 of Subsection 2.11.4, and defined as `vi` in the input file `ex2-32.sp` of Example 2.32.
5. Sampling and amplification phase durations: $T_S = T_A = 52$ ns. Clock waveforms to be used in verification are similar to those displayed in Fig. 2.50 of Subsection 2.11.4, and defined as `vphi1` and `vphi2` in the input file `ex2-32.sp` of Example 2.32.
6. Input capacitance of the circuit driven by the amplifier: $C_I = 0.5$ pF.
7. Absolute value of the worst-case 4.5-σ dc error: $|\Delta v_{O(dc)}| = 0.9$ mV.

8. Absolute value of the worst-case settling error: $|\Delta v_{O(\mathrm{set})}| = 0.62$ mV.
9. MIM capacitor parameters: Pelgrom coefficient $K_C = 0.0095$ μm, areal coefficient $C_A = 1$ fF/μm^2, fringe coefficient $C_F = 0.2$ fF/μm.

This example involves the very same configuration and specification set defined in Example 2.32 of Subsection 2.11.4, where verification is performed with abstract models for switches and OTA. Example 3.18 of Subsection 3.9.2 also involves the same configuration and specification set but it relies on transmission-gate switches instead of abstract switches.

Work Shown in Fig. 4.49(a) is an annotated electrical schematic of the OTA. Part b of the same figure depicts the top-level schematic. As usual, we begin with the derivation of open-loop constraints from closed-loop specifications and expressing them in terms of device metrics.

- $A_{cl} = 3$ V/V, $v_i = \pm 0.15$ V, and $V_B = 0.9$ V imply $V_{O(\min)} = 0.45$ V and $V_{O(\max)} = 1.35$ V for the limits of output swing. Assuming identical dimensions for M8 and M8c and also for M7 and M7c, the corresponding output-range limitations described by (4.172) and (4.173) yield

$$2V_{DS\mathrm{sat}(8,8c)} + \Delta V_{DS8} \;>\; -0.45\,\mathrm{V}, \tag{4.175}$$

$$2V_{DS\mathrm{sat}(7,7c)} + \Delta V_{DS7} \;<\; 0.45\,\mathrm{V}. \tag{4.176}$$

- Since virtual short in dc steady state keeps the common-mode voltage constant at $V_{CM} = 0.9$ V, (4.139) yields

$$V_{GS(3,4)} - V_{DS\mathrm{sat}(1,2)} + V_{GS(1,2)} < 0.9\,\mathrm{V}. \tag{4.177}$$

  By the same token, (4.140) yields

$$V_{DS\mathrm{sat}9} + V_{GS(1,2)} > -0.9\,\mathrm{V}. \tag{4.178}$$

- The dc error of this amplifier is generally modeled by (2.184) in Example 2.31 of Subsection 2.11.3. Its offset-error component is intrinsically negligible due to the suppression of V_{os} by βA_0. The charge-error component can also be reduced by supplementing the feedback switch with a dummy, which leaves us with the gain and external-network error components. As long as C_1 and C_2 are laid out with proper physical-design techniques, the mean value of the external-network error component will also be negligible. We therefore consider only the distribution of this error component. Gain error, on the other hand, has no distribution; it is a systematic error. The worst-case dc error is observed when these

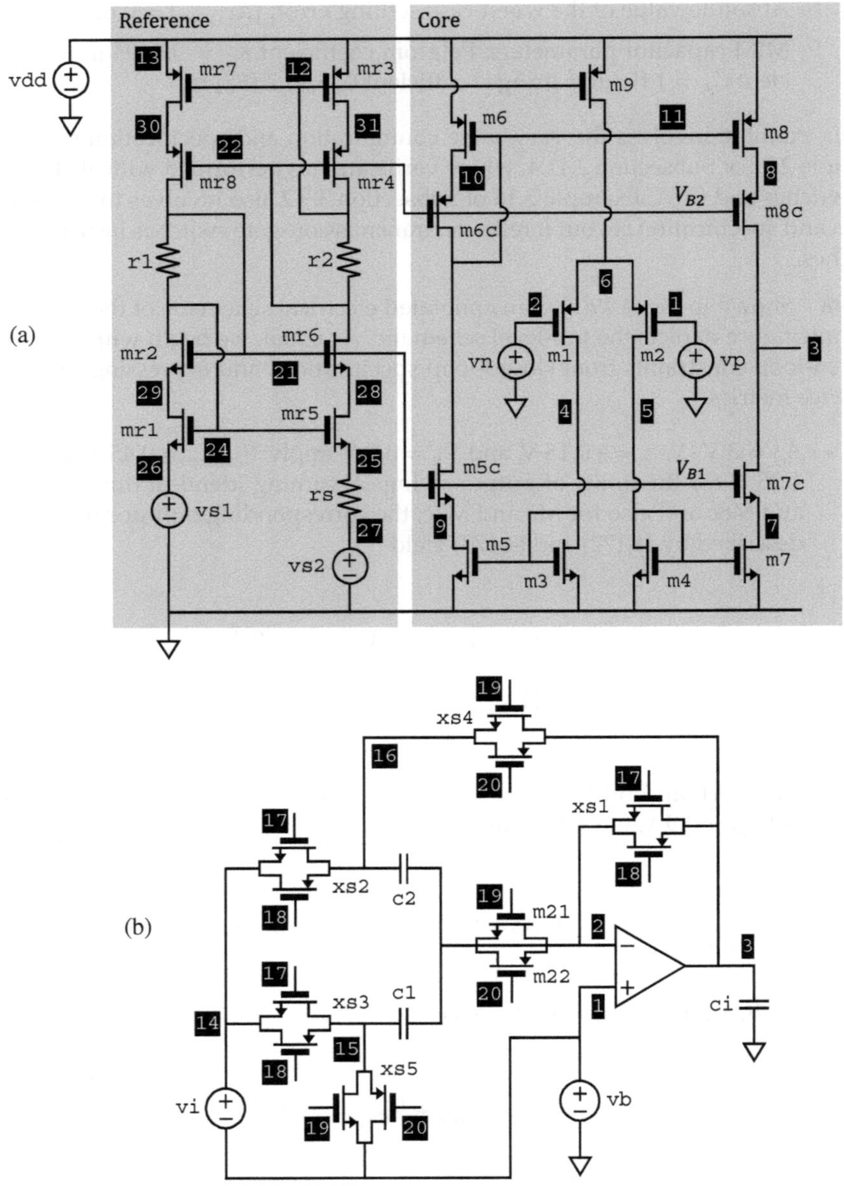

FIGURE 4.49 The discrete-time amplifier designed in Example 4.22. (a) Reference and core sections of the OTA. (b) Top-level schematic including the transmission gates.

two components have the same polarity. According to (2.184), its magnitude can therefore be expressed as follows:

$$|\Delta v_{O(\text{dc})}| = 4.5\sigma\left[\frac{C_1}{C_2}\right]|v_i(t_2)| + \frac{|v_o(t_4)|}{\beta A_0}. \tag{4.179}$$

Since the dc closed-loop gain is specified as $A_{cl} = 3$ V/V, (2.171) implies $C_1/C_2 = 2$. If we build C_2 with k copies of a unit capacitance C_u, it will take $2k$ copies to build C_1. According to (2.171) and (2.28), the standard deviation of the ratio C_1/C_2 is therefore given by

$$\sigma\left[\frac{C_1}{C_2}\right] = \sqrt{\frac{6}{k}}\frac{\sigma[C_u]}{C_u}.$$

Assuming a common-centroid layout pattern with $k = 12$, and using (3.138) together with the specified Pelgrom coefficient for $\sigma[C_u]/C_u$, we turn this equation into

$$\sigma\left[C_1/C_2\right] = \frac{6.72\times10^{-3}}{\sqrt{A_u}},$$

where $\sqrt{A_u}$ is the side length of a square unit-capacitor in units of µm. We now substitute into (4.179) this result for $\sigma\left[C_1/C_2\right]$ in addition to $|\Delta v_{O(\text{dc})}| = 0.9$ mV for the specified worst-case dc error, $\beta = 1/3$ for the feedback factor, $|v_i(t_2)| = 0.15$ V for the magnitude of the largest input signal, and $|v_o(t_4)| = 0.45$ V for the magnitude of the largest output signal. Although the outcome

$$\sqrt{A_u} = \frac{4.53\times10^{-3}}{0.9\times10^{-3} - \frac{1.35}{A_0}}$$

indicates the necessity of a large A_0 for keeping A_u small, the benefit of doing so diminishes roughly beyond $A_0 = 3500$ V/V as the gain error becomes negligible in comparison with the external-network error. We therefore specify the dc differential gain as $A_0 = 3500$ V/V, which implies $\sqrt{A_u} = 8.8$ µm, and therefore an area $A_u = 77.4$ µm^2 and a perimeter $P_u = 35.2$ µm for each unit capacitor. Substituting these numbers into (3.137) together with the specified values of the areal and fringe coefficients, we obtain $C_u = 84$ fF, hence approximately $C_2 = 1$ pF and $C_1 = 2$ pF.

- Next, we consider the settling-error specification. A detailed model of this type of error is presented in Example 2.32 for a slew-free transient throughout the entire cycle. In order to evaluate the viability of a slew-free transient in both phases of the present application, we have to determine the differential-input range V_{DIR} needed for accommodating the extreme values of the differential-input voltage observed at $t = t_1$ and $t = t_3$ in Fig. 2.50 of Example 2.32. To this end, we adopt the following expressions derived in Example 2.32 for these extreme values:

$$v_{ID}(t_1) = V_B - v_O(t_1) = -\frac{C_1 + C_2}{C_1 + C_2 + C_I}\left[v_i - \left(1 - \frac{C_I}{C_2}\right) v_{i(\mathrm{pre})}\right],$$

$$v_{ID}(t_3) = v_i.$$

According to the former of the two, $|v_{ID}(t_1)|$ is maximized in the case of v_i and $v_{i(\mathrm{pre})}$ having opposite polarities and maximum magnitudes. Considering the largest specified input signal magnitude of 0.15 V, this maximum is calculated as $|v_{ID}(t_1)| = 0.193$ V. The maximum of $|v_{ID}(t_3)|$, on the other hand, equals 0.15 V for the same input signal peak. These are sufficiently small to design for slew-free transients in both phases. Of course, the larger the V_{DIR} with respect to these maximum voltages, the better the quality of slew-free transient. We target $V_{DIR} = 0.22$ V, which is larger than both maxima but not too large to cause a low transconductance efficiency.

After determining the transconductance efficiency $g_{m(1,2)}/(I_B/2) = 9.1\ \mathrm{V}^{-1}$ from (4.85), we move to the calculation of $g_{m(1,2)}$ itself from the settling-error expression (2.196). Considering $G_m = g_{m(1,2)}$ for the symmetrical OTA, and substituting the peak input signal values $v_i = 0.15$ V and $v_{i(\mathrm{pre})} = -0.15$ V together with the specified values of $|\Delta v_{O(\mathrm{set})}|$, T_S, T_A, C_I, and the calculated values of C_1 and C_2, we arrive at $g_{m(1,2)} = 500\ \mu\mathrm{A/V}$. This result and the calculated value of the transconductance efficiency yield $I_B = 110\ \mu\mathrm{A}$ for the bias current.

We now move to the steps of device layer of design and verification.

1. *Sizing M1 and M2* For driver devices, we decide to exercise the low-threshold option in order to leave a large saturation margin for the bias-current source. We also set the drawn channel length $L_{DR(1,2)} = 0.4\ \mu\mathrm{m}$ as justified in step 1 of the design flow given in Example 4.19. Next, $V_{GS(1,2)} - V_{TP} = -0.2$ V is determined from Fig. 4.14 for the transconductance efficiency $g_{m(1,2)}/(I_B/2) = 9.1\ \mathrm{V}^{-1}$ calculated above. This much overdrive voltage leads to $I_{\square(1,2)} = 1.4\ \mu\mathrm{A/sq}$ from Fig. 4.12. Since driver devices conduct $I_B/2 = 55\ \mu\mathrm{A}$ each, their aspect ratio should therefore be $(W/L)_{1,2} = 39.3$. Relying on this result as well as (3.1) and (3.2), we arrive at $W_{DR(1,2)} = 14\ \mu\mathrm{m}$.
2. *Evaluating the Common-Mode Range Constraints Represented by (4.177) and (4.178)* We apply the procedure introduced in step 3 of the design flow of Example 4.19. Both constraints contain $V_{GS(1,2)}$. For determining this voltage, we reexpress the fixed overdrive voltage $V_{GS(1,2)} - V_{TP} = -0.2$ V of the drivers as

$$V_{GB(1,2)} - V_{SB(1,2)} - V_{TP(1,2)} = -0.2\ \mathrm{V},$$

which, for

$$V_{GB(1,2)} = V_B - V_{DD} = 0.9 - 1.8 = -0.9\text{ V}$$

yields

$$V_{SB(1,2)} + V_{TP(1,2)} = -0.7\text{ V}.$$

Referring to Fig. 4.10 for this sum, we read $V_{TP(1,2)} = -0.215$ V and $V_{SB(1,2)} = -0.485$ V from the plot belonging to $L_{DR(1,2)} = 0.4$ μm. This leads to

$$V_{GS(1,2)} = V_{GB(1,2)} - V_{SB(1,2)} = -0.9 - (-0.485) = -0.415\text{ V}.$$

Constraint (4.177) contains $V_{DSsat(1,2)}$ too. For $V_{GS(1,2)} - V_{TP} = -0.2$ V, Fig. 4.11 indicates

$$V_{DSsat(1,2)} = -0.175\text{ V}.$$

Finally, by substituting the calculated values of $V_{GS(1,2)}$ and $V_{DSsat(1,2)}$ into (4.177) and (4.178), we arrive at $V_{GS(3,4)} < 1.14$ V and $V_{DSsat9} > -0.485$ V.

3. *Sizing M3, M4, M5, M5c, M7, and M7c* These identical devices should be sized in accordance with the more restrictive between (a) (4.176) concerning the drain-source saturation voltage of M7, M7c, M5, and M5c; and (b) the upper limit found in the preceding step for the gate-source voltage of M3 and M4. Let us consider the former of these two constraints first. In order to decide how to partition the available 0.45 V into $V_{DSsat(7,7c)}$ and ΔV_{DS7}, recall that V_{DIR} has been selected only marginally higher than the extreme value expected for v_{ID}. The OTA will therefore come close to slewing at times in transient state, and therefore force M7 and M7c to conduct almost a full I_B. The resulting increase in the overdrive voltages of these two devices will threaten the saturation condition of M7 by raising V_{DSsat7} while lowering V_{DS7}. If allowed to exit saturation, M7 will reduce its current, and thus slow down the transient. Driven by this concern, we allow for a sizeable saturation margin $\Delta V_{DS7} = 0.15$ V, and settle for $V_{DSsat(7,7c)} = 0.15$ V.

   Regardless of the threshold option exercised, the drain-source saturation voltage just determined can be realized with a sufficiently small $V_{GS(3,4)}$ complying with the upper limit 1.14 V found in step 2. What limits the threshold option is the current-reference section of the OTA. If we select the low-threshold option for M3, M4, M5, M5c, M7, and M7c, we should do the same for the corresponding devices MR1 and MR2 of the reference section. These two devices, however, are interconnected in a low-voltage cascode topology. As stated earlier in Subsection 4.2.4, saturation is not unconditional in that topology if assembled with low-threshold devices. For evaluating the present case in this regard, we first note that the aspect ratio and current of M7 are to be scaled by the same factor with respect to those of MR1. This makes their sheet currents and therefore overdrive voltages equal, which, in turn, translates into equal gate-source voltages

$V_{GSR1} = V_{GS7}$ for these two devices. For an estimate of this voltage in the low-threshold option, consider (a) the 0.2-V overdrive voltage indicated by Fig. 4.6 for $V_{DSsat(7,7c)} = 0.15$ V, and (b) the 0.18-V maximum possible zero-bias threshold voltage indicated by Fig. 4.5. Since the sum $V_{GSR1} = 0.2 + 0.18 = 0.33$ V is less than $2V_{DSsat(7,7c)} + \Delta V_{DS7} = 0.45$ V, the condition of saturation as expressed originally by (4.29) cannot be satisfied with low-threshold devices. We therefore decide to implement M3, M4, M5, M5c, M7, and M7c with standard-threshold devices.

Sizing of M7 and M7c begins by determining the overdrive voltage of these devices from Fig. 3.8 for $V_{DSsat(7,7c)} = 0.15$ V. The outcome is 0.195 V. Moving to Fig. 3.10, we find $I_{\square(7,7c)} = 4.5$ μA/sq for this much overdrive. Considering $I_{D(7,7c)} = I_B/2 = 55$ μA, the aspect ratio of these devices is determined as $(W/L)_{7,7c} = 12.3$. Next, their channel length is determined with an iterative simulation similar to that of Example 4.6. This is done by assuming $r_{id7c} = r_{id8c}$, for which, (4.170) yields the following target value for the current/resistance product of M7c:

$$r_{id7c} \times (I_B/2) = 2\frac{A_0}{g_{m(1,2)}}\frac{I_B}{2} = 2\frac{3500}{500 \times 10^{-6}}\frac{110 \times 10^{-6}}{2} = 770 \text{ V}.$$

Adding a 10% margin, we target $r_{id7c} \times (I_B/2) = 850$ V. The outcome of iterative simulation is $L_{DR(7,7c)} = 0.9$ μm, which also applies to M3, M4, M5, and M5c. Iterative simulation also yields $V_{B1} = 1.015$ V for the bias voltage of the cascoding devices M5c and M7c, and $V_{GS7} = 0.625$ V for M7, M3, M4, and M5. Since V_{GS7} is larger than 0.45 V, the condition set by (4.29) for saturation is indeed satisfied by the standard-threshold option. The drawn channel width of M3, M4, M5, M5c, M7, and M7c will be determined in step 5 after deciding on how to partition these devices into copies of unit devices.

4. *Sizing M6, M6c, M8, and M8c* These devices are sized for satisfying the upper limit of the output range as represented by (4.175). We do this with $V_{DSsat(8,8c)} = -0.15$ V and $\Delta V_{DS8} = -0.15$ V. As the reader can verify with an analysis similar to the one performed in step 3 for M7 and M7c, implementation with low-threshold devices is not viable for the low-voltage cascode topology of M6 and M6c. As a matter of fact, a large threshold voltage is particularly desirable as it will bring the drain voltage of M6c closer to the output bias voltage $V_O = 0.9$ V, and thus reduce the systematic offset arising from the mismatch between these two voltages.

   Figure 3.17 yields $(V_{GS} - V_{TP})_{8,8c} = -0.175$ V for $V_{DSsat(8,8c)} = -0.15$ V. According to Fig. 3.19, this leads to $I_{\square(8,8c)} = 0.9$ μA/sq. Since these devices are supposed to conduct $I_B/2 = 55$ μA, their aspect ratio should therefore be $(W/L)_{8,8c} = 62$. Next, targeting $r_{id8c} \times (I_B/2) = 850$ V, we determine their channel length with iterative simulation. It is left to the reader to put together an input file for this purpose. It is important to maintain throughout the simulation (a) an overdrive voltage of −0.175 V for both devices, and (b) $V_{SB8c} = V_{DSsat8} + \Delta V_{DS8} = -0.3$ V for the source-bulk voltage of M8c. The outcome is $L_{DR(8,8c)} = 0.65$ μm, which also applies to M6 and M6c. The corresponding channel width will be determined in step 5 after partitioning into unit devices. Also note that the iterative simulation performed for sizing yields $V_{B2} - V_{DD} = -0.99$ V for the gate voltage of the cascode devices M6c and M8c with respect to V_{DD}, and $V_{GS(6,8)} = -0.6$ V for the gate-source voltage of M6 and M8.

5. *Designing the Reference Section* Having determined the aspect ratios and channel lengths of core devices, we now move to the reference section, which generates I_B, V_{B1}, and V_{B2}.

We begin by testing the condition $V_{DSsat9} > -0.485$ V imposed on M9 in step 2. This device has the same overdrive voltage as MR3, and, therefore, as M6 and M8, which implies $V_{DSsat9} = V_{DSsat8} = -0.15$ V. The condition is therefore satisfied with a sizeable margin.

In order to keep the current flowing in the reference block small, we decide to build each of M8, M8c, M6, and M6c with five copies of a PMOS unit device while using only one unit for each of MR3, MR4, MR7, and MR8. Such a unit device is supposed to conduct 11 μA. Its aspect ratio, being one-fifth of what is derived in step 4, should be $W/L = 62/5 = 12.4$. Considering also $L_{DR} =$ 0.65 μm, the drawn width of a unit device is calculated as $W_{DR} = 7.35$ μm. Since M9 is supposed to conduct twice as much current as each of M8, M8c, M6, and M6c does, it has to be constructed with 10 unit devices. Also note that the gate-source voltage V_{GS9} of this device equals $V_{GS(6,8)}$ hence −0.6 V as calculated in step 4. Considering also $V_{B2} - V_{DD} = -0.99$ V, we calculate the voltage across R_2 as 0.39 V, which implies $R_2 = 35.5$ kΩ for 11-μA reference current.

We construct each of M3, M4, M5, M5c, M7, and M7c with five copies of an NMOS unit device while using one copy of such device for each of MR1, MR2, and MR6, and four copies for MR5. The aspect ratio of such a unit device is one-fifth of $(W/L)_{7,7c} = 12.3$, hence 2.46. For $L_{DR} = 0.9$ μm, the drawn width of the NMOS unit devices is to be $W_{DR} = 2.15$ μm.

Finally, note that the 0.39-V difference between the voltages $V_{GSR1} = V_{GS7} =$ 0.625 V and $V_{B1} = 1.015$ V determined in step 3 is absorbed by R_1. For the 11-μA current flowing in each reference branch, this leads to $R_1 = 35.5$ kΩ, whose equality to R_2 is purely coincidental.

As to R_S, iterative operation-point simulation is the most accurate way of determining its value. The input file `ex4-22a.sp` presented below is used for this purpose. It defines an operating-point analysis only for the reference section of the schematic shown in Fig. 4.49(a). Iteration is done by comparing the current measured by `vs1` or `vs2` with the target 11 μA, and trimming R_S until the target is reached. Once R_S has thus been determined, it is worth comparing the voltages of nodes 12, 21, 22, and 24 with their target values to make sure that the reference section works as intended.

```
filename: ex4-22a.sp
.incl bsim.sp
mr1 29 24 26 0 cmosn w=2.15u l=0.9u nrd=0.1 nrs=0.1
mr2 24 21 29 0 cmosn w=2.15u l=0.9u nrd=0.1 nrs=0.1
mr3 31 12 13 13 cmosp w=7.35u l=0.65u nrd=0.03 nrs=0.03
mr4 12 22 31 13 cmosp w=7.35u l=0.65u nrd=0.03 nrs=0.03
mr5 28 24 25 0 cmosn w=2.15u l=0.9u nrd=0.1 nrs=0.1 m=4
mr6 22 21 28 0 cmosn w=2.15u l=0.9u nrd=0.1 nrs=0.1
mr7 30 12 13 13 cmosp w=7.35u l=0.65u nrd=0.03 nrs=0.03
mr8 21 22 30 13 cmosp w=7.35u l=0.65u nrd=0.03 nrs=0.03
```

```
vs1 26 0 0
vs2 27 0 0
rs 25 27 7.4k
r1 21 24 35.5k
r2 12 22 35.5k
vdd 13 0 1.8
.op
.end
```

Iterative simulation yields $R_S = 7.4\ \text{k}\Omega$.

6. *Verifying the Open-Loop DC Response* We combine the reference block with the OTA and run a dc analysis for verifying the overall open-loop dc performance in terms of bias currents, voltages, dc differential gain, output range, and systematic offset. We do this for the schematic shown in Fig. 4.49(a) and netlisted in the self-plotting input file `ex4-22b.sp` presented below. Note that the inverting-input voltage is kept constant at 0.9 V whereas the noninverting-input is swept between 0.85 and 0.95 V.

```
filename: ex4-22b.sp
.incl bsimlvt.sp
.incl bsim.sp
**********************ota section*********************
m1 4 2 6 13 cmosplvt w=14u l=0.4u nrd=0.014 nrs=0.014
+  ad=7.7p pd=29.1u as=7.7p ps=29.1u
m2 5 1 6 13 cmosplvt w=14u l=0.4u nrd=0.014 nrs=0.014
+  ad=7.7p pd=29.1u as=7.7p ps=29.1u
m3 4 4 0 0 cmosn w=2.15u l=0.9u nrd=0.1 nrs=0.1
+  ad=1.2p pd=5.4u m=5
m4 5 5 0 0 cmosn w=2.15u l=0.9u nrd=0.1 nrs=0.1
+  ad=1.2p pd=5.4u m=5
m5 9 4 0 0 cmosn w=2.15u l=0.9u nrd=0.1 nrs=0.1
+  ad=1.2p pd=5.4u m=5
m5c 11 21 9 0 cmosn w=2.15u l=0.9u nrd=0.1 nrs=0.1
+  ad=1.21p pd=5.4u as=1.21p ps=5.4u m=5
m6 10 11 13 13 cmosp w=7.35u l=0.65u nrd=0.03 nrs=0.03
+  ad=4.1p pd=15.8u m=5
m6c 11 22 10 13 cmosp w=7.35u l=0.65u nrd=0.03 nrs=0.03
+  ad=4.1p pd=15.8u as=4.1p ps=15.8u m=5
m7 7 5 0 0 cmosn w=2.15u l=0.9u nrd=0.1 nrs=0.1
+  ad=1.2p pd=5.4u m=5
m7c 3 21 7 0 cmosn w=2.15u l=0.9u nrd=0.1 nrs=0.1
+  ad=1.2p pd=5.4u as=1.2p ps=5.4u m=5
m8 8 11 13 13 cmosp w=7.35u l=0.65u nrd=0.03 nrs=0.03
+  ad=4.1p pd=15.8u m=5
m8c 3 22 8 13 cmosp w=7.35u l=0.65u nrd=0.03 nrs=0.03
+  ad=4.1p pd=15.8u as=4.1p ps=15.8u m=5
m9 6 12 13 13 cmosp w=7.35u l=0.65u nrd=0.03 nrs=0.03
+  ad=4.1p pd=15.8u m=10
******************reference section******************
mr1 29 24 26 0 cmosn w=2.15u l=0.9u nrd=0.1 nrs=0.1
```

```
mr2 24 21 29 0 cmosn w=2.15u l=0.9u nrd=0.1 nrs=0.1
mr3 31 12 13 13 cmosp w=7.35u l=0.65u nrd=0.03 nrs=0.03
mr4 12 22 31 13 cmosp w=7.35u l=0.65u nrd=0.03 nrs=0.03
mr5 28 24 25 0 cmosn w=2.15u l=0.9u nrd=0.1 nrs=0.1 m=4
mr6 22 21 28 0 cmosn w=2.15u l=0.9u nrd=0.1 nrs=0.1
mr7 30 12 13 13 cmosp w=7.35u l=0.65u nrd=0.03 nrs=0.03
mr8 21 22 30 13 cmosp w=7.35u l=0.65u nrd=0.03 nrs=0.03
vs1 26 0 0
vs2 27 0 0
rs 25 27 7.4k
r1 21 24 35.5k
r2 12 22 35.5k
*************supply, inputs and analysis*************
vdd 13 0 1.8
vn 2 0 0.9
vp 1 0 0.9
.save v(3) v(4) v(5) v(11) i(vs1) i(vs2) @m8[id] @m6[id]
.control
dc vp 0.85 0.95 10u
plot v(3) v(4) v(5) v(11)
plot deriv(v(3)) vs v(3)
plot i(vs1) i(vs2) @m8[id] @m6[id]
.endc
.end
```

7. *Switch Design and Verification of Transient Response* We decide to deploy low-threshold devices for transmission gates in order to achieve the necessary switch conductance with a narrower channel. This will help keep the capacitive loading of the closed-loop configuration light. However, the leakage current of a low-threshold device in cutoff is orders of magnitude higher than what a standard-threshold device leaks. The reader can ascertain this claim from (3.21) or (3.49) by inspecting the variation of the device current with the threshold voltage in subthreshold. Such a leakage can be tolerated only in the case of a sufficiently high clock frequency. Also worth noting is that the leakage is a very strongly increasing function of the temperature of operation. Quite generally, therefore, the qualification of a given threshold option for switch use should be evaluated at the highest expected temperature.

   We minimize the drawn channel length of switching devices to 0.2 μm, and leave the drawn width as a variable to be determined with an iterative transient simulation. The objective of this simulation is to minimize the adverse effect of switch resistance on the settling performance. The following input file `ex4-22c.sp` netlisting the top-level schematic of Fig. 4.49(b) is used for this purpose. Clock and input waveforms are the same as those used in Example 2.32 and shown in Fig. 2.50. Line 4 defines the width as a parameter. Also note the adoption of half width for the devices of the dummy switch. The reader is invited to rerun the simulation for different values of the width, and determine the best for the smallest settling error. Running the simulation will also enable the reader to compare the differential-input voltage and output voltage waveforms with those presented in Fig. 2.50. It is worth mentioning that only the gain-error component will be observable because no Monte Carlo analysis is included for evaluating the external-network error.

```
filename: ex4-22c.sp
.incl bsimlvt.sp
.incl bsim.sp
.param width=5u
**********************ota core***********************
m1 4 2 6 13 cmosplvt w=14u l=0.4u nrd=0.014 nrs=0.014
+  ad=7.7p pd=29.1u as=7.7p ps=29.1u
m2 5 1 6 13 cmosplvt w=14u l=0.4u nrd=0.014 nrs=0.014
+  ad=7.7p pd=29.1u as=7.7p ps=29.1u
m3 4 4 0 0 cmosn w=2.15u l=0.9u nrd=0.1 nrs=0.1
+  ad=1.2p pd=5.4u m=5
m4 5 5 0 0 cmosn w=2.15u l=0.9u nrd=0.1 nrs=0.1
+  ad=1.2p pd=5.4u m=5
m5 9 4 0 0 cmosn w=2.15u l=0.9u nrd=0.1 nrs=0.1
+  ad=1.2p pd=5.4u m=5
m5c 11 21 9 0 cmosn w=2.15u l=0.9u nrd=0.1 nrs=0.1
+  ad=1.2p pd=5.4u as=1.21p ps=5.4u m=5
m6 10 11 13 13 cmosp w=7.35u l=0.65u nrd=0.03 nrs=0.03
+  ad=4.1p pd=15.8u m=5
m6c 11 22 10 13 cmosp w=7.35u l=0.65u nrd=0.03 nrs=0.03
+  ad=4.1p pd=15.8u as=4.1p ps=15.8u m=5
m7 7 5 0 0 cmosn w=2.15u l=0.9u nrd=0.1 nrs=0.1
+  ad=1.2p pd=5.4u m=5
m7c 3 21 7 0 cmosn w=2.15u l=0.9u nrd=0.1 nrs=0.1
+  ad=1.2p pd=5.4u as=1.2p ps=5.4u m=5
m8 8 11 13 13 cmosp w=7.35u l=0.65u nrd=0.03 nrs=0.03
+  ad=4.1p pd=15.8u m=5
m8c 3 22 8 13 cmosp w=7.35u l=0.65u nrd=0.03 nrs=0.03
+  ad=4.1p pd=15.8u as=4.1p ps=15.8u m=5
m9 6 12 13 13 cmosp w=7.35u l=0.65u nrd=0.03 nrs=0.03
+  ad=4.1p pd=15.8u m=10
*****************reference section******************
mr1 29 24 26 0 cmosn w=2.15u l=0.9u nrd=0.1 nrs=0.1
mr2 24 21 29 0 cmosn w=2.15u l=0.9u nrd=0.1 nrs=0.1
mr3 31 12 13 13 cmosp w=7.35u l=0.65u nrd=0.03 nrs=0.03
mr4 12 22 31 13 cmosp w=7.35u l=0.65u nrd=0.03 nrs=0.03
mr5 28 24 25 0 cmosn w=2.15u l=0.9u nrd=0.1 nrs=0.1 m=4
mr6 22 21 28 0 cmosn w=2.15u l=0.9u nrd=0.1 nrs=0.1
mr7 30 12 13 13 cmosp w=7.35u l=0.65u nrd=0.03 nrs=0.03
mr8 21 22 30 13 cmosp w=7.35u l=0.65u nrd=0.03 nrs=0.03
vs1 26 0 0
vs2 27 0 0
rs 25 27 7.4k
r1 21 24 35.5k
r2 12 22 35.5k
************transmission gate subcircuit*************
```

```
.subckt tg in out ngate pgate power
mn in ngate out 0 cmosnlvt w=width l=0.2u
+  nrd={0.2e-6/width} nrs={0.2e-6/width}
+  ad={width*0.55e-6} pd={1.1e-6+2*width}
+  as={width*0.55e-6} ps={1.1e-6+2*width}
mp in pgate out power cmosplvt w=width l=0.2u
+  nrd={0.2e-6/width} nrs={0.2e-6/width}
+  ad={width*0.55e-6} pd={1.1e-6+2*width}
+  as={width*0.55e-6} ps={1.1e-6+2*width}
.ends
*******************dummy switch**********************
m21 2 19 2 0 cmosnlvt w={width/2} l=0.2u
+  nrd={0.2e-6/width} nrs={0.2e-6/width}
+  ad={width*0.27e-6} pd={1.1e-6+width}
+  as={width*0.27e-6} ps={1.1e-6+width}
m22 2 20 2 13 cmosplvt w={width/2} l=0.2u
+  nrd={0.2e-6/width} nrs={0.2e-6/width}
+  ad={width*0.27e-6} pd={1.1e-6+width}
+  as={width*0.27e-6} ps={1.1e-6+width}
*****************external network********************
xs1 2 3 17 18 13 tg
xs2 14 16 17 18 13 tg
xs3 14 15 17 18 13 tg
xs4 16 3 19 20 13 tg
xs5 1 15 19 20 13 tg
c1 15 2 2p
c2 16 2 1p
ci 3 0 0.5p
*******input signal, clocks, supply and analysis*******
vi 14 1 pwl 0 -0.15 74n -0.15 76n 0.15 189n 0.15
+  191n 0 300n 0 302n -0.15 441n -0.15
vphi1 17 0 pulse 1.8 0 47n 2n 2n 58n 112n
vphi1c 18 0 pulse 0 1.8 47n 2n 2n 58n 112n
vphi2 19 0 pulse 0 1.8 51n 2n 2n 50n 112n
vphi2c 20 0 pulse 1.8 0 51n 2n 2n 50n 112n
vdd 13 0 1.8
vb 1 0 0.9
.control
tran 10p 441n
plot v(3) v(14) v(17) v(19)
.endc
.end ▲
```

4.7 Folded-Cascode OTA

One very popular OTA topology is the so-called *folded cascode,* whose performance is comparable to that of the cascoded-symmetrical OTA. Its properties are modeled in the first three subsections of this section. The final subsection is devoted to an even more popular version offering a rail-to-rail common-mode range.

4.7.1 Topology and DC Transfer Characteristic

Elementary folded-cascode OTAs with NMOS and PMOS drivers are presented in Figs. 4.50(a) and (b), respectively. The matching drivers M1 and M2 and the current source/sink I_B constitute a differential input stage. v_{IP} and v_{IN} are the voltages of the non-inverting input and inverting input, respectively. Output is at the interconnected drain terminals of M4c and M6c. v_O denotes the output voltage. M3 and M4 are two matching current sources or sinks biased commonly by V_{B1}. M3c and M4c are also matching devices that cascode M1, M2, M3, and M4. The remaining devices M5, M5c, M6, and M6c constitute a low-voltage cascode current mirror.

Before we start discussing the dc operation, note that V_{B2} and V_{B3} are properly set by design to hold M3 and M5 unconditionally in saturation. Furthermore, M3 and M4 are properly sized to source $I_B + \Delta I$, where ΔI is a safety margin on the order of a few percent of I_B. The purpose of this margin will be clarified later.

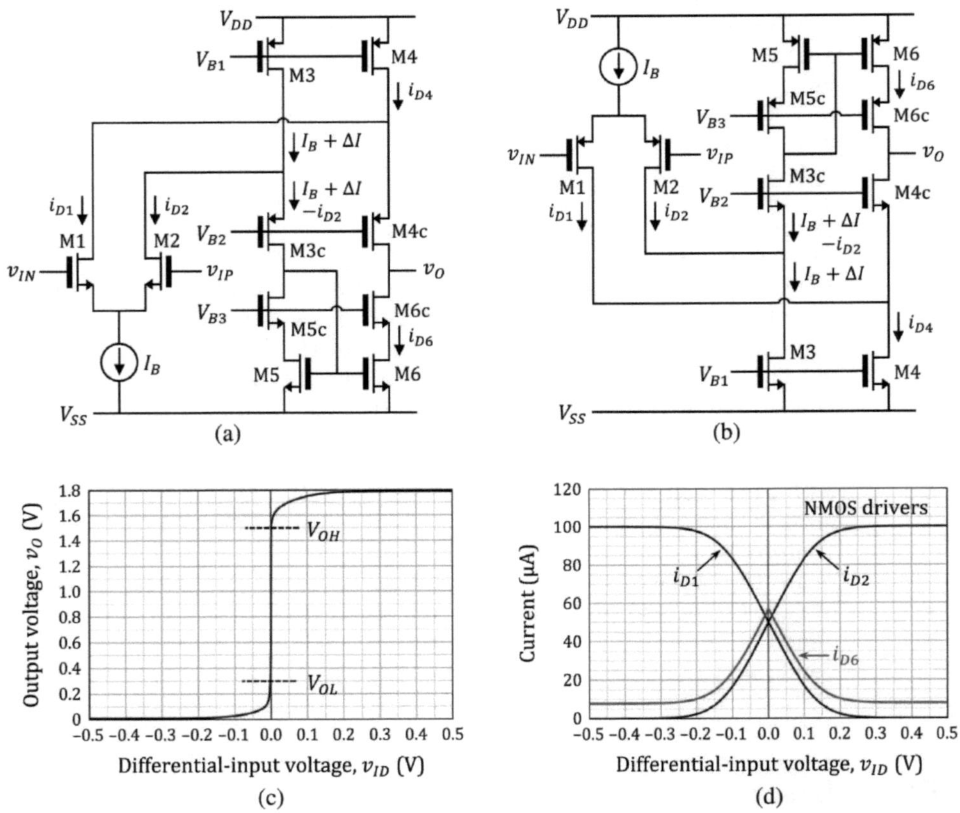

FIGURE 4.50 (a) Folded-cascode OTA with NMOS drivers. (b) Folded-cascode OTA with PMOS drivers. (c) DC transfer characteristic of a single-supply folded-cascode OTA. (d) Transconductance characteristics of the folded-cascode OTA whose dc transfer characteristic is presented in part c.

The dc operation is now discussed with reference to the transfer and transconductance characteristics shown in Figs. 4.49(c) and (d) for a single-supply OTA of NMOS drivers operating with $I_B = 100\ \mu A$ and $\Delta I = 8\ \mu A$.

First, consider the case of a large negative v_{ID} in excess of the differential-input range so that $i_{D2} = 0$ and $i_{D1} = I_B = 100\ \mu A$. As we will shortly justify, M4 is saturated in this case. It therefore conducts $i_{D4} = I_B + \Delta I = 108\ \mu A$ exactly like its unconditionally saturated pair M3. However, since i_{D1} steals one full I_B away from this current, what is left to conduct for M4c, M6c, and M6 is a mere $i_{D6} = \Delta I = 8\ \mu A$. On the other hand, the reference device M5 of the low-voltage cascode current mirror conducts a full $I_B + \Delta I = 108\ \mu A$ supplied by M3. The only way in which M6 can prevent itself from conducting a much smaller current is to force itself into nonsaturation with a vanishingly small drain-source voltage. For large negative values of v_{ID}, we indeed observe a very small v_O in Fig. 4.49(c). What ensures saturation for M4 and M4c as assumed in the foregoing description is this small v_O.

M2 starts conducting as the negative v_{ID} is brought sufficiently close to zero. As a result, i_{D2} rises and i_{D1} falls. This leads to a decreasing current in M5 and an increasing current in M6. The shrinking difference between these two currents necessitates an increasingly lighter nonsaturation for M6, hence a higher v_O, which is why v_O rises as v_{ID} is made increasingly less negative.

For $v_{ID} = 0$, the currents of M5 and M6 are equalized at $I_B/2 + \Delta I = 58\ \mu A$. M6, which is no more obliged to limit its current, escapes into saturation. As we make v_{ID} positive, hence $i_{D2} > I_B/2$ and $i_{D1} < I_B/2$, the current flowing in M5 is lowered below $I_B/2 + \Delta I$ and is mirrored by the saturated M6 to i_{D6}. Since both i_{D1} and i_{D6} are now lower than their respective levels observed for $v_{ID} = 0$, the current $i_{D4} = i_{D1} + i_{D6}$ of M4 should also be lower. A very steeply rising v_O lowers it by taking M4 out of saturation and thus preventing it from mirroring a full $I_B + \Delta I$ from M3.

A sufficiently large positive v_{ID} causes $i_{D1} = 0$ and $i_{D2} = I_B$. The current flowing in M5 and therefore in M6 are minimized to ΔI when this happens. This calls for $i_{D4} = \Delta I$ for M4 also. As a result, v_O is maximized at a level close to V_{DD}. This is indeed what we observe in Fig. 4.49(c) for large positive values of v_{ID}.

It is obvious from the foregoing analysis and Fig. 4.49(d) that, for any given v_{ID}, i_{D6} equals ΔI plus the smaller between i_{D1} and i_{D2}.

Ideally, a folded-cascode OTA should operate properly even with $\Delta I = 0$. The problem with such an implementation is that if the current supplied by any of M3 and M4 falls short of the bias current I_B of the drivers due to inevitable mismatch, then M5 or M6 will be forced into current starvation whenever v_{ID} reaches the limit of the differential-input range. This is not an issue for dc response, in which, these limits are not reached anyway but it will adversely affect the frequency and step responses if slewing is involved.

4.7.2 DC Differential Gain and Range Limitations

As in all single-stage OTAs, the dc differential gain of a folded-cascode OTA is the product of the effective transconductance G_m and the resistance r_o seen into the OTA at the output node. For both types of drivers, it is described by

$$A_0 = \underbrace{g_{m(1,2)}}_{G_m} \underbrace{(r_{id6c} \parallel r_{id4c})}_{r_o}, \tag{4.180}$$

where

$$r_{id6c} \approx A_{\max 6c} r_{ds6} = r_{ds6c} \left(g_{m6c} + g_{mb6c} \right) r_{ds6}, \tag{4.181}$$

and

$$r_{id4c} \approx A_{\max 4c} \left(r_{ds4} \parallel r_{ds(1,2)} \right) = r_{ds4c} \left(g_{m4c} + g_{mb4c} \right) \left(r_{ds4} \parallel r_{ds(1,2)} \right). \tag{4.182}$$

As usual, the bulk transconductance of M4c or M6c is to be ignored in the case of operation with a source-bulk short.

In (4.181), $A_{\max 6c}$ multiplies r_{ds6} because M6c cascodes M6. Why $A_{\max 4c}$ multiplies $r_{ds4} \parallel r_{ds(1,2)}$ in (4.182), however, may need a more elaborate explanation. To this end, let us drive the output node of the OTA with an independent voltage source, and trace the paths through which the resulting current flows to ground. One path goes through M4c and M4, and therefore exhibits an equivalent resistance $A_{\max 4c} r_{ds4}$. Another path goes through M4c, M1, M2, M3c, M5c, and M5. The equivalent resistance of this path, as determined from (4.83), equals $2r_{ds(1,2)}$ to a very good approximation because the resistance loading the drain of M2 is due to the very small $1/(g_{m3c} + g_{mb3c})$ seen into the source terminal of M3c. Therefore, the equivalent resistance of this second path equals $2A_{\max 4c} r_{ds(1,2)}$. However, a copy of the current flowing in this path is mirrored from M5 to M6, which creates a third path of equivalent resistance $2A_{\max 4c} r_{ds(1,2)}$ from output to ground. What (4.182) represents is the parallel equivalent of the resistances of these three paths.

The sensitivity of A_0 to the output voltage v_O is mainly due to the Early effect on r_{ds4c} and r_{ds6c} because the drain-source voltages of the rest of the devices are virtually independent of v_O. Equalizing $A_{\max 6c} r_{ds6}$ and $A_{\max 4c} \left(r_{ds4} \parallel r_{ds(1,2)} \right)$ helps reduce this sensitivity for the reason put forward in Subsection 4.5.3 for the basic OTA.

In a folded-cascode OTA with NMOS drivers, the lower limit of the common-mode range is given by

$$V_{IL} = V_{SS} + V_C + V_{GS(1,2)} \quad \text{(NMOS drivers)}, \tag{4.183}$$

where V_C is the compliance limit of the biasing current sink, and $V_{GS(1,2)}$ is the driver gate-source voltage specific to the lower limit of the common-mode range.

The upper limit of the common-mode range in the NMOS-driver version is imposed by the saturation condition of M1 and M2. This is because an increasing common-mode voltage reduces the drain-source voltage of these devices by pulling their common-source voltage up while their drain voltages are held constant by M3c and M4c at $V_{B2} - V_{GS(3c,4c)}$. Therefore,

$$V_{IH} = V_{B2} - V_{GS(3c,4c)} - V_{DSsat(1,2)} + V_{GS(1,2)} \quad \text{(NMOS drivers)}, \tag{4.184}$$

where $V_{GS(1,2)}$ is specific to the upper limit.

The lower limit of the output range in the NMOS-driver version is dictated by the saturation condition of M6c. Assuming a saturation margin ΔV_{DS6} for M6, it is described by

$$V_{OL} = V_{SS} + V_{DSsat6} + V_{DSsat6c} + \Delta V_{DS6} \quad \text{(NMOS drivers).} \tag{4.185}$$

The upper limit of the output range in the NMOS-driver version is dictated by the saturation condition of M4c, and is expressed as

$$V_{OH} = V_{DD} + V_{DSsat4} + V_{DSsat4c} + \Delta V_{DS4} \quad \text{(NMOS drivers),} \tag{4.186}$$

where ΔV_{DS4} is the saturation margin of M4.

The PMOS-driver counterparts of (4.183) to (4.186) are as follows:

$$V_{IL} = V_{B2} - V_{GS(3c,4c)} - V_{DSsat(1,2)} + V_{GS(1,2)} \quad \text{(PMOS drivers),} \tag{4.187}$$

where $V_{GS(1,2)}$ is specific to the lower limit of the common-mode range;

$$V_{IH} = V_{DD} + V_C + V_{GS(1,2)} \quad \text{(PMOS drivers),} \tag{4.188}$$

where V_C is the compliance limit of the biasing current source, and $V_{GS(1,2)}$ is specific to the higher limit of the common-mode range;

$$V_{OL} = V_{SS} + V_{DSsat4} + V_{DSsat4c} + \Delta V_{DS4} \quad \text{(PMOS drivers),} \tag{4.189}$$

where ΔV_{DS4} is the saturation margin of M4; and

$$V_{OH} = V_{DD} + V_{DSsat6} + V_{DSsat6c} + \Delta V_{DS6} \quad \text{(PMOS drivers),} \tag{4.190}$$

where ΔV_{DS6} is the saturation margin of M6.

4.7.3 Frequency Response, Step Response, and Noise-Related Properties

As stated in Subsection 4.5.4, the frequency response of an OTA can be extracted with adequate accuracy from a dc small-signal equivalent circuit supplemented with lumped node capacitances. The schematic shown in Fig. 4.51 serves this purpose for a folded-cascode OTA driving a capacitive load C_L. Although it is constructed with NMOS drivers, the conclusions reached from its analysis are also applicable to the PMOS-driver version.

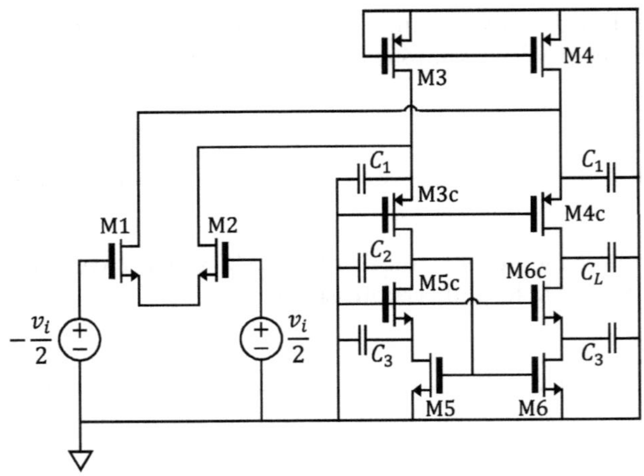

FIGURE 4.51 Circuit schematic for extracting the frequency response of a folded-cascode OTA.

The open-loop transfer function $A \doteq v_o/v_i$ as determined from Fig. 4.51 is described by

$$A = g_{m(1,2)} r_o \frac{A_Z(s)}{A_P(s)}, \tag{4.191a}$$

$$A_Z(s) \doteq 1 + s\frac{C_2}{2g_{m(5,6)}} + s^2 \frac{C_2C_3}{2g_{m(5,6)}g'_{m(5c,6c)}}, \tag{4.191b}$$

$$A_P(s) \doteq \left(1 + sr_oC_L\right)\left(1 + s\frac{C_1}{g'_{m(3c,4c)}}\right)\left(1 + s\frac{C_2}{g_{m(5,6)}} + s^2\frac{C_2C_3}{g_{m(5,6)}g'_{m(5c,6c)}}\right), \tag{4.191c}$$

where the primed transconductance symbol generally represents the sum of gate and bulk-transconductance parameters but the bulk-transconductance parameter is to be ignored in the special case of a device operating with a source-bulk short.

The dominant pole of the transfer function usually belongs to the output node which features by far the largest small-signal resistance among all nodes. Open-loop bandwidth is therefore given by

$$f_o = \frac{1}{2\pi \underbrace{\left(r_{id4c} \parallel r_{id6c}\right)}_{r_o} C_L}. \tag{4.192}$$

The remaining zeros and nondominant poles are associated mainly with the cascoding and mirroring devices. Their frequencies, being proportional to transconductance to capacitance ratios, can be raised by decreasing the channel length and/or increasing the overdrive voltage as explained in Subsection 4.5.4.

As to the gain/bandwidth product f_{GBW} of a folded-cascode OTA, we obtain the very same (4.145) by substituting (4.180) and (4.192) into (2.43). There is nothing special about the differential-input range V_{DIR} either. Depending on the type of the differential-pair topology, it is modeled by (4.85), (4.98), or (4.100).

The noise analysis and modeling equations presented in Subsection 4.5.5 for the basic symmetrical OTA are also valid for the folded-cascode OTA with the only exception that the number of noise-contributing devices is 6 instead of 8. The mean-square output noise voltage attributable to the folded-cascode OTA of a closed-loop configuration is therefore described by (4.168) or (4.169) provided that the upper limit of summation is changed to 6. Cascoding devices M3c, M4c, M5c, and M6c are to be ignored because their contribution to output noise is negligible as justified in Subsection 4.6.6 for the cascoded-symmetrical OTA.

Example 4.23 The discrete-time amplifier specified in Example 4.22 is to be implemented with a folded-cascode OTA of PMOS drivers.

Work An OTA schematic including the reference section is shown in Fig. 4.52(a). For the closed-loop configuration, the reader is referred to Fig. 4.49(b) in Example 4.22. Open-loop constraints as derived from closed-loop specifications in that example are also valid. They are itemized below.

- Output-range constraints:
  Equation (4.190) leads to the following counterpart of (4.175):

$$2V_{DSsat(6,6c)} + \Delta V_{DS6} > -0.45\text{ V}. \tag{4.193}$$

  Equation (4.189) leads to the following counterpart of (4.176):

$$2V_{DSsat(4,4c)} + \Delta V_{DS4} < 0.45\text{ V}. \tag{4.194}$$

- Common-mode range constraints:
  Equations (2.7) and (4.187) lead to the following counterpart of (4.177):

$$V_{B2} - V_{GS(3c,4c)} - V_{DSsat(1,2)} + V_{GS(1,2)} < 0.9\text{ V}. \tag{4.195}$$

  Equations (2.8) and (4.188) lead to the following counterpart of (4.178):

$$V_C + V_{GS(1,2)} > -0.9\text{ V}. \tag{4.196}$$

- Implications of dc-error specifications: $A_0 = 3500$ V/V, $C_2 = 1$ pF and $C_1 = 2$ pF.
- Implications of settling-error specifications: $g_{m(1,2)} = 500\ \mu$A/V and $I_B = 110\ \mu$A. ΔI is set to one-tenth of I_B to 11 μA.

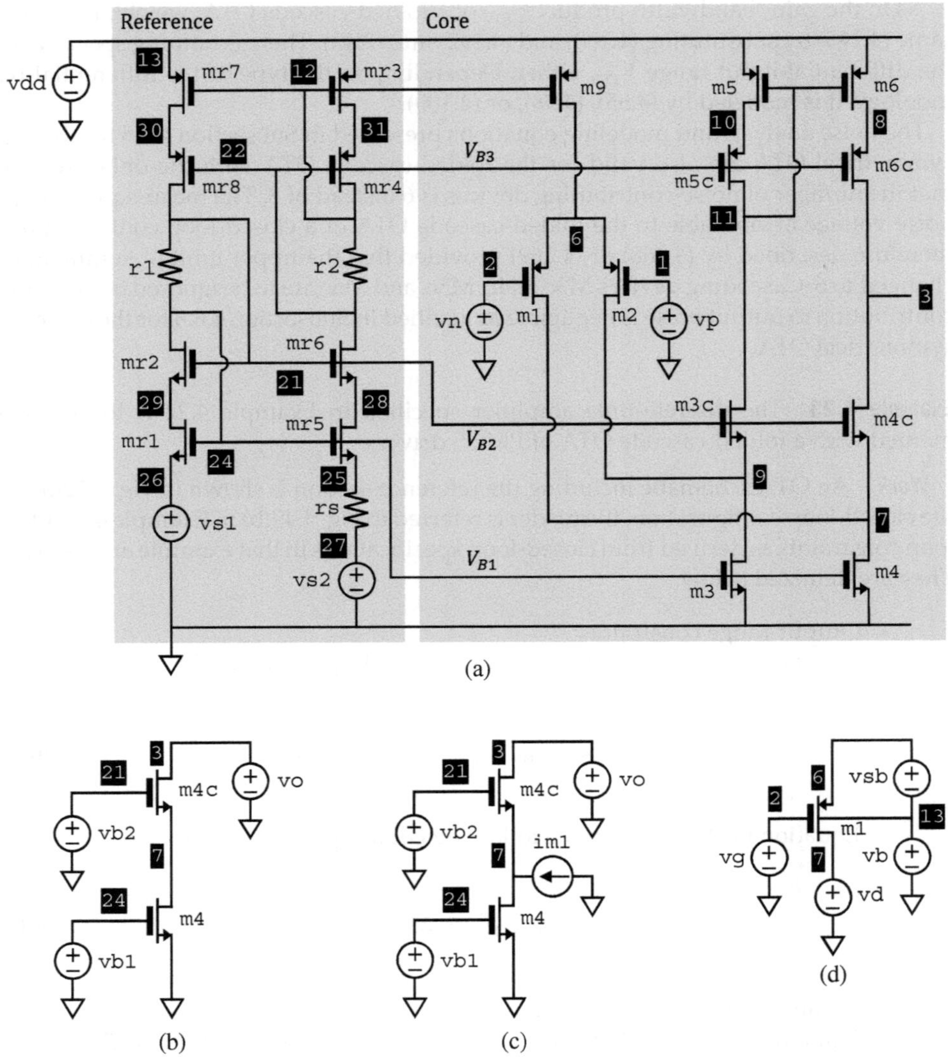

FIGURE 4.52 (a) Schematic of the folded-cascode OTA designed in Example 4.23. Top-level schematic of the discrete-time closed-loop configuration is identical to the one presented in Fig. 4.49(b). (b) Schematic used for determining the channel length of M3, M3c, M4, and M4c with iterative simulation. (c) Schematic used for extracting A_{max4c}. (d) Schematic used for determining the channel length of M1 and M2 with iterative simulation.

Device-sizing and verification steps are as follows:

1. *Partitioning the Output Resistance* r_o Considering $A_0 = 3500$ V/V and $g_{m(1,2)} = 500\ \mu\text{A/V}$, (4.180) yields $r_o = 7\ \text{M}\Omega$. We partition this resistance equally between $r_{id6c} = A_{\text{max6c}} r_{ds6} = 14\ \text{M}\Omega$, $r_{id4c} = A_{\text{max4c}}(r_{ds4} \parallel r_{ds(1,2)}) = 14\ \text{M}\Omega$.

2. *Sizing M5, M5c, M6, and M6c* Following the argument presented in step 4 of the design flow of Example 4.22, we select the standard-threshold option, and partition the constraint defined by (4.193) into $\Delta V_{DS6} = -0.15$ V and $V_{DSsat(6,6c)} = -0.15$ V. The design tool of Fig. 3.17 indicates −0.175 V for the overdrive voltage of these devices. The corresponding sheet current is determined as 0.9 μA/sq from Fig. 3.19. Since the bias current of these devices equals $(I_B/2) + \Delta I = 55 + 11 = 66$ μA, their effective aspect ratio is obtained as $W/L = 73.5$.

   Next, by targeting $I_{D6c}A_{max6c}r_{ds6} = 66 \times 10^{-6} \times 14 \times 10^{6} = 924$ V, we determine the channel length of M6 and M6c (and therefore M5 and M5c) with an iterative simulation similar to what is done for M8 and M8c in step 4 of Example 4.22. In all iterations, we maintain −0.175 V for the overdrive voltage of M6 and M6c, and keep the source-bulk voltage of M6c at $V_{SB6c} = V_{DSsat6} + \Delta V_{DS6} = -0.3$ V. The outcome of this step is $L_{DR(6,6c)} = 0.75$ μm, which also applies to M5 and M5c. Also note that the iterative simulation performed for sizing yields $V_{B3} - V_{DD} = -0.983$ V for the bias voltage of the cascode devices M6c and M5c with respect to V_{DD}, and $V_{GS(5,6)} = -0.595$ V for the gate-source voltage of M6 and M5.

   With a view to keeping the current of the reference section low, we decide to implement each of M5, M5c, M6 and M6c with six parallel copies of a unit device conducting 11 μA. The aspect ratio of this PMOS unit device should therefore be $73.5/6 = 12.25$, which implies $W_{DR} = 8.5$ μm drawn channel width for the 0.75-μm drawn channel length extracted with simulation.

3. *Sizing M3, M3c, M4, and M4c* Threshold option is not constrained for these devices. We select the standard-threshold option. Targeting $V_{DSsat4c} = V_{DSsat4}$, we partition (4.194) into $\Delta V_{DS4} = 0.15$ V and $V_{DSsat(4,4c)} = 0.15$ V, and refer to the design tool of Fig. 3.8 to obtain $V_{GS(4,4c)} - V_{TN} = 0.195$ V. The corresponding sheet current is determined from Fig. 3.11 as 4.5 μA/sq. Since the bias current of M3c and of M4c equals $(I_B/2) + \Delta I = 55 + 11 = 66$ μA, the aspect ratio of these devices is set to $(W/L)_{3c,4c} = 14.7$. M3 and M4, on the other hand, conduct $I_B + \Delta I = 110 + 11 = 121$ μA each. Their aspect ratio is calculated as $(W/L)_{3,4} = 26.9$.

   For determining the channel length of these devices, we need to partition $A_{max4c}(r_{ds4} \parallel r_{ds(1,2)}) = 14$ MΩ first. If we do so by allocating comparable resistances to M4 and M1, we will most probably end up with an extremely long channel for M4 because, compared to M1, this device conducts a factor of 2.2 higher current while operating under a considerably smaller drain-source voltage. It is therefore reasonable to allocate a smaller resistance to M4. $r_{ds4}/r_{ds(1,2)} = 1/3$ is probably appropriate because a higher ratio would lower $A_{max4c}r_{ds4}$ only marginally while necessitating a disproportionately larger $A_{max4c}r_{ds(1,2)}$. This selection leads us to $A_{max4c}r_{ds(1,2)} = 56$ MΩ and $A_{max4c}r_{ds4} = 18.7$ MΩ.

   The channel length of M4 and M4c is extracted by targeting

$$A_{max4c}r_{ds4}I_{D4} = 18.7\,(\text{M}\Omega) \times 121\,(\mu\text{A}) = 2263\text{ V}$$

in an iterative simulation, in which, the branch of M1 is omitted in order not to corrupt the outcome with the finite output resistance of this device. Simulation is performed on the schematic shown in Fig. 4.52(b) and netlisted in the following input file `ex4-23a.sp`:

```
filename: ex4-23a.sp
.incl bsim.sp
m4 7 24 0 0 cmosn l=1.8u w=10u
vb1 24 0 0.600
m4c 3 21 7 0 cmosn l=1.8u w=10u
vb2 21 0 0.985
vo 3 0 0.9
.dc vo 0.8 1.0 0.1m
.end
```

The reader may have noticed that this schematic does not reflect the differences that exist in the actual circuit between M4 and M4c in terms of currents and aspect ratios. Exactly like the cascode arrangements we have previously sized with iterative simulation, these two devices conduct the same current and feature the same arbitrarily selected channel width. For an explanation of how this arrangement works despite the seemingly unrealistic features, let us reexpress $A_{\max 4c} r_{ds4} I_{D4}$ quite generally as

$$A_{\max 4c} r_{ds4} I_{D4} = \underbrace{[(g_{m4c} + g_{mb4c})/I_{D4c}]}_{\substack{\text{Transconductance}\\ \text{efficiency}\\ \text{of M4c}}} \; \underbrace{r_{ds4c} I_{D4c}}_{\substack{\text{Early}\\ \text{voltage}\\ \text{of M4c}}} \; \underbrace{r_{ds4} I_{D4}}_{\substack{\text{Early}\\ \text{voltage}\\ \text{of M4}}} .$$

According to this expression, $A_{\max 4c} r_{ds4} I_{D4}$ is independent of W_4 and W_{4c} because neither the transconductance efficiency nor any of the two Early voltages is dependent on channel width. This fact enables us to construct the netlist of iterative simulation with arbitrarily selected widths. However, the widths must be kept identical because otherwise we cannot maintain identical overdrive voltages in these devices while they conduct the same current as an inevitable result of the omission of M1.

The outcome of each simulation is plotted with

```
plot i(vo)/deriv(i(vo))
```

and its value for $v_O = 0.9$ V is compared with 2263 V. This procedure eventually yields $L_{DR(4,4c)} = 1.8$ µm which, of course, also applies to M3 and M3c. $V_{B1} = 0.600$ V and $V_{B2} = 0.985$ V are additional outcomes of the iterative simulation.

We implement each of M3 and M4 with 11 unit devices each conducting 11-µA current, and use 6 such unit devices to implement M3c and also M4c. The aspect ratio of each unit device will be 2.45 as obtained from the ratio 26.9/11. The corresponding drawn channel width is $W_{DR} = 4.35$ µm.

4. *Sizing M1 and M2* As explained in step 1 of the design flow in Example 4.22, the low-threshold option is appropriate for these devices. As also concluded in that example, design should target $V_{GS(1,2)} - V_{TP} = -0.200$ V for the overdrive voltage and $(W/L)_{1,2} = 39.3$ for the aspect ratio. However, the constraint imposed on the channel length in the present example is fundamentally different

from those applied in Example 4.22. Unlike the case of symmetrical OTA, A_0 of a folded-cascode OTA is highly affected by $L_{DR(1,2)}$ through $r_{ds(1,2)}$. This is why we need to determine the channel length with an iterative simulation in accordance with the specification set for $A_{\max 4c} r_{ds(1,2)}$ in step 3. Since the cascoding device M4c has already been sized in step 3, the netlist of the simulation will contain M1 alone. Firstly, however, $A_{\max 4c}$ should be determined in order to define the target value of $r_{ds(1,2)}$ from $A_{\max 4c} r_{ds(1,2)}$. This is done with a separate simulation based on the schematic shown in Fig. 4.52(c) and netlisted in the following input file `ex4-23b.sp`:

```
filename: ex4-23b.sp
.incl bsim.sp
m4 7 24 0 0 cmosn l=1.8u w=4.35u m=11
vb1 24 0 0.6
m4c 3 21 7 0 cmosn l=1.8u w=4.35u m=6
vb2 21 0 0.985
vo 3 0 0.9
im1 0 7 55u
.control
op
print (@m4c[gm]+@m4c[gmbs])/@m4c[gds]
.endc
.end
```

Notice the representation of M4 and M4c with their exact dimensions as well as the inclusion of an independent current source `im1` imitating the current conducted by M1 in the original circuit. Simulation yields $A_{\max 4c} = 272$ V/V, for which, $A_{\max 4c} r_{ds(1,2)} = 56$ MΩ implies $r_{ds(1,2)} = 206$ kΩ. Considering the bias current $I_B/2 = 55$ μA, we target $V_{A(1,2)} = r_{ds(1,2)}(I_B/2) = 11.3$ V for the Early voltage to be supported by the channel length of M1. Iterative simulation is performed with the following input file `ex4-23c.sp` on the schematic shown in Fig. 4.52(d):

```
filename: ex4-23c.sp
.incl bsimlvt.sp
m1 7 2 6 13 cmosplvt l=0.50u w=10u
vg 2 0 0.9
vd 7 0 0.3
vdd 13 0 1.8
vsb 6 13 -0.490
.dc vd 0.2 0.4 0.1m
.end
```

Gate voltage is kept constant at 0.9 V as defined for the operation point. Drain voltage is swept around 0.3 V, which is what the sum $V_{DSsat(4,4c)} + \Delta V_{DS4}$ implies as determined in step 3. Source-bulk voltage `vsb` is adjusted in each iteration to maintain the specified overdrive voltage $V_{GS(1,2)} - V_{TP} = -0.200$ V. Considering the fixed 0.9-V gate voltage, this overdrive voltage implies $V_{S(1,2)} + V_{TP} = 1.1$ V. By referring the source voltage to the bulk rather than ground, we turn this equation into

$$V_{SB(1,2)} + V_{TP} = 1.1 - V_{DD} = -0.7\,\text{V}.$$

In each iteration, we read from Fig. 4.10 the specific value of $V_{SB(1,2)}$ that satisfies this equation, and set the voltage source `vsb` of the netlist accordingly. After running each simulation, we plot the Early voltage with

```
plot i(vd)/deriv(i(vd))
```

and read its value for 0.3-V drain voltage. This procedure eventually yields $L_{DR(1,2)} = 0.5\ \mu m$, for which, the previously determined aspect ratio implies $W_{DR(1,2)} = 17.4\ \mu m$. Yet another outcome of the simulation is $V_{SB(1,2)} = -0.490$ V, which, considering the already specified overdrive voltage, implies $V_{GS(1,2)} = -0.410$ V.

5. *Evaluating the Constraints (4.195) and (4.196) Imposed by the Common-Mode Range* The condition stated by (4.195) concerns the lower limit of the common-mode range. $V_{B2} - V_{GS(3c,4c)}$ equals $V_{DSsat(4,4c)} + \Delta V_{DS4}$, hence 0.3 V. Also considering $V_{GS(1,2)} = -0.410$ V as determined in step 4, and $V_{DSsat(1,2)} = -0.175$ V as read out from Fig. 4.11 for $V_{GS(1,2)} - V_{TP} = -0.200$ V, the left-hand side of (4.195) turns out to be 65 mV, which is much smaller than the right-hand side. This condition is obviously met.

   Condition (4.196) is related to the upper limit of the common-mode range. For $V_{GS(1,2)} = -0.410$ V, it sets the constraint $V_C > -0.490$ V on the compliance limit of the bias-current source.

6. *Designing the Reference Section* Repeating the design flow presented in Example 4.22 for this purpose, we arrive at $R_1 = 35\ k\Omega$, $R_2 = 35.3\ k\Omega$, and $R_S = 7.5\ k\Omega$. M9 is implemented with 10 copies of the PMOS unit device sized in step 2 for M5, M6, M5c, and M6c. One copy of the same unit device is used for implementing MR3, MR4, MR7, and MR8. One copy of the NMOS unit device sized in step 3 for M3, M4, M3c, and M4c is used for implementing each of MR1, MR2, and M6. MR5 comprises four copies of the same unit device.

7. *Verification of Open-Loop Response* It is left to the reader to verify the open-loop response along the lines of step 6 of Example 4.22. The OTA core and reference section of the input file `ex4-23d.sp` presented next in step 8 can be used for this purpose.

8. *Switch Design and Verification of Transient Response* Design considerations presented in step 7 of Example 4.22 apply also to the present case. Verification is done with the following input file `ex4-23d.sp`, which netlists the closed-loop configuration shown in Fig. 4.49(b).

```
filename: ex4-23d.sp
.incl bsimlvt.sp
.incl bsim.sp
.param width=5u
***********************ota core************************
m1  7 2 6 13 cmosplvt w=17.4u l=0.5u nrd=0.01 nrs=0.01
+  ad=9.6p pd=35.9u as=9.6p ps=35.9u
m2  9 1 6 13 cmosplvt w=17.4u l=0.5u nrd=0.01 nrs=0.01
+  ad=9.6p pd=35.9u as=9.6p ps=35.9u
m3  9 24 0 0 cmosn w=4.35u l=1.8u nrd=0.05 nrs=0.05
+  ad=2.4p pd=9.8u m=11
m3c 11 21 9 0 cmosn w=4.35u l=1.8u nrd=0.05 nrs=0.05
```

```
+  ad=2.4p pd=9.8u as=2.4p ps=9.8u m=6
m4 7 24 0 0 cmosn w=4.35u l=1.8u nrd=0.05 nrs=0.05
+  ad=2.4p pd=9.8u m=11
m4c 3 21 7 0 cmosn w=4.35u l=1.8u nrd=0.05 nrs=0.05
+  ad=2.4p pd=9.8u as=2.4p ps=9.8u m=6
m5 10 11 13 13 cmosp w=8.5u l=0.75u nrd=0.023 nrs=0.023
+  ad=4.7p pd=18.1u m=6
m5c 11 22 10 13 cmosp w=8.5u l=0.75u nrd=0.023 nrs=0.023
+  ad=4.7p pd=18.1u as=4.7p ps=18.1u m=6
m6 8 11 13 13 cmosp w=8.5u l=0.75u nrd=0.023 nrs=0.023
+  ad=4.7p pd=18.1u m=6
m6c 3 22 8 13 cmosp w=8.5u l=0.75u nrd=0.023 nrs=0.023
+  ad=4.7p pd=18.1u as=4.7p ps=18.1u m=6
m9 6 12 13 13 cmosp w=8.5u l=0.75u nrd=0.023 nrs=0.023
+  ad=4.7p pd=18.1u m=10
******************reference section*******************
mr1 29 24 26 0 cmosn w=4.35u l=1.8u nrd=0.05 nrs=0.05
mr2 24 21 29 0 cmosn w=4.35u l=1.8u nrd=0.05 nrs=0.05
mr3 31 12 13 13 cmosp w=8.5u l=0.75u nrd=0.023 nrs=0.023
mr4 12 22 31 13 cmosp w=8.5u l=0.75u nrd=0.023 nrs=0.023
mr5 28 24 25 0 cmosn w=4.35u l=1.8u nrd=0.05 nrs=0.05 m=4
mr6 22 21 28 0 cmosn w=4.35u l=1.8u nrd=0.05 nrs=0.05
mr7 30 12 13 13 cmosp w=8.5u l=0.75u nrd=0.023 nrs=0.023
mr8 21 22 30 13 cmosp w=8.5u l=0.75u nrd=0.023 nrs=0.023
vs1 26 0 0
vs2 27 0 0
rs 25 27 7.5k
r1 21 24 35k
r2 12 22 35.3k
***************transmission gate subcircuit**************
.subckt tg in out ngate pgate power
mn in ngate out 0 cmosn w=width l=0.2u
+  nrd={width/0.2e-6} nrs={width/0.2e-6}
+  ad={width*0.55e-6} pd={1.1e-6+2*width}
+  as={width*0.55e-6} ps={1.1e-6+2*width}
mp in pgate out power cmosp w=width l=0.2u
+  nrd={width/0.2e-6} nrs={width/0.2e-6}
+  ad={width*0.55e-6} pd={1.1e-6+2*width}
+  as={width*0.55e-6} ps={1.1e-6+2*width}
.ends
********************dummy switch*********************
m21 2 19 2 0 cmosn w={width/2} l=0.2u
+  nrd={width/0.4e-6} nrs={width/0.4e-6}
+  ad={width*0.27e-6} pd={1.1e-6+width}
+  as={width*0.27e-6} ps={1.1e-6+width}
m22 2 20 2 13 cmosp w={width/2} l=0.2u
+  nrd={width/0.4e-6} nrs={width/0.4e-6}
+  ad={width*0.27e-6} pd={1.1e-6+width}
+  as={width*0.27e-6} ps={1.1e-6+width}
*******************external network*******************
```

```
xs1 2 3 17 18 13 tg
xs2 14 16 17 18 13 tg
xs3 14 15 17 18 13 tg
xs4 16 3 19 20 13 tg
xs5 1 15 19 20 13 tg
c1 15 2 2p
c2 16 2 1p
ci 3 0 0.5p
********input signal, clocks, supply and analysis********
vi 14 1 pwl 0 -0.15 74n -0.15 76n 0.15 189n 0.15
+  191n 0 300n 0 302n -0.15 441n -0.15
vphi1 17 0 pulse 1.8 0 47n 2n 2n 58n 112n
vphi1c 18 0 pulse 0 1.8 47n 2n 2n 58n 112n
vphi2 19 0 pulse 0 1.8 51n 2n 2n 50n 112n
vphi2c 20 0 pulse 1.8 0 51n 2n 2n 50n 112n
vdd 13 0 1.8
vb 1 0 0.9
.control
tran 10p 441n
plot v(3) v(14) v(17) v(19)
.endc
*********************************************************
.end ▲
```

4.7.4 Rail-to-Rail Folded-Cascode OTA

In a folded-cascode OTA, the limits of the output range can be set very close to the power-supply rails. As to the limits of the common-mode range, we can bring V_{IH} of the NMOS version above V_{DD} and V_{IL} of the PMOS version below V_{SS}. However, V_{IL} of the NMOS version and V_{IH} of the PMOS version have to be separated from the corresponding rail by at least the sum of the driver gate-source voltage and current source/sink compliance limit. Although deploying low-threshold drivers as done in Example 4.23 alleviates this problem, a true rail-to-rail common-mode range is impossible in the fundamental topologies shown in Figs. 4.50(a) and (b). A radical solution to this problem is to use both NMOS and PMOS differential pairs in the same OTA so that at least one of the two is active for any given value of the common-mode voltage. This can be done by adding a PMOS differential pair to the internal nodes of the low-voltage current mirror of a folded-cascode OTA with NMOS drivers as shown in Fig. 4.53(a), or by adding an NMOS differential pair to the corresponding nodes of a folded-cascode OTA with PMOS drivers. We study only the former topology in this book but, if needed, the reader can easily extend the conclusions to the latter.

DC Metrics

Referring to Fig. 4.53(a), notice $I_{BP} = I_{BN} = I_B$ for the bias currents of the differential pairs, and $I_B + \Delta I$ for the bias currents supplied by M3 and M4. These currents imply $(I_B/2) + \Delta I$ for M3c, M4c, M5c, and M6c, and $I_B + \Delta I$ for M5 and M6. Also note that the PMOS- and NMOS-driver devices are properly sized to equalize their transconductances g_{mp} and g_{mn} when conducting $I_B/2$. Let us denote this particular transconductance with g_m. As shown

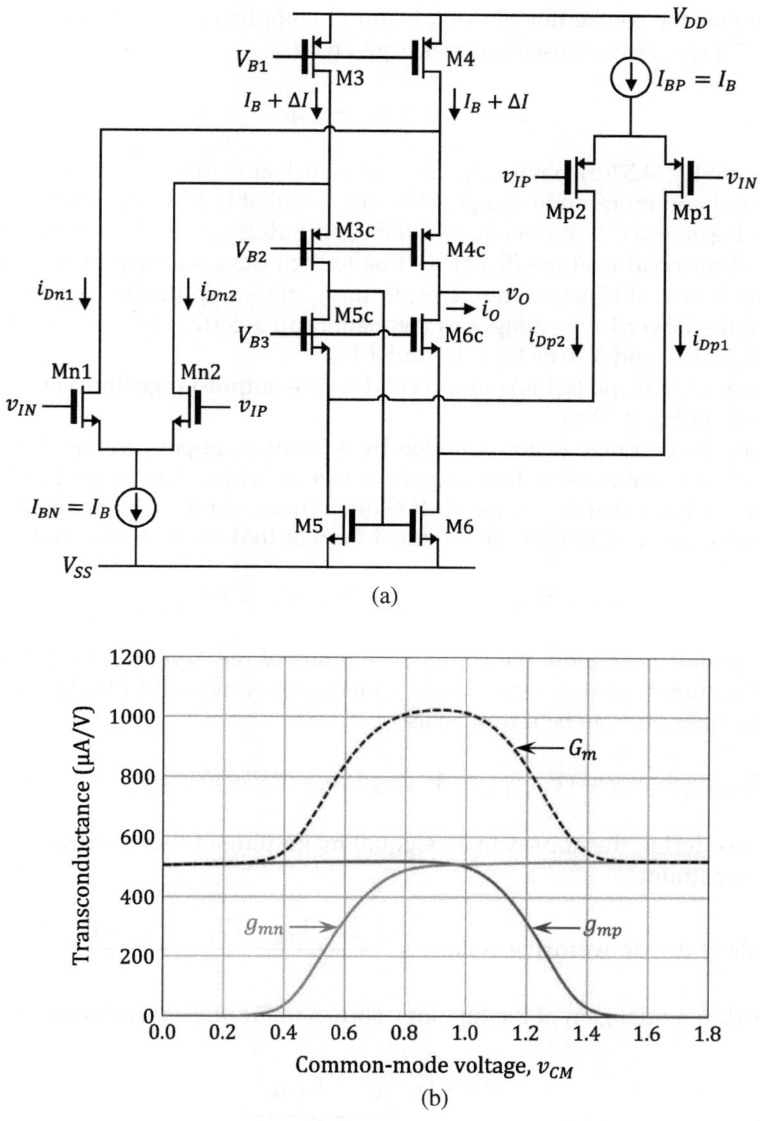

FIGURE 4.53 (a) Rail-to-rail folded-cascode OTA. (b) Variation of PMOS-driver transconductance g_{mp}, NMOS-driver transconductance g_{mn}, and effective transconductance G_m with common-mode voltage v_{CM}.

in Fig. 4.53(b), g_{mp} and g_{mn} vary with the common-mode voltage v_{CM} in a complementary and overlapping fashion throughout the rail-to-rail common-mode range between V_{SS} and V_{DD}. At the low end, g_{mn} vanishes as the compliance limit of the current-sink supplying I_{BN} is breached, but g_{mp} remains equal to g_m because the current-source I_{BP} is fully compliant. An opposite observation is made at the high end where g_{mp} vanishes as the current-source I_{BP} moves out of compliance but g_{mn} remains equal to g_m. In midrange, where neither the source nor the sink is out of compliance, we observe $g_{mp} = g_m$ and $g_{mn} = g_m$. The effective transconductance, given by

$$G_m = g_{mp} + g_{mn}, \tag{4.197}$$

and traced in Fig. 4.53(b), obviously does not vanish anywhere in between V_{SS} and V_{DD}. A rail-to-rail common-mode range is therefore available from the OTA topology presented in Fig. 4.53(a). Also note that NMOS- and PMOS-driver devices exhibit identical midrange transconductance efficiencies due to their design for equalized transconductances and identical bias currents. This, in turn, implies approximately the same magnitude for their overdrive voltages as the reader can ascertain by comparing the design tools of Figs. 3.12 and 3.20 or Figs. 4.9 and 4.14.

Equations (4.185) and (4.186) are still valid for the output-range limits of the rail-to-rail OTA shown in Fig. 4.53(a).

Common-mode range is not bounded by a lower or upper limit, and therefore can accommodate a common-mode swing of any size provided that V_{IH} of the PMOS differential pair is higher than V_{IL} of the NMOS differential pair in order to avoid a midrange gap. This necessary condition can be stated from (4.188) and (4.183) as follows:

$$V_{SS} + V_{CN} + V_{GS(n1,n2)} \leq V_{DD} + V_{CP} + V_{GS(p1,p2)},$$

where V_{CN} and V_{CP} denote the compliance limits of the NMOS bias-current sink and PMOS bias-current source, respectively. In terms of NMOS and PMOS overdrive voltages, this expression can be rewritten as

$$V_{GS(n1,n2)} - V_{TN} - (V_{GS(p1,p2)} - V_{TP}) \leq V_{DD} - V_{SS} + V_{CP} - V_{CN} - V_{TN} + V_{TP},$$

which, considering the approximately equal magnitude of the overdrive voltages, implies the constraint

$$\text{Magnitude of driver overdrive voltage} \leq \frac{V_{DD} - V_{SS} + V_{CP} - V_{CN} - V_{TN} + V_{TP}}{2}. \tag{4.198}$$

The dc differential gain of the topology shown in Fig. 4.53(a) is described by

$$A_0 = G_m \underbrace{\left(r_{id6c} \parallel r_{id4c}\right)}_{r_o}, \tag{4.199}$$

where

$$r_{id6c} = A_{\max 6c}(r_{ds6} \parallel r_{ds(p1,p2)}) = r_{ds6c}(g_{m6c} + g_{mb6c})(r_{ds6} \parallel r_{ds(p1,p2)}), \tag{4.200}$$

and

$$r_{id4c} = A_{\max 4c}(r_{ds4} \parallel r_{ds(n1,n2)}) = r_{ds4c}(g_{m4c} + g_{mb4c})(r_{ds4} \parallel r_{ds(n1,n2)}). \tag{4.201}$$

As usual, the bulk transconductance parameter is to be ignored if the device operates with a source-bulk short.

Properties Related to Frequency Response and Step Response

Among the metrics related to frequency response, open-loop bandwidth is still described by (4.192) provided that a single-pole response is dictated by the output pole. Considering also (4.199), the gain-bandwidth product is given by

$$f_{GBW} = \frac{G_m}{2\pi C_L}, \tag{4.202}$$

which is a function of the common-mode voltage through G_m. As observed in Fig. 4.53(b), a factor-of-two variation is possible over the rail-to-rail common-mode range.

Nondominant poles are due to the cascoding and mirroring devices. As usual, their frequencies are increasing functions of the overdrive voltage and decreasing functions of the channel length.

One frequency-response metric that needs special attention is the differential-input range V_{DIR}, which, unlike that of the basic folded-cascode OTA, exhibits asymmetry and varies strongly with the common-mode voltage. These properties are analyzed in the following three paragraphs, and are illustrated in Fig. 4.54 with simulated transconductance characteristics. These characteristics relate the output current i_O of the OTA topology of Fig. 4.53(a) to the noninverting-input voltage v_{IP} for a constant inverting-input voltage V_{IN} and a constant output voltage V_O.

First, consider the characteristic shown in part a of Fig. 4.54. It represents the case of such a high V_{IN} that, in the common-mode state of $v_{IP} = V_{IN}$, the current source I_{BP} is completely out of compliance, and therefore the PMOS differential pair is disabled. In this case, the slope of the plot around the common-mode operation point equals g_m as contributed by the NMOS differential pair alone. In order to determine the extreme values of the output current i_O for this case, let us first consider $v_{IP} \ll V_{IN}$. This will turn on Mp2, and thus bring I_{BP} into compliance. However, Mn2 will be forced into cutoff. Driver currents are therefore described by $I_{Dp1} = 0$, $I_{Dp2} = I_B$, $I_{Dn1} = I_B$, and $I_{Dn2} = 0$. As the reader can ascertain from Fig. 4.53(a), the output current in this state attains its most negative value $i_{O(\min)} = -2I_B$, and therefore defines a differential input range

$$V_{DIR}^{-} = -\frac{2I_B}{g_m} \tag{4.203}$$

for the negative excursion of the differential input voltage. For determining the most positive value of the output current for the case of a high V_{IN}, consider $v_{IP} \gg V_{IN}$. This will keep I_{BP} out of compliance, and therefore maintain $I_{Dp1} = 0$, $I_{Dp2} = 0$. In the NMOS differential pair, however, it leads to $I_{Dn1} = 0$ and $I_{Dn2} = I_B$. An inspection of Fig. 4.53(a) reveals $i_{O(\max)} = I_B$ for this extreme. Since the slope equals g_m, the corresponding differential-input range is determined as

$$V_{DIR}^{+} = \frac{I_B}{g_m}. \tag{4.204}$$

The asymmetry of the differential-input range is obvious from (4.203) and (4.204).

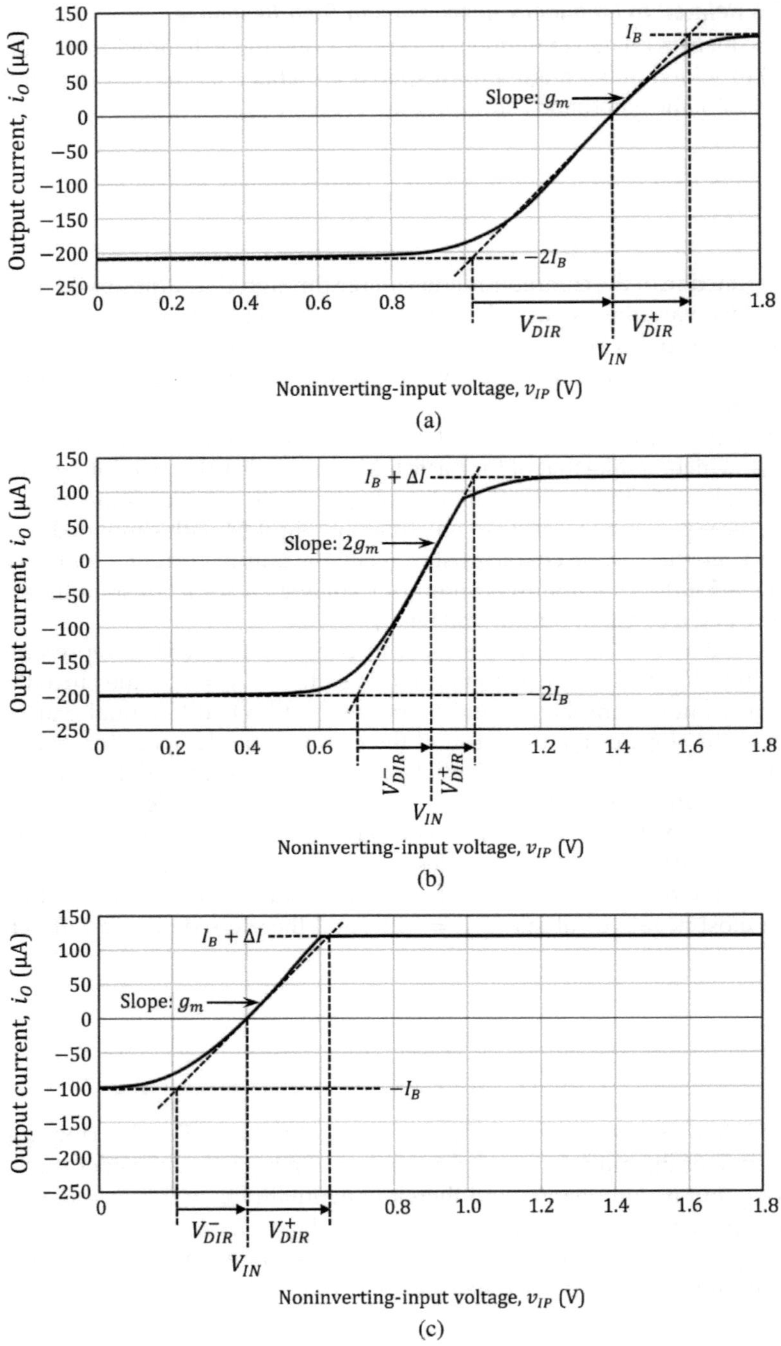

FIGURE 4.54 Transconductance characteristics relating the output current i_O to the noninverting-input voltage v_{IP} for a constant V_{IN} in the OTA topology of Fig. 4.53(a). (a) V_{IN} close to upper rail. (b) V_{IN} midway between the rails. (c) V_{IN} close to lower rail.

Next, consider the characteristic shown in Fig. 4.54(b), which belongs to the case of a V_{IN} midway between the rails, hence a slope $2g_m$. An analysis conducted along the lines of the preceding paragraph again indicates $i_{O(\min)} = -2I_B$ for the negative peak output current, and thus leads to

$$V_{DIR}^{-} = -\frac{I_B}{g_m} \tag{4.205}$$

for the negative differential-input range. A little more intricate analysis is needed for the positive differential-input range. Raising v_{IP} far above V_{IN} in this case implies $I_{Dn1} = 0$, $I_{Dn2} = I_B$, and $I_{Dp2} = 0$, which leave M5 and therefore M6 with only ΔI to sink. Since the current flowing in Mp1 cannot be larger than what M6 sinks, and Mp2 cannot conduct any current, the current source I_{BP} should be forced partially out of compliance in order to keep its current well below I_B. This is accomplished by the movement of the common-source voltage of the PMOS differential pair sufficiently close to V_{DD}. In order to limit the current of Mp1 despite the very large negative gate-source voltage thus created, the drain voltage of this device also rises close to V_{DD}. Carrying the same large voltage at its source but a rather low V_{B3} at its gate, M6c responds by moving into cutoff. Since M6c is thus prevented from sinking any current while M4c sources $I_B + \Delta I$, the output current in this case is given by $i_{O(\max)} = I_B + \Delta I$. Considering the effective transconductance $2g_m$, the positive differential-input range is therefore described by

$$V_{DIR}^{+} = \frac{I_B + \Delta I}{2g_m}. \tag{4.206}$$

Finally, consider the characteristic shown in Fig. 4.54(c), which belongs to the case of a sufficiently low V_{IN} keeping the current sink I_{BN} completely out of compliance in common mode. Since NMOS differential pair is disabled in this case, the slope of the characteristic equals g_m. A v_{IP} much lower than V_{IN} does not change the disabled state of the NMOS differential pair but diverts the current I_B of the current source I_{BP} to Mp2 in its entirety. Therefore, $I_{Dp1} = 0$, $I_{Dp2} = I_B$, $I_{Dn1} = 0$, and $I_{Dn2} = 0$. An analysis based on these implications indicates $i_{O(\min)} = -I_B$. Also considering the slope, the negative differential-input range is determined exactly as described by (4.205). As to the positive differential-input range, the analysis presented in the preceding paragraph is applicable with the exception that the slope of the characteristic is halved. This leads to

$$V_{DIR}^{+} = \frac{I_B + \Delta I}{g_m}. \tag{4.207}$$

The foregoing analysis of V_{DIR} is based on the assumption of a v_{IP} varying around a constant V_{IN}. If v_{IN} varies around a constant midrange V_{IP}, V_{DIR}^{-}, and V_{DIR}^{+} are still described by (4.205) and (4.206), respectively. In the case of a large V_{IP} that disables the PMOS differential pair in common mode, (4.205) and (4.207) are applicable. If V_{IP} is sufficiently low to disable the NMOS differential pair in common mode, then, (4.203) and (4.204) are the valid expressions.

The asymmetry and variability observed in the differential-input range greatly complicate the application of the analytical techniques presented in Subsections 2.5.2 and

2.6.2. A simplified way in which these techniques can still be applied for worst-case modeling is (a) to determine the most relevant of the three common-mode scenarios analyzed above, and (b) to assume a constant V_{DIR} equaling the most restrictive between V_{DIR}^- and V_{DIR}^+.

Constant-G_m Implementation

One particular concern regarding the dc differential gain A_0 is the distorting effect of its variation with common-mode voltage v_{CM}. The root cause of this variation is the dependence of bias currents and therefore G_m and r_o on v_{CM}. This dependence makes the gain-bandwidth product f_{GBW} and nondominant pole-frequencies also functions of v_{CM}.

There exist several techniques by which the effect of v_{CM} on the small-signal parameters of a rail-to-rail folded-cascode OTA can be prevented.[9] Supplementing the driver devices with bypass devices as labeled Mn3, Mn4, Mp3, and Mp4 in Fig. 4.55(a) is one such technique. These additional devices are biased by a dc gate voltage V_{ref} set midway between V_{SS} and V_{DD}. Their sizing is such that, for $v_{CM} = V_{ref}$, they divert sufficient current away from all four driver devices to reduce each of g_{mp} and g_{mn} to one half of the nominal g_m a driver exhibits when conducting full $I_B/2$. The effective transconductance given by (4.197) is therefore reduced from $2g_m$ to g_m in the middle of the rail-to-rail common-mode range.

For $v_{CM} \ll V_{\text{ref}}$, the PMOS bypass devices Mp3 and Mp4 are cut off, and each of driver devices Mp1 and Mp2 conducts $I_B/2$ and offers $g_{mp} = g_m$. g_{mn}, on the other hand, is negligible because Mn3 and Mn4 divert virtually all I_B away from Mn1 and Mn2. This leads to $G_m = g_m$ exactly like what is available in the middle of the common-mode range.

For $v_{CM} \gg V_{\text{ref}}$, M3p and M4p conduct $I_B/2$ each, whereas M1p and M2p are cut off with $g_{mp} = 0$. In contrast, the NMOS drivers Mn1 and Mn2 conduct $I_B/2$ each, whereas Mn3 and Mn4 are cut off. This leads to $g_{mn} = g_m$, and therefore $G_m = g_m$ again. As shown in Fig. 4.55(b), a reasonably constant G_m is thus maintained throughout the rail-to-rail common-mode range. Bias currents of all cascoding and mirroring devices also remain constant, which ensures nearly constant overdrive voltages, transconductances, and output resistances for all. The dc differential gain of the OTA and nondominant-pole frequencies are thus made independent of the common-mode voltage.

As stated above, bypass devices are sized to divert sufficient current from the driver devices to reduce the midrange values of g_{mp} and g_{mn} to $g_m/2$. Referring to (3.16) and (3.28), note that the transconductance of a MOSFET in strong inversion is proportional to the square root of the bias current. In order to cut the transconductance by half, the current should therefore be reduced by a factor of four. Assuming identical channel lengths for driver and bypass devices, this can be accomplished by making the bypass device three times wider than the driver device. Next, referring to (3.29), note that the transconductance of a MOSFET in subthreshold is proportional to the current itself. Halving the current by setting equal widths for the bypass and driver devices will therefore be appropriate for this case. Since the operation point is usually somewhere in between strong inversion and subthreshold, the scaling factor of the width of the bypass device with respect to the driver device is expected to be in between 1 and 3. The most accurate way of determining the proper scaling factor is to perform an iterative simulation.

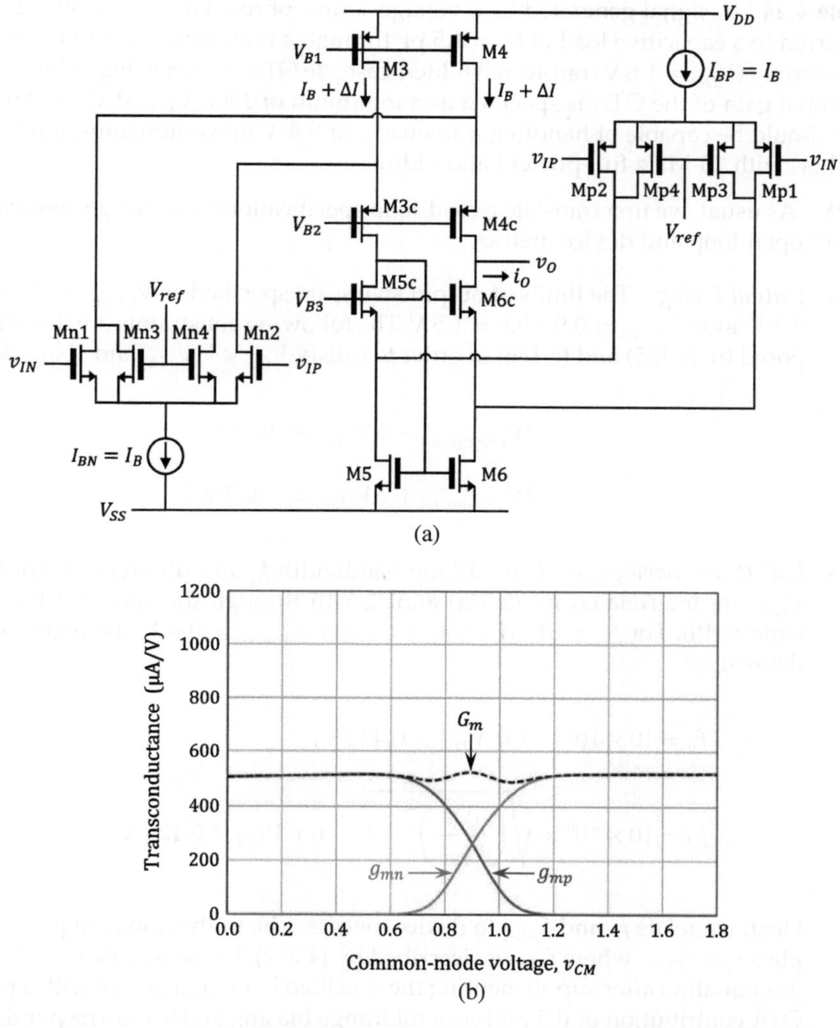

FIGURE 4.55 (a) Constant-g_m rail-to-rail folded-cascode OTA. (b) Variation of PMOS-driver transconductance g_{mp}, NMOS-driver transconductance g_{mn}, and effective transconductance G_m with common-mode voltage v_{CM}.

Since the scaling factor needed for bypassing is greater than unity, bypass devices are to be wider than driver devices. They add considerable amount of parasitic capacitance to internal nodes and thus lower the frequencies of nondominant poles. If this happens to be a critical issue, we can implement bypassing by adjusting the gate voltages rather than the widths of bypass devices. This calls for two separate reference voltages: one close to V_{SS} and biasing the PMOS bypass devices, and another close to V_{DD} and biasing the NMOS bypass devices. As a matter of fact, the original proposal of the constant-g_m technique described in this subsection is based on this type of implementation.[10]

Example 4.24 A signal generated by a voltage source of resistance $R_S = 30\ \text{k}\Omega$ is to be transferred to a capacitive load of $C_L = 4.5$ pF through a unity-gain buffer built with the single-supply ($V_{DD} = 1.8$ V) rail-to-rail folded-cascode OTA shown in Fig. 4.56(a). The dc differential gain of the OTA is specified as a minimum of 2000 V/V at $V_O = 0.9$ V. The buffer should be capable of handling a sinewave of 0.6-V maximum amplitude around 0.9-V bias with 10-MHz full-power bandwidth.

Work As usual, we first translate closed-loop specifications into design constraints in terms of open-loop and device metrics.

- *Output Swing* The limits of output swing are specified as $V_{O(\text{min})} = 0.9 - 0.6 = 0.3$ V and $V_{O(\text{max})} = 0.9 + 0.6 = 1.5$ V. The following constraints are therefore imposed by (4.185) and (4.186) in order to satisfy $V_{OL} < V_{O(\text{min})}$ and $V_{OH} > V_{O(\text{max})}$:

$$2V_{DSsat(6,6c)} + \Delta V_{DS6} < 0.3\,\text{V}, \tag{4.208}$$

$$2V_{DSsat(4,4c)} + \Delta V_{DS4} > -0.3\,\text{V}. \tag{4.209}$$

- *Full-Power Bandwidth* Closed-loop bandwidth f_{cl} and differential-input range V_{DIR} are interrelated by (2.59a) and (2.59b) through the specified full-power bandwidth. For $f_{fp} = 10$ MHz, $\alpha_i = 1$, and $X_{i(\text{max})} = 0.6$ V, the relationship is defined by

$$f_{cl} = 10 \times 10^6 \qquad \text{for } V_{DIR} > 0.424\ \text{V}, \tag{4.210a}$$

$$f_{cl} = 10 \times 10^6 \times \sqrt{\left(\frac{0.6}{V_{DIR}}\right)^2 - 1} \qquad \text{for } V_{DIR} < 0.424\ \text{V}. \tag{4.210b}$$

Next, we relate f_{cl} and V_{DIR} to device metrics. The unity-gain configuration implies $f_{cl} = f_{GBW}$, where f_{GBW} is described by (4.202). We substitute $C_L = 5$ pF into this equation after supplementing the specified load capacitance with a possible OTA contribution of 0.5 pF. Since midrange biasing at 0.9 V corresponds to the case of Fig. 4.54(b), we assume $G_m = 2g_m$ in (4.202), where g_m denotes driver transconductance. Replacing g_m with the product of driver transconductance efficiency $g_m/(I_B/2)$ and driver bias current $I_B/2$, we finally arrive at

$$f_{cl} = \frac{g_m/(I_B/2)}{1.57 \times 10^{-11}} \frac{I_B}{2}. \tag{4.211}$$

For V_{DIR}, midrange biasing at 0.9 V implies (4.205) for negative swing and (4.206) for positive swing. We use the more restrictive (4.206), and assume $\Delta I = I_B/10$. The outcome is

$$V_{DIR} = \frac{1.1}{g_m/\left(I_B/2\right)}. \tag{4.212}$$

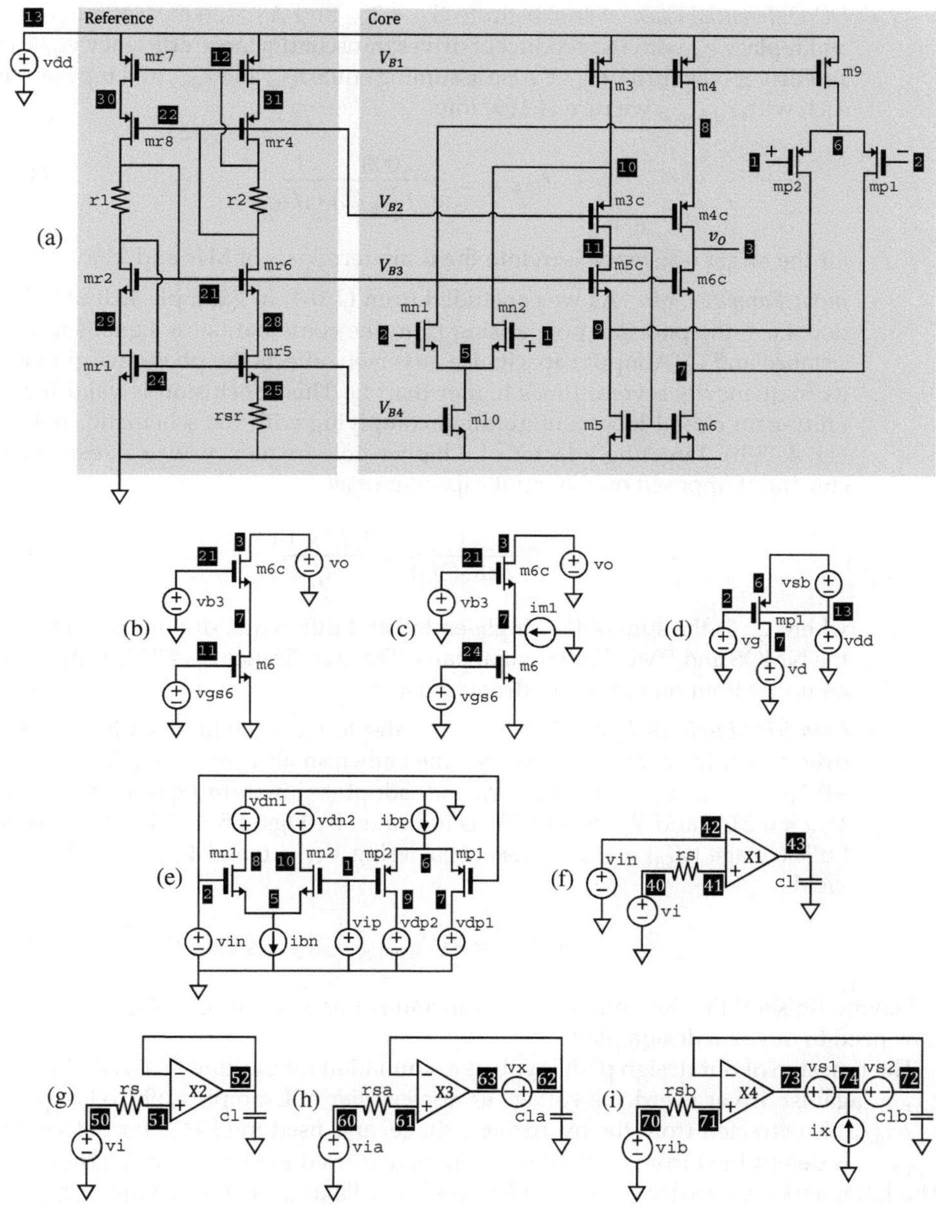

FIGURE 4.56 Simulation schematics used in Example 4.24. (a) OTA subcircuit netlisted in `ex4-24e.sp`. (b) Schematic netlisted in `ex4-24a.sp` for determining the channel length of M6 and M6c. (c) Schematic netlisted in `ex4-24b.sp` for determining A_{max6c} and r_{ds6}. (d) Schematic netlisted in `ex4-24c.sp` for determining the channel length of Mp1 and Mp2. (e) Schematic netlisted in `ex4-24d.sp` for determining the input capacitance of the OTA. (f) Schematic netlisted in `ex4-24e.sp` for open-loop dc and ac verification. (g) Schematic netlisted in `ex4-24e.sp` for closed-loop dc and transient verification. (h) and (i) Schematics netlisted in `ex4-24e.sp` for loop-gain verification.

- *DC Differential Gain* We substitute $G_m = 2g_m$ and $A_0 \geq 2000$ V/V into (4.199), and replace g_m with the product of driver transconductance efficiency $g_m/(I_B/2)$ and driver bias current $I_B/2$. Also assuming equal r_{id4c} and r_{id6c}, and representing each with $r_{id(4c,6c)}$, we turn (4.199) into

$$r_{id(4c,6c)} \geq \frac{2000}{g_m/\left(I_B/2\right)} \frac{1}{I_B/2} \tag{4.213}$$

for the target resistance seen into the drain terminals of M4c and M6c.

- *Input Parasitic Pole* As we concluded from (4.164) in Example 4.20 of Subsection 4.6.4, the parasitic pole arising from the combination of signal-source resistance and OTA input capacitance adversely affects the phase margin unless its frequency is several times higher than f_{cl}. This conclusion is valid for any unity-gain closed-loop configuration complying with the schematic shown in Fig. 4.45(b). Targeting a factor of 3 higher pole frequency, we can express the constraint imposed on the input capacitance as

$$C_{in} \leq \frac{1}{2\pi R_S(3f_{cl})} = \frac{1.77\times 10^{-6}}{f_{cl}}, \tag{4.214}$$

where C_{in} is the sum of the single-ended and differential-input capacitances of the NMOS and PMOS differential pairs. The significance of (4.214) is that it sets an upper limit on the size of driver devices.

- *Overdrive Limit for Driver Devices* In order to relax the limit set by (4.198) on driver overdrive voltages, we assume rather small compliance limits $V_{CP} = -0.20$ V and $V_{CN} = 0.20$ V, and adopt a low-threshold option, hence $V_{TN} < 0.21$ V and $V_{TP} > -0.15$ V as indicated by Figs. 4.5 and 4.10 for a source-bulk voltage $|V_{SB}| = 0.2$ V. Also considering $V_{SS} = 0$ and $V_{DD} = 1.8$ V, (4.198) yields

$$V_{GS(n1,n2)} - V_{TN} = -(V_{GS(p1,p2)} - V_{TP}) \leq 0.52 \text{ V}. \tag{4.215}$$

Having finished the derivation of the constraints imposed on open-loop metrics, we now need to devise a design plan.

The first step of our design plan involves a simulation for plotting f_{cl}, I_B, $(W/L)_{1,2}$, and $r_{id(4c,6c)}$ against driver overdrive voltage as done earlier in Example 4.19. To be specific, $g_m/(I_B/2)$ is extracted from the overdrive voltage, and used in (4.212) for determining V_{DIR}. f_{cl} is determined from (4.210a) or (4.210b), and used in (4.211) for determining I_B. The latter and the overdrive voltage yield $(W/L)_{1,2}$. Finally, the lower limit of $r_{id(4c,6c)}$ is determined from (4.213). These plots will enable us to determine an optimum overdrive voltage within the limit defined by (4.215) for driver devices. The target values of the four plotted metrics will therefore be fixed at that stage.

Next, we will rely on (4.208) and (4.209) for determining $V_{DSsat(6,6c)}$ and $V_{DSsat(4,4c)}$, hence the overdrive voltages and sheet currents of the standard-threshold devices M3, M3c, M4, M4c, M5, M5c, M6, and M6c. Since the bias currents of these devices are also known, we will be able to calculate their aspect ratios. An iterative simulation will be performed subsequently for determining the channel lengths in accordance with (4.213). We will finish by designing a bias-current reference.

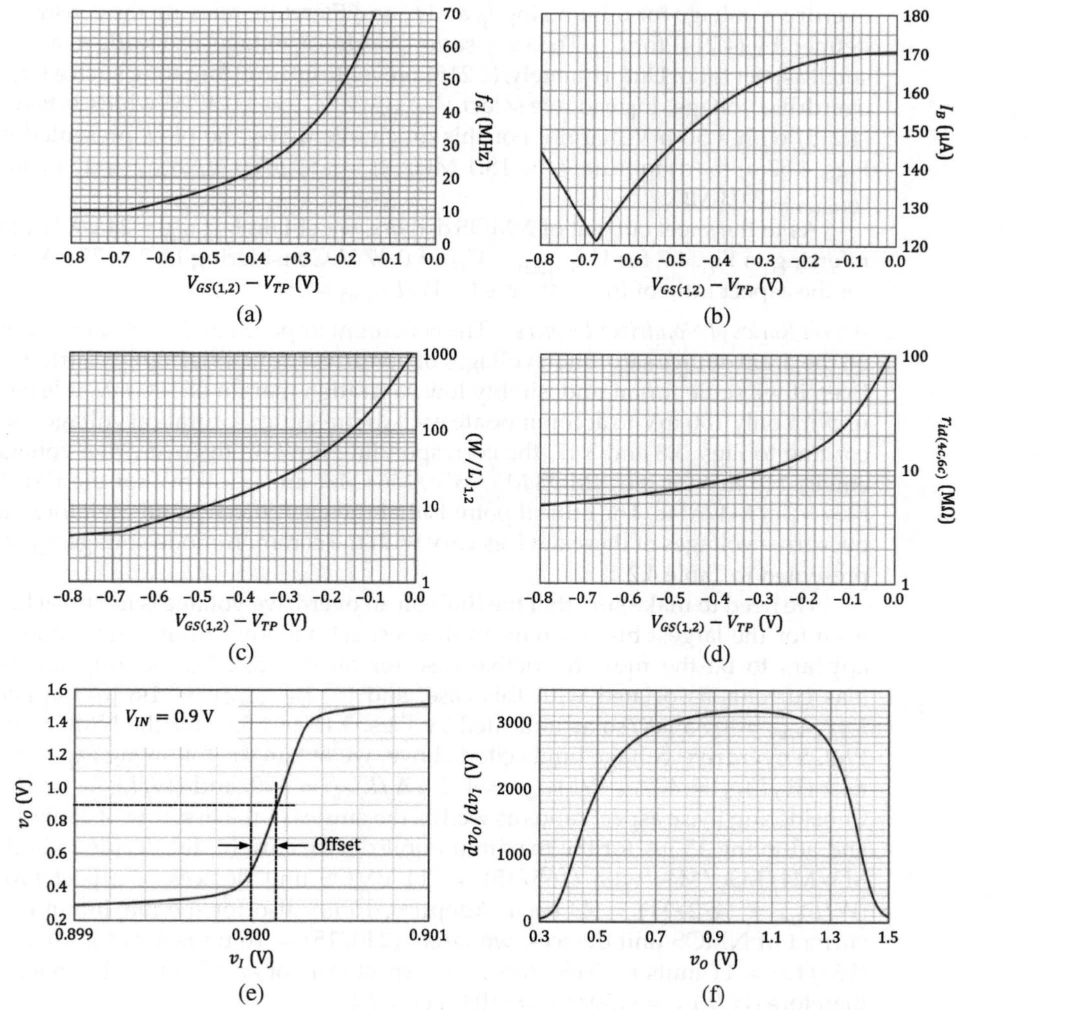

FIGURE 4.57 (a) Closed-loop bandwidth against driver overdrive from (4.210a) and (4.210b). (b) Bias current against driver overdrive from (4.211). (c) Driver aspect ratio against driver overdrive from (4.158). (d) Minimum resistance seen into the drain of M4c or M6c against driver overdrive from (4.213). (e) Simulated dc transfer characteristic v_O against v_{IP} for $v_{IN} = 0.9$ V. (f) Simulated slope of the dc transfer characteristic against output voltage.

Steps of design flow are described next.

1. *Driver Overdrive Voltage and Aspect Ratios* Simulated plots of f_{cl}, I_B, $(W/L)_{p1,p2}$, and $r_{id(4c,6c)}$ are displayed in Figs. 4.57(a) to (d) as functions of PMOS-driver overdrive voltage $V_{GS(p1,p2)} - V_{TP}$.[6] These plots indicate −0.675 V as the optimum

[6] The reader is invited to verify these plots by writing and running an input file along the lines of `ex4-19a.sp`.

overdrive voltage for minimizing I_B and f_{cl} and therefore relaxing the constraint defined by (4.214) for C_{in}. The very same overdrive voltage also leads to a very small aspect ratio. Unfortunately, (4.215) prevents us from targeting such a large overdrive voltage. Instead, we select $V_{GS(p1,p2)} - V_{TP} = -0.47$ V, which satisfies (4.215) with a 50-mV margin. For this much overdrive, the plots presented in Figs. 4.57(a) to (d) indicate f_{cl} = 18.1 MHz, I_B = 150 μA, $(W/L)_{p1,p2}$ = 10.6, and $r_{id(4c,6c)}$ = 7.03 MΩ.

As to the sheet current of NMOS drivers, we obtain $I_{\square(n1,n2)}$ = 24 μA/sq by referring to Fig. 4.7 for $V_{GS(n1,n2)} - V_{TN} = 0.47$ V. Considering $I_B/2$ = 75 μA, we set the aspect ratio of these drivers to $(W/L)_{n1,n2}$ = 3.1.

2. *Aspect Ratios of Nondriver Devices* The constraint imposed by (4.208) and (4.209) on the drain-source saturation voltages of nondriver devices is highly restrictive. Even if we settle for an undesirably low saturation margin of 100 mV, it leaves us with only 100 mV to accommodate each drain-source saturation voltage. According to Figs. 3.8 and 3.17, the corresponding limit on the overdrive voltage equals 115 mV for the NMOS M5/M5c/M6/M6c and −100 mV for the PMOS M3/M3c/M4/M4c. The critical point is that the bias currents and therefore the overdrive voltages of these devices vary with the common-mode voltage v_{CM} as presented in Table 4.2.

   We need to make sure that the limit on an overdrive voltage is not breached even for the largest bias current. In this respect, the low common-mode level appears to be the most restrictive case for all devices. Considering (a) the bias currents associated with this case, and (b) $I_{\square(5,5c,6,6c)}$ = 1.6 μA/sq and $I_{\square(3,3c,4,4c)}$ = 0.35 μA/sq as indicated by Figs. 3.11 and 3.19 for the NMOS and PMOS overdrive voltage limits cited above, we obtain the following aspect ratios: $(W/L)_{3,4}$ = 472, $(W/L)_{3c,4c}$ = 472, $(W/L)_{5,6}$ = 150, and $(W/L)_{5c,6c}$ = 103. Considering these aspect ratios as well as the entries of the first row of Table 4.2, and adopting 15 μA for the maximum unit-device current, we decide to build M3/M4/M3c/M4c with (165/15) = 11 PMOS unit devices of aspect ratio $(W/L)_{uP}$ = (472/11) = 43 each. Adopting 15 μA also for the maximum bias current of NMOS unit devices, we target (240/15) = 16 units for M5/M6 and (165/15) = 11 units for M5c/M6c. The aspect ratio of an NMOS unit device is therefore $(W/L)_{uN}$ = (150/16) = (103/11) = 9.4.

   The very same unit devices are also to be used for M9 and M10. Each of the two will contain 10 units to conduct I_B = 150 μA. Also note that the 100-mV

Common mode	M3/M4	M3c/M4c	M5/M6	M5c/M6c
Low	$I_B + \Delta I$ (165 μA)	$I_B + \Delta I$ (165 μA)	$(3I_B/2) + \Delta I$ (240 μA)	$I_B + \Delta I$ (165 μA)
Medium	$I_B + \Delta I$ (165 μA)	$(I_B/2) + \Delta I$ (90 μA)	$I_B + \Delta I$ (165 μA)	$(I_B/2) + \Delta I$ (90 μA)
High	$I_B + \Delta I$ (165 μA)	$(I_B/2) + \Delta I$ (90 μA)	$(I_B/2) + \Delta I$ (90 μA)	$(I_B/2) + \Delta I$ (90 μA)

TABLE 4.2 Bias Currents in Rail-to-Rail Folded-Cascode OTA

drain-source saturation voltage selected for nondriver devices also applies to M9 and M10 due to the common use of these unit devices. Considering our earlier assumption of $V_{CP} = -0.20$ V and $V_{CN} = 0.20$ V, this much drain-source saturation voltage will obviously provide each of M9 and M10 with 100-mV saturation margin.

3. *Bias Voltages of Nondriver Devices* We need the estimates of the overdrive voltages of M4, M4c, M6, and M6c around the operation point $V_O = 0.9$ V for determining the channel lengths of these devices by iterative simulation. We already derived these voltages in step 2 for low levels of common mode. We now need to determine them for medium levels of common mode where the operation point is defined. To this end, we first refer to Table 4.2 for determining the relative change in bias currents hence sheet currents as the common mode changes from low to medium. Since we already know the low-level values of sheet currents, the relative change will enable us to determine sheet currents and therefore overdrive voltages in medium-level common mode. The result is $I_{\square 4} = 0.35$ μA/sq, $I_{\square 4c} = 0.19$ μA/sq, $I_{\square 6} = 1.1$ μA/sq, and $I_{\square 6c} = 0.87$ μA/sq. Next, relying on Figs. 3.11 and 3.19, we determine the corresponding overdrive voltages as $V_{GS4} - V_{TP} = -100$ mV, $V_{GS4c} - V_{TP} = -65$ mV, $V_{GS6} - V_{TN} = 92$ mV, and $V_{GS6c} - V_{TN} = 75$ mV.

   Also required while performing the iterative simulation procedure are the source-bulk voltages of M4c and M6c at the operation point. For low-level common mode, these two voltages are set by design to $V_{SB4c} = V_{DSsat4} + \Delta V_{DS4} = -0.1 - 0.1 = -0.2$ V and $V_{SB6c} = V_{DSsat6} + \Delta V_{DS6} = 0.1 + 0.1 = 0.2$ V. Considering the variation of $V_{GS4c} - V_{TP}$ from −100 to −65 mV as the level of common mode changes from low to medium, and ignoring the corresponding variation in threshold voltages, we expect V_{SB4c} to change from −0.2 to −0.235 V. In the case of M6c, low-to-medium change in common mode brings $V_{GS6c} - V_{TN}$ from 0.115 V down to 0.075 V. Ignoring the variation of the threshold voltage, we therefore expect V_{SB6c} to change from 0.2 to 0.24 V.

4. *Sizing M5/M5c/M6/M6c and Mp1/Mp2* Channel lengths of these devices are determined for a target resistance $r_{id(4c,6c)} = 7.03$ MΩ by applying the procedure described in step 3 of Example 4.23. Considering (4.200) first, we partition r_{id6c} assuming $r_{ds(p1,p2)} = 3r_{ds6}$. The purpose of this unequal partitioning is to avoid the excessively long channel that would otherwise be needed for M6 in order to compensate for the effect of its low drain-source voltage. This leads us to $A_{\max 6c} r_{ds6} = 9.4$ MΩ and $A_{\max 6c} r_{ds(p1,p2)} = 28.2$ MΩ.

   Next, we determine the channel length of M6 and M6c by performing an iterative simulation on these two devices in accordance with the schematic shown in Fig. 4.56(b). This schematic is similar to the one presented in Fig. 4.52(b). The reader is referred to Example 4.23 for its construction. Simulation is targeted for

$$I_{D6} A_{\max 6c} r_{ds6} = 165 \times 10^{-6} \times 9.4 \times 10^{6} = 1551 \text{ V},$$

   and is run with the following input file `ex4-24a.sp`:

```
filename: ex4-24a.sp
.incl bsim.sp
m6 7 11 0 0 cmosn l=1.3u w=10u
vgs6 11 0 0.500
```

```
m6c 3 21 7 0 cmosn l=1.3u w=10u
vb3 21 0 0.800
vo 3 0 0.9
.dc vo 0.8 1 0.1m
.end
```

In each iteration, after selecting the channel length, we determine V_{TN6} from Fig. 3.7 for $V_{SB6} = 0$, and add it to $V_{GS6} - V_{TN} = 92$ mV to obtain V_{GS6}. Similarly, we determine V_{TN6c} from Fig. 3.7 for $V_{SB6c} = 0.240$ V, and add it to $V_{GS6c} - V_{TN} = 75$ mV to obtain V_{GS6c}. V_{B3} is then calculated by adding V_{SB6c} to V_{GS6c}. The outcome is $L_{DR(6,6c)} = 1.3$ µm, which also applies to M5 and M5c. Also note the additional outcomes $V_{GS6} = 0.5$ V and $V_{B3} = 0.8$ V.

We can now calculate the drawn channel width of the NMOS unit devices to be used for implementing M5, M5c, M6, and M6c. Considering $(W/L)_{uN} = 9.4$ and $L_{DR(6,6c)} = 1.3$ µm, the width turns out to be 12 µm.

A subsequent simulation run on the schematic shown in Fig. 4.56(c) enables us to determine the actual values of $A_{\max 6c}$ and r_{ds6}. Its input file `ex4-24b.sp` is constructed along the lines of `ex4-23b.sp` of Example 4.23, and is presented below.

```
filename: ex4-24b.sp
.incl bsim.sp
m6 7 11 0 0 cmosn l=1.8u w=16.6u m=11
vgs6 11 0 0.515
m6c 3 21 7 0 cmosn l=1.8u w=16.6u m=6
vb3 21 0 0.77
vo 3 0 0.9
iml 0 7 75u
.control
op
print (@m6c[gm]+@m6c[gmbs])/@m6c[gds] 1/@m6[gds]
.endc
.end
```

This simulation yields $r_{ds6} = 38.7$ kΩ and $A_{\max 6c} = 281$ V. The target $A_{\max 6c} r_{ds(p1,p2)} = 28.2$ MΩ therefore implies $r_{ds(p1,p2)} = 100$ kΩ for PMOS-driver devices. Considering the bias current $I_B/2 = 75$ µA of these devices, the target Early voltage is calculated as $V_{A(p1,p2)} = 7.53$ V. For determining their channel length, we run an iterative simulation on the schematic shown in Fig. 4.56(d). The following input file `ex4-24c.sp` is used for this purpose.

```
filename: ex4-24c.sp
.incl bsimlvt.sp
mp1 7 2 6 13 cmosplvt l=0.2u w=10u
vg 2 0 0.9
vd 7 0 0.24
vdd 13 0 1.8
vsb 6 13 -0.25
.dc vd 0.1 0.3 0.1m
.end
```

For the construction of this file, the reader is referred to the description of the input file `ex4-23c.sp` in Example 4.23.

Even for a minimum channel length $L_{DR(p1,p2)} = 0.2$ μm, this simulation yields an Early voltage about twice as large as the target value. This doesn't come as a surprise because these devices operate with a large drain-source voltage. While an Early voltage larger than the target is welcome since its sole implication is to elevate A_0 beyond the specified limit, a minimum channel length is definitely undesirable as it may lead to excessive mismatch between the two driver devices. We therefore decide to set $L_{DR(p1,p2)} = 0.4$ μm. Rerunning `ex4-24c.sp` with this channel length and the corresponding source-bulk voltage $V_{SB} = -0.28$ V yields $V_{A(p1,p2)} = 19.9$ V. For $I_B/2 = 75$ μA, the corresponding value of $r_{ds(p1,p2)}$ is 265 kΩ. Considering also $r_{ds6} = 38.7$ kΩ and $A_{max6c} = 281$ V, the revised value of r_{id6c} is determined from (4.200) as 9.49 MΩ. Since we prefer $r_{id4c} = r_{id6c}$, the target value of r_{id4c} is also revised to 9.49 MΩ. For an estimate of the resulting dc differential gain, we first refer to Figs. 4.9 and 4.14 for reading the transconductance efficiency corresponding to the overdrive voltage ±0.47 V targeted for the drivers. The outcome $g_m/I_D \approx 4$ μA/V together with the target bias current 75 μA lead us to $g_m = 300$ μA/V hence $G_m = 2g_m = 600$ μA/V. For $r_{id4c} = r_{id6c} = 9.49$ MΩ, we therefore estimate $A_0 = 2847$ V/V for the dc differential gain.

The sizing of PMOS drivers is finalized with $W_{DR(p1,p2)} = 3.7$ μm as implied by $(W/L)_{p1,p2} = 10.6$ and $L_{DR(p1,p2)} = 0.4$ μm.

5. *Sizing M3/M3c/M4/M4c and Mn1/Mn2* Channel lengths of these devices are determined by following the procedure of step 4 after partitioning the target $r_{id4c} = 9.49$ MΩ into $A_{max4c} r_{ds4} = 12.7$ MΩ and $A_{max4c} r_{ds(n1,n2)} = 38$ MΩ. Its execution is left to the reader as an exercise. It leads to $L_{3,3c,4,4c} = 1.8$ μm. Considering the aspect ratio determined in step 2, we calculate 75 μm for the corresponding unit-device drawn channel width. Iterative simulation also yields $V_{B1} - V_{DD} = -0.5$ V, $V_{B2} - V_{DD} = -0.8$ V, $L_{DR(n1,n2)} = 1$ μm, and therefore $W_{DR(n1,n2)} = 3$ μm.

6. *Input Capacitance* Having determined device dimensions, we are now ready to check the actual input capacitance against the condition imposed by (4.214). For the target small-signal bandwidth $f_{cl} = 18.1$ MHz, this condition sets an upper limit $C_{in} = C_{se} + C_d = 98$ fF. The sum of $C_{se} + C_d$ can be extracted with simulation on the schematic shown in Fig. 4.56(e). It is a derivative of the netlist defined in `ex4-14.sp` of Example 4.14 and is simulated with the following input file `ex4-24d.sp`.

```
filename: ex4-24d.sp
.incl bsimlvt.sp
mp1 7 2 6 13 cmosplvt w=3.7u l=0.4u nrd=0.06 nrs=0.06
+ ad=2.03p pd=8.5u as=2.03p ps=8.5u
mp2 9 1 6 13 cmosplvt w=3.7u l=0.4u nrd=0.06 nrs=0.06
+ ad=2.03p pd=8.5u as=2.03p ps=8.5u
mn1 8 2 5 0 cmosnlvt w=3u l=1u nrd=0.07 nrs=0.07
+ ad=1.65p pd=7.1u as=1.65p ps=7.1u
mn2 10 1 5 0 cmosnlvt w=3u l=1u nrd=0.07 nrs=0.07
+ ad=1.65p pd=7.1u as=1.65p ps=7.1u
```

```
vdp1 7 0 0.24
vdp2 9 0 0.24
vdn1 8 0 1.565
vdn2 10 0 1.565
vip 1 0 dc 0.9 ac 1
vin 2 0 0.9
ibp 0 6 150u
ibn 5 0 150u
.control
ac dec 100 10k 100meg
plot imag(-i(vip)/v(1))/(2*pi*frequency) ; plotting cse+cd
.endc
.end
```

The outcome C_{in} = 25.6 fF obviously satisfies the limit by a large margin.

7. *Current Reference* All devices except MR5 are implemented with one unit device conducting 15-μA reference current. MR5 consists of four unit devices. Based on the values cited for V_{GS6} and V_{B3} in step 4 and for V_{B1} and V_{B2} in step 5, we calculate 20 kΩ for both R_1 and R_2. As usual, R_{SR} is determined with iterative simulation. It turns out to be 4 kΩ. As a final note, the reader is reminded that the current reference, being a self-biased circuit, must be supplemented with a start-up unit. No such unit is included in Fig. 4.56(a).

8. *Verification* Verification is performed with the input file `ex4-24e.sp` presented below.

```
filename: ex4-24e.sp
.incl bsimlvt.sp
.incl bsim.sp
**************ota subckt****Fig.4.56(a)***************
.subckt ota 1 2 3
mp1 7 2 6 13 cmosplvt w=3.7u l=0.4u nrd=0.06 nrs=0.06
+  ad=2.03p pd=8.5u as=2.03p ps=8.5u
mp2 9 1 6 13 cmosplvt w=3.7u l=0.4u nrd=0.06 nrs=0.06
+  ad=2.03p pd=8.5u as=2.03p ps=8.5u
mn1 8 2 5 0 cmosnlvt w=3u l=1u nrd=0.07 nrs=0.07
+  ad=1.65p pd=7.1u as=1.65p ps=7.1u
mn2 10 1 5 0 cmosnlvt w=3u l=1u nrd=0.07 nrs=0.07
+  ad=1.65p pd=7.1u as=1.65p ps=7.1u
m5 9 11 0 0 cmosn w=12u l=1.3u nrd=0.017 nrs=0.017
+  ad=6.6p pd=25.1u m=16
m5c 11 21 9 0 cmosn w=12u l=1.3u nrd=0.017 nrs=0.017
+ ad=6.6p pd=25.1u as=6.6p ps=25.1u m=11
m6 7 11 0 0 cmosn w=12u l=1.3u nrd=0.017 nrs=0.017
+  ad=6.6p pd=25.1u m=16
m6c 3 21 7 0 cmosn w=12u l=1.3u nrd=0.017 nrs=0.017
+  ad=6.6p pd=25.1u as=6.6p ps=25.1u m=11
m3 10 12 13 13 cmosp w=75u l=1.8u nrd=0.003 nrs=0.003
+  ad=41.25p pd=151.1u m=11
m3c 11 22 10 13 cmosp w=75u l=1.8u nrd=0.003 nrs=0.003
```

```
+  ad=41.25p pd=151.1u as=41.25p ps=151.1u m=6
m4 8 12 13 13 cmosp w=75u l=1.8u nrd=0.003 nrs=0.003
+ ad=41.25p pd=151.1u m=11
m4c 3 22 8 13 cmosp w=75u l=1.8u nrd=0.003 nrs=0.003
+  ad=41.25p pd=151.1u as=41.25p ps=151.1u m=6
m9 6 12 13 13 cmosp w=75u l=1.8u nrd=0.003 nrs=0.003
+  ad=41.25p pd=151.1u m=10
m10 5 24 0 0 cmosn w=12u l=1.3u nrd=0.017 nrs=0.017
+  ad=6.6p pd=25.1u m=10
mr1 29 24 0 0 cmosn w=12u l=1.3u nrd=0.017 nrs=0.017
+  ad=6.6p pd=25.1u
mr2 24 21 29 0 cmosn w=12u l=1.3u nrd=0.017 nrs=0.017
+  ad=6.6p pd=25.1u as=6.6p ps=25.1u
mr3 31 12 13 13 cmosp w=75u l=1.8u nrd=0.003 nrs=0.003
+  ad=41.25p pd=151.1u
mr4 12 22 31 13 cmosp w=75u l=1.8u nrd=0.003 nrs=0.003
+  ad=41.25p pd=151.1u as=41.25p ps=151.1u
mr5 28 24 25 0 cmosn w=12u l=1.3u nrd=0.017 nrs=0.017
+  ad=6.6p pd=25.1u as=6.6p ps=25.1u m=4
mr6 22 21 28 0 cmosn w=12u l=1.3u nrd=0.017 nrs=0.017
+  ad=6.6p pd=25.1u as=6.6p ps=25.1u
mr7 30 12 13 13 cmosp w=75u l=1.8u nrd=0.003 nrs=0.003
+  ad=41.25p pd=151.1u
mr8 21 22 30 13 cmosp w=75u l=1.8u nrd=0.003 nrs=0.003
+  ad=41.25p pd=151.1u as=41.25p ps=151.1u
rsr 25 0 4k
r1 21 24 20k
r2 12 22 20k
vdd 13 0 1.8
.ends ota
*************open-loop dc****Fig.4.56(f)**************
x1 41 42 43 ota
vin 42 0 0.9
vi 40 0 dc
rs 40 41 30k
cl 43 0 4.5p
.dc vi 0.899 0.901 10u
*************open-loop ac****Fig.4.56(f)**************
*x1 41 42 43 ota
*vin 42 0 0.9
*vi 40 0 dc 0.900143 ac 1
*rs 40 41 30k
*cl 43 0 4.5p
*.ac dec 10 100 100meg
************closed-loop dc****Fig.4.56(g)*************
*x2 51 52 52 ota
*vi 50 0 dc
*rs 50 51 30k
*cl 52 0 4.5p
*.dc vi 0 1.8 1m
```

```
*.save all @m.x2.m4[id] @m.x2.m4c[id]
*+  @m.x2.m6[id] @m.x2.m6c[id] @m.x2.m9[id]
*+  @m.x2.m10[id] @m.x2.mp1[gm] @m.x2.mn1[gm]
**********loop gain****Fig.4.56(h) and (i)************
*x3 61 62 63 ota     ; for voltage return ratio
*cla 62 0 4.5p       ; for voltage return ratio
*via 60 0 dc 1.5     ; for voltage return ratio
*rsa 60 61 30k       ; for voltage return ratio
*vx 62 63 dc 0 ac 1 ; for voltage return ratio
*x4 71 72 73 ota     ; for current return ratio
*clb 72 0 4.5p       ; for current return ratio
*vib 70 0 dc 1.5     ; for current return ratio
*rsb 70 71 30k       ; for voltage return ratio
*ix 0 74 dc 0 ac 1  ; for current return ratio
*vs1 73 74 0         ; for current return ratio
*vs2 74 72 0         ; for current return ratio
*.control
*ac dec 100 1k 100meg
*let rv=-v(63)/v(62)
*let ri=-i(vs1)/i(vs2)
*set units=degrees
*plot db((rv*ri-1)/(2+rv+ri)) ph((rv*ri-1)/(2+rv+ri))
*let magnitude=mag((rv*ri-1)/(2+rv+ri))
*let phasedif=ph((rv*ri-1)/(2+rv+ri))+180
*meas ac fu when magnitude=1
*meas ac pm find phasedif at=fu
*.endc
*********transient and fourier****Fig.4.56(g)*********
*x2 51 52 52 ota
*vi 50 0 sin 0.9 0.6 1k
*rs 50 51 30k
*cl 52 0 4.5p
*.control
*tran 1u 20m
*fourier 1k v(52)
*plot v(52)
*.endc
*******************************************************
.end
```

The OTA and its current reference are netlisted as a subcircuit, and is used in five different simulations. An open-loop dc simulation is performed on the schematic shown in Fig. 4.56(f) with a view to determine the systematic offset and dc differential gain of the dc transfer characteristic. Although not needed in a dc simulation, R_S and C_L are included for later use in an ac simulation. Plotting the output voltage yields the characteristic shown in Fig. 4.57(e), which indicates 143-μV systematic offset. This offset arises from the difference between the drain voltages of M5c and M6c at the operation point. A plot of the derivative of the output voltage against output voltage itself is presented in Fig. 4.57(f). It indicates dc differential gain $A_0 = 3153$ V/V around $V_O = 0.9$ V.

An open-loop ac simulation is included for verifying small-signal frequency response. It is run on the very same schematic defined for the open-loop dc simulation. However, the dc offset determined in that simulation is added to the noninverting-input voltage in order to bring the OTA to its operation point. The reader is invited to plot the magnitude response with the command

```
plot vdb(43)-vdb(40)
```

A closed-loop dc simulation is performed on the schematic shown in Fig. 4.56(g), where R_S and C_L are retained for later use in a transient simulation. This closed-loop analysis has two purposes. One is to quantify the systematic part of the dc error on a plot of $v_O - v_I$ against v_I as obtained with the command

```
plot v(52,50)
```

The other is to verify the expected variation of key internal variables with common-mode voltage. v_I is qualified as the common-mode voltage due to the virtual short existing between the two input terminals of the OTA in a closed-loop dc operation. Notice the inclusion of a line for saving critical device currents and transconductances of the OTA. The reader may add more. Also note that, if needed, the voltage of an internal node k can be plotted with command

```
plot v(x2.k)
```

The reader will greatly benefit from comparing as many OTA variables as possible with the corresponding targets set in design. Entries of Table 4.2 are probably the most significant of all.

Loop-gain extraction is based on the schematics presented in Figs. 4.56(h) and (i). Its netlist is similar to that of the input file `ex4-19b.sp` of Example 4.19. The simulator directly plots the magnitude and phase of βA and also prints the unity-gain frequency and phase margin. Since phase margin may change as the common-mode moves away from the operation point, it is advisable to repeat this simulation for different values of `via` and `vib` in between the specified limits 0.3 and 1.5 V.

Transient response is simulated on the schematic shown in Fig. 4.56(g). v_I is specified as a sinewave of 0.9-V bias, 0.6-V amplitude, and 1-kHz frequency. Simulator directly plots the waveform of v_O, and prints the THD. The reader may wish to check the consistency of the output voltage peaks with the systematic dc-error data obtained in the closed-loop dc simulation. Also of interest is the transient response at higher frequencies as it will enable the reader to observe the extent of deterioration in THD as the frequency approaches the specified full-power bandwidth at 10 MHz. However, when resetting the frequency, it is important to also reset the simulation parameters T_{step} and T_{stop} and the fundamental-frequency setting of Fourier analysis as advised in Example 2.15. ▲

4.8 Miller OTA

The OTA topologies we have so far studied have a single gain stage and therefore a rather low dc differential gain unless supported by cascoding. The Miller OTA, in contrast, deploys two cascaded gain stages for providing a large gain. Its topological properties and dc response are rather straightforward, and will be covered in the first subsection. The relatively more complicated frequency and step responses will follow in two separate subsections.

4.8.1 Topology and DC Response

Two Miller OTA schematics differing in the type of driver devices are presented in Fig. 4.58. These schematics are intended for explaining the dc response only. In Subsection 4.8.2, we will supplement them with additional components for establishing a stable frequency response.

Both versions shown in Fig. 4.58 feature an input stage in the form of a basic OTA. This stage is built with (a) a differential pair of drivers M1 and M2 biased by the current I_{B1} of M6, and (b) a current-mirror load comprising M3 and M4. A common-source amplifier built with the driver M5 and the current source/sink M7 forms the second stage of the cascade topology. Since the dc voltage gain of this second stage is typically around 100 V/V whereas the output swing of the OTA is around a volt at most, the dc signal observed at the input of the second stage is no larger than a few tens of millivolts. As far as range limitations are concerned, we therefore can assume the drain-source voltage of M4 to be nearly constant at its bias level V_{DS4}. Also note that V_{DS4} is usually equated to $V_{GS(3,4)}$ in order to establish identical bias conditions for M3 and M4. This is how we avoid a systematic offset that would otherwise arise from asymmetry. Equalization is done by scaling the aspect ratio of M5 with respect to M3 by the ratio $I_{B2}/(I_{B1}/2)$.

Common-mode range limits of a Miller OTA are not different from those of a basic OTA. Assuming $V_{DS4} = V_{GS(3,4)}$ as just explained, these limits are described for the PMOS input-driver version by

$$V_{IL} = V_{SS} + V_{GS(3,4)} - V_{DSsat(1,2)} + V_{GS(1,2)} \quad \text{(PMOS input drivers),} \qquad (4.216)$$

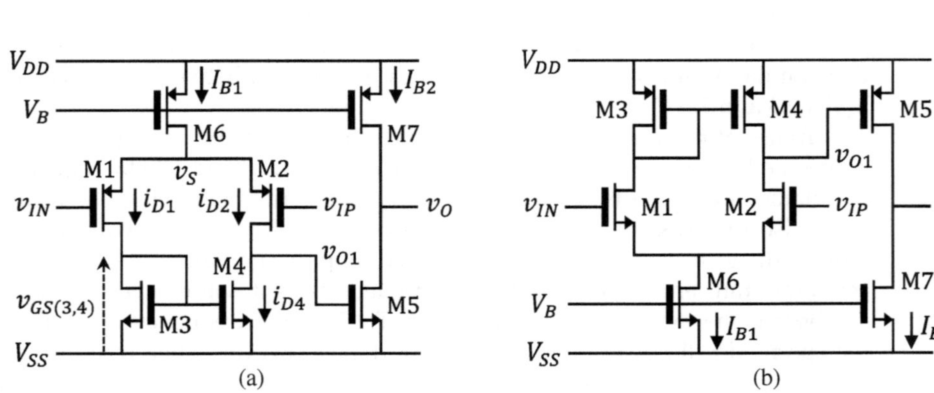

FIGURE 4.58 Miller OTA with (a) PMOS input drivers, and (b) NMOS input drivers.

where $V_{GS(1,2)}$ is specific to the lower limit if input drivers operate without source-bulk short, and

$$V_{IH} = V_{DD} + V_{DSsat6} + V_{GS(1,2)} \quad \text{(PMOS input drivers)}, \tag{4.217}$$

where $V_{GS(1,2)}$ is specific to the upper limit if input drivers operate without source-bulk short.

The NMOS counterparts of these expressions are as follows:

$$V_{IL} = V_{SS} + V_{DSsat6} + V_{GS(1,2)} \quad \text{(NMOS input drivers)}, \tag{4.218}$$

where $V_{GS(1,2)}$ is specific to the lower limit if input drivers operate without source-bulk short, and

$$V_{IH} = V_{DD} + V_{GS(3,4)} - V_{DSsat(1,2)} + V_{GS(1,2)} \quad \text{(NMOS input drivers)}, \tag{4.219}$$

where $V_{GS(1,2)}$ is specific to the upper limit if input drivers operate without source-bulk short.

The output range of a Miller OTA is limited by the saturation conditions of M5 and M7. For the PMOS input-driver version, these limits are described by

$$V_{OH} = V_{DD} + V_{DSsat7} \quad \text{(PMOS input drivers)}, \tag{4.220}$$

and

$$V_{OL} = V_{SS} + V_{DSsat5} \quad \text{(PMOS input drivers)}. \tag{4.221}$$

The NMOS input-driver counterparts of these expressions are as follows:

$$V_{OH} = V_{DD} + V_{DSsat5} \quad \text{(NMOS input drivers)}, \tag{4.222}$$

and

$$V_{OL} = V_{SS} + V_{DSsat7} \quad \text{(NMOS input drivers)}. \tag{4.223}$$

It is obvious from these expressions that a wide output range calls for a small drain-source saturation voltage and therefore a small overdrive voltage for M5 and M7.

The dc differential gain of a Miller OTA is the product of the individual gains of its two stages. Regardless of the type of drivers, it is described by

$$A_0 = \underbrace{g_{m(1,2)}\left(r_{ds2} \parallel r_{ds4}\right)}_{\text{First stage}} \times \underbrace{g_{m5}\left(r_{ds5} \parallel r_{ds7}\right)}_{\text{Second stage}} . \tag{4.224}$$

4.8.2 Frequency Response and Noise-Related Properties

The frequency response of the cascaded topology of a Miller OTA is highly unfavorable in regard to stability unless treated with what we call *frequency compensation*. This problem arises from the fact that nondominant-pole frequencies are very close to the dominant-pole frequency in an uncompensated Miller OTA. The simplest technique available for solving this problem is called *pole-splitting compensation*. It is applied by placing a single capacitor between the input and output nodes of the second stage. A derivative, known as *lead compensation*, supplements the capacitor with a series resistor. We will study both techniques in this subsection. For other compensation techniques, the reader is referred to [11], [12], and [13].

The topology shown in Fig. 4.59(a) belongs to a lead-compensated Miller OTA. Replacing R_C with a short circuit results in pole-splitting compensation. Finally, removing both R_C and C_C reduces the topology to an uncompensated OTA. The frequency response of each of these three topologies can be analyzed on the simplified equivalent circuit presented in Fig. 4.59(b) for both PMOS and NMOS input-driver versions. Notice the representation of the first stage with driver transconductance $g_{m(1,2)}$ and equivalent output resistance $r_1 \doteq r_{ds(1,2)} \parallel r_{ds4}$. v_{id} denotes the differential-input signal $v_{id} \doteq v_{IP} - v_{IN}$. The second stage is represented also with the driver transconductance g_{m5} and the equivalent output resistance $r_2 \doteq r_{ds5} \parallel r_{ds7}$. As to the capacitances, only three of them influence the transfer function to any significant extent. Namely, the parasitic capacitance C_1 at the output node of the first stage; the load capacitance C_L, which also includes the parasitic output capacitance of the second stage; and the compensation capacitance C_C. Assuming

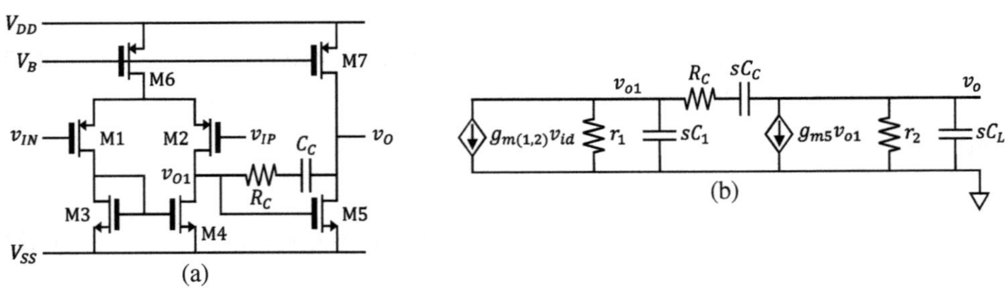

FIGURE 4.59 Frequency-compensated Miller OTA. (a) Circuit topology in the general case of lead compensation. (b) Small-signal equivalent circuit.

$r_1 \gg R_C, r_2 \gg R_C, g_{m5}r_2 \gg 1$, and widely separated poles $f_{P1} \ll f_{P2} \ll f_{P3}$, this general equivalent circuit yields the transfer function

$$A \doteq \frac{v_o}{v_{id}} = A_0 \frac{1 \pm \frac{s}{2\pi f_Z}}{\left(1+\frac{s}{2\pi f_{P1}}\right)\left(1+\frac{s}{2\pi f_{P2}}\right)\left(1+\frac{s}{2\pi f_{P3}}\right)}, \tag{4.225}$$

where

$$f_{P1} \approx \frac{1}{2\pi\left[r_1\left(C_1+g_{m5}r_2C_C\right)+r_2\left(C_L+C_C\right)\right]}, \tag{4.226}$$

$$f_{P2} \approx \frac{r_1\left(C_1+g_{m5}r_2C_C\right)+r_2\left(C_L+C_C\right)}{2\pi r_1 r_2\left(C_1C_L+C_1C_C+C_LC_C\right)}, \tag{4.227}$$

$$f_{P3} \approx \frac{C_1C_L+C_1C_C+C_LC_C}{2\pi R_C C_C C_1 C_L}, \tag{4.228}$$

$$f_Z = \frac{1}{2\pi C_C\left|R_C-\frac{1}{g_{m5}}\right|}. \tag{4.229}$$

Note that the zero is located in the left half plane if $R_C > 1/g_{m5}$. Otherwise it is in the right half plane.

Frequency Response without Compensation

For an uncompensated Miller OTA, the expressions of the poles and zero of (4.225) are modified by (a) substituting a zero for R_C, and (b) replacing C_C with C_{gd5} of M5 in (4.226) and (4.227), and with C_{dg5} of the same device in (4.229). Assuming $C_{gd5} \ll C_L$, this modification turns (4.226) and (4.227) into

$$f_{P1} \approx \frac{1}{2\pi\left[r_1\left(C_1+g_{m5}r_2C_{gd5}\right)+r_2C_L\right]}, \tag{4.230}$$

$$f_{P2} \approx \frac{r_1\left(C_1+g_{m5}r_2C_{gd5}\right)+r_2C_L}{2\pi r_1 r_2 C_1 C_L}. \tag{4.231}$$

It also leads to $f_{P3} = \infty$, and moves the zero to the right half plane. Since f_{P1} is lower than f_{P2}, the dominant- and nondominant-pole frequencies are given by $f_o = f_{P1}$ and $f_{nd} = f_{P2}$, respectively. For an evaluation of these pole frequencies from stability point of view, we now consider a unity-gain buffer amplifier, which happens to be the most critical application in this respect.

As the reader will recall from Subsection 2.7.2, the extent of stability is quantified in terms of phase margin, which is well approximated by (2.94) for any amplifier application with acceptable damping. Considering also $f_{cl} \approx f_{GBW} = A_0 f_o$ in a unity-gain buffer amplifier, we conclude from (2.94) that the ratio f_{nd}/f_o hence f_{P2}/f_{P1} should be larger than A_0 for a phase margin better than 45°. Now, consider the typical case of $r_2 C_L$ being the dominant term of the denominator of (4.230) and the numerator of (4.231) so that $f_{P2}/f_{P1} \approx r_2 C_L / r_1 C_1$. Since r_2 and r_1 are comparable resistances, a phase margin greater than 45° obviously calls for $C_L > A_0 C_1$. Considering the very large A_0 of the Miller OTA, the lower limit thus imposed on C_L turns out to be impractically large. This problem is exacerbated by the presence of the right half-plane zero. A right half-plane zero lifts the roll-off rate of the magnitude by 20 dB/dec while rotating the phase in the negative direction. Its impact on the phase margin is extremely unfavorable because it not only raises the unity-gain frequency but also lowers the frequency at which the phase drops to −180°.

Pole-Splitting Compensation

Notice in (4.230) and (4.231) the product of C_{gd5} and second-stage dc voltage gain $g_{m5} r_2$. Since this gain is typically around 100 V/V, the effective capacitance represented by this product can be made very large by supplementing C_{gd5} with an external capacitor placed between the gate and drain terminals of M5. The sum of the two capacitances, represented by C_C in Fig. 4.59(a), moves f_{P1} to a lower frequency and f_{P2} to a higher frequency. The ratio f_{P2}/f_{P1} can thus be raised as much as needed for achieving a desired phase margin. This is what we call pole-splitting frequency compensation.

Assuming $g_{m5} r_1 r_2 C_C$ to be the dominant term in the denominator of (4.226) and the numerator of (4.227), pole and zero frequencies of the compensated OTA are described by

$$f_{P1} \approx \frac{1}{2\pi g_{m5} r_1 r_2 C_C} = \frac{g_{m(1,2)}}{2\pi A_0 C_C}, \tag{4.232}$$

$$f_{P2} \approx \frac{g_{m5} C_C}{2\pi \left(C_1 C_L + C_1 C_C + C_L C_C\right)}, \tag{4.233}$$

$$f_Z \approx \frac{g_{m5}}{2\pi C_C}, \tag{4.234}$$

and $f_{P3} = \infty$.

The pole at f_{P1} acts as the dominant pole, and therefore sets the open-loop bandwidth to $f_o = f_{P1}$. Considering (4.232) together with (2.43), we can express the gain-bandwidth product resulting from pole-splitting compensation as

$$f_{GBW} = \frac{g_{m(1,2)}}{2\pi C_C}. \tag{4.235}$$

This expression differs from (4.115) of the symmetrical-OTA and folded-OTA topologies in that C_C replaces the load capacitance C_L.

Any phase margin greater than about 45° in an amplifier compensated with pole splitting is determined by the proximity of the closed-loop bandwidth f_{cl} to (a) nondominant-pole frequency $f_{nd} = f_{P2}$ and (b) right half-plane zero frequency f_Z. We illustrate this relationship in Fig. 4.60 with the magnitude and phase plots of loop gain for various

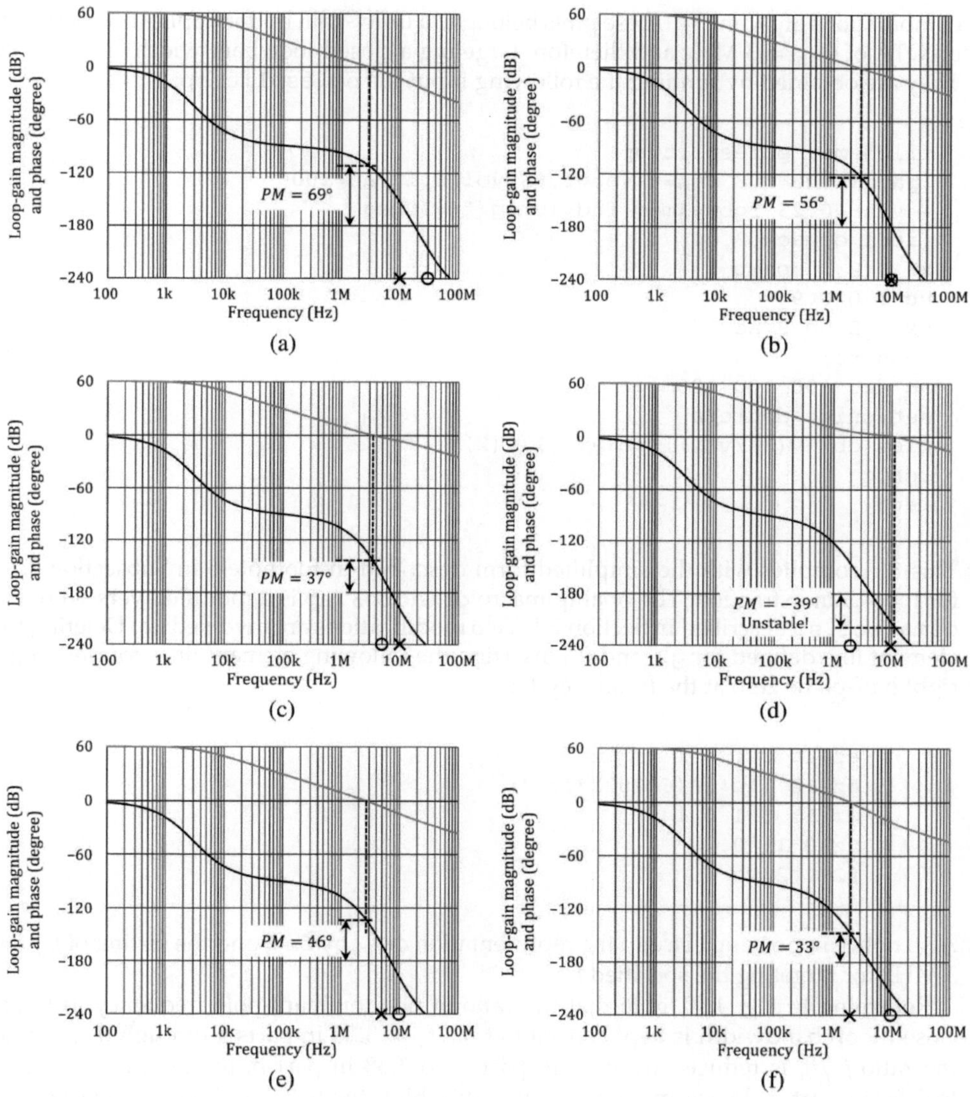

FIGURE 4.60 Simulated magnitude (gray) and phase (black) characteristics of loop gain in a unity-gain buffer application compensated by pole splitting. Gain-bandwidth product equals 3 MHz. Nondominant-pole frequency f_{nd} and right half-plane zero frequency f_Z are identified with cross and circle, respectively. (a) $f_{nd} = 10$ MHz, $f_Z = 30$ MHz. (b) $f_{nd} = 10$ MHz, $f_Z = 10$ MHz. (c) $f_{nd} = 10$ MHz, $f_Z = 5$ MHz. (d) $f_{nd} = 10$ MHz, $f_Z = 2$ MHz. (e) $f_{nd} = 5$ MHz, $f_Z = 10$ MHz. (f) $f_{nd} = 2$ MHz, $f_Z = 10$ MHz.

combinations of f_{P2} and f_Z. These plots belong to a unity-gain buffer application based on an OTA of f_{GBW} = 3 MHz and therefore targeting a closed-loop bandwidth f_{cl} = 3 MHz. They are obtained by running the following input file `polesplit.sp`:

```
filename: polesplit.sp
.param vdd=1.8 vss=0 voh=1.65 vol=0.15 a0=1000
+ vdir=0.25 fgbw=3meg fnd=10meg fz=30meg
.incl opmacro3.sp
x1 1 2 3 opamp
vb 1 0 0.9
vx 3 2 dc 0 ac 1
.control
ac dec 100 100 100meg
set units=degrees
plot db(-v(3)/v(2)) ph(-v(3)/v(2))
.endc
.end
```

This file complies with the simplified form described in footnote 8 in Subsection 2.7.2 for loop-gain extraction. The opamp macro `opmacro3.sp` is a modified version of the `opmacro1.sp` described in Section 2.1. Two modifications are involved: (a) Deleting the element line defined for `g2`, and (b) inserting the following element lines for defining a right half-plane zero at the frequency `fz`:

```
ez 10 0 1 0 1
cz 10 11 '1/(6.2832*fz)'
gz 11 0 1 0 1
vz 0 11 0
f2 0 2 vz 1
```

Also notice in `polesplit.sp` the representation of f_{P2} by `fnd`, and the setting of f_{GBW} = 3 MHz for targeting the specified f_{cl}.

Returning to Fig. 4.60, note that the ratio of nondominant-pole frequency to target closed-loop bandwidth is kept constant at f_{P2}/f_{cl} = 3.33 in parts a through d, whereas the ratio f_Z/f_{cl} is reduced from 10 in part a, to 3.33 in part b, to 1.66 in part c, and to 0.66 in part d. Phase margin is reasonably high for $f_Z/f_{cl} \geq 3.33$ but deteriorates rapidly as f_Z is lowered below this level. Too low an f_Z may even lead to a negative phase margin and therefore to instability as observed in part d. Notice how such a low right half-plane zero elevates the unity-gain frequency to around 10 MHz while lowering the frequency of −180° phase rotation to around 4.5 MHz. It is obvious from the first four parts of Fig. 4.60 that f_Z/f_{cl} should be kept above 3 for a reasonable phase margin.

Displayed in parts e and f of Fig. 4.60 are the magnitude and phase characteristics obtained after transposing the zero and nondominant-pole frequencies associated with parts c and d, respectively. They still indicate low levels of phase margin but not as low as those observed in parts c and d. This is because the nondominant pole accelerates the rotation of phase but, unlike a right half-plane zero, it lowers the unity-gain frequency f_u rather than raising it. Obviously, the threat posed by the nondominant pole on stability is not as severe as that of the right half-plane zero but still a reasonable phase margin calls for a ratio f_{P2}/f_{cl} larger than about 2.

In the case of pole-splitting frequency compensation, (2.94) can accurately describe the phase margin only in those rare cases of f_Z being at least an order of magnitude higher than f_{P2}. A more general description taking the right half-plane zero into account is

$$\text{PM} \approx \arctan\left(\frac{\frac{f_{P2}}{f_{cl}}\frac{f_Z}{f_{cl}} - 1}{\frac{f_{P2}}{f_{cl}} + \frac{f_Z}{f_{cl}}}\right). \tag{4.236}$$

Note that this expression is subject to the following assumptions: $f_{P2} \gg f_{P1}, f_Z \gg f_{P1}$, and $f_u \approx f_{cl}$. The validity of the last of these three assumptions rapidly deteriorates for $f_Z/f_{cl} < 3$ or $f_{P1}/f_{cl} < 2$ as previously observed in Fig. 4.60. This is why the lower limit of reliable phase-margin estimates obtained with (4.236) is around 45°.

For relating the variables f_Z/f_{cl} and f_{P2}/f_{cl} of (4.236) to device performance metrics, we consider (a) (4.233) and (4.234) for f_{P2} and f_Z, (b) $f_{cl} = \beta A_0 f_{P1}$ as obtained from (2.48) and (2.43), and (c) (4.232) for f_{P1}. The following two expressions are thus obtained:

$$\frac{f_Z}{f_{cl}} = \frac{g_{m5}}{\beta g_{m(1,2)}}, \tag{4.237}$$

$$\frac{f_{P2}}{f_{cl}} = \frac{C_C^2}{C_1 C_L + C_1 C_C + C_L C_C}\frac{g_{m5}}{\beta g_{m(1,2)}}. \tag{4.238}$$

It is important to note that the assumption $f_u = f_{cl}$ underlying (4.236) restricts the validity of this equation to amplifier applications. Same restriction applies also to (4.237) and (4.238). In filter applications, the best way of evaluating the effect of f_{P2} and f_Z on PM before embarking on device design is to simulate the application with an opamp macro like `opmacro3.sp`.

Lead Compensation

In lead compensation, C_C is supplemented with R_C. Pole and zero frequencies are therefore described generally by (4.226) to (4.229). However, R_C is made sufficiently larger than $1/g_{m5}$ so that the zero moves to the left half plane and cancels out the second pole by establishing $f_Z = f_{P2}$. The resistance needed for lead compensation can be determined from this equation as

$$R_C = \frac{C_1 C_L + C_1 C_C + C_L C_C}{g_{m5} C_C^2}, \tag{4.239}$$

provided that $g_{m5} r_1 r_2 C_C$ is the dominant term in the numerator of (4.227), and $1/g_{m5}$ is negligible in comparison with R_C in (4.229).

What results from pole-zero cancelation is a two-pole transfer function whose dominant pole is still at f_{P1} as described by (4.232). As a consequence, (4.235) retains its validity for f_{GBW}. The third pole, whose frequency is now finite as described by (4.228), acts as the nondominant pole. By substituting the right-hand side of (4.239) for R_C in (4.228), its frequency is obtained as

$$f_{P3} = \frac{C_C g_{m5}}{2\pi C_1 C_L}. \tag{4.240}$$

The phase margin of an amplifier application with lead compensation can be related to device metrics through (2.94). By substituting (4.240) and (4.232) for f_{nd} and f_o, respectively, and using $f_{cl} = \beta A_0 f_{P1}$, we obtain the following phase-margin expression:

$$\text{PM} \approx \arctan\left(\frac{C_C^2}{C_1 C_L}\frac{g_{m5}}{\beta g_{m(1,2)}}\right). \tag{4.241}$$

The target value determined for R_C from (4.239) may be significantly different from the actual value needed for exact cancelation because C_1 and C_L cannot be estimated accurately. Furthermore, the fabricated value of R_C itself will differ from the target value due to the inaccuracy and imprecision involved in resistor implementation. A varying temperature will also invalidate (4.239) unless the temperature sensitivity of R_C does not perfectly match the temperature sensitivity of the right-hand side of this equation. Yet another deviation will be observed in those applications where C_L fluctuates due to dynamic loading. These effects preclude an exact cancelation of the pole/zero pair. Instead, we end up with a *pole/zero doublet*, which is how we name a pair of pole and zero whose frequencies have a small difference. Although the effect of this imperfection on closed-loop frequency response is insignificant, step response may be affected significantly as will be explained in Subsection 4.8.3.

One way of minimizing the separation between the pole and zero of the doublet is to implement the resistor with a MOSFET whose conductance is scaled by a constant factor with respect to the transconductance g_{m5} of M5 so that both are affected similarly by process and temperature-related variations. Shown in Fig. 4.61 is such an arrangement, where MC functions as R_C.[14] Note that the aspect ratio of M10 is scaled by a factor S with respect to M5. Same factor applies to the scaling of M8 with respect to M7,

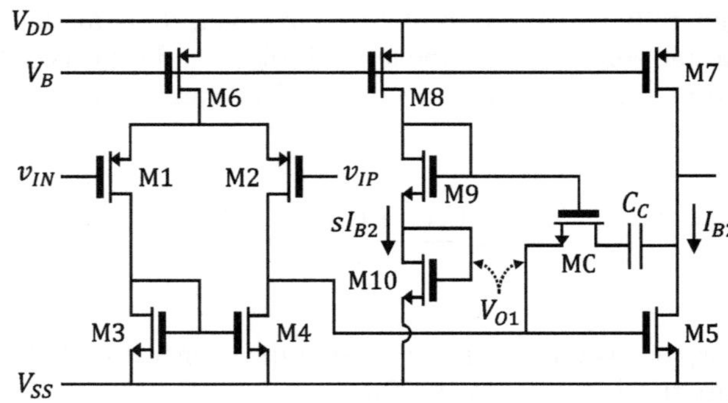

FIGURE 4.61 Implementing lead compensation with a MOSFET resistor MC. M8, M9, and M10 are used for biasing MC.

and therefore implies SI_{B2} for the current flowing in M10. Having identically scaled aspect ratios and currents, M10 and M5 operate with identical sheet currents hence identical overdrive voltages. Considering their identical channel lengths and zero source-bulk voltages, we expect them to have the same threshold voltage, hence the same gate-source voltage marked V_{O1} in Fig. 4.61. Now moving to M9 and MC, we observe the same source-bulk voltage V_{O1} for these two devices. Assuming identical channel lengths, we therefore expect them to have the same threshold voltage, which, considering the interconnection of their gate terminals and common source-bulk voltages, implies identical overdrive voltages for these two devices. Finally, by matching M10 to M9 we make the overdrive voltages of these two, and therefore those of MC and M5 identical too. The key conclusion here is the identity of the overdrive voltages of MC and M5. Let us represent this voltage with $V_{O1} - V_{TN}$ and use it in (3.131) to model R_C under the assumption of strong inversion. Since MC doesn't conduct any dc current, we also substitute a zero for V_{DS} in (3.131). The outcome is

$$R_C = \left(\frac{L}{W}\right)_C \frac{1}{\mu_N C_{OX} \left(V_{O1} - V_{TN}\right)}. \tag{4.242}$$

Next, consider (3.26) for g_{m5} in strong inversion; that is,

$$g_{m5} = \left(\frac{W}{L}\right)_5 \frac{\mu_N C_{OX}}{A_N} \left(V_{O1} - V_{TN}\right). \tag{4.243}$$

These two expressions turn (4.239) into

$$\left(\frac{W}{L}\right)_C = \frac{1}{A_N} \frac{C_C^2}{C_1 C_L + C_1 C_C + C_L C_C} \left(\frac{W}{L}\right)_5, \tag{4.244}$$

which obviously establishes a relationship to be satisfied between the aspect ratios of MC and M5 for implementing lead compensation.[7] Since the matching needed for implementation involves two MOSFETs instead of a MOSFET and a resistor, the accuracy of pole/zero cancelation is much improved.

Differential-Input Range

Referring back to Subsection 2.5.2, recall that an opamp starts slewing whenever one of its amplifier stages ceases to function as a transconductor. All OTA topologies presented prior to the Miller OTA have a single amplifier stage in the form of a differential pair, whose transconductor functionality collapses when the magnitude of the differential-input voltage exceeds the differential-input range described by (4.85). Miller OTA, on the other hand, has two amplifier stages. Slewing may be due to one or the other depending on the polarity of the differential-input signal $v_{id} \doteq v_{IP} - v_{IN}$.

Considering the schematic shown in Fig. 4.58(a) for a Miller OTA with PMOS input drivers, note that a negative v_{id} enhances the transconductance of M5 by raising v_{O1}. In

---

[7] Although obtained for strong inversion, the validity of (4.244) is general. The reader can verify this claim for subthreshold by (a) extracting the counterpart of (3.131) by differentiating (3.21) with respect to v_{DS}, (b) taking the reciprocal after substituting $v_{DS} = 0$, and (c) comparing the outcome with (3.27) for g_m.

the NMOS input-driver topology depicted in Fig. 4.58(b), a positive v_{id} does the same by lowering v_{O1}. Since g_{m5} is not reduced in these cases, slewing cannot originate from the second stage regardless of the magnitude of v_{id}. Instead, the differential pair of the first stage will slew under the condition $|v_{id}| > V_{DIR1}$, where

$$V_{DIR1} \doteq \frac{I_{B1}}{g_{m(1,2)}} \tag{4.245}$$

is adopted from (4.84).

A positive v_{id} in PMOS input-driver version causes v_{O1} to fall. Since M5 is thus driven toward cutoff, the second stage may also cease to function as a transconductor. Slewing will occur either in the first stage or in the second stage depending on whose transconductance collapses for a smaller v_{id}. We know that the differential pair sustains its functionality for any positive v_{id} smaller than V_{DIR1} as defined by (4.245). For determining the limiting value of v_{id} for the second stage, consider Fig. 4.62, which shows the transconductance characteristic of M5 relating i_{D5} to v_{O1}. By extrapolating the tangential line of slope g_{m5} at the operation point, we can determine the most negative swing v_{o1} for which this driver still exhibits a reasonable transconductance. This leads to

$$v_{o1} = -\frac{I_{B2}}{g_{m5}}. \tag{4.246}$$

The specific value of the OTA differential-input signal v_{id} causing this much swing at the input of the second stage can be regarded as the differential-input range imposed by the second stage. We denote it with V_{DIR2}, and relate it to device performance metrics by determining the magnitude of the transfer function from v_{id} to v_{o1}, and subsequently replacing v_{id} with V_{DIR2} and substituting the absolute value of the right-hand side of (4.246) for v_{o1}. This transfer function can be extracted from Fig. 4.59(b) after simplifying the equivalent circuit by removing r_1 and r_2. What justifies this simplification is that these

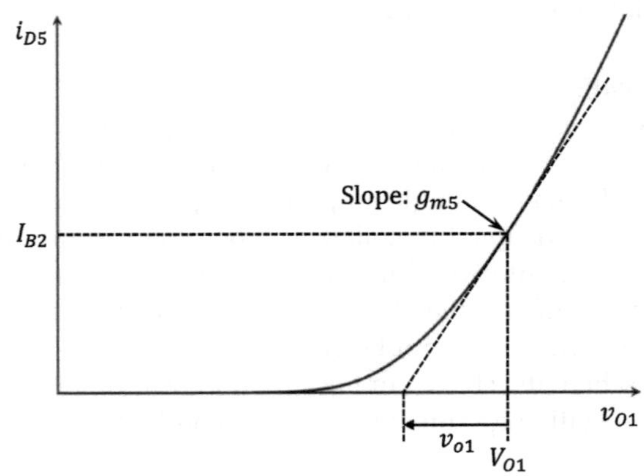

FIGURE 4.62 Typical transconductance characteristic of the second-stage driver M5 in the Miller OTA shown in Fig. 4.58(a). v_{o1} denotes the most negative input signal for which the stage is regarded to be at the limit of its input range.

two resistances are effective only at frequencies lower than the open-loop bandwidth f_o, whereas V_{DIR2} is relevant only at the much higher frequency f_{cl} of closed-loop bandwidth. For $R_C = 0$ of pole-splitting compensation the transfer function turns out to be

$$\frac{v_{o1}}{v_{id}} = -\frac{g_{m(1,2)}\left(C_L + C_C\right)}{g_{m5}C_C}\frac{1}{1+\frac{s}{2\pi f_{P2}}},$$

in which, f_{P2} described by (4.233) is the nondominant-pole frequency as explained previously for pole-splitting compensation. Since the nondominant pole must in any case be placed above f_{cl} for an acceptable phase margin, we can safely ignore this pole in modeling the differential-input range, and thus approximate the magnitude of the transfer function with

$$\left|\frac{v_{o1}}{v_{id}}\right| \approx \frac{g_{m(1,2)}\left(C_L + C_C\right)}{g_{m5}C_C},$$

which yields

$$V_{DIR2} = \frac{C_C}{C_L + C_C}\frac{I_{B2}}{g_{m(1,2)}} \tag{4.247}$$

after replacing v_{id} with V_{DIR2}, and v_{o1} with the absolute value of the right-hand side of (4.246). Considering (4.245), we can express this result also as

$$V_{DIR2} = \frac{C_C}{C_L + C_C}\frac{I_{B2}}{I_{B1}}V_{DIR1}. \tag{4.248}$$

Although (4.247) and therefore (4.248) are derived for pole-splitting compensation, both are also valid to a good approximation for lead compensation particularly if C_C is selected to be smaller than C_L.

As pointed out above, the smaller between V_{DIR1} and V_{DIR2} should be adopted as V_{DIR} in determining the slewing condition for a positive v_{id} in a Miller OTA of PMOS input drivers. In the case of NMOS input drivers, second-stage transconductance is threatened by the negative values of v_{id}, and slewing prevails whenever the magnitude of v_{id} exceeds the smaller between V_{DIR1} and V_{DIR2}.

Noise-Related Properties

The noise contribution of a Miller OTA in a closed-loop configuration is primarily due to first-stage devices because the magnitude $|A_k(f)|$ of noise transfer function of second-stage devices is much smaller. Closed-loop mean-square output noise voltage is therefore well approximated by (4.123) of the basic OTA. Equation (4.124) of the special case of amplifier applications is also valid but C_L has to be replaced by C_C as implied by the substitution of (4.235) for the gain-bandwidth product in $f_{cl} = \beta f_{GBW}$.

4.8.3 Step Response

The step response of a properly compensated Miller OTA complies with the general step-response models presented in Sections 2.6 and 2.7 depending on whether the differential gain A is classified as a single-pole transfer function or a two-pole transfer function, respectively. However, the general model may grossly underestimate the settling time in the case of lead compensation with imperfect pole-zero cancelation. For a detailed description of this effect, we will consider a lead-compensated Miller OTA whose third pole is located at a much higher frequency than f_{cl}. If the frequencies of zero and second pole are perfectly coincident, the OTA will exhibit a single-pole response. Assuming slew-free conditions, such a response and the associated settling time will be described by (2.77) and (2.78), respectively.

Now, suppose that the zero and second-pole frequencies are not perfectly coincident, and therefore the differential gain is given by

$$A \doteq \frac{v_o}{v_{id}} = A_0 \frac{1 + \frac{s}{\omega_Z}}{\left(1 + \frac{s}{\omega_{P1}}\right)\left(1 + \frac{s}{\omega_{P2}}\right)}, \tag{4.249}$$

where, besides the dominant pole at ω_{P1}, a pole/zero doublet appears at frequencies ω_Z and ω_{P2}. Slew-free output signal in this case is given approximately by

$$v_o(t) = \left\{ 1 - \underbrace{\left(1 - \frac{f_{P2} - f_Z}{f_{cl}}\right) \exp\left(-2\pi \frac{f_{P2}}{f_Z} f_{cl} t\right)}_{\text{Fast exponential}} \right.$$

$$\left. - \underbrace{\frac{f_{P2} - f_Z}{f_{cl}} \exp\left[-2\pi \left(1 - \frac{f_{P2} - f_Z}{f_{cl}}\right) f_Z t\right]}_{\text{Slow exponential}} \right\} \frac{\alpha_i}{\beta} X_i, \tag{4.250}$$

where the original definition (2.48) is still valid for f_{cl} but f_{GBW} appearing in that definition is to be expressed as $f_{GBW} = A_0 f_{P1}$ because the open-loop bandwidth is denoted by f_{P1} in the present context. The approximation involved in (4.250) is due to the assumption that the pole and zero frequencies of the doublet are (a) closely separated ($|f_{P2} - f_Z|/f_Z \ll 1$), and (b) located well inside the closed-loop bandwidth ($f_{P2} \ll f_{cl}$ and $f_Z \ll f_{cl}$).

Notice in (4.250) the presence of two exponential components, one identified as *fast* and the other as *slow*. The fast component decays at a rate determined by the frequency $(f_{P2}/f_Z)f_{cl}$, which is not much different from f_{cl} due to the assumption of closely separated f_{P2} and f_Z. Also obvious is that the relative swing of this component may be larger or smaller than 1 depending on whether f_{P2} is lower or higher than f_Z but its deviation from 1 is expected to be small as a result of the doublet condition $|f_{P2} - f_Z| \ll f_{cl}$. In the case of perfect cancelation, this component alone shapes the step response with a frequency f_{cl}

and a relative swing 1, as the reader can ascertain by comparing (4.250) with (2.77) for $f_{P2} = f_Z$.

As to the slow component of step response, (4.250) indicates a frequency close to f_Z due to the condition $|f_{P2} - f_Z| \ll f_c$. This, of course, is much lower than the frequency of the fast component. Also indicated is an equally lower relative swing as compared to the fast component.

How the two components shape up the step response is now exemplified with the waveforms of an unbiased ($V_B = 0$) split-supply unity-gain buffer amplifier ($\alpha_i = 1$, $\beta = 1$) application shown in Fig. 4.63. The OTA is specified by $A_0 = 10^4$ V/V and $f_{P1} = 1$ kHz, which collectively imply $f_{GBW} = 10$ MHz. Considering $\beta = 1$ of the unity-gain application, the corresponding closed-loop bandwidth is determined as $f_{cl} =$ 10 MHz. Input voltage is stepped from −0.5 to 0.5 V at $t = 10$ ns as depicted by the waveform marked v_I in Fig. 4.63(a). Since the peak value of the output signal is 0.5 V, (2.26) indicates $E_{\text{gain}} = 50$ μV for the gain error. Output voltage v_O is therefore expected to reach 499.95 mV in the steady state. The waveforms shown in Fig. 4.63(a) for v_O are extracted for a fixed second-pole frequency $f_{P2} = 100$ kHz and three different zero frequencies $f_Z = 90$, 100, and 110 kHz. Apparently, the waveforms rise at different rates,

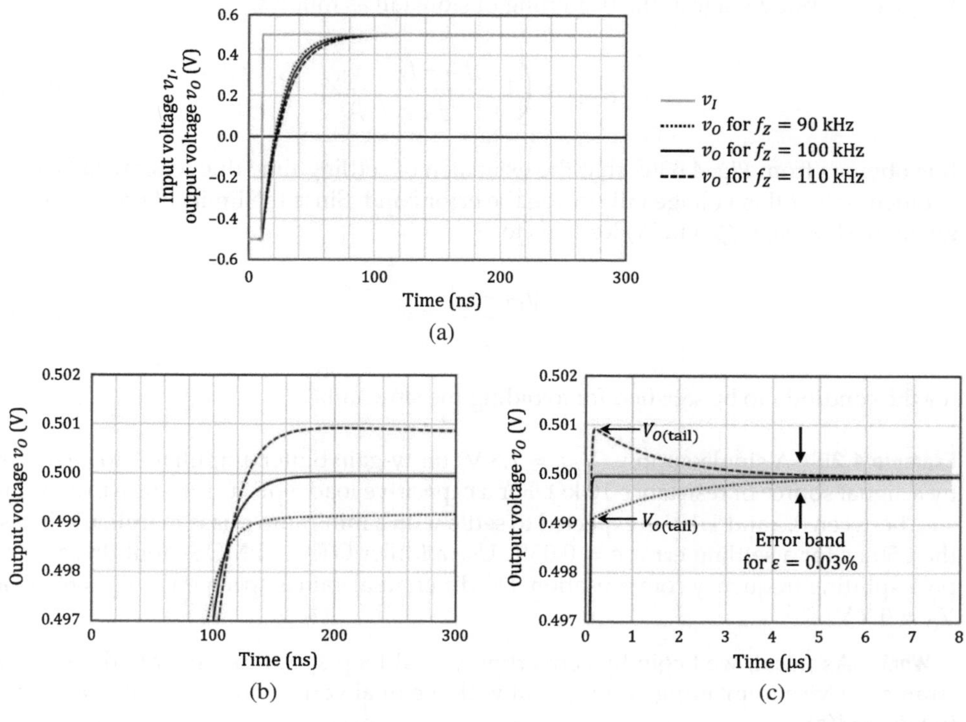

FIGURE 4.63 The effect of a pole-zero doublet on the step response of a unity-gain buffer amplifier built with a lead-compensated Miller OTA. Input voltage steps from −0.5 to 0.5 V. Gain error: 50 μV. Closed-loop bandwidth: 10 MHz. Doublet-pole frequency: 90, 100, and 110 kHz. (a) Waveforms of input voltage v_I and output voltage v_O. (b) Waveform of v_O on an expanded voltage scale. (c) Long-term variation of v_O on an expanded voltage scale.

which can be attributed to the differences in the frequency $(f_{P2}/f_Z)f_{cl}$ of the fast exponential components. However, zooming in on the final level as shown in Fig. 4.63(b), we discover a much more significant disparity among the waveforms. While the waveform observed for perfect pole-zero cancelation ($f_Z = f_{P2} = 100$ kHz) obviously settles at the expected final level within the displayed timescale, the waveforms for $f_Z = 90$ kHz and $f_Z = 110$ kHz are far from settling within the same time frame. To be precise, the former of the two waveforms falls short of the final level, whereas the latter overshoots it. Expanding the timescale as in Fig. 4.63(c) reveals the rest of the picture: The progress toward the final value is slowed down as the slow exponential component starts dominating the waveform beyond the initial 150 ns or so of the transient. Any settling-error specification that can be met only on this slow tail of the waveform will inevitably elongate the settling time beyond what is predicted by (2.78). Suppose, for example, $\epsilon = 3 \times 10^{-4}$ is specified for the settling error. The error band is therefore limited by 499.95 ± 0.30 mV. As observed in Fig. 4.63(c), the corresponding settling time is around 139 ns for perfect pole-zero cancelation with $f_Z = 100$ kHz but is extended to 2 μs for $f_Z = 90$ kHz or $f_Z = 110$ kHz.

Notice in Fig. 4.63(c) that the slow exponential is still close to its initial value when fast exponential vanishes early in transient response. By ignoring the fast exponential and replacing the slow exponential with 1 in (4.250) we determine the specific value $V_{O(\text{tail})}$ of output voltage at the beginning of slow tail as follows:

$$V_{O(\text{tail})} \approx \left(1 - \frac{f_{P2} - f_Z}{f_{cl}}\right)\frac{\alpha_i}{\beta}X_i. \tag{4.251}$$

It is obvious from Fig. 4.63(c) that the extension of settling time due to slow tail can be avoided only if this voltage falls inside the error band. Since the limits of error band are given by $(1 \pm \epsilon)\alpha_i X_i/\beta$, (4.251) leads us to

$$\frac{|f_{P2} - f_Z|}{f_{cl}} < \epsilon \tag{4.252}$$

for the condition to be satisfied for avoiding the slow tail.

Example 4.25 A single-supply ($V_{DD} = 1.8$ V) unity-gain buffer amplifier is to be driven by a signal source of resistance 10 kΩ. For a capacitive load 5 pF, its response to an input step between 0.3 and 1.1 V is expected to settle with a minimum phase margin 65° in less than 50 ns for a settling error $\epsilon = 0.03\%$. Use a Miller OTA of PMOS input drivers and pole-splitting frequency compensation. DC differential gain is specified as $A_0 = 65$ dB at $V_O = 0.7$ V.

Work As usual, we begin by converting closed-loop specifications into design constraints. Device numbering is compliant with the final verification schematic presented in Fig. 4.64(a).

- *Output Range* Signal specifications dictate $V_{OH} > 1.1$ V and $V_{OL} < 0.3$ V. For $V_{DD} = 1.8$ V and $V_{SS} = 0$, (4.220) and (4.221) yield

$$V_{DSsat7} > -0.7\,\text{V} \tag{4.253}$$

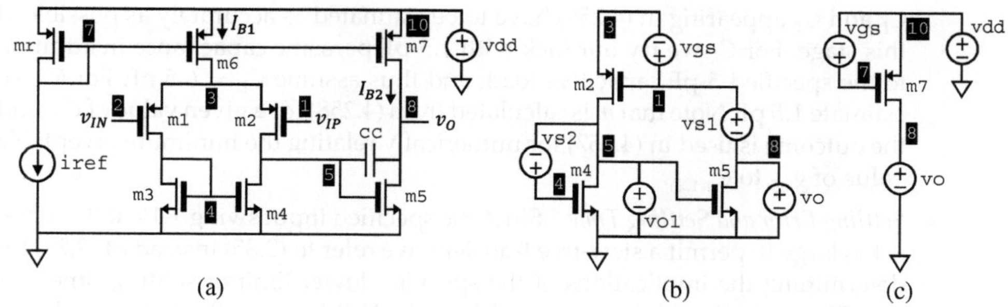

FIGURE 4.64 Simulation schematics used in Example 4.25. (a) OTA subcircuit of `ex4-25.sp`. (b) `ex4-25b.sp`. (c) `ex4-25c.sp`.

and

$$V_{DSsat5} < 0.3\text{ V}. \tag{4.254}$$

- *Common-Mode Range* Signal specifications dictate $V_{IL} < 0.3$ V and $V_{IH} > 1.1$ V, which, according to (4.216) and (4.217), translate into

$$V_{GS(3,4)} - V_{DSsat(1,2)} + V_{GS(1,2)} < 0.3\text{ V} \tag{4.255}$$

and

$$V_{DSsat6} + V_{GS(1,2)} > -0.7\text{ V}. \tag{4.256}$$

As explained later in step 1 of design flow, M1 and M2 will be operated with source-bulk shorts. This is why no distinction is made between the gate-source voltages represented in (4.255) and (4.256).

- *Phase Margin* Equation (4.236) describes the phase margin for pole-splitting compensation. It takes into consideration the second pole and right half-plane zero but not the pole associated with the current-mirror load of the first stage or the pole arising from the combination of signal-source resistance and input capacitance. In order to compensate for this deficiency we raise the target to PM = 70°. By substituting into (4.236) the right-hand sides of (4.237) and (4.238) for f_Z/f_{cl} and f_{P2}/f_{cl}, and considering tan 70° = 2.75 as well as $\beta = 1$, we arrive at

$$g_{m5} \geq \frac{2.75\,(m+1) + \sqrt{7.56\,(m+1)^2 + 4m}}{2m} g_{m(1,2)}, \tag{4.257}$$

where m is defined by

$$m \doteq \frac{C_C^2}{C_1 C_L + C_1 C_C + C_L C_C}. \tag{4.258}$$

C_L and C_1 appearing in (4.258) have to be estimated as accurately as possible at this stage. For C_L we try our luck with 1.5-pF parasitic capacitance in addition to the specified 5-pF capacitive load, and thus assume $C_L = 6.5$ pF. For C_1 we estimate 1.5 pF. Note that m is calculated from (4.258) for a given value of C_C, and the outcome is used in (4.257) for numerically relating the minimum acceptable value of g_{m5} to $g_{m(1,2)}$.

- *Settling Error and Settling Time* Since the specified input swing $1.1 - 0.3 = 0.8$ V is too large to permit a slew-free transient, we refer to (2.85) instead of (2.78) for determining the implications of the specified lower limit of settling time. For $t_S = 50$ ns, $\epsilon = 3 \times 10^{-4}$, $\alpha_i = 1$, and $X_i = 0.8$ V, it imposes the following lower limit on the closed-loop bandwidth f_{cl}:

$$f_{cl} \geq \frac{1}{2\pi \times 50 \times 10^{-9}} \left[\frac{0.8}{V_{DIR}} + \ln\left(\frac{V_{DIR}}{0.8}\right) - 1 + \ln\left(\frac{1}{3 \times 10^{-4}}\right) \right]. \tag{4.259}$$

As explained in Subsection 4.8.2 for PMOS first-stage drivers, the differential-input range is given by $V_{DIR} = V_{DIR1}$ when v_{id} is negative, and by $V_{DIR} = \min\left(V_{DIR1}, V_{DIR2}\right)$ when v_{id} is positive. As a matter of fact it is preferable to target $V_{DIR2} = V_{DIR1}$ because a comparison between (4.245) and (4.247) indicates that otherwise either I_{B1} or I_{B2} will be unnecessarily large. According to (4.248), the optimum implementation $V_{DIR} = V_{DIR1} = V_{DIR2}$ necessitates

$$I_{B2} = \frac{C_L + C_C}{C_C} I_{B1}. \tag{4.260}$$

- *Input Parasitic Pole* As explained in Example 4.20, the adverse effect of an input pole on phase margin can be reduced by moving the frequency of this pole sufficiently above the closed-loop bandwidth. In Example 4.24, we practiced this rule on a folded-cascode OTA by setting the pole frequency to $3f_{cl}$. In a Miller OTA compensated by pole splitting we may need to push the input pole further away from f_{cl} because the phase margin is already severely restricted by the presence of a right half-plane zero. In the present application we target $10f_{cl}$ for its frequency. Considering the specified source resistance $R_S = 10$ kΩ, the maximum permissible value of the input capacitance permitted by this target is as follows:

$$C_{in(\max)} \leq \frac{1}{2\pi R_S (10 f_{cl})} = \frac{1.59 \times 10^{-6}}{f_{cl}}. \tag{4.261}$$

- *DC Differential Gain* It is reasonable to reexpress (4.224) in terms of transconductance efficiencies and Early voltages since these are the preferred metrics of design. The result is

$$A_0 = \frac{g_{m(1,2)}}{I_{B1}/2} \frac{V_{A2} V_{A4}}{V_{A2} + V_{A4}} \frac{g_{m5}}{I_{B2}} \frac{V_{A5} V_{A7}}{V_{A5} + V_{A7}}, \tag{4.262}$$

where $V_{A2} \doteq r_{ds2} I_{B1}/2$, $V_{A4} \doteq r_{ds4} I_{B1}/2$, $V_{A5} \doteq r_{ds5} I_{B2}$, and $V_{A7} \doteq r_{ds7} I_{B2}$. The target set for A_0 equals 1780 V/V as implied by the specified 65-dB gain.

- *Systematic Offset* As explained in Subsection 4.8.1, systematic offset can be minimized by scaling the aspect ratio $(W/L)_5$ of M5 with respect to the aspect ratio $(W/L)_{3,4}$ of M3 and M4 by the ratio of I_{B2} to $I_{B1}/2$, that is,

$$\left(\frac{W}{L}\right)_5 = \frac{I_{B2}}{I_{B1}/2}\left(\frac{W}{L}\right)_{3,4}. \tag{4.263}$$

We now put together a design plan.

An inspection of the constraints reveals the overdrive voltage $V_{GS(1,2)} - V_{TP}$ of first-stage drivers and the compensation capacitance C_C as two independent variables of design. Once the target values of these two have been decided upon, we can determine $g_{m(1,2)}/(I_{B1}/2)$ from the relevant device-design tool, and use the outcome in (4.245) for V_{DIR}. The latter enables us to calculate f_{cl} from (4.259). By substituting f_{cl} for f_{GBW} in (4.235) and using the target value of C_C, we determine $g_{m(1,2)}$. The calculated value of f_{cl} also enables us to determine the maximum acceptable value of $C_{in(\max)}$ from (4.261). Next, we determine I_{B1} from (4.245) by using the calculated values of V_{DIR} and $g_{m(1,2)}$. Since $I_{\square(1,2)}$ is available from $V_{GS(1,2)} - V_{TP}$, and I_{B1} is already quantified, we also determine $(W/L)_{1,2}$ at this stage.

For the second-stage driver M5, we first determine the lower limit of g_{m5} from (4.257). Since I_{B2} is available from (4.260), we calculate the corresponding lower limit of g_{m5}/I_{B2} which sets an upper limit on $V_{GS5} - V_{TN}$. Note that this is the highest overdrive voltage of M5 that can satisfy the phase-margin specification. Two more specification items impose upper limits on this overdrive voltage. One is (4.254) and the other is (4.255), which actually limits $V_{GS(3,4)}$ but of course V_{GS5} and $V_{GS(3,4)}$ are to be equalized in order not to cause a systematic offset. Obviously, the most restrictive among the three limits will define the target overdrive voltage of M5. After quantifying the corresponding $I_{\square 5}$ from the relevant design tool, we will obtain $(W/L)_5$ from I_{B2}. This will be followed by the calculation of $(W/L)_{3,4}$ from (4.263). M3, M4, and M5 will be built with copies of unit devices in order to scale them precisely.

M6 and M7 share the same overdrive voltage which is constrained indirectly by (4.253) and (4.256). The latter is definitely more stringent due to the negative value of $V_{GS(1,2)}$. We determine $V_{GS(6,7)} - V_{TP}$ and therefore $I_{\square(6,7)}$ after setting a reasonable V_{DSsat6} in accordance with (4.256). The aspect ratio of M6 is derived from $I_{\square 6}$ and I_{B1}. The aspect ratio M7, on the other hand, is scaled with respect to $(W/L)_6$ by the factor I_{B2}/I_{B1}. The practice of scaling with copies of a unit device is meaningful also for these two devices.

The channel length of M1 and M2 is our main design variable for setting V_{A2} of (4.262). However, it is subject to an upper limit because (4.261) restricts the gate area of these devices by imposing an upper limit on C_{in}.

V_{A4} and V_{A5} of (4.262) are controlled by the common channel length of M3, M4, and M5 but we need to be aware of the fact that C_1 and C_L are increasing functions of this channel length. If these two capacitances significantly exceed the estimates assumed up front, all steps of design up to this step should be revised.

A long channel for M7 helps raise V_{A7} of (4.262). However, we need to keep in mind that it also increases C_L by implying a wide channel.

Steps of execution are presented next.

1. *$V_{GS(1,2)} - V_{TP}$ and C_C, and therefore I_{B1}, $(W/L)_{1,2}$ and I_{B2}* Targeting $V_{DSsat6} = -0.15$ V and allocating a small saturation margin $\Delta V_{DS6} = -0.1$ V to M6, we determine the limit imposed by (4.256) on the gate-source voltage of M1 and

M2 as $V_{GS(1,2)} > -0.45$ V, which is too small to yield an adequate overdrive voltage in standard-threshold option. We select the low-threshold option, and also decide to operate these devices with source-bulk shorts for avoiding the body effect. According to Fig. 4.10, V_{TP} for a typical channel length is just below −0.1 V in low-threshold PMOS devices. The constraint defined by (4.256) is thus translated into $V_{GS(1,2)} - V_{TP} > -0.45 - (-0.1) = -0.35$ V for the overdrive voltage of M1 and M2.

The plots of I_{B1}, $(W/L)_{1,2}$, $C_{in(max)}$, I_{B2}, and g_{m5}/I_{B2} versus $V_{GS(1,2)} - V_{TP}$ as obtained by simulation for six different values of C_C are presented in Fig. 4.65. The input file `ex4-25a.sp` used for this purpose is copied below.

```
filename: ex4-25a.sp
.incl bsimlvt.sp
m1 2 1 0 0 cmosplvt w=6u l=0.8u
vds1 2 0 -0.9
vgs1 1 0 dc
**********************parameters**********************
vcl 3 0 6.5p      ; cl
vc1 4 0 1.5p      ; c1
vcc 5 0 0.5p      ; cc
vpm 6 0 2.75      ; tan(pm)
vxi 7 0 0.8       ; step height
vepsilon 8 0 3e-4 ; settling error
vts 9 0 50n       ; settling time
***********************analysis***********************
.save @m1[vth] @m1[id] @m1[gm] v(1) v(3) v(4)
+  v(5) v(6) v(7) v(8) v(9)
.control
destroy all
foreach comp 0.5p 1p 1.5p 2p 2.5p 3p
alter vcc=$comp
dc vgs1 -0.45 -0.15 10m
****************calculations for M1,M2****************
let gmeff12=@m1[gm]/@m1[id]           ; gm/ID of M1,M2
let vdir=2/gmeff12                    ; (4.245)
let fcl=((v(7)/vdir)+ln(vdir/v(7))
+ -1+ln(1/v(8)))/(6.283*v(9))        ; (4.259)
let gm12=6.283*fcl*v(5)               ; (4.235)
let cinmax=1.59e-6/fcl                ; (4.261)
let ib1=vdir*gm12                     ; (4.245)
let asp12=(ib1/2)/(@m1[id]/8.08)     ; W/L of M1,M2 (3.25)
*****************calculations for M5******************
let m=v(5)^2/(v(4)*v(3)
+ +v(4)*v(5)+v(3)*v(5))              ; (4.258)
let gm5=((v(6)*(m+1)+sqrt(v(6)^2*(m+1)^2+4*m))
+ /(2*m))*gm12                       ; (4.257)
let ib2=(v(3)+v(5))*ib1/v(5)         ; (4.260)
let gmeff5=gm5/ib2                   ; gm/ID of M5
end
************************plots*************************
```

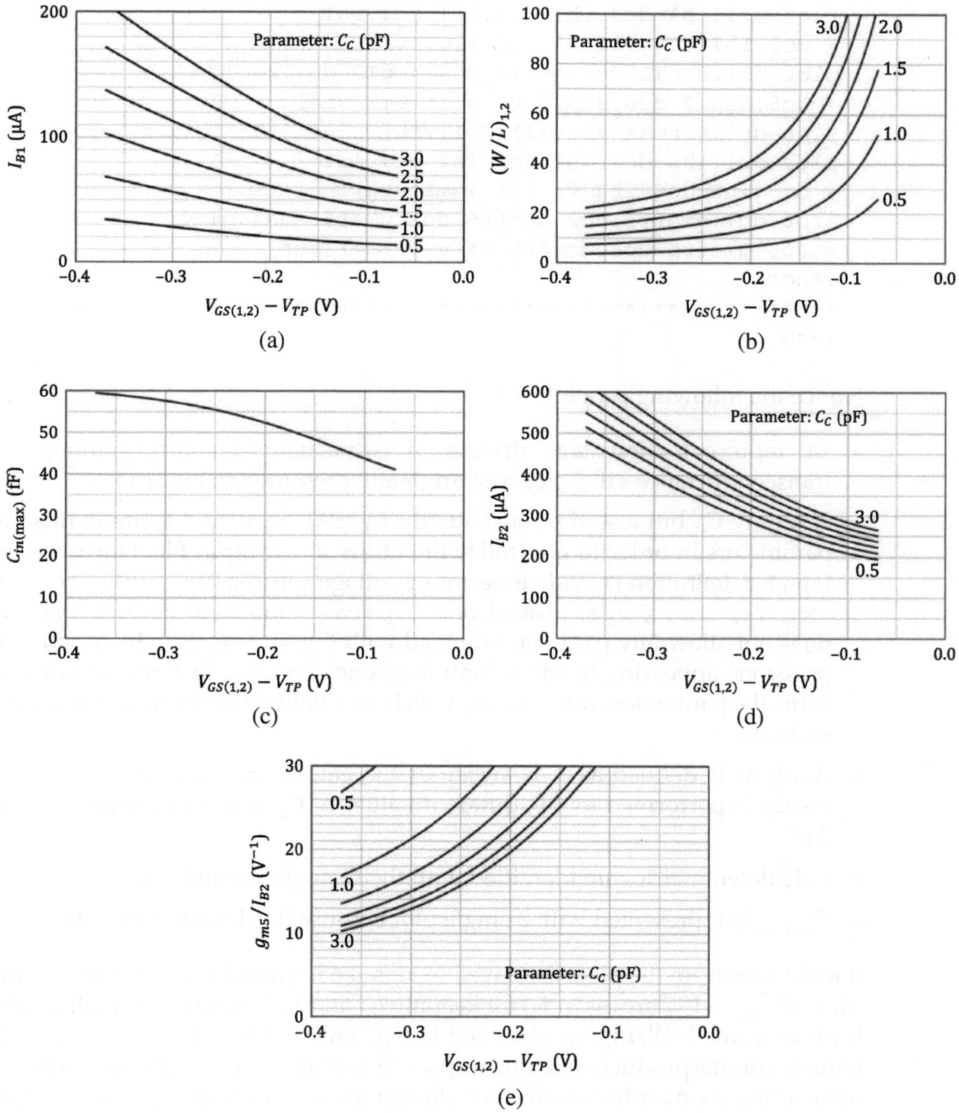

FIGURE 4.65 Data obtained by simulation with `ex4-25a.sp` and plotted against the overdrive voltage $V_{GS(1,2)} - V_{TP}$ of first-stage drivers for six different values of compensation capacitance C_C. (a) First-stage bias current I_{B1}. (b) Aspect ratio $(W/L)_{1,2}$ of first-stage drivers. (c) Maximum tolerable input capacitance $C_{in(max)}$. (d) Second-stage bias current I_{B2}. (e) Transconductance efficiency g_{m5}/I_{B2} of second-stage driver.

```
plot dc1.ib1 dc2.ib1 dc3.ib1 dc4.ib1
+ dc5.ib1 dc6.ib1 vs v(1)+@m1[vth]
plot dc1.asp12 dc2.asp12 dc3.asp12 dc4.asp12
+ dc5.asp12 dc6.asp12 vs v(1)+@m1[vth]
plot dc1.cinmax vs v(1)+@m1[vth]
plot dc1.ib2 dc2.ib2 dc3.ib2 dc4.ib2
+ dc5.ib2 dc6.ib2 vs v(1)+@m1[vth]
plot dc1.gmeff5 dc2.gmeff5 dc3.gmeff5 dc4.gmeff5
+ dc5.gmeff5 dc6.gmeff5 vs v(1)+@m1[vth]
.endc
*******************************************************
.end
```

Notice the following properties:

- `m1` represents first-stage drivers. It is included for incorporating the transconductance-efficiency tool originally presented in Fig. 4.14.
- Not only C_C but also the constants C_1, C_L, PM, X_i, ϵ, and t_S are defined as parameters in order to generalize the utility of the input file. However, parameter definition is made in terms of voltage sources (`vcl`, `vc1`, `vcc`, `vpm`, `vxi`, `vepsilon`, `vts`) instead of a "`.param`" statement because Ngspice does not allow any parameter defined with this statement to be used in expressions appearing inside a control section. Defining in terms of voltages turns the parameters into vectors, which are qualified for inclusion in control sections.
- Analysis is defined as a dc sweep of the gate-source voltage `vgs1` of `m1`. Sweep is performed for six different values of C_C ranging between 0.5 and 3 pF.
- Calculated metrics are identified with their equation numbers.
- $C_{in(\max)}$ is represented with a single plot because it is independent of C_C.

It is obvious from the plots displayed in Figs. 4.65(a) and (d) that a small C_C and a low $V_{GS(1,2)} - V_{TP}$ are essential for keeping I_{B1} and I_{B2} low. While a small C_C also leads to a small $(W/L)_{1,2}$ as observed in Fig. 4.65(b), a low $V_{GS(1,2)} - V_{TP}$ is obviously counterproductive in this respect. Yet another undesirable consequence of lowering the overdrive is the reduction of the upper limit $C_{in(\max)}$ of the input capacitance as observed in Fig. 4.65(c). As a matter of fact, too low a first-stage overdrive is prohibitive as indicated by the transconductance-efficiency plots of M5 shown in Fig. 4.65(e). As we concluded in Subsection 3.3.5, g_m/I_D cannot be made larger than about 30 V^{-1} even under deep subthreshold conditions. For M5, it needs to be considerably smaller than the maximum because otherwise the overdrive voltage would be too small to support the range of I_{B2} indicated in Fig. 4.65(d) without necessitating a very large aspect ratio. A large aspect ratio for M5 is of course undesirable as it would increase the footprint of this device and therefore enlarge C_1 and C_L. One final observation made in Fig. 4.65(e) is that the necessity of a low g_{m5}/I_{B2} can be fulfilled only with a large C_C.

Based on the foregoing observations, we decide to push $V_{GS(1,2)} - V_{TP}$ to the limit −0.35 V while striking a balance between the benefits and drawbacks of

C_C by setting it to 2 pF. According to the data of Fig. 4.65, this selection implies the following targets: $I_{B1} = 130$ μA, $(W/L)_{1,2} = 15.6$, $C_{in(\max)} = 59$ fF, $I_{B2} = 554$ μA, and $g_{m5}/I_{B2} = 12.7\ V^{-1}$. Although not shown in Fig. 4.65, the rest of the variables defined in `ex4-25a.sp` can also be plotted after the completion of the simulation. It is important not to forget plotting specifically against $V_{GS(1,2)} - V_{TP}$ with a command extension

```
vs v(1)+@m1[vth]
```

in order to be able to read the correct values corresponding to our first-stage overdrive selection. Values thus obtained include $g_{m(1,2)}/(I_{B1}/2) = 5.19\ \text{V}^{-1}$, $f_{cl} = 26.9$ MHz, $f_Z/f_{P2} = 6.44$, $f_{P2}/f_{cl} = 3.22$, and $V_{DIR} = 0.385$ V.

2. *$(W/L)_{3,4}$ and $(W/L)_5$* In order to quantify the upper limit imposed on the common overdrive voltage of these three devices, we first consider (4.255). Since $V_{GS(1,2)} - V_{TP} = -0.35$ V is already fixed, and a PMOS threshold voltage around -0.1 V is indicated by Fig. 4.10 for the case of source-bulk short for M1 and M2, we estimate $V_{GS(1,2)} \approx -0.45$ V. Also reading $V_{DSsat(1,2)} \approx -0.290$ V from Fig. 4.11 for $V_{GS(1,2)} - V_{TP} = -0.35$ V, we turn (4.255) into $V_{GS(3,4)} \leq 0.47$ V. Judging from Figs. 3.7 and 4.5 for a channel length not exceeding 1.2 μm, the corresponding overdrive voltage $V_{GS(3,4)} - V_{TN}$ is not larger than about 60 mV for the standard-threshold option but close to 400 mV for the low-threshold option.

   Next, we consider (4.254) together with Figs. 3.8 and 4.6, and arrive at $V_{GS5} - V_{TN} \leq 0.45$ V for both threshold options. This limit is obviously less restrictive than the previous one regardless of the threshold option.

   Finally, we consider the limit imposed on $V_{GS5} - V_{TN}$ by $g_{m5}/I_{B2} = 12.7\ V^{-1}$ as determined in step 1. According to Figs. 3.12 and 4.9, the corresponding overdrive voltage equals 0.15 V for both threshold options. Since we need to adopt the smallest of the three limits while preferring a large overdrive voltage for minimizing the device footprint, we decide to select the low-threshold option and therefore adopt $V_{GS(3,4,5)} - V_{TN} = 0.15$ V.

   Reading $I_{\square 5} = 2.5$ μA/sq from Fig. 4.7 or Fig. 4.8, and considering the target $I_{B2} = 554$ μA selected in step 1, we determine the aspect ratio of M5 as $(W/L)_5 = 222$. For $I_{B1} = 130$ μA, this leads to $(W/L)_{3,4} = 26$ from (4.263). Since the ratio between the two aspect ratios is close to 8.5, we decide to construct each of M3 and M4 with two unit devices of $(W/L)_{u(3,4,5)} = 13$, and combine 17 such unit devices for M5.

3. *$(W/L)_6$ and $(W/L)_7$* Threshold option is not restricted for these devices. We select the standard-threshold option, and read $V_{GS(6,7)} - V_{TP} = -0.175$ V from Fig. 3.17 for the target $V_{DSsat6} = -0.150$ V set in step 1. The corresponding $I_{\square(6,7)} = 0.9$ μA/sq indicated by Fig. 3.19 leads to $(W/L)_6 = 144$ for $I_{B1} = 130$ μA. Since the factor of scaling from M6 to M7 should be one half of the scale 17/2 from the pair M3/M4 to M5, we construct M6 with four copies of a unit device of $(W/L)_{u(6,7)} = 36$, and use 17 copies of this unit device for constructing M7. Current-reference design is not a part of this example. We simply put together four copies of the unit device as MR for mirroring I_{B1} and I_{B2} from $I_{\text{REF}} = 130$ μA.

4. *$L_{DR(1,2)}$ and $W_{DR(1,2)}$* These dimensions must satisfy the aspect ratio determined in step 1 with a maximum input capacitance of 59 fF. As concluded in Example 4.20, the input capacitance of a unity-gain buffer configuration is given by

$C_{in} = C_{se} + C_d$. Referring to (4.92) and (4.93) for $g_{mb} = 0$, we can express it in terms of first-stage driver capacitances as follows:

$$C_{in} = \frac{C_{gg} + C_{gd} + C_{gb}}{2},$$

where C_{gb} is generally negligible in comparison with the remaining two components. In order to express this equation in terms of $L_{DR(1,2)}$ and $W_{DR(1,2)}$ we substitute (3.105) and (3.106) for C_{gg} and C_{gd}, and manipulate the resulting expression by (a) replacing C_{gge} with the right-hand side of (3.99) but ignoring the negligible C_{gbe}, and (b) adopting (3.90) for the areal density of C_{ggi}, (3.92) for C_{gde}, and (3.93) for C_{gse}. The outcome

$$C_{in} = \frac{1}{3}C_{OX}W_{1,2}L_{1,2} + \left(\frac{\mathrm{CGSO}}{2} + \mathrm{CGDO}\right)W_{1,2}$$

evaluated together with $C_{in} \leq 59$ fF and $(W/L)_{1,2} = 15.6$ yields $L_{1,2} \leq 0.96$ µm. We prefer staying on the safe side by setting $L_{DR(1,2)} = 0.5$ µm. Whether such a short channel can still sustain an adequate output resistance r_{ds2} in M2 remains to be seen. We cross our fingers and calculate the corresponding channel width as $W_{DR(1,2)} = 6.9$ µm.

5. *$L_{DR(3,4,5)}$, L_{DR7}, $W_{DRu(3,4,5)}$, and $W_{DRu(6,7)}$* Referring to (4.262), note that the two transconductance-efficiency factors it contains have already been determined. V_{A2} is not a free variable either because the overdrive voltage and channel length of M2 have been quantified while the drain-source voltage may vary only very weakly depending on the selection of $L_{DR(3,4,5)}$.[8] The remaining three factors V_{A4}, V_{A5}, and V_{A7} are controlled by two independent variables, namely, $L_{DR(3,4,5)}$ and L_{DR7}. We set the target values of these channel lengths by applying the following procedure:
   a. By repetitively running a parametric simulation with different values of $L_{DR(3,4,5)}$, determine the corresponding values of V_{A2}, V_{A4}, and V_{A5}. Substitute this data into (4.262) together with the specified $A_0 = 65$ dB, and calculate the necessary V_{A7} for each value of $L_{DR(3,4,5)}$ used in this parametric simulation.
   b. By repetitively running a parametric simulation with different values of L_{DR7}, determine the corresponding values of V_{A7}.
   c. By matching the values of V_{A7} in the data sets extracted in steps a and b, determine the corresponding L_{DR7} for each $L_{DR(3,4,5)}$ involved in step a.
   d. Identify the pair of $L_{DR(3,4,5)}$ and L_{DR7} that comes closest to satisfying $V_{A7} = V_{A5}$ while minimizing the parasitic capacitances of M5 and M7. Target this pair in design.

   The parametric simulation mentioned in step a of this procedure is performed with the input file `ex4-25b.sp` presented below and based on the schematic shown in Fig. 4.64(b).

[8] Selection of $L_{DR(3,4,5)}$ affects V_{DS2} in the following way: A longer $L_{DR(3,4,5)}$ leads to a smaller threshold voltage for M3, M4, and M5. Since the overdrive voltage of these devices has already been fixed, a smaller threshold voltage implies a larger V_{GS5}, which, in turn, makes V_{DS2} less negative because the drain of M2 is tied to the gate of M5. This leads to a smaller V_{A2} but the effect is weak because the relative variation of V_{DS2} with $L_{DR(3,4,5)}$ is small.

```
filename: ex4-25b.sp
.incl bsimlvt.sp
.param length=0.4u vol=0.305
m2 5 1 3 3 cmosplvt w=6.9u l=0.5u
m4 5 4 0 0 cmosnlvt w=10u l=length
m5 8 5 0 0 cmosnlvt w=10u l=length
vs1 1 8 0
vs2 4 5 0
vgs 1 3 -0.445
vo 8 0 0.7
vol 5 0 vol
.save @m2[id] @m2[gds] @m4[id] @m4[gds] @m5[id] @m5[gds]
.control
op
print @m2[id]/@m2[gds] @m4[id]/@m4[gds] @m5[id]/@m5[gds]
.endc
.end
```

The netlist includes M2, M4, and M5 and retains the node numbers of the main schematic shown in Fig. 4.64(a). M2 is specified with its exact dimensions, whereas M4 and M5 are specified with an arbitrarily selected channel width and a parametric channel length. `vs1` is a short-circuit element maintaining the same fixed bias voltage $V_O = 0.7$ V at the gate of M2 and the drain of M5 as implied by the unity-gain application. `vs2` is a short-circuit element too. It interconnects the gate and drain of M4 to establish the virtual equality of the voltages of these two terminals as maintained in actual operation. Notice the presence of a source `vol` supplying this voltage to the drain of M2 and gate of M5 to complete the connectivity of the first-stage output of the OTA. The gate-source voltage of M2 is supplied by voltage source `vgs`. This voltage is fixed at −0.445 V as calculated from $V_{GS(1,2)} - V_{TP} = -0.350$ V and $V_{TP} = -0.095$ V. The threshold voltage used in this calculation is obtained from Fig. 4.10 for $L_{DR(1,2)} = 0.5$ μm and $V_{SB(1,2)} = 0$. Note that $L_{DR(3,4,5)}$ is defined by the parameter `length`. In each iteration, we also enter the value of parameter `vol` before running the simulation. Being equal to V_{GS5}, it is calculated by determining V_{TN5} from Fig. 4.5 for the selected value of `length`, and using the outcome in $V_{GS5} - V_{TN5} = 0.15$ V. The control section automatically puts out the simulated values of V_{A2}, V_{A4}, and V_{A5}. Results are presented in Table 4.3.

$L_{DR(3,4,5)}$ (μm)	0.4	0.6	0.8	1.0	1.2	1.4	2.0
V_{A2} (V)	15.37	15.75	15.96	16.09	16.24	16.31	16.44
V_{A4} (V)	3.14	3.73	4.16	4.52	4.72	4.94	5.38
V_{A5} (V)	7.80	10.04	12.08	14.20	16.07	18.11	23.78
V_{A7} (V)	–	82.12	25.30	16.57	13.64	11.72	9.24

TABLE 4.3 Early Voltages of M2, M4, M5, and M7 as Functions of Channel in Length Example 4.25

L_{DR7} (μm)	0.2	0.4	0.6	0.8	1.0	1.2	1.4	2.0
V_{A7} (V)	9.83	12.02	13.53	14.64	15.76	16.62	17.41	20.13

TABLE 4.4 Early Voltage of M7 as a Function of Channel Length in Example 4.25

The parametric simulation mentioned in step b of the procedure is run with the following input file `ex4-25c.sp`, which is based on the schematic shown in Fig. 4.64(c).

```
filename: ex4-25c.sp
.incl bsim.sp
m7 8 7 10 10 cmosp l=0.2u w=10u
vgs 7 10 -0.650
vo 8 0 0.7
vdd 10 0 1.8
.save @m7[gds] @m7[id]
.control
op
print @m7[id]/@m7[gds]
.endc
.end
```

Each iteration is performed by setting the channel length `l` and the corresponding gate-source voltage `vgs`. The latter is calculated from (a) the fixed overdrive voltage $V_{GS7} - V_{TP} = -0.175$ V, and (b) the threshold voltage indicated by Fig. 3.16 for $V_{SB} = 0$. Results are presented in Table 4.4.

According to Table 4.3, the desirable condition $V_{A7} \approx V_{A5}$ prevails around $V_{A7} \approx 15$ V and necessitates $L_{DR(3,4,5)} = 1.1$ μm. As indicated by Table 4.4, this much V_{A7} can be achieved with $L_{DR7} = 0.9$ μm.

Considering the aspect ratios determined in step 2 for the unit devices of M3, M4, and M5, and in step 3 for those of M6 and M7, we calculate the following channel widths for unit devices: $W_{DRu(3,4,5)} = 13.9$ μm, $W_{DRu(6,7)} = 30.3$ μm.

6. *Verification* Performed with the input file `ex4-25.sp` presented below.

```
filename: ex4-25.sp
.incl bsimlvt.sp
.incl bsim.sp
***************ota subckt****fig.4.64(a)****************
.subckt ota 1 2 8
m1 4 2 3 3 cmosplvt l=0.5u w=6.9u ad=3.8p pd=14.9u
m2 5 1 3 3 cmosplvt l=0.5u w=6.9u ad=3.8p pd=14.9u
m3 4 4 0 0 cmosnlvt l=1.1u w=13.9u ad=7.7p pd=28.9u m=2
m4 5 4 0 0 cmosnlvt l=1.1u w=13.9u ad=7.7p pd=28.9u m=2
m5 8 5 0 0 cmosnlvt l=1.1u w=13.9u ad=7.7p pd=28.9u m=17
m6 3 7 10 10 cmosp l=0.9u w=30.3u ad=16.7p pd=61.7u m=4
m7 8 7 10 10 cmosp l=0.9u w=30.3u ad=16.7p pd=61.7u m=17
mr 7 7 10 10 cmosp l=0.9u w=30.3u ad=16.7p pd=61.7u m=4
iref 7 0 130u
cc 8 5 2p
```

```
vdd 10 0 1.8
.ends ota
**************open-loop dc****fig.4.56(f)***************
x1 41 42 43 ota
vin 42 0 0.7
vi 40 0 dc
rs 40 41 10k
cl 43 0 5p
.dc vi 0.69 0.71 1u
************loop gain****fig.4.56(h) and (i)************
*x3 61 62 63 ota     ; for voltage return ratio
*cla 62 0 5p         ; for voltage return ratio
*via 60 0 dc 0.7     ; for voltage return ratio
*rsa 60 61 10k       ; for voltage return ratio
*vx 62 63 dc 0 ac 1  ; for voltage return ratio
*x4 71 72 73 ota     ; for current return ratio
*clb 72 0 5p         ; for current return ratio
*vib 70 0 dc 0.7     ; for current return ratio
*rsb 70 71 10k       ; for voltage return ratio
*ix 0 74 dc 0 ac 1   ; for current return ratio
*vs1 73 74 0         ; for current return ratio
*vs2 74 72 0         ; for current return ratio
*.control
*ac dec 100 1k 1g
*let rv=-v(63)/v(62)
*let ri=-i(vs1)/i(vs2)
*set units=degrees
*plot db((rv*ri-1)/(2+rv+ri)) ph((rv*ri-1)/(2+rv+ri))
*let magnitude=mag((rv*ri-1)/(2+rv+ri))
*let phasedif=ph((rv*ri-1)/(2+rv+ri))+180
*meas ac fu when magnitude=1
*meas ac pm find phasedif at=fu
*.endc
**************open-loop ac****fig.4.56(f)***************
*x1 41 42 43 ota
*vin 42 0 0.7
*vi 40 0 dc 0.700057 ac 1
*rs 40 41 10k
*cl 43 0 5p
*.ac dec 100 1k 1g
*************closed-loop ac****Fig.4.56(g)**************
*x2 51 52 52 ota
*vi 50 0 dc 0.7 ac 1
*rs 50 51 10k
*cl 52 0 5p
*.ac dec 100 1k 1g
***************transient****Fig.4.56(g)****************
*x2 51 52 52 ota
*vi 50 0 pulse 0.3 1.1 10n 0.1n 0.1n 100n
*rs 50 51 10k
```

```
*cl 52 0 5p
*.tran 10p 200n
*******************************************************
.end
```

Notice the definition of the OTA as a subcircuit complying with the schematic given in Fig. 4.64(a). First, a dc simulation is run for determining the dc gain and systematic offset. It is based on the schematic presented earlier in Fig. 4.56(f) of Example 4.24. Systematic offset turns out to be 57 μV. The dc gain, as plotted in dB units with the command

```
plot 20*log(deriv(v(43))) vs v(43)
```

confirms the 65-dB target at $V_O = 0.7$ V.

Next comes a loop-gain simulation, whose netlist complies with the schematic presented in Figs. 4.56(h) and (i) of Example 4.24. It prints out 67.1° for the phase margin.

What follows is an ac simulation by which the open-loop small-signal frequency response is evaluated on the schematic given in Fig. 4.56(f) of Example 4.24. It yields $f_u = 22.0$ MHz, which falls short of the target value 26.9 MHz set in design for $f_{cl} = f_{GBW}$ in accordance with (4.235). This difference is partially due to the fact that the expectation $f_u = f_{cl} = f_{GBW}$ in a unity-gain configuration is not perfectly justified unless a perfect single-pole frequency response prevails. For an explanation of this nonideality, the reader is referred to Subsection 2.7.2 in general and to the comments on (2.93) in particular. Yet another reason is that the actual value of f_{GBW} and therefore f_{cl} is smaller than the target value due to the contribution of the gate-to-drain extrinsic capacitance C_{gde} of M5 to C_C. The reader can evaluate the significance of this contribution by estimating the value of C_{gde} from (3.92) without ignoring the number of unit devices paralleled for M5, and repeating the ac simulation after resetting the capacitance of the compensation capacitor to $C_C = 2 \times 10^{-12} - C_{gde}$.

A closed-loop ac simulation is run next on the schematic given in Fig. 4.56(g) of Example 4.24. The aim is to determine the 3-dB frequency, which is expected to move considerably above f_{cl} due to the underdamped response implied by $f_{P2}/f_{cl} = 3.22 < 4$ and confirmed by the phase margin determined earlier. The reader is referred to Subsection 2.7.1 for the consequences of an underdamped response. This simulation indicates a 3-dB frequency at 35.0 MHz, which is indeed significantly higher than the target 26.9-MHz set for f_{cl} in design.

The final simulation involves a transient analysis of the step response. It is run on the schematic used earlier for closed-loop ac simulation. Shown in Fig. 4.66(a) are the waveforms in one full period. An underdamped response is obvious from the overshoot and undershoot observed in settling. Considering $X_i = 0.8$ V and $\epsilon = 0.03\%$, the error band of settling is calculated as ± 240 μV and marked on the expanded waveforms of falling-edge settling and rising-edge settling in Figs. 4.66(b) and (c), respectively. Since the high-to-low transition of v_I occurs at $t = 110$ ns, and v_O enters the error band at $t = 154.5$ ns, the falling-edge settling time is around 44.5 ns, which satisfies the specification. The rising edge

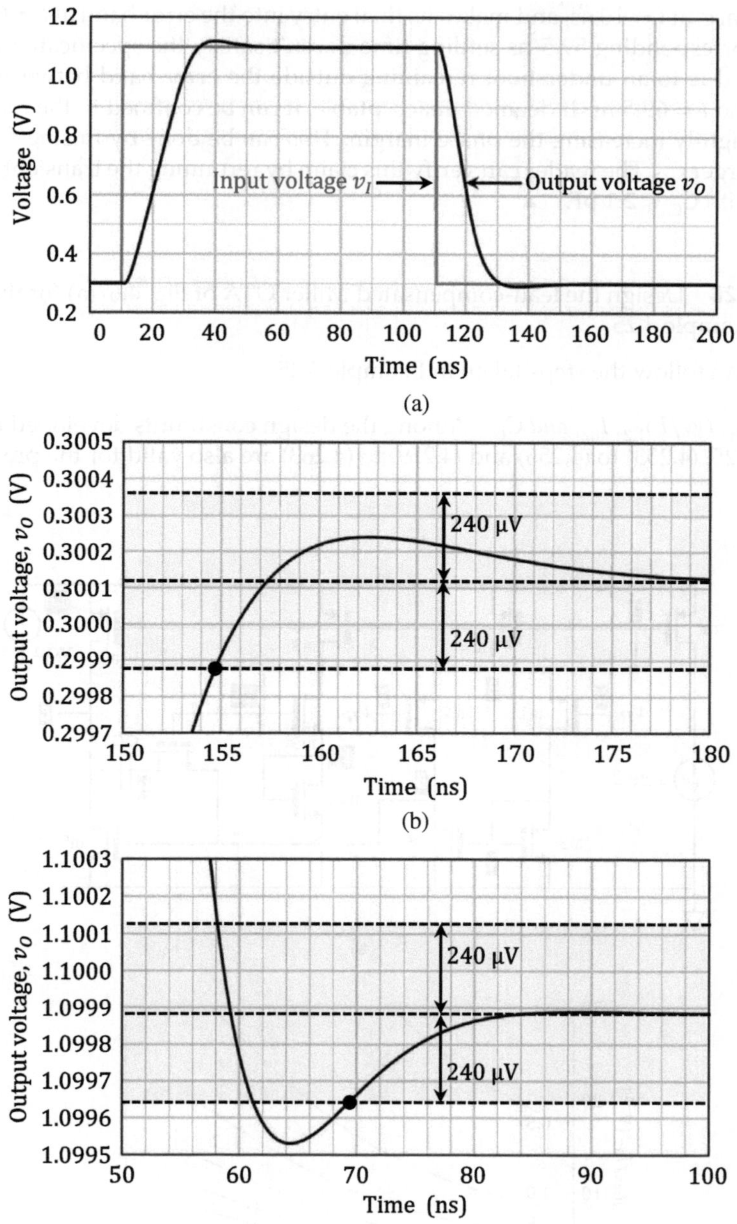

FIGURE 4.66 Step response obtained with `ex4-25.sp`. (a) Input and output waveforms in one period. (b) Zoomed-in output waveform in falling-edge settling, where the dot marks the final entry into the error band. (c) Zoomed-in output waveform in rising-edge settling, where the dot marks the final entry into the error band.

starts at t = 10 ns, and makes its final entry into the error band at t = 69.5 ns. The corresponding 59.5-ns settling time doesn't satisfy the specification. The delay is due to an undershoot remaining outside the error band between t = 61 ns and t = 69.5 ns. If deemed unacceptable, it can be confined to the error band by slightly increasing the phase margin. This can be done by raising f_{P2}/f_{cl} with a larger C_C. The reader can verify this claim by rerunning the transient simulation with C_C = 2.1 pF. ▲

Example 4.26 Design the lead-compensated Miller OTA of Fig. 4.67(a) for the specifications of Example 4.25.

Work We follow the steps taken in Example 4.25.

1. I_{B1}, $(W/L)_{1,2}$, I_{B2}, *and* C_C Among the design constraints developed in Example 4.25, (4.253) to (4.256) and (4.259) to (4.263) are also valid for the present design

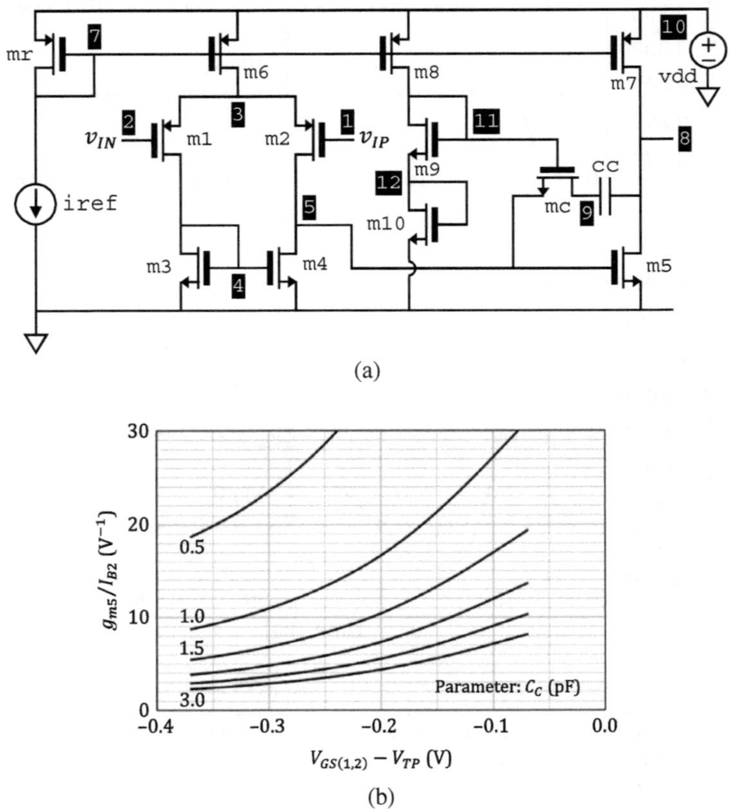

FIGURE 4.67 (a) OTA subcircuit of Example 4.26. (b) Transconductance efficiency g_{m5}/I_{B2} of second-stage driver.

but (4.257) and (4.258) are replaced by the following equation obtained from (4.241):

$$g_{m5} \geq \frac{C_1 C_L g_{m(1,2)}}{C_C^2} \tan(\text{PM}). \tag{4.264}$$

The plots presented in Fig. 4.65 for I_{B1}, $(W/L)_{1,2}$, $C_{in(\max)}$, and I_{B2} are still valid. The plot of g_{m5}/I_{B2} should now be extracted in compliance with (4.264). To this end, use the input file `ex4-25a.sp` after modifying its "calculations for M5" section as follows:

```
****calculations for M5****
let gm5=v(4)*v(3)*v(6)*gm12/v(5)^2       ; (4.264)
let ib2=(v(3)+v(5))*ib1/v(5)             ; (4.260)
let gmeff5=gm5/ib2      ; transconductance efficiency of M5
end
```

The plot thus obtained for g_{m5}/I_{B2} is shown in Fig. 4.67(b). A comparison with the pole-splitting counterpart shown in Fig. 4.65(e) indicates much smaller lower limits for this transconductance efficiency, which justifies the selection of a smaller $V_{GS(1,2)} - V_{TP}$ for the present case. We adopt $V_{GS(1,2)} - V_{TP} = -0.25$ V alongside $C_C = 2$ pF, and read $g_{m5}/I_{B2} = 5.9\ V^{-1}$ from Fig. 4.67(b). Other targets indicated by Figs. 4.65(a) to (d) are as follows: $I_{B1} = 96.5$ μA, $(W/L)_{1,2} = 21.5$, $C_{in(\max)} = 55.5$ fF, and $I_{B2} = 410$ μA. Simulation with the modified `ex4-25a.sp` also yields $g_{m(1,2)}/(I_{B1}/2) = 7.5\ \text{V}^{-1}$, $f_{cl} = 28.6$ MHz, and $V_{DIR} = 0.268$ V.[9]

2. $(W/L)_{3,4}$ *and* $(W/L)_5$ For determining the upper limit imposed on $V_{GS(3,4,5)} - V_{TN}$, we first consider (4.255). For $V_{GS(1,2)} - V_{TP} = -0.25$ V and $V_{TP} \approx -0.1$ V of the low-threshold M1 and M2 operating with $V_{SB(1,2)} = 0$, we expect $V_{GS(1,2)} \approx -0.35$ V. Furthermore, Fig. 4.11 indicates $V_{DSsat(1,2)} \approx -0.21$ V for $V_{GS(1,2)} - V_{TP} = -0.25$ V. These numbers imply $V_{GS(3,4)} \leq 0.44$ V from (4.255). Considering the threshold voltages indicated by Figs. 3.7 and 4.5 for a channel length not exceeding 1.2 μm, we conclude that $V_{GS(3,4)} - V_{TN}$ should be less than about 40 mV for the standard-threshold option and close to 0.37 V for the low-threshold option.

   The upper limit imposed on $V_{GS5} - V_{TN}$ by (4.254) is around 0.45 V for both threshold options as indicated by Figs. 3.8 and 4.6. This is obviously less restrictive than the limits identified in the preceding paragraph.

   As to the constraint imposed on $V_{GS5} - V_{TN}$ by $g_{m5}/I_{B2} = 5.9\ V^{-1}$, Figs. 3.12 and 4.9 indicate 0.33 V for both threshold options. We select the low-threshold option in order to avoid the 40-mV limit, and therefore target $V_{GS(3,4,5)} - V_{TN} = 0.33$ V. The corresponding sheet current $I_{\square 5} = 12$ μA/sq indicated by Fig. 4.7 implies $(W/L)_{3,4} = 4$ and $(W/L)_5 = 34.5$ for the bias currents $I_{B1} = 96.5$ μA and $I_{B2} = 410$ μA. We decide to implement M3 and M4 with two unit devices of $(W/L)_{u(3,4,5)} = 2$, and combine 17 such unit devices for M5.

3. $(W/L)_6$ *and* $(W/L)_7$ Considering $V_{GS(1,2)} - V_{TP} = -0.25$ V and assuming $V_{TP(1,2)} \approx -0.1$ V as indicated by Fig. 4.10 for typical channel lengths, we predict

---

[9] As explained in Example 4.25, it is important to plot these additional variables specifically against $V_{GS(1,2)} - V_{TP}$ in order to be able to read their values corresponding to $V_{GS(1,2)} - V_{TP} = -0.25$ V.

$L_{DR(3,4,5)}$ (µm)	0.4	0.6	0.8	1.0	1.2	1.4	2.0
V_{A2} (V)	10.18	10.59	10.86	10.95	11.17	11.27	11.42
V_{A4} (V)	5.41	6.40	7.21	8.07	8.62	9.26	10.87
V_{A5} (V)	10.2	13.42	16.29	19.12	21.87	24.56	32.02
V_{A7} (V)	–	49.1	21.6	15.8	13.3	11.7	9.32

TABLE 4.5 Early Voltages of M2, M4, M5, and M7 as Functions of Channel Length in Example 4.26

$V_{GS(1,2)} \approx -0.35$ V, which leads to $V_{DSsat6} > -0.35$ V from (4.256). We decide to target $V_{DSsat6} = -0.25$ V, and therefore allow $\Delta V_{DS6} = -0.1$ V for the saturation margin of M6. We select the standard-threshold option, and read $V_{GS(6,7)} - V_{TP} = -0.31$ V from Fig. 3.17. For this much overdrive, Fig. 3.19 indicates $I_{\square(6,7)} = 2.7$ µA/sq. Aspect ratios of M6 and M7 are therefore calculated as $(W/L)_6 = 35.7$ and $(W/L)_7 = 152$. We decide to construct M6 with 4 copies of a unit device of aspect ratio $(W/L)_{u(6,7)} = 9$, and use 17 copies for M7. Four copies of the same unit device are used for MR, and I_{REF} is set to 96.5 µA.

4. *$L_{DR(1,2)}$ and $W_{DR(1,2)}$* Following the path of step 4 of Example 4.25 we arrive at $L_{max(1,2)} = 0.8$ µm, and decide to set $L_{DR(1,2)} = 0.5$ µm, which implies $W_{DR(1,2)} = 11$ µm for $(W/L)_{1,2} = 21.5$.

5. *$L_{DR(3,4,5)}$, L_{DR7}, $W_{DRu(3,4,5)}$, and $W_{DRu(6,7)}$* The procedure explained in step 5 of Example 4.25 is applied. Only two modifications are needed in the input file `ex4-25b.sp`: (a) changing the channel width of M2 to 11 µm, and (b) replacing the value of the voltage source `vgs` with –0.345 V as calculated from $V_{GS(1,2)} - V_{TP} = -0.25$ V and $V_{TP} = -0.095$ V. Otherwise the file is used as instructed in step 5(a) of Example 4.25. The outcome of the simulations performed with this file is presented in Table 4.5.

   The parametric simulation mentioned in step 5(b) of Example 4.25 is performed with the input file `ex4-25c.sp`, in which, each iteration involves resetting of the channel length `l` and the corresponding gate-source voltage `vgs`. The latter is calculated from (a) the fixed overdrive voltage $V_{GS7} - V_{TP} = -0.31$ V and (b) the threshold voltage indicated by Fig. 3.16 for $V_{SB} = 0$. Results are presented in Table 4.6.

   According to Table 4.5, the desirable condition of comparable V_{A7} and V_{A5} is met around $V_{A7} \approx 18$ V and $L_{DR(3,4,5)} \approx 0.9$ µm. The corresponding $L_{DR(6,7)}$, as determined from Table 4.6, is around 0.5 µm. Adopting these channel lengths and considering the aspect ratios determined in steps 2 and 3, we arrive at $W_{DRu(3,4,5)} = 1.75$ µm and $W_{DRu(6,7)} = 4$ µm for unit-device widths.

L_{DR7} (µm)	0.2	0.4	0.6	0.8	1.0	1.2	1.4	2.0
V_{A7} (V)	11.8	16.2	18.8	20.7	22.4	23.8	25.0	28.7

TABLE 4.6 Early Voltage of M7 as a Function of Channel Length in Example 4.26

6. *M8, M9, M10, and MC* M8 is implemented with one copy of the PMOS unit device sized for M6, M7, and MR in step 5. Each of M9 and M10 is implemented with one copy of the NMOS unit device sized for M3, M4, and M5 in step 5.

   Compensation resistance R_C, as calculated from (4.239), equals 2.66 kΩ. However, the estimates $C_1 = 1.5$ pF and $C_L = 6.5$ pF used in this calculation are likely to be larger than the actual values of these capacitances because the dimensions of M2, M4, M5, and M7 are much smaller than what would justify them. We therefore expect the optimum value of R_C to be smaller. Instead of trying to calculate it accurately, we install M8, M9, M10, and MC together with $C_C = 2$ pF into the final schematic, and iteratively simulate the step response by varying the number of parallel unit devices for MC until the desired settling performance is achieved. The unit device to be used for constructing MC is the one defined for M9 and M10.

7. *Verification* Verification is performed with the input file `ex4-26.sp`, which differs from `ex4-25.sp` only in the netlist of the OTA subcircuit, which now reflects the OTA schematic presented in Fig. 4.67 as follows:

```
***************ota subckt****fig.4.67(a)***************
.subckt ota 1 2 8
m1 4 2 3 3 cmosplvt l=0.5u w=11u ad=6.1p pd=23u
m2 5 1 3 3 cmosplvt l=0.5u w=11u ad=6.1p pd=23u
m3 4 4 0 0 cmosnlvt l=0.9u w=1.75u ad=0.96p pd=4.6u m=2
m4 5 4 0 0 cmosnlvt l=0.9u w=1.75u ad=0.96p pd=4.6u m=2
m5 8 5 0 0 cmosnlvt l=0.9u w=1.75u ad=0.96p pd=4.6u m=17
m6 3 7 10 10 cmosp l=0.5u w=4u ad=2.2p pd=9.1u m=4
m7 8 7 10 10 cmosp l=0.5u w=4u ad=2.2p pd=9.1u m=17
m8 11 7 10 10 cmosp l=0.5u w=4u ad=2.2p pd=9.1u m=1
m9 11 11 12 0 cmosnlvt l=0.9u w=1.75u ad=0.96p pd=4.6u
+ as=0.96p ps=4.6u m=1
m10 12 12 0 0 cmosnlvt l=0.9u w=1.75u ad=0.96p pd=4.6u m=1
mc 9 11 5 0 cmosnlvt l=0.9u w=1.75u ad=0.96p pd=4.6u
+ as=0.96p ps=4.6u m=3
mr 7 7 10 10 cmosp l=0.5u w=4u ad=2.2p pd=9.1u m=4
iref 7 0 96.5u
cc 8 9 2p
vdd 10 0 1.8
.ends ota
```

As a final word, the iterative step-response simulation needed for determining the number of unit devices of MC is performed with the transient section of `ex4-26.sp`. ▲

4.9 Opamp with a Push-Pull Source-Follower Output Stage

As stated in Section 2.8, any continuous-time amplifier configuration whose load or external network is resistive necessitates an opamp of low output resistance for maintaining a high dc differential gain. Such an opamp can be constructed simply by adding a buffering stage to the output of an OTA. The most basic among such buffering stages is the source follower studied in Subsection 4.4.2 and applied to the conversion of an OTA into an opamp in Example 4.12. However, the basic source follower is very inefficient in terms of power consumption. This deficiency arises from the fixed setting of its bias current above the maximum load current. Although this setting prevents the driver from current starvation, it also proves to be unnecessarily large most of the time because load current rarely rises to its maximum. This efficiency problem is alleviated by buffering with *push-pull* versions of source-follower and common-source stages. Opamps with a push-pull source-follower buffer stage are the subject of this section. Buffering with a push-pull common-source stage is discussed in Section 4.10.

4.9.1 Topologies and Operation

Shown in Fig. 4.68 are examples of buffering with a push-pull source follower in cascoded-symmetrical OTA, folded-cascode OTA, and Miller OTA, whose original unconverted topologies are presented in Figs. 4.48, 4.50, and 4.58, respectively. Only high-gain OTA topologies are considered because buffered opamp applications generally necessitate a large dc differential gain. In all three examples, the push-pull source-follower stage is built with M10 and M11, which are driven by two additional devices, M12 and M13, which are embedded into the original output stage of the OTA. Although these conversion examples involve OTAs with PMOS input drivers, the very same topology of M10, M11, M12, and M13 can also be used for converting OTAs with NMOS input drivers.

Before going into the details of push-pull operation, note that M12 and M13 operate with virtually constant gate-source voltages V_{GS12} and V_{GS13} because the signal component of the current flowing in these devices is much smaller than the bias component I_{B2}. As a consequence, the difference $\Delta V \doteq v_{G10} - v_{G11} = V_{GS12} - V_{GS13}$ between the gate voltages of M10 and M11 remains constant, which implies identical voltage signals at these two gates. These signal components are closely approximated by the voltage signal that would be observed at the output of the unconverted OTA because the small-signal resistances of these diode-connected devices are much smaller than the output resistance of original OTA output stage.

Push-pull operation is now explained with the aid of Fig. 4.68 while referring also to Fig. 4.69, where typical plots of v_{GS10}, v_{GS11}, i_{D10}, i_{D11}, and v_O are presented as functions of v_{G11}.

Let us begin with the bias condition $v_O = V_B$ as indicated in Fig. 4.69(a). It is obvious from Fig. 4.68 that in this condition no voltage is imposed onto the load resistance R_L, which implies $i_L = 0$ hence $i_{D10} = i_{D11}$. This common bias current of M10 and M11 is marked I_Q in Fig. 4.69(b), and the corresponding gate-source voltages are identified as V_{GS10} and V_{GS11} in Fig. 4.69(c).

Now suppose that v_{G10} and v_{G11}, which are separated by the constant ΔV, are pulled upward by a signal component. Since the voltage gain from gate to source is positive but lesser than 1 V/V, v_O rises above V_B as shown in Fig. 4.69(a) but its variation remains smaller than that of v_{G10} and v_{G11}. As a consequence, v_{GS10} becomes more positive than the bias level V_{GS10} while v_{GS11} becomes less negative than the bias level V_{GS11} as indicated in Fig. 4.69(c). This, in turn, leads to $i_{D10} > I_Q$ and $i_{D11} < I_Q$ as observed in Fig. 4.69(b). The difference $i_{D10} - i_{D11}$ flowing into R_L as i_L is what supports the signal component $v_O - V_B = R_L i_L$ observed in the output voltage.

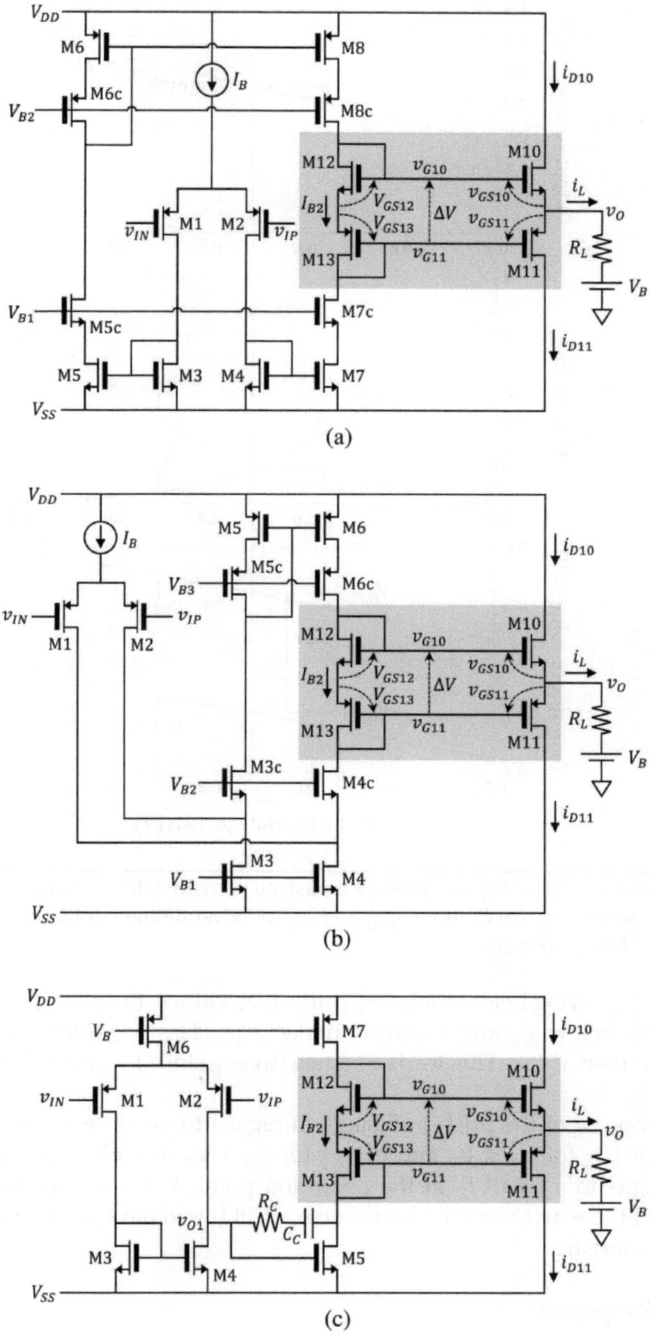

FIGURE 4.68 Converting an OTA into an opamp with a push-pull source-follower buffer stage. The part added to the OTA is highlighted in gray. (a) Cascoded-symmetrical opamp. (b) Folded-cascode opamp. (c) Miller opamp.

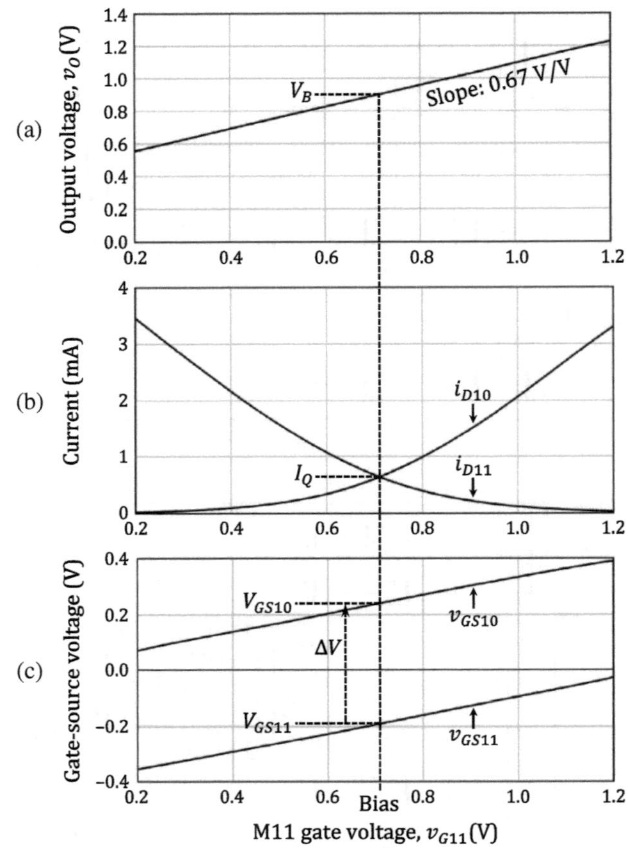

FIGURE 4.69 Plots of some key variables of a push-pull emitter-follower stage as functions of v_{G11}. (a) Output voltage v_O. (b) Currents i_{D10} and i_{D11} of driver devices. (c) Gate-source voltages v_{GS10} and v_{GS11} of driver devices.

If v_{G10} and v_{G11} swing below their respective bias values, the push-pull stage responds by lowering v_O below V_B and therefore makes v_{GS10} less positive than V_{GS10} but v_{GS11} more negative than V_{GS11}. This leads to a negative i_L since i_{D11} rises above I_Q while i_{D10} drops below I_Q.

The key property of the push-pull stage in regard to power efficiency is the gradual suppression of i_{D10} for $v_O < V_B$ and of i_{D11} for $v_O > V_B$ as well as the lowering of both of these currents to a small I_Q at the operation point. What usually prevents us from lowering I_Q all the way to zero is the adverse effect it will have on the linearity of the dc transfer characteristic.

4.9.2 DC Response

Output Range

The lower limit of the output range of an opamp built with a push-pull source-follower buffer stage is given by

$$V_{OL} = V_{OL(\text{ota})} - V_{GS11L}, \tag{4.265}$$

where $V_{OL(\text{ota})}$ is the lower limit of the output range in the unconverted OTA, and V_{GS11L} is the gate-source voltage of M11 for $v_O = V_{OL}$. For an OTA core with PMOS input drivers, $V_{OL(\text{ota})}$ is described by (4.173) or (4.189) or (4.221) depending on the type of the OTA core.

The upper limit of the output range is described by

$$V_{OH} = V_{OH(\text{ota})} - V_{GS10U}, \tag{4.266}$$

where $V_{OH(\text{ota})}$ is the upper limit of the output range in the unconverted OTA, and V_{GS10U} is the gate-source voltage of M10 for $v_O = V_{OH}$. For an OTA core with PMOS drivers, $V_{OH(\text{ota})}$ is described by (4.172) or (4.190) or (4.220) depending on the type of the OTA core.

Equations (4.265) and (4.266) are also valid for an opamp having an OTA core of NMOS input drivers provided that $V_{OL(\text{ota})}$ and $V_{OH(\text{ota})}$ are represented by appropriate expressions.

Since V_{GS11L} is the most negative value of v_{GS11}, the lower limit V_{OL} indicated by (4.265) will be considerably above $V_{OL(\text{ota})}$ unless M11 is implemented with a low-threshold PMOS device operating with a source-bulk short. The upper limit V_{OH} is also highly restrictive unless the most positive value V_{GS10U} of v_{GS10} is minimized by using a low-threshold DNW NMOS device for M10 and short-circuiting its source and bulk. Even if such device options are exercised, output range will still be narrower than the output range of the unconverted OTA. Opamps with a push-pull source-follower output stage are definitely unsuitable for integration in low-voltage technologies.

DC Differential Gain, Output Resistance, and Effective Transconductance

The dc differential gain of any of the opamps shown in Fig. 4.68 is described by

$$A_0 = A_{0(\text{ota})} A_{0(\text{buf})}, \tag{4.267}$$

where $A_{0(\text{ota})}$ and $A_{0(\text{buf})}$ represent the dc gains of the OTA core and the buffer, respectively. These two components are defined by

$$A_{0(\text{ota})} \doteq \frac{v_{g10}}{v_{id}} = \frac{v_{g11}}{v_{id}}, \tag{4.268}$$

and

$$A_{0(\text{buf})} \doteq \frac{v_o}{v_{g10}} = \frac{v_o}{v_{g11}}, \tag{4.269}$$

where voltage symbols represent dc small-signal components. $A_{0(\text{ota})}$ is described by (4.170) for cascoded-symmetrical OTA, (4.180) for folded-cascode OTA, and (4.224) for

Miller OTA. $A_{0(\mathrm{buf})}$ is the sum of the gate-to-source voltage gains of M10 and M11. Assuming source-bulk shorts for both devices, it is given by

$$A_{0(\mathrm{buf})} = \frac{R_L\left(g_{m10} + g_{m11}\right)}{1 + R_L\left(g_{m10} + g_{m11}\right)}. \tag{4.270}$$

Substituting the right-hand side of (4.270) for $A_{0(\mathrm{buf})}$ in (4.267), and comparing the outcome with (2.96) lead us to the following descriptions of the output resistance r_o and effective transconductance G_m of the opamp:

$$r_o = \frac{1}{g_{m10} + g_{m11}}, \tag{4.271}$$

$$G_m = A_{0(\mathrm{ota})}\left(g_{m10} + g_{m11}\right). \tag{4.272}$$

4.9.3 Frequency Response, Step Response, and Noise

The overall transfer function of an opamp is described generally by

$$A = A_{\mathrm{ota}} A_{\mathrm{buf}}, \tag{4.273}$$

where A_{ota} and A_{buf} denote the transfer functions defined by the voltage gains of the OTA core and the buffer stage, respectively.

For a push-pull source-follower output stage, A_{buf} is characterized by a transfer function similar to (4.71), that is

$$A_{\mathrm{buf}} = A_{0(\mathrm{buf})} \frac{1 + \frac{s}{2\pi f_Z}}{1 + \frac{s}{2\pi f_P}}, \tag{4.274}$$

where f_Z and f_P are determined collectively by M10 and M11 as follows:

$$f_Z = \frac{g_{m10} + g_{m11}}{2\pi\left(C_{gse10} + C_{gse11} + C_{ggi10} + C_{ggi11} + C_{dgi10} + C_{dgi11}\right)}, \tag{4.275}$$

$$f_P = \frac{g_{m10} + g_{m11} + (1/R_L)}{2\pi\left(C_{gse10} + C_{gse11} + C_{ggi10} + C_{ggi11} + C_{dgi10} + C_{dgi11} + C_{dbe10} + C_{dbe11} + C_L\right)}. \tag{4.276}$$

Equations (4.275) and (4.276) are subject to the assumption that both devices operate with source-bulk shorts and C_{gbe}, C_{gdi}, and C_{ddi} are negligible. C_L appearing in (4.276) represents the total capacitance loading the output port of the opamp. In most opamp applications, f_Z and f_P are too high to have a significant effect on the frequency response. Equations (4.275) and (4.276) will help the reader decide whether to ignore them or not in a given application.

A_{ota} is given by (4.143) in a cascoded-symmetrical opamp, (4.191a) to (4.191c) in a folded-cascode opamp, and (4.225) to (4.229) in a Miller opamp. C_L appearing in some of these OTA transfer functions denotes the load capacitance of the OTA core. From now on, we will represent it with $C_{L(\text{ota})}$ in order not to confuse it with the symbol C_L used in (4.276) for the load capacitance of the buffer.

Referring to Fig. 4.68, we identify the following three components for $C_{L(\text{ota})}$:

1. Capacitances seen into the drain terminals of M7c and M8c in cascoded-symmetrical opamp, M4c and M6c in folded-cascode opamp, and M5 and M7 in Miller opamp.
2. n-well/substrate capacitance in the PMOS M13 and p-well/DNW capacitance in the NMOS M12.
3. Capacitances seen into the gate terminals of M10 and M11.

The last of these three components is described generally by (4.76) for each of M10 and M11. Assuming source-bulk short and negligible C_{gbe}, C_{gdi}, C_{gbi}, and C_{ddi} in these devices, we can express its total with

$$C_{i10} + C_{i11} = \left(C_{gse10} + C_{gse11} + C_{ggi10} + C_{ggi11}\right)\left(1 - A_{0(\text{buf})}\right) + C_{gde10} + C_{gde11} \tag{4.277}$$

for frequencies lower than the pole frequency f_P of the buffer.

The open-loop bandwidth f_o is described by (4.174) in the case of a cascoded-symmetrical core and by (4.192) in the case of a folded-cascode core. However, C_L should be interpreted as $C_{L(\text{ota})}$ in these equations. For a Miller core, (4.232) is valid but A_0 should be interpreted as $A_{0(\text{ota})}$.

The gain-bandwidth product, as determined from (4.267) and the relevant f_o expression, is given by

$$f_{GBW} = \frac{g_{m(1,2)}A_{0(\text{buf})}}{2\pi C_{L(\text{ota})}} \tag{4.278}$$

for the cascoded-symmetrical and folded-cascode opamps. For the Miller opamp, it is described by

$$f_{GBW} = \frac{g_{m(1,2)}A_{0(\text{buf})}}{2\pi C_C}. \tag{4.279}$$

There is no difference between an opamp and its core OTA as regard to differential-input range. For an analytical description of V_{DIR} in a cascoded-symmetrical opamp or a folded-cascode opamp, the reader is referred to Subsection 4.4.3. For a Miller opamp, the contents of Subsection 4.8.2 are relevant. The step-response analysis presented in Subsection 4.8.3 is also valid for a Miller opamp.

The closed-loop output noise contributed by an opamp with a push-pull source follower is primarily due to the OTA core because the magnitude $|A_k(f)|$ of the noise transfer function of a buffer-stage device is a very small fraction of the magnitude of open-loop differential gain. The noise contribution of M12 and M13 is also small because the noise current of each of these two devices mainly loops inside the device itself due to the very high resistance of the external path between the drain terminals of these devices and ground. The reader is therefore referred to the appropriate OTA subsection for an analytical description of $V_{no(\mathrm{opamp})}$.

Example 4.27 The transresistance amplifier shown in Fig. 4.70(a) is to be implemented with a single-supply (V_{DD} = 1.8 V) Miller opamp built with PMOS drivers in its input stage and compensated by pole splitting. It is supposed to convert a sinusoidal current signal i_i of amplitude I_i = 200 μA into an output-voltage signal v_o of amplitude V_o = 300 mV. Bias voltage and load resistance are specified as V_B = 0.9 V and R_L = 300 Ω. Full-power bandwidth is specified as 20 MHz. Design should target a minimum of 67° for phase margin, and emphasize linearity while keeping the total power-supply current below 150 μA at the operation point. The dc differential gain defined at the operation point is to be larger than 2000 V/V. Assuming that the design tools presented in Figs. 4.5 to 4.9 are also valid for low-threshold DNW NMOS devices, build the opamp with a push-pull source-follower output stage in compliance with the schematic shown in Fig. 4.70(b).

During verification also evaluate the *total harmonic distortion plus noise* (THD+N) of the transresistance amplifier as a function of frequency for peak input-signal amplitude. This is a popular metric used for assessing the dynamic performance of amplifiers. It is defined by

$$\mathrm{THD{+}N} \doteq \sqrt{\mathrm{THD}^2 + \frac{V_{no}^2}{V_{o1}^2}}, \tag{4.280}$$

where THD is the total harmonic distortion of the amplifier as defined by (2.66), V_{no}^2 represents mean-square output noise, and V_{o1}^2 is the mean-square value of the fundamental component of the output signal.

Work As usual, we begin by converting closed-loop specifications into open-loop and external-network design constraints.

- *Feedback Resistor* Resistance is specified as $R = V_o/I_i = 300\,\mathrm{mV}/200\,\mu\mathrm{A} = 1500\,\Omega$.
- *External-Network Parameters* $\alpha_i = -R = -1500\,\Omega$ and $\beta = 1$.
- *Common-Mode Range* Common-mode voltage is fixed at 0.9 V. Any range including 0.9 V is therefore acceptable.
- *Output Range* Since we use a source-follower buffer it is reasonable not to target an output range wider than the specified output swing. Therefore $V_{OL} = V_B - V_o = 0.9 - 0.3 = 0.6$ V and $V_{OH} = V_B + V_o = 0.9 + 0.3 = 1.2$ V.
- *Full-Power Bandwidth* f_{fp} is related to closed-loop small-signal bandwidth f_{cl} through (2.59a) or (2.59b) depending on whether a slew-free or slew-limited

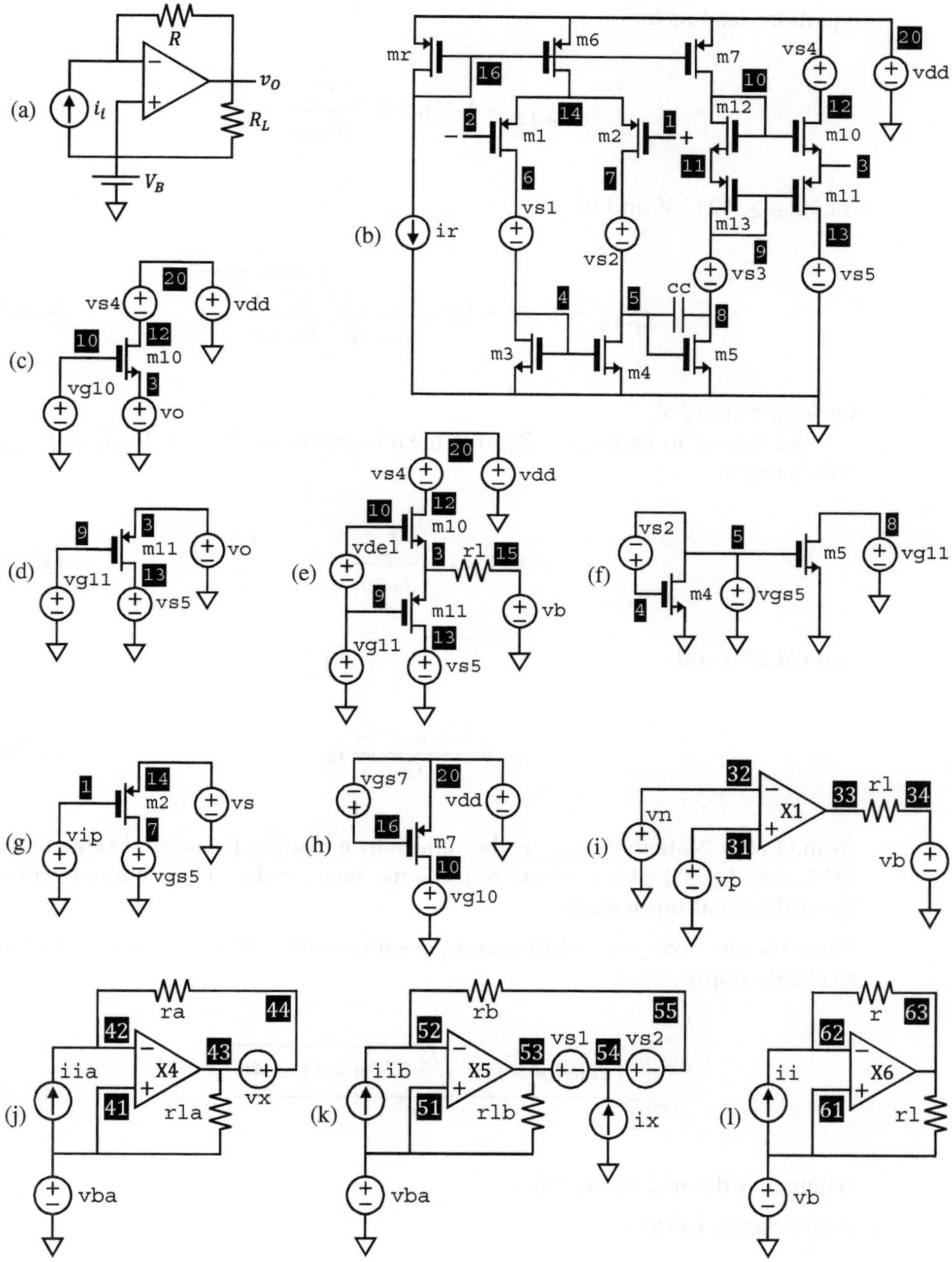

FIGURE 4.70 Top-level and simulation schematics of Example 4.27. (a) Top level. (b) Opamp. (c) ex4-27a.sp. (d) ex4-27b.sp. (e) ex4-27c.sp. (f) ex4-27e.sp. (g) ex4-27f.sp. (h) ex4-27g.sp. (i) Open-loop dc verification in ex4-27h.sp. (j) and (k) Loop-gain verification in ex4-27h.sp. (l) Transient and Fourier verification in ex4-27h.sp.

response is involved. Considering (2.48) for f_{cl} and (4.279) for f_{GBW}, these two equations lead us to

$$g_{m(1,2)} = 1.256 \times 10^8 \frac{C_C}{A_{0(\text{buf})}} \tag{4.281}$$

for $V_{DIR} \geq 0.212$ V, and to

$$g_{m(1,2)} = 1.256 \times 10^8 \frac{C_C}{A_{0(\text{buf})}} \sqrt{\left(\frac{0.3}{V_{DIR}}\right)^2 - 1} \tag{4.282}$$

for $V_{DIR} \leq 0.212$ V.

As argued in Example 4.25, it is meaningful to set $V_{DIR} = V_{DIR1} = V_{DIR2}$, which implies

$$V_{DIR} = \frac{I_{B1}}{g_{m(1,2)}} \tag{4.283}$$

from (4.245) and

$$I_{B2} = \frac{C_{L(\text{ota})} + C_C}{C_C} I_{B1} \tag{4.284}$$

from (4.260). Note that $C_{L(\text{ota})}$ is the capacitance loading the second stage of the OTA core, I_{B2} is the bias current of the same stage, and I_{B1} is the bias current of the differential-input stage.

- *Phase Margin* For pole-splitting compensation with a 67° phase margin, (4.236) to (4.238) imply

$$g_{m5} = \frac{2.36\,(m+1) + \sqrt{5.57\,(m+1)^2 + 4m}}{2m} g_{m(1,2)}, \tag{4.285}$$

where m is defined by (4.258).

- *Power-Supply Current*

$$I_{B1} + I_{B2} + I_Q \leq 150\ \mu\text{A}, \tag{4.286}$$

where I_Q is the bias-current component contributed by the buffer stage. Since linearity is emphasized, this component should be made as large as possible.

- *DC Differential Gain* For the specified 2000-V/V dc differential gain, (4.267) implies

$$A_{0(\text{ota})}A_{0(\text{buf})} \geq 2000 \text{ V/V}, \tag{4.287}$$

where $A_{0(\text{ota})}$ is described by (4.262).

Design commences with the sizing and biasing of buffer devices M10 and M11, and moves to the OTA core after the capacitive loading and bias conditions imposed by these two devices have been determined. The OTA core is designed largely along the lines of Example 4.25.

1. L_{DR10} *and* W_{DR10} M10 is a low-threshold DNW device. At the upper limit of the output range, the voltage of its interconnected source and bulk terminals rises to $V_{OH} = 1.2$ V while the gate terminal receives $V_{OH(\text{ota})}$. Assuming $i_{D11} = 0$ at this limit, the current flowing in M10 is maximized to $i_{D10(\text{max})} = I_i + (V_o/R_L) = 1.2$ mA. It is meaningful to bring $V_{OH(\text{ota})}$ as close to V_{DD} as possible so that the maximized current is conducted with a large overdrive voltage. This will lead to a small footprint and a small capacitive contribution to $C_{L(\text{ota})}$. A small $C_{L(\text{ota})}$, in turn, will translate into a small second-stage bias current I_{B2} as implied by (4.284). We assign −0.2 V to M7 as a worst-case drain-source voltage, and thus set $V_{OH(\text{ota})} = 1.6$ V. For M10, we also select the shortest permissible channel length $L_{DR10} = 0.2$ μm because the output resistance is not a critical metric in a source follower.

   Now that the channel length as well as the maximum current and the corresponding terminal voltages have been fixed for M10, we can determine the channel width of this device with the aid of Figs. 4.5 and 4.7. Alternatively, we can run an iterative operation-point simulation for this purpose. The width is iterated until the target value of the peak current is reached. The schematic used in this simulation is shown in Fig. 4.70(c). Its input file `ex4-27a.sp` is copied below.

```
filename: ex4-27a.sp
.incl bsimlvt.sp
m10 12 10 3 3 cmosnlvt w=36u l=0.2u
vg10 10 0 1.6
vo 3 0 1.2
vs4 20 12 0
vdd 20 0 1.8
.control
op
print i(vs4)
.endc
.end
```

   The outcome of this iterative simulation is $W_{DR10} = 36$ μm.

2. L_{DR11} *and* W_{DR11} The procedure applied to M10 in step 1 is repeated for M11, which is a low-threshold PMOS device also operating with a source-bulk short. We consider the lower limit $V_{OL} = 0.6$ V of the output range as the common voltage of source and bulk terminals, and 1.2 mA as the maximized drain current. Setting $L_{DR11} = 0.2$ μm as the channel length, and $V_{OL(\text{ota})} = 0.2$ V as the

gate voltage, we determine $W_{DR11} = 62\ \mu\text{m}$ with an iterative simulation on the schematic presented in Fig. 4.70(d). Input file `ex4-27b.sp` used in this simulation is as follows:

```
filename: ex4-27b.sp
.incl bsimlvt.sp
m11 13 9 3 3 cmosplvt w=62u l=0.2u
vg11 9 0 0.2
vo 3 0 0.6
vs5 13 0 0
.control
op
print i(vs5)
.endc
.end
```

3. $\Delta V, A_{0(buf)}, C_{i10} + C_{i11}, V_{G10},$ *and* V_{G11} Consider the operation point, at which, both M10 and M11 conduct the same drain current I_Q. Hoping that one-third of the maximum total power-supply current specified by (4.286) suffices for $I_{B1} + I_{B2}$, we allocate 100 μA to I_Q, and run an iterative dc simulation on M10 and M11 for determining the necessary gate-to-gate bias voltage ΔV for generating this much bias current. The schematic used in this simulation is presented in Fig. 4.70(e). It is netlisted in the following input file `ex4-27c.sp`.

```
filename: ex4-27c.sp
.incl bsimlvt.sp
m10 12 10 3 3 cmosnlvt w=36u l=0.2u
m11 13 9 3 3 cmosplvt w=62u l=0.2u
rl 3 15 300
vb 15 0 0.9
vg11 9 0 dc
vdd 20 0 1.8
vdel 10 9 0.37
vs4 20 12 0
vs5 13 0 0
.save i(vs4) i(vs5) @m10[cgg] @m11[cgg] v(9) v(10) v(3)
.control
dc vg11 0.2 1.3 1m
plot i(vs4) i(vs5) vs v(3)
.endc
.end
```

Notice in Fig. 4.70(e) the setting of open-loop loading conditions by placing the load resistance `rl` to the output of the buffer, which is necessary for accurately extracting the open-loop gain $A_{0(\mathrm{buf})}$ of this stage. Feedback resistance R is not included because disabling the input current source in accordance with the procedure explained in Section 2.8 leaves the left-hand terminal of the feedback resistor floating. Also notice the placement of a dc voltage source `vdel` between the gates of M10 and M11 to represent ΔV. Another voltage source, named `vg11`, defines the gate voltage of M11. In each iteration `vdel` is kept constant but `vg11` is swept, and the currents of M10 and M11 are plotted against the output voltage v_O. The common value of the two device currents indicated for $v_O = 0.9$ V is of course I_Q. If I_Q turns out to be higher than 100 μA, we lower ΔV, and rerun the simulation. If I_Q happens to be lower than the target, then, we raise ΔV, and rerun. This procedure yields $\Delta V = 0.37$ V for $I_Q = 95$ μA to be precise. After convergence, we also determine $A_{0(\text{buf})} = 0.516$ V/V from this simulation by plotting the derivative of v_O against v_O itself, and reading the outcome for $v_O = 0.9$ V. We do this with the command

```
plot deriv(v(3)) vs v(3)
```

By default, the derivative is taken with respect to the sweeping variable v_{G11}, which happens to be the input voltage of the buffer stage.

The iterative simulation run with ex4-27c.sp has two additional utilities. One is the characterization of the contribution of the buffer stage to $C_{L(\text{ota})}$, and the other is the evaluation of the bias values V_{G10} and V_{G11} of gate voltages, which will be needed later for determining the channel lengths of M5 and M7. C_{ggi10} and C_{ggi11} are plotted with

```
plot @m10[cgg] @m11[cgg] vs v(3)
```

and characterized for $v_O = 0.9$ V as 32 fF and 52.5 fF, respectively. Other parasitic capacitances appearing in (4.277) are calculated from (3.92) and (3.93) as $C_{gse10} = C_{gde10} = 31$ fF and $C_{gse11} = C_{gde11} = 48.5$ fF. According to (4.277), the capacitance contributed by M10 and M11 to $C_{L(\text{ota})}$ is therefore 167 fF.

As to the gate voltages of M10 and M11 at the operation point, we plot them against output voltage with

```
plot v(10) v(9) vs v(3)
```

and read $V_{G10} = 1.107$ V and $V_{G11} = 0.737$ V for $v_O = 0.9$ V.

4. I_{B1}, $(W/L)_{1,2}$, I_{B2}, $(W/L)_5$, *and* C_C The first three of these variables and transconductance efficiencies g_{m5}/I_{B2} and $g_{m(1,2)}/(I_{B1}/2)$ are characterized as functions of $V_{GS(1,2)} - V_{TP}$ with a dc simulation, in which, C_C is treated as a parameter. The input file ex4-27d.sp used in this simulation is presented below.

```
filename: ex4-27d.sp
.incl bsimlvt.sp
m1 2 1 0 0 cmosplvt w=6u l=0.8u
vds1 2 0 -0.9
vgs1 1 0 dc
vcl 3 0 250f          ;cl(ota)
vcl 4 0 50f           ;cl
vcc 5 0 100f          ;cc
vpm 6 0 2.36          ;tan(pm)
.save @m1[vth] @m1[id] @m1[gm] v(1) v(3) v(4) v(5) v(6)
.control
destroy all
foreach comp 100f 200f 300f 400f 500f
alter vcc=$comp
dc vgs1 -0.45 -0.15 0.1m
******************calculations for M1,M2******************
let vdir=2*@m1[id]/@m1[gm]                              ; (4.283)
```

```
let gm12='(vdir>=0.213)*2.43e8*V(5)+(vdir<0.213)*
+ sqrt((0.3/vdir)^2-1)*2.43e8*v(5)'          ; (4.281),(4.282)
let ib1=vdir*gm12                                      ; (4.283)
let asp12=(ib1/2)/(@m1[id]/8.08)         ; W/L of M1,M2 (3.25)
*******************calculations for M5*******************
let m=v(5)^2/(v(4)*v(3)+v(4)*v(5)+v(3)*v(5))          ; (4.258)
let gm5=((v(6)*(m+1)+sqrt(v(6)^2*(m+1)^2+4*m))
+  /(2*m))*gm12                                        ; (4.285)
let ib2=(v(3)+v(5))*ib1/v(5)                           ; (4.284)
let gmeff5=gm5/ib2                                   ; gm/ID of M5
end
**************************plots**************************
plot dc1.ib1 dc2.ib1 dc3.ib1 dc4.ib1
+  dc5.ib1 vs v(1)+@m1[vth]
plot dc1.asp12 dc2.asp12 dc3.asp12 dc4.asp12
+  dc5.asp12 vs v(1)+@m1[vth]
plot dc1.ib2 dc2.ib2 dc3.ib2 dc4.ib2 dc5.ib2
+  vs v(1)+@m1[vth]
plot dc1.gmeff5 dc2.gmeff5 dc3.gmeff5 dc4.gmeff5
+  dc5.gmeff5 vs v(1)+@m1[vth]
.endc
*********************************************************
.end
```

This file is a modified version of `ex4-25a.sp` used in Example 4.25. `m1` is included for extracting the transconductance-efficiency tool needed for designing M1 and M2. As usual in design-tool extraction, device dimensions are set arbitrarily but reasonably. The drain-source voltage setting is also arbitrary as long as saturation is guaranteed. Notice the selection of low-threshold option for M1 and M2, which provides a large headroom to current-source M6. Also notice the dc sweeping of the gate-source voltage in 0.1-mV steps within the range between −0.45 and −0.15 V. $C_{L(\text{ota})}$, C_1, C_C, and tan(PM) of phase margin are represented by voltage sources `vcl`, `vc1`, `vcc`, and `vpm`, respectively. This is how we can represent parameters inside a control section.

$C_{L(\text{ota})}$ = 250 fF is an estimate for the capacitance loading the OTA. It is determined by combining (a) 167 fF for the contribution of M10 and M11, and (b) half as much for the contribution of M5, M7, M12, and M13. C_1 is estimated to be 50 fF. The reader may prefer experimenting with different settings for these parasitic capacitance parameters. As to tan(PM), we set it to 2.35 as specified. Simulation is run for five different values of C_C between 100 and 500 fF. It automatically generates four plots, namely, I_{B1}, $(W/L)_{1,2}$, I_{B2}, and g_{m5}/I_{B2}, which are displayed in Figs. 4.71(a) to (d).

According to the plots shown in Figs. 4.71(a) and (c), I_{B1} and I_{B2} are minimized for a given C_C if $V_{GS(1,2)} - V_{TP}$ is set to the borderline between slew-free and slew-limited operations at −0.199 V. Also note that the sum $I_{B1} + I_{B2}$ has an upper limit 55 μA imposed by (4.286) for I_Q = 95 μA. We move the operation point slightly into the slew-free zone by adopting $V_{GS(1,2)} - V_{TP} = -0.22$ V, for which, the condition $I_{B1} + I_{B2} < 55$ μA is satisfied by any C_C less than approximately 300 fF. We decide to set C_C to this upper limit because Fig. 4.71(d)

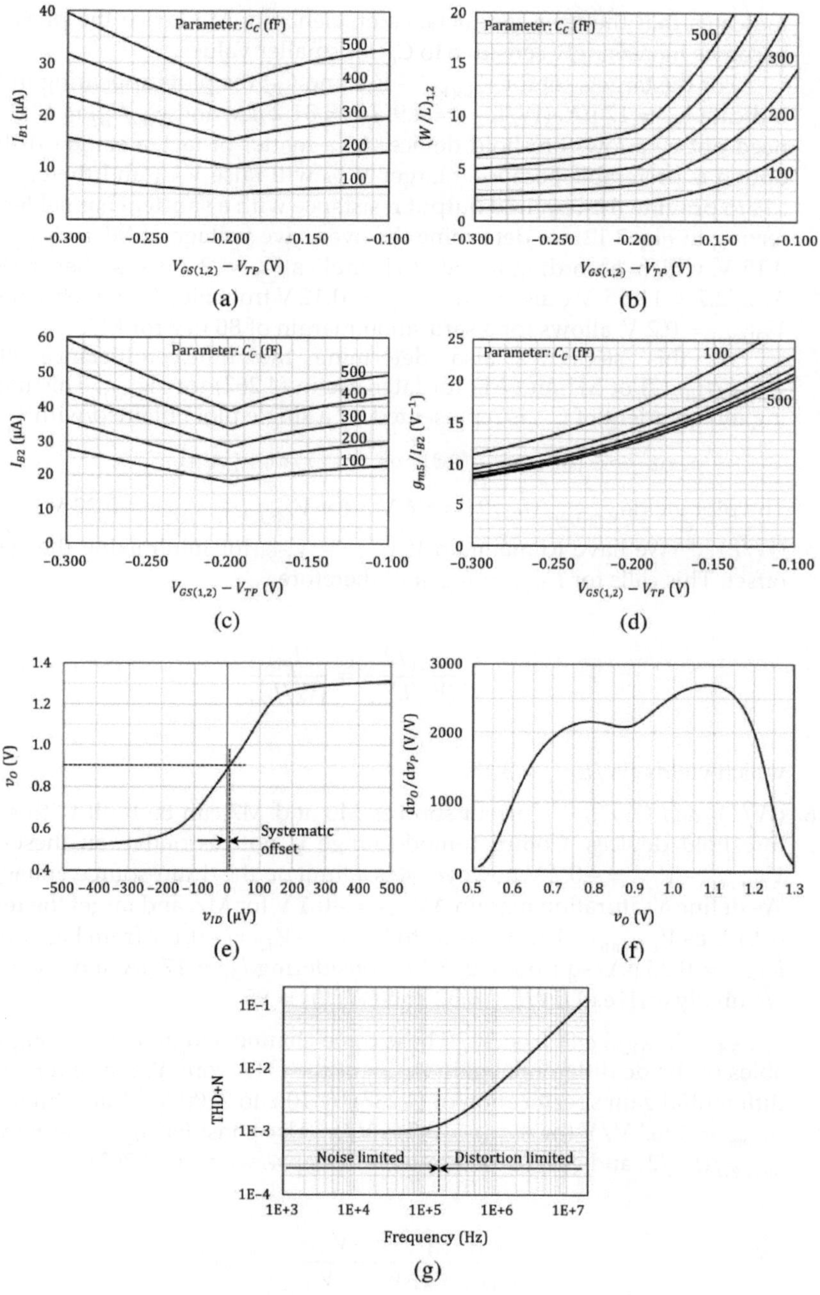

FIGURE 4.71 Simulated plots of Example 4.27. (a) First-stage bias current I_{B1} as obtained with `ex4-27d.sp`. (b) Aspect ratio $(W/L)_{1,2}$ of first-stage drivers as obtained with `ex4-27d.sp`. (c) Second-stage bias current I_{B2} as obtained with `ex4-27d.sp`. (d) Transconductance efficiency g_{m5}/I_{B2} of second-stage driver as obtained with `ex4-27d.sp`. (e) DC transfer characteristic obtained with `ex4-27h.sp`. (f) Derivative of the dc transfer characteristic obtained with `ex4-27h.sp`. (g) THD+N versus frequency as obtained with `ex4-27h.sp`.

indicates that the transconductance efficiency of M5 hence the phase margin becomes increasingly sensitive to C_C for smaller values.

For the targets set for $V_{GS(1,2)} - V_{TP}$ and C_C, the plots appearing in Fig. 4.71 indicate $I_{B1} = 17\ \mu A$, $(W/L)_{1,2} = 4.9$, $I_{B2} = 31.2\ \mu A$, and $g_{m5}/I_{B2} = 12.3\ V^{-1}$. M5 need not be a low-threshold device. As a matter of fact, a standard-threshold option is preferable because a larger V_{GS5} will raise V_{DS4} and therefore enable M4 to provide the required output resistance with a shorter channel length. Referring to Fig. 3.12, we determine the overdrive voltage of M5 as $V_{GS5} - V_{TN} = 0.15$ V, which, according to Fig. 3.11 implies $I_{\square 5} = 2.7\ \mu A/sq$ hence $(W/L)_5 = 31.2/2.7 = 11.55$. We also read $V_{DSsat5} = 0.12$ V from Fig. 3.8, which, considering $V_{OL(ota)} = 0.2$ V, allows for a saturation margin of 80 mV for M5.

At this step, we also determine the transconductance efficiency $g_{m(1,2)}/(I_{B1}/2)$ of M1 and M2 for later use in (4.262) for $A_{0(ota)}$. This metric, being independent of C_C, is represented by a single plot obtained with

```
plot @m1[gm]/@m1[id] vs v(1)+@m1[vth]
```

The plot yields $g_{m(1,2)}/(I_{B1}/2) = 8.6\ V^{-1}$ for $V_{GS(1,2)} - V_{TP} = -0.22$ V.

5. $(W/L)_{3,4}$ We have to maintain $V_{GS(3,4)} = V_{GS5}$ for minimizing the systematic offset. This calls for $I_{\square(3,4)} = I_{\square 5}$, and therefore

$$\frac{I_{B1}/2}{(W/L)_{3,4}} = \frac{I_{B2}}{(W/L)_5},$$

which yields $(W/L)_{3,4} = 3.15$.

6. $(W/L)_6$ *and* $(W/L)_7$ Current sources M6 and M7 can be built with standard-threshold devices. Common-mode range is not restrictive on these devices. $V_{OH(ota)} - V_{DD} = -0.2$ V however sets a limit on the drain-source voltage of M7. We define a saturation margin $\Delta V_{DS7} = -0.1$ V for M7, and target the remaining -0.1 V as $V_{DSsat(6,7)}$. This leads us to $V_{GS(6,7)} - V_{TP} = -0.1$ V from Fig. 3.16, and to $I_{\square(6,7)} = 0.33\ \mu A/sq$ from Fig. 3.19. Considering $I_{B1} = 17\ \mu A$ and $I_{B2} = 31.2\ \mu A$, we finally arrive at $(W/L)_6 = 52$ and $(W/L)_7 = 95$.

7. $L_{DR(3,4,5)}$, $L_{DR(1,2)}$, *and* $L_{DR(6,7)}$ These three channel lengths are among the variables of the dc differential gain $A_{0(ota)}$ of the OTA core. We raise the opamp dc differential gain specification of (4.287) by 10% to 2200 V/V, and thus calculate $A_{0(ota)} = 4263$ V/V for $A_{0(buf)} = 0.516$ V/V. Also considering the target values of $g_{m(1,2)}/(I_{B1}/2)$ and g_{m5}/I_{B2} determined in step 4, we turn (4.262) into

$$\frac{V_{A2}V_{A4}}{V_{A2}+V_{A4}}\frac{V_{A5}V_{A7}}{V_{A5}+V_{A7}} = 40.3\ V^2.$$

We target $V_{A2} = V_{A4}$ and $V_{A7} = V_{A5}$ for minimizing the sensitivity of A_0 to v_O, and thus turn this equation further into

$$V_{A4}V_{A5} = 162\ V^2. \tag{4.288}$$

Both V_{A4} and V_{A5} depend on the same $L_{DR(3,4,5)}$. We determine this common channel length with an iterative simulation targeting the constraint defined by (4.288). The very same iterative simulation also yields V_{A4} and V_{A5} individually. We set these two as targets to be satisfied for V_{A2} and V_{A7}, respectively, and perform two additional iterative simulations for determining $L_{DR(1,2)}$ and $L_{DR(6,7)}$ from these target values.

The schematic used for determining $L_{DR(3,4,5)}$ is shown in Fig. 4.70(f). Iterative simulation is performed with the following input file `ex4-27e.sp`:

```
filename: ex4-27e.sp
.incl bsim.sp
.param length=0.9u vgs5=0.580
m4 5 4 0 0 cmosn w=10u l=length
m5 8 5 0 0 cmosn w=10u l=length
vs2 4 5 0
vg11 8 0 0.737
vgs5 5 0 vgs5
.control
op
print @m4[id]*@m5[id]/(@m5[gds]*@m4[gds])
+  @m4[id]/@m4[gds] @m5[id]/@m5[gds]
.endc
.end
```

As usual, channel width is assigned arbitrarily for M4 and M5. The parameter `length` defines $L_{DR(3,4,5)}$. The virtual diode connection of M4 is established by the zero-voltage source `vs2`. Voltage source `vg11` sets the drain-source voltage of M5 to 0.737 V as determined in step 3. Another voltage source, `vgs5`, defines the voltage of node 5 but it is retained as a simulation parameter. In each iteration, after estimating the channel length, we determine V_{TN} from Fig. 3.7, and calculate the corresponding V_{GS5} from $V_{GS5} - V_{TN} = 0.15$ V as derived in step 4. This is how we update `vgs5`. Also notice the control section added for directly printing $V_{A4}V_{A5}$, V_{A4}, and V_{A5}. This iterative simulation yields $L_{DR(3,4,5)} = 0.9$ µm, $V_{A4}V_{A5} = 161$ V^2, $V_{A4} = 11.9$ V, and $V_{A5} = 13.6$ V. $V_{GS5} = 0.58$ V is also an outcome of this simulation. It is needed for determining $L_{DR(1,2)}$.

$L_{DR(1,2)}$ is extracted with the schematic shown in Fig. 4.70(g). The input file `ex4-27f.sp` used for this purpose is copied below.

```
filename: ex4-27f.sp
.incl bsimlvt.sp
.param length=0.6u vs=1.206
m2 7 1 14 14 cmosplvt w=10u l=length
vs 14 0 vs
vgs5 7 0 0.58
vip 1 0 0.9
.control
op
print @m2[id]/@m2[gds]
.endc
.end
```

The netlist is arranged for properly biasing M2. The fixed 0.9-V gate voltage of this device is supplied by the voltage source `vip`. Drain terminal is biased by `vgs5` to 0.58 V as determined in the preceding simulation. Channel length and source voltage are defined as simulation parameters. In each iteration, source voltage is calculated from $V_S = 1.119 - V_{TP}$, which is implied by the overdrive voltage $V_{GS(1,2)} - V_{TP} = -0.22$ V determined in step 4. The value of V_{TP} needed in this calculation is obtained from Fig. 4.10 for the selected channel length and zero source-bulk voltage. The target $V_{A2} = V_{A4} = 11.9$ V of this iterative simulation is reached for $L_{1,2} = 0.6$ μm.

The schematic used for extracting $L_{DR(6,7)}$ is shown in Fig. 4.70(h). It is netlisted by the following input file `ex4-27g.sp`.

```
filename: ex4-27g.sp
.incl bsim.sp
.param length=1.4u vgs7=-0.505
m7 10 16 20 20 cmosp w=10u l=length
vgs7 16 20 vgs7
vg10 10 0 1.107
vdd 20 0 1.8
.control
op
print @m7[id]/@m7[gds]
.endc
.end
```

This iterative simulation is run on M7, whose drain voltage is kept constant at 1.107 V by `vg10` as determined in step 3. The gate-source voltage of this device is treated as a simulation parameter `vgs7` whose value is calculated in each iteration from $V_{GS7} - V_{TP} = -0.1$ V. We determine this overdrive voltage from Fig. 3.17 after deciding on $V_{DSsat7} = -0.1$ V in view of the –0.2-V drain-source voltage allocated to M7 in step 1. The value of V_{TP} needed for determining V_{GS7} is obtained from Fig. 3.16 for the selected channel length and the zero source-bulk voltage. We target $V_{A7} = V_{A5} = 13.6$ V for the Early voltage, and arrive at $L_{DR(6,7)} = 1.4$ μm for the channel length.

8. $W_{DR(1,2)}$, $W_{DRu(3,4,5)}$, $W_{DRu(6,7)}$
   a. $(W/L)_{1,2} = 4.9$ and $L_{DR(1,2)} = 0.6$ μm imply $W_{DR(1,2)} = 2.7$ μm.
   b. $(W/L)_{3,4} = 3.15$, $(W/L)_5 = 11.55$, and $L_{DR(3,4,5)} = 0.9$ μm can be implemented with unit devices of width $W_{DRu(3,4,5)} = 0.9$ μm. Each of M3 and M4 is built with three unit devices. Eleven unit devices are paralleled for M5.
   c. $(W/L)_6 = 52$, $(W/L)_7 = 95$, and $L_{DR(6,7)} = 1.4$ μm can be implemented with unit devices of width $W_{DRu(6,7)} = 12$ μm. M6 is implemented with six unit devices. M7 needs 11.

9. L_{DR12}, W_{DR12}, L_{DR13}, W_{DR13} M12 and M13 are supposed to establish a gate-to-gate bias voltage $\Delta V = 0.37$ V for M10 and M11 while conducting $I_{B2} = 21.7$ μA. Considering the bias current $I_Q = 95$ μA of M10 and M11, we can fulfill this requirement by setting $L_{DR(12,13)} = L_{DR(10,11)} = 0.2$ μm and scaling the widths of M12 and M13 by a factor of 31.2/95 = 0.328 with respect to the widths of M10 and M11. However, the significant difference between the drain-source voltages of these minimum-length devices severely invalidates the assumption

that the scaling factor between their currents equals the scaling factor between their channel widths. It is therefore meaningful to begin with the scaling factor 0.325 as an initial estimate, and trim W_{DR12} and W_{DR13} with an iterative simulation until the bias current of the buffer reaches the target $I_Q = 95\ \mu A$. Iteration can be performed on the final schematic as a part of open-loop dc verification.

10. *Verification* Performed with the input file `ex4-27h.sp` presented below.

```
filename: ex4-27h.sp
.incl bsimlvt.sp
.incl bsim.sp
************op-amp subcircuit  fig.4.70(b)*************
.subckt opamp1 1 2 3
.incl bsimlvt.sp
.incl bsim.sp
m1 6 2 14 14 cmosplvt l=0.6u w=2.7u ad=1.49p pd=6.5u
m2 7 1 14 14 cmosplvt l=0.6u w=2.7u ad=1.49p pd=6.5u
m3 4 4 0 0 cmosn l=0.9u w=0.9u ad=0.5p pd=2.9u m=3
m4 5 4 0 0 cmosn l=0.9u w=0.9u ad=0.5p pd=2.9u m=3
m5 8 5 0 0 cmosn l=0.9u w=0.9u ad=0.5p pd=2.9u m=11
m6 14 16 20 20 cmosp l=1.4u w=12u ad=6.6p pd=25.1u m=6
m7 10 16 20 20 cmosp l=1.4u w=12u ad=6.6p pd=25.1u m=11
m10 12 10 3 3 cmosnlvt l=0.2u w=36u ad=19.8p pd=73.1u
m11 13 9 3 3 cmosplvt l=0.2u w=62u ad=34.1p pd=125u
m12 10 10 11 11 cmosnlvt l=0.2u w=15u ad=5.94p pd=22.7u
m13 9 9 11 11 cmosplvt l=0.2u w=25.8u ad=10.2p pd=38.9u
mr 16 16 20 20 cmosp l=1.4u w=12u ad=6.6p pd=25.1u m=6
ir 16 0 17u
cc 8 5 300f
vdd 20 0 1.8
vs1 6 4 0       ; ammeter
vs2 7 5 0       ; ammeter
vs3 9 8 0       ; ammeter
vs4 20 12 0     ; ammeter
vs5 13 0 0      ; ammeter
.ends
***************open-loop dc  fig.4.70(i)***************
x1 31 32 33 opamp1
rl 33 34 300
vb 34 0 0.9
vn 32 0 0.9
vp 31 0 dc
.control
dc vp 0.8995 0.9005 1u
plot v.x1.vs4#branch v.x1.vs5#branch vs v(33)       ; IQ
plot v(33) vs v(31,32)      ; dc transfer characteristic
plot deriv(v(33)) vs v(33)        ; dc differential gain
.endc
***************loop gain  fig.4.70(j),(k)***************
*x4 41 42 43 opamp1      ; for voltage return ratio
```

```
*rla 43 41 300           ; for voltage return ratio
*iia 41 42 0u            ; for voltage return ratio
*vba 41 0 0.9            ; for voltage return ratio
*ra 42 44 1500           ; for voltage return ratio
*vx 44 43 dc 0 ac 1      ; for voltage return ratio
*x5 51 52 53 opamp1      ; for current return ratio
*rlb 53 51 300           ; for current return ratio
*iib 51 52 0u            ; for current return ratio
*vbb 51 0 0.9            ; for current return ratio
*rb 52 55 1500           ; for voltage return ratio
*ix 0 54 dc 0 ac 1       ; for current return ratio
*vs1 53 54 0             ; for current return ratio
*vs2 54 55 0             ; for current return ratio
*.control
*set units=degrees
*ac dec 100 1k 1g
*let rv=-v(43)/v(44)
*let ri=-i(vs1)/i(vs2)
*plot db((rv*ri-1)/(2+rv+ri)) ph((rv*ri-1)/(2+rv+ri))
*let magnitude=mag((rv*ri-1)/(2+rv+ri))
*let phasedif=ph((rv*ri-1)/(2+rv+ri))+180
*meas ac fu when magnitude=1
*meas ac pm find phasedif at=fu
*.endc
*****************noise  fig.4.70(l)******************
*x6 61 62 63 opamp1
*r 62 63 1500
*rl 63 61 300
*vb 61 0 0.9
*ii 0 62 dc 0 ac 1
*.control
*noise v(63) ii dec 100 1 100meg
*print (onoise_total)^2
*.endc
*********transient and fourier  fig.4.70(l)**********
*x6 61 62 63 opamp1
*r 62 63 1500
*rl 63 61 300
*vb 61 0 0.9
*ii 0 62 sin 0 200u 1meg
*.control
*tran 0.5n 10u
*plot v(63)
*fourier 1meg v(63)
*.endc
*****************************************************
.end
```

Opamp is represented by the subcircuit `opamp1`, whose schematic is given in Fig. 4.70(b).

An open-loop dc analysis is performed first. It is based on the schematic shown in Fig. 4.70(i). Notice the placement of a load resistance R_L in series with the bias source V_B across the output port for establishing proper open-loop loading conditions. In this simulation, we keep the inverting-input voltage constant at 0.9 V, and sweep the noninverting-input voltage around 0.9 V. The first plot generated by this simulation displays i_{D10} and i_{D11} as `v.x1.vs4#branch` and `v.x1.vs5#branch` against the output voltage v_O. The common value of these currents measured for $v_O = 0.9$ V is the standby current I_Q. We iterate this simulation after trimming the scaling factor between the widths of M12/M13 and M10/M11. The existing width setting yields $I_Q = 99$ μA.

Another plot generated by the open-loop dc simulation is the dc transfer characteristic presented in Fig. 4.71(e). It indicates a systematic offset around 8 μV, and appears to be compliant with the output range specified by the limits $V_{OL} = 0.6$ V and $V_{OH} = 1.2$ V. The derivative of this characteristic is plotted in Fig. 4.71(f) against the output voltage. It exhibits a dc gain $A_0 = 2111$ V/V at $v_O = 0.9$ V. Its twin-peak formation is typical of a push-pull operation with a low buffer-stage bias current I_Q.

The reader may also utilize the open-loop dc simulation for verifying the operation-point value of any internal variable or device parameter by plotting against v_O and reading at $v_O = 0.9$ V.

A loop-gain simulation is run next for automatically extracting the phase margin of the specified application. It involves the schematics presented in Figs. 4.70(j) and (k), and yields PM = 72° at a unity-gain frequency f_u = 17.9 MHz at the operation point, where $i_i = 0$. It is advisable to repeat this simulation for different values of the input signal within the specified range -200 μA $\leq i_i \leq 200$ μA. This is done by resetting both `iia` and `iib` to the desired value of i_i. Any stability issue that may be prevailing outside the operation point will thus be disclosed.

The final step of verification involves transient and noise simulations for evaluating THD+N for the transresistance amplifier. V_{no}^2 appearing in the equation of definition (4.280) of this metric is obtained with a noise simulation performed on the schematic shown in Fig. 4.70(l). It is printed out as $V_{no}^2 = 4.20 \times 10^{-8}$ V^2 in response to the command

```
print (onoise_total)^2
```

THD and V_{o1}^2 are characterized with the Fourier analysis extension of a transient simulation. The reader is referred to Examples 2.15 and 4.24 for its details. This simulation is run on the schematic presented in Fig. 4.70(l). The value of THD is printed directly after the completion of the transient simulation. For the existing fundamental-frequency setting of 1 MHz, for example, THD equals 0.00518.

For V_{o1}^2, the reader is referred to the table printed directly by the simulator. The second row identified by a 1 in the first column belongs to the fundamental component of the output signal. The number cited for `Magnitude` in this row is the amplitude. V_{o1}^2 is calculated by taking the square of this number and dividing the outcome by 2. For the existing 1-MHz frequency setting, this calculation yields $V_{o1}^2 = 0.045$ V^2.

The transient simulation is rerun for different frequencies within the bandwidth of interest. This is done by resetting the frequency in those lines where the signal source i_i and Fourier analysis are defined. The time step and duration of transient analysis are also adjusted in each run to cover at least 10 periods with adequate time resolution. The THD+N thus characterized for peak input signal $i_i = 200\ \mu\text{A}$ is plotted in Fig. 4.71(g) as a function of the fundamental frequency. Note that the dynamic response is limited by noise at low frequencies and by distortion at high frequencies. The borderline between the two appears to be around 150 kHz in the present application. ▲

4.10 Opamp with a Push-Pull Common-Source Output Stage

The source-follower buffer stage studied in Section 4.9 necessitates low-threshold and DNW device options for a reasonably wide output range. This technological restriction can be removed if the buffer is constructed as a common-source stage. Converting an OTA into an opamp with such an output stage in push-pull topology is the subject of this section.

Topology and Operation

Shown in Fig. 4.72 are three opamp topologies featuring a push-pull common-source buffer stage as highlighted in gray. They differ only in the type of the OTA core. In all three, M10 and M11 act as output drivers whereas M12 and M13 are responsible for push-pull control. Although the opamps are depicted with PMOS input drivers, the same push-pull stage can also be used for converting an OTA of NMOS input drivers.

The buffer operates with constant voltages V_{G12} and V_{G13} at the gate terminals of M12 and M13 with respect to ground. These two devices, when conducting $I_{B2}/2$ each, generate the correct gate voltages V_{G10} and V_{G11} needed by M10 and M11 to conduct the bias current I_Q. How this relationship is established will be explained later. Presently, suppose that the OTA forces the buffer out of its operation point by raising v_{G11} above the bias level V_{G11}. M11 responds by raising i_{D11} above I_Q, and M13 responds by lowering i_{D3} below $I_{B2}/2$. Since the sum of i_{D12} and i_{D13} is a constant I_{B2}, i_{D12} has to increase in response. This calls for a more negative gate-source voltage for M12. Since V_{G12} is fixed, the gate-source voltage is made more negative by v_{G10} rising above its bias level V_{G10}. This, in turn, lowers i_{D10} below I_Q. The difference between the larger i_{D11} and the smaller I_{D12} flows into the buffer stage as a negative i_L through R_L, and thus lowers the output voltage v_O below V_B. All these actions are reversed whenever v_{G11} is lowered below its bias level V_{G11} by the OTA. To be specific, $v_{G11} < V_{G11}$ leads to $i_{D13} > I_{B2}/2$, $i_{D12} < I_{B2}/2$, $v_{G10} < V_{G10}$, $i_{D10} > I_Q$, $i_{D11} < I_Q$, $i_L > 0$, and $v_O > V_B$.

The constant gate voltages V_{G12} and V_{G13} are supplied by the biasing circuit highlighted in gray in Fig. 4.73. This circuit works on the basis of the following scaling relationships between pairs M12/M15, M10/M14, M13/M16, and M11/M17:

$$\left(\frac{W}{L}\right)_{15} = \frac{I_{\text{ref1}}}{I_{B2}/2}\left(\frac{W}{L}\right)_{12}, \tag{4.289}$$

$$\left(\frac{W}{L}\right)_{10} = \frac{I_Q}{I_{\text{ref1}}}\left(\frac{W}{L}\right)_{14}, \tag{4.290}$$

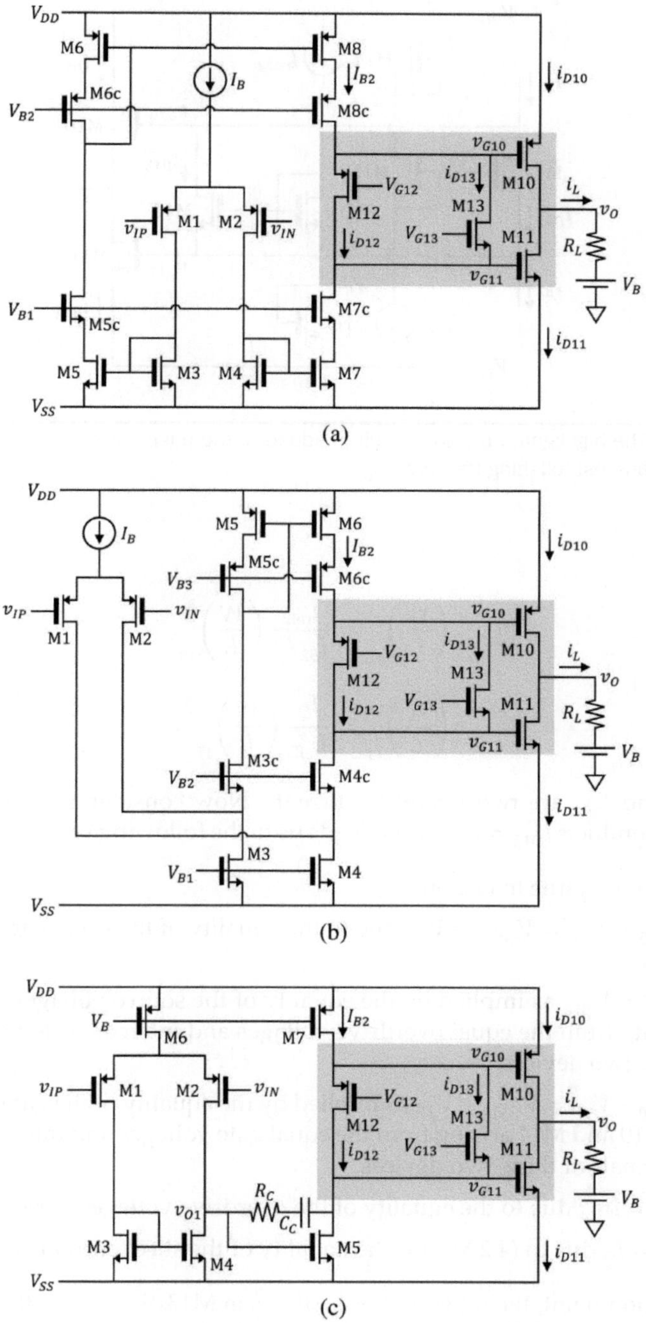

FIGURE 4.72 Converting an OTA into an opamp with a push-pull common-source buffer stage. The part added to the OTA is highlighted in gray. (a) Cascoded-symmetrical opamp. (b) Folded-cascode opamp. (c) Miller opamp.

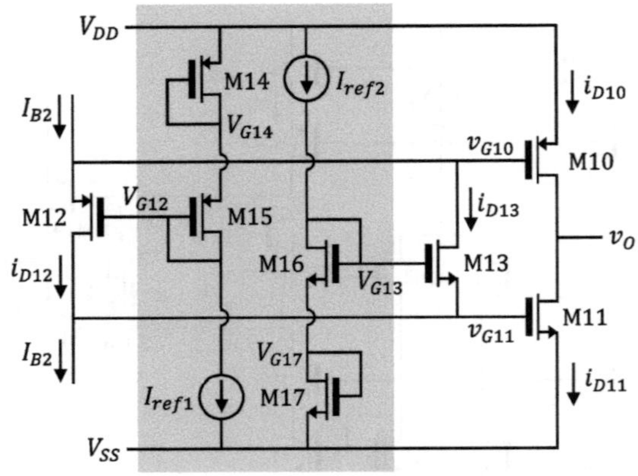

FIGURE 4.73 The highlighted biasing circuit is added to the push-pull controller for setting V_{G12} and V_{G13} and thus establishing the target I_Q.

and

$$\left(\frac{W}{L}\right)_{16} = \frac{I_{\text{ref2}}}{I_{B2}/2}\left(\frac{W}{L}\right)_{13}, \tag{4.291}$$

$$\left(\frac{W}{L}\right)_{11} = \frac{I_Q}{I_{\text{ref2}}}\left(\frac{W}{L}\right)_{17}, \tag{4.292}$$

where I_{ref1} and I_{ref2} are two reference currents. Now consider the operation point, at which, M12 conducts $i_{D12} = I_{B2}/2$, and leads us to the following sequence of conclusions:

- $I_{\square 15} = I_{\square 12}$ due to (4.289).
- $V_{GS15} - V_{TP} = V_{GS12} - V_{TP}$ due to the equality of the sheet currents of M12 and M15.
- $V_{G10} = V_{G14}$ as implied by the equality of the source voltages of M12 and M15 arising from the equal overdrive voltages and interconnected gate terminals of these two devices.
- $V_{GS10} - V_{TP} = V_{GS14} - V_{TP}$ as implied by the equality of the gate-source voltages of M10 and M14 arising from the equal gate voltages and interconnected source terminals of these two devices.
- $I_{\square 10} = I_{\square 14}$ due to the equality of the overdrive voltages of M10 and M14.
- $i_{D10} = I_Q$ due to (4.290) and the equality of the sheet currents of M10 and M14.

At the operation point, the other half of I_{B2} flows in M13. Starting with $i_{D13} = I_{B2}/2$, and repeating the analysis for M11, M13, M16, and M17 while considering (4.291) and (4.292), we also arrive at $i_{D11} = I_Q$ for the current conducted by M11 at the operation point. It is therefore obvious that, as long as (4.289) to (4.292) are satisfied by design, V_{G12} and V_{G13} will assume proper values to ensure each of M10 and M11 to conduct I_Q at the operation point.

Output Range

The output range offered by a common-source buffer stage is limited by the saturation conditions of M10 and M11 as described by

$$V_{OL} = V_{SS} + V_{DSsat11}, \tag{4.293}$$

$$V_{OH} = V_{DD} + V_{DSsat10}. \tag{4.294}$$

DC Differential Gain, Output Resistance, and Effective Transconductance

For the dc differential gain, the reader is referred to (4.267) to (4.269), where $A_{0(\text{ota})}$ and $A_{0(\text{buf})}$ are negative valued for the signal definition in Fig. 4.72. $A_{0(\text{buf})}$ is given by

$$A_{0(\text{buf})} = -(R_L \parallel r_{ds10} \parallel r_{ds11})(g_{m10} + g_{m11}), \tag{4.295}$$

which is the sum of the gate-to-drain voltage gains of M10 and M11. Comparing (4.267) with (2.96) while considering (4.295) for $A_{0(\text{buf})}$ leads us to

$$r_o = r_{ds10} \parallel r_{ds11} \tag{4.296}$$

for the output resistance, and to

$$G_m = -A_{0(\text{ota})}(g_{m10} + g_{m11}) \tag{4.297}$$

for the effective transconductance. It is therefore obvious that the common-source buffer stage boosts the effective transconductance while maintaining a large output resistance.

Frequency Response, Step Response, and Noise

Assuming a small R_L as justified in any application that necessitates buffering, and excluding the cases involving an unusually large opamp load capacitance C_L, the time constant associated with the output port of the common-source buffer is expected to be too small to have a significant effect on the frequency response of the opamp. The push-pull controller however can significantly alter the response by introducing a low-frequency zero as a result of the time delay between the signal components of v_{G10} and v_{G11}. If necessary, this zero can be removed by enabling simultaneous arrival of signal at both gates by inserting a bypass capacitor between the gates of M10 and M11. Provided that the zero is located at a frequency well above the closed-loop bandwidth, or is removed by bypassing, the frequency response will be dictated primarily by the OTA core as in the case of source-follower buffering. The expressions cited or presented in Subsection 4.9.3 for the open-loop bandwidth f_o and gain-bandwidth product f_{GBW} are therefore also valid for an opamp deploying a common-source buffer.

One unique implication of common-source buffering on frequency response is the enhancement of the load capacitance $C_{L(\mathrm{ota})}$ of the OTA core by the so-called Miller effect. This capacitance is described by

$$C_{L(\mathrm{ota})} = C_1 + (1 - A_{0(\mathrm{buf})})\left(C_{gd10} + C_{gd11}\right), \tag{4.298}$$

where C_1 is the load capacitance excluding the gate-drain capacitances C_{gd10} and C_{gd11} of M10 and M11. Note that the contribution of these two gate-drain capacitances is enhanced by a factor $1 - A_{0(\mathrm{buf})}$. For an explanation of this effect, refer to any of the buffers shown in Fig. 4.72, and consider the small-signal dc relationship $v_o = A_{0(\mathrm{buf})} v_{g(10,11)}$ between the drain signal v_o and the common gate signal $v_{g(10,11)}$. The small-signal current flowing through the gate-drain capacitance C_{gd} of M10 or M11 is related to these voltages by

$$i_{gd} = C_{gd}\left[\frac{dv_{g(10,11)}}{dt} - \frac{dv_o}{dt}\right] = \left(1 - A_{0(\mathrm{buf})}\right) C_{gd}\frac{dv_{g(10,11)}}{dt},$$

which obviously implies an effective capacitance $(1 - A_{0(\mathrm{buf})})C_{gd}$ between the gate nodes and ground. Since this effective capacitance shunts C_1, the total load capacitance of the OTA core is indeed the sum of the two as expressed in (4.298). Incidentally, the Miller effect is what we exploited in Subsection 4.8.2 for the frequency compensation of a Miller OTA. Although not mentioned specifically, the multiplication of C_C by A_0 in (4.232) is due to the Miller effect.

As to the differential-input range, there is no difference between an opamp and its OTA core. The reader is referred to Subsection 4.4.3 for the cascoded-symmetrical opamp and the folded-cascode opamp, and to Subsection 4.8.2 for the Miller opamp. Noise properties are also approximately the same as the OTA core as long as the gain of the buffer stage is much smaller than that of the OTA.

Example 4.28 A single-supply ($V_{DD} = 1.8$ V) folded-cascode opamp buffered by a common-source push-pull stage is shown in Fig. 4.74(a). It is to be used in the noninverting amplifier shown in Fig. 4.74(b). The amplifier is supposed to operate with $V_B = 0.9$ V and to drive $R_L = 250\ \Omega$ and $C_L = 4$ pF. A closed-loop dc gain of 7 V/V is targeted with $R_1 = 10$ kΩ and $R_2 = 60$ kΩ. A full-power bandwidth of 4.5 MHz is specified for a sinusoidal input signal of maximum amplitude $V_i = 100$ mV. At the operation point, the opamp should exhibit $A_0 \geq 4000$ V/V while conducting $I_Q \leq 250$ µA in its buffer stage.

Work Converting the closed-loop specifications into open-loop specifications is what we do first.

- *External Network Parameters* $\alpha_i = 1$ and $\beta = 1/7$.
- *Common-Mode Range* $V_{IL} \leq V_{CM(\mathrm{min})} = V_B - V_i = 0.9 - 0.1 = 0.8$ V and $V_{IH} \geq V_{CM(\mathrm{max})} = V_B + V_i = 0.9 + 0.1 = 1.1$ V.
- *Output Range* $V_{OL} \leq V_{O(\mathrm{min})} = -7V_i + V_B = -0.7 + 0.9 = 0.2$ V and $V_{OH} \geq V_{O(\mathrm{max})} = 7V_i + V_B = 0.7 + 0.9 = 1.6$ V.

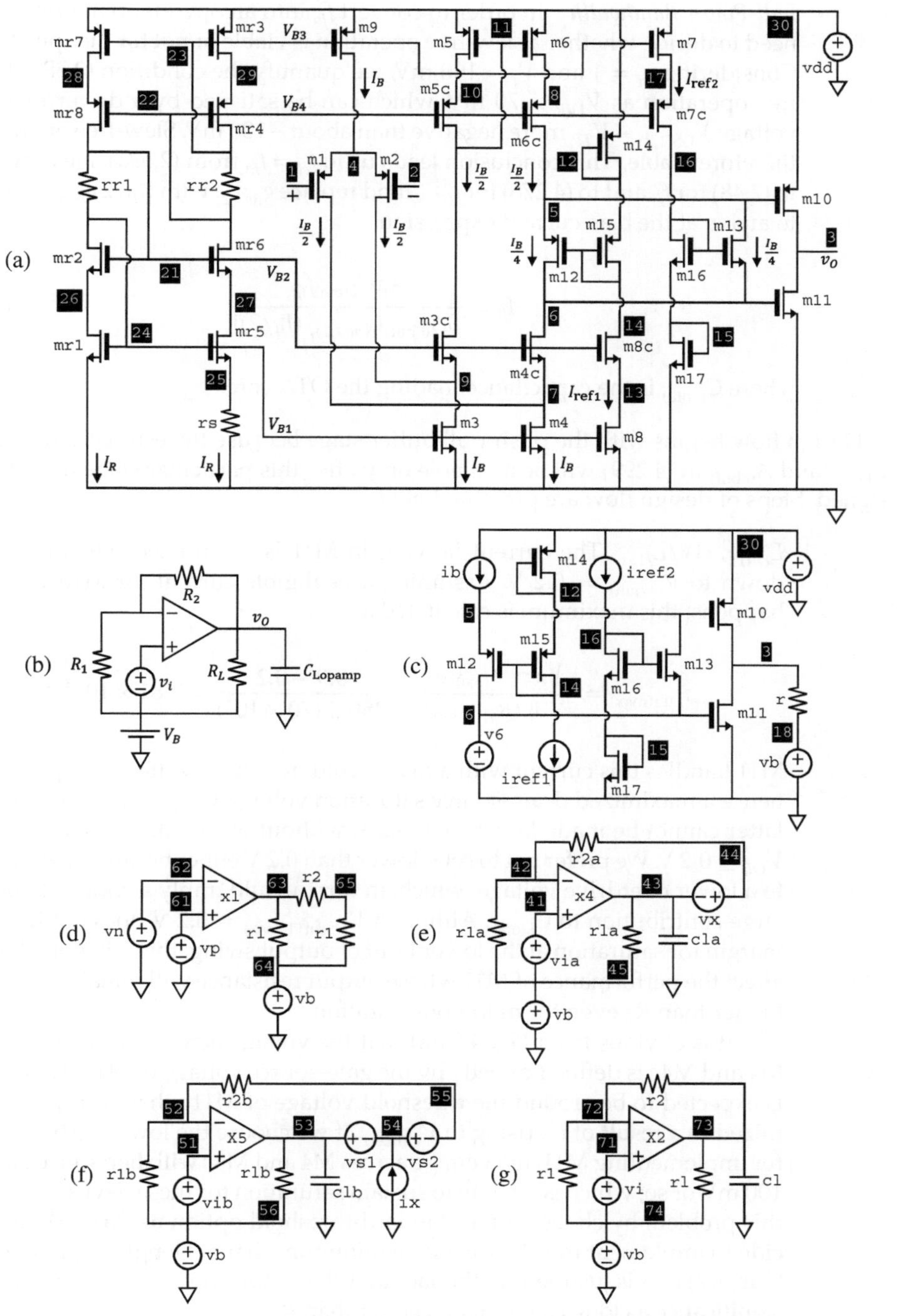

FIGURE 4.74 Top-level and simulation schematics of Example 4.28. (a) Opamp. (b) Top level. (c) `ex4-28c.sp`. (d) Open-loop dc verification in `ex4-28d.sp`. (e) and (f) Loop-gain verification in `ex4-28d.sp`. (g) Closed-loop ac and transient verifications in `ex4-28d.sp`.

- *Full-Power Bandwidth* In order to convert f_{fp} into an open-loop metric, we first need to decide whether a slew-free operation is viable or not for the specified V_i. Considering $\alpha_i = 1$ and $V_i = 100$ mV, we quantify the condition (2.58) of slew-free operation as $V_{DIR} > 70$ mV, which can be satisfied by a driver overdrive voltage $V_{GS(1,2)} - V_{TP}$ more negative than about −100 mV. Slew-free operation is therefore viable. This conclusion leads us to $f_{cl} = f_{fp}$ from (2.59a). Next, we refer to (2.48) for f_{cl} and to (4.278) for f_{GBW}, and replace $g_{m(1,2)}$ with $(I_B/2)[g_{m(1,2)}/(I_B/2)]$ to arrive at the bias-current expression

$$I_B = \frac{4\pi C_{L(\text{ota})} f_{fp}}{\beta A_{0(\text{buf})}[g_{m(1,2)}/(I_B/2)]}, \tag{4.299}$$

where $C_{L(\text{ota})}$ is the capacitance loading the OTA core.

Design flow begins with the push-pull buffer stage because the estimates needed for $C_{L(\text{ota})}$ and $A_{0(\text{buf})}$ in (4.299) will be available only after this particular stage has been designed. Steps of design flow are presented next.

1. L_{DR11}, $(W/L)_{11}$ The current flowing in M11 is maximized when v_O swings down to $V_{O(\min)} = 0.2$ V. Assuming a negligible current for M10 when this happens, this maximum is calculated as

$$i_{D11(\max)} = \frac{V_B - V_{O(\min)}}{R_L \parallel (R_1 + R_2)} = \frac{0.9 - 0.2}{250 \parallel (70 \times 10^3)} = 2.81 \times 10^{-3}\ \text{A}.$$

M11 handles this current with a maximized overdrive voltage $v_{GS11(\max)} - V_{TN}$ hence a maximized drain-source saturation voltage $V_{DSsat11(\max)}$. However, the latter cannot be made larger than 0.2 V without breaching the specified limit $V_{OL} \leq 0.2$ V. We prefer not to set it lower than 0.2 V either because it would lead to a lower overdrive voltage, which, in turn, would imply a wide M11 hence a large contribution to $C_{L(\text{ota})}$. Although $V_{DSsat11(\max)} = 0.2$ V does not leave any margin for saturation at the lower limit of output swing, this will not adversely affect the performance of M11 whose output resistance will remain sufficiently higher than R_L even for mild nonsaturation.

It is obvious from Fig. 4.74(a) that the voltage across the cascode pair of M4 and M4c is defined directly by the gate-source voltage of M11. This voltage is expected to be around the threshold voltage of M11 whenever i_{D11} is minimized as a result of v_O rising to $V_{O(\max)}$. If we choose the low-threshold option for implementing M11, the voltage across M4 and M4c will therefore be around 100 mV or so, which is too low to sustain saturation for these devices. We avoid this problem by choosing the standard-threshold option for M11. We also decide to implement this device with a minimum channel length $L_{DR11} = 0.2$ μm. Our decision is justified by the fact that the output resistance of M11 is of no significance as long as it remains larger than R_L.

At this stage, we quantify the overdrive voltage $v_{GS11(\max)} - V_{TN}$ corresponding to $V_{DSsat11(\max)} = 0.2$ V. We need it for determining the maximum sheet current $I_{\square 11(\max)}$ of M11. Since the expected value of $i_{D11(\max)}$ is also available, the sheet current will enable us to determine the aspect ratio $(W/L)_{11}$.

Unfortunately, the design tools given in Fig. 3.8 for V_{DSsat} and in Fig. 3.10 or Fig. 3.11 for $I_{\square}$ are not very accurate for a device having a minimum channel length and operating at the borderline between saturation and nonsaturation. We extract these tools specifically for M11 by running a simulation with the input file `nmos2.sp` for $L_{DR} = 0.2$ μm and $V_{DS} = 0.2$ V. They yield $v_{GS11(\max)} - V_{TN} = 0.424$ V and $I_{\square 11(\max)} = 13.7$ μA/sq. The aspect ratio of M11 is therefore determined as

$$\left(\frac{W}{L}\right)_{11} = \frac{i_{D11(\max)}}{I_{\square 11(\max)}} = \frac{2.81 \times 10^{-3}}{13.7 \times 10^{-6}} = 205,$$

which implies $W_{11} = 34$ μm. We also read $V_{TN} = 0.494$ V for the threshold voltage of M11, and therefore calculate the maximum gate-source voltage $v_{GS11(\max)} = 0.918$ V for later use in sizing M13. As will be clarified in step 3 of design flow, sizing of M13 also necessitates the operation-point values of the gate-source voltage, transconductance, output resistance, and intrinsic gate capacitance of M11. We obtain these four with an iterative operation-point simulation, in which, M11 is specified with the channel dimensions determined above and operated with fixed 0.9-V drain-source voltage. Gate-source voltage is iterated until the device current converges to $I_Q = 250$ μA. The following input file `ex4-28a.sp` is used for this purpose:

```
filename: ex4-28a.sp
.incl bsim.sp
m11 3 6 0 0 cmosn w=34u l=0.2u
vds11 3 0 0.9
vgs11 6 0 0.567
.control
op
print @m11[id] @m11[gm] 1/@m11[gds] @m11[cgg]
.endc
.end
```

The outcomes of this simulation are $V_{GS11} = 0.567$ V, $g_{m11} = 4.23$ mA/V, $r_{ds11} = 10.92$ kΩ, and $C_{ggi11} = 37.2$ fF.

2. L_{DR10}, $(W/L)_{10}$ The approach taken in step 1 for sizing M11 is also appropriate for sizing M10. Again, the standard-threshold option and minimum channel length are chosen. This device conducts the largest current $i_{D11(\max)} = 2.81$ mA when v_O swings up to $V_{O(\max)} = 1.6$ V. By targeting −0.2 V for the drain-source saturation voltage, and running `pmos2.sp` with $L_{DR} = 0.2$ μm and $V_{DS} = -0.2$ V, we arrive at $v_{GS10(\min)} - V_{TP} = -0.240$ V for the most negative overdrive voltage and $I_{\square 10(\max)} = 1.55$ μA/sq for the largest sheet current. The aspect ratio of M10 is therefore calculated as

$$\left(\frac{W}{L}\right)_{10} = \frac{i_{D10(\max)}}{I_{\square 10(\max)}} = \frac{2.81 \times 10^{-3}}{1.55 \times 10^{-6}} = 1813,$$

which leads to $W_{10} = 256$ μm. Simulation also yields $V_{TP} = -0.473$ V. The most negative gate-source voltage of M10 is therefore estimated to be $v_{GS10(\text{min})} = -0.713$ V. Finally, an iterative simulation is run for extracting the operation-point parameters of M10. It is run with the following input file `ex4-28b.sp`:

```
filename: ex4-28b.sp
.incl bsim.sp
m10 3 5 0 0 cmosp w=256u l=0.2u
vds11 3 0 -0.9
vgs11 5 0 -0.510
.control
op
print @m10[id] @m10[gm] 1/@m10[gds] @m10[cgg]
.endc
.end
```

Iterative simulation yields $V_{GS10} = -0.510$ V, $g_{m10} = 4.82$ mA/V, $r_{ds10} = 23.01$ kΩ, and $C_{ggi10} = 206$ fF.

3. I_B, $(W/L)_{1,2}$, L_{DR13}, W_{DR13} The reader may have noticed in Fig. 4.74(a) the setting of the bias currents of M3 and M4 to I_B without supplementing them with the safety margin ΔI indicated in the original folded-cascode OTA schematic of Fig. 4.50. As explained in the last paragraph of Subsection 4.7.1, such a margin is not indispensable if the OTA is expected not to slew. In its absence the sum of the currents of M12 and M13 is fixed to $I_B/2$. At the operation point, each of these two devices will therefore conduct $I_B/4$. I_B itself is described by (4.299), where f_{fp} is specified as 5 MHz, and β equals 1/7. As to $g_{m(1,2)}/(I_B/2)$, we determine it as 15 V^{-1} from Fig. 4.14 assuming low-threshold option with an overdrive voltage $V_{GS(1,2)} - V_{TP} = -0.100$ V to ensure slew-free operation. Between the remaining two variables of (4.299) $A_{0(\text{buf})}$ is calculated from (4.295) for the parameter values extracted in steps 1 and 2. The outcome is $A_{0(\text{buf})} = -2.18$ V/V. As to $C_{L(\text{ota})}$, it is almost certain that M10 and M11 will be its dominant contributors. We estimate it from the following specific form of (4.298)

$$C_{L(\text{ota})} = \underbrace{1.1\left(C_{ggi10} + C_{gse10} + C_{ggi11} + C_{gse11}\right)}_{C_1} + \left(1 - A_{0(\text{buf})}\right)\left(C_{gde10} + C_{gde11}\right), \tag{4.300}$$

where the extrinsic components are assumed to be dominant in C_{gd10} and C_{gd11}. Also note the assumption that C_1 is 10% higher than the total gate capacitance of M10 and M11. This is a ballpark estimate for the contribution of the rest of the devices attached to the output of the OTA. Considering the estimates of C_{ggi10} and C_{ggi11} obtained in steps 1 and 2, and calculating the extrinsic capacitances from (3.92) and (3.93), we obtain $C_{L(\text{ota})} = 1.25$ pF from (4.300). Out of this total, 520 fF is due to C_1. Having quantified all variables of (4.299), we estimate the bias current as $I_B = 15$ μA.

According to Fig. 4.13, the overdrive voltage −0.100 V selected for M1 and M2 implies $I_{\square(1,2)} = 430$ nA/sq for these devices. Since each conducts 7.5 μA, their aspect ratio should be $(W/L)_{1,2} = 7.5/0.430 = 17.5$.

As explained above, M13 should be designed to conduct $I_B/4$ at the operation point. We now know that this current amounts to 3.75 μA. In order to determine the sheet current and therefore the aspect ratio of this device, we also need to know its overdrive voltage. Notice from Fig. 4.74(a) the identity of its source voltage to the gate-source voltage of M11, hence $V_{S13} = V_{GS11} = 0.567$ V as determined in step 1. For determining its fixed gate voltage V_{G13}, we consider the lower limit of the output swing, at which, the source voltage is maximized to $v_{S13(\text{max})} = v_{GS11(\text{max})} = 0.918$ V as obtained in step 1, whereas the current i_{D13} practically vanishes. Assuming a zero overdrive voltage in this state, the gate voltage is given by $V_{G13} = 0.918 + V_{TN13(\text{max})}$. The overdrive voltage at the operation point is therefore described by

$$V_{GS13} - V_{TN13} = V_{G13} - V_{S13} - V_{TN13} = 0.351 + V_{TN13(\text{max})} - V_{TN13},$$

where $V_{TN13(\text{max})}$ and V_{TN13} are to be evaluated for $V_{S13(\text{max})} = 0.918$ V and $V_{S13} = 0.567$ V, respectively. We opt for a low-threshold device for M13 in order not to overly restrict the voltage across the cascode current source M7/M7b. According to Fig. 4.5, the difference $V_{TN13(\text{max})} - V_{TN13}$ is approximately constant and equal to 0.085 V regardless of the channel length. This leads us to $v_{GS13} - V_{TN13} = 0.436$ V, for which, Fig. 4.7 indicates $I_{\square 13} = 20.5$ μA/sq, and therefore $(W/L)_{13} = 0.183$ for the 3.75-μA standby current. Since this aspect ratio is indicative of a narrow device whose channel length is not critical in any respect, it is meaningful to select a small but not minimum width first, and then calculate the corresponding channel length. We select $W_{DR13} = 1$ μm, and calculate $L_{DR13} = 5.4$ μm.

4. L_{DR12}, W_{DR12} The path we followed in step 3 is also qualitatively valid for sizing the low-threshold M12. At the operation point, its source voltage with respect to ground is given by $V_{S12} = V_{GS10} + V_{DD} = -0.510 + 1.8 = 1.290$ V, where V_{GS10} is transferred from step 2. The constant gate voltage V_{G12} with respect to ground is determined for the upper limit of the output swing, at which, the source voltage with respect to ground attains its minimum value $v_{S12(\text{min})} = v_{GS10(\text{min})} + V_{DD} = -0.713 + 1.8 = 1.087$ V, where $v_{GS10(\text{min})}$ is transferred from step 2. Assuming a zero overdrive voltage for M12 at this limit, we write $V_{G12} = 1.087 + V_{TP12(\text{min})}$ for the gate voltage, and therefore arrive at

$$V_{GS12} - V_{TP12} = V_{G12} - V_{S12} - V_{TP12} = -0.203 + V_{TP12(\text{min})} - V_{TP12},$$

where $V_{TP12(\text{min})}$ and V_{TP12} are to be evaluated for $V_{SB12} = -0.713$ V and $V_{SB12} = -0.510$ V, respectively. Since Fig. 4.10 indicates $V_{TP12(\text{min})} - V_{TP12} = -0.045$ V irrespective of the channel length, we obtain $V_{GS12} - V_{TP12} = -0.248$ V for the overdrive voltage. $I_{\square 10} = 2.2$ μA/sq follows from Fig. 4.12, and leads to

$(W/L)_{12} = 1.70$. Selecting $L_{DR12} = 1.2$ μm for the noncritical channel length, we therefore arrive at $W_{DR12} = 2.0$ μm.

5. *I_{ref2}, M16, and $W_{DRu(11,17)}$* Selection of I_{ref2} involves a trade-off between (a) power efficiency, which favors a low current; and (b) precision of mirroring from M17 to M11 and from M16 to M13, which favors large channel areas for M16 and M17 hence a high current. We select $I_{\text{ref2}} = 8 \times (I_B/4) = 30$ μA, and thus satisfy (4.291) by implementing M16 with eight copies of M13. The reader is referred to step 3 for the channel dimensions of M13.

   In (4.292), we round down the ratio $I_Q/I_{\text{ref2}} = 250/30 = 8.33$ to the nearest integer 8, and considering $W_{11} = 34$ μm obtained in step 1, arrive at $W_{17} = 4.25$ μm. For increasing dispersion in layout, we divide this width into four, and thus set the unit-device width as $W_{DRu(11,17)} = 1.1$ μm. The number of copies to be used is 4 for M17 and 32 for M11.

6. *I_{ref1}, M15, and $W_{DRu(10,14)}$* I_{ref1} is subject to the very same trade-off mentioned in step 5. We select $I_{\text{ref1}} = I_{\text{ref2}} = 30$ μA, and satisfy (4.289) by implementing M15 with eight copies of M12, whose dimensions are made available in step 4. Setting $I_Q/I_{\text{ref1}} = 8$ and considering $W_{10} = 256$ μm lead us to $W_{14} = 31.8$ μm, which is divided into eight unit devices of drawn width $W_{DRu(10,14)} = 4$ μm. M10 is to be implemented with 64 such unit devices as implied by (4.290).

7. *Verification of the Buffer Stage* In this step, we verify the dc operation of the buffer stage, and trim device sizes if necessary. We also take the opportunity provided by the stage-wide simulation to accurately determine the compliance limits of the cascode pairs M4/M4c and M6/M6c.

   The schematic used for verification is shown in Fig. 4.74(c). It is netlisted in the following input file:

```
filename: ex4-28c.sp
.incl bsimlvt.sp
.incl bsim.sp
m10 3 5 30 30 cmosp w=4u l=0.2u m=64
m11 3 6 0 0 cmosn w=1.1u l=0.2u m=32
m12 6 14 5 30 cmosplvt w=2u l=1.2u
m13 5 16 6 0 cmosnlvt w=1u l=5.4u
m14 12 12 30 30 cmosp w=4u l=0.2u m=8
m15 14 14 12 30 cmosplvt w=2u l=1.2u m=8 ; m=9 after trim
m16 16 16 15 0 cmosnlvt w=1u l=5.4u m=8 ; m=9 after trim
m17 15 15 0 0 cmosn w=1.1u l=0.2u m=4
r 3 18 243 ; combination of rl, rl and r2
vb 18 0 0.9
iref1 14 0 30u
iref2 30 16 30u
ib 30 5 7.5u
vdd 30 0 1.8
v6 6 0
.save all @m10[id] @m11[id]
.control
dc v6 0 1.8 1m
```

```
plot @m10[id] @m11[id] vs v(3)
plot v(6) v(5)-v(30) vs v(3)
.endc
.end
```

Notice the use of (a) independent current sources for applying I_{B2}, I_{ref1}, and I_{ref2}; and (b) an independent voltage source v6 to sweep the voltage of node 6 to mimic the dc operation of the OTA core. This simulation generates two plots. One of them displays i_{D10} and i_{D11} against v_O, which indicates 275 μA for I_Q. The reason for this higher-than-expected bias current is that M12 and M13 operate with larger drain-source voltages in comparison with their respective reference devices M15 and M16. Increasing the number of unit devices of M15 and M16 from 8 to 9 will raise V_{G12} and lower V_{G13}, and thus reduce the standby current down to $I_Q = 210$ μA. The second plot generated by this simulation displays two voltages: one between node 6 and ground and another between node 5 and V_{DD}. These voltages are needed for defining compliance limits for the cascode pairs M4/M4c and M6/M6c. Their most restrictive values are $V_5 - V_{DD} = -0.404$ V for $v_O = 0.2$ V and $V_6 = 0.402$ V for $v_O = 1.6$ V.

8. $(W/L)_{u(3,3c,4,4c)}$ Since these four devices conduct rather small currents, they can operate with small overdrive voltages hence small drain-source saturation voltages without necessitating excessive aspect ratios. Based on this argument, we partition $V_6 = 0.402$ V into $V_{DS\text{sat}(3,3c,4,4c)} = 0.100$ V, $\Delta V_{DS(3,4)} = 0.150$ V, and $\Delta V_{DS4c} = 0.052$ V. Selecting the standard-threshold option, we read $V_{GS(3,3c,4,4c)} - V_{TN} = 0.115$ V from Fig. 3.8, and $I_{\square(3,3c,4,4c)} = 1.7$ μA/sq from Fig. 3.10. Considering $I_{D(3,4)} = I_B = 15$ μA, the aspect ratio of M3 and M4 is determined as $(W/L)_{3,4} = 15/1.7 = 8.82$. We divide this into four unit devices, and use two copies for implementing M3c and M4c. $(W/L)_{u(3,3c,4,4c)} = 8.82/4 = 2.20$ is the aspect ratio of the common unit device of M3, M3c, M4, and M4c.

9. $(W/L)_{5,5c,6,6c}$ Following the lines of step 8, we partition $V_5 - V_{DD} = -0.404$ V into $V_{DS\text{sat}(5,5c,6,6c)} = -0.100$ V, $\Delta V_{DS(5,6)} = -0.150$ V, and $\Delta V_{DS6c} = -0.054$ V. The low-voltage cascode arrangement necessitates standard-threshold option for these devices, for which, Fig. 3.17 indicates $V_{GS(5,5c,6,6c)} - V_{TP} = -0.100$ V. The corresponding sheet current as indicated by Fig. 3.19 is 330 nA/sq. Since these devices conduct $I_B/2 = 7.5$ μA each, their aspect ratio should be $(W/L)_{5,5c,6,6c} = 22.7$.

10. $L_{DR(3,3c,4,4c)}$, $W_{DRu(3,3c,4,4c)}$, $L_{DR(5,5c,6,6c)}$, $W_{DR(5,5c,6,6c)}$, $L_{DR(1,2)}$, *and* $W_{DR(1,2)}$ Since the dc differential gain at the operation point is specified as $A_0 \geq 4000$ V/V, and $|A_{0(\text{buf})}|$ equals 2.16 V/V, the dc differential gain to be targeted for the OTA core should be no less than 1850 V/V. We target $|A_{0(\text{ota})}| = 2100$ V/V. Considering (4.180) together with $g_{m(1,2)}/(I_B/2) = 15\ \text{V}^{-1}$ and $I_B/2 = 7.5$ μA, the output resistance r_o of the OTA corresponding to this much gain is calculated as 18.7 MΩ. Targeting equal contributions by M4/M4c and M6/M6c, we define $r_{id4c} = r_{id6c} = 37.4$ MΩ. As argued in step 3 of Example 4.23, it is reasonable to target $r_{ds(1,2)} = 3r_{ds4}$ for partitioning $r_{id4c} = A_{\max 4c}(r_{ds(1,2)} \parallel r_{ds4})$, hence $A_{\max 4c} r_{ds(1,2)} = 150$ MΩ and $A_{\max 4c} r_{ds4} = 50$ MΩ.
An iterative simulation is performed for determining $L_{DR(3,3c,4,4c)}$. It is run on a schematic similar to Fig. 4.52(b) with an input file similar to `ex4-23a.sp`.

Considering $I_{D4} = I_B = 15\ \mu A$, the target of simulation is $A_{max4c} r_{ds4} I_{D4} = 750$ V. Since the overdrive of these devices equals 0.115 V, V_{B1} is reset in each iteration to $V_{TN4} + 0.115$ V after reading V_{TN4} from Fig. 3.7. V_{B2} equals the sum of the source, overdrive, and threshold voltages of M4. The first of these voltages is given by $V_{S4c} = V_{DSsat4} + \Delta V_{DS4} = 0.250$ V. The overdrive voltage is constant at 0.115 V, and V_{TN4c} is determined from Fig. 3.7. The outcome of the iterative simulation is $L_{DR(3,3c,4,4c)} = 1.1\ \mu m$, which yields $W_{DRu(3,3c,4,4c)} = 2.4\ \mu m$ from the aspect ratio calculated in step 8. This simulation also yields $V_{B1} = 0.530$ V and $V_{B2} = 0.855$ V. A_{max4c}, as extracted with an input file similar to `ex4-23b.sp`, equals 138 V/V.

For $L_{DR(1,2)}$, we run an iterative simulation on a schematic similar to Fig. 4.52(d) with an input file similar to `ex4-23c.sp`. Assuming that the bulk terminals of M1 and M2 are connected to the rail, `vb` is set to 1.8 V. `vd` equals $V_{DS(3,4)}$, hence 0.250 V. `vg` is kept at 0.9 V. Source-bulk voltage `vsb` is determined in each iteration from Fig. 4.10 considering

$$V_{SB(1,2)} + V_{TP(1,2)} = V_{GB(1,2)} - (V_{GS(1,2)} - V_{TP(1,2)}) = -0.9 - (-0.100) = -0.8 \text{ V}.$$

Iterative simulation yields $L_{DR(1,2)} = 0.6\ \mu m$, for which, we obtain $W_{DR(1,2)} = 9.5\ \mu m$ for the aspect ratio determined in step 3.

$L_{DR(5,5c,6,6c)}$ is determined with an iterative simulation similar to the one described in step 2 of Example 4.23 in Section 4.7. It targets $r_{id6c} = 37.4\ M\Omega$, and maintains a constant overdrive voltage −0.100 V for M6 and M6c. The source-bulk voltage of M6c is also kept constant at $V_{SB6c} = V_{DSsat6} + \Delta V_{DS6} = -0.100 - 0.150 = -0.250$ V. This simulation yields $L_{DR(5,5c,6,6c)} = 0.4\ \mu m$, for which, the aspect ratio determined in step 9 implies $W_{DR(5,5c,6,6c)} = 7.8\ \mu m$. Additional outcomes are $V_{B3} - V_{DD} = -0.540$ V and $V_{B4} - V_{DD} = -0.865$ V.

11. *Reference Section* Designed for $I_R = 7.5\ \mu A$. The unit device defined in step 8 for conducting $I_B/4 = 3.75\ \mu A$ is doubled for MR1, MR2, and MR6, and octupled for MR5, M8, and M8c. PMOS devices are built in units of the device defined in step 9. MR3, MR4, MR7, and MR8 contain one unit each. M9 has two units. M7 and M7c have four units each. Considering the specified I_B and the values of V_{B1}, V_{B2}, V_{B3}, and V_{B4} estimated in step 10, R_{R1} and R_{R2} are set to 43.3 kΩ each. As usual, R_S is determined with iterative simulation.
12. *Verification* The opamp schematic shown in Fig. 4.74(a) is netlisted as a subcircuit in the following input file `ex4-28d.sp` for verification purposes:

```
filename: ex4-28d.sp
.incl bsimlvt.sp
.incl bsim.sp
******op-amp****************************************
.subckt opamp 1 2 3
m1 7 1 4 4 cmosplvt w=9.5u l=0.6u ad=5.3p pd=20.1u
m2 9 2 4 4 cmosplvt w=9.5u l=0.6u ad=5.3p pd=20.1u
m3 9 24 0 0 cmosn w=2.4u l=1.1u ad=1.32p pd=5.9u m=4
m3c 11 21 9 0 cmosn w=2.4u l=1.1u ad=1.32p pd=5.9u
+  as=1.32p ps=5.9u m=2
m4 7 24 0 0 cmosn w=2.4u l=1.1u ad=1.32p pd=5.9u m=4
m4c 6 21 7 0 cmosn w=2.4u l=1.1u ad=1.32p pd=5.9u
```

```
+  as=1.32p ps=5.9u m=2
m5 10 11 30 30 cmosp w=7.8u l=0.40u ad=4.29p pd=16.7u
m5c 11 22 10 30 cmosp w=7.8u l=0.40u ad=4.29p pd=16.7u
+  as=4.29p ps=16.7u
m6 8 11 30 30 cmosp w=7.8u l=0.40u ad=4.29p pd=16.7u
m6c 5 22 8 30 cmosp w=7.8u l=0.40u ad=4.29p pd=16.7u
+  as=4.29p ps=16.7u
m7 17 23 30 30 cmosp w=7.8u l=0.40u ad=4.29p pd=16.7u
+  m=4
m7c 16 22 17 30 cmosp w=7.8u l=0.40u ad=4.29p pd=16.7u
+  as=4.29p ps=16.7u m=4
m8 13 24 0 0 cmosn w=2.400u l=1.1u ad=1.32p pd=5.9u
+  m=8
m8c 14 21 13 0 cmosn w=2.4u l=1.1u ad=1.32p pd=5.9u
+  as=1.32p ps=5.9u m=8
m9 4 23 30 30 cmosp w=7.8u l=0.40u ad=4.29p pd=16.7u
+  m=2
m10 3 5 30 30 cmosp w=4u l=0.2u ad=2.2p pd=9.1u
+  m=64
m11 3 6 0 0 cmosn w=1.1u l=0.2u ad=0.3p pd=2.2u
+  m=32
m12 6 14 5 30 cmosplvt w=2u l=1.2u ad=1.1p pd=5.1u
+  as=1.1p ps=5.1u
m13 5 16 6 0 cmosnlvt w=1u l=5.4u ad=0.55p pd=3.1u
+  as=0.55p ps=3.1u
m14 12 12 30 30 cmosp w=4u l=0.2u ad=2.2p pd=9.1u
+  m=8
m15 14 14 12 30 cmosplvt w=2u l=1.2u ad=1.1p pd=5.1u
+  as=1.1p ps=5.1u m=9
m16 16 16 15 0 cmosnlvt w=1u l=5.4u ad=0.55p pd=3.1u
+  as=0.55p ps=3.1u m=9
m17 15 15 0 0 cmosn w=1.1u l=0.2u ad=0.3p pd=2.2u
+  m=4
******reference section*****************************
mr1 26 24 0 0 cmosn w=2.4u l=1.1u ad=1.32p pd=5.9u
+  m=2
mr2 24 21 26 0 cmosn w=2.4u l=1.1u ad=1.32p pd=5.9u
+  as=1.32p ps=5.9u m=2
mr3 29 23 30 30 cmosp w=7.8u l=0.40u ad=4.29p pd=16.7u
mr4 23 22 29 30 cmosp w=7.8u l=0.40u ad=4.29p pd=16.7u
+  as=4.29p ps=16.7u
mr5 27 24 25 0 cmosn w=2.4u l=1.1u ad=1.32p pd=5.9u
+  as=1.32p ps=5.9u m=8
mr6 22 21 27 0 cmosn w=2.4u l=1.1u ad=1.32p pd=5.9u
+  as=1.32p ps=5.9u m=2
mr7 28 23 30 30 cmosp w=7.8u l=0.40u ad=4.29p pd=16.7u
mr8 21 22 28 30 cmosp w=7.8u l=0.40u ad=4.29p pd=16.7u
+  as=4.29p ps=16.7u
rs 25 0 8k
rr1 21 24 43.3k
```

```
rr2 23 22 43.3k
vdd 30 0 1.8
.ends opamp
*******open-loop*************************************
x1 61 62 63 opamp
rl 63 64 250
r1 64 65 10k
r2 63 65 60k
vb 64 0 0.9
vn 62 0 0.9
vp 61 0
.dc vp 0.895 0.905 1u
.save all @m.x1.mr1[id] @m.x1.mr3[id] @m.x1.m3[id]
+  @m.x1.m4[id] @m.x1.m5[id] @m.x1.m6[id] @m.x1.m7[id]
+  @m.x1.m8[id] @m.x1.m10[id] @m.x1.m11[id]
*******loop gain**************************************
*x4 41 42 43 opamp      ; for voltage return ratio
*r1a 42 45 10k          ; for voltage return ratio
*r2a 42 44 60k          ; for voltage return ratio
*rla 43 45 250          ; for voltage return ratio
*cla 43 0 4p            ; for voltage return ratio
*via 41 45 0            ; for voltage return ratio
*vba 45 0 0.9           ; for voltage return ratio
*vx 44 43 dc 0 ac 1     ; for voltage return ratio
*x5 51 52 53 opamp      ; for current return ratio
*r1b 52 56 10k          ; for current return ratio
*r2b 52 55 60k          ; for current return ratio
*rlb 53 56 250          ; for current return ratio
*clb 53 0 4p            ; for current return ratio
*ix 0 54 dc 0 ac 1      ; for current return ratio
*vs1 53 54 0            ; for current return ratio
*vs2 54 55 0            ; for current return ratio
*vib 51 56 0            ; for current return ratio
*vbb 56 0 0.9           ; for current return ratio
*.control
*set units=degrees
*ac dec 100 100 100meg
*let rv=-v(43)/v(44)
*let ri=-i(vs1)/i(vs2)
*plot db((rv*ri-1)/(2+rv+ri)) ph((rv*ri-1)/(2+rv+ri))
*let magnitude=mag((rv*ri-1)/(2+rv+ri))
*let phasedif=ph((rv*ri-1)/(2+rv+ri))+180
*meas ac fu when magnitude=1
*meas ac pm find phasedif at=fu
*.endc
*******closed-loop ac*********************************
*x2 71 72 73 opamp
*r1 72 74 10k
*r2 72 73 60k
*rl 73 74 250
```

```
*cl 73 0 4p
*vb 74 0 dc 0.9
*vi 71 74 dc 0 ac 1
*.control
*ac dec 100 1k 100meg
*plot vdb(73)-vdb(71)
*.endc
*******closed-loop transient************************
*x2 71 72 73 opamp
*r1 72 74 10k
*r2 72 73 60k
*rl 73 74 250
*cl 73 0 4p
*vb 74 0 dc 0.9
*vi 71 74 sin 0 0.1 4.5meg
*.control
*tran 1n 3u
*fourier 4.5meg v(73)
*plot v(73)
*.endc
*****************************************************
.end
```

An iterative open-loop dc simulation is performed on the top-level schematic of Fig. 4.74(d) for determining the reference resistance R_S. In each iteration, the currents of MR1 and MR3 are plotted with

```
plot @m.x1.mr1[id] @m.x1.mr3[id] vs v(63)
```

against the output voltage at node 63, and compared with the target 7.5 μA for $v_O = 0.9$ V. This effort eventually leads to $R_S = 8$ kΩ, for which, all other branch currents also closely match their target values. Next, i_{D10} and i_{D11} are plotted against the output voltage for determining the standby current. The outcome, shown in Fig. 4.75(a), indicates $I_Q = 249$ μA, which is somewhat higher than what we determined in Fig. 4.74(c) after trimming in step 7, but it is still in compliance with the specification.

Yet another outcome of the open-loop dc simulation is the dc transfer characteristic displayed in Fig. 4.75(b). It shows the variation of the output voltage with the differential-input voltage. It exhibits a systematic offset around 93 μV.

The final outcome of the open-loop dc simulation is the derivative of the output voltage plotted against the output voltage itself in Fig. 4.75(c). It enables us to verify the dc differential gain. The plot exhibits the usual twin-peak formation of low-I_Q push-pull opamps and indicates $A_0 = 4700$ V/V for $v_O = 0.9$ V.

We also run a set of loop-gain simulations based on the schematics displayed in Figs. 4.74(e) and (f). These simulations are intended for determining the phase margin over the entire range of input signal. The reason why we usually need multiple loop-gain simulations in the case of push-pull amplifiers is that the loop gain and therefore the phase margin may depend significantly on the signal level due to the highly nonlinear transfer function as exemplified by Fig. 4.75(b). Repeating the loop-gain simulation in 20-mV increments of `via` and `vib` within the specified input-signal range indicates

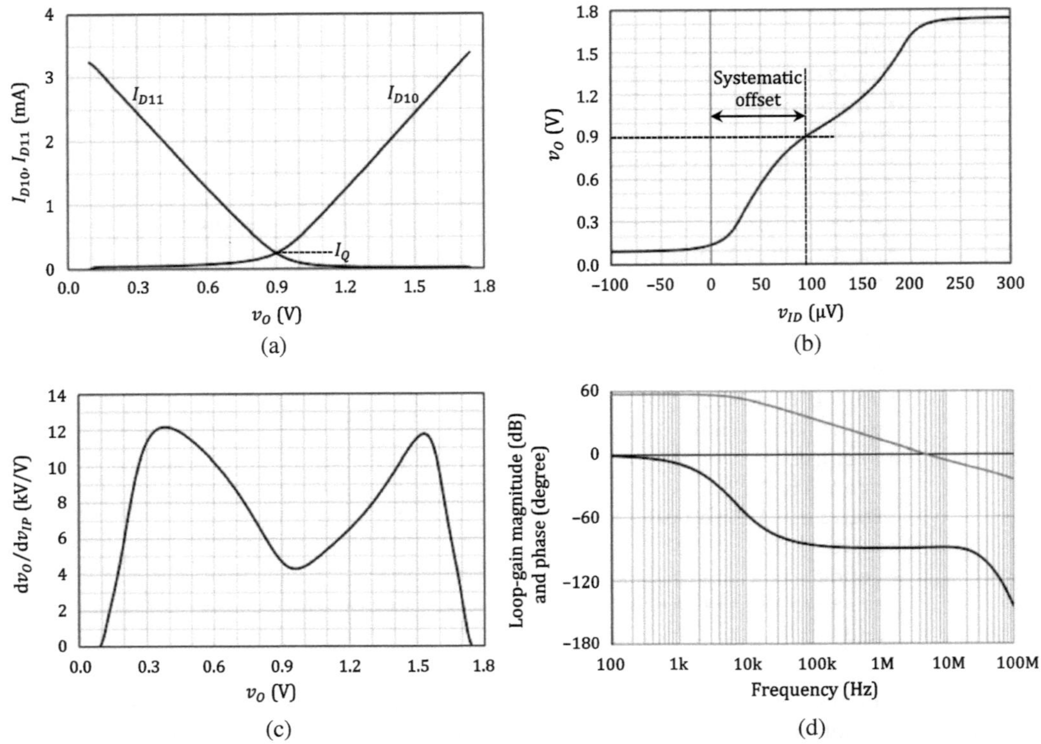

FIGURE 4.75 Simulated plots of Example 4.28 as obtained with `ex4-28d.sp`. (a) Currents of the push-pull driver devices M10 and M11 against the output voltage. (b) DC transfer characteristic. (c) Derivative of the dc transfer characteristic against the noninverting-input voltage for a constant 0.9-V inverting-input voltage. (d) Magnitude (gray) and phase (black) of loop gain for $v_i = 0$.

a phase margin better than 82°. As a matter of fact, for -80 mV $< v_i < 0$ it exceeds 90° due to the presence of a low-frequency zero arising from the time difference between the gate signals of push-pull buffer drivers. The magnitude and phase plots belonging to the case of $v_i = 0$ are presented in Fig. 4.75(d) as an example. The reader is invited to rerun this simulation after removing the zero by placing a 1-pF capacitor between nodes 5 and 6, and to observe the consequences on the magnitude and phase plots and on the phase margin. It is obvious from these loop-gain simulations that stability is ensured for the present application with a large phase margin regardless of whether the zero is removed or not.

The final two simulations are performed on the closed-loop schematic given in Fig. 4.74(g). The ac simulation yields $f_{cl} = 4.9$ MHz for closed-loop bandwidth at the operation point. The transient simulation is for verifying the integrity of the output-signal waveform for maximum input signal strength at the specified full-power bandwidth. It yields a peak output signal amplitude around 0.58 V, which is somewhat higher than the nominal peak amplitude $0.7/\sqrt{2} = 0.49$ V. The transient simulation also returns a THD around 8%. The amplitude and frequency conditions set for this simulation are of course the most unfavorable in this respect. ▲

4.11 Fully-Differential OTAs and Opamps

Application aspects of electrical design with fully-differential OTAs or opamps are the subject of Sections 2.10 and 2.12. The circuit-layer properties of such OTAs and opamps are covered in the present section.

The first subsection introduces the topologies and properties of fully-differential OTAs, and describes their conversion into opamps. The second subsection covers common-mode feedback (CMFB) circuits, which are essential for properly biasing a fully-differential OTA or opamp. The coverage culminates in the third subsection with design examples involving fully-differential, continuous-time, and discrete-time applications.

4.11.1 Core Topologies and Properties

Topologies and properties of fully-differential basic OTA, symmetrical OTA, folded-cascode OTA, and Miller OTA are presented in this subsection. Their conversion into opamps is also described.

Basic OTA

Shown in Figs. 4.76(a) and (b) are fully-differential versions of basic OTAs with NMOS and PMOS drivers, respectively. Differential output voltage is defined by $v_O \doteq v_{OP} - v_{ON}$. These OTAs differ from their single-ended counterparts in that M4 is configured not as a diode-connected current reference but as a current source/sink exactly like its matching pair M3. Notice the perfectly balanced topology as well as the presence of a common-mode feedback circuit marked CMFB, whose implementation is the subject of Subsection 4.11.2. As explained in Subsection 2.10.1, CMFB keeps the output common-mode voltage $v_{OCM} \doteq (v_{OP} + v_{ON})/2$ at an externally defined reference voltage V_{OCM}. It functions by comparing v_{OCM} with V_{OCM} and adjusting the bias current $i_{D(3,4)}$ of M3 and M4 by means of the gate voltage v_B in a negative feedback loop. For an explanation of how this kind of adjustment equates v_{OCM} to V_{OCM}, consider the PMOS-driver version displayed in Fig. 4.76(b). For $v_{OCM} < V_{OCM}$, CMFB lowers $i_{D(3,4)}$ below $I_B/2$ by pulling v_B down. The positive difference between $I_B/2$ and $i_{D(3,4)}$ raises v_{OP} and v_{ON} and therefore v_{OCM} by charging the parasitic capacitances of the two output nodes. If v_{OCM} exceeds V_{OCM}, then, CMFB pulls v_B up, and thus forces $i_{D(3,4)}$ to exceed $I_B/2$. The difference between these two currents discharges the parasitic capacitances and therefore forces v_{OP} and v_{ON} down toward V_{OCM}. If necessary, CMFB can be applied only to a subset of the unit devices by which M3 and M4 are constructed. In such a partial control, the rest of the unit devices are dc biased as constant-current sinks.

The foregoing description is also valid for the NMOS-driver version with the exception that v_{OCM} is stabilized at V_{OCM} by raising $i_{D(3,4)}$ for $v_{OCM} < V_{OCM}$, and lowering it otherwise.

It is worth mentioning that CMFB regulation can also be implemented by controlling the current of M5 instead of M3 and M4. Partial control is possible in this case too.

Equations (4.102) to (4.105) are still valid for the range limits of each single-ended output. The reader is referred to (2.125) to (2.130) for the limits of the differential-output range.

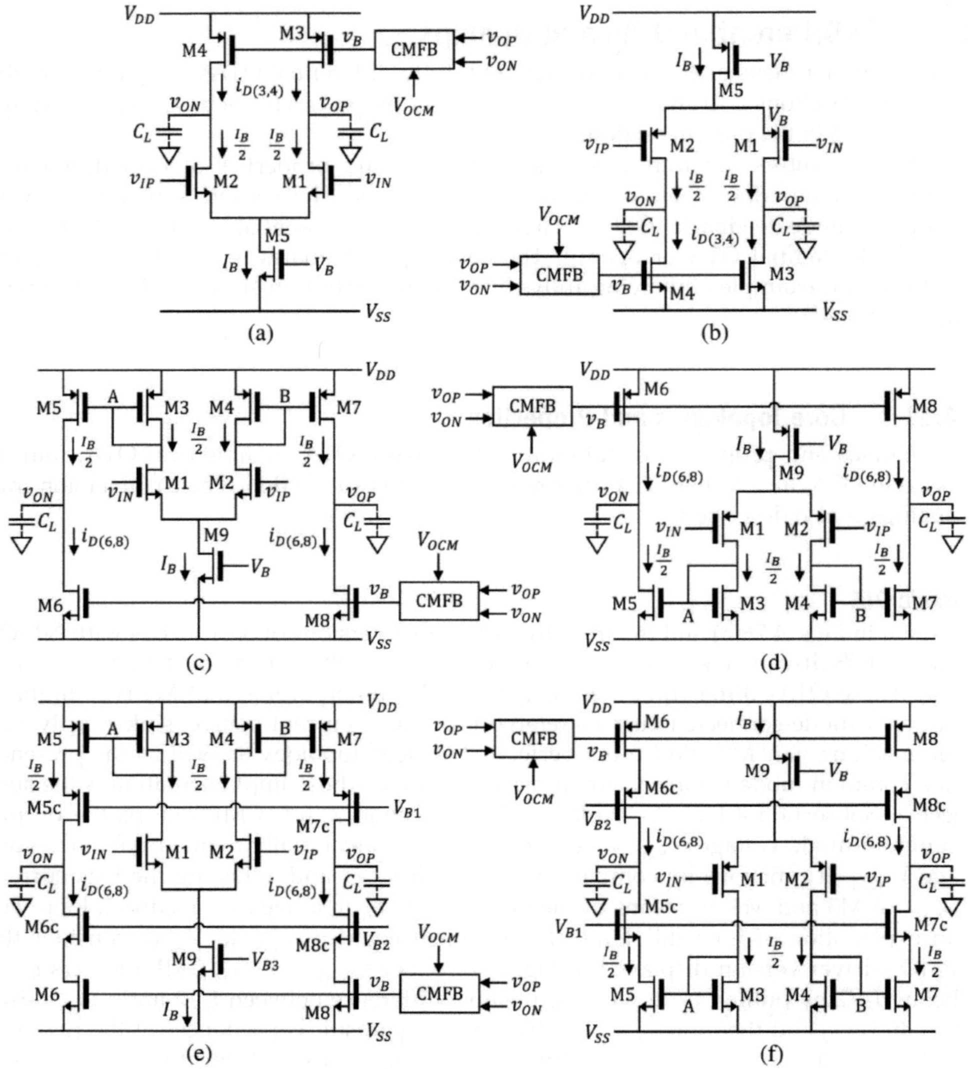

FIGURE 4.76 Fully-differential OTA topologies. (a) Basic OTA with NMOS drivers. (b) Basic OTA with PMOS drivers. (c) Symmetrical OTA with NMOS drivers. (d) Symmetrical OTA with PMOS drivers. (e) Cascoded-symmetrical OTA with NMOS drivers. (f) Cascoded-symmetrical OTA with PMOS drivers.

As to the limits of the common-mode range, V_{IL} of the NMOS-driver version and V_{IH} of the PMOS-driver version are still given by (4.106) and (4.109), respectively. For V_{IH} in NMOS driver and V_{IL} in PMOS driver, (4.107) and (4.108) are replaced by

$$V_{IH} = V_{DD} + V_{DS\text{sat}(3,4)} - V_{DS\text{sat}(1,2)} + V_{GS(1,2)} \quad \text{(NMOS drivers),} \qquad (4.301)$$

where $V_{GS(1,2)}$ is specific to the upper limit, and

$$V_{IL} = V_{SS} + V_{DSsat(3,4)} - V_{DSsat(1,2)} + V_{GS(1,2)} \quad \text{(PMOS drivers)}, \tag{4.302}$$

where $V_{GS(1,2)}$ is specific to the lower limit.

DC differential gain from differential input to any of the single-ended outputs is exactly one half of the dc differential gain A_0 of the corresponding single-ended basic OTA topology. The dc gain from differential input to differential output, as denoted by A_{dif0} in (2.145), doubles the gain back to A_0. Equations (4.110) and (4.111) are therefore valid for A_{dif0}.

Small-signal frequency and step responses are governed by the transfer function

$$A_{\text{dif}} = \frac{g_{m(1,2)} r_o}{1 + sC_L r_o}, \tag{4.303}$$

where C_L denotes the capacitance loading each single-ended output, and r_o represents the single-ended output resistance defined by $r_o \doteq r_{ds(1,2)} \parallel r_{ds(3,4)}$. Note that the high-frequency pole-zero pair appearing in (4.113) of the single-ended counterpart is absent due to the removal of the diode connection of M4 but (4.114) and (4.115) are still valid for the open-loop bandwidth and gain-bandwidth products.

The differential-input range V_{DIR} of a fully-differential basic OTA is modeled by the very same expressions presented in Subsection 4.4.3 for the basic differential pair, in Subsection 4.4.4 for the source-degenerated differential pair, and in Subsection 4.4.5 for the super-gm differential pair.

Inherent noise properties of a fully-differential basic OTA are not different from those of the single-ended counterpart. For details, the reader is referred to Subsection 4.5.5.

Symmetrical OTA

The NMOS-driver and PMOS-driver versions of the fully-differential symmetrical OTAs are shown in Figs. 4.76(c) and (d). They differ from their single-ended counterparts in that M6 functions as a current source or current sink exactly like M8. A symmetrical and balanced topology is thus created. As usual in all fully-differential OTAs, a CMFB circuit is also present. This circuit fully or partially controls the current $i_{D(6,8)}$ of M6 and M8 for the purpose of equating the common-mode output voltage v_{OCM} to an externally defined reference voltage V_{OCM}.

Range limits of each single-ended output are still described by (4.133) to (4.136). Limits of the differential-output range are given by (2.125) to (2.130). As to common-mode range limits, (4.137) to (4.140) retain their validity.

The dc gain A_{dif0} defined from differential input to differential output is given by (4.141) or (4.142) provided that the factor of scaling from M3 to M5 and from M4 to M7 is 1.

Small-signal frequency response and step response are governed by the transfer function

$$A_{\text{dif}} = \frac{g_{m(1,2)} r_o}{(1 + sr_o C_L)\left(1 + s\frac{C_1}{g_{m(3,4)}}\right)}, \tag{4.304}$$

where r_o represents the single-ended output resistance $r_{ds(5,7)} \parallel r_{ds(6,8)}$, C_L is the capacitance loading each single-ended output, and C_1 is the parasitic capacitance associated with each of the nodes marked A and B in Figs. 4.76(c) and (d). A comparison between (4.143) and (4.304) shows that the pole-zero pair associated with the diode-connected M6 of the single-ended symmetrical OTA is removed in the fully-differential version. However, the dominant-pole frequency and gain-bandwidth product of the resulting two-pole response are still described by (4.144) and (4.145), respectively.

Inherent noise properties of a fully-differential symmetrical OTA are similar to those of the single-ended counterpart. The contents of Subsection 4.6.5 therefore continue to be applicable.

Cascoded versions of fully-differential symmetrical OTA are shown in Figs. 4.76(e) and (f) with NMOS drivers and PMOS drivers, respectively. Contents of Subsection 4.6.6 concerning their single-ended counterparts are also valid for these two.

Folded-Cascode OTA

Fully-differential folded-cascode OTA topologies are shown in Figs. 4.77(a) and (b). The CMFB loop can be established by controlling the current $i_{D(5,6)}$ of M5 and M6 as shown in these schematics, or alternatively, the current of M3 and M4, or even the bias current I_B of driver devices. Control can be full or partial.

DC gain A_{dif0} from differential input to differential output is described by (4.180) to (4.182). As usual in a fully-differential OTA, dc gain to each single-ended output is one half of A_{dif0}.

Equations (4.183) to (4.190) are valid for common-mode range limits and single-ended output range limits. For differential-output range limits, the reader is referred to (2.125) to (2.130).

Open-loop transfer function is given approximately by

$$A_{\text{dif}} = \frac{g_{m(1,2)} r_o}{\left(1 + s r_o C_L\right)\left(1 + s \frac{C_1}{g_{m(3c,4c)} + g_{mb(3c,4c)}}\right)}, \tag{4.305}$$

where r_o is the single-ended output resistance $r_{id(3c,4c)} \parallel r_{id(5c,6c)}$, and C_1 is the parasitic capacitance associated with each of the nodes marked A and B in Figs. 4.77(a) and (b). Note that $g_{mb(3c,4c)}$ is to be ignored if M3c and M4c operate with source-bulk shorts.

For open-loop bandwidth and gain-bandwidth products, the reader is referred to Subsection 4.7.3 of single-ended folded-cascode OTA. The very same subsection is also relevant for noise properties.

Miller OTA

Fully-differential Miller OTA topologies are presented in Figs. 4.77(c) and (d). Notice the duplication of the second stage for establishing a fully-differential output. Frequency compensation network is also duplicated. The CMFB circuit may control the current $i_{D(3,4)}$ of M3 and M4 as depicted in these schematics, or alternatively, the bias current I_{B1} of driver devices. Again, control may be full or partial. Equations (4.220) to (4.223) are valid for the range limits of each single-ended output. Limits of differential-output range are described by (2.125) to (2.130).

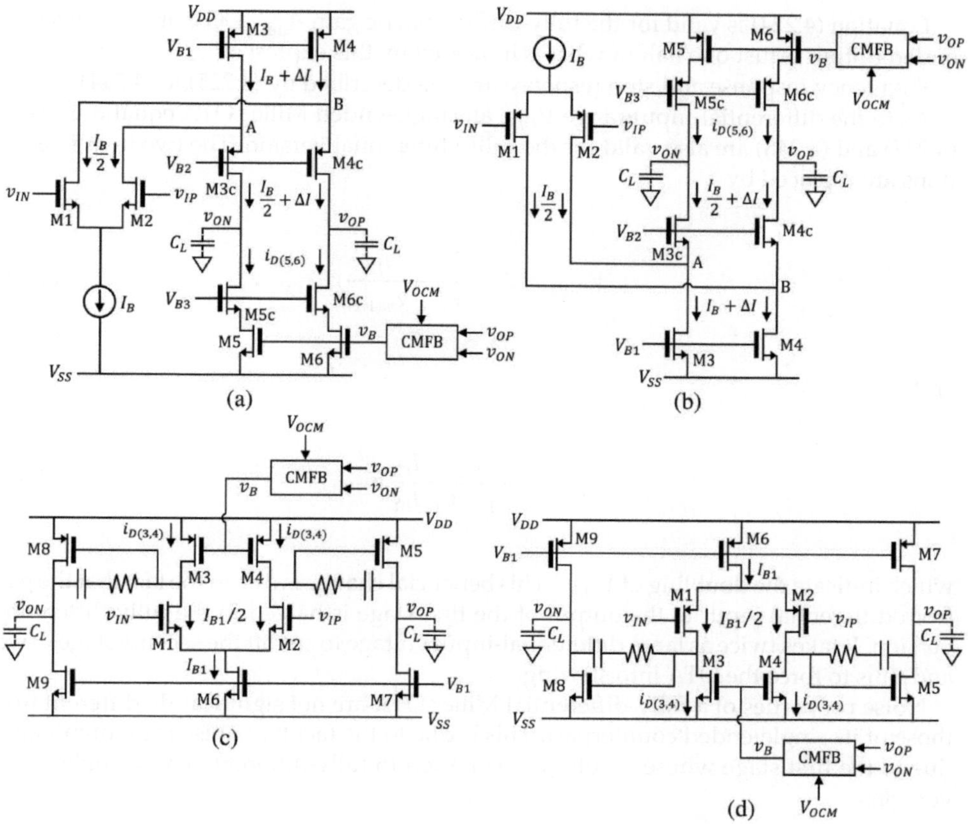

FIGURE 4.77 Fully-differential OTA topologies. (a) Folded-cascode OTA with NMOS drivers. (b) Folded-cascode OTA with PMOS drivers. (c) Miller OTA with NMOS drivers. (d) Miller OTA with PMOS drivers.

As to the common-mode range, the upper limit V_{IH} of the PMOS-driver version and the lower limit V_{IL} of the NMOS-driver version are still given by (4.217) and (4.218), respectively. V_{IL} for PMOS drivers and V_{IH} for NMOS drivers are described by

$$V_{IL} = V_{SS} + V_{DSsat(3,4)} - V_{DSsat(1,2)} + V_{GS(1,2)} \quad \text{(PMOS drivers)}, \qquad (4.306)$$

where $V_{GS(1,2)}$ is specific to the lower limit, and

$$V_{IH} = V_{DD} + V_{DSsat(3,4)} - V_{DSsat(1,2)} + V_{GS(1,2)} \quad \text{(NMOS drivers)}, \qquad (4.307)$$

where $V_{GS(1,2)}$ is specific to the upper limit.

Equation (4.224) is valid for the fully-differential dc gain A_{dif0}. DC gain to each single-ended output is just one half of what is indicated by this expression.

Frequency response and step response are still described by (4.225) to (4.244).

As to the differential-input range V_{DIR}, all single-ended Miller OTA equations except (4.247) and (4.248) are also valid for the fully-differential version. The two invalid equations are replaced by

$$V_{DIR2} = 2\frac{C_C}{C_L + C_C}\frac{I_{B2}}{g_{m(1,2)}}, \tag{4.308}$$

and

$$V_{DIR2} = 2\frac{C_C}{C_L + C_C}\frac{I_{B2}}{I_{B1}}V_{DIR1}, \tag{4.309}$$

which indicate the doubling of V_{DIR2}. This beneficial change is due to the fact that the gain from differential input to the output of the first stage is halved in the fully-differential version. It takes twice as large differential-input voltage to cut off the second-stage driver and thus to force the OTA into slewing.

Noise properties of a fully-differential Miller OTA are not significantly different from those of its single-ended counterpart. This is due to the fact that noise is predominantly due to the first stage whose topology is common to fully-differential and single-ended versions.

Fully-Differential Opamps

A fully-differential OTA is transformed into a fully-differential opamp by adding a buffer stage to each of the two single-ended outputs and feeding the buffered outputs into a CMFB circuit for regulating the output common-mode level.

Limits of the differential-output range are described by (2.125) to (2.130). For the single-ended limits appearing as V_{OL} and V_{OH} in these equations, the reader is referred to Section 4.9 in the case of a push-pull source-follower buffer stage, and to Section 4.10 for a push-pull common-source buffer stage. As to the limits of the common-mode range, there is no difference between the fully-differential and single-ended versions of an opamp. The dc gain to differential output is also not different from the dc differential gain A_0 of the corresponding single-ended opamp.

Properties related to frequency response and step response are well approximated by those of the OTA core provided that the frequency of the pole associated with the output ports of the fully-differential opamp is sufficiently higher than the closed-loop bandwidth. Noise properties are also similar to those of the OTA core.

4.11.2 Common-Mode Feedback Circuits

The common-mode feedback block marked CMFB in Figs. 4.76 and 4.77 regulates the common-mode output voltage $v_{OCM} \doteq (v_{OP} + v_{ON})/2$ of a fully-differential circuit around an external reference voltage V_{OCM}. It does this by sensing v_{OCM}, comparing it with V_{OCM}, and adjusting the current sources or sinks inside the OTA. It is expected to keep the difference between v_{OCM} and V_{OCM} small, and to respond quickly to any perturbation in v_{OCM} with a reasonable settling behavior. Needless to say, it must function without degrading the performance of the OTA or opamp.

There exist a wide variety of CMFB circuit topologies. In this subsection we study three examples, one suitable for continuous-time opamp applications, another for continuous-time OTA applications, and a third for discrete-time OTA applications. Their circuit schematics are presented in Fig. 4.78.

CMFB for Opamp-Based Continuous-Time Configurations

Two versions of a CMFB circuit suitable for continuous-time opamp applications are shown in Figs. 4.78(a) and (b). The former of the two is good for controlling the current sources of the OTA core such as M6 and M8 of Fig. 4.76(f), M5 and M6 of Fig. 4.77(b), or M3 and M4 of Fig. 4.77(c). The version shown in Fig. 4.78(b) is deployed if the current sinks of the OTA core are to be controlled as exemplified in Fig. 4.76(e) and in Figs. 4.77(a) and (d). In both versions, v_{OCM} is generated by averaging v_{OP} and v_{ON} with resistors RF1 and RF2. The outcome is compared with the reference voltage V_{OCM} in a basic OTA whose drivers are MF1 and MF2. v_{B1} and v_{B2} denote the voltages of the two complementary output nodes of the CMFB circuit. For $v_{OCM} = V_{OCM}$, these two output voltages stay at their common operation-point value. For $v_{OCM} > V_{OCM}$, most of I_{BF} is routed to MF4 in Fig. 4.78(a) and to MF3 in Fig. 4.78(b). In response, v_{B1} moves up and v_{B2} moves down from the common operation-point value in both topologies. For $v_{OCM} < V_{OCM}$, the directions of their movement are transposed. Between these two outputs, the one that defines a negative CMFB is used for closing the loop over the opamp.

The issue of how close this CMFB can bring v_{OCM} to V_{OCM} is addressed later in Example 4.29 on a fully-differential opamp of folded-cascode OTA core. The very same example includes an analysis of the bandwidth hence speed of the loop gain too. It presently suffices to note that the loop gain of this CMFB topology features two additional poles on top of those belonging to the differential-mode loop gain. One of these poles is associated with the gate of MF2. Between MF3 and MF4, the one that resides inside the CMFB loop contributes the other pole. But as long as R_{F1} and R_{F2} are sufficiently small, and the transconductance of MF3 and MF4 is sufficiently high, these additional poles don't threaten the stability of the CMFB loop.

CMFB for OTA-Based Continuous-Time Configurations

The CMFB circuit shown in Fig. 4.78(a) or (b) is not suitable for OTA applications because the unbuffered outputs of an OTA cannot handle the resistive loading of R_{F1} and R_{F2}. Two versions of a CMFB circuit that can be used in continuous-time fully-differential OTA applications are presented in Figs. 4.78(c) and (d). The version shown in part c is applicable to current-source control of an OTA, whereas the version shown in part d is good for current-sink control. The so-called *differential difference amplifier* (DDA) topology of this CMFB circuit features two source-degenerated differential pairs of matching devices MF1/MF2 and MF3/MF4. The branch currents flowing in MF2 and MF3 are combined by the diode-connected MF5 and mirrored to the source or sink device of a fully-differential OTA for common-mode control.

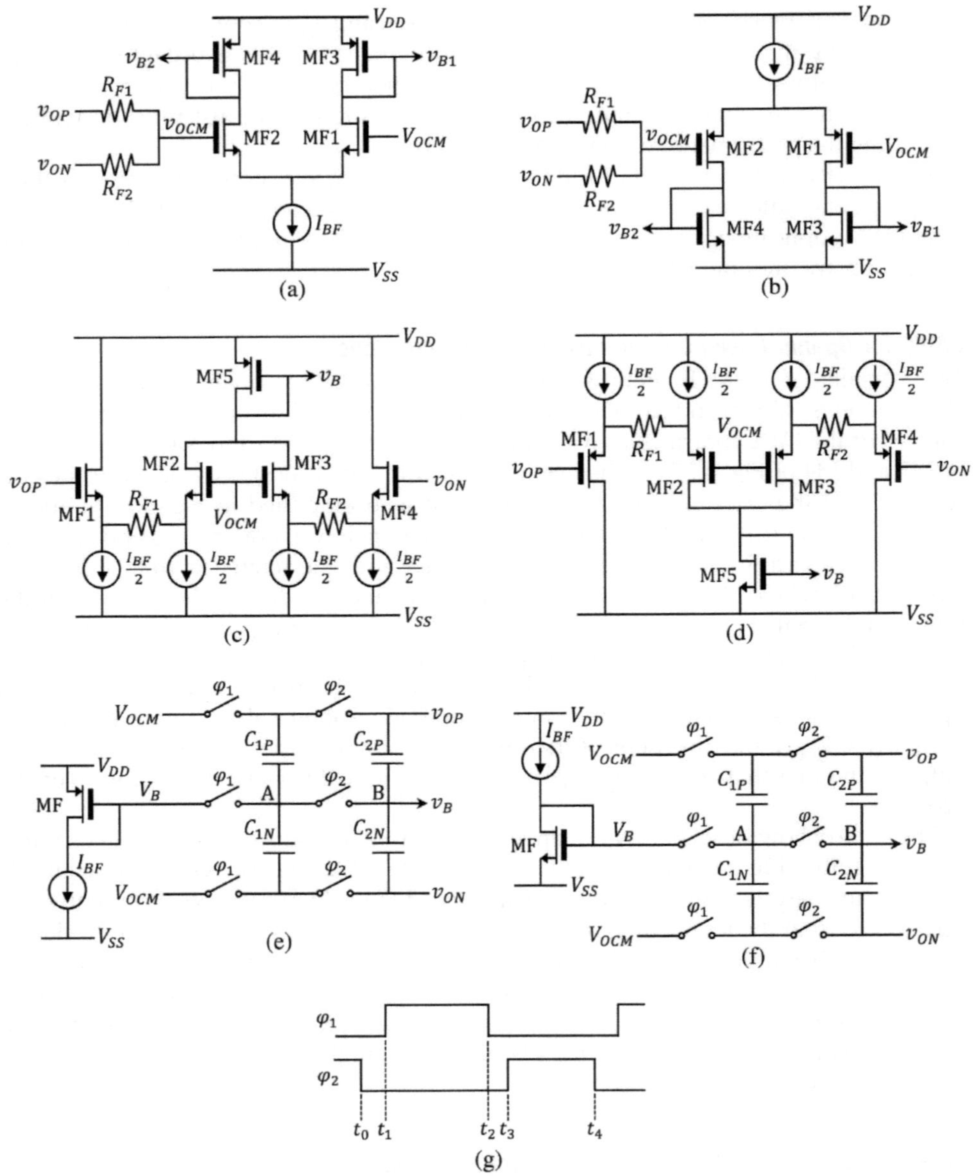

FIGURE 4.78 Common-mode feedback topologies. (a) For continuous-time opamp applications with current-source control. (b) For continuous-time opamp applications with current-sink control. (c) For continuous-time OTA applications with current-source control. (d) For continuous-time OTA applications with current-sink control. (e) For discrete-time OTA applications with current-source control. (f) For discrete-time OTA applications with current-sink control. (g) Clock timing diagram for the topologies shown in parts e and f.

For an explanation of how the DDA works as a CMFB circuit, consider the schematic presented in Fig. 4.78(c). Representing the effective transconductance of each differential pair with G_m, we can express the currents of MF2 and MF3 as

$$i_{DF2} = \frac{I_{BF}}{2} + G_m \left(V_{OCM} - v_{OP}\right)$$

$$i_{DF3} = \frac{I_{BF}}{2} + G_m \left(V_{OCM} - v_{ON}\right).$$

The current $i_{DF5} = i_{DF2} + i_{DF3}$ to be mirrored to the OTA is therefore described by

$$i_{DF5} = I_{BF} + 2G_m \left(V_{OCM} - \frac{v_{OP} + v_{ON}}{2}\right),$$

whose variable part is obviously proportional to $V_{OCM} - v_{OCM}$ as needed for CMFB control.

The use of source degeneration provides the differential pairs with a wide differential-input range V_{DIR}, which is essential for handling the large differences that may exist between the reference voltage V_{OCM} and the OTA output voltages v_{OP} and v_{ON}. To be specific, the extreme values of these single-ended output voltages that can be processed by the CMFB are related to V_{DIR} by

$$v_{OP(\min)} = v_{ON(\min)} = V_{OCM} - V_{DIR}, \tag{4.310}$$

and

$$v_{OP(\max)} = v_{ON(\max)} = V_{OCM} + V_{DIR}. \tag{4.311}$$

According to (4.98), one way of widening V_{DIR} is to decrease the transconductance efficiency of the driver devices, which calls for a large overdrive voltage. Another way to the same end is to increase $R_F I_{BF}$, where R_F represents the degeneration resistance. However, the extreme values of v_{OP} and v_{ON} that can be handled by the CMFB are defined ultimately by the compliance limit V_C of the current sources or sinks of the CMFB circuit. This limitation is described by

$$v_{OP(\min)} = v_{ON(\min)} \geq V_C + V_{GSL} \tag{4.312}$$

for the schematic shown in Fig. 4.78(c), and by

$$v_{OP(\max)} = v_{ON(\max)} \leq V_C + V_{GSU} \tag{4.313}$$

for the schematic shown in Fig. 4.78(d). Here, V_{GSL} denotes the gate-source voltage of the NMOS-driver MF1 or MF4 receiving $v_{OP(\min)}$ or $v_{ON(\min)}$ at its gate, and V_{GSU} denotes the gate-source voltage of the PMOS driver MF1 or MF4 receiving $v_{OP(\max)}$ or $v_{ON(\max)}$ at its gate.

The polarity of voltage gain from $v_{OCM} \doteq (v_{OP} + v_{ON})/2$ to v_B is positive in both of the CMFB circuits shown in Figs. 4.78(c) and (d). In any of the fully-differential OTAs depicted in Fig. 4.76 or in the folded-cascode OTAs depicted in Figs. 4.77(a) and (b), this will result in a negative feedback loop as expected from a proper CMFB circuit. If applied to any of the Miller-OTA topologies shown in Figs. 4.77(c) and (d), however, it will create a positive feedback loop, which, of course, is unacceptable. The polarity of the loop can be reversed in this case by applying V_{OCM} to the gates of MF1 and MF4, and using the gates of MF2 and MF3 for applying v_{OP} and v_{ON}. The loop can also be made negative by retaining the CMFB topology of Fig. 4.78(c) or (d) but applying v_B to the gate of M6 instead of M3 and M4 inside the Miller OTA.

The frequency-related properties of the common-mode loop gain are usually dominated by those of the differential-mode loop gain as long as the poles associated with the CMFB circuit are at sufficiently high frequencies. For details, the reader is referred to Example 4.30.

CMFB for Discrete-Time Configurations

A CMFB circuit that neither loads the outputs of an OTA by resistive sensing nor restricts them with the limitation of (4.312) or (4.313) is presented in Fig. 4.78(e) for current-source control and in Fig. 4.78(f) for current-sink control. However, this circuit is suitable only for discrete-time applications. It is built with (a) matching capacitors C1p and C1n of capacitance C_1, (b) matching capacitors C2p and C2n of capacitance C_2, (c) a current-reference device MF generating a bias voltage V_B while conducting a reference current I_{BF}, and (d) six switches controlled by two-phase nonoverlapping clocks φ_1 and φ_2 whose timing diagram is shown in Fig. 4.78(g). Similar to the CMFB circuits discussed previously, v_B acts as the controlling voltage of the current sources or sinks of an OTA core.

For an analysis of the common-mode feedback action, consider node A in any of the two versions presented in Figs. 4.78(e) and (f). The charge of this node is set in phase $\varphi_1 = 1$ when it is connected to MF. This charge is described by

$$Q_A\big|_{\varphi_1=1} = 2C_1(V_B - V_{OCM}). \tag{4.314}$$

In phase $\varphi_2 = 1$, node A is disconnected from MF but connected to node B. Its charge is now described by

$$Q_A\big|_{\varphi_2=1} = C_1(v_B - v_{OP}) + C_1(v_B - v_{ON}). \tag{4.315}$$

Now consider the periodic steady state, in which, v_B, v_{OP}, and v_{ON} are time invariant. Since no charge exchange can take place between nodes A and B in that state, we can assume $Q_A\big|_{\varphi_2=1} = Q_A\big|_{\varphi_1=1}$, which, according to (4.314) and (4.315), implies

$$v_B = V_B + \frac{v_{OP} + v_{ON}}{2} - V_{OCM} \tag{4.316}$$

for the voltage applied to the OTA core in phase $\varphi_2 = 1$. However, the same voltage is retained at node B also during $\varphi_2 = 0$ because C2p and C2n ensure uninterrupted averaging of v_{OP} and v_{ON}. This is how the common-mode feedback loop responds to the variation of v_{OCM} at any instant of the clock period.

Equation (4.316) indicates a positive gain from $v_{OCM} \doteq (v_{OP} + v_{ON})/2$ to v_B. In order to establish a negative feedback when regulating any of the Miller OTAs depicted in Figs. 4.77(c) and (d), v_B should therefore be applied to the gate of M6 instead of the gates of M3 and M4.

An application of the discrete-time CMFB circuit is presented in Example 4.31.

4.11.3 Design Examples

This subsection contains three examples of fully-differential design. Example 4.29 involves a continuous-time opamp application. A continuous-time OTA application is the subject of Example 4.30. Finally, a discrete-time OTA application is presented in Example 4.31.

Example 4.29 In this example, we design the fully-differential version of the continuous-time opamp of Example 4.28. Its schematic, shown in Fig. 4.79, is just an extension of the original schematic shown in Fig. 4.74(a). Circuit blocks that are common to both halves of the opamp are highlighted in gray. They include a beta-multiplier current reference, a push-pull biasing circuit for defining the bias current of output buffers, and a CMFB circuit. Notice the addition of M5a and M6a to the folded-cascode OTA core for CMFB control. Dimensions determined in Example 4.28 for all devices other than M5 and M6 are retained. The aspect ratio of M5 and M6 is reduced for accommodating M5a and M6a. Besides these four, only the devices of the CMFB circuit are to be sized in the present work. Notice the adoption of the CMFB technique presented originally in Fig. 4.78(a). It is slightly modified by applying low-voltage cascoding to MF3 and MF4.

This opamp is to operate in the voltage amplifier shown in Fig. 4.80(a), which complies with the generic fully-differential configuration shown earlier in Fig. 2.37(a). However, the reference polarity of the input signal v_i is transposed in the present schematic. Closed-loop specifications are similar to those of Example 4.28. The sinusoidal input signal specified with a 100-mV maximum amplitude in that example is now applied around an input common-mode voltage $V_{ICM} = 0.9$ V. V_{OCM} also equals 0.9 V. With regard to external-network resistors, we set R_1 to 10 kΩ as in Example 4.28 but, as implied by (2.131), R_2 needs to be 70 kΩ in order to satisfy the dc gain specification of 7 V/V. Output is loaded differentially by a $R_L = 500$-Ω resistance. Additionally, each single-ended output is loaded by a capacitance $C_L = 4$ pF. This resistance/capacitance combination represents exactly the same loading condition defined in Example 4.28.

We are supposed to design the CMFB circuit, and verify the opamp and amplifier performances with simulation using the following input file `ex4-29.sp`:

```
filename: ex4-29.sp
.incl bsimlvt.sp
.incl bsim.sp
**************************opamp****************************
.subckt opamp 1 2 3 33 45 46
m1 7 1 4 4 cmosplvt w=9.5u l=0.6u ad=5.3p pd=20.1u
m2 9 2 4 4 cmosplvt w=9.5u l=0.6u ad=5.3p pd=20.1u
```

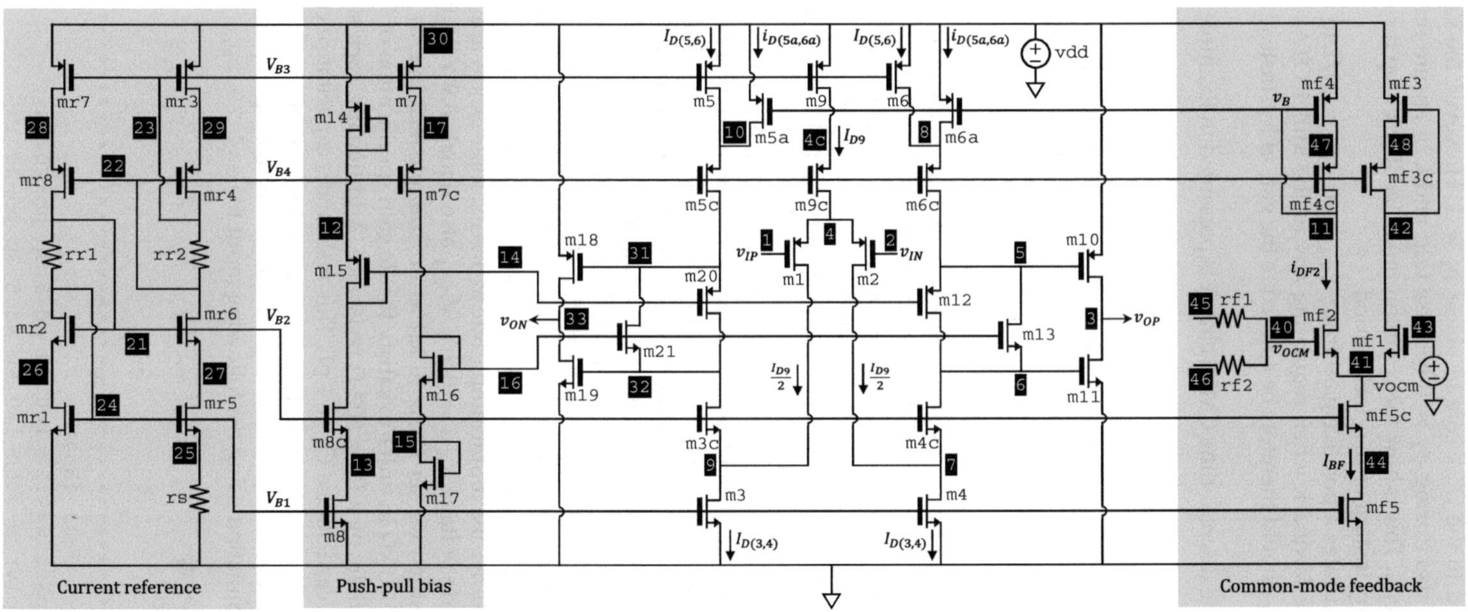

FIGURE 4.79 Circuit schematic of the fully-differential opamp simulated in Example 4.29. Node numbers are 1 and 2 for opamp input terminals and 3 and 39 for opamp output terminals. Input terminals of the CMFB circuit at nodes 45 and 46 are separated from nodes 3 and 33 in order to enable test-source insertion for return-ratio extraction in loop-gain characterization.

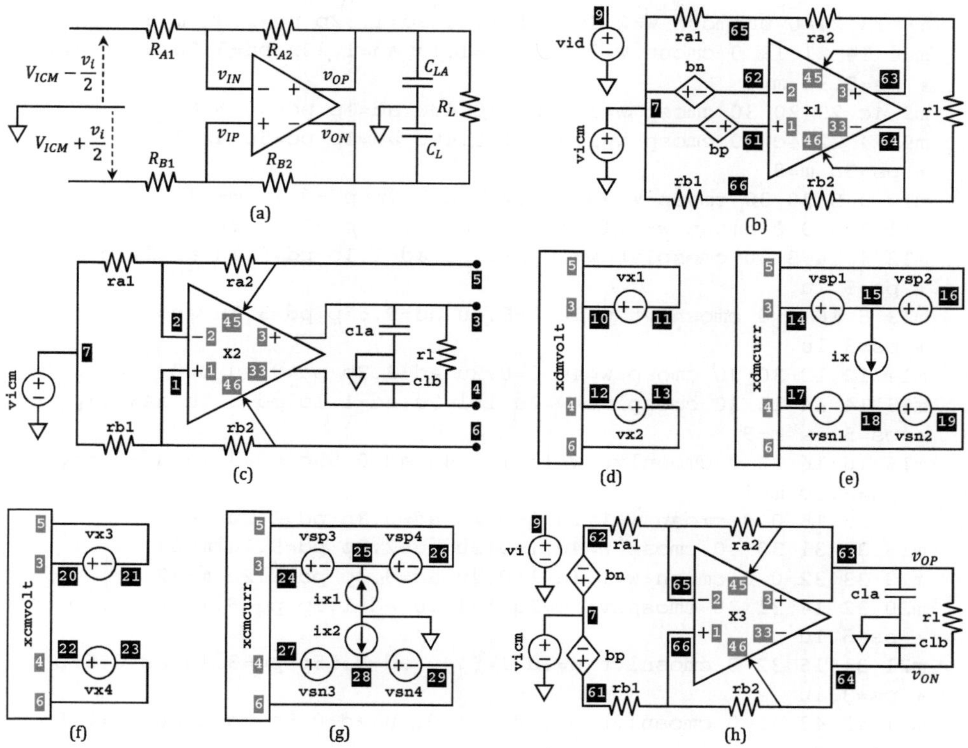

FIGURE 4.80 Circuit schematics of Example 4.29. (a) Top level. (b) Open-loop dc simulation. (c) Subcircuit used in loop-gain simulations. (d) Differential-mode voltage return ratio simulation. (e) Differential-mode current return ratio simulation. (f) Common-mode voltage return ratio simulation. (g) Common-mode current return ratio simulation. (h) Closed-loop transient simulation.

```
m3 9 24 0 0 cmosn w=2.4u l=1.1u ad=1.32p pd=5.9u m=4
m3c 32 21 9 0 cmosn w=2.4u l=1.1u ad=1.32p pd=5.9u as=1.32p
+ ps=5.9u m=2
m4 7 24 0 0 cmosn w=2.4u l=1.1u ad=1.32p pd=5.9u m=4
m4c 6 21 7 0 cmosn w=2.4u l=1.1u ad=1.32p pd=5.9u as=1.32p
+ ps=5.9u m=2
m5 10 23 30 30 cmosp w=1.95u l=0.40u ad=1p pd=5u m=3
m5a 10 11 30 30 cmosp w=1.95u l=0.40u ad=1p pd=5u
m5c 31 22 10 30 cmosp w=1.95u l=0.40u ad=1p pd=5u as=1p
+ ps=5u m=4
m6 8 23 30 30 cmosp w=1.95u l=0.40u ad=1p pd=5u m=3
m6a 8 11 30 30 cmosp w=1.95u l=0.40u ad=1p pd=5u
m6c 5 22 8 30 cmosp w=1.95u l=0.40u ad=1p pd=5u as=1p
+ ps=5u m=4
m7 17 23 30 30 cmosp w=1.95u l=0.40u ad=1p pd=5u m=16
m7c 16 22 17 30 cmosp w=1.95u l=0.40u ad=1p pd=5u as=1p
+ ps=5u m=16
```

```
m8 13 24 0 0 cmosn w=2.400u l=1.1u ad=1.32p pd=5.9u m=8
m8c 14 21 13 0 cmosn w=2.400u l=1.1u ad=1.32p pd=5.9u as=1.32p
+ ps=5.9u m=8
m9 4c 23 30 30 cmosp w=1.95u l=0.40u ad=1p pd=5u m=8
m9c 4 22 4c 30 cmosp w=1.95u l=0.40u ad=1p pd=5u as=1p
+ ps=5u m=8
m10 3 5 30 30 cmosp w=4u l=0.2u ad=2.2p pd=9.1u m=64
m11 3 6 0 0 cmosn w=1.1u l=0.2u ad=0.3p pd=2.2u m=32
m12 6 14 5 30 cmosplvt w=2u l=1.2u ad=1.1p pd=5.1u as=1.1p
+ ps=5.1u
m13 5 16 6 0 cmosnlvt w=1u l=5.4u ad=0.55p pd=3.1u as=0.55p
+ ps=3.1u
m14 12 12 30 30 cmosp w=4u l=0.2u ad=2.2p pd=9.1u m=8
m15 14 14 12 30 cmosplvt w=2u l=1.2u ad=1.1p pd=5.1u as=1.1p
+ ps=5.1u m=9
m16 16 16 15 0 cmosnlvt w=1u l=5.4u ad=0.55p pd=3.1u as=0.55p
+ ps=3.1u m=9
m17 15 15 0 0 cmosn w=1.1u l=0.2u ad=0.3p pd=2.2u m=4
m18 33 31 30 30 cmosp w=4u l=0.2u ad=2.2p pd=9.1u m=64
m19 33 32 0 0 cmosn w=1.1u l=0.2u ad=0.3p pd=2.2u m=32
m20 32 14 31 30 cmosplvt w=2u l=1.2u ad=1.1p pd=5.1u as=1.1p
+ ps=5.1u
m21 31 16 32 0 cmosnlvt w=1u l=5.4u ad=0.55p pd=3.1u as=0.55p
+ ps=3.1u
mf1 42 43 41 0 cmosnlvt w=0.85u l=0.6u ad=0.5p pd=2.8u as=0.5p
+ ps=2.8u
mf2 11 40 41 0 cmosnlvt w=0.85u l=0.6u ad=0.5p pd=2.8u as=0.5p
+ ps=2.8u
mf3 48 42 30 30 cmosp w=1.95u l=0.40u ad=1p pd=5u
mf3c 42 22 48 30 cmosp w=1.95u l=0.40u ad=1p pd=5u as=1p ps=5u
mf4 47 11 30 30 cmosp w=1.95u l=0.40u ad=1p pd=5u
mf4c 11 22 47 30 cmosp w=1.95u l=0.40u ad=1p pd=5u as=1p ps=5u
mf5 44 24 0 0 cmosn w=2.4u l=1.1u ad=1.32p pd=5.9u
mf5c 41 21 44 0 cmosn w=2.4u l=1.1u ad=1.32p pd=5.9u as=1.32p
+ ps=5.9u
rf1 45 40 10k
rf2 46 40 10k
vocm 43 0 0.900
mr1 26 24 0 0 cmosn w=2.4u l=1.1u ad=1.32p pd=5.9u m=2
mr2 24 21 26 0 cmosn w=2.4u l=1.1u ad=1.32p pd=5.9u as=1.32p
+ ps=5.9u m=2
mr3 29 23 30 30 cmosp w=1.95u l=0.40u ad=1p pd=5u m=4
mr4 23 22 29 30 cmosp w=1.95u l=0.40u ad=1p pd=5u as=1p
+ ps=5u m=4
mr5 27 24 25 0 cmosn w=2.4u l=1.1u ad=1.32p pd=5.9u as=1.32p
+ ps=5.9u m=8
mr6 22 21 27 0 cmosn w=2.4u l=1.1u ad=1.32p pd=5.9u as=1.32p
+ ps=5.9u m=2
mr7 28 23 30 30 cmosp w=1.95u l=0.40u ad=1p pd=5u m=4
mr8 21 22 28 30 cmosp w=1.95u l=0.40u ad=1p pd=5u as=1p
```

```
+ ps=5u m=4
rs 25 0 8k
rr1 21 24 43.3k
rr2 23 22 43.3k
vdd 30 0 1.8
.ends opamp
*************************open loop**************************
x1 61 62 63 64 63 64 opamp
rl 63 64 500
ra1 65 7 10k
ra2 63 65 70k
rb1 66 7 10k
rb2 64 66 70k
vicm 7 0 0.900
bp 61 7 v=v(9)/2
bn 7 62 v=v(9)/2
vid 9 0
.save all @m.x1.m10[id] @m.x1.m11[id] @m.x1.m19[id]
+  @m.x1.m18[id]
.control
dc vid -1m 1m 0.2u
plot v(63) v(64)
plot v(63,64)
plot x1.40-x1.43 vs v(63,64)
plot deriv(v(63,64)) vs v(63,64)
plot @m.x1.m10[id] @m.x1.m11[id] @m.x1.m19[id]
+  @m.x1.m18[id] vs v(63,64)
.endc
****************amplifier subckt for loop gain***************
*.subckt amplifier 3 4 5 6
*x2 1 2 3 4 5 6 opamp
*rl 3 4 500
*cla 3 0 4p
*clb 4 0 4p
*ra1 2 7 10k
*ra2 2 5 70k
*rb1 1 7 10k
*rb2 1 6 70k
*vicm 7 0 0.9
*.ends amplifier
*****************differential-mode loop gain****************
*xdmvolt 10 12 11 13 amplifier
*vx1 10 11 dc 0 ac 1
*vx2 13 12 dc 0 ac 1
*xdmcurr 14 17 16 19 amplifier
*vsp1 14 15 0
*vsp2 15 16 0
*vsn1 17 18 0
*vsn2 18 19 0
*ix 15 18 dc 0 ac 1
```

```
*.control
*set units=degrees
*ac dec 100 100 100meg
*let rv=-(v(12,10)/v(13,11)) ; voltage return ratio
*let ri=-(i(vsn1)-i(vsp1))
*+  /(i(vsn2)-i(vsp2)) ; current return ratio
*plot db((rv*ri-1)/(2+rv+ri)) ph((rv*ri-1)/(2+rv+ri))
*let magnitude=mag((rv*ri-1)/(2+rv+ri))
*let phasedif=ph((rv*ri-1)/(2+rv+ri))+180
*meas ac fu when magnitude=1
*meas ac pm find phasedif at=fu
*.endc
*******************common-mode loop gain*******************
*xcmvolt 20 22 21 23 amplifier
*vx3 20 21 dc 0 ac 1
*vx4 22 23 dc 0 ac 1
*xcmcurr 24 27 26 29 amplifier
*vsp3 24 25 0
*vsp4 25 26 0
*vsn3 27 28 0
*vsn4 28 29 0
*ix1 0 25 dc 0 ac 1
*ix2 0 28 dc 0 ac 1
*.save all
*.control
*set units=degrees
*ac dec 100 10 1g
*let rv=-(v(22)+v(20))/(v(23)+v(21)) ; voltage return ratio
*let ri=-(i(vsn3)+i(vsp3))
*+  /(i(vsn4)+i(vsp4)) ; current return ratio
*plot db((rv*ri-1)/(2+rv+ri)) ph((rv*ri-1)/(2+rv+ri))
*let magnitude=mag((rv*ri-1)/(2+rv+ri))
*let phasedif=ph((rv*ri-1)/(2+rv+ri))+180
*meas ac fu when magnitude=1
*meas ac pm find phasedif at=fu
*.endc
******************closed-loop transient*******************
*x3 66 65 63 64 63 64 opamp
*rl 63 64 500
*cla 63 0 4p
*clb 64 0 4p
*ra1 65 62 10k
*ra2 63 65 70k
*rb1 66 61 10k
*rb2 64 66 70k
*vicm 7 0 pwl 0 0.9 1u 0.9 1.01u 1.1 2u 1.1 2.01u 0.7
*+  3u 0.7 3.01u 0.9
*bp 61 7 v=v(9)/2
*bn 7 62 v=v(9)/2
*vi 9 0 sin 0 0.1 4meg
```

```
*.control
*tran 1n 8u
*fourier 4meg v(63,64)
*plot v(7) v(63,64)
*.endc
***********************************************************
.end
```

Work

1. *CMFB Design* CMFB devices and M5a/M6a are sized for achieving a sufficiently fast common-mode response while minimizing the error $v_{OCM} - V_{OCM}$. We begin by modeling the common-mode error $v_{OCM} - V_{OCM}$. As a first step, notice in Fig. 4.79 the mirroring of $i_{D(5a,6a)}$ from i_{DF2}. The former of these two currents is related to the currents flowing in M3/M4, M9, and M5/M6 through

$$i_{D(5a,6a)} = I_{D(3,4)} - \frac{I_{D9}}{2} - I_{D(5,6)}. \tag{4.317}$$

i_{DF2}, on the other hand, is described by

$$i_{DF2} = \frac{g_{m(F1,F2)}}{2}\left(v_{OCM} - V_{OCM}\right) + \frac{I_{BF}}{2}, \tag{4.318}$$

where $g_{m(F1,F2)}$ is the gate transconductance of MF1 and MF2, and I_{BF} is the bias current of the CMFB circuit. Assuming unity scaling in mirroring from MF4 to M5a/M6a, hence $i_{D(5a,6a)} = i_{DF2}$, and solving this equation with (4.317) and (4.318), we obtain the following expression for the common-mode error:

$$v_{OCM} - V_{OCM} = 2\frac{I_{D(3,4)} - I_{D9}/2 - I_{BF}/2 - I_{D(5,6)}}{g_{m(F1,F2)}}. \tag{4.319}$$

Adopting I_B as the symbol of total bias current flowing in input drivers M1 and M2, we can represent the nominal values of the currents appearing in (4.319) with

$$I_{D(3,4)} = I_B,$$

$$I_{D9} = I_B,$$

$$I_{BF} = kI_B,$$

$$I_{D(5,6)} = (1-k)\frac{I_B}{2},$$

where k is the fraction of I_B selected for I_{BF}. As the reader can ascertain from (4.319), no common-mode error would be observed if all four of these currents matched their nominal values perfectly. In practice they mismatch. However, the systematic component of mismatch due to disparate drain-source voltages is almost perfectly eliminated by application of cascoding to all current sources and sinks. For layout-related remedies of mismatch, the reader is referred to Subsection 4.2.2. It is also obvious from (4.319) that the common-mode error arising from any residual mismatch can be lowered by setting a large $g_{m(F1,F2)}$.

For identifying the parameters of the dc gain and bandwidth of the CMFB loop, we break the loop at node 40 in Fig. 4.79, and apply a voltage signal to the gate of MF2. This signal is transformed into a current signal by the effective transconductance $g_{m(F1,F2)}/2$ of the CMFB circuit and subsequently mirrored to M5a and M6a. These two devices source this current onto the impedances associated with nodes 5 and 6 on the right-hand side of the OTA, and nodes 31 and 32 on the left-hand side. Due to the symmetrical topology of the OTA core, we can trace the rest of the loop on one-half only. Selecting the right half, we identify three parallel components for the total impedance of nodes 5 and 6 with respect to ground. Namely, r_{id4c}, r_{id6c}, and $C_{L(\text{ota})}$ as described by (4.298). The dominant pole of CMFB loop gain is usually associated with the time constant of this combination. The common-mode signal experiences a further voltage gain $A'_{0(\text{buf})}$ as it moves to node 3 through the buffer stage of M10 and M11, and completes the loop over RF1. The common-mode gain of the buffer stage is given by

$$A'_{0(\text{buf})} = -(g_{m10} + g_{m11})(r_{ds10} \parallel r_{ds11}). \tag{4.320}$$

Note that, unlike (4.295), this equation does not contain R_L. The reason is that no common-mode current signal flows inside R_L or $R_{F1} + R_{F2}$ because a common-mode voltage signal is present at both terminals of these resistances at nodes 3 and 33. Since $R_L \ll r_{ds10} \parallel r_{ds11}$, a comparison between (4.320) and (4.295) leads us to the conclusion that the gain $A'_{0(\text{buf})}$ experienced by the common-mode signal through the buffer is much larger than the gain $A_{0(\text{buf})}$ experienced by the differential-mode signal. This is why the overall dc gain of the CMFB loop, as described by

$$A'_0 = \underbrace{-(g_{m10} + g_{m11})(r_{ds10} \parallel r_{ds11})}_{A'_{0(\text{buf})}} \frac{g_{m(F1,F2)}}{2}(r_{id4c} \parallel r_{id6c}), \tag{4.321}$$

significantly exceeds its differential counterpart described by (4.267).

Considering the load impedance explained above for the OTA core, the dominant-pole frequency of the CMFB loop is given approximately by

$$f_P = \frac{1}{2\pi\left(r_{id4c} \parallel r_{id6c}\right)\underbrace{\left[C_1 + \left(1 + \left|A'_{0(\text{buf})}\right|\right)\left(C_{gd10} + C_{gd11}\right)\right]}_{C_{L(\text{ota})}}}, \tag{4.322}$$

where (4.298) is adopted for $C_{L(\text{ota})}$. Assuming a single-pole frequency response, the bandwidth f'_{cl} of the CMFB loop equals the gain-bandwidth product $A'_0 f_P$. Referring to (4.321) and (4.322), we formulate it as

$$f'_{cl} = \frac{g_{m(F1,F2)}}{4\pi\left[\frac{C_1}{|A'_{0(\text{buf})}|} + \frac{1+|A'_{0(\text{buf})}|}{|A'_{0(\text{buf})}|}\left(C_{gd10} + C_{gd11}\right)\right]}, \qquad (4.323)$$

where $A'_{0(\text{buf})}$ is given by (4.320). On the other hand, differential-mode closed-loop bandwidth is given by $f_{cl} = \beta f_{GBW}$, where f_{GBW} is the gain-bandwidth product described by (4.278), and β represents the feedback factor $\beta = R_1/(R_1 + R_2)$. Therefore

$$f_{cl} = \frac{\beta g_{m(1,2)}}{2\pi\left[\frac{C_1}{|A_{0(\text{buf})}|} + \frac{1+|A_{0(\text{buf})}|}{|A_{0(\text{buf})}|}\left(C_{gd10} + C_{gd11}\right)\right]}. \qquad (4.324)$$

Since $|A_{0(\text{buf})}| < |A'_{0(\text{buf})}|$ and $\beta = 1/8$ in the present application, we expect from (4.323) and (4.324) to move f'_{cl} above f_{cl} by setting $g_{m(F1,F2)} \approx g_{m(1,2)}/4$. This will ensure adequate speed for the transient response of the CMFB loop.

Having modeled the relevant properties of CMFB, we are now ready for biasing and sizing its devices. To this end, we rely on the following list of parameters as quantified in Example 4.28: $g_{m(1,2)} = 112$ μA/V from the data of step 3; $C_{gd10} \approx C_{gde10} = 200$ fF as calculated from (3.92); $C_{gd11} \approx C_{gde11} = 30$ fF as calculated from (3.92); and $C_1 = 0.52$ pF from step 3.

Targeting $g_{m(F1,F2)} = g_{m(1,2)}/4 = 28$ μA/V in line with the bandwidth argument presented earlier, and assuming $|A'_{0(\text{buf})}| \gg 1$ as well as a negligible $C_1/|A'_{0(\text{buf})}|$ in (4.323), we estimate the common-mode bandwidth as $f'_{cl} =$ 9.7 MHz. Differential-mode bandwidth is expected to be 87.5% of the single-ended counterpart f_{cl} targeted in Example 4.28 because β is reduced from 1/7 to 1/8 in the fully-differential version in order to maintain the same 7-V/V dc closed-loop gain. Hence $f_{cl} = 3.94$ MHz.

Selecting $k = 1/4$ and therefore $I_{BF} = I_B/4 = 3.75$ μA enable us to implement MF5 and MF5c with one copy of the very same unit device sized in step 11 of Example 4.28 for implementing the NMOS devices of the current-reference section. This selection implies $g_{m(F1,F2)}/(kI_B/2) = 28/1.875 = 15\ \text{V}^{-1}$ for the transconductance efficiency of MF1 and MF2.

We select low-threshold devices for MF1 and MF2 in order not to restrict the headroom of the cascode current sink MF5/MF5c of the CMFB circuit. For this selection, Fig. 4.9 indicates $V_{GS(F1,F2)} - V_{TN} \approx 116$ mV, which, in turn, implies $I_{\square(F1,F2)} \approx 1.3$ μA/sq from Fig. 4.8. Since these devices are supposed to conduct 1.875 μA, their aspect ratio should be 1.45. Channel length is not critical for CMFB operation but in order to avoid excessive random offset we prefer the larger-than-minimum setting $L_{DR(F1,F2)} = 0.6$ μm, which, in turn, implies $W_{DR(F1,F2)} = 0.85$ μm for drawn channel width.

Considering $k = 1/4$ as well as the matching of M5a and M6a to MF4, we need to set the aspect ratio of these three devices to one-third of the aspect ratio of M5 and M6. We do this by defining a PMOS unit device of one-fourth the original aspect ratio of MR3 and MR7, and using three copies of this unit device for each of M5 and M6, and one copy for each of M5a, M6a, MF3, MF3c, MF4, and MF4c. The rest of the current-sourcing PMOS devices and their cascode pairs are also built with copies of this unit device without altering the overall aspect ratios assigned in Example 4.28.

Each of MF5 and MF5c is built with one copy of the NMOS unit device defined in Example 4.28 for conducting $I_B/4$ in current sinks.

The averaging resistances R_{F1} and R_{F2} must be large enough not to load the opamp, yet small enough not to introduce a low-frequency pole at node 40. Striking a balance between matching and footprint is also important. 10 kΩ appears to be a reasonable selection.

2. *Open-Loop DC Simulation* The opamp is represented in the input file by a subcircuit named `opamp`. It is accessible by the following six pins: (a) Differential input nodes 1 and 2, (b) differential output nodes 3 and 33, and (c) CMFB input nodes 45 and 46. The open-loop dc simulation is performed on the schematic presented in Fig. 4.80(b). Notice the short-circuiting of pins 3 and 45 as well as 33 and 46 for closing the common-mode feedback loop. Differential-mode feedback is left open by not short-circuiting any of the node pairs 62/65 or 61/66. Also notice the inclusion of all resistive external components that load the output of the opamp in the original application. An independent voltage source `vicm` supplies the input common-mode voltage, around which, the differential-input signal is applied by the complementary dependent voltage sources `bp` and `bn`. Differential-input voltage is swept by the independent voltage source `vid`, and the following five plots are generated automatically: (a) Single-ended output voltages v_{OP} and v_{ON} as functions of v_{ID}, (b) differential-output voltage $v_{OP} - v_{ON}$ as a function of v_{ID}, (c) common-mode error $v_{OCM} - V_{OCM}$ as a function of $v_{OP} - v_{ON}$, (d) dc differential gain $A_{0\text{dif}}$ as a function of $v_{OP} - v_{ON}$, and (e) currents of buffer devices as functions of $v_{OP} - v_{ON}$.

First of these plots, shown in Fig. 4.81(a), verifies the range limits $V_{OL} \leq 0.2$ V and $V_{OH} \geq 1.6$ V as specified earlier in Example 4.28 for single-ended outputs. Second plot, presented in Fig. 4.81(b), shows how the fully-differential topology doubles the output range and also removes the systematic offset of the single-ended opamp previously observed in Fig. 4.75(b). Third plot, shown in Fig. 4.81(c), indicates 971 μV for the common-mode error at the operation point $v_{OP} - v_{ON} = 0$. The plot obtained for A_{dif0} and presented in Fig. 4.81(d) exhibits the symmetry expected from a fully-differential opamp. This is the feature by which the second-order harmonic distortion is eliminated. However, the gain at the operation point for $v_{OP} - v_{ON} = 0$ is somewhat smaller than its single-ended counterpart observed earlier in Fig. 4.75(c) for $v_O = 0.9$ V. This is attributable to the cascoding of M9 by M9c, which slightly lowers I_{D9} below its single-ended counterpart and therefore leads to a lower $g_{m(1,2)}$. The reader can check the credibility of this claim by repeating the simulation after replacing M9c with a short circuit, and comparing the new gain plot with the one displayed in Fig. 4.75(c). Incidentally, if M9 is not cascoded, the common-mode error will worsen to −35.6 mV.

Plots of buffer currents are presented in Fig. 4.81(e). They indicate approximately 209 μA for the operation-point current I_Q, which is lower than

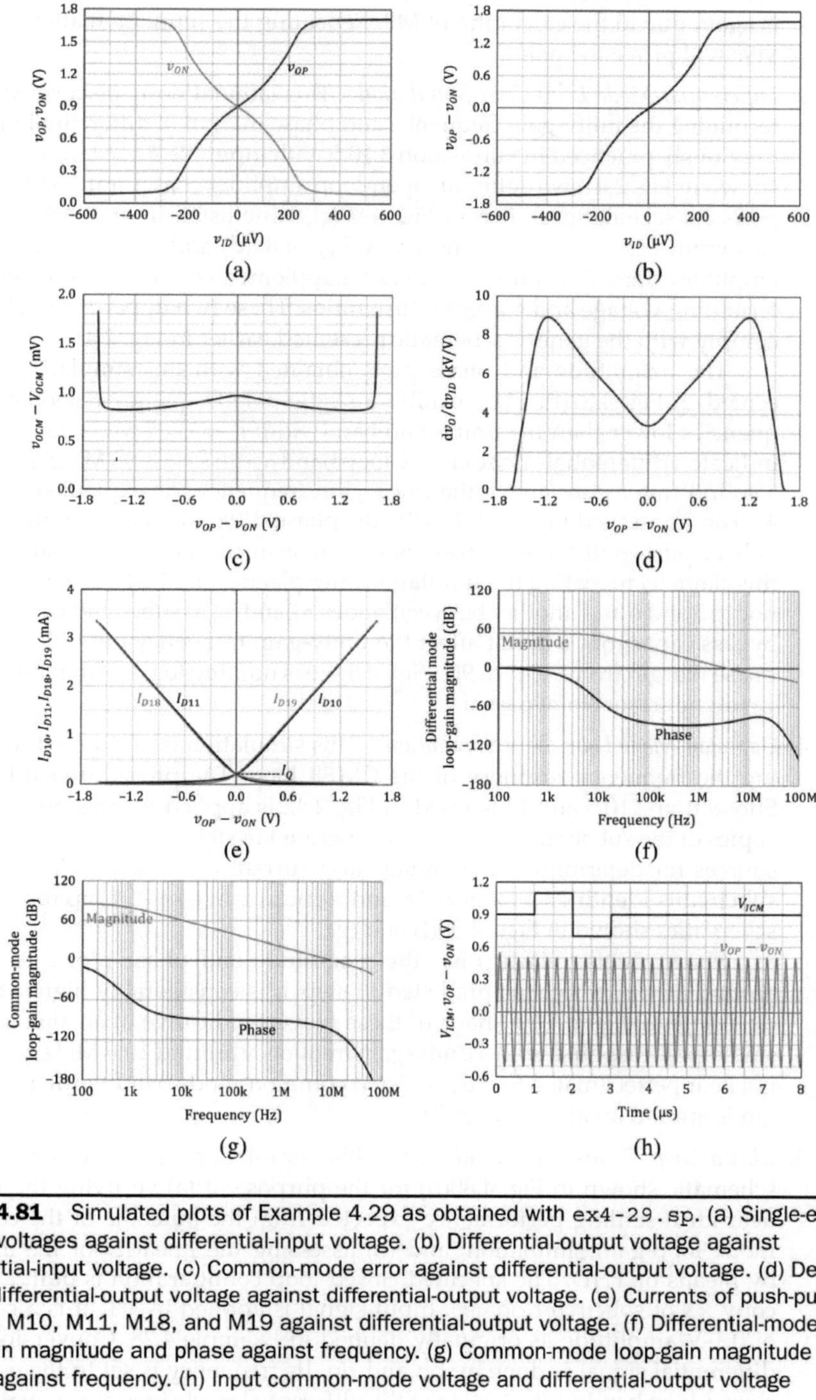

FIGURE 4.81 Simulated plots of Example 4.29 as obtained with `ex4-29.sp`. (a) Single-ended output voltages against differential-input voltage. (b) Differential-output voltage against differential-input voltage. (c) Common-mode error against differential-output voltage. (d) Derivative of the differential-output voltage against differential-output voltage. (e) Currents of push-pull driver devices M10, M11, M18, and M19 against differential-output voltage. (f) Differential-mode loop-gain magnitude and phase against frequency. (g) Common-mode loop-gain magnitude and phase against frequency. (h) Input common-mode voltage and differential-output voltage waveforms.

the single-ended counterpart 249 μA observed in Example 4.28. The difference is again due to the cascoding of M9. Repeating the simulation after removing M9c will prove the point.

3. *Differential-Mode Loop-Gain Simulation* This simulation is performed for determining the unity-gain frequency and phase margin in differential mode. As previously practiced in Subsection 2.10.5 with input file `fdloopgain.sp` and shown in Fig. 2.41, we define an open-loop amplifier configuration for this purpose. Its schematic, shown in Fig. 4.80(c), is netlisted in `ex4-29.sp` by the subcircuit `amplifier`. As shown in Figs. 4.80(d) and (e), two copies of this amplifier, `xdmvolt` and `xdmcurr`, are supplemented with test sources for determining voltage and current return ratios. These two upper-level schematics comply with the generic schematic presented earlier in Fig. 2.41.

   The magnitude and phase plots obtained with this simulation are presented in Fig. 4.81(f). They exhibit a single-pole frequency response for frequencies lower than the unity-gain bandwidth f_u = 3.85 MHz, and therefore indicate a differential-mode closed-loop bandwidth f_{cl} = 3.85 MHz. These plots also indicate a zero above the unity-gain frequency. As explained in Section 4.10 and observed in Example 4.28, the phase difference between the gate signals of push-pull buffer drivers is the cause of this zero. The reader can test this claim by repeating the simulation after placing one 1-pF capacitor between nodes 5 and 6 and another between nodes 31 and 32 in subcircuit `opamp`. These bypass capacitors will not affect the unity-gain frequency but will reduce the phase margin from 96.4° to 90°. Since there is nothing to gain from this change, we do not resort to bypassing.

4. *Common-Mode Loop-Gain Simulation* This simulation is performed for assessing the frequency response of the CMFB loop. The procedure explained in Subsection 2.10.5 and illustrated in Fig. 2.42 is applied for this purpose. Two copies of the subcircuit `amplifier` described in step 3 are combined with test sources for determining the voltage and current return ratios. The resulting subcircuits identified as `xcmvolt` and `xcmcurr` in `ex4-29.sp` represent the schematics shown in Figs. 4.80(f) and (g).

   Shown in Fig. 4.81(g) are the magnitude and phase plots. DC gain is around 85 dB, which, as predicted in step 1, is considerably higher than the dc differential gain. The shape of these plots is indicative of an almost perfect single-pole response with a unity-gain frequency around 10.4 MHz, which also is a near perfect match to the predicted common-mode bandwidth. Phase margin is around an almost ideal 74°.

5. *Closed-Loop Transient Simulation* This simulation is performed on the schematic shown in Fig. 4.80(h) for the purpose of (a) verifying the fast and near-ideal settling performance expected from the outcome of the common-mode loop-gain simulation, and (b) assessing the linearity of the amplifier by means of THD. The specified closed-loop configuration is built around a copy `x3` of subcircuit `opamp`. Input signal is applied by `vi`. It is a sinewave of 0.1-V amplitude as originally defined in Example 4.28. Conversion into a differential signal is done by `bp` and `bn`. Its frequency is set to the estimated closed-loop bandwidth 4 MHz of the differential mode in order to evaluate the worst-case distortion. Also notice the definition of the input common-mode

voltage by the piecewise linear source `vicm`. As observed in Fig. 4.81(h), initially this voltage is pulsed up and down from its specified value 0.9 V for perturbing the common-mode state and observing the CMFB response. During the final 5-μs, however, it is kept constant in order to ensure a reliable distortion measurement.

The waveform of the differential-output voltage $v_{OP} - v_{ON}$ obtained with this simulation is presented in Fig. 4.81(h). The effect of common-mode perturbation on the amplitude of the waveform is on the order of millivolts, and can be observed only by zooming in on the peaks. Furthermore, the effect is observed only on the very first peak that follows the perturbation. Settling behavior is therefore in compliance with the conclusion of the loop-gain simulation.

THD is printed as 4.08%, which is roughly one half of what the single-ended amplifier exhibits in Example 4.28. Notice in the table of Fourier analysis the suppression of all even-order harmonics as expected from a fully-differential application. ▲

Example 4.30 The open-loop amplifier/filter designed in Example 4.18 is to be converted into the fully-differential configuration shown in Fig. 4.82(a). We begin with a description of the ideal dc response. Notice the application of the input signal v_i directly to the differential input of OTA-1. Assuming an ideal OTA of output resistance $r_{o1} = \infty$, the output current of this OTA is described by $i_{o1} = G_{m1}v_i$, where G_{m1} is the effective transconductance. OTA-2, configured with external input-output short circuits on both halves, operates with a differential-input signal $-v_o$, where v_o is the output signal. Assuming $r_{o2} = \infty$ also for this OTA, we represent its output current with $i_{o2} = -G_{m2}v_o$, where G_{m2} denotes its effective transconductance. The configuration also implies $i_{o2} = -i_{o1}$ since C_{L1} and C_{L2} are open circuit in a dc excitation. This equation together with the previous two lead us to $v_o = (G_{m1}/G_{m2})v_i$, which is identical to the ideal dc gain expression of the single-ended version of the amplifier.

Shown in Fig. 4.82(b) is a small-signal equivalent circuit taking into account the finite output resistances r_{o1} and r_{o2} as well as the load capacitances $C_{L1} = C_{L2} = C_L$. Its analysis yields the very same transfer function (4.128) of the single-ended version. The dc gain and bandwidth are described by

$$A_0 = \frac{2G_{m2}r_o}{1 + 2G_{m2}r_o}\frac{G_{m1}}{G_{m2}}, \tag{4.325}$$

$$f_o = \frac{1 + 2G_{m2}r_o}{2G_{m2}r_o}\frac{2G_{m2}}{2\pi C_L}. \tag{4.326}$$

Since effective transconductances are one half of their single-ended counterparts, these two expressions are not different from their single-ended counterparts (4.129) and (4.130).

Work A device-level schematic of the application excluding the CMFB circuit is presented in Fig. 4.82(c). Notice (a) two fully-differential basic OTAs, (b) input-output

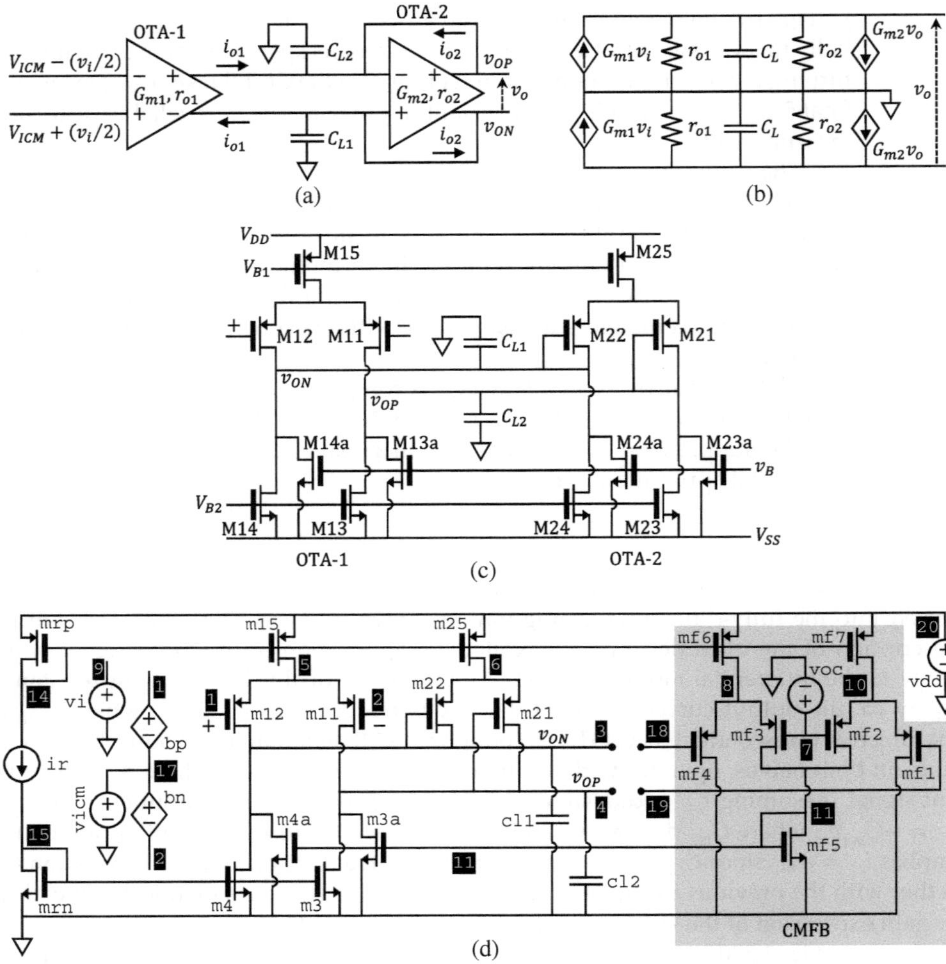

FIGURE 4.82 Fully-differential open-loop amplifier/filter application of Example 4.30. (a) Top-level schematic. (b) Small-signal equivalent circuit. (c) Circuit schematic, where V_{B1} and V_{B2} are bias voltages, and v_B is the control voltage applied by the CMFB circuit. (d) Simulation schematic.

shorts in OTA-2, which turn the drivers M21 and M22 into diode-connected devices, and (c) the addition of M13a, M14a, M23a, and M24a for common-mode control. The schematic contains four pairs of parallel-connected devices: M13/M23, M13a/M23a, M14/M24, and M14a/M24a. Each of these pairs can be implemented with a single device as represented by M3, M3a, M4, and M4a in the simulation schematic shown in Fig. 4.82(d).

Since differential outputs in both OTAs are interconnected, a single CMFB circuit can serve both. Figure 4.82(d) includes such a common CMFB circuit, which is a simplified version of the DDA presented earlier in Fig. 4.78(d) for continuous-time OTA applications. Notice the omission of source degeneration resistors thanks to the rather small maximum differential voltage expected at the inputs of DDA differential pairs. To be specific, the maximum output signal amplitude specified as 90 mV in Example 4.18 implies at most 45 mV for the differential-input voltage of each differential pair in the DDA. In contrast, the differential-input range of these differential pairs, as estimated from (4.85) and Fig. 4.14, will be around 0.525 V if we build the drivers of the DDA as scaled replicas of OTA drivers whose overdrive voltage equals −0.460 V.

Also note in regard to the schematic shown in Fig. 4.82(d) that input nodes 18 and 19 of the CMFB circuit are normally connected to amplifier output nodes 3 and 4, respectively. They are separated only for the purpose of inserting test sources when performing a common-mode loop-gain simulation.

The dc value A_0' and the dominant-pole frequency f_P of the common-mode loop gain are described by

$$A_0' = \frac{g_{m(F1,F2,F3,F4)}}{g_{mF5}} g_{m(3a,4a)} r_o, \tag{4.327}$$

$$f_P = \frac{1}{2\pi r_o C_L}, \tag{4.328}$$

where $r_o = r_{o1} \parallel r_{o2}$ denotes the small-signal resistance associated with the output terminals of OTA-1 and OTA-2 at nodes 3 and 4, and C_L is the load capacitance between each of these nodes and ground. Among the contents of (4.327), the product $g_{m(3a,4a)} r_o$ represents the magnitude of the dc voltage gain from node 11 to nodes 3 and 4. The rest represents the magnitude of the dc voltage gain from nodes 18 and 19 to node 11. The bandwidth $f_{cl}' = A_0' f_P$ of the common-mode loop gain is therefore given by

$$f_{cl}' = \frac{g_{m(F1,F2,F3,F4)}}{g_{mF5}} \frac{g_{m(3a,4a)}}{2\pi C_L}. \tag{4.329}$$

As to the differential-mode bandwidth, we substitute $g_{m(21,22)}/2$ for G_{m2} in (4.326), and assume $G_{m2} r_o \gg 1$ to obtain

$$f_o = \frac{g_{m(21,22)}}{2\pi C_L}. \tag{4.330}$$

We now start sizing the devices of CMFB circuit. To this end, recall that M23 of Example 4.18 is implemented with one unit NMOS device of aspect ratio 1.33 whereas three copies of the same unit device are used for implementing M13. In Fig. 4.82(c) of the present fully-differential version, M23 is represented by the combination of M23 and M23a, and M13 is represented by the combination of M13 and M13a. These four transistors are further combined into M3 and M3a in Fig. 4.82(c). It is therefore obvious that the combined aspect ratio of M3 and M3a will be $4 \times 1.33 = 5.32$. What is not obvious is how this aspect ratio should be partitioned between the two transistors. The answer comes from the optimum setting $f_{cl}' \approx f_o$ of the common-mode and differential-mode bandwidths, and applies also to the combination of M4 and M4a. Referring to (4.329) and (4.330), the condition of comparable bandwidths can be expressed as

$$\frac{f'_{cl}}{f_o} = \frac{g_{m(3a,4a)}}{g_{mF5}} \frac{g_{m(F1,F2,F3,F4)}}{g_{m(21,22)}} = 1,$$

which can be turned into the following form in terms of transconductance efficiencies:

$$\frac{g_{m(3a,4a)}/I_{D(3a,4a)}}{g_{mF5}/I_{DF5}} \frac{g_{m(F1,F2,F3,F4)}/I_{D(F1,F2,F3,F4)}}{g_{m(21,22)}/I_{D(21,22)}} \frac{I_{D(3a,4a)}}{I_{DF5}} \frac{I_{D(F1,F2,F3,F4)}}{I_{D(21,22)}} = 1, \tag{4.331}$$

where the current symbols represent bias values. Note that the very first ratio of transconductance efficiencies appearing on the left-hand side of this equation equals 1 because M3a, M4a, and MF5 operate with identical overdrive voltages. If we also target identical overdrive voltages for CMFB drivers and OTA-2 drivers, the second ratio of transconductance efficiencies becomes 1 as well. Next, assume a scaling factor S from MF5 to M3a and M4a, which replaces the ratio $I_{D(3a,4a)}/I_{DF5}$ with S. Finally, consider the following four equations:

$$I_{D(21,22)} = I_{B2}/2,$$

$$I_{D(F1,F2,F3,F4)} = I_{DF5}/2,$$

$$I_{DF5} = I_{D(3a,4a)}/S,$$

$$k \doteq \frac{I_{D(3a,4a)}}{I_{D(3a,4a)} + I_{D(3,4)}} = \frac{I_{D(3a,4a)}}{(I_{B1} + I_{B2})/2} = \frac{I_{D(3a,4a)}}{2I_{B2}}.$$

These four equations imply $I_{D(F1,F2,F3,F4)}/I_{D(21,22)} = 2k/S$, whose substitution into (4.331) finally leads us to the surprisingly simple result $k = 1/2$. We therefore partition the aspect ratio 5.32 equally between M3 and M3a and also between M4 and M4a. We also target $S = 4$ with a view to keeping the current consumption of the CMFB circuit low. As a consequence of these decisions, aspect ratio of MF5 is finalized as $(W/L)_{F5} = 0.66$. The channel length $L_{DR(23,24)} = 2$ μm determined in step 4 of Example 4.18 is appropriate for the standard-threshold devices MF5, M3, M3a, M4, and M4a. The channel width of MF5 is therefore obtained as $W_{DR(F5)} = 1.3$ μm. Each of M3, M3a, M4, and M4a is to be implemented with four copies of this device.

The outcome $(W/L)_{21,22} = 4.56$ of step 3 of design flow in Example 4.18, combined with

$$\frac{I_{D(F1,F2,F3,F4)}}{I_{D(21,22)}} = \frac{2k}{S} = \frac{1}{4}$$

of the present implementation, leads us to

$$(W/L)_{F1,F2,F3,F4} = \frac{4.56}{4} = 1.14$$

for the aspect ratio of DDA drivers. Adopting $L_{DR(F1,F2,F3,F4)} = 2.0$ μm, as originally selected for all PMOS drivers in Example 4.18, we therefore arrive at $W_{DR(F1,F2,F3,F4)} = 2.2$ μm for their channel width. Each of M21 and M22 is to be built with four copies of these drivers. M11 and M12 will take 12 copies each.

The final step of device sizing involves PMOS current sources. In Example 4.18, a single device of effective aspect ratio $(W/L)_{25} = 86$ implements M25. In the present application, each of MF6 and MF7 is supposed to conduct one-fourth of the current sourced by M25. We therefore divide this aspect ratio by 4, and set $(W/L)_{F6,F7} = 21.5$ for DDA current sources. For the channel length $L_{DR(F6,F7)} = 1.2$ μm defined in Example 4.18, the width is determined as $W_{DR(F6,F7)} = 24.5$ μm. Four and twelve copies of this unit device are needed for implementing M25 and M15, respectively.

For the sake of brevity, no current-reference circuit is included in the schematic presented in Fig. 4.82(d). Instead, bias currents are generated by an independent current source `ir` of 16.15 μA and distributed by unit reference devices `mrp` and `mrn`.

Verification is done with the following input file `ex4-30.sp`:

```
filename: ex4-30.sp
.incl bsimlvt.sp
.incl bsim.sp
m3 4 15 0 0 cmosn w=1.3u l=2u nrd=0.154 nrs=0.154
+  ad=0.72p pd=3.7u m=4
m3a 4 11 0 0 cmosn w=1.3u l=2u nrd=0.154 nrs=0.154
+  ad=0.72p pd=3.7u m=4
m4 3 15 0 0 cmosn w=1.3u l=2u nrd=0.154 nrs=0.154
+  ad=0.72p pd=3.7u m=4
m4a 3 11 0 0 cmosn w=1.3u l=2u nrd=0.154 nrs=0.154
+  ad=0.72p pd=3.7u m=4
m11 4 2 5 5 cmosplvt w=2.2u l=2u nrd=0.09 nrs=0.09
+  as=1.21p ps=5.5u ad=1.21p pd=5.5u m=12
m12 3 1 5 5 cmosplvt w=2.2u l=2u nrd=0.09 nrs=0.09
+  as=1.21p ps=5.5u ad=1.21p pd=5.5u m=12
m15 5 14 20 20 cmosp w=24.5u l=1.2u nrd=0.01 nrs=0.01
+  ad=13.5p pd=50u m=12
m21 4 4 6 6 cmosplvt w=2.2u l=2u nrd=0.09 nrs=0.09
+  as=1.21p ps=5.5u ad=1.21p pd=5.5u m=4
m22 3 3 6 6 cmosplvt w=2.2u l=2u nrd=0.09 nrs=0.09
+  as=1.21p ps=5.5u ad=1.21p pd=5.5u m=4
m25 6 14 20 20 cmosp w=24.5u l=1.2u nrd=0.01 nrs=0.01
+  ad=13.5p pd=50u m=4
mf1 0 19 10 10 cmosplvt w=2.2u l=2u nrd=0.09 nrs=0.09
+  as=1.21p ps=5.5u
mf2 11 7 10 10 cmosplvt w=2.2u l=2u nrd=0.09 nrs=0.09
+  as=1.21p ps=5.5u ad=1.21p pd=5.5u
mf3 11 7 8 8 cmosplvt w=2.2u l=2u nrd=0.09 nrs=0.09
+  as=1.21p ps=5.5u ad=1.21p pd=5.5u
mf4 0 18 8 8 cmosplvt w=2.2u l=2u nrd=0.09 nrs=0.09
+  as=1.21p ps=5.5u
mf5 11 11 0 0 cmosn w=1.3u l=2u nrd=0.154 nrs=0.154
+  ad=0.72p pd=3.7u
mf6 8 14 20 20 cmosp w=24.5u l=1.2u nrd=0.01 nrs=0.01
```

```
+  ad=13.5p pd=50u
mf7 10 14 20 20 cmosp w=24.5u l=1.2u nrd=0.01 nrs=0.01
+  ad=13.5p pd=50u
mrp 14 14 20 20 cmosp w=24.5u l=1.2u nrd=0.01 nrs=0.01
+  ad=13.5p pd=50u
mrn 15 15 0 0 cmosn w=1.3u l=2u nrd=0.154 nrs=0.154
+  ad=0.72p pd=3.7u
ir 14 15 16.15u
cl1 3 0 20p
cl2 4 0 20p
vdd 20 0 1.8
vocm 7 0 0.9
vicm 17 0 0.9
bp 1 17 v=v(9)/2
bn 17 2 v=v(9)/2
***************************dc******************************
vs1 3 18 0
vs2 4 19 0
vi 9 0
.save all @m3[id] @m3a[id] @m4[id] @m4a[id] @m11[id] @m12[id]
+ @m21[id] @m22[id] @mf1[id] @mf2[id] @mf3[id] @mf4[id]
.control
dc vi -30m 30m 0.1m
plot v(3) v(4)
plot v(4,3)
plot deriv(v(4,3))
.endc
***************************ac******************************
*vs1 3 18 0
*vs2 4 19 0
*vi 9 0 dc 0 ac 1
*.control
*ac dec 100 100 10meg
*plot vdb(4,3)
*.endc
***************************tran****************************
*vs1 3 18 0
*vs2 4 19 0
*vi 9 0 sin 0 30m 1meg
*.control
*tran 1n 15u
*plot v(4,3)
*fourier 1meg v(4,3)
*.endc
******************common mode loop gain*******************
*vi 9 0 0
*vs1 3 18 dc 0 ac 1
*vs2 4 19 dc 0 ac 1
*.control
*set units=degrees
```

```
*ac dec 100 100 100meg
*let rv=-(v(3)+v(4))/(v(18)+v(19)) ; voltage return ratio
*plot db(rv) ph(rv)
*.endc
***********************************************************
.end
```

The dc section of the simulation is intended mainly as a sanity check but it is also used for evaluating the common-mode error and voltage gain. Reading the common value of the single-ended output voltages at nodes 3 and 4 for $v_i = 0$, and subtracting it from $V_{OCM} = 0.9$ V yield -2.25 mV for the systematic common-mode error. The voltage gain indicated by the derivative of the differential-output voltage is close to 2.75 V/V, which is a near perfect fit to what the single-ended version offered in Example 4.18.

The purpose of the ac simulation is to determine the 3-dB frequency for evaluating the differential bandwidth. It yields $f_{cl} = 1.04$ MHz, which closely fits the specification.

The transient simulation is run mainly for evaluating the linearity of the amplifier at the upper limit 1 MHz of the frequency band. The reader is invited to compare the THD with that of the single-ended version simulated with the input file `ex4-18d.sp`.

In all three simulations just described, common-mode loop is closed by two short-circuit sources named `vs1` and `vs2`. In the final part, where a common-mode loop-gain analysis is defined, these two are deployed as test sources for extracting the voltage return ratio. Considering the argument presented in footnote 8 of Subsection 2.7.2, it is unnecessary to extract the current return ratio because the impedance seen into the amplifier through nodes 3 and 4 is definitely much smaller than the impedance seen into the CMFB circuit through nodes 18 and 19. The reader is invited to run this simulation, and compare the indicated unity-gain frequency with the common-mode bandwidth specification on which the design is based. ▲

Example 4.31 The discrete-time amplifier configuration shown in Fig. 4.49 and designed in Example 4.22 is to be converted into a fully-differential configuration complying with the top-level schematic presented in Fig. 2.54.

Work An electrical schematic of the fully-differential OTA including CMFB and current-reference circuits is given in Fig. 4.83(a). Shown in part b of the same figure is the top-level closed-loop configuration.

The original single-ended amplifier configuration is based on a cascoded-symmetrical OTA with PMOS drivers. Here, we adopt the fully-differential version introduced in Fig. 4.76(f) for this type of OTA, and supplement its current reference with a start-up circuit. For CMFB, we prefer the switched-capacitor topology presented in Fig. 4.78(e) for current-source control. However, C_{2p} and C_{2n} are removed because C_{A1}, C_{A2}, C_{B1}, and C_{B2} of the external network readily perform the task of averaging v_{OP} and v_{ON} in sampling phases. Device dimensions including those of transmission gate switches are identical to those determined in Example 4.22. V_{ICM} and V_{OCM} are set to 0.9 V as needed for replicating the original bias conditions. OTA output nodes 3 and 11 are normally connected to CMFB input nodes 33 and 34. They are left open in Fig. 4.83(a) as a provision for driving the CMFB circuit with voltage-dependent voltage sources without capacitively loading the OTA. This will help us assess the effect of CMFB loading on the OTA.

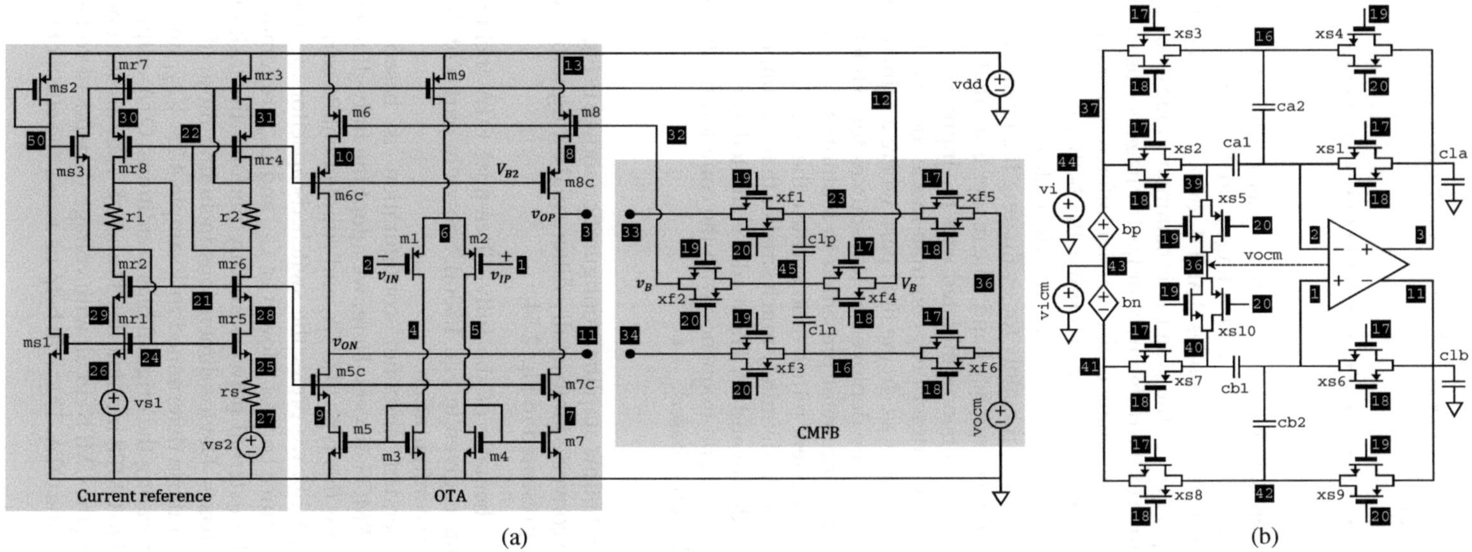

FIGURE 4.83 Fully-differential discrete-time amplifier application of Example 4.31. (a) Electrical schematic of the OTA. (b) Top-level schematic, where the OTA symbol represents the electrical schematic shown in part a.

Verification is performed with a transient analysis performed with the following input file ex4-31.sp. The netlist complies with the schematics shown in Fig. 4.83.

```
filename: ex4-31.sp
.incl bsimlvt.sp
.incl bsim.sp
************************ota core**************************
m1 4 2 6 13 cmosplvt w=14u l=0.4u nrd=0.014 nrs=0.014
+ ad=7.7p pd=29.1u as=7.7p ps=29.1u
m2 5 1 6 13 cmosplvt w=14u l=0.4u nrd=0.014 nrs=0.014
+ ad=7.7p pd=29.1u as=7.7p ps=29.1u
m3 4 4 0 0 cmosn w=2.15u l=0.9u nrd=0.1 nrs=0.1
+ ad=1.2p pd=5.4u m=5
m4 5 5 0 0 cmosn w=2.15u l=0.9u nrd=0.1 nrs=0.1
+ ad=1.2p pd=5.4u m=5
m5 9 4 0 0 cmosn w=2.15u l=0.9u nrd=0.1 nrs=0.1
+ ad=1.2p pd=5.4u m=5

m5c 11 21 9 0 cmosn w=2.15u l=0.9u nrd=0.1 nrs=0.1
+ ad=1.2p pd=5.4u as=1.21p ps=5.4u m=5
m6c 11 22 10 13 cmosp w=7.35u l=0.65u nrd=0.03 nrs=0.03
+ ad=4.1p pd=15.8u as=4.1p ps=15.8u m=5
m6 10 32 13 13 cmosp w=7.35u l=0.65u nrd=0.03 nrs=0.03
+ ad=4.1p pd=15.8u m=5
m7 7 5 0 0 cmosn w=2.15u l=0.9u nrd=0.1 nrs=0.1
+ ad=1.2p pd=5.4u m=5
m7c 3 21 7 0 cmosn w=2.15u l=0.9u nrd=0.1 nrs=0.1
+ ad=1.2p pd=5.4u as=1.2p ps=5.4u m=5
m8c 3 22 8 13 cmosp w=7.35u l=0.65u nrd=0.03 nrs=0.03
+ ad=4.1p pd=15.8u as=4.1p ps=15.8u m=5
m8 8 32 13 13 cmosp w=7.35u l=0.65u nrd=0.03 nrs=0.03
+ ad=4.1p pd=15.8u m=5
m9 6 12 13 13 cmosp w=7.35u l=0.65u nrd=0.03 nrs=0.03
+ ad=4.1p pd=15.8u m=10
vdd 13 0 1.8
********************current reference*********************
mr1 29 24 26 0 cmosn w=2.15u l=0.9u nrd=0.1 nrs=0.1
mr2 24 21 29 0 cmosn w=2.15u l=0.9u nrd=0.1 nrs=0.1
mr3 31 12 13 13 cmosp w=7.35u l=0.65u nrd=0.03 nrs=0.03
mr4 12 22 31 13 cmosp w=7.35u l=0.65u nrd=0.03 nrs=0.03
mr5 28 24 25 0 cmosn w=2.15u l=0.9u nrd=0.1 nrs=0.1 m=4
mr6 22 21 28 0 cmosn w=2.15u l=0.9u nrd=0.1 nrs=0.1
mr7 30 12 13 13 cmosp w=7.35u l=0.65u nrd=0.03 nrs=0.03
mr8 21 22 30 13 cmosp w=7.35u l=0.65u nrd=0.03 nrs=0.03
vs1 26 0 0
vs2 27 0 0
rs 25 27 7.4k
r1 21 24 35.5k
r2 12 22 35.5k
ms1 50 24 0 0 cmosn w=2u l=1u nrd=0.1 nrs=0.1
```

```
ms2 50 50 13 13 cmosp w=1u l=3u nrd=0.03 nrs=0.03
ms3 12 50 24 0 cmosn w=2u l=1u nrd=0.1 nrs=0.1
********************switch subcircuit*********************
.subckt tg in out ngate pgate power
mn in ngate out 0 cmosnlvt w=5u l=0.2u nrd=0.04 nrs=0.04
+ ad=2.75p pd=11.1u as=2.75p ps=11.1u
mp in pgate out power cmosplvt w=5u l=0.2u nrd=0.04 nrs=0.04
+ ad=2.75p pd=11.1u as=2.75p ps=11.1u
.ends tg
******************common-mode feedback********************
xf1 23 33 19 20 13 tg
xf2 45 32 19 20 13 tg
xf3 16 34 19 20 13 tg
xf4 12 45 17 18 13 tg
xf5 36 23 17 18 13 tg
xf6 36 16 17 18 13 tg
c1p 23 45 0.25p
c1n 16 45 0.25p
vocm 36 0 0.9
*********************external network*********************
xs1 2 3 17 18 13 tg
xs2 37 39 17 18 13 tg
xs3 37 38 17 18 13 tg
xs4 38 3 19 20 13 tg
xs5 39 36 19 20 13 tg
xs6 1 11 17 18 13 tg
xs7 40 41 17 18 13 tg
xs8 41 42 17 18 13 tg
xs9 42 11 19 20 13 tg
xs10 36 40 19 20 13 tg
ca1 39 2 2p
ca2 38 2 1p
cb1 40 1 2p
cb2 42 1 1p
cla 3 0 0.5p
clb 11 0 0.5p
********************transient analysis*********************
vs3 3 33 0
vs4 11 34 0
*b1 33 0 v=v(3)  ; connects CMFB without loading OTA
*b2 34 0 v=v(11) ; connects CMFB without loading OTA
vi 44 0 pwl 0 -0.15 410n -0.15 412n 0.15 525n 0.15
+ 527n 0 636n 0 638n -0.15 777n -0.15
vicm 43 0 0.9
bp 37 43 v=v(44)/2
bn 43 41 v=v(44)/2
vphi1 17 0 pulse 1.8 0 47n 2n 2n 58n 112n
vphi1c 18 0 pulse 0 1.8 47n 2n 2n 58n 112n
vphi2 19 0 pulse 0 1.8 51n 2n 2n 50n 112n
vphi2c 20 0 pulse 1.8 0 51n 2n 2n 50n 112n
```

```
.tran 100p 777n
.options trtol=1
.end
```

Simulation is run initially with the voltage-controlled voltage sources `b1` and `b2` disabled but the zero-voltage sources `vs3` and `vs4` enabled, so that the CMFB circuit is driven directly by the OTA as usual. Clock waveform is similar to the one defined in `ex4-22c.sp` of Example 4.22. However, the duration of simulation is extended in order to allow time for the common-mode control voltage v_B to develop into a steady state following start-up. Input voltage `vi` is also similar to what is defined in `ex4-22c.sp` except for the extension of the initial period, in which, it stays at −0.15 V. The reader is invited to run the simulation, and compare the differential output-voltage waveform `v(3,11)` with its single-ended counterpart obtained with `ex4-22c.sp`. A relatively slower response will be evident. This is a consequence of the capacitive loading imposed by the CMFB circuit onto the output nodes of the OTA. Disabling `vs3` and `vs4` and enabling `b1` and `b2` will remove this additional load while maintaining the common-mode feedback. Repeating the transient simulation with this modification and comparing the waveform with that of `ex4-22c.sp` will help the reader appreciate the extent of capacitive loading. Yet another interesting aspect of the transient simulation is the evolution of the CMFB control voltage v_B of node 32 following start-up. ▲

4.12 References

1. P. G. Drennan, M. L. Kniffin, and D. R. Locascio, "Implications of proximity effects for analog design," in *Proceedings of IEEE Custom Integrated Circuits Conference*, pp. 169–176, 2006.
2. A. Hastings, *The Art of Analog Layout*, 2nd Edition. Pearson: Upper Saddle River, NJ, 2005.
3. M. J. M. Pelgrom, A. C, J. Duinmaijer, and A. P. G. Welbers, "Matching properties of MOS transistors," *IEEE Journal of Solid-State Circuits*, vol. 24, pp. 1433–1440, 1989.
4. T.-H. Yeh, J. C. H. Lin, S.-C. Wong, H. Huang, and J. Y. C. Sun, "Mis-match characterization of 1.8V and 3.3V devices in 0.18μm mixed signal CMOS technology," *Proceedings of IEEE International Conference on Microelectronic Test Structures*, vol. 14, pp. 77–82, 2001.
5. H. Banba, H. Shiga, A. Umezawa, T. Miyaba, T. Tanzawa, S. Atsumi, and K. Sakui, "A CMOS bandgap reference circuit with sub-1-V operation," *IEEE Journal of Solid-State Circuits*, vol. 34, pp. 670–674, 1999.
6. G. A. Rincon-Mora, *Voltage References: From Diodes to Precision High-Order Bandgap Circuits*, Hoboken, NJ: Wiley-IEEE Press, 2001.
7. C.-W. Kok, and W.-S. Tam, *CMOS Voltage References: An Analytical and Practical Perspective*, Hoboken, NJ: Wiley-IEEE Press, 2013.
8. A. Boni, "Op-amps and startup circuits for CMOS bandgap references with near 1-V supply," *IEEE Journal of Solid-State Circuits*, vol. 37, pp. 1339–1343, 2002.
9. S. Yan, J. Hu, T. Song, and E. Sanchez-Sinencio, "Constant-g_m techniques for rail-to-rail CMOS amplifier input stages: A comparative study," in *Proceedings of IEEE International Symposium on Circuits and Systems*, pp. 2571–2574, 2005.

10. K.-J. de Langen, and J. H. Huijsing, "Compact low-voltage power-efficient operational amplifier cells for VLSI," *IEEE Journal of Solid-State Circuits*, vol. 33, pp. 1482–1496, 1998.
11. Y. P. Tsividis, and P. R. Gray, "An integrated NMOS operational amplifier with internal compensation," *IEEE Journal of Solid-State Circuits*, vol. SC-11, pp. 748–753, 1976.
12. B. K. Ahuja, "An improved frequency compensation technique for CMOS operational amplifiers," *IEEE Journal of Solid-State Circuits*, vol. SC-18, pp. 629–633, 1983.
13. D. B. Ribner, and M. A. Copeland, "Design techniques for cascoded CMOS op amps with improved PSRR and common-mode input range," *IEEE Journal of Solid-State Circuits*, vol. SC-19, pp. 919–925, 1984.
14. W. C. Black, Jr., D. J. Allstot, and R. A. Reed, "A high performance low power CMOS channel filter," *IEEE Journal of Solid-State Circuits*, vol. SC-19, pp. 929–938, 1980.

4.13 Exercises

Assume 0.18-μm CMOS for the manufacturing technology unless design specifications stipulate 90-nm CMOS. Refer to Exercise 3.1 for the details of the 90-nm CMOS technology adopted in this book.

4.1 Change the compliance-limit and output bias-voltage specifications of Example 4.1 to $V_C = 0.125$ V and $V_O = 0.6$ V, and redesign in the 90-nm CMOS technology adopted in this book.

4.2 Change the power-supply voltage, compliance-limit, and output bias-voltage specifications of Example 4.4 to $V_{DD} = 1.2$ V, $V_C = -0.125$ V, and $V_O = 0.6$ V, and redesign in the 90-nm CMOS technology adopted in this book.

4.3 Design a single-supply ($V_{DD} = 1.2$ V) low-voltage cascode sink in the 90-nm CMOS technology adopted in this book. It is supposed to comply with the following specifications:

1. Topology: Fig. 4.15(b).
2. Reference current: $I_R = 5$ μA.
3. Scaling: One unit-device for each of MR1 and MR2. Five unit devices for each of M1 and M2.
4. Output resistance: $r_o = 25$ MΩ for $V_O = 0.6$ V.
5. Compliance limit: $V_C = 0.2$ V.

4.4 Design a single-supply ($V_{DD} = 1.8$ V) self-cascode current source in accordance with the following specifications:

1. Nominal output current and reference current: $I_O = I_R = 1$ μA.
2. Output resistance: $r_o = 200$ MΩ for $V_O = 0.9$ V.
3. Compliance limit: $V_C = -120$ mV.

4.5 Design the PMOS-driver version of the source-degenerated differential pair of Fig. 4.32(a) in accordance with the following specifications:

1. Supply voltage: Single supply $V_{DD} = 1.8$ V.
2. Effective transconductance: $G_m = 150\ \mu\text{A/V}$.
3. Output resistance: $r_{o1} = 500\ \text{k}\Omega$ for $R_{D2} = 0$.
4. Total harmonic distortion: THD $= 0.5\%$ on i_{D1} for 1-kHz sinewave of 100-mV input-signal amplitude.
5. Common-mode operation point: $V_{CM} = 0.9$ V and $V_{D1} = V_{D2} = 0.3$ V.
6. Overdrive voltage: $V_{GS} - V_{TP} = -200$ mV.

4.6 Rework Example 4.17 after changing specification items 1, 4, and 5 to 5-pF external load, 35-MHz closed-loop bandwidth, and 100-fF closed-loop input capacitance, respectively.

4.7 Design the unity-gain amplifier of Example 4.19 with a symmetrical OTA of twin-well NMOS drivers. Retain all specifications with the exception of bias voltage. Target $V_B = 1$ V for the latter.

4.8 Rework Example 4.22 for the discrete-time closed-loop configuration shown in Fig. 2.59 of Exercise 2.19 after deleting item 3 of specifications and changing item 7 to $|\Delta v_{O(dc)}| = 0.3$ mV.

4.9 Rework Example 4.24 for 20-MHz full-power bandwidth while retaining the rest of the original specifications.

4.10 Assuming $V_{DD} = 1.2$ V, $V_O = 0.4$ V, and an input voltage stepping between 0.2 and 0.6 V but retaining the rest of the specifications, will it be possible to design the buffer amplifier of Example 4.25 in the 90-nm CMOS technology adopted in this book?

4.11 Design a cascoded-symmetrical opamp for the inverting amplifier specified in Example 4.28.

4.12 Convert the OTA designed in Exercise 4.8 into a fully-differential OTA complying with the topology shown in Fig. 4.83(a), and use it for implementing the fully-differential version of the closed-loop configuration described in Exercise 4.8. Compare the performances of the two versions of the closed-loop configuration as done in Example 4.31 for the amplifier of Fig. 4.83(b).

Index

Note: Page numbers followed by *f* denote figures; page numbers followed by *t* denote tables; page numbers followed by *n* denote footnotes.

A

B

C

D

E

O

P

9 781260 441451

Printed by Libri Plureos GmbH in Hamburg,
Germany